ICM '76
Part II

Proceedings of the

International Conference on Magnetism

ICM '76

Part II
(Chapters 4-6)

Amsterdam, The Netherlands
September 6-10, 1976

Editors:

P.F. de Châtel and J.J.M. Franse
Natuurkundig Laboratorium, University of Amsterdam

1977

NORTH-HOLLAND PUBLISHING COMPANY
AMSTERDAM · NEW YORK · OXFORD

© North-Holland publishing Company, 1977

All Rights Reserved. No part of this publication may be reproduced, stored in a retrieval system, or transmitted, in any form or by any means, electronic, mechanical, photocopying, recording or otherwise, without the prior permission of the Copyright owner

NORTH-HOLLAND ISBN 07204 0731 1

REPRINTED FROM PHYSICA VOLUMES 86–88 B+C – PART II

PUBLISHERS:
NORTH-HOLLAND PUBLISHING COMPANY
AMSTERDAM–NEW YORK–OXFORD

SOLE DISTRIBUTORS FOR THE U.S.A. AND CANADA:
ELSEVIER NORTH-HOLLAND, INC.
52 VANDERBILT AVENUE, NEW YORK, N.Y. 10017

PRINTED IN THE NETHERLANDS

CHAPTER 4

CRITICAL PHENOMENA
DIMENSIONALITY AND IMPURITY EFFECTS

Recent developments in critical magnetic behaviour (*Invited paper 5AI1*)	545
Experimental studies of multicritical points in magnetic systems (*Invited paper 5AI2*)	550
Critical properties of the *XY* model (*Invited paper 6BI2*)	556
Critical behaviour in magnetic insulators (*Session 8E*)	562
Field-dependent critical behaviour (*Session 7E*)	572
Multicritical behaviour (*Session 5E*)	590
Theory; Critical behaviour (*Session 5V*)	602
Theory; Renormalization group (*Session 4D*)	618
Magnetic ordering in $CuSO_4 \cdot 5H_2O$ and related linear chain compounds in external field (*Invited paper 3BI2*)	634
Neutron scattering studies of low dimensional magnetic systems (*Invited paper 2BI1*)	639
One- and two-dimensional magnetism I (*Session 2E*)	647
One- and two-dimensional magnetism II (*Session 3X*)	660
One- and two-dimensional magnetism III (*Session 9X*)	680
Impurity effects in magnetic chains (exp.) (*Session 1F*)	705
Magnetic insulators; Disorder effects (*Session 2X*)	710
Magnetic insulators; Impurity effects (*Session 3F*)	725

RECENT DEVELOPMENTS IN CRITICAL MAGNETIC BEHAVIOUR

A. AHARONY

Department of Physics and Astronomy, Tel-Aviv University, Ramat-Aviv, Tel-Aviv, Israel

(Invited paper)

Competition between various types of interaction leads to complex phase diagrams, exhibiting multicritical points. Bicritical points appear due to uniaxial assymmetry, resulting e.g. from uniaxial stress. These may become tetracritical when cubic symmetry exists. Tricritical and bicritical points appear in metamagnets. Tetracritical points exist for randomly mixed ferromagnets and antiferromagnets. Competing interactions also lead to new types of critical behaviour, e.g. "dipolar", "cubic", etc. All of these may be affected by interactions with elastic degrees of freedom, random impurities, etc. Present theoretical understanding of these phenomena, based mainly on renormalization group arguments, is reviewed, and relevant experiments are mentioned.

1. Introduction

Renormalization Group has had a very important impact on our theoretical understanding of critical behaviour of magnets [1–5]. Since the last Conference in this series, much more is now known about crossover phenomena, competing interactions and multicritical points. Since much of this knowledge is published, I shall only list here the results which to my feeling had, or should have, the most important impact on experimental work. Detailed references are included, and the most important features of each problem are summarized.

In general, we consider a Hamiltonian of the form

$$\mathcal{H} = \mathcal{H}_0 + g\mathcal{H}_1, \tag{1}$$

where $\mathcal{H}_0$ represents some known critical behavior of the system, and $\mathcal{H}_1$ is some additional interaction. In many cases, g is a *relevant* variable, and the *asymptotic* critical behavior of $\mathcal{H}$ is distinct from that of $\mathcal{H}_0$. One then speaks of a *crossover* from one behavior to another. Each such behaviour is represented by a *fixed point* in Hamiltonian space, and the crossover is represented by a *flow* of the effective, coarse grained, Hamiltonian among the possible fixed points [2]. This flow yields a crossover scaling function, e.g. for the free energy density,

$$F(T, g, H) = |t|^{2-\alpha} f(H/|t|^{\Delta}, g/|t|^{\phi}), \tag{2}$$

where $t = [T - T_c(g=0)]/T_c(g=0)$, H is the ordering (magnetic) field and α, Δ and ϕ are critical exponents associated with $\mathcal{H}_0$. If g is relevant, then the "crossover exponent" ϕ is positive. The function $f(0, y)$ then has a singular point for some value of y, of the form $(y - y_0)^{2-\dot{\alpha}}$, which then yields the new critical behaviour, with a specific heat exponent $\dot{\alpha}$.

There are various types of parameters g to consider. One type involves a microscopic interaction coupling constant, e.g. that of dipole-dipole interactions. This variable cannot be varied experimentally. Thus, one only varies t, and follows the crossover as T approaches $T_c(g)$, where $T_c(g)$ is the value of T for which $g/|t|^{\phi}$ reaches the singular point y_0 of f. Different values of g may be observed for different materials.

A more interesting type involves *non-ordering fields*, which can be varied experimentally. These include a uniform magnetic field on an antiferromagnet, isotropic and anisotropic stress, the chemical potential of impurities, etc. For any fixed g, one observes a transition at $T_c(g)$. One thus finds *lines of critical points*, all of which are characterized by the same universal exponents, normalized scaling functions, etc. Two or more such lines may meet at the point $g = 0$, which is then called "*multicritical*".

In the next two sections we thus review a few examples in each of these categories. Again, emphasis is put on results and on references, and not on any derivation.

2. Crossover effects

In this section we shall list a few typical cases, in which crossover occurs from one critical behavior to another due to some microscopic interaction. Many such cases are listed and discussed at length in ref. 6.

2.1. Isotropic dipolar systems

Dipole–dipole interactions are highly relevant for ferromagnets with low T_c [7]. These were discussed in detail in a series of papers [8–10], and we shall therefore skip all details.

One finds, that in spite of the long range of the dipolar interactions, their critical exponents are still quite different from those of mean field theory. In fact, calculations to order ϵ^2 ($\epsilon = 4 - d$, where d is the dimensionality of the system) of the exponents [9] give results which are numerically quite close to those found in the short range case! Similarly, the equation of state scaling function is also found [10] to be barely distinguishable from that of non-dipolar materials.

One might therefore expect, that the crossover from isotropic short-range behaviour (observed relatively far from T_c) to the dipolar one (observed as $T \to T_c$) will not be detected. It turns out, however, that the function $f(0, g/|t|^{\phi})$ [eq. (1)] is not as "smooth" as naively expected. Recent ϵ-expansions of this function, and of its second derivatives corresponding to the specific heat and to the magnetic susceptibility [11, 12] show, that for some range of values of t one may observe an effective exponent, e.g.

$$\gamma_{\text{eff}} = -\frac{\partial \ln \chi(T)}{\partial \ln t} \tag{3}$$

[$\chi(T)$ is the magnetic susceptibility], which is significantly smaller than both its short range and dipolar counterparts. This may explain the low values of γ observed in EuO and EuS [13], and probably also some of the specific heat data on these materials [14].

2.2. Uniaxial dipolar systems

The case of uniaxial dipolar systems at $d = 3$ is most interesting, as it may be solved *exactly* using the renormalization group technique [15–17]. The result is, that the asymptotic uniaxial behavior exhibits mean-field like exponents, modified by logarithmic corrections. For example, the singular term in the specific heat behaves as

$$C = A |\ln |t||^{\frac{1}{3}}, \tag{4}$$

the susceptibility behaves as

$$\chi = \Gamma |t|^{-1} |\ln |t||^{\frac{1}{3}}, \tag{5}$$

etc. These predictions have been recently confirmed by experiments on LiTbF$_4$ [18, 19]. *This is the first clear experimental verification of renormalization group theory.*

In addition to the power-law forms, e.g. (4) and (5), one can also derive universal relations among the *amplitudes* [20]. These are relatively less sensitive to experimental errors, and are also confirmed by recent experiments [21].

2.3. Cubic systems

The simplest single ion interaction of cubic symmetry has the form

$$\mathcal{H}_c = v \sum_X \sum_{\alpha=1}^n S^\alpha(X)^4, \tag{6}$$

where S^α ($\alpha = 1, 2, \ldots, n$) is the α-component of the (n-component) spin vector at site X. The sign of v determines whether the spins tend to align along a cubic axis ($v < 0$) or along a diagonal ($v > 0$).

Renormalization group study near $d = 4$ shows, that the critical behavior of the cubic system is characterized mainly by a close competition of two fixed points, i.e. the isotropic (Heisenberg-like) one, with $v^* = 0$, and the "cubic" one ($v^* \neq 0$) [22]. The isotropic fixed point is stable if the cubic crossover exponent, ϕ_v, is positive [see eq. (2)]. If ν is the correlation length exponent, then $\phi_v = \lambda_v / \nu$, and the exponent λ_v has been calculated both to order ϵ^3, with $\epsilon = 4 - d$ [23], and to order $(\epsilon')^2$, with $\epsilon' = d - 2$ [24]. Clearly, λ_v becomes positive for $n > n_c(d)$, where $n_c(4) = 4$ and $n_c(2) = 2$. Extrapolation of λ_v at $n = 3$ towards $d = 3$ from both ends seems to indicate that $n_c(3) > 3$, so that the usual three component Heisenberg model is "safe" from crossover to cubic behaviour. However, the cubic terms yield *corrections* to scaling, which lead to important experimental consequences. For example, the spins will align along *easy axes* (cubic axes or diagonals) for *any* $|t| \neq 0$, and the transverse susceptibility below T_c will not be infinite [25].

When v is negative and large in magnitude, the transition becomes *first order* [26, 27]. In the limit $v \to -\infty$, the spins align exactly only along the cubic axes. This case may then be studied using a discrete spin model at $d = 2$ [28]. It turns out that in this limit the first order transition for large n is related to a fixed point of a Potts model. It should be mentioned here that quite generally any first order transition may be related to a fixed point of the renormalization group [29].

The exponent λ_v also determines the stability of the isotropic fixed point with respect to many other quartic "spin" interactions. It has recently been found [30], that many antiferromagnets have order parameters with $n > 4$, for which indeed $\lambda_v > 0$ and crossover occurs to new types of behavior.

3. Multicritical points

We now turn to the cases in which g is a non-ordering field, which can be varied experimentally.

3.1. Tricritical points

A tricritical point separates between a second order and a first order transition region on the critical line. A spin-1 Ising model, due to Blume, Emery and Griffiths [31], constructed for ^{3}He–^{4}He mixtures, indeed exhibits a tricritical point. The same model is directly applicable also to spin-$\frac{1}{2}$ Ising systems with *annealed nonmagnetic impurities* [32]. The continuous spin version of this model is also exactly soluble by renormalization group techniques at $d = 3$, yielding mean field behavior plus logarithmic corrections [33]. These corrections have not yet been observed in experiments. The same model was shown to apply, near $d = 4$, to metamagnets (uniaxial antiferromagnets in a uniform magnetic field) [34].

The two models are quite distinct for discrete spins at $d = 2$. However, discrete spin renormalization group studies of the metamagnet [29] and of the Blume–Emery–Griffiths model [35] seem to yield quite close exponents, e.g. $\nu \simeq 0.5$, $\phi \simeq 0.3$.

In addition to the behavior at the tricritical point itself, the full crossover function from tricritical to critical behavior has also been recently calculated to order ϵ [36].

It should be noted, that the cubic system discussed above exhibits a different type of tricritical point; for $n > n_c(d)$, the isotropic point ($v^* = 0$) is a tricritical point on the $T_c(v)$ line. For $v < 0$ one observes a first order transition, while for $v > 0$ one observes cubic second order behavior [6].

3.2. Bicritical and tetracritical points

These occur in general if one adds a quadratic symmetry breaking term, of the form

$$\mathcal{H}_g = \tfrac{1}{2} g \sum_X \left[S^1(X)^2 - \frac{1}{n-1} \sum_{\alpha=2}^{n} S^\alpha(X)^2 \right]. \tag{7}$$

Such terms arise, e.g. due to a uniaxially *anisotropic external stress* along the 1-axis [25]. This will lead to an alignment of the spins along the 1-axis ($g < 0$) or perpendicular to it ($g > 0$). If $\mathcal{H}_g$ appears together with $\mathcal{H}_v$ of eq. (6), then there are two possible phase diagrams in the g–T plane, as shown in fig. 1 [25]. For $v \leq 0$ one observes a *bicritical* point, as shown in case (a), while for $v > 0$ one observes a *tetracritical* point [fig. 1(b)].

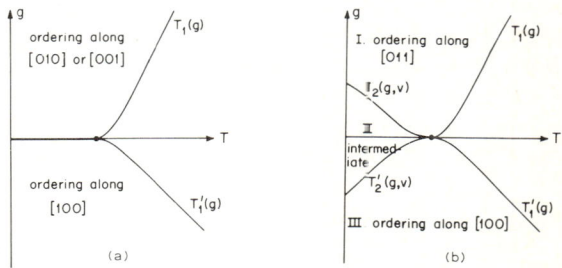

Fig. 1. (a) Bicritical point ($v \leq 0$); (b) tetracritical point ($v > 0$).

It turns out that similar Hamiltonians, and therefore similar phase diagrams, occur in many other cases. The mostly known one is that of the *spin flop bicritical point* in anisotropic antiferromagnets [37–38], where g represents the uniform magnetic field. Less investigated, but probably richer in structure, is the tetracritical point in *randomly mixed magnets* [39], where g stands for the concentration of antiferromagnetically bonded ions in the ferromagnetic material. Another interesting case has to do with isotropic antiferromagnets; when a uniform magnetic field is applied along a symmetry axis, this leads to a bicritical or a tetracritical spin flop point [40].

In all these cases, the multicritical point is described by the isotropic Heisenberg (or XY) fixed point of the renormalization group, and the crossover function (from Heisenberg to Ising or to XY) is also available [12, 41].

3.3. Other interactions

Much progress has been recently also made on the effects of *elastic degrees of freedom*. Here, g represents the coupling of the spin to the strain variables. It turns out, that whenever the specific heat exponent α is positive, these will lead to a crossover to a first order transition due to a mechanical instability [42], which can be avoided only if the boundaries are pinned. In this case, the exponents become renormalized [43]. If the material is anisotropic [42, 44], the transition is always first order.

It may be interesting to point out, that for the uniaxial dipolar case, discussed above, renormalized behavior means that eqs. (4) and (5) are replaced by [45]

$$C = \tilde{A} |\ln |t||^{-\frac{1}{3}}, \quad \chi = \tilde{\Gamma} |t|^{-1} |\ln |t||^{\frac{2}{3}}. \tag{8}$$

Much progress has also been made in the study of *random systems*, where g stands for an impurity concentration. Since this is reviewed elsewhere in this conference [46], let me only mention again the random uniaxial ferromagnet, for which eqs. (4) and (5) become [47]

$$C = -\tilde{A} |\ln |t||^{\frac{1}{2}} \exp[-2(D|\ln|t||)^{\frac{1}{2}}],$$

$$\chi = \tilde{\Gamma} |t|^{-1} \exp[(D|\ln|t||)^{\frac{1}{2}}], \tag{9}$$

where $D = 9/(81 \ln \frac{4}{3} + 53)$ is universal.

4. Conclusion

In conclusion, renormalization group theory has enabled us to produce a very large amount of explicit theoretical predictions on the various possible types of critical behavior and on the crossover phenomena among them. Although some experiments already verify some of these predictions, theory seems to be far ahead of experiment, and many experiments should be done in order to verify some of the predictions and in order to obtain actual quantitative results (Most of the calculations use the ϵ-expansion, near $d = 4$, and therefore give only qualitative results at $d = 3$).

I have benefitted from many discussions with a large majority of the people whose names appear in the list of references. Most of these are also acknowledged for sending me preprints of their work prior to publication. This work was supported by a grant from the United States – Israel Binational Foundation (BSF), Jerusalem, Israel.

References

[1] K.G. Wilson and J. Kogut, Phys. Reports 12C (1974) 75.
[2] M.E. Fisher, paper presented at the 1973 Int. Conf. on Magnetism (Moscow, August 1973) and Rev. Mod. Phys. 46 (1974) 597.
[3] S.K. Ma, Rev. Mod. Phys. 45 (1973) 589.
[4] S.K. Ma, Modern Theory of Critical Phenomena (Benjamin, Reading, Mass., 1976).
[5] C. Domb and M.S. Green, eds., Phase Transitions and Critical Phenomena, Vol. 6, (Academic, New York, 1976).
[6] A. Aharony, in ref. 5.
[7] M.E. Fisher and A. Aharony, Phys. Rev. Lett. 30 (1973) 559.
[8] A. Aharony and M.E. Fisher, Phys. Rev. B8 (1973) 3323.
A. Aharony, Phys. Rev. B8 (1973) 3342, 3349, 3358.
[9] A.D. Bruce and A. Aharony, Phys. Rev. B10 (1974) 2078.
[10] A. Aharony and A.D. Bruce, Phys. Rev. B10 (1974) 2973.
[11] A.D. Bruce, J.M. Kosterlitz and D.R. Nelson, J. Phys. C9 (1976) 825.
Th. Natterman and S. Trimper, J. Phys. C9 (1976) 2589.
[12] D.R. Nelson, paper presented at the 21st Conf. on Magnetism and Magnetic Materials, Dec. 1975, A.I.P. Conf. Proc. (in press).
[13] N. Menyuk, K. Dwight and T.B. Reed, Phys. Rev. B3 (1971) 1689.
J. Høg and J. Johanssen, Int. J. Mag. 4 (1973) 11.
[14] A. Kornblit and G. Ahlers, Phys. Rev. B11 (1975) 2678.
G. Ahlers and A. Kornblit, Phys. Rev. B12 (1975) 1938.
[15] A.I. Larkin and Khmel'nitzkii, Zh. Eksp. Teor. Fiz. 56, (1969) 2087 [Sov. Phys. – JETP 29 (1969) 1123].
[16] A. Aharony, Phys. Rev. B8 (1973) 3363; B9 (1974) 3946(E).
[17] E. Brézin and J. Zinn-Justin, Phys. Rev. B13 (1976) 251.
[18] G. Ahlers, A. Kornblit and H.J. Guggenheim, Phys. Rev. Lett. 34 (1975) 1227.
[19] J. Als-Nielsen, L.M. Holmes and H.J. Guggenheim, Phys. Rev. B12 (1975) 180.
J. Als-Nielsen, L.M. Holmes, F. Krebs Larsen and H.J. Guggenheim, Phys. Rev. B12 (1975) 191 and to be published.
[20] A. Aharony and B.I. Halperin, Phys. Rev. Lett. 35 (1975) 1308.
[21] J. Als-Nielsen, Phys. Rev. Lett. 37 (1976) 1161.
[22] A. Aharony, Phys. Rev. B8 (1973) 4270.

[23] I.J. Ketley and D.J. Wallace, J. Phys. A6 (1973) 1667.
[24] R.A. Pelcovits and D.R. Nelson, Phys. Lett. 57A (1976) 23.
E. Brézin, J. Zinn-Justin and J.C. Le Guillou, Phys. Rev. B, to appear.
[25] A.D. Bruce and A. Aharony, Phys. Rev. B11 (1975) 478.
A. Aharony and A.D. Bruce, A.I.P. Conf. Proc. 24 (1975) 296.
[26] D.J. Wallace, J. Phys. C6 (1973) 1390.
[27] J. Rudnick, unpublished.
[28] A. Aharony, J. Phys. A (in press).
[29] B. Nienhuis and M. Nauenberg, Phys. Rev. B13 (1976) 2021.
[30] R. Alben, Comp. Rend. Acad. Sci., Paris 279 (1974) B-111.
D. Mukamel, Phys. Rev. Lett. 34 (1975) 481.
D. Mukamel and S. Krinsky, Phys. Rev. B13 (1976) 5065, 5078.
P. Bak and D. Mukamel, Phys. Rev. B13 (1976) 5086.
M. Droz and M.D. Continko-Filho, paper presented at the 21st Conf. on Magnetism and Magnetic Materials, A.I.P. Conf. Proc. (in press).
T. Garel and P. Pfeuty, J. Phys. C9 (1976) L245.
[31] M. Blume, V.J. Emery and R.B. Griffiths, Phys. Rev. A4 (1971) 1071.
[32] M. Wortis, Phys. Lett. 47A (1974) 445.
[33] F.J. Wegner and E.K. Riedel, Phys. Rev. B7 (1973) 248.
[34] D.R. Nelson and M.E. Fisher, Phys. Rev. B11 (1975) 1030.
M.E. Fisher and D.R. Nelson, Phys. Rev. B12 (1975) 263.
[35] J. Adler and A. Aharony, to be published.
[36] D.R. Nelson and J. Rudnick, Phys. Rev. Lett. 35 (1975) 178.
[37] M.E. Fisher, A.I.P. Conf. Prof. 24 (1975) 273 and references therein; Phys. Rev. Lett. 34 (1975) 1634.
[38] H. Rohrer, Phys. Rev. Lett. 34 (1975) 1638; A.I.P. Conf. Proc. 24 (1975) 268.
[39] A. Aharony, Phys. Rev. Lett. 34 (1975) 590.
A. Aharony and S. Fishman, Phys. Rev. Lett. 37 (1976) 1587.
[40] D. Mukamel, Phys. Rev. B14 (1976) 1303.
[41] D.R. Nelson and E. Domany, Phys. Rev. B13 (1976) 236.
J.M. Kosterlitz, J. Phys. C9 (1976) 497.
A.D. Bruce and D.J. Wallace, J. Phys. A9 (1976) 1117.
[42] D.J. Bergman and B.I. Halperin, Phys. Rev. B13 (1976) 2145.
[43] M.E. Fisher, Phys. Rev. 176 (1968) 257.
[44] M. de Moura, T.C. Lubensky, Y. Imry and A. Aharony, Phys. Rev. B13 (1976) 2176.
[45] A. Aharony, to be published.
[46] A.B. Harris, See Contents of these Proceedings under Chapter 5.
[47] A. Aharony, Phys. Rev. B13 (1976) 2092.

EXPERIMENTAL STUDIES OF MULTICRITICAL POINTS IN MAGNETIC SYSTEMS†

W.P. WOLF

Department of Engineering and Applied Science, Yale University, New Haven, CT. 06520, USA

(Invited paper)

A brief review is given of recent experimental studies of magnetic systems displaying either tricritical or bicritical point behavior. The tricritical systems include the highly anisotropic antiferromagnets $FeCl_2$, $CsCoCl_3 \cdot 2D_2O$ and $Dy_3Al_5O_{12}$, while the bicritical systems include the weakly anisotropic antiferromagnets MnF_2, $GdAlO_3$ and $NiCl_2 \cdot 6H_2O$. There have also been Monte-Carlo computer studies on both kinds of systems. The results provide significant tests for a number of theoretical predictions and in general good agreement is found. Most of the remaining discrepancies can probably be attributed to experimental difficulties, which are quite severe in studies of this kind.

1. Introduction

A multicritical point (MCP) is a special point on phase boundary at which competitition between two or more types of order leads to the new kind of cooperative behavior [1, 2]. They are found in many different physical systems including mixtures of liquid ^3He-^4He [3], multicomponent fluid mixtures [4], structural phase transitions [5], liquid crystals [6] and ferroelectrics [7], as well as in various magnetic materials, and one reason for their interest is the general similarity in behavior of such widely differing systems. The special appeal of magnetic materials for studies of this kind is the ready access to appropriate fields and temperatures to explore the phase diagrams in the region of possible multicritical points. The direct relation of magnetic spin systems to simple microscopic models is also useful for theoretical comparisons.

There have been a number of interesting theoretical predictions [1, 2, 8–19] for multicritical behavior, but we shall here restrict ourselves to tricritical and bicritical systems, since these are the only kinds of magnetic multicritical systems for which detailed experimental results have been reported. The predictions may be discussed under several headings, even though they mostly stem from the same basic concepts.

One class of predictions concerns the *shapes* of the phase boundaries as the multicritical point is approached [10, 13, 18]. Using renormalization group techniques together with the results of series expansions, one finds phase boundaries such as those shown in fig. 1. The

†Supported in part by AROD and by NSF grant DMR 76-23102.

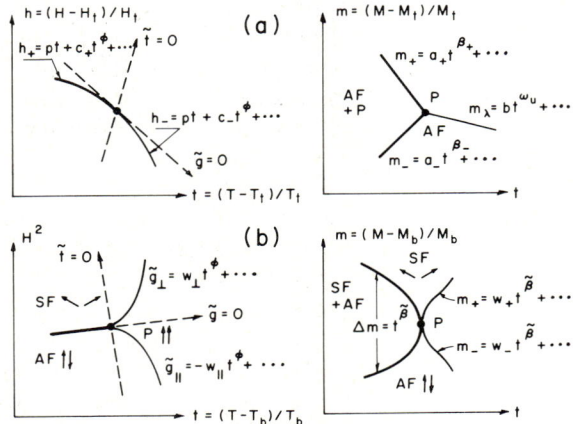

Fig. 1. Theoretical predictions for phase boundaries of (a) tricritical and (b) bicritical systems. Heavy lines denote first order transitions. $\bar{g} = 0$ and $\tilde{t} = 0$ denote the axes of the appropriate scaling fields. The orientation of $\bar{g} = 0$ is fixed by the geometry of the figure but the slopes of the $\tilde{t} = 0$ axes are nonuniversal. For tricritical systems, the exponents defining the phase boundaries are given by $\phi = 2$, $\beta_\pm = \omega_u = 1$, as in classical theory [1], but the amplitudes a_+ and b are not equal [16, 18]. For bicritical systems, ϕ and $\tilde{\beta}$ take on nonclassical but universal values (see section 4) and the phase boundaries must be expressed in terms of the appropriate scaling variables $\bar{g} = H^2 - H_b^2 - pt$ and $\tilde{t} = t + q(H^2 - H_b^2)$ [14].

differences from classical mean field (Landau) theory are quite striking and they can readily be tested.

The theory also predicts the variation of thermodynamic functions along different paths approaching a multicritical point [9, 13]. These paths are classified according to their relation to the *scaling axes* which are in turn determined by the form of the phase diagram [3, 9, 14, 15]. In general, one expects singular behavior which can be characterized by one or more *critical exponents*, though there may also be singular terms involving logarithms which cannot be represented by a simple exponent [17]. At this time there is still no direct evidence for any such logarithmic terms and we shall discuss the experimental results in terms of the more usual exponent laws. It should be noted, however, that even though the log terms are not observed directly, they may well affect the values of fitted critical exponents, which might account for some apparent discrepancies in the analysis of experimental results [20]. Values for various predicted exponents will be given in tables I and II together with the available experimental results.

As in the case of ordinary critical points, it is believed that the exponents are not all independent, and by extending the usual scaling hypothesis one can derive a number of relations between them [9, 11]. Scaling theory also provides a basis for constructing *equations of state* which allow the combination of different experimental measurements into a single curve describing the behavior around the multicritical point [8, 9, 11, 19]. There have recently also been some attempts to calculate the detailed form of these scaling functions [18, 19], but so far there have been no detailed comparisons of these results with any magnetic experiments.

The theory also makes some general predictions concerning the *universality* of different types of behavior which can be tested [15]. In particular, one expects to find similar behavior at all multicritical points of a given kind so that, as in the case of ordinary critical points, one should find similar exponents in widely differing materials.

We can clearly not discuss all of these predictions here, and further details may be found in the references noted and in the references cited in these papers. However, it may be useful to comment briefly on one limitation of the theory which may affect the interpretation of experiments. As in the case of ordinary critical point theory, the predictions relate principally to the *asymptotic* limit as the multicritical point is approached and there are presently few indications for the size of higher order terms. The region over which the asymptotic predictions are expected to be valid is therefore undefined and some of the observed discrepancies may be due to nonasymptotic behavior rather than fundamental inadequacies in the theory or the experiments. This problem is particularly serious in the case of multicritical points since experimental difficulties make it very difficult to obtain unambiguous results very close to the MCP and one is, as always, faced with fitting several parameters over a finite range of variables.

2. Experimental problems

From the phase diagrams shown in fig. 1 it can be seen that the experimental location of either a tricritical or a bicritical point involves finding the point at which a first order discontinuity goes over into a second order singularity. In practice, this can be quite difficult, since the corresponding susceptibility close to the first order transition diverges as the MCP is approached, thereby masking the vanishing discontinuity. This problem is complicated by demagnetizing effects, especially in irregularly shaped samples, and by the inevitable "rounding" which is found in all real crystals.

There have been several different approaches to overcome this problem, and we shall mention these in the discussion of specific materials in sections 3 and 4. However, it is probably important to remark at this point that none of the presently used techniques is entirely satisfactory and that one may expect a significant improvement in the overall comparison of theory and experiment if this problem could be resolved.

The second major problem, which affects particularly bicritical point studies, concerns the alignment of the field relative to the preferred easy-axis of the magnetic system. This has been discussed by Rohrer [21] and by Fisher [15] and it remains as a major source of uncertainty in present day studies, even though alignments as close as 10^{-4} deg. have been achieved [22]. The solution to this problem is probably to find materials in which the alignment is not so critical.

A third, and perhaps trivial difficulty which confronts the experimentalist, especially in the study of tricritical systems, is the bewildering variety of notations which have been used by theorists to express their predictions. An attempt was made by Griffiths [12] to propose a unified notation for tricritical points, but unfortunately he did not explicitly consider all of the cases encountered experimentally and a different notation based on Riedel's original scaling ideas [9] is still in use [3]. We shall here use Griffiths' notation supplemented by Riedel's where necessary. Useful summaries of definitions have also been given in refs. 23 and 24. For bicritical points the notation problem is not so acute, perhaps because there has not yet been so much work in this field.

3. Experimental results for tricritical systems

Detailed experiments have been reported for three materials and there have also been computer studies using Monte-Carlo techniques.

a) *FeCl$_2$*. This is a well known and simple metamagnet which unfortunately suffers from the disadvantage of poor mechanical and chemical properties. This makes it hard to control demagnetizing effects and the tricritical studies have therefore used neutron scattering [25] and optical techniques [26] to measure the magnetization, since these methods permit shielding of the sample edges. In the neutron experiments it is of course possible to measure also the antiferromagnetic order parameter and the critical fluctuations close to the phase boundaries. The published results are summarized in table I. It can be seen that there is general agreement, but there is also some disagreement with theory and in one case, between the different experiments. The reasons for these differences are not known at this time. In addition to the measurements in the immediate vicinity of the tricritical point, the neutron experiments also yielded values for the critical exponent at different points along the second order phase boundary [25]. The results were found to be consistent with the hypothesis of smoothness, which predicts that the critical exponents should be the same for different paths of approach towards the critical line.

b) *CsCoCl$_3$·2D$_2$O*. This is a somewhat more complicated four sublattice antiferromagnet of which only relatively small irregular crystals are available [24]. The only tricritical point studies so far have used neutron scattering [24, 27, 28], which again allows masking of sample edges. The results in table I can be seen to be in relatively poor quantitative agreement with theory, though the general behavior was as expected. It is not clear at this time whether the disagreement lies outside the experimental uncertainties, which are quite large for this material. The measurements for CsCoCl$_3$·2D$_2$O have also been tested for scaling and good data collapsing was found. See also ref. 28. As in the case of FeCl$_2$, the results along the critical line above the tricritical point were found to be consistent with the smoothness hypothesis.

c) *Dy$_3$Al$_5$O$_{12}$*. Dysprosium aluminum garnet (DAG) is an Ising-like six sublattice cubic an-

Table I. Experimental and theoretical tricritical exponents[a]

	α_t	β_+	β_-	β_u	γ_u	δ_u	ϕ	β_1	β_t	β_t^*
FeCl$_2$ Neutrons[b]	—	1	0.36 ±0.04	—	—	—	~2	0.19 ±0.02	—	0.36 ±0.04
Optical[c]	—	1.03 ±0.05	1.13 ±0.14	1.11 ±0.11	—	—	~2	—	—	—
CsCoCl$_3$·2D$_2$O[d]	0.65 ±0.05	0.7 ±0.4	0.7 ±0.4	0.65 ±0.20	~1.3	3±1	2±0.2	0.3 ±0.1	0.15 ±0.02	0.36 ±0.05
Dy$_3$Al$_5$O$_{12}$: H∥{110}[e]	—	~1	~1	0.98 ±0.05	1.01 ±0.07	2.1 ±0.25	1.95 ±0.11	—	—	—
Theory[f]	1/2	1	1	1	1	2	2	1/2	1/4	1/2

[a] Notation based on ref. 12. See also refs. 23 and 24.
[b] Ref. 25.
[c] Ref. 26.
[d] Refs. 24 and 28.
[e] Refs. 31 and 32.
[f] After ref. 10. See also refs. 3 and 23.

tiferromagnet, and early experiments with fields applied along {111} indicated significant discrepancies from tricritical point theory [29]. It was subsequently recognized that the symmetry of this material allows an "induced staggered field" if $H_x H_y H_z \neq 0$, so that for $H\|\{111\}$ one actually observes a "wing critical point" [30]. Later experiments with $H\|\{110\}$, located a true tricritical point and the results [31, 32] in table I can be seen to be in very good agreement with the theory. The measurements were also found to be consistent with scaling [33].

The method used for these studies was based on the difference in the transient response of domain states in the first order region from that of the homogeneous antiferromagnetic or paramagnetic states. So far, only measurements of the magnetization have been reported and it would now be interesting to study the antiferromagnetic order parameter.

d) *Computer studies*. These are free of the usual experimental problems of demagnetizing effects and impurities, but they suffer severe limitations due to finite sample size. The largest system studied so far had $20 \times 20 \times 20$ spins, for which extrapolation to the $N = \infty$ limit is still quite hard. However, the results [34] were generally consistent with the theory and with larger computers one should be able to make quite detailed tricritical studies in this way.

One particular advantage of computer studies is that it is easy to apply "staggered" as well as uniform fields, so that the entire $H-H_s-T$ phase diagram can be investigated [34].

4. Experimental results for bicritical systems

Results for bicritical systems are much more limited and in particular there have not yet been any neutron studies of the order parameter. There are again measurements on three materials and one computer study.

a) MnF_2. This is one of the best known examples of a Heisenberg-like ($n = 3$) antiferromagnet with weak anisotropy. Its bicritical point at $H_b \approx 11.8$ T and $T_b \approx 65$ K is somewhat hard to study, especially since extremely accurate alignment of the sample relative to the field is required. In the experiments of King and Rohrer [22] the alignment was estimated to be better than 10^{-4} deg., a nontrivial achievement. The phase boundaries were located using both differential susceptibility and NMR measurements, but in both cases there were some questions of interpretation which increase the overall uncertainty of the analysis. In addition to the cross-over exponent ϕ, the results give values for $Q = \omega_\perp/\omega_\|$, the ratio of two amplitudes describing the second order phase boundaries and q, the slope of the optimal scaling axis $\tilde{t} = 0$ (see fig. 1b). The fitted values shown in table II can be seen to be in quite good agreement with the theory, though the overall uncertainties are still quite large.

Table II. Experimental and theoretical bicritical exponents

	ϕ	$Q = \omega_\perp/\omega_\|$	q	$\tilde{\beta}$	$\tilde{\gamma}$
$n = 3$: $MnF_2^{(a)}$	1.29 1.26	1.25 1.75	1.38 1.06	—	—
Theory[b]	1.25 ±0.015	2.5	~1.35	0.85 ±0.07	0.40 ±0.09
$n = 2$: $GdAlO_3^{(c)}$	1.15 ± 0.08 1.25 ± 0.07	—[d]		0.92 ±0.03	~0.15
$NiCl_2 \cdot 6H_2O^{(e)}$	~1.18	—[d]		—	—
Theory[b]	1.175 ±0.015	1		0.84 ±0.05	0.33 ±0.06

[a] Ref. 22. Two estimates for phase boundaries.
[b] Refs. 13–15.
[c] Ref. 21. Two alignments ~0.14 deg. and ~0.08 deg.
[d] Fits assuming $\omega_\perp/\omega_\| \equiv 1$.
[e] Ref. 36.

b) *GdAlO₃*. This is an orthorhombic antiferromagnet [21] and one would expect its bicritical point to be described by a two-component order parameter ($n = 2$). The theory for $n = 2$ bicritical points predicts $\omega_\perp/\omega_\parallel \equiv 1$ and this was used in fitting the data. From previous work one expects GdAlO₃ to be not as sensitive to orientation as MnF₂, but experiments with estimated misalignments of only 0.14 deg. and 0.08 deg. still showed significant differences [21]. However, the results in table II are in generally good agreement with the theory, and they certainly confirm that $\phi > 1$, the classical value. Analysis of the magnetization leads to a value of $\tilde{\beta}$ (corrected for misalignment) which is in good agreement with theory, and an estimate of the susceptibility exponent $\tilde{\gamma}$ which appears to be rather low, though the uncertainties are quite large. More recent experiments on GdAlO₃ [35] have studied the variation of the phase boundaries as a function of the orientation of the applied field and found evidence for a predicted *line* of bicritical points for one particular plane. The corresponding bicritical exponents have not yet been reported.

c) *NiCl₂·6H₂O*. This is a monoclinic crystal which has rather more complicated properties and so far only the phase boundary has been studied near the bicritical point [36]. The results were found to be consistent with the theory for an $n = 2$ bicritical point, but no detailed analysis has been reported.

d) *Computer studies*. Monte-Carlo studies of bicritical points are even more difficult than those on tricritical points because the three dimensional nature of the spins leads to much slower convergence of the statistical iteration process. So far, only preliminary results have been reported [37]. These are consistent with the theory, but more work is clearly needed.

5. Other systems

There are many other magnetic materials which show field induced phase transitions and for some of these one would expect to find multicritical point behavior similar to that discussed above. In some of these cases one might hope to find situations which are less sensitive to some of the experimental problems which have limited experiments so far.

In some cases one may expect to find new kinds of behavior. For example, results which have been reported for DyPO₄ [38–40], FeBr₂ [41] and MnO under pressure [42] all show features which do not fit obviously, at this time, into the simple types of multicritical points illustrated in fig. 1. There are also predictions of tetracritical behavior [13, 42–44], and the onset of spiral order (Lifshitz point) [45–47] which have not yet been studied experimentally in any detail.

It would seem clear that there is still a considerable amount of experimental and theoretical work to be done in this field.

It is a pleasure to thank R.J. Birgeneau, R.B. Griffiths, N. Giordano, D. Jasnow, D.P. Landau, H. Meyer and H. Rohrer for their help in preparing this paper. I am also grateful to A.L.M. Bongaarts for sending me preprints of his work.

References

[1] L.D. Landau, Phys. Z. Sowjetunion 11 (1937) 26. Reprinted in Collected Papers of L.D. Landau, D. ter Haar, ed. (Pergamon, London, 1965) p. 193.
[2] See for example, A. Aharony, Physica 86–88B (1977) 545, Proceedings of International Conference on Magnetism, Amsterdam, 1976 and references cited therein.
[3] See for example, E.K. Riedel, H. Meyer and R.P. Behringer, J. Low Temp. Physics 22 (1976) 369.
G. Ahlers in *The Physics of Liquid and Solid Helium*, Part I, K.H. Bennemann and J.B. Ketterson, eds. (Wiley, New York, 1976) p. 85 and references cited in these papers.
[4] See for example, B. Widom, J. Phys. Chem. 77 (1973) 2196.
[5] See for example, C.W. Garland, D.E. Bruins and T.J. Greytak, Phys. Rev. B12 (1975) 2759.
[6] P.H. Keyes, H.T. Weston and W.B. Daniels, Phys. Rev. Lett. 31 (1973) 628.
[7] P.S. Peercy, Phys. Rev. Lett. 35 (1975) 1581.
[8] R.B. Griffiths, Phys. Rev. Lett. 24 (1970) 715.
[9] E.K. Riedel, Phys. Rev. Lett. 28 (1972) 675.
[10] E.K. Riedel and F.J. Wegner, Phys. Rev. Lett. 29 (1972) 349.
[11] A. Hankey, H.E. Stanley and T.S. Chang, Phys. Rev. Lett. 29 (1972) 278.
[12] R.B. Griffiths, Phys. Rev. B7 (1973) 545.
[13] M.E. Fisher and D.R. Nelson, Phys. Rev. Lett. 32 (1974) 1350.
[14] M.E. Fisher, Phys. Rev. Lett. 34 (1975) 1634.
[15] M.E. Fisher, Proceedings of 20th Conference on Magnetism and Magnetic Materials, AIP Conf. Proc. 24 (1975) 273, and references cited therein.
[16] V.J. Emery, Phys. Rev. B11 (1975) 3397.
[17] M.J. Stephen, E. Abrahams and J.P. Straley, Phys. Rev. B12 (1975) 256, and references cited therein.

[18] D.R. Nelson and J. Rudnick, Phys. Rev. Lett. 35 (1975) 178.
[19] D.R. Nelson, Proceedings of 21st Conference on Magnetism and Magnetic Materials, AIP Conf. Proc. 29 (1976) 450, and references cited therein.
[20] See for example, W.B. Yelon, D.E. Cox, P.J. Kortman and W.B. Daniels, Phys. Rev. B9 (1974) 4843.
[21] H. Rohrer, Phys. Rev. Lett. 34 (1975) 1638, and references cited therein.
[22] A.R. King and H. Rohrer, Proceedings of 21st Conference on Magnetism and Magnetic Materials, AIP Conf. Proc. 29 (1975) 420; ibid p. 456.
[23] J.M. Kincaid and E.G.D. Cohen, Physics Reports 22C (1975) 57.
[24] A.L.M. Bongaarts, thesis, University of Eindhoven (1975), unpublished.
[25] R.J. Birgeneau, G. Shirane, M. Blume and W.C. Koehler, Phys. Rev. Lett. 33 (1974) 1098, Proceedings of the 20th Conference on Magnetism and Magnetic Materials, AIP Conf. Proc. 24 (1975) 258, and to be published.
[26] J.A. Griffin and S.E. Schnatterly, Phys. Rev. Lett. 33 (1974) 1576 and AIP Conf. Proc. 24 (1975) 195.
[27] A.L.M. Bongaarts, W.J.M. de Jonge and P. van der Leeden, Phys. Rev. Lett. 37 (1976) 1007.
[28] A.L.M. Bongaarts and W.J.M. de Jonge, Proceedings of International Conference on Magnetism, Amsterdam, 1976, Physica 86–88B (1977) 595, 671.
[29] W.P. Wolf, D.P. Landau, B.E. Keen and B. Schneider, Phys. Rev. B5 (1972) 4472.
[30] M. Blume, L.M. Corliss, J.M. Hastings and E. Schiller, Phys. Rev. Lett. 32 (1974) 544.
[31] N. Giordano and W.P. Wolf, Phys. Rev. Lett. 35 (1975) 799.
[32] N. Giordano, Phys. Rev. B14 (1976) 2927.
[33] N. Giordano and W.P. Wolf, Proceedings of 21st Conference on Magnetism and Magnetic Materials, AIP Conf. Proc. 29 (1976) 459.
[34] D.P. Landau, Phys. Rev. B14 (1976) 4054.
[35] H. Rohrer, B. Derighetti and Ch. Gerber, Proceedings of International Conference on Magnetism, Amsterdam, 1976, Physica 86–88B (1977) 597.
[36] N.F. Oliveira, Jr., A. Paduhan Filho and S.F. Salinas, Phys. Lett. 55A (1975) 293.
[37] D.P. Landau and K. Binder, Proceedings of 21st Conference on Magnetism and Magnetic Materials, AIP Conf. Proc. 29 (1976) 461.
[38] J.E. Battison, A. Kasten, M.J.M. Leask and J.B. Lowry, Solid State Comm. 17 (1975) 1363.
[39] P.J. Becker and I.R. Jahn, J. Phys. C: Solid State Phys. 9 (1976) L505.
[40] M. Régis, J. Ferré, Y. Farge and I.R. Jahn, Proceedings International Conference on Magnetism, Amsterdam, 1976, Physica 86–88B (1977) 599.
[41] J.M. Kincaid and E.G.D. Cohen, Phys. Lett. 50A (1974) 317, and references cited therein.
[42] P. Bak, S. Krinsky and D. Mukamel, Phys. Rev. Lett. 36 (1976) 829, and references cited therein.
[43] D. Mukamel, Phys. Rev. B14 (1976) 1303.
[44] A.D. Bruce and A. Aharony, Phys. Rev. B11 (1975) 478.
[45] R.M. Hornreich, M. Luban and S. Shtrikman, Phys. Rev. Lett. 35 (1975) 1678.
[46] M. Droz and M.D. Coutinho-Filho, AIP Conf. Proc. 29 (1976) 465.
[47] J.F. Nicoll, G.F. Tuthill, T.S. Chang and H.E. Stanley, Proceedings of the Joint MMM-Intermag Conference, 1976, to be published.

CRITICAL PROPERTIES OF THE XY MODEL

D.D. BETTS
Theoretical Physics Institute, University of Alberta, Edmonton, Canada

(Invited paper)

The spin one half XY model is of interest as the probably simplest quantum mechanical many body system, as a model for liquid ^{4}He near the lambda transition, as a model for a class of antiferromagnetic insulators near T_N and as a model for granular superconductors. The latest estimates of the critical properties of the three dimensional spin one half XY model are presented. The evidence for and nature of a phase transition in the two dimensional XY model are discussed. Recent experimental measurements on the compounds $CoCl_2 \cdot 6H_2O$ and $CoBr_2 \cdot 6H_2O$ and the compounds $CO(C_5H_5NO)_6 (ClO_4)_2$ and $Co(C_5H_5NO)_6 (BF_4)_2$ are compared with theoretical predictions for the spin 1/2 XY model on the square and simple cubic lattices, respectively. Renormalization group calculations are touched on.

1. Introduction

A rather general model for the interaction of magnetic ions localized on the sites of a lattice is specified by the interaction Hamiltonian of the anisotropic Heisenberg model expressed in terms of pseudospin variables S^α,

$$\mathcal{H}_0 = -\frac{3}{2S(S+1)} \sum_{i,j} [J_{ij}^\perp (S_i^x S_j^x + S_i^y S_j^y) + J_{ij}^\parallel S_i^z S_j^z]. \quad (1)$$

The indices i and j label the sites of the lattice while $J_{ij}^\perp$ and $J_{ij}^\parallel$ are "exchange" energy parameters. In theoretical studies calculational simplicity is achieved by studying the special cases of the Ising model, $J^\perp = 0$, the XY model, $J^\parallel = 0$, or the isotropic Heisenberg model, $J^\perp = J^\parallel$.

The above restrictions are not so severe as they might seem because of the *principle of universality* [1]. According to this principle all systems which undergo second order phase transitions fall into a small number of *universality classes* distinguished by a very few parameters notably the dimensionality, d, of the system and the dimensionality, n, of the order parameter. In the critical region the thermodynamic properties of all systems of the same universality class are identical apart from typically two scale factors. The universality principle has been well verified so far [2], so we concentrate on the nearest neighbour, spin 1/2 XY model as a simple example of an $n = 2$ system.

As the properties of the XY model have been reviewed [3] recently, we emphasize subsequent developments. In section 2 we outline the relation of the spin $1/2\ XY$ model to a class of magnetic insulators, to liquid helium and to granular superconductors. The dimensionality of the lattice is of supreme importance with regard to the existence and nature of a phase transition for the XY model. The one dimensional model is paramagnetic at all temperatures, while there seems little doubt that the three dimensional model undergoes a second order phase transition at a finite temperature to a phase with conventional long range order. The existence of a phase transition, the nature of a low temperature phase and the type of transition are all controversial questions for the two dimensional model, which we consider last.

A thorough review has been published [4] on magnetic systems as realizations of the $d = 1, 2$ and 3 Ising, XY and Heisenberg models. A very recent exciting experimental development, discussed below, has been the investigation of the properties of good examples of systems approximating both the two and the three dimensional spin one half XY models.

2. Experimental realizations of the XY model

A magnetic system of pseudo spins governed by the Hamiltonian (1) will belong to the $n = 2$ universality class of the XY model provided $J^\perp > J^\parallel$. The examples recently studied in detail include [5] the antiferromagnetic layer ($d = 2$) compounds, $CoCl_2 \cdot 6H_2O$ and $CoBr_2 6H_2O$, and [6] the antiferromagnetic pyradine compounds $Co(C_5H_5NO)_6 (ClO_4)_2$ and $Co(C_5H_5NO)_6(BF_4)_2$ in which the Co^{2+} ions form a nearly simple cubic magnetic lattice.

In terms of the true spin, $\tilde{S} = 3/2$, of the Co^{2+}, we may assume [6] a Hamiltonian containing the basic isotropic Heisenberg exchange term and a single ion anisotropy due to a local axial crystalline field,

$$\tilde{\mathcal{H}} = -2\tilde{J}\sum_{\langle i,j\rangle} \tilde{S}_i \cdot \tilde{S}_j - \tilde{D}\sum_i [(\tilde{S}_i^z)^2 - \tilde{S}(\tilde{S}+1)/3)]. \quad (2)$$

The effect of the single ion term is to split in energy the Kramers doublet, only the lowest doublet being appreciably populated, so that there is an effective spin, $S = 1/2$, in the interesting temperature range. The components of the effective spin are related to those of the true spin by $S^\alpha = 2\tilde{S}^\alpha/g^\alpha$. On comparing (1) and (2) we see that $J^\perp = (g^\perp/2)^2 \tilde{J}$ and $J^\| = (g^\|/2)^2 \tilde{J}$. In the above compounds $J^\| \approx 0.3 J^\perp$. In the halogen compounds there is a slight easy plane anisotropy, $J^y/J^x \approx 0.96$, important only as $T \to T_c$.

The $S = 1/2$ anisotropic Heisenberg model was derived [7] as a model of liquid ^{4}He in the vicinity of the lambda transition. It is necessary that $J^\perp > J^\|$ for the order parameter to have the appropriate dimensionality, $n = 2$. The superfluid order parameter, ψ, of liquid ^{4}He is the analogue of the perpendicular magnetization, $M^\perp$, of an XY like ferromagnet. The conjugate field, η, which is physically unrealizable, is the analogue of $H^\perp$. The axial field, H^z, of the ferromagnet is analogous to the pressure or chemical potential in ^{4}He while the axial magnetization, M^z, is linearly related to the density of ^{4}He [3].

For a quite different realization of the $S = 1/2$ XY model [8] consider a regular lattice of spherical grains of superconducting material embedded in a nonconducting matrix. As the temperature is lowered each grain will undergo, at T_c, a transition to a superconducting state which should become fully coherent for grain size $a \gg \xi(T)$, the coherence length. The phases of the superconducting order parameter in different grains should remain at first completely random. All electrons are coupled into spin zero Cooper pairs which can tunnel from one grain to another governed by a tunneling Hamiltonian,

$$\mathcal{H} = -\sum_{i,j} v_{i,j}(S_i^+ S_j^- + S_i^- S_j^+), \quad (3)$$

where $S_i^\pm = \Sigma_k S_{ik}^\pm$, $S_{ik}^\pm$ being the Cooper pair creation and annihilation operators for momentum k in grain i. This is precisely the $S = 1/2$ XY model, and thus a further phase transition is expected at a temperature $T_0(<T_c)$ below which there will be long range order in the sense that phases of the superconducting order parameters in the individual grains achieve a long range correlation.

Finally certain ferrodistortive lattice dynamical models with quartic anisotropy [9], though not quite XY models, nevertheless seem to belong to the same universality class.

3. Critical properties of the three dimensional XY model

The order parameter of the XY model is the perpendicular or transverse magnetization $M^\perp = (M^x, M^y)$ for ferromagnets or the staggered transverse magnetization, $N^\perp$ for antiferromagnets where $M^\alpha = \mu_B g^\alpha \Sigma S_i^\alpha$. For an XY ferromagnet quantities of particular interest and their expected critical behaviour include:

(i) the reduced specific heat,

$$C_H/Nk_B \approx \begin{cases} A(1 - T_c/T)^{-\alpha} & T > T_c \\ A'(1 - T/T_c)^{-\alpha'} & T < T_c, \end{cases} \quad (4a)$$

(ii) the fluctuation in the long range order,

$$Y = \langle (M^x)^2 \rangle / N(\mu_B g^\perp)^2 \approx \begin{cases} C(1 - T_c/T)^{-\gamma} & T > T_c \\ C'(1 - T/T_c)^{-\gamma} & T < T_c. \end{cases}$$
(4b)

The perpendicular susceptibility $\chi^\perp = \partial M^x/\partial H^x$ will behave in the critical region like Y. The values of the amplitudes and critical temperatures depend on the specific properties of the model such as the underlying lattice and the spin value and likewise for experimental systems vary from substance to substance, but the critical exponents should have the values of the simple XY model for all models and systems of the same universality class.

The methods of deriving high temperature series expansions and the actual coefficients as far as they were then known have already been reviewed [3]. Dekeyser and Rogiers [10] have recently extended the high temperature series for the free energy and the transverse magnetization fluctuation, Y, to include arbitrary longitudinal field, H^z. Their analysis of the fluctuation series in zero field, which contains one additional coefficient, confirming the previous estimate, yields $\gamma = 1.333 \pm 0.01$. It is attractive to assume that $\gamma = 4/3$ exactly.

The high temperature specific heat series is an even function of T for loose packed lattices, so the only existing series in three dimensions which can be analyzed directly is that for the f.c.c. lattice [3], for which $\alpha = 0.02 \pm 0.05$. It is attractive to assume a logarithmic singularity ($\alpha = 0$). Scaling laws then give for the other exponents $\beta = 1/3$, $\delta = 5$, $\Delta = 5/3$, $\nu = 2/3$ and $\eta = 0$. The amplitude, A^F, and additive constant, B_0^F have also been estimated [3] in

$$C_H^F/Nk_B \approx -A^F \ln(1 - T_c^F/T) + B_0^F. \quad (5)$$

It is now of experimental interest to have the specific heat curve for the simple cubic (S) lattice. Assuming the same form, (5), for the s.c. lattice and adopting the above values for the critical exponents and estimates for C^F, A^F and C^S one finds [11] with the aid of one scale factor universality [12] that $A_S = 0.319 \pm 0.005$.

The additive constant, B_0^S, can next be found by evaluating at K_C Padé approximants to the series for the specific heat with the logarithmic singularity subtracted off. Unfortunately there seem to be errors in the last two coefficients of the original series [13] so we reproduce the revised series in its entirety.

$$C_H^S/Nk_B = (3/4)K^2 + (1/32)K^4 + (51/180)K^6 \\ + 8.60052077\,K^8 + 65.7598958\,K^{10} + \cdots \quad (6)$$

From the revised series $B_0^S = -0.39 \pm 0.05$, only slightly larger than the previous estimate [11].

Although the critical properties of the XY model for $T < T_c$ cannot yet be predicted, the simple spin wave theory result that $C_H \propto T^3$ at low temperature is useful.

4. Nature of the two dimensional XY model

A conventional type of phase transition would involve a transition at a temperature T_c between a low temperature phase having a nonzero thermal expectation value of some long range order parameter, O, and a high temperature phase for which $\langle O \rangle = 0$. For a square lattice some possible types of long range order are illustrated in fig. 1. Fig. 1(a) illustrates the Néel state of longitudinal staggered magnetization, N^z, with spins pointing alternatively up and down as for an Ising antiferromagnet. Fig. 1(b) illustrates a state of nonzero transverse magnetization, $M^\perp$, while fig. 1(c) shows a state of non-zero transverse staggered magnetization, $N^\perp$. Finally fig. 1(d) illustrates a vorticity state in which clockwise and counter clockwise vertices form a lattice of spacing 2δ.

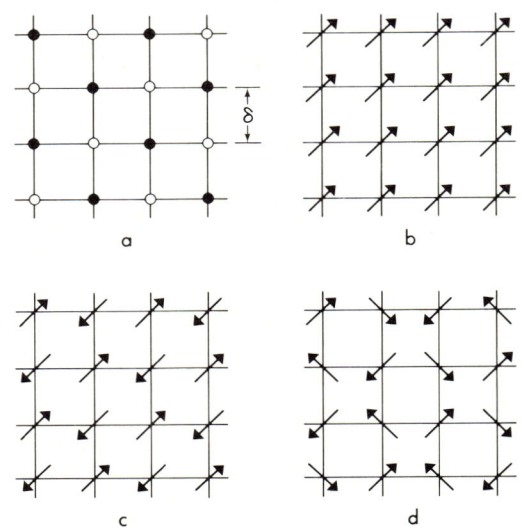

Fig. 1. Possible ordered states of the two dimensional XY model. (a) staggered longitudinal magnetization, N^z, (Néel state), (b) transverse magnetization, M^x, (c) staggered transverse magnetization, N^x, (d) longitudinal vorticity, V^z.

For a class of two dimensional spin models including the XY and isotropic Heisenberg models no phase of nonzero magnetization or staggered magnetization exists for any zero temperature [14]. Meanwhile Stanley and Kaplan [15] conjectured, on the basis of 6 term high temperature susceptibility series [16], that some sort of phase transition takes place at a finite T_c for the isotropic Heisenberg model for $S > 1/2$. The low temperature phase would be characterized by correlations which decay with distance sufficiently slowly to yield an infinite susceptibility without a finite spontaneous magnetization. On the basis of longer series [17] a Stanley–Kaplan transition in the two dimensional isotropic Heisenberg model ($n = 3$) now seems unlikely for any S.

Models with $n = 2$ which have been investigated by high temperature series expansion include the plane rotator model [18], the classical XY model [19] and the $S = 1/2\ XY$ model [20]. All of these models show stronger evidence of a divergent susceptibility at a finite T_c than does the isotropic Heisenberg model. One puzzling feature however is that the *same* methods of analysis yield $\gamma \simeq 3.0$ for the $S = \infty$ models and $\gamma \simeq 1.5$ for the $S = 1/2$ model.

For the $d = 2$ classical XY model Berezinskii [21] has argued that for spins separated sufficiently far apart the relative phase may greatly exceed 2π. Kosterlitz and Thouless [22] introduced the concept of a low temperature phase of "topological long range order" in which point dislocations or vortices of opposite Burgers vector are bound together. Above a transition temperature $T_c \simeq \pi J^\perp / k_B$ the dislocations become unbound and isolated vortices will be found. Fig. 1(d) illustrates a simple form of topological long range order. Kosterlitz [23] then predicted that the susceptibility would have the unusual critical behaviour,

$$\chi \approx \chi_0 \exp A(1 - T/T_c)^{-\nu} \quad (7)$$

rather than the conventional power law divergence. The series for the classical, $d = 2$, $n = 2$ models favours (7) over a power law [24], but for $S = 1/2$ we favour a power law divergence.

Contrary to general belief the Bethe–Peierls approximation to the solution of the $S = 1/2$ Ising model has been found *not* to become exact on the Cayley tree [25] but rather the free energy is of the form [26].

$$F(T, H) = F_{reg}(T, H^2) + G(T)|H|^{\kappa(T)} \quad (8)$$

to lowest order in the field H. The exponent κ varies from 1 at $T = 0$ to ∞ at $T = T_\infty$. For $T > T'$ where $T' \geqslant T_\infty$ $F(T, 0)$ is analytic. Thus there is a line of phase transitions varying from first order at $T = 0$ to infinite order at $T = T_\infty$.

Zittartz [27, 28] has shown that the $d = 2$ plane rotator model undergoes a *continuous phase transition* of the above type. As a consequence there are the following properties all of which are in accord with previous information about the two dimensional XY model: (i) the spontaneous magnetization vanishes identically [14]; (ii) the specific heat is analytic at all temperatures [29]; (iii) at low temperatures the singular exponents should depend on temperature [21]; (iv) the initial susceptibility should diverge at a temperature T_2 such that $\kappa(T_2) = 2$ [18–20].

We have made a detailed study of the ground state properties of clusters of 4, 8 and 16 spins coordinated as in a square lattice with periodic boundary conditions [30]. By extrapolation we estimate the ground state energy per bond for the infinite square lattice, $-E_0/2NJ^\perp \approx 0.54$ compared with the mean field value for the completely aligned state of 0.50. For the 16 spin "lattice" we have computed exactly all inequivalent two spin correlations. The correlations $\langle \sigma_0^x \sigma_r^x \rangle$ for five inequivalent values of r are all of the order of $+0.5$ and fall very slowly with distance. The nearest neighbour correlation, $\langle \sigma_0^z \sigma_\delta^z \rangle \approx -0.18$ and others are negligible. The ground state is more complex than any of the states of fig. 1 but closest to fig. 1a.

The renormalization group method [31, 32] provides an alternative to series expansion methods for investigating phase transitions and critical phenomena. The approach of Niemeijer and van Leeuwen [33] in real space is most convenient for $d = 2$, finite S, and can be adapted to $n > 1$ systems such as the $S = 1/2\ XY$ model.

For the plane rotator model Lublin [34] finds a T_c 30% lower than the series estimate [18] and critical exponents $\gamma = 2.4 \pm 0.5$ and $\alpha = -0.7 \pm 0.3$. The estimate of γ is not inconsistent with the series estimate. $\alpha = -1$ would indicate an analytic specific heat in agreement with Zittartz [28].

In the renormalization group calculations for the $S = 1/2\ XY$ model [35, 36, 37] the chosen lattice is divided into identical cells each containing a small number, l^d, of sites and in such a way that the cells themselves form the same lattice (with lattice spacing $l\delta$). All properties of the system within the cells can be calculated exactly while interactions between spins in different cells are treated as a perturbation. In order to make the system of cells look like the original system some of the degrees of freedom are summed over in a partial trace operation. The partial trace must be performed in such a way as to preserve the free energy,

$$\exp(-\beta \mathcal{H}_{cell}) = \text{tr}' \exp(-\beta \mathcal{H}_{site}). \quad (9)$$

By equating corresponding matrix elements on both sides of (9) a set of nonlinear recurrence relations can be obtained for the various interaction parameters in the Hamiltonian. These relations can then be solved numerically to obtain fixed points, critical temperatures, exponents etc. in the usual way. Calculations have been carried out for symmetrical three [35], five [36] and seven [37] spin cells on the triangular, square and triangular lattices to second order in the cumulant expansion. Starting with nearest neighbour XY interactions the renormalization operation generates second and third neighbour XY interactions and first neighbour Ising interactions. Solution of the recurrence relations for the four interaction strengths gave a finite critical temperature for the triangular lattice about 30% lower than the series estimate and a negative value for δ! However the zero field free energy curves for both lattices are in excellent agreement with the curves from high temperature series expansions.

Extensions of these calculations either by increasing the cell size or by going to third order in the cumulant expansion would be very laborious and are not being contemplated at present. However it would be desirable to have longer high temperature series and very shortly two additional coefficients will be available* [38].

5. Comparison with experiment

The $S = 1/2\ XY$ model in three dimensions behaves in the critical region in remarkably good agreement with liquid ^{4}He near the lambda transition [3], but until recently good magnetic examples of the XY model have not been studied.

Algra et al. [6] have measured the specific heats of $Co(C_5H_5NO)_6(ClO_4)_2$ and $Co(C_5H_5NO)_6(BF_4)_2$ below 1 K in an adiabatic demagnetization apparatus using the heat pulse technique with a cerium magnesium nitrate thermometer. Phase transitions, marked by lambda anomalies in the specific heat, were observed at $T_c = 0.428$ K and $T_c = 0.357$ K, respectively. The two specific heat curves can be brought into very near coincidence by plotting against T/T_c.

To compare theory with experiment, for $T > T_c$, Algra et al. constructed a theoretical curve by smoothly joining the critical part, (6), to a high temperature part obtained more directly from the series. (Since $K_c^S \approx 0.50$ the errors noted in the coefficients of K^8 and K^{10} make no detectable contribution to the curve.) In fig. 2 we have plotted the dimensionless specific heat,

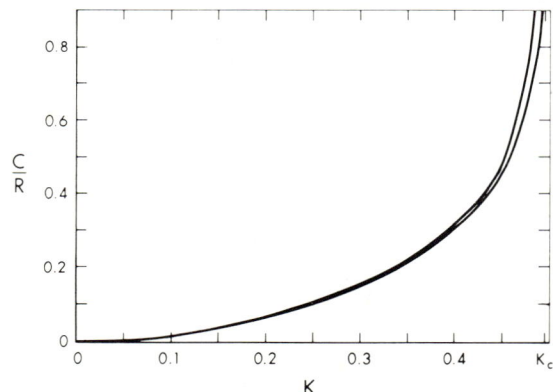

Fig. 2. Smoothed experimental specific heat data for $Co(PyNO)_6(BF_4)_2$ (upper curve) compared with the specific heat of $S = 1/2\ XY$ model on the s.c. lattice (lower curve) versus reduced inverse temperature, $K = J^\perp/k_BT$.

C/Nk_B, for the $S = 1/2\ XY$ model on the simple cubic lattice using the corrected coefficients in (5) and (6) and for comparison the experimental curve [6] for $Co(PyNO)_6(BF_4)_2$. Possible reasons for the small discrepancy between theory and experiment are: (i) in the experimental system $J^\parallel \approx 0.3\ J^\perp$ (ii) some anisotropy in the easy plane is to be expected (iii) in the calculation there are significant uncertainties in estimate of K_c^S, A^S and particularly B_0^S.

Table I
Critical properties of the $S = 1/2$ Ising, XY and Heisenberg models on the s.c. lattice compared with experiment [6]

System	kT_C/J	$(S_\infty - S_C)/Nk_B$	$-E_C/Nk_BT_C$
Ising	2.255	0.135	0.220
XY	2.02	0.20	0.41
Heisenberg	1.68	0.26	0.63
$Co(PyNO)_6(BF_4)_2$	1.98	0.25	0.47
$Co(PyNO)_6(ClO_4)_2$	2.02	0.24	0.43

* In the course of these investigations it has been discovered that the original series for the zero field partition function for the square and triangular lattices [29] contain errors in coefficients beyond the sixth degree. The errors do not effect the previous conclusion [29] that the free energy is analytic.

De Jongh et al. [39] have compared the experimental specific heat and the initial susceptibility of $CoCl_2 \cdot 6H_2O$ and $CoBr_2 \cdot 6H_2O$ with theoretical predictions for the $S = 1/2$ XY antiferromagnet on the square lattice. The theoretical specific heat curve is obtained from the high temperature series for $T \geq 1.5 \, J^\perp/k_B$ (the errors in the last two coefficients as originally reported have no effect on this portion of the curve). For $T \leq 0.3 \, J^\perp/k_B$ the simple spin wave result [40] $C/Nk_B = 0.28 \, (k_B T/J^\perp)^2$ is used, and in the intermediate region a smooth curve is drawn so that the area under the curve yields the correct entropy [5, 40].

The exchange constants for the chlorine and bromine salts, $J^\perp/k_B = -2.05 K$ and $J^\perp/k_B = -2.45 K$ respectively, are obtained by fitting experimental data to the high temperature end of the theoretical curve. Upon fixing $J^\perp$ the specific heat curve for $CoCl_2 \cdot 6H_2O$ is in excellent agreement with the theoretical curve for all $T > 1.5 |J^\perp|/k_B$. There is however a sharp spike in the experimental curve just below this temperature, which could be due either to easy plane anisotropy causing a $d = 2$ Ising logarithmic singularity or to the effect of interplanar interactions causing a $d = 3$ XY nearly logarithmic singularity.

The initial susceptibility of both salts is also in excellent agreement with the theoretical for the XY model for $T > T_c$ [39].

References

[1] L.P. Kadanoff in Critical Phenomena, M.S. Green, ed. (Academic Press, New York, 1971).
[2] See e.g. D.D. Betts, A.J. Guttmann and G.S. Joyce, J. Phys. C4 (1971) 1994.
M. Ferer and M. Wortis, Phys. Rev. B6 (1972) 3426.
P.C. Hohenberg, A. Aharony, B.I. Halperin and E.D. Siggia, Phys. Rev. B13 (1976) 2986, where other references may be found.
[3] D.D. Betts in Phase Transitions and Critical Phenomena, Vol. 3, C. Domb and M.S. Green, eds. (Academic Press, London, 1974).
[4] L.J. de Jongh and A.R. Miedema, Advances in Phys. 23 (1974) 1.
[5] J.W. Metselaar, L.J. de Jongh and D. de Klerk, Physica 79B (1975) 53.
[6] H.A. Algra, L.J. de Jongh, W.J. Huiskamp and R.L. Carlin, Physica 83B (1976) 71.
[7] T. Matsubara and H. Matsuda, Prog. Theor. Phys. 16 (1956) 416.
[8] J. Rosenblatt, A. Raboutou and R. Pellan in Proc. LT 14, Vol. 2, M. Krusius and M. Vuorio, eds. (North-Holland, Amsterdam, 1975).
[9] T. Schneider and E. Stoll, Phys. Rev. Lett. 36 (1976) 1501.
[10] R. Dekeyser and J. Rogiers, Physica 81A (1975) 72.
[11] D.J. Austen and D.D. Betts, Phys. Lett. 53A (1975) 313.
[12] D.D. Betts and D.S. Ritchie, Phys. Rev. Lett. 34 (1975) 788.
[13] J.T. Tsai and C.J. Elliott, Phys. Lett. 45A (1973) 295.
[14] N.D. Mermin and H. Wagner, Phys. Rev. Lett. 17 (1966) 1133.
[15] H.E. Stanley and T.A. Kaplan, Phys. Rev. Lett. 17 (1966) 913.
[16] G.S. Rushbrooke and P.J. Wood, Mol. Phys. 1 (1958) 257.
[17] K. Yamaji and J. Kondo, J. Phys. Soc. Japan 35 (1973) 25.
[18] M.A. Moore, Phys. Rev. Lett. 23 (1969) 861.
[19] M. Ferer, M.A. Moore and M. Wortis, Phys. Rev. B8 (1973) 5205.
[20] D.D. Betts, C.J. Elliott and R.V. Ditzian, Can. J. Phys. 49 (1971) 1327.
[21] V.L. Berezinskii, Soviet Phys. JETP 32 (1971) 493.
[22] J.M. Kosterlitz and D.J. Thouless, J. Phys. C6 (1973) 1181.
[23] J.M. Kosterlitz, J. Phys. C7 (1974) 1046.
[24] W.J. Camp and J.P. Van Dyke, J. Phys. C8 (1975) 336.
[25] T.P. Eggarter, Phys. Rev. B9 (1974) 2989.
[26] E. Müller-Hartmann and J. Zittartz, Phys. Rev. Lett. 33 (1974) 893.
[27] J. Zittartz, Z. Physik B23 (1976) 55.
[28] J. Zittartz, Z. Physik B23 (1976) 63.
[29] D.D. Betts, J.T. Tsai and C.J. Elliott, in Proc. Int. Conf. on Magnetism, Moscow, 1973, Vol. IV (Nauka, Moscow, 1974).
[30] J. Oitmaa and D.D. Betts, to be published.
[31] K.G. Wilson, Phys. Rev. B4 (1971) 3174.
[32] K.G. Wilson and J. Kogut, Phys. Reports 12C (1974) 75. M.E. Fisher, Rev. Mod. Phys. 46 (1974) 597.
[33] Th. Niemeijer and J.M.J. van Leeuwen, Phys. Rev. Lett. 31 (1973) 1411.
[34] D.M. Lublin, Phys. Rev. Lett. 34 (1975) 568.
[35] J. Rogiers and R. Dekeyser, Phys. Rev. B13 (1976) 4886.
[36] D.D. Betts and M. Plischke, Can. J. Phys. 54 (1976) 1553.
[37] J. Rogiers and D.D. Betts, Physica, in press.
[38] J. Rogiers and D.D. Betts, to be published.
[39] L.J. de Jongh, D.D. Betts and D.J. Austen, Solid State Commun. 15 (1974) 1711.
[40] P. Bloembergen, Physica 79B (1975) 467.

MAGNETO-ELECTRIC MEASUREMENTS ON GdAlO$_3$

A.H. COOKE, N.J. ENGLAND, S.J. SWITHENBY and M.R. WELLS

The Clarendon Laboratory, Oxford, England

The magnetic structure and part of the magnetic phase diagram have been investigated using magneto-electric techniques. Measurements of the magneto-electric susceptibility close to T_N have enabled the critical coefficients β and B to be evaluated.

1. Introduction

Gadolinium aluminate, GdAlO$_3$, which has a distorted perovskite structure, space group Pbnm, orders antiferromagnetically at $T_N = 3.87$ K [1]. The magnetic behaviour may be described by a simple Heisenberg model in which the dominant term in the Hamiltonian is the exchange interaction between nearest neighbour Gd^{3+} ions. We have used magneto electric (ME) measurements to examine its magnetic structure, the critical behaviour just below T_N, and the spin-flop magnetic phase transition.

2. The magnetic structure

Previous measurements have shown that the diagonal elements of the ME susceptibility tensor are finite [2], identifying the magnetic point group as *mmm*. Fig. 1 illustrates the temperature dependence of these elements α_{xx}, α_{yy} and α_{zz}. Proposed mechanisms [3] for the ME effect lead to the magnitude of α being approximately proportional to the product of the sublattice magnetization and the magnetic susceptibility. Thus the ME susceptibility along the ordering direction reaches a maximum at 2.7 K, while the ME susceptibility in the perpendicular direction reaches a maximum at $T = 0$. From these results the ordering direction is identified as the crystalline a axis, and the magnetically ordered structure as predominantly G_x in Bertaut's notation [4].

3. Critical behaviour

In a simple antiferromagnet the magnetic susceptibility along an axis perpendicular to the ordering direction is constant, and indeed this is a good approximation for GdAlO$_3$ [1], so that the dominant term in the temperature dependence of α is the variation of the sublattice magnetisation. With this assumption we have used the variation of α_{zz} with temperature just below T_N to evaluate the critical coefficients B and β in the expression $\alpha/\alpha_0 = B(1 - T/T_N)^\beta$, which was shown theoretically and experimentally by Rado [5] to be applicable in another system. From the results shown in Fig. 2 the values $\beta = 0.36 \pm 0.01$ and $B = 1.11 \pm 0.04$ have been obtained. These are in excellent agreement not only with values from NMR experiments on the critical behaviour of the sublattice magnetisation of 3-dimensional spin 7/2 Heisenberg

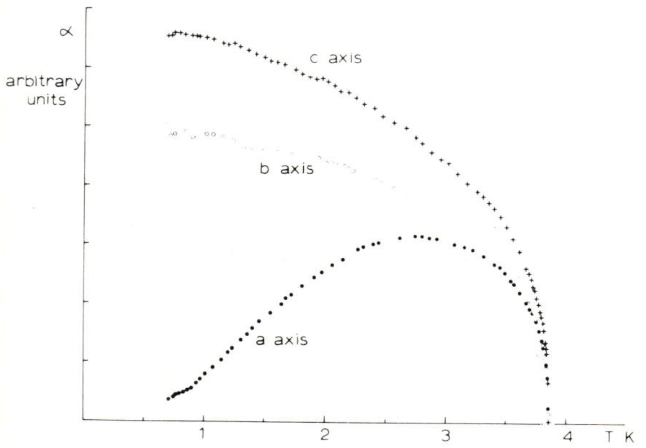

Fig. 1. The temperature dependence of the components α_{xx}, α_{yy}, α_{zz} of the ME susceptibility in GdAlO$_3$.

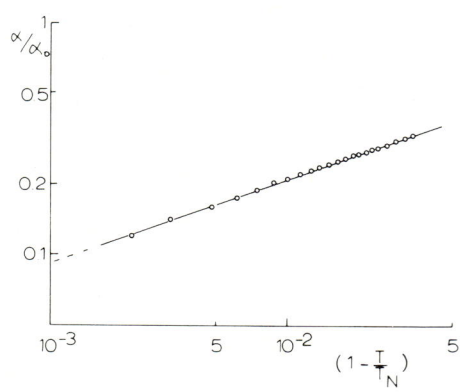

Fig. 2. The ME susceptibility along the c axis as a function of reduced temperature.

systems but also with the theoretical predictions of $\beta = 0.36$ [6]. Our results contrast with those of Gorodetsky et al. [7], who from ME experiments on a similar system, $GdVO_4$, found a critical exponent $\beta = 0.50$, in agreement with classical molecular field theory.

An advantage of the ME technique is that measurements can be taken very close to T_N. By experimentation at lower values of reduced temperature ($<2 \times 10^{-3}$), we intend to look for the change from Heisenberg to Ising critical behaviour.

4. Magnetic Phase Transitions

Since the form of the ME susceptibility tensor is governed by the symmetry of the crystal, the effect provides a powerful tool for the examination of the magnetic phase transitions. We have mentioned that in zero magnetic field the ME susceptibility tensor elements in $GdAlO_3$ are diagonal corresponding to a magnetic point group mmm. Any other ordering mode would either exclude the ME effect entirely, or change the form of the tensor, leaving only off-diagonal elements. We have observed the spin-flop phase transition in $GdAlO_3$ by slowly sweeping a magnetic field along the a axis while monitoring α_{xx}. The most prominent features of the curves are the enhanced effects in the region of H_c, the spin-flop field. Fig. 3 shows the results of a magnetic moment measurement, uncorrected for demagnetising effects, and a ME suscep-

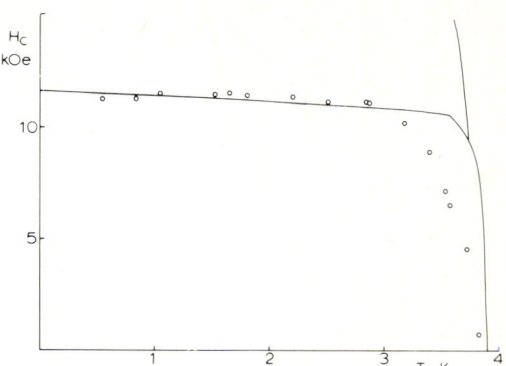

Fig. 4. The magnetic phase diagram for $GdAlO_3$ in low fields.

tibility measurement, both made on the same crystal at 0.5 K. Comparing the two measurements we see that the peaks in the trace of α_{xx} coincide with the beginning and end of the spin-flop transition. With this assumption we have used the ME effect to plot the antiferromagnetic phase diagram for $GdAlO_3$ over the range $0.5 < T < 3.8$ K. The results are shown in Fig. 4 together with the phase diagram derived by Blazey and Rohrer [8] from magnetisation measurements using a pulsed field technique. It is not surprising that the two curves are at variance with each other close to T_N. In this region our identification of the critical field becomes difficult and the character of the phase transition is different.

One of the authors (NJE) would like to acknowledge the support of an SRC grant.

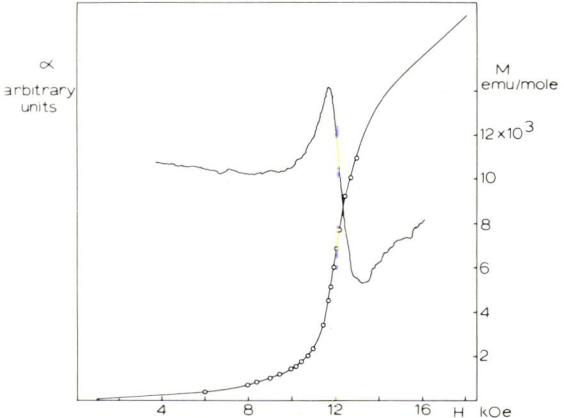

Fig. 3. The magnetisation and ME susceptibility as a function of applied magnetic field parallel to the a axis.

References

[1] J.D. Cashion, A.H. Cooke, T.L. Thorp and M.R. Wells, Proc. Roy. Soc. A318 (1970) 473.
[2] A.H. Cooke, N.J. England, N.F. Preston, S.J. Swithenby, and M.R. Wells, Sol. State Comm. 18 (1976) 545.
[3] G.T. Rado, Int. J. Magn. 6 (1974) 121.
[4] E.F. Bertaut, Magnetism, Vol. 3 (Academic Press, New York, 1963) p. 149.
[5] G.T. Rado, Sol. St. Comm. 8 1349 (1970). Erratum: 9 No2 vii (1971).
[6] L.J. de Jongh, and A.R. Miedema, Experiments on Simple Magnetic Model Systems, Taylor and Francis, London, 1974).
[7] G. Gorodetsky, R.M. Hornreich, and B.M. Wanklyn, Phys. Rev. B8 (1973) 2263.
[8] K.W. Blazey, and H. Rohrer, Phys. Rev. 173 (1968) 574.

DETERMINATION OF CRITICAL SUSCEPTIBILITY EXPONENTS ABOVE AND BELOW T_c WITH RELAXATION TECHNIQUES IN 3-DIMENSIONAL FERROMAGNETIC $CuRb_2Br_4 \cdot 2H_2O$

W.L.C. RUTTEN and J.C. VERSTELLE

Kamerlingh Onnes Laboratory, State University, Leiden, The Netherlands

Dynamic susceptibility measurements on $CuRb_2Br_4 \cdot 2H_2O \mathbin{/\mkern-4mu/} c$-axis yield single relaxation above T_c and a double process below T_c. The single process intensity diverges according to a power-law $\chi(0) \sim \epsilon^{-\gamma}$ with $\gamma = 1.26$. The intensity χ_I of the fast process below T_c can be interpreted as the susceptibility of paramagnetic spins in a "predomain" or "superparamagnetic" phase. It has critical exponent $\gamma' = 1.32$.

The frequency-dependent susceptibility $\chi(\omega) \equiv \chi'(\omega) - i\chi''(\omega)$ has been measured in the critical temperature region above and below T_c on an ellipsoïdal sample of 3-dimensional ferromagnetic $CuRb_2Br_4 \cdot 2H_2O$ parallel to the c-axis, which is the easy axis of magnetization below $T_c = 1.874$ K. The measurements were performed in zero external magnetic field in the frequency-range 0–62 MHz using bridge methods.

Above T_c the relaxation is Debye-like with a single relaxation time (fig. 1) and yields a static susceptibility $\chi(0)$, corrected for demagnetizing effects with an exponent $\gamma = 1.26 \mp 0.01$ (fig. 2). The correction has been made according to $\chi(0) = \chi(0)_e / (1 - D\chi(0)_e)$ in which χ_e is the experimental susceptibility and D is the demagnetizing factor. This value of D, determined from $\chi(0)$ below T_c agreed within a few percent

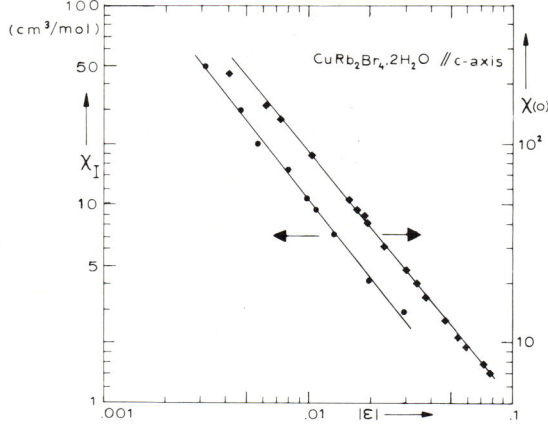

Fig. 2. Susceptibility χ_I below T_c (●) and $\chi(0)$ above T_c (♦) vs. $|\epsilon| = |(T - 1.874)/1.874]$ on a log-log scale.

with that estimated from the ratios of the ellipsoid semi-axes [1].

Below T_c the relaxation process is more complicated. From T_c down to $|\epsilon| \equiv |(T - T_c)/T_c| = 0.03$ the relaxation is split up in two separated processes, each Debye-like (fig. 1), with increasing intensity for the low-frequency process and decreasing intensity for the high-frequency part. The latter intensity χ_I is also reported in fig. 2, and shows a critical behavior with a value of 1.32 ∓ 0.05 for the exponent. For higher $|\epsilon|$-values below T_c the high-frequency process becomes resonant-like with frequencies outside our range. An explanation in terms of the formation of a domain-structure below T_c seems plausible. The strongly coupled spins within the domains will relax rather slowly as a whole, while the spins in the walls, being in a state of subtle equilibrium between exchange, anisotropy and demagnetization, will behave much more like a paramagnetic system and relax at much higher frequencies.

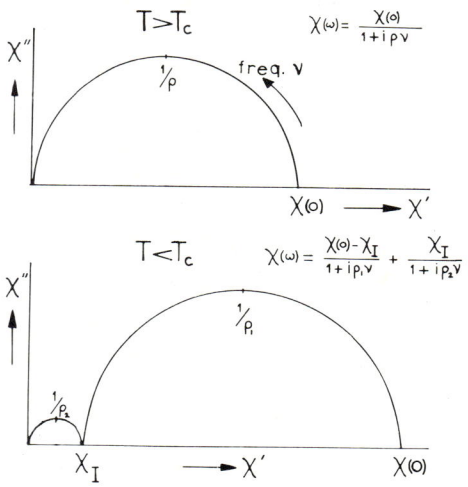

Fig. 1. Cole–Cole plot (Argand-diagram) illustrating the precedure of finding $\chi(0)$ and χ_I.

It is important to note that it is not the relaxation or resonance of walls as a whole we measure from T_c until $|\epsilon| = 0.03$. Much work has been done on this subject (for example see refs. 2 and 3) and one can crudely calculate the resonance or relaxation frequency for a 180° Bloch-wall. From the equation of motion $m_w \ddot{x} + \beta_f \dot{x} + ax = 2M_s H e^{j\omega t}$ (m_w: mass of the wall, β_f: friction coefficient, a: stiffness constant, M_s: saturation magnetization, $H e^{j\omega t}$: alternating magnetic field) it follows that resonance will occur if $\beta_f^2 \ll m_w a$, or relaxation if $\beta_f^2 \gg m_w a$. The quantities β_f, m_w and a can be expressed in terms of exchange J, uniaxial anisotropy K_u and M_s, with K_u and M_s functions of temperature. With the values $K_u = 1.62$ erg/cm^3 ($T < 0.3 T_c$), $J/k = 0.63$ K, $M_s = 12$ Oe ($T < 0.3 T_c$), β (magnetization exponent) $= 0.367$ [4] and the temperature dependence of K_u as calculated by van Vleck and Zener [5] the condition $\beta_f^2 \ll m_w a$ is fulfilled for all $T < T_c$. So the relaxations we measure cannot be ascribed to domain-wall movements.

Also because of the low $|\epsilon|$-values involved, the above suggested domain picture does not seem to apply. A complete domain formation probably does not yet occur in this immediate neighbourhood of T_c. According to Landau and Semenchenko [6] one has to deal in this region with the strong development of fluctuations, which lead to the formation of so-called "clusters" of ordered spins. The system is in the "pre-domain" phase. The spins outside the clusters should still behave like paramagnetic spins. A support for this point of view is found in the measured relaxation times (fig. 3). The relaxation rate $1/\rho = 1/2\pi\tau$ (τ: relaxation time) of the fast process below T_c shows a similar behavior as the rate above T_c as a function of the relative temperature $|\epsilon|$. Moreover, theory about "clusters" or "superparamagnets" yields us a quantitative estimate of the cluster radius. According to [7] the inverse relaxation time is given by $\tau^{-1} = 10^9 \times e^{-K_u V/kT}$ in which k is Boltzmann's constant and V is the cluster volume. For example, at $T = 1.863$ K ($|\epsilon| = 0.0059$) the low-frequency relaxation has a measured time $\tau = 2 \times 10^{-6}$ s, which would imply a radius of about a hundred lattice constants.

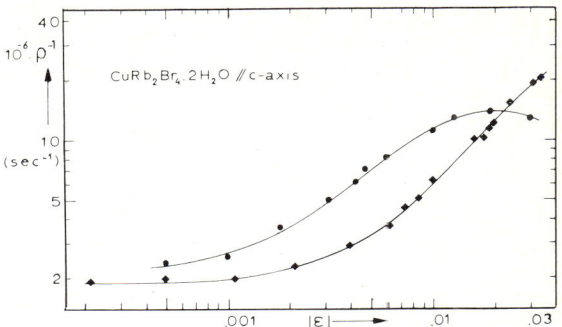

Fig. 3. Relaxation rates above T_c(◆) and of the fast process below T_c(●) vs. $|\epsilon|$ on a log-log scale.

This is indeed much less than the usual domain-width values.

Resuming, this leads us to the conclusion that the intermediate susceptibility χ_I, which determines the onset of the high-frequency process, is the paramagnetic susceptibility which should follow the critical behavior with exponent $\gamma' = \gamma$. The thus measured γ' (see fig. 2) has the value 1.32 ∓ 0.05, like γ too small for a 3-dimensional Heisenberg system. The scaling result $\gamma' = \gamma$ is, however, satisfied within the error margins. For the amplitudes C_0 and C'_0 of the susceptibilities above and below T_c we find the ratio $C_0/C'_0 = 8.45$ at $|\epsilon| = 0.02$. If our interpretation would be correct, these measurements would constitute the first experimental determination of γ' in a ferromagnetic system.

Acknowledgements

We wish to thank Drs. H.A. Groenendijk and Dr. A.J. van Duyneveldt of our laboratory for enabling us to do some additional measurements at very low frequencies.

References

[1] J.A. Osborn, Phys. Rev. 67 (1945) 351.
[2] J. Smit and H.P.J. Wijn, Ferrites (Philips Technical Library, Eindhoven, 1959).
[3] A. Hubert, Theorie der Domänenwände in geordneten Medien (Springer Verlag, New York, 1974).
[4] W.J. Looyestijn, T.O. Klaassen and N.J. Poulis, Physica 83B (1976) 169.
[5] F. Keffer, Phys. Rev. 100 (1955) 1692.
[6] See K.P. Belov, Magnetic transitions (Consultants Bureau, New York, 1961).
[7] W.F. Brown Jr., J. Appl. Phys. 30, Suppl., 130 S (1959).

ANISOTROPY OF NEUTRON CRITICAL SCATTERING IN IRON MONOCRYSTAL

A. JABŁONKA, R. CISZEWSKI, D. SIKORSKA and A. SOBASZEK

Institute of Physics, Warsaw Technical University, Koszykowa 75, 00-662 Warsaw, Poland

It was found that the intensity of neutron critical scattering in an iron monocrystal is smaller for the scattering vector q along the [100] direction, than for q along the [110] direction. The difference in these intensities of scattering decreases with increasing temperature.

In the last years, X-ray and neutron critical scattering experiments were performed in ferroelectrics, dielectrics and binary alloys. An anisotropy of scattering was found in BaTiO$_3$ [1] TGS [3], AuCu$_3$ [3], NaNO$_3$ [2], [4].

The aim of this experiment was to investigate whether anisotropy of scattering exists for ferromagnets. For this purpose an iron monocrystal with 4% Si was used. The sample was cut in the shape of a cylinder with its vertical axis parallel to the [100] direction; the diameter of the crystal was equal to 14 mm and the height-40 mm. The sample was heated in high-temperature vacuum furnace. The temperature inside the furnace was automatically stabilized within ± 0.1 K and its gradient along the vertical axis of the sample did not exceed ± 0.1 K. For temperature measurements two chromel-alumel thermocouples were mounted at the basis and the top of the sample. The Curie temperature was determined from neutron critical scattering measurements for very small q. It was found to be equal to $T_c = (28.74 \pm 0.01)$ mV. Measurements were performed on DN-501 double-axis neutron spectrometer, situated at the "Ewa" reactor at Swierk. A monochromatic neutron beam of wavelength $\lambda = 1.53$ Å was extracted from the reactor beam by Bragg diffraction in the focusing Zn monochromator crystal. The collimation in front of the sample was 10 min. of arc and in front of the BF$_3$ counter 13 min. of arc.

Critical scattering measurements were performed for different temperatures above the Curie point and for the scattering vector parallel to [100] or [110] directions.

Typical curves are represented in fig. 1 a,b,c. The curves in fig. 1 were obtained after the subtraction of the background intensity, measured for $T_c + 75$ K and represented in fig. 2. All measurements were performed for the reciprocal lattice vector $\tau = 0$ in the range 0.09 Å^{-1} $< q <$ 0.20 Å^{-1}. Another set of measurements was performed around the (110)

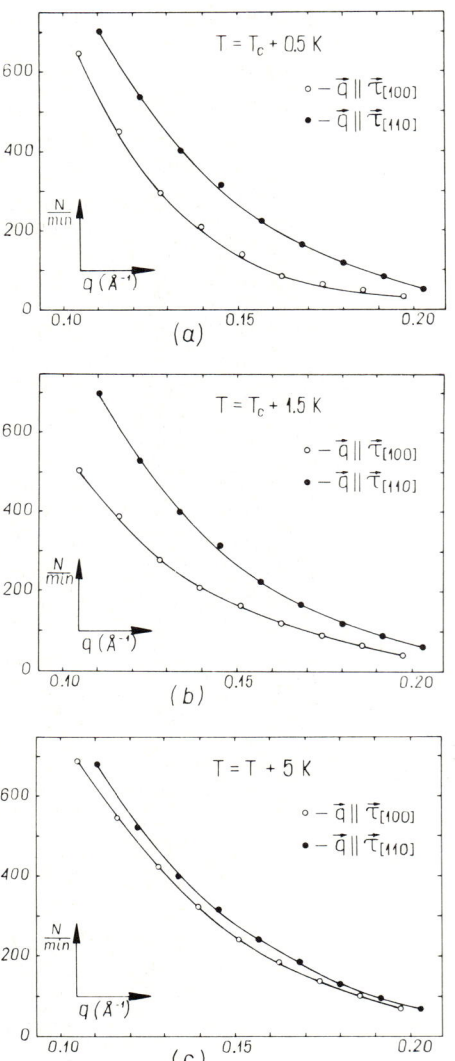

Fig. 1 a,b,c. Intensity of small angle critical scattering versus scattering vector, for q parallel to [100] or to [110] directions.

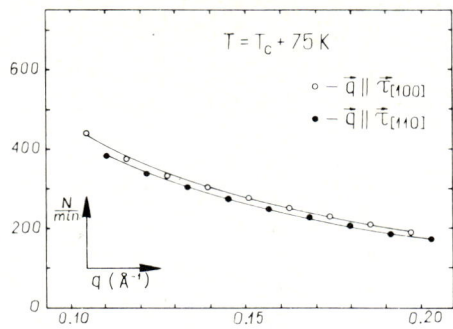

Fig. 2. Intensity of scattering versus scattering vector q in [100] or [110] directions for $T_c + 75$ K, accepted as the background intensity.

reciprocal lattice point. These results will be discussed in a separate paper [5].

The statistical counting error did not exceed 5%. The angular position of the spectrometer arm and sample table was known with the accuracy of 1%.

The results of measurements for various temperatures show that the intensity of scattering is always larger for $q\|\tau_{[110]}$ than for $q\|\tau_{[100]}$ and that this difference decreases with increasing temperature. The detected anisotropy of scattering cannot be explained by a spherically symmetric cross-section. This anisotropy is in agreement with the cross-section derived by Kociński and Marzec [6] for a body centered cubic monocrystal. The connection between crystalline symmetry and anisotropy of critical scattering was first discussed in the papers [7]–[10].

The influence of the resolution function on the obtained results will be discussed in detail in paper [5]. For $\tau = 0$ the resolution corrections are nearly identical for both crystal orientations with respect to the scattering vector.

Measurements of the temperature shift of the maximum of neutron critical scattering performed for different crystallographic directions [11] show also the existence of the anisotropy of neutron critical scattering.

We wish to thank prof. J. Kociński for suggesting the problem and inspiring discussions, as well as Mr. A. Czajkowski and Mr. S. Zduńczyk for technical assistance.

References

[1] Y. Yamada, G. Shirane and A. Linz, Phys. Rev. 177 (1969) 848.
[2] Y. Fuji and Y. Yamada, J. Phys. Soc. Japan 30 (1971) 1676.
[3] B. Pura, and J. Przedmojski, Phys. Lett. 43A (1973) 217, 48A (1974) 82.
[4] J. Przedmojski and B. Pura, Phys. Stat. Sol. (b) 69, K 37 (1975).
[5] R. Ciszewski et al., to be published.
[6] J. Kociński and W. Marzec, to be published.
[7] L. Wojtczak and J. Kociński, Phys. Lett. 32A (1970) 389, 34A (1971) 306.
[8] J. Kociński and L. Wojtczak, Phys. Lett. 36A (1971) 171, 43A (1973) 215.
[9] J. Kociński, L. Wojtczak and B. Mrygoń, Acta Phys. Polon. A 43 (1973) 425.
[10] J. Milczarek and J. Kociński, Phys Lett. 47A (1974) 11.
[11] R. Ciszewski et al., to be published.

CORRECTIONS TO SCALING IN THE SPECIFIC HEAT OF Cu(NH$_4$)$_2$Br$_4$·2H$_2$O NEAR T_c[†]

J.E. RIVES, D.P. LANDAU, H.P. LUNG* and H.S. CHANDRASEKHAR

Department of Physics and Astronomy, University of Georgia, Athens, GA 30602 U.S.A.

The specific heat of the Heisenberg ferromagnet Cu(NH$_4$)$_2$Br$_4$·2H$_2$O has been measured with high resolution near T_c. The data outside the 'rounded' region are consistent with power law behavior including first order corrections to scaling: $C/R = (A/\alpha)t^{-\alpha}(1 + Dt^x) + B$. The fitted parameters agree well with predictions for a 3-dim Heisenberg model with short range interactions.

1. Introduction

The critical behavior of the insulating ferromagnet Cu(NH$_4$)$_2$Br$_4$·2H$_2$O is expected to closely reflect the behavior of an ideal $S = \frac{1}{2}$ 3 dim. Heisenberg model with short range interactions. The tetragonal structure [1], $a = 7.98$ Å, $c = 8.41$ Å, is almost BCC and an isotropic exchange interaction [2, 3] is primarily between nearest-neighbors with $J_1/kT_c \approx 0.35$. A smaller exchange coupling exists between next-nearest-neighbors $J_2/J_1 \approx 0.25$ and there is a quite small exchange anisotropy [4, 5] of $\approx 0.015 J_1$. The long range dipole interactions are quite small leading to a calculated dipole field [5] of less than 1 Oe.

Existing experimental estimates for critical exponents are not totally satisfactory. The susceptibility exponent γ obtained from NMR [6] ($\gamma = 1.26$) and low frequency induction techniques [3] ($\gamma = 1.31$) is below the present theoretical prediction of $\gamma \approx 1.38$ for the 3-dim Heisenberg model [7]. In addition bulk magnetization studies [5, 6, 8] yield $\beta \approx 0.38$–0.40 and $\delta \approx 3.9$–4.3 as opposed to the theoretical estimates of $\beta \approx 0.36$, $\delta = 4.4$, respectively. Previous specific heat measurements [2] yielded $\alpha = -0.14 \pm 0.20$ and $\alpha' = 0.06 \pm 0.18$. Within the rather large error limits these values are not inconsistent with predictions but hardly confirm them. It has recently been demonstrated [9] that corrections to scaling must be included in the analysis of experimental data if the correct asymptotic behavior is to be extracted. With this in mind we hope to obtain a definitive result which may then clarify the critical behavior in general.

[†] Research supported in part by the National Science Foundation under Grant No. DMR71-01778 A02.
* Present address: Dept. of Computer Science, State University of New York at Buffalo, Buffalo, NY, U.S.A.

2. Experimental

A standard heat pulse technique was used to measure the specific heat from 1.56 to 1.86 K. Temperature was measured using an Allen–Bradley $\frac{1}{2}$W carbon resistor calibrated against the vapor pressure of liquid helium-4. Specific heat data were taken on four different samples, obtained from D. Walton and G. Dixon.

3. Results and analysis

Data near T_c on samples II and IV showed a broad, almost flat region ≈ 20 mK wide. This maximum could be composed of two weak unresolved peaks (possibly due to multiple crystal formation during growth). Sample III exhibited similar behavior but the flattened region was at most 5 mK. Data from sample I, while showing considerable rounding near T_c, showed a single peak and were analyzed in detail.

We first fitted the data to a simple power law

$$C/R = (A/\alpha)t^{-\alpha} + B, \qquad (1)$$

where $t = |1 - T/T_c|$ and the scaling theory constraints [9] $\alpha = \alpha'$ and $B = B'$ were adopted. (These constraints were not used in the earlier study [2].) The agreement with the simple power law may be tested by plotting $\ln(B - C/R)$ vs $\ln t$ for various choices of T_c and B. For the proper choice of constants such plots should show data lying on straight lines (for $T > T_c$, $T < T_c$) with slope $-\alpha$. No good agreement was obtained in any of the fits if the constraints $\alpha = \alpha'$, and $B = B'$ were retained. It was *not* possible to improve the agreement with eq. (1) by shifting T_c or B!

Of course the anisotropy should cause a crossover to Ising-like behavior very close to T_c, but the analysis showed no evidence for such behavior and we therefore believe that crossover occurs in the rounded region. We therefore reanalyzed the data using the predicted form for lowest order corrections to scaling so that

$$C/R = (A/\alpha)t^{-\alpha}(1 + Dt^x) + B. \qquad (2)$$

The exponent, x, in the correction term has been estimated from renormalization group theory to be 0.5 ± 0.2 and experimental measurements suggest $x = 0.5 \pm 0.1$ [9] although there is some evidence [10] that for $S = \frac{1}{2}$ the lowest order correction has zero amplitude and the next order term with $x = 1.0$ is appropriate. We therefore considered both possibilities in fitting the data. We wish to emphasize that scaling conditions do not assume specific values for exponents or amplitudes, and except for x all parameters were free to vary.

The solid curves shown in fig. 1 correspond to the best fit using eq. (2) (with $x = 0.5$) and yielded the parameters given in table I. The quality of the fit did not vary appreciably for the different choices of x. The *asymptotic* form [i.e. eq. (1)] obtained with these parameters is shown by the dashed line. We estimate that even without rounding this form would fit the data only over perhaps half a decade. It is therefore easy to understand why the simple power law fit was so unsuccessful. The fit is quite good over a very wide range of t with evidence of rounding near T_c. (The obviously rounded data were not included in the analysis, and the best estimate $T_c = 1.675$ was almost 10 mK above the maximum [11].) Both the values for α and the amplitude ratio A/A' are consistent with the best estimates for the $S = \frac{1}{2}$ 3-dim Heisenberg model with short range interactions [9]. Qualitatively similar results were obtained on the other three samples, and although the broadened nature of the peak made any truly independent analysis impossible, the parameters given in table I described the specific heat away from T_c in these samples as well.

Although these data do not allow us to determine experimentally whether or not the lowest order correction vanishes, it is clear that corrections to scaling are very important. Clearly then the extraction of the correct asymptotic parameters in this range of t is possible only if corrections to scaling are included in the analysis. We suspect that the small discrepancies between theoretical and experimental estimates for β, γ, δ might similarly be due to the omission of corrections to scaling in the analyses.

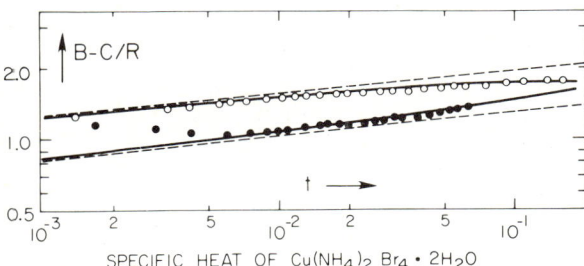

Fig. 1. $(B - C/R)$ vs t for Cu(NH$_4$)$_2$Br$_4$·2H$_2$O. Simple power law, ----; Corrected power law ———; Present data for $T < T_c$, ●, and $T > T_c$, ○.

Table I.

	Present Results		Theory [9]
	$x = 0.5$	$x = 1.0$	
$\alpha = \alpha'$	-0.10 ± 0.02	-0.08 ± 0.01	-0.1
$B = B'$	2.2 ± 0.1	2.15 ± 0.1	—
$-D = D'$	0.37 ± 0.05	1.1 ± 0.10	—
A/A'	1.5 ± 0.1	1.4 ± 0.1	1.52
$P = \alpha^{-1}(1 - A/A')$	5.0	3.7	5.92

References

[1] R.W.G. Wyckoff, Crystal Structures, Vol. III, 2nd edition (Interscience, New York, 1948) p. 618.
[2] A.R. Miedema, R.F. Wielinga and W.J. Huiskamp, Physica 31 (1965) 1585.
[3] L.J. DeJongh, A.R. Miedema and R.F. Wielinga, Physica 46 (1970) 44.
[4] H. Suzuki and T. Watanabe, Phys. Lett. 26A (1967) 103.
[5] R.F. Wielinga and W.J. Huiskamp, Physica 40 (1969) 602.
[6] J.-P. Renard and E. Velu, J. de Physique 32 (1971) Cl-1154.
[7] G.S. Rushbrooke, G.A. Baker, Jr. and P.J. Wood, Phase Transitions and Critical Phenomena, Vol. 3, C. Domb and M.S. Green, eds. (Academic Press, New York, 1974) p. 246.
[8] E. Velu, D. Cadoul, B. Lecuyer and J.-P. Renard, Phys. Lett. 36A (1971) 443.
[9] See e.g. M. Barmatz, P.C. Hohenberg and A. Kornblit, Phys. Rev. B12 (1975) 1947, G. Ahlers and A. Kornblit, Phys. Rev. B12 (1975) 1938; and references therein.
[10] W.J. Camp and J.P. Van Dyke, J. Phys. A9 (1976) 731.
[11] A difficulty in this compound is the wide variation of T_c from crystal to crystal. Our other three samples center around $T_c = 1.71$ K, and other workers [3–7] have reported values of T_c ranging from 1.77 to 1.82 K.

CRITICAL AND ELASTIC BEHAVIOUR OF PARAMAGNETIC MANGANESE OXIDE

D. HERRMANN-RONZAUD* and A.S. PAVLOVIC[†]
Laboratoire de Magnétisme, C.N.R.S.

and

A. WAINTAL[‡]
Service National des Champs Intenses, C.N.R.S., 166X, 38042-Grenoble-Cedex, France

Acoustical velocity measurements of the longitudinal and transverse modes have been performed on a multidomain MnO single crystal, and on a monodomain crystal obtained by applying uniaxial stress, from T_N to 300 K. Critical behaviour of the $\Delta v/v$ is described by a logarithmic law. The Debye temperature, extrapolated to 0 K, is 435 K.

1. Introduction

The elastic coefficients C_{11}, C_{12} and C_{44} have been determined between $T = 300$ K and T_N from acoustical velocity measurements on MnO single crystals without and with an applied uniaxial stress of 200 bar along the [111] direction. The uniaxial stress was applied to de-twin the crystal [1] in order to obtain a near single T-domain crystal.

Acoustical longitudinal and transverse waves of 15 MHz frequency and wave vector along the [1̄10] direction were sent through the crystal and the velocities determined by the method of "superposition of echoes".

2. Results

The elastic stiffness constants C_{11}, C_{12} and C_{44} of a monodomain MnO single crystal, from 118.5 K to room temperature, are given in fig. 1.

These results are in good agreement with the measurements performed by Oliver [2] and by Uchida and Saito [3] at room temperature as well as those of Cracknell and Evans [4] just above T_N.

On a stressed crystal, transducer bond breaking ("Nonaq" as bonding agent) made the determination of the Néel temperature impossible. Therefore strain-gauge measurements were used to determine $T_N = (118. \pm 0.1)$ K.

The C_{44} has also been measured on an unstressed sample. At most temperatures, both stressed and unstressed values were the same within 0.5%, but as T_N was approached, the values became 5% higher for the unstressed crystal. Although the region very near T_N was explored very thoroughly, the data was scattered and difficult to reproduce (cf. [4]). Approaching T_N from the paramagnetic side a new family of echoes appeared at T_N which was very close to the old family of echoes. Below T_N the amplitude of the old echoes decreased while the amplitude of the new echoes increased down to 116 to 114 K, where all the echoes were lost because of bond breaking. The phenomenon was reproducible over many temperature cycles. The only interpretation which seems plausible is that the new family of echoes is created by the formation of antiferromagnetic T-domains, which have slightly altered cell dimensions. From this, it appears that in this temperature region both paramagnetic and antiferromagnetic regions co-exist. This is what would be expected for a first-order phase transition. Apparently, the transducer bond does not break immediately at T_N for an unstressed crystal because at zero stress, the dilatation is small.

The Debye temperature θ_D, represented in fig. 2, has been calculated as a function of temperature from the elastic stiffness constants. Extrapolation of the high temperature elastic stiffness constants to 0 K gave values which were used to calculate $\theta_D(0\ K) = 435$ K; this value is in fair agreement with that determined from specific heat data [5].

The critical behaviour of the transverse and longitudinal $\Delta v/v$ along [1̄10] (fig. 3) is represented by the relation $\Delta v/v = A \ln \epsilon + B$, where $A = 2.67 \times 10^{-2}$, $B = 0.25 \times 10^{-2}$ and $A = 0.78 \times$

* and: Institut Laue-Langevin, 156X, 38042-Grenoble-Cedex.
† Permanent address: Dept. of Physics, West Virginia University, Morgantown, W. Va. 26506, U.S.A.
‡ and: Laboratoire des Rayons X, C.N.R.S., 38042-Grenoble.

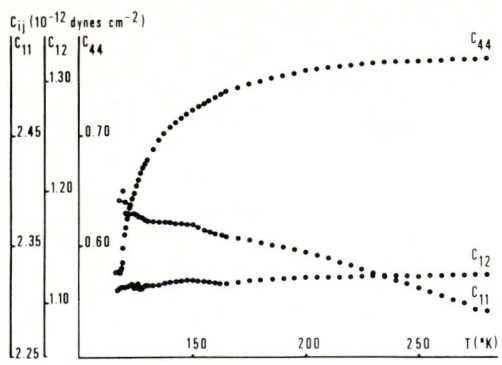

Fig. 1. Elastic stiffness constants of MnO (applied stress of 200 bar).

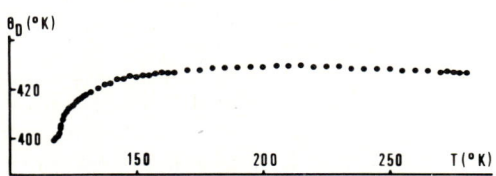

Fig. 2. Debye temperature of MnO (applied stress of 200 bar).

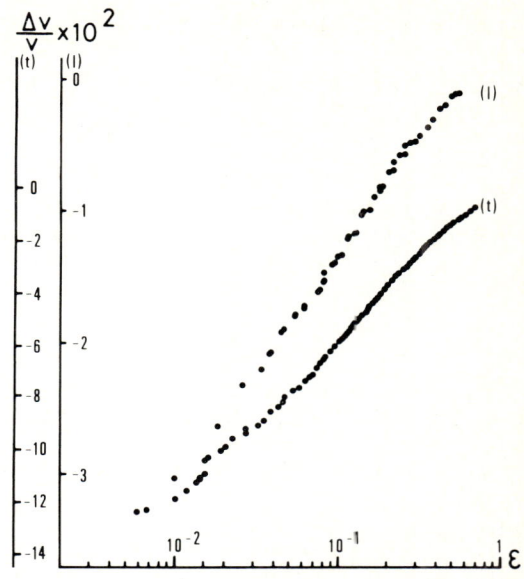

Fig. 3. Variation of longitudinal (l) and transverse (t) velocities (ϵ: reduced temperature; $T_N = 118.5$ K).

10^{-2}, $B = 0.47 \times 10^{-2}$, respectively, in the range $10^{-2} < \epsilon < 2 \times 10^{-1}$, ϵ is the reduced temperature. A similar logarithmic law has been reported for RbMnF$_3$ [6], for the rare earths metals Ho, Gd, Dy, Tb [6], and for several other materials. Bennett [7] has shown that the velocity shift should follow either a power law ($\Delta v/v \sim \omega^o \epsilon^{-\xi}$) or a logarithmic law ($\Delta v/v \sim -\omega^o \ln \epsilon$). These arise in the theory from phonons interacting with the critical fluctuations of the spin system near the transition temperature.

References

[1] D. Bloch, D. Herrmann–Ronzaud, C. Vettier, , W.B. Yelon and R. Alben, Phys. Rev. Letters 35 (1975) 963.
[2] D.W. Oliver, J. Appl. Phys. 40 (1969) 893.
[3] N. Uchida and S. Saito, J. Acoust. Soc. Am. 51 (1972) 1602.
[4] M.F. Cracknell and R.G. Evans, Sol. Stat. Commun. 8 (1970) 359.
[5] J.A. Hofmann, A. Paskin, K.J. Tauer and R.J. Weiss, J. Phys. Chem. Solids 1 (1956) 45.
[6] B. Lüthi, T.J. Moran and R.J. Pollina, J. Phys. Chem. Solids 31 (1970) 1741.
[7] H.S. Bennett, Phys. Rev. 185 (1969) 801.

CRITICAL BEHAVIOR OF CUBIC AND TETRAGONAL FERROMAGNETS IN A FIELD

D. MUKAMEL, E. DOMANY and M.E. FISHER

Baker Laboratory and Laboratory of Atomic and Solid State Physics, Cornell University, Ithaca, New York 14853, U.S.A.

The phase diagrams and critical behavior of cubic and tetragonal ferromagnets in a field are discussed using mean field, scaling, and renormalization group arguments. In particular we find that for cubic ferromagnets with three easy axes and $H\|[111]$ there is a multicritical phase transition region for $T < T_c$ which is described by the three-component Potts model. Competing predictions for the multicritical behavior can be tested experimentally.

The thermodynamic behavior of cubic and tetragonal ferromagnets in a field below the Curie point is strongly influenced by the nature of the small anisotropic interactions. We find that a variety of phase transitions, critical and multicritical points should occur in low magnetic fields (of order H_{Aniso}). In particular, cubic ferromagnets such as Fe, $PrAl_2$, $NdAl_2$, etc. [1], provide experimentally accessible realizations of the ($q = 3$)-component Potts model, which is a system of appreciable theoretical interest [2, 3] because of competing predictions for its multicritical behavior in $d = 3$ dimensions.

The simplest model which describes a cubic ferromagnet is a continuous spin Hamiltonian

$$\bar{\mathcal{H}} = -\mathcal{H}/k_BT = \int d\mathbf{R}\{-\tfrac{1}{2}|\nabla s|^2 + \mathbf{h}\cdot\mathbf{s} - U[s(\mathbf{R})]\}, \tag{1}$$

where $s = (s_x, s_y, s_z)$ is an ($n = 3$)-component spin field, $\mathbf{h} = m\mathbf{H}/k_BT$ is the reduced magnetic field, and

$$U(s) = \tfrac{1}{2}r|s|^2 + u|s|^4 + v(s_x^4 + s_y^4 + s_z^4), \tag{2}$$

with $r \sim T - T_c$. For stability we require $u + v > 0$ and $3u + v > 0$. To describe a tetragonal crystal with hard [001] axis one may set $s_z \equiv 0$ in (1) and (2). The phase diagrams associated with this Hamiltonian depend on the sign of the cubic coefficient v.

(A) For $v < 0$ we may show that the Hamiltonian (1) with $\mathbf{h}$ close to a cubic diagonal is equivalent to a continuous ($q = 3$)-component Potts model. An orthogonal transformation $s \Rightarrow \sigma = (\sigma_0, \sigma_1, \sigma_2)$ with $\sigma_0 \propto (s_x + s_y + s_z)$ and $\sigma_1 \propto (s_x - s_y)$, takes U into the same form except that the last term becomes

$$\tfrac{1}{3}v[6\sigma_0^2(\sigma_1^2 + \sigma_2^2) - 2\sqrt{2}\sigma_0(\sigma_2^3 - 3\sigma_1^2\sigma_2)$$
$$+ \tfrac{3}{2}(\sigma_1^2 + \sigma_2^2)^2 + \sigma_0^4]. \tag{3}$$

Now fix $T < T_c$ so that $r < 0$, and vary the diagonal field $\mathbf{h} = (h_0, h_1 = h_2 = 0)$. For large $\mathbf{h}$ the spin component σ_0 is noncritical and may hence be replaced by its mean value $\bar{\sigma}_0(h_0)$, while, by symmetry, $\langle\sigma_1\rangle = \langle\sigma_2\rangle = 0$. However, as h_0 is reduced, the transverse components σ_1 and σ_2 become critical when $\tilde{r}(\bar{\sigma}_0) = r + 4(u + v)\bar{\sigma}_0^2 \approx ah_0 - b|v|$ vanishes. The system may then be described in terms of the reduced, *two*-component order parameter, $\mathbf{\Psi} = (\sigma_1, \sigma_2)$, with a reduced Hamiltonian of the same form (1) but with last term $\tilde{v}(\bar{\sigma}_0)[\Psi_2^3 - 3\Psi_1^2\Psi_2]$; but such a coupling merely describes the continuous spin version of the $q = 3$ Potts model [2].

High temperature series extrapolation studies [2] and exact analytic calculations of the latent heat [3] have shown that the ($d = 2$)-dimensional Potts model for $q = 3$ exhibits a *continuous* phase transition in zero symmetry breaking field, $h_1 = h_2 = 0$. The (H_x, H_y, H_z) phase diagram, based on the calculations in $d = 2$ dimensions [3] is shown schematically in fig. 1. Here the space $\mathbf{H}$ is divided, at small fields, into six regions by twelve first order planes. On the principal diagonals [111], etc., three distinct magnetic phases, related to one another by a three-fold rotation, are in equilibrium. The first order surfaces terminate in critical edges. Three critical lines meet on each diagonal at an *anomalous* tricritical point or Potts point, labeled P, P' etc., in fig. 1.

A quite distinct phase diagram is predicted by mean field theory [2, 4] and is presumably correct for $d \geq 4$. Now the critical edges of the original first order planes terminate in sets of

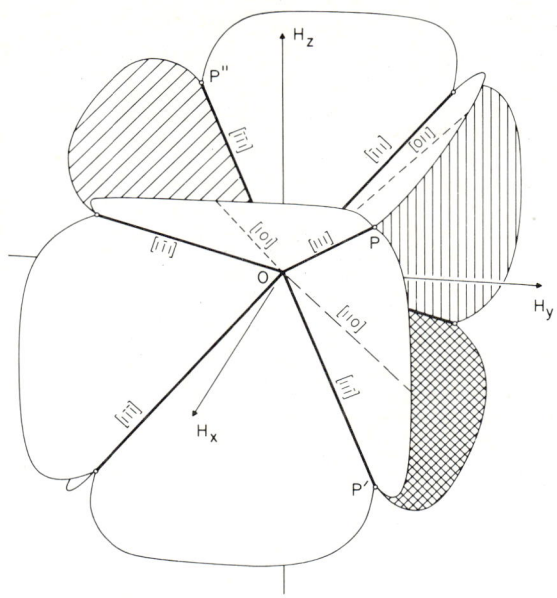

Fig. 1. View in (H_x, H_y, H_z) space of the schematic phase diagram of a cubic ferromagnet with three easy axes at fixed T below T_c, drawn in accord with results for the two-dimensional $q = 3$ Potts model. Bold lines denote lines of magnetic triple points; thin curves represent critical lines; the open circles marked P, P', etc., mark the "anomalous tricritical" or Potts points.

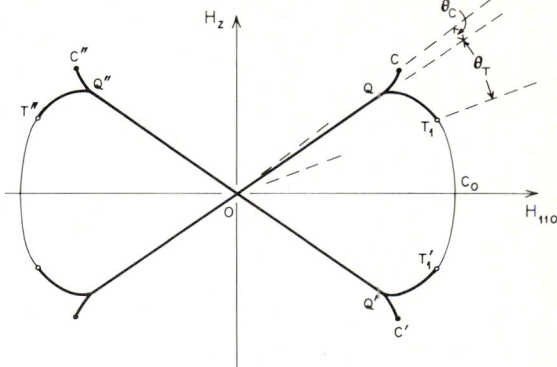

Fig. 2. Schematic section of the mean-field phase diagram analogous to fig. 1 in the (H_{110}, H_z)-plane, showing critical points C, C_0, tricritical points T_1, etc., first order phase boundaries (bold), and critical lines (thin). At $T \ll T_c$ the opening angles are $\theta_C \simeq 0.36°$ and $\theta_T \simeq 1.68°$.

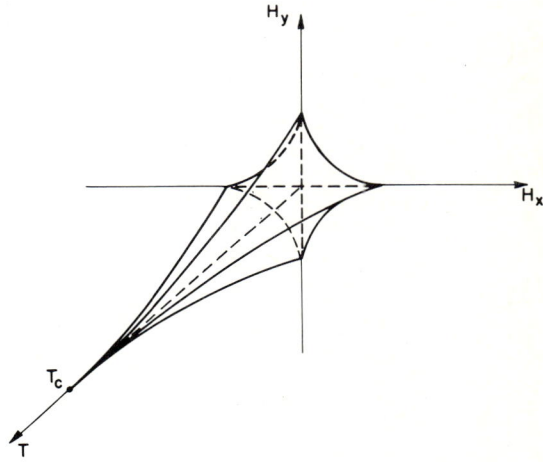

Fig. 3. Schematic view in (H_x, H_y, T) space of the phase diagram of a cubic ferromagnet with four easy axes (i.e. $v > 0$).

three symmetrically arranged *tri*critical points, T_1, T_2, and T_3, which are connected by three critical "wing edges" bounding new coexistence surfaces which converge on a four-phase or quadruple point on the diagonal axis. The presence of the wing surfaces ensures a first order transition as a function of field at fixed orientation *even if* H is somewhat misaligned from the three-fold axis. Fig. 2 represents schematically a section of the mean field phase diagram in the (H_{110}, H_z) plane.

(B) When $v > 0$ a *tetragonal* ferromagnet with a hard z-axis is found to display Ising-like *critical lines* $H_c(T) \sim (T_c - T)^\psi$ when H lies along the x or y axes. An explicit, general ϵ-expansion renormalization group calculation yields $\psi = 2 - \alpha - \beta - \phi_v$ which, with appropriate XY (or $n = 2$) exponents, yields $\psi \simeq 1.7$–1.8. In *cubic* systems fluted *critical surfaces* appear when H lies in the (x, y), (y, z) or (z, x) planes: see fig. 3. The critical behavior is Ising-like unless H lies along the x, y, or z axes in which case it is XY-like. (Sznajd [5] has studied the mean field phase diagram for this case.)

The predictions of general scaling theory [6] are informative: for the magnetization of a cubic ferromagnet as $t = (T - T_c)/T_c \to 0$, one may write

$$M(H, T) \approx |t|^\beta Y(h/|t|^{\Delta^*}|v|^\zeta, v|t|^{-\phi_v}), \quad (4)$$

where $\Delta^* = 2 - \alpha - \beta - \zeta\phi_v$, in which $\alpha \simeq -0.14$ and $\beta \simeq 0.36$ are Heisenberg ($n = 3$) exponents while $\phi_v \simeq -0.03 < 0$ is the Heisenberg crossover exponent for cubic anisotropy [7]; finally ζ is an *arbitrary* exponent. Above T_c one may, as usual, choose $\zeta = 0$ and v is then a standard "irrelevant" variable [7], which can be neglected as $t \to 0$. Below T_c, however, v is a "dangerous irrelevant variable", which cannot merely be equated to zero. Nevertheless, in mean field theory (where $\phi_v \equiv 0$) the unique choice $\zeta = 1$ allows one to set $v = 0$ in the *second* argument of $Y(x, y)$, so that $M/|t|^\beta$ becomes a function only of $h/|v|^\zeta |t|^{\Delta^*}$. We expect this result to hold more generally for appropriate, $\zeta(d)$. The result $\psi = 2 - \alpha - \beta - \phi_v$ mentioned above [4], then implies via extended scaling [6] that $\zeta(d) = 1$. Consequently the whole H phase diagrams of figs. 1 and 2 should scale as $v|t|^{\Delta^*}$ with $\Delta^* \simeq 1.8$. Experimental tests of our qualitative and quantitative predictions are highly desirable!

The support of the National Science Foundation and the Materials Science Center at Cornell University is gratefully acknowledged.

References

[1] W.J. Carr, Jr., *Encyclopedia of Physics*, H.P.J. Wijn, ed. (Springer-Verlag, Berlin, 1966) vol. XVIII/2, p. 284.
H.G. Purwins, E. Walker, B. Barbara, M.F. Rossignol and P. Bak, J. Phys. C7 (1974) 3573.
P. Bak, J. Phys. C7 (1974) 4097.
[2] J.P. Straley and M.E. Fisher, J. Phys. A6 (1973) 1310.
[3] R.J. Baxter, J. Phys. C6 (1973) L445.
[4] D. Mukamel, E. Domany and M.E. Fisher, Phys. Rev. Lett. 37 (1976) 565. Previous work (e.g. ref. 1) has been restricted to the symmetry axes.
[5] J. Sznjd, Acta Physica Polonica. A47 (1975) 61.
[6] See M.E. Fisher, Proc. Nobel Symposium 24 (1973) 16.
[7] M.E. Fisher, Rev. Mod. Phys. 46 (1974) 597.
A.D. Bruce and A. Aharony, Phys. Rev. B11 (1975) 428.

PROPERTIES OF A ONE-DIMENSIONAL MAGNET IN A MAGNETIC FIELD

J.M. LOVELUCK and S.W. LOVESEY

Institut Laue-Langevin, 156X Centre de Tri, 38042 GRENOBLE, France

The exact static, and dynamical properties of a one-dimensional, quasi-classical spin-wave model are presented, and discussed as functions of wavevector and applied magnetic field.

We calculate exactly the dynamical properties of a one-dimensional quasi-classical ferromagnetic spin-wave model subject to an applied field. This model permits the study of damping and renormalisation of spin-waves as a function of long-range disorder, which is reduced by application of the magnetic field.

It is well known that the one-dimensional, isotropic, Heisenberg magnet does not order at a finite temperature. However, the model possesses strong short-range order at temperatures small compared with the, nearest neighbour, exchange interaction J. The magnetic order is enhanced by the application of a magnetic field, which we choose to be parallel to the z-direction. In view of the substantial short-range order, the static properties of a one-dimensional magnet in an applied field are given to a good approximation, for low temperatures, by the corresponding properties of a system of non-interacting excitations with energies

$$\omega_k = h + 2J(1 - \cos k),$$

where h is the applied field (in units of $g\mu_B$) and lengths are measured in units of the lattice spacing. Assuming that the excitations form a gas of quasi-classical harmonic oscillators, with frequencies ω_k, the magnetization, for example, is readily shown to be

$$\langle S^z \rangle = 1 - T^*/K, \qquad K^2 = h^*(h^* + 4), \qquad (1)$$

where the reduced temperature, T^*, and reduced magnetic field, h^*, are defined by $k_B T/J$ and h/J, respectively. Clearly, for the validity of the model, a weak lower bound on h, for a given T, is obtained by requiring that $\langle S^z \rangle$ be positive. The pair correlation functions

$$\langle S^z_0 S^z_l \rangle - \langle S^z \rangle^2 = (T^*/K)^2 \{\tfrac{1}{2}(2 + h^* - K)\}^{2l}, \qquad l = 0, 1, 2, \ldots \qquad (2)$$

and the right-hand side reduces, in the limit of large fields, to the result

$$(T^*/h^*)^2 (h^*)^{-2l}.$$

From (2), we can obtain readily the wave-vector dependent susceptibility defined by

$$\chi(k) = (1/k_B T)\langle S^z_k S^z_{-k}\rangle$$

$$\equiv (1/k_B T) \sum_l \exp(ikl)(\langle S^z_0 S^z_l \rangle - \langle S^z \rangle^2)$$

with the result,

$$J\chi(k) = (T^*/K)\left\{\frac{(2 + h^*)}{K^2 + 2(1 - \cos k)}\right\}. \qquad (3)$$

This result shows that K is the inverse correlation length, in reduced units. We conclude the summary of the static properties by noting that the specific heats at constant field, C_h, and constant magnetization, C_m, are

$$C_h = k_B, \qquad C_m = k_B\{1 - K/(2 + h^*)\}. \qquad (4)$$

The deviation from unity of the ratio C_h/C_m is a measure of the coupling between fluctuations in the energy and magnetization induced by the applied field, and from the result (4) we see that the ratio increases as the square of the field for large magnetic fields.

The results (1), (3) and (4) have been compared successfully, for modest fields and low temperatures (for example, for $T^* = 0.1$, $h^* \geq 0.5$), with exact numerical results for a classical Heisenberg magnet by Blume et al. [1] and Lovesey and Loveluck [2].

We turn now to the study of the normalised spectral density

$$F(k, \omega) = \int_{-\infty}^{\infty} \frac{dt}{2\pi} \exp(i\omega t)\{\langle S^z_k(t) S^z_{-k}\rangle/\langle S^z_k S^z_{-k}\rangle\}. \qquad (5)$$

All the moments of $F(k, \omega)$ can be calculated analytically, and the time dependent correlation function can be obtained by summing its power series in t. We find that $F(k, \omega)$ is non-zero only for frequencies which satisfy

$$|\omega| \leq \theta_k; \qquad \theta_k^2 = 8J^2(1 - \cos k). \tag{6}$$

It is convenient to express the result for $F(k, \omega)$ in terms of a frequency Ω_k, and a damping function Δ_k defined by

$$\Omega_k^2 = 4J^2(1 - \cos k)(1 - \cos k - \tfrac{1}{2}K^2 \cos k), \tag{7}$$

$$\Delta_k^2 = 2J^2(2 + h^*)K \sin k (1 - \cos k), \tag{8}$$

and the result is, for $|\omega| < \theta_k$,

$$\pi J^2 F(k, \omega) = (T^*/\chi(k))$$
$$\times \{2(1 - \cos k)[2(1 - \cos k)(6 + K^2 + 2 \cos k) - \omega^2]\}$$
$$\times \{[\theta_k^2 - \omega^2]^{1/2}|\omega^2 - \Omega_k^2 - i\Delta_k^2|^2\}^{-1}. \tag{9}$$

The singularity at $\omega = \pm \theta_k$ is an anomaly of our model, and it is akin to the singularity found by Kim and Nelkin [3] in the velocity autocorrelation function for a disordered harmonic solid.

The frequency Ω_k coincides with the spin-wave frequency in the limit of zero applied field, and for finite fields it is real for wavevectors $k > k_c$, where

$$k_c = \cos^{-1}\{2/(2 + K^2)\}. \tag{10}$$

The critical wavevector increases monotonically with h^*, and achieves 92% of its saturation value, $\pi/2$, for $h^* = 1.0$. It is important to note also that Ω_k exceeds θ_k for wavevectors $k > (\pi - k_c)$. Consequently, $F(k, \omega)$, when displayed as a function of ω for a fixed k, can possess a peak at non-zero ω (other than the physically uninteresting singularity at $\omega = \theta_k$) only for wavevectors in the range $k_c < k < (\pi - k_c)$, and this range shrinks rapidly to a negligible value with increasing applied field. Moreover, the ratio Δ_k/Ω_k is small only near the zero boundary. We conclude from this discussion that a well defined collective mode exists in the quasi-classical spin-wave model for a small range of wavevectors, for small fields.

References

[1] M. Blume, P. Heller and N.A. Lurie, Phys. Rev. B11 (1975) 4483.
[2] S.W. Lovesey and J.M. Loveluck, J. Phys. C9 (1976) 3639.
[3] K. Kim and M. Nelkin, Phys. Rev. B7 (1973) 2762.

CRITICAL BEHAVIOUR OF THE ISING MODEL IN A TRANSVERSE FIELD

J. OITMAA

School of Physics, The University of New South Wales, Kensington, N.S.W. 2033, Australia

and

M. PLISCHKE

Department of Physics, Simon Fraser University, Burnaby, B.C., Canada

High temperature series have been obtained for the fluctuation in the order parameter of the Ising model with a transverse field on the three common cubic lattices. Accurate estimates of the position of the critical line in the temperature-field plane have been obtained.

A simple lattice spin model which has been used to describe a wide variety of physical systems is the "Ising model in a transverse field", described by the Hamiltonian

$$\mathcal{H} = -J \sum_{\langle ij \rangle} S_i^z S_j^z - \Gamma \sum_i S_i^x \qquad (1)$$

in which S^z, S^x are quantum-mechanical spin $\frac{1}{2}$ operators and the sums are over nearest neighbour pairs and over all sites, respectively. An extensive discussion of the application of this model to real physical systems has been given by Stinchcombe [1]. An important application is to induced moment magnetism in systems with singlet ground and first excited crystal field states (Birgeneau [2]).

In two and three dimensions no exact results are known but the following qualitative behaviour is to be expected. The transverse field Γ effectively hinders the S^z ordering and the critical temperature $T_c(\Gamma)$ will decrease monotonically from the Ising value at $\Gamma = 0$ to zero at some value $\Gamma = \Gamma_c$; for $\Gamma > \Gamma_c$ no transition occurs. There will thus be a critical line in the (T, Γ) plane separating ordered and disordered phases.

We have used the technique of high temperature series expansions to study the critical behaviour of this model for the simple cubic, body-centred cubic, and face-centred cubic lattices. In particular we have obtained accurate estimates of the position of the critical line, the value of the critical field Γ_c, and the value of the critical exponent γ. The results for the simple cubic lattice extend earlier work of Elliott and Wood [3]; the other results are new.

The derivation of high temperature series for this model is complicated by the fact that the two terms in (1) do not commute and thus cannot be simultaneously diagonalized. Furthermore the order parameter $M^z = (1/N) \sum_i S_i^z$ does not commute with the total Hamiltonian and thus the square of the order parameter fluctuation $\langle (M^z)^2 \rangle$ is not directly related to the susceptibility as it is, for example, in the ordinary Ising model. In the critical region the asymptotic behaviour of both quantities is the same. We have obtained a high temperature expansion for the latter quantity, in particular for the quantity

$$Y = 4N \langle (M^z)^2 \rangle$$

$$= 1 + 4 \sum_{r=0}^{\infty} \frac{1}{r!} \langle \sum_{m,n}{}' S_m^z S_n^z$$

$$\times \{4K \sum_{\langle ij \rangle} S_i^z S_j^z + 2h \sum_i S_i^x\}^r \rangle_N, \qquad (2)$$

where $4K = J/kT$, $2h = \Gamma/kT$, and

$$\langle A \rangle_N \equiv \text{coeff. of } N \text{ in } [2^{-N} \text{Tr}\{A\}].$$

Terms in the expansion (2) can be associated with graphs on the lattice. These graphs can be enumerated and the associated factors calculated by computer. The final result can be conveniently written in the form

$$Y = 1 + \sum_{r=1}^{\infty} y_r(h) K^r. \qquad (3)$$

where $y_r(h)$ are infinite polynomials in h^2. We have computed the leading terms in these polynomials for the three common cubic lattices, up to y_{10} inclusive.

A detailed discussion of the derivation and explicit expressions for the polynomials are given in a recent paper [4].

To analyse the series we have written $h = \eta K$, obtained a series in the single variable K for a number of values of the parameter η, and used standard Padé approximant and ratio methods to obtain estimates of the critical "temperature" K_c and exponent γ. These results are shown in Fig. 1 as plots of $4kT_c/zJ$ versus $2\Gamma/zJ$, where z is the coordination number. Our estimates of the quantity Γ_c/J are 5.33 ± 0.3 for the FCC lattice, 3.48 ± 0.02 for the BCC latice, 2.54 ± 0.02 for the SC lattice. Analysis of the series also shows convincingly that the critical exponent γ is constant at 1.25 along the critical line.

It has recently been proved by Suzuki [5] that at $\Gamma = \Gamma_c$ and $T = 0$ the exponent γ should change discontinuously to the value corresponding to an Ising model of one higher dimensionality. In this region our series become irregular and we are unable to see any evidence for crossover to the expected four dimensional behaviour.

An interesting result is found by plotting $T_c(\Gamma)/T_c(0)$ versus Γ/Γ_c [4]. The points for all three lattices fall with remarkable accuracy on a single curve, indicating that the law of corresponding states is obeyed. This result is of importance in applying these results to real systems in which the exchange interaction will not normally be of the nearest neighbour type.

This work was carried out while both authors were at the University of Alberta. They thank the Theoretical Physics Institute and the National Research Council of Canada for support.

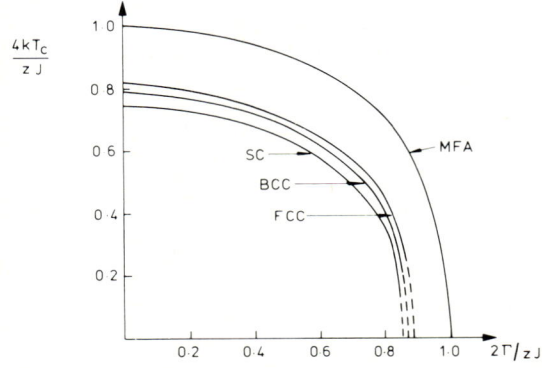

Fig. 1. Phase boundary in the T, Γ plane for the three cubic lattices, and the mean field approximation.

References

[1] R.B. Stinchombe, J. Phys. C: Solid St. Phys. 6 (1973) 2459.
[2] R.J. Birgeneau, in: AIP Conference Proceedings No. 10, Magnetism and Magnetic Materials, C.D. Graham and J.J. Rhyne, eds. (1972) p. 1664.
[3] R.J. Elliott and C. Wood, J. Phys. C: Solid St. Phys. 4 (1971) 2359.
[4] J. Oitmaa and M. Plischke, J. Phys. C: Solid St. Phys. 9 (1976) 2093.
[5] M. Suzuki, Prog. Theor. Phys. 56 (1976), in press.

QUANTUM CLASSICAL CROSSOVER CRITICAL BEHAVIOR OF THE ISING MODEL IN A TRANSVERSE FIELD

P. PFEUTY

Laboratorie de Physique des Solides*, Université Paris-Sud, 91405 Orsay, France

The Ising model (spin 1/2) in a transverse field has many applications in the field of cooperative phenomena (magnetism, Jahn–Teller effects, ferroelectricity). The quantum classical crossover critical behavior is studied for a d-dimensional system with exact results when $d = 1$. The crossover behavior depends strongly on the relative magnitude of the crossover exponent ϕ and of the critical line shift exponent ψ.

The spin 1/2 Ising model in a transverse field is defined by the Hamiltonian

$$H = -\Gamma \sum_i \sigma_i^x - \frac{1}{2} \sum_{i \neq j} J_{ij} \sigma_i^z \sigma_j^z, \qquad (1)$$

where the σ's are Pauli spin matrices, Γ is the transverse field and J_{ij} is the interaction between spins on sites i and j. This quantum model has applications in a wide range of cooperative phenomena: magnetic and ferroelectric phase transitions [1], structural phase transitions [2], coupled chain-systems [3] and more recently meson field theory [4].

If the transverse field Γ is less than a critical value Γ_c, the model exhibits the usual Ising model like transition at a finite critical temperature $T_c(\Gamma)$ and with the critical exponents unaltered by the presence of a non-commuting transverse field [5]. The transition at $T = 0$ when Γ approaches Γ_c has different critical exponents; the exponents of the d-dimensional Ising model in a transverse field at $T = 0$ are the same as those of the $(d+1)$-dimensional Ising model ($\Gamma = 0$), but with Γ playing the role of T [6]. For $d = 1$ the previous statement is exact [7, 8]. At low temperature when Γ is close to Γ_c, there is a competition between the classical and the "quantum" critical behavior leading to a crossover behavior [9].

The strong equivalence between the $d = 1$ Ising model in a transverse field at $T = 0$ and the $d = 2$ Ising model at $\Gamma = 0$ [8, 10], leads to a natural extension to higher dimensionalities with the following consequence [11]: The quantum-classical-crossover critical behavior of the d-dimensional Ising model in a transverse field is assumed to be "analagous" to the $d + 1$ to d-dimensional crossover critical behavior for the usual Ising model when one length of the system becomes finite of order N with *periodic boundary conditions*. The temperature and the transverse field of the d-dimensional quantum model correspond respectively to $1/N$ and the temperature of the equivalent $(d+1)$-dimensional classical model [12].

The analogy with the scaling theory of finite size effects [13] can then be used to study the quantum-classical-crossover behavior of this model. If we introduce the shift exponent ψ defined by $T_c(\Gamma) \sim (\Gamma_c - \Gamma)^\psi$ and the crossover exponent Φ defined by the crossover temperature $T^\times \sim (\Gamma_c - \Gamma)^\Phi$, then usually because of the periodic boundary conditions in the finite size direction, Φ and ψ are not equal. From simple arguments [13, 14], the crossover exponent Φ is equal to ν_{d+1} (the exponent ν describing the divergence of the correlation length ξ of the $d + 1$ dimensional Ising model). The crossover behavior is rather different when $\Phi < \psi$ and when $\Phi > \psi$. Lets consider the correlation length ξ. In the region $(\Gamma - \Gamma_c)$ and T small ξ behaves in a scaled fashion:

$$\xi \sim \{\Gamma - \Gamma_c(T)\}^{-\nu_{d+1}} f(x), \quad x = T/\{\Gamma - \Gamma_c(T)\}^{\Phi = \nu_{d+1}}. \qquad (2)$$

When $\Phi = \psi$ the scaling is extended (15) and ξ becomes

$$\xi \sim |\Gamma_c - \Gamma|^{-\nu_{d+1}} g(x),$$
$$x = T \cdot \mathrm{sign}\,(\Gamma_c - \Gamma)/|\Gamma_c - \Gamma|^{\Phi = \nu_{d+1}}. \qquad (3)$$

The main differences between the two cases a): $\Phi \leq \psi$ and b): $\Phi > \psi$ appear when we consider the behavior at $\Gamma = \Gamma_c$ lowering the tem-

*Laboratoire associé au C.N.R.S.

perature. From eq. (2) we get

$$\xi \sim T^{-A}$$

with

$$A = \nu_{d+1}/\psi \quad \text{if} \quad \Phi \leq \psi \quad \text{and} \quad (4.\text{a})$$

$$\alpha = \nu_{d+1}/\psi + (\nu_{d+1} - \nu_d)(1/\Phi - 1/\psi) \quad \text{if} \quad \Phi > \psi. \quad (4.\text{b})$$

For $d > 3$ we expect $\Phi = \psi = 1/2$ (for $d > 3$, $\nu_{d+1} = 1/2$ and $\psi = 1/2$ for all d [13] when periodic boundary conditions are imposed). The case $d = 3$ is marginal and logarithmic terms are expected. For d less than 3, ψ being fixed to 1/2 and Φ being larger than 1/2, eq. (4.b) should be valid (for $d = 2$ the difference between the values of A calculated from eq. (4.a) and from eq. (4.b) is of the order of 0.1 when A is of the order of 1).

For $d = 1$ the situation is an interesting limit case: the critical line $T_c(\Gamma)$ is pushed towards the axis $T = 0$; Φ is equal to $\nu_2 = 1$ and ψ is not defined. The equivalence with the finite size crossover behavior is exact and the crossover behavior can be obtained exactly. For instance the inverse correlation length is equal to

$$\xi^{-1} \sim T \cdot F(x), \quad x = T/(\Gamma - \Gamma_c), \quad (5)$$

where the scaling function $F(x)$ can be determined analytically in certain limits [11]. For instance $F^-(|x|) = F(x < 0)$ behaves like

$$F^-_{x \to 0}(|x|) \sim e^{-1/|x|} |x|^{-1/2} \left(\sum_n a_n |x|^n \right). \quad (6)$$

To conclude we suggest to pursue a more detailed study both numerically and experimentally to test the assumptions and the predictions proposed in this short communication. The main point is that a physical system represented by the Ising model in a transverse field is a good candidate to study finite size crossover effects with *periodic boundary conditions*, by varying easily accessible parameters such as temperature and transverse field at sufficiently low temperature and close to the critical transverse field.

References

[1] R.B. Stinchcombe, J. Phys. C (Solid St. Phys.) 6 (1973) 2459.
[2] R.A. Bishop and J.A. Krumhansl, Phys. Rev. B (1975).
[3] J. Lajzerowicz and P. Pfeuty, J. de Physique (supplement) 32 C 5a (1971) 193.
[4] D. Amati, M. Le Bellac, G. Marchesini and M. Ciafolini, unpublished (1976).
[5] R.J. Elliott and C. Wood, J. Phys. C (Solid St. Phys.) 4 (1971) 2359.
[6] P Pfeuty and R.J. Elliott, J. Phys. C (Solid St. Phys.) 4 (1971) 2370.
[7] P. Pfeuty, Ann. Phys. N.Y. 57 (1970) 79.
[8] M. Suzuki, Progr. Theor. Phys. 46 (1971) 1337.
[9] E. Riedel and F. Wegner, Z. Physik 225 (1969) 195.
[10] J. Lajzerowicz and P. Pfeuty, Phys. Rev. B 11 (1975) 4560.
[11] P. Pfeuty, J. Phys. C (Solid St. Phys.) (1976).
[12] R. P. Feynman, Phys. Rev. 84 (1951) 108.
[13] M.E. Fisher, Critical Phenomena, M.S. Green, ed Acad. Press, New York (1971) p. 73.
[14] J. Hertz Proceedings of the San Francisco Magnetism Conference (1974).
[15] M.E. Fisher and D. Jasnow, Theory of correlations in the critical region, to be published.

DYNAMICS OF THE ONE-DIMENSIONAL TRANSVERSE ISING MODEL

L.L. GONCALVES*† and R.J. ELLIOTT
Department of Theoretical Physics, 12 Parks Road, Oxford, OX1 3PQ, England

An approximate expression for the time-dependent longitudinal correlation function $\langle S_1^z(t) S_{1+n}^z(0) \rangle$ is presented. It gives the exact results in the limiting cases, where the transverse field, or the exchange interaction, is zero. Some results are given for the frequency transform of the self and nearest neighbour correlation functions.

The one-dimensional Ising model ($S = 1/2$) in a transverse field can be considered as a particular case of the XY-model with anisotropy and has been studied by many authors. Katsura [1] evaluated the free energy and Pfeuty [2] studied the static longitudinal correlation function at $T = 0$. The transverse time-dependent correlation was calculated by Niemeyer [3], and the longitudinal one was studied at $T = 0$ and large separation between spins by McCoy et al. [4]. Recently Capel et al. [5] presented the high temperature expansion of the longitudinal susceptibility and discussed the difficulties in its calculation. In this paper we propose an approximation which allows one to overcome some of these difficulties.

The hamiltonian of the system can be written as:

$$H = -J \sum_j S_j^z S_{j+1}^z - \Gamma \sum_j S_j^x. \quad (1)$$

Introducing the transformation,

$$a_j^+ = S_j^x + i S_j^y = \exp\left[i\pi \sum_{l=1}^{j-1} c_l^+ c_l\right] c_j^+, \quad (2)$$

where c's are fermion operators, we get:

$$H = H^+ P^+ + H^- P^-, \quad (3)$$

where $P^+(P^-)$ is a projection operator for states of an even (odd) number of c_j excitations, and $H^+(H^-)$ are the c-anticyclic (c-cyclic) hamiltonian, given by:

$$H^\pm = -\tfrac{1}{4} J \sum_{j=1}^{N-1} (c_j^+ c_{j+1} + c_j^+ c_{j+1}^+ + \text{h.c.})$$

$$-\Gamma \sum_{j=1}^N (\tfrac{1}{2} - c_j^+ c_j) \pm \tfrac{1}{4} J (c_N^+ c_1 - c_N^+ c_1^+ + \text{h.c.}). \quad (4)$$

The hamiltonian (4) can be put into diagonal form, as shown by McCoy et al. [4], as:

$$H^\pm = \sum_k \pm E_k (\beta_k^+ \beta_k - \tfrac{1}{2}), \quad (5)$$

where

$$E_k = (\tfrac{1}{4} J^2 - J\Gamma \cos ka + \Gamma^2)^{\tfrac{1}{2}}. \quad (6)$$

The $+$ sign in (5) holds for values

$$k^+ = \pm 2\pi(n + \tfrac{1}{2})/Na \quad (7)$$

and the $-$ sign for

$$k^- = \pm 2\pi n / Na \quad (8)$$

and n runs over integers from 0 to $N/2$. a is the lattice parameter.

The longitudinal time-dependent correlation function $\langle S_j^z(t) S_{j+n}^z(0) \rangle$ in the limit $N \to \infty$ is written as [5]:

$$\langle S_1^z(t) S_{1+n}^z(0) \rangle = \langle e^{iH^-t} S_1^z e^{-iH^+t} S_{1+n}^z \rangle, \quad (9)$$

where the average on the right-hand side is calculated over the eigenstates of H^-, and the density operator is $e^{-\beta H^-}$. We introduce an approximation by adding and substracting cyclic terms so that

$$\langle S_1^z(t) S_{1+n}^z(0) \rangle = \langle e^{iH^-t} S_1^z e^{-iH^-t} \cdot e^{i(H^- - H^+)t} S_{1+n}^z \rangle. \quad (10)$$

Then, after some algebraic manipulations we get:

---

*On leave from Departamento de Fisica da Universidade Federal do Ceara, Brazil.
† Supported by CNPq (Brazilian Government) and by Universidade Federal do Ceara, Brazil.

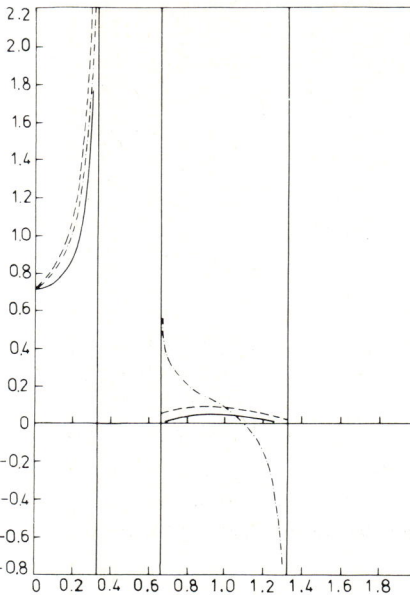

Fig. 1. The frequency transforms of the correlation functions at $\lambda = 1.5$. Full line—self-correlation function ($n = 0$) at high temperature $\beta = 0$. Dashed line—self-correlation function at low temperature $\beta = 100$. Dot–dash line—near neighbour correlation function ($n = 1$) at $\beta = 100$. [N.B. This function tends to zero as $\beta \to 0$]. Note that the response is confined to the frequency ranges $0 < \omega/J < 1/2\lambda$; $1 - 1/2\lambda < \omega/J < 1 + 1/2\lambda$ if $\lambda > 1$.

$$\langle S_1^z(t) S_{1+n}^z(0) \rangle = \frac{1}{4N} \sum_k \sum_{j=1}^{n+1}$$
$$\times [\{M_{n+1}^j \cos(j-1)ka \cos \tfrac{1}{2}Jt$$
$$\times (\cos E_k t - i\tau_k \sin E_k t)\} \times (-1)^{j-1}$$
$$+ i\{N_{n+1}^j \sin \tfrac{1}{2} Jt(\tau_k \cos E_k t - i \sin E_k t)$$
$$\times (-1)^j \times [(\Gamma/E_k) \cos(j-2)ka$$
$$- (J/2E_k) \cos(j-1)ka]\}], \quad (11)$$

where $\tau_k = \tanh \tfrac{1}{2} \beta E_k$ and M_{n+1}^j and N_{n+1}^j are respectively minors of a normal and shifted Toeplitz determinants [6] calculated with respect to the $(n+1)$th row jth column, in first case and the 1st column, jth row in the second. Toelitz determinants are generated by the function:

$$G(l) = -\frac{1}{N} \sum_k \tau_k [(\Gamma/E_k) \cos lka$$
$$- (J/2E_k) \cos(l+1)ka]. \quad (12)$$

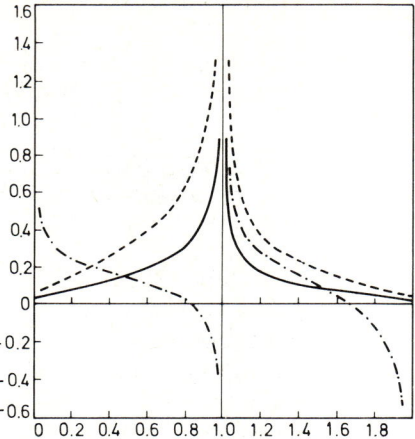

Fig. 2. As fig. 1 for $\lambda = 0.5$. The response now covers the range $0 < \omega/J < 2$ but there are singular points at $\omega/J = 1/2\lambda$, $1 \pm (1/2\lambda)$.

The approximate result shown in eq. (11) gives the known exact results in the limiting cases, where the field is zero, or the exchange interaction is zero. Besides, the first two moments of the auto-correlation function are exact at any temperature, and only the auto-correlation function is different from zero at $T = \infty$. Although the introduction of the change shown in eq. (10) implies the loss of the translational symmetry we believe it gives a satisfactory approximation and that the variation of $\langle S_j^z(t) S_{j+n}^z(0) \rangle$ with j will be small.

The auto-correlation function and the near neighbour correlation function are shown in the figs. 1 and 2, respectively, for various

$$\lambda = (J/2\Gamma) \quad \text{and} \quad \beta = (J/kT).$$

It is planned to extend these results and use them in the study of spin-phonon systems.

References

[1] S. Katsura, Phys. Rev. 127 (1962) 1508.
[2] P. Pfeuty, Annals of Physics 57 (1970) 79.
[3] T.H. Niemeyer, Physica 36 (1967) 377.
[4] B.M. McCoy, E. Barouch and D.B. Abraham, Phys. Rev. A4 (1971) 2331.
[5] H.W. Capel, E.J. Van Dongen and T.J. Siskens, Physica 76 (1974) 445.
[6] R.E. Hartwig and M.E. Fisher, Archive for Rational Mechanics and Analysis 32 (1969) 190.

SCALING BEHAVIOR OF THE SPECIFIC HEAT OF Gd NEAR T_c*

M.B. SALAMON, D.S. SIMONS† and C.C. HUANG

Department of Physics and Materials Research Laboratory, University of Illinois at Urbana-Champaign, 61801, USA

The specific heat of Gd near T_c is shown to scale with magnetic field in accordance with scaling law predictions. A surprisingly close agreement is found between the scaled data and the predictions of the spin-1/2 Heisenberg model.

In a recent paper [1] we presented data on the magnetic field dependence of the specific heat of gadolinium in the vicinity of the Curie temperature. There, we tested the scaling-theory result that the quantity [2]

$$t'(y) = (C_H - C_0)H^{\alpha/\beta\delta} \quad (1)$$

depends on magnetic field and temperature only in the combination

$$y = (1 - T/T_c)H^{-1/\beta\delta}. \quad (2)$$

The specific heat at constant internal field C_H is measured in units of $R = 8.31$ J/K mole and H is in units of $k_B T_c/7.55\mu_B = 5.74 \times 10^5$ Oe. The Curie temperature T_c has been chosen to coincide with the peak in the zero-field specific heat curve. The exponents have their usual meaning [3]. In our earlier analysis, we used published values [4, 5] of $\beta = 0.37$ and $\delta = 4.39$, obtaining α from the scaling equation $\alpha = 2 - \beta(\delta + 1)$, which must hold if scaling theory is to be valid. These values of the exponents yield a positive value of α and, as may be seen in fig. 10 of ref. 1, lead to a failure of the scaling law (1) and (2). Similar results follow from the values of β and δ reported more recently by Deschizeaux and Develey [6].

In this paper we employ the value $\alpha = -0.20 \pm 0.02$, obtained through a power-law fit of both our own specific heat data for $H = 0$ and Lewis' earlier data [7]. Using this value of α and $\beta = 0.38$ [6], we obtain $\delta = 4.78$ from the above scaling law. With these values of the exponents, the scaling law result (1) and (2) is satisfied by our data as may be seen in fig. 1.

In order to compare the experimental func-

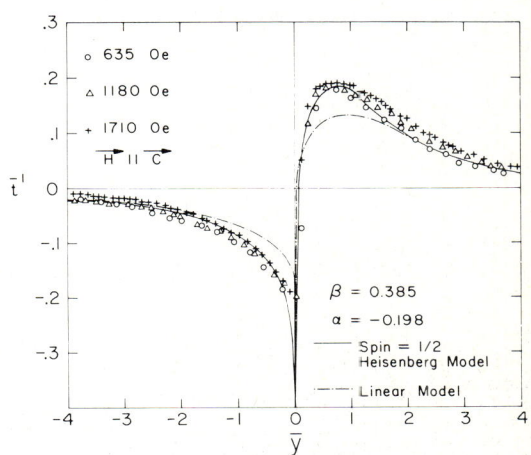

Fig. 1. Scaled specific heat in universal units with exponents and corresponding amplitudes as shown in table I. Theoretical predictions of the spin-½ Heisenberg model and the linear model with spin-½ Heisenberg exponents are shown for comparison.

tion $t'(y)$ with theoretical equations of state, the quantities t' and y must be redefined to be independent of the amplitudes B and D associated with the exponents β and δ, respectively [2]. The scale factors required are

$$t_0 = B^{2/\beta}D^{1+\alpha/\beta\delta} \quad \text{and} \quad y_0 = B^{-1/\beta}D^{-1/\beta\delta}. \quad (4)$$

The quantity B is known from the experimental determination of β [6], but the value of D must be regarded as unknown, owing to our use of a value of δ obtained from the scaling law. To find D, we have three possibilities: i) treat it as an adjustable parameter in the final comparison with theoretical results, ii) choose it such that in combination with B it gives the correct specific heat amplitude, or iii) obtain a fit to experimental magnetization data using an assumed value of δ. All three approaches lead to $D \cong 4$. For the last method, we have combined the data of refs. 5 and 6 and forced a fit with $\delta = 4.78$; the

---

* Supported in part by National Science Foundation Grant DMR 72-03026 and DMR 76-10379.
† Present address: Knolls Atomic Power Laboratory, Schenectady, New York 12301, USA.

resulting value $D = 3.95$ is used in (4). The parameters are summarized in table I.

Table I
Critical exponents and amplitudes for gadolinium. ( ) denotes a value obtained from the equation $2 - \alpha = \beta(\delta + 1)$.

	δ	D	β	B	α
Ref. 4	4.39	2.35	0.37	1.16	(0.01)
Ref. 5	3.6	1.01	0.38	1.25	(0.25)
Used here	(4.78)	3.95	0.38	1.25	−0.20
Spin-1/2 Heisenberg	4.71		0.385		−0.198

For our test of (1) and (2), we express t' and y in universal units as noted above,

$$\bar{t}' = t'/t_0 \quad \text{and} \quad \bar{y} = y/y_0.$$

In fig. 1, we have plotted the experimental values of $\bar{t}'$ vs $\bar{y}$ for three different applied fields, using the parameters of table I. The dot-dash curve gives the result of the universalized linear model [8] using the same exponents. We note in passing that the values $a = 7.01$ and $k = 1.20$ obtained here differ significantly from those reported earlier [8]. This discrepancy may result from the use in ref. 8 of incorrectly scaled data from ref. 4.

As may be seen in fig. 1, this set of exponents and amplitudes leads to a verification of the scaling result (1) and (2). There is a significant deviation, however, of the form of the experimental curve from the predictions of the linear model, which deviation cannot be reduced by reasonable (±30%) changes in D. Somewhat surprisingly, Gd appears to behave very similarly to a spin-1/2 Heisenberg model. The closeness of the exponents is paralleled by the close agreement of the function $\bar{t}'(\bar{y})$ [2], shown as a solid curve in fig. 1, with the experimental data. In light of the large spin (7/2) and the complexity of Gd metal, this close agreement should probably be regarded as fortuitous. Further, as pointed out by Barmatz et al. curves of very similar form can be obtained from quite varied equations of state [9].

The failure of the experimentally determined values of α, β and δ to satisfy the scaling law equation remains a serious problem. In view of the possible cross-over in Gd to dipolar or Ising-like behavior close to T_c it is not possible to draw any strong conclusions regarding the Heisenberg-like nature of the material. It is noteworthy, however, that the scaling of the specific heat with external magnetic fields follows the simple predictions of (1) and (2) through the observed critical region.

We wish to acknowledge the assistance of F.L. Lederman in obtaining exponents for the specific heat. One of us (MBS) acknowledges the hospitality of the Technical University of Munich and the support of the Alexander von Humboldt Foundation during the early stages of this work. We have benefited from helpful comments from P. Hohenberg.

References

[1] D.S. Simons and M.B. Salamon, Phys. Rev. B 10 (1974) 4680.
[2] R. Krasnow and H.E. Stanley, Phys. Rev. B 8 (1971) 332.
[3] M.E. Fisher, Repts. Prog. Phys. 30 (1967) 615.
[4] M. Vincentini-Missoni, R.I. Joseph, M.S. Green and J.M.H. Levelt-Sengers, Phys. Rev. B 1 (1970) 2312.
[5] C.D. Graham, Jr., J. Appl. Phys. 36 (1965) 1135.
P. Heller, Repts. Prog. Phys. 30 (1967) 731.
[6] M.N. Deschizeaux and G. Develey, J. de Phys. 32 (1971) 319.
[7] E.A.S. Lewis, Phys. Rev. B 1 (1970) 4368.
[8] J.T. Ho, Phys. Rev. Lett. 26 (1971) 1485.
[9] M. Barmatz, P.C. Hohenberg and A. Kornblitt, Phys. Rev. B 12 (1975) 1947.

COMMENSURABILITY EFFECTS ON THE CRITICAL BEHAVIOR OF SYSTEMS WITH HELICAL ORDERING

T. GAREL* and P. PFEUTY

Laboratoire de Physique des Solides, Université Paris-Sud, 91405 Orsay, France

For systems with a phase transition to an ordered phase with a periodic structure, it is of interest to understand the effect of the commensurability of this structure with the underlying lattice. We consider magnetic systems with helical ordering and show that the commensurability effect may change the order of the transition.

It is well known that the critical behavior of physical systems depends on a small number of parameters: (i) space dimension d, (ii) number of components N of the order parameter, and symmetry of the Hamiltonian (iii) range of interaction. From now on, we shall only be concerned with systems having short-range forces, and we wish to emphasize point (ii). Of particular interest are the systems whose transitions are associated with the instability, in reciprocal space, of a star of wave vectors. A few examples are the charge density waves in quasi-1d and -2d systems, the spin density waves in some rare earth and transition metals, soft phonons in structural transitions, etc. Naturally, the question arises as to whether the unstable wave vector (s) is commensurate or not with the underlying reciprocal lattice. The commensurability problem has been studied in various situations and can have interesting consequences on the critical behavior (Peierls transition [1, 2], adsorption [3, 4]).

We shall first focus on the critical behavior of rare earth helimagnets [5, 6]. Two wave vectors $(\pm k_0)$ belong to the critical star, k_0 being parallel to the hexagonal axis. The polarization of the spin density waves can be longitudinal (Er, Tm) or transverse (Tb, Dy, Ho), corresponding to an $N = 2$ or $N = 4$ order parameter. In the quartic terms of a Landau–Ginzburg–Wilson Hamiltonian, there is one commensurability invariant, non-zero if, and only if, $k_0 = \frac{1}{4}\tau$ (τ = reciprocal lattice vector). We have studied, with renormalization group (R G) techniques the critical behavior of helimagnets to order ϵ^2 ($\epsilon = 4 - d$). The results are the following [6]. The commensurability invariant plays no role for the $N = 2$ case, which is found to have an isotropic stable fixed point. However, for the $N = 4$ case, two situations are to be considered: a) If $k_0 \neq \frac{1}{4}\tau$, there is a stable fixed point, yielding *to order ϵ^2* the critical exponents of an isotropic vector model. b) If $k_0 = \frac{1}{4}\tau$, the commensurability invariant is relevant with respect to the a) fixed point. We therefore expect a "first order transition". The influence of the commensurability of $4k_0$ is of course not a general feature. In the structural transition of NbO_2 [7], for instance, the critical behavior is the same for $4k_0 = \tau$ and $4k_0 \neq \tau$. The previous results should be compared with experiment; the most reliable experiments have been made on Dy. There is no agreement with our predictions, which could be due to the presence of dipolar interactions. Nevertheless, it should be interesting to obtain a situation where $4k_0 = \tau$. We notice that for Dy and Ho, one has $k_0 \simeq 0.23\,\tau$ and $k_0 \simeq 0.28\,\tau$ [10]. However, the alloying or pressure experiments seem hard to perform.

Another case where commensurability effects are important, concerns the number of components N of the order parameter. To illustrate this, let us consider the magnetic transition in chromium. What we have here is a sinusoidally modulated structure with an incommensurate wave vector, close, but not equal to $\frac{1}{2}\tau$. Due to the cubic symmetry, the star is composed of six wave vectors. The polarization is transverse, yielding an $N = 12$ order parameter. No fixed point is found in the R G analysis [8] and this was interpreted as a first order transition, in agreement with experiment. If one adds small quantities of manganese to chromium, one can lock the wave vector to $\frac{1}{2}\tau$ (simple antiferromagnet phase). It is easy to check that the paramagnetic to simple antiferromagnetic phase transition is described by an $N = 2$ order parameter. One then knows that this transition is second order, if the effects of disorder are neglected. A similar

---

* Permanent address: Service de Physique du Solide et de résonance magnétique, C.E.N. Saclay B.P. n° 2, 91 – Gif sur Yvette, France.

effect occurs in the trigonal lattice. If the star of the instability is associated with the six corners of the Brillouin zone, one has two independent wave vectors ($\pm k_0$, $k_0 = \frac{1}{3}\tau$). If the wave vectors lie in the same directions, but inside the Brillouin zone, one has six independent wave vectors (as in neodymium). N can therefore be divided by three (we consider here only magnetic transitions to avoid the presence of cubic terms).

It is known [9] that the universality of critical behavior is not the same for $N \leq 3$ and $N \geq 4$. For $N \leq 3$, the anisotropies in the order parameter space are irrelevant, whereas they are relevant for $N \geq 4$. This has been illustrated in the cases $N = 2$ and $N = 4$ for helimagnets, as well as $N = 12$ and $N = 2$ for chromium and chromium + manganese. The phase transitions with a star of wave vectors are a convenient way to study this change of universality as a function of N (magnetic field, pressure, alloying). To conclude, one can point out several interesting problems. The case of disordered helimagnets can be treated with an Nl order parameter and the usual $l = 0$ limit. In the same line, it would be of interest to know if the magnetic phase transition of solid helium 3 is describable by a star, and if this could explain the "abrupt transition" experimentally observed. In crystals, the star of wave vectors is of course discrete; on the contrary, the nematic–smectic C transition provides an unusual case of continuous star [11].

References

[1] S.A. Brazovsky and I.E. Dzyaloshinsky, Proceedings of LT 14, Helsinki (1975) p. 338.
P. Bak and V.J. Emery, Phys. Rev. Lett. 36 (1976) 978.
[2] G. Toulouse, Nuovo Cimento 23B (1974) 234.
[3] S.C. Ying, Phys. Rev. B3 (1971) 4160.
[4] S. Alexander, Phys. Lett. 54A (1975) 353.
[5] D. Mukamel, S. Krinsky and P. Bak preprint.
P. Bak and D. Mukamel, Phys. Rev. B13 (1976) 5086.
[6] T. Garel and P. Pfeuty, J. Phys. C9 (1976) L245; this article is part of the 3rd Cycle Thesis of T. Garel, Orsay (1976).
[7] D. Mukamel, Phys. Rev. Lett. 34 (1975) 481.
[8] P. Bak, S. Krinsky and D. Mukamel, Phys. Rev. Lett. 36 (1976) 52.
[9] E. Brézin, J.C. Le Guillou and J. Zinn-Justin, Phys. Rev. B10 (1974) 892.
[10] A. Herpin, Théorie du Magnétisme, P.U.F., Paris, 1968.
[11] J.B. Swift to be published in Phys. Rev.
M. Gabay and A.T. Garel, unpublished.

CRITICAL BEHAVIOUR AT A FIELD-INDUCED MAGNETIC PHASE TRANSITION AT $T = 0$*

H. BECK and H. THOMAS

Institut für Physik der Universität Basel, Basel, Switzerland

We study the critical behaviour at the field-induced phase transition in an anisotropic ferromagnet at the magnetic displacive limit $T = 0$, $H = H_K$. The critical behaviour of a classical magnet is shown to be equivalent to that of a classical lattice-dynamical system at the displacive limit of a structural phase transition. We also report exact results for a quantum magnet with infinite-range interactions, for the transverse Ising model and the anisotropic XY-chain, and for the spin-wave theory of a uniaxial ferromagnet.

1. Introduction

An anisotropic ferromagnet with the field applied in a hard direction has a line of second-order phase transitions which intersects the line $T = 0$ at the anisotropy field $H = H_K$ (magnetic displacive limit (DL)). In contrast to phase transitions at $T_c \neq 0$, the critical behaviour at the DL is different for classical and for quantum-mechanical systems. We are interested in this difference, and in the relation to the DL of structural phase transition [1–3].

2. Classical spins

For a classical spin model, there exists always a correspondence to a classical lattice-dynamical model. We consider a model of $(n+1)$-component spins $S_l = (s_{l1}, \ldots, s_{ln}, z_l)$; $\sum_\alpha s_{l\alpha}^2 + z_l^2 = S^2$, with Hamiltonian

$$\mathcal{H} = -\frac{1}{2} \sum_{ll',\alpha} J_{ll'}^{(\alpha)} S_{l\alpha} S_{l'\alpha} - \frac{1}{2} \sum_{l,\alpha} L^{(\alpha)} S_{l\alpha}^2 - H \sum_l z_l. \quad (1)$$

We eliminate the hard-direction spin component z_l and expand the Hamiltonian in powers of the $s_{l\alpha}$,

$$\mathcal{H} = \sum_l \left\{ \frac{1}{2S} \sum_\alpha (H - H_K^{(\alpha)}) s_{l\alpha}^2 + \frac{1}{8S^3} H \left(\sum_\alpha s_{l\alpha}^2 \right)^2 \right\}$$
$$+ \sum_{ll'} \left\{ \frac{1}{4} \sum_\alpha J_{ll'}^{(\alpha)} (s_{l\alpha} - s_{l'\alpha})^2 \right. \quad (2)$$
$$\left. + \frac{1}{16 S^2} J_{ll'}^{(z)} \left[\sum_\alpha (s_{l\alpha}^2 - s_{l'\alpha}^2) \right]^2 \right\} + \cdots,$$

where $H_K^{(\alpha)} = S(J_0^{(\alpha)} - J_0^{(z)} + L^{(\alpha)} - L^{(z)})$. The DL is given by the largest H_K, and the m directions $\alpha = 1, 2, \ldots, m$ belonging to this H_K form the m-dimensional easy subspace.

* Work supported by the Swiss National Science Foundation.

Since according to usual renormalization group (RG) arguments the q-dependence of the terms of higher than second order is irrelevant, this model is equivalent to the anisotropic n-vector model. Moreover, all components except the easiest ones are irrelevant. Therefore, the critical behaviour is equivalent to that of a classical m-vector model. The results derived in refs. [1, 2] are given in column 3 of table I.

3. Quantum-mechanical spins

3.1. Infinite-range interaction

In the case of infinite-range interactions, $J_{ll'}^{(\alpha)} = I^{(\alpha)}/N$, and $L = 0$, the Hamiltonian can be expressed in terms of the components of total spin $\mathcal{S}_\alpha = \sum_l S_{l\alpha}$,

$$\mathcal{H} = -\sum_\alpha \left(\frac{I^{(\alpha)}}{2N} \mathcal{S}_\alpha^2 + H_\alpha \mathcal{S}_\alpha \right) + \text{const} + \mathcal{O}(1). \quad (3)$$

In this case, the system behaves according to mean-field theory. This can be seen by explicit calculation of the partition function

$$Z = \sum_{\mathcal{S}} g_S(\mathcal{S}) \sum_\lambda e^{-\beta E(\mathcal{S}, \lambda)}, \quad (4)$$

where $\lambda = 1, \ldots, 2\mathcal{S} + 1$ numbers the states belonging to given $\mathcal{S}$, and the statistical weight $g_S(\mathcal{S})$ of state $(\mathcal{S}, \lambda)$ is in the thermodynamic limit given by $g_S(\mathcal{S}) \sim \exp[N\eta_S(\mathcal{S}/NS)]$ with η_S = single-spin entropy. The sum over states λ is replaced by the integral of a classical spin vector $\mathcal{S} = \mathcal{S}(\sin\theta\cos\phi, \sin\theta\sin\phi, \cos\theta)$ over solid angle, and the sum over $\mathcal{S}$ is performed by saddle-point integration. The saddle point condition then gives the usual mean-field equations. Analysis at the DL leads to the results given in column 4 of table I.

Table I

Exp.	Definition of exponent	Classical spin model	Quantum spin with inf. range interact. $(S<\infty)$*	Transverse Ising model	Anisotropic XY-chain $(d=1)$	Spin wave approx. uniaxial magnet $(d>2, S<\infty)$*
$\bar{\nu}$	$\xi_x(0, H) \propto |H - H_K|^{-\bar{\nu}}$	$\frac{1}{2}$	—	$\bar{\nu}(d) = \nu_{cl}(d+1)$	1	$\frac{1}{2}$
$\bar{\gamma}$	$\chi_x(0, H) \propto |H - H_K|^{-\bar{\gamma}}$	1	1	$\bar{\gamma}(d) = \gamma_{cl}(d+1)$	$\frac{7}{4}$	
η_C	$C_{xx}(0, H_K, R) \propto R^{-d+2-\eta_C}$	$0(C_{xx}(R) \propto TR^{-d+2})$	—	$\eta_C(d) = 1 + \eta_{cl}(d+1)$	$1+\frac{1}{4}$	$\begin{cases} C_{xx} \propto \delta_{R,0} + \bar{C}_{xx} \\ \bar{C}_{xx} \propto TR^{-d+2} \end{cases}$
$\bar{\beta}$	$M_x(0, H) \propto (H_K - H)^{\bar{\beta}}$ $(H < H_K)$	$\frac{1}{2}$	$\frac{1}{2}$	$\bar{\beta}(d) = \beta_{cl}(d+1)$	$\frac{1}{8}$	
δ	$M_x(0, H) \propto h_x^{1/\delta}$	3	3	$\delta(d) = \delta_{cl}(d+1)$		
$\bar{\alpha}$	$\Delta M_z(0, H) \propto |H - H_K|^{-\bar{\alpha}+1}$	$0(H < H_K)$	0	$\bar{\alpha}(d) = \alpha_{cl}(d+1)$	$0(\log)$	
ν	$\xi_x(T, H_K) \propto T^{-\nu}$	$\frac{1}{2}$ $(d>2)$; $1/(4-d)$ $(d<2)$	—	$\nu(d) = (d-1)\nu_{cl}(d+1)$ (for $d>1$) from scaling at quantum dis-placive limit	$\frac{1}{2}$?	$\nu = \frac{1}{2}\psi$
γ	$\chi_x(T, H_K) \propto T^{-\gamma}$	1 $(d>2)$; $2/(4-d)$ $(d<2)$	$\chi_x \propto e^{Sf(1)}/kT$			$\gamma = \psi$
α	$\Delta C_H(T, H_K) \propto T^{-\alpha+1}$	0 $(d>2)$; $(4-2d)/(4-d)$ $(d<2)$	$C_H \propto \dfrac{1}{T^2} e^{-Sf(1)/kT}$			$1 - d/2$
ψ	$T_c \propto (H_K - H)^{1/\psi}$	1 $(d>2)$	$H_K - H \propto e^{-Sf(1)/kT}$			

*The results for $S = \infty$ are consistent with the classical model, column 3.

3.2. Transverse Ising model

The Ising model in a transverse magnetic field

$$\mathcal{H}_{TI} = -\tfrac{1}{2}\sum_{ll'} J_{ll'} S_{lx}S_{l'x} - H\sum_{l} S_{lz} \quad (5)$$

can be related to a lattice-dynamical model. Young [4] has shown that the partition function of $\mathcal{H}_{TI}$ can be replaced by a functional integral over classical fields $s(q,m)$ depending on wave-vector q and Matsubara frequency $z_m = m \cdot 2\pi kT$:

$$Z_{TI} \to \bar{Z} \propto \int \delta s(qm) e^{-\beta\bar{\mathcal{H}}} \quad (6)$$

with

$$\bar{\mathcal{H}} = \sum_{r=1}^{\infty}\sum_{\{m_i\}} \int \prod_{i=1}^{2r}[d^d q_i s(q_i, m_i)]$$
$$\times u^{(2r)}(\{q_i m_i\})\delta(\Sigma q_i)\delta(\Sigma m_i). \quad (7)$$

The coefficient $u^{(2)}$ is given by

$$u^{(2)}(qm) = r_0 + bz_m^2 + cq^2 + \mathcal{O}(q^4) \quad (8)$$

with $r_0 \propto H - \tfrac{1}{2}J_0 \text{th}\,\beta H$ for $S = \tfrac{1}{2}$.

At $T = 0$ the z_m form a continuum [4] and $\bar{Z}$ is the Landau–Ginzburg–Wilson functional of a classical $(d+1)$-dimensional system [4,5], after introduction of an irrelevant frequency cut-off. Since the original field H plays the role of the temperature in the classical system, S_z is mapped onto the energy density. These facts lead to the exponents with respect to $H - H_K$ listed in column 5 of table I.

In order to discuss the critical behaviour for $H = H_K$ and $T \to 0$ we first note that usual RG arguments imply that the dependence of $u^{(2n)}$ on q and z_m is irrelevant for $n > 2$. Thus $\bar{\mathcal{H}}$ can be replaced by $\tilde{\mathcal{H}}$ with constant $u^{(4)}$, $u^{(6)}$ etc. But such a simplified partition function $\tilde{Z}$ is also the functional integral version of the partition function Z_L for a lattice Hamiltonian

$$\mathcal{H}_L = \int d^d q \left[\frac{1}{2M}\Pi^+(q)\Pi(q)\right.$$
$$+ \tfrac{1}{2}(r_0 + cq^2)Q^+(q)Q(q)\right]$$
$$+ \sum_{r=2}^{\infty} u^{(2r)} \int \prod_{i=1}^{2r}[d^d q_i Q(q_i)]\delta(\Sigma q_i) \quad (9)$$

for particles with mass $M = b$ and displacement and momentum operators Q and Π, respectively (see Appendix of ref. 3). The critical behaviour of $\mathcal{H}_{TI}$ for $T \to T_c = 0$ is thus identical with the DL of the quantum lattice system $\mathcal{H}_L$. An RG calculation [5] relates the exponent $\nu(d)$ for $d > 1$ to the classical value $\nu_{cl}(d+1)$, see table I. $\gamma(d)$ is then obtained through the scaling relation $\gamma = (2 - \eta_\chi)\nu$ where η_χ determines the long-distance behaviour of the response function χ. (The exponent η_C for the correlation function obeys $\eta_C = 1 + \eta_\chi$.)

As column 6 of table I shows, these results are consistent with exact calculations for anisotropic XY-chains [6].

3.3 Spin-wave approximation

For a uniaxial Heisenberg model with an easy plane

$$\mathcal{H} = -\frac{1}{2}\sum_{ll'}[J^{(1)}_{ll'}(S_{lx}S_{l'x} + S_{ly}S_{l'y})$$
$$+ J^{(3)}_{ll'}S_{lz}S_{l'z}] - HS_{lz'} \quad (10)$$

spin-wave theory is expected to give exact results for $H > H_K$, $T \to 0$ for $d > 2$. The results are included in column 7 of table I.

4. Conclusion

The magnetic DL, $T = 0$, $H = H_K$, is an isolated point on the line of second-order phase transitions in the (H, T)-plane with its own set of critical exponents. Critical behaviour at this point may be easier to study experimentally than at the DL of a structural transition, because the parameter H is easy to control.

Similar results hold for the antiferromagnet at the spin-flop/paramagnetic transition at $T = 0$.

References

[1] H. Beck, T. Schneider and E. Stoll, Phys. Rev. B12 (1975) 5198.
[2] R. Oppermann and H. Thomas, Z. Phys. B22 (1975) 887.
[3] H. Beck, T. Schneider and E. Stoll, Phys. Rev. B13 (1976) 1123.
[4] A.P. Young, J. of Physics C8 (1975) L309.
[5] H. Beck and R. Schäfer, Phys. Lett. 58A (1976) 73.
[6] E. Barouch and B.M. McCoy, Phys. Rev. A3 (1971) 786.

NOVEL TWO-VARIABLE APPROXIMANTS FOR STUDYING MAGNETIC MULTICRITICAL BEHAVIOR

M.E. FISHER
Baker Laboratory, Cornell University, Ithaca, New York, 14853, USA

Series expansions in two variables, e.g. $\chi(T, H) = \Sigma_{jk} a_{jk} (J/k_B T)^j (mH/k_B T)^k$, are important in studying magnetism but established extrapolation methods (e.g. ratio, Padé, and Canterbury techniques) fail near multicritical points. A novel method of series extrapolation is explained which can exactly represent the expected two-variable, two-exponent (γ, ϕ), asymptotic scaling behavior.

In the study of critical phenomena an important role has been played by the extrapolation of power series expansions for functions of one variable, such as $f(x) \equiv \chi(T)$, with $x = J/k_B T$. The Padé approximant technique [1], has been particularly fruitful: a direct Padé approximant of the form

$$f(x) = \sum_{k=0}^{\infty} a_k x^k \simeq [L/M]_f = P_L(x)/Q_M(x), \quad (1)$$

where $P_L(x)$ and $Q_M(x)$ are polynomials of degrees L and M with coefficients p_l and q_m chosen to match the expansion of $f(x)$ to order $k = L + M$, can accurately represent functions whose only singularities are poles. However, it is not suitable for functions displaying critical-type singularities of the form $f(x) \approx Z_0 (x_c - x)^{-\gamma}$. Baker overcame this difficulty by studying instead the logarithmic derivative

$$\frac{d}{dx} \ln f(x) = \frac{(df/dx)}{f(x)} \simeq [L/M]_{D \log f} = \frac{P_L(x)}{Q_M(x)}. \quad (2)$$

A zero of $Q_M(x)$ should approximate the critical value x_c and the corresponding residue of $[L/M]_{D \log f}$ yields estimates for the critical exponent γ.

In the study of multicritical phenomena, functions of *two* variables play a crucial part and an effective method of extrapolating the corresponding series expansions

$$f(x, y) = \sum_{k, k'=0} a_{kk'} x^k y^{k'}, \quad (3)$$

would be useful. A typical physical situation is a bicritical point as in an antiferromagnet [2], where, for example, f might again be a susceptibility $\chi(T, H)$, with $x = J/k_B T$, but the second variable would be the field, $y = H/k_B T$. Some success has been achieved [3] in cases where the multicritical point is symmetrically located, by examining the set of *single*-variable functions $f(x, 0)$, $df(x, 0)/dy, \ldots d^j f(x, 0)/dy^j$. Nonetheless, the need for handling both variables in the multicritical region is clearly felt; this is especially so near tricritical points occurring at points of low symmetry.

Recently the standard Padé approximants (1) have been generalized by approximating $f(x, y)$ by a ratio of polynomials $P_L(x, y)$ and $Q_M(x, y)$ of two variables [4]. But these Canterbury approximants, cannot represent branch point singularities of critical type and the trick of looking at $d(\ln f)/dx$ is unsatisfactory near a multicritical point, the essential feature being that the nature of the critical singularity changes abruptly (e.g., from Ising, to Heisenberg, to XY type) as the multicritical point is traversed. Canterbury-type approximants can only handle exactly a simple pole moving analytically in the (x, y) plane [5]. More specifically, in the neighborhood of a multicritical point at (x_c, y_c) one physically expects [2, 3] scaling behavior of the form

$$f(x, y) \approx (x - x_c)^{-\gamma} Z \left(\frac{y - y_c}{(x - x_c)^\phi} \right). \quad (4)$$

where $Z(z)$ is the single-variable crossover scaling function, while ϕ is the crossover exponent. (For simplicity we assume, here and below, that the multicritical scaling axes are parallel to the x and y axes.) A really effective method of extrapolation should yield estimates for (x_c, y_c), for both exponents γ and ϕ, and for the scaling function $Z(z)$. The additional singularities implied by nonintegral ϕ and (necessarily) nonanalytic $Z(z)$, appear to rule out Canterbury approximants.

An alternative route generalizes (2) in a different direction. Thus, one may regard a

D log Padé approximant, $F_{LM}(x) \simeq f(x)$, as defined by the solution of the approximating differential equation [6], $P_L(x)F = Q_M(x)\,df/dx$, in which the coefficients p_l and q_m are chosen so that the power series solution of the equation agrees with the expansion (1) to optimal order. In the critical region we expect $Q(x) \simeq Q_c'(x - x_c)$ and $P(x_c) \simeq -\gamma Q_c'$.

In this vein, now note that if a function $F(x, y)$ obeyed a strict equality in the scaling expression (4), it would satisfy the *partial differential equation*

$$P_c F = Q_c'(x - x_c)\frac{\partial F}{\partial x} + R_c'(y - y_c)\frac{\partial F}{\partial y}, \quad (5)$$

with constants

$$P_c = -\gamma Q_c' \quad \text{and} \quad R_c' = \phi Q_c'. \quad (6)$$

related to the critical exponents. (This partial differential equation resembles the Callan–Symanzik renormalization equation in field theory.)

On this basis, a novel class of approximants, $F_{LMN}(x, y)$, for a function of two variables $f(x, y)$, may be defined as the solution of the partial differential equation

$$P_L(x,y)F(x,y) = Q_M(x,y)\frac{\partial F}{\partial x} + R_N(x,y)\frac{\partial F}{\partial y}, \quad (7)$$

satisfying appropriate boundary conditions (imposed say at small x). The two-variable polynominals $P_L(x, y)$, $Q_M(x, y)$, and $R_N(x, y)$ are to be chosen so that the power series solution for $F(x, y)$ agrees with the expansion (3) as far as possible. As in ordinary Padé approximants, this simply leads to a set of simultaneous linear equations which may be solved by standard methods (provided they are not singular). These new approximants might be termed "partial differential approximants" (PDA's) or "PQR approximants." Evidently they have the potentiality of representing crossover scaling behavior like (4) *exactly*: in the region of the multicritical point one expects to find $Q(x, y) \approx Q_c'(x - x_c)$, $R(x, y) \approx R_c'(y - y_c)$ and $P(x_c, y_c) = P_c$ with the coefficients satisfying (6). This should thus yield estimates for both exponents γ and ϕ as desired.

To solve the defining partial differential equation various numerical techniques may be used, but a trajectory-integral or characteristic method will probably be most convenient. Thus one introduces a time-like variable, τ, with $x = x(\tau)$, $y = y(\tau)$, and $F = F[x(\tau), y(\tau)]$. The partial differential equation (7) is then equivalent to the set of *ordinary* differential equations

$$\dot F = \frac{dF}{d\tau} = P(x, y)F,$$

$$\dot x = \frac{dx}{d\tau} = Q(x, y), \quad \dot y = \frac{dy}{d\tau} = R(x, y). \quad (8)$$

In the multicritical region the trajectories $[x(\tau), y(\tau)]$ may be expected to have the scaling form $y - y_c \propto (x - x_c)^\phi$. The generalizations of (7) and (8) to more variables are obvious.

Note that the new approximants can exactly represent [i.e., with *finite* polynomials P, Q, and R] functions $f(x, y) = A(x, y) + B(x, y)$ where A and B are of the form

$$A(x, y) = A_0 \exp\left[\frac{a_0(x, y)}{a_{-1}(x, y)}\right]\prod_{j=1}^{J}[a_j(x, y)]^{\alpha_j}, \quad (9)$$

in which the $a_j(x, y)$, for $j = -1, 0, 1, \ldots, J$, are polynomials in x and y while the α_j are arbitrary exponents. These forms include $f = (1 - 2x + y)^{-\gamma} + e^{x - 2y}$ which was used by Roberts et al. [5] to test the Canterbury approximants (which cannot represent this f exactly).

Any numerical technique must prove its value by tests which may reveal difficulties of various sorts (ill-conditioned equations, etc.) which must be recognized and dealt with by appropriate devices before the method can be judged practical and reliable. Initial trials of our approximants are promising but experience will show how far their potential to elucidate multicritical phenomena can actually be realized.

I am happy to acknowledge fruitful comments and essential programming work by Dr. M. Springgate and Mr. Robert M. Kerr, and the support of the National Science Foundation and the Materials Science Center at Cornell University. The first conception of the new approximants arose at a stimulating workshop on Padé approximants organized by Dr. D. Bessis and held at CNRS, Marseille in May 1975.

References

[1] See G.A. Baker, Jr., Phys. Rev. 124 (1961) 768 and e.g. M.E. Fisher, Rocky Mtn. J. Math. 4 (1975) 181.
[2] M.E. Fisher and D.R. Nelson, Phys. Rev. Lett. 32 (1974) 1350.
    M.E. Fisher, ibid 34 (1975) 1634.
    H. Rohrer, ibid 34 (1975) 1638.
[3] P. Pfeuty, D. Jasnow and M.E. Fisher, Phys. Rev. B 10 (1974) 2088.
[4] J.S.R. Chisholm, Math. Comput. 27 (1973) 841.
    J.S.R. Chisholm and J. McEwan, Proc. Roy. Soc. A 336 (1974) 421.
[5] D. Roberts, H.P. Griffiths and D.W. Wood, J. Phys. A 8 (1975) 1365.
[6] A.J. Guttmann and G.S. Joyce, J. Phys. A 5 (1972) L81.
    J. Gammel and D.S. Gaunt (private communication).

LIGHT SCATTERING FROM METAMAGNETIC DOMAINS IN DYSPROSIUM ALUMINUM GARNET*

N. GIORDANO† and W.P. WOLF
Department of Engineering and Applied Science, Yale University, New Haven, CT. 06520, USA

Light scattering techniques have been used to study the behavior of dysprosium aluminum garnet near the "wing" critical point which is found for magnetic fields applied along a {111} direction. The extent of the mixed phase region and the critical exponent β have been determined.

Recent work on metamagnetic [1, 2] and spin-flop systems [3] has shown that the uncertainties in the location of the first-order phase boundaries can be a major experimental problem in the study of multicritical points. This is because large pre-transition susceptibilities coupled with unavoidable "rounding" effects often make it difficult to isolate the discontinuities associated with the first-order transition. There has therefore been increasing interest in developing new experimental techniques for locating first-order phase boundaries [4]. One approach to this problem is to use the fact that demagnetizing effects cause the first-order transition to be spread over a range of applied fields in which domains of the two phases are in coexistence [5]. Any property which is sensitive to the presence or absence of domains can therefore be used to locate the phase boundaries.

Since the coexisting domains have different optical properties, the sample in the mixed phase region will act as a grating and thus scatter light. By monitoring the scattering as a function of applied field, H, and temperature, T, the first-order phase boundaries in the H–T plane can be determined. Light scattering from domains has been previously observed in several ferromagnets [6], in several antiferromagnets [7] and has very recently been used by Dillon et al. [7] to study the tricritical behavior of $FeCl_2$.

In this paper we describe the application of the light scattering method to the study of the "wing" critical point [8] found in the Ising-like antiferromagnet [1] dysprosium aluminum garnet (DAG) for fields applied along a {111} direction. This point is of particular interest as two previous studies [1] have reported values for several of the critical exponents which appear to be inconsistent with the theoretical predictions.

In our experiments the sample was a platelet of DAG oriented with a {111} direction perpendicular to the largest face, and the light source was a He–Ne laser ($\lambda = 632.8$ nm). Details of the experimental arrangement will be given elsewhere [9].

Most of the measurements were made with the field swept continuously. The scattering was found to be a strong function of both sweep speed and direction, and showed large hysteresis effects [9]. This is not surprising since from the work of Dillon et al. [10] we expect the domain configuration and hence the scattering to be extremely sensitive to the magnetic history of the sample. By studying the systematics of the hysteresis as a function of sweep speed and direction, we found it possible to use the scattering to locate the equilibrium phase boundaries quite accurately. A detailed description of the hysteresis effects will be given elsewhere [9]. Fig. 1 shows typical results for the scattering very close to the phase boundaries. Here the onsets of the scattering locate the equilibrium phase boundaries. The boundaries could be located to within about ±10 Oe at 1.3 K and ±30 Oe near the critical point ($T_c = 1.596$ K).

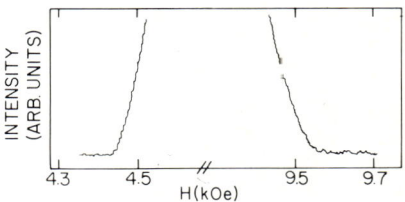

Fig. 1. Scattering near the phase boundaries at 1.30 K and a scattering angle of 2.5°. In both traces the field was swept out of the mixed phase at 2 Oe/s (see ref. 9).

---

* Supported in part by N.S.F. grant DMR 76-23102 and the U.S. Army Research Office.
† N.S.F. Graduate Fellow.

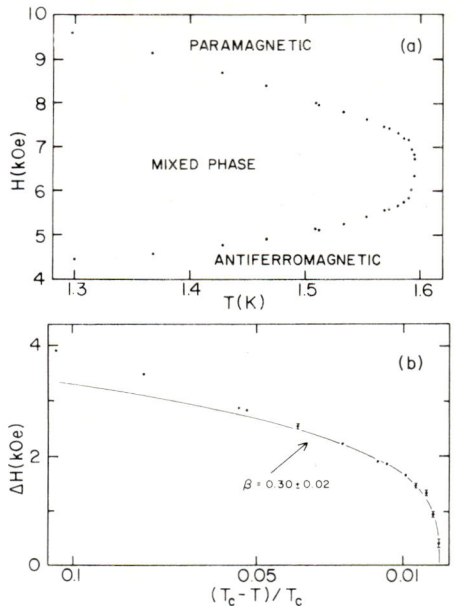

Fig. 2. (a) H–T Phase diagram. (b) ΔH vs $(T_c - T)/T_c$ near the critical point. Note that both scales are linear. The solid line is a fit to a power law (see text).

Our results for the H–T phase diagram are shown in fig. 2a. We see that the phase boundaries appear to approach the critical point with an infinite gradient, as expected [11]. Since the extent of the mixed phase region at constant temperature, ΔH, is proportional to the discontinuity in the magnetization [5], the data of fig. 2a can be used to determine the critical exponent [11] β. Fig. 2b shows the data for ΔH near the critical point. The solid line is a fit to the expected power law $\Delta H \sim t^\beta$ ($t \equiv (T_c - T)/T_c$) which gave $\beta = 0.30 \pm 0.02$ for $t < 0.04$, in good agreement with the present theory, which predicts $\beta = 0.3125 {}^{+0.003}_{-0.006}$ for Ising-like systems [11]. An interesting feature of fig. 2b is that the deviations from the power law as t increases, are towards *larger* values of ΔH. This is opposite to the direction usually found [12], and may indicate a crossover towards tricritical behavior [13] corresponding to a larger value of β. This could also account for the higher values of β found in previous work [1] which was limited to larger values of t than the present experiments.

The techniques described here should prove useful in the study of many other systems in which first order phase boundaries are found. Results of a light scattering study of the tricritical point found in DAG for fields along {110}, and of the behavior near the tricritical wings in DAG for fields near {110} will be reported elsewhere [9].

We are greatly indebted to J.F. Dillon, Jr. for much advice and encouragement. We also thank R. Alben, R.K. Chang, E. Yi Chen and A.R. Tanguay, Jr. for many helpful discussions.

References

[1] W.P. Wolf, B. Schneider, D.P. Landau and B.E. Keen, Phys. Rev. B5 (1972) 4472.
A.T. Skjeltorp, R. Alben and W.P. Wolf, AIP Conf. Proc. 18 (1974) 770.
[2] R.J. Birgeneau, G. Shirane, M. Blume and W.C. Koehler, Phys. Rev. Lett. 33 (1974) 1098.
[3] H. Rohrer, Phys. Rev. Lett. 34 (1975) 1638.
[4] See for example, J.A. Griffin, S.E. Schnatterly, Y. Farge, M. Regis and M.P. Fontana, Phys. Rev. B10 (1974) 1960.
N. Giordano and W.P. Wolf, Phys. Rev. Lett. 35 (1975) 799.
[5] A.F.G. Wyatt, J. Phys. C1 (1968) 684.
[6] See for example, J.F. Dillon, Jr. and J. P. Remeika, J. Appl. Phys. 34 (1963) 637.
J.C. Suits, J. Appl. Phys. 38 (1967) 1498.
R.V. Telesnin, A.G. Shishkov, E.N. Ilicheva, N.G. Kanavina and N.A. Ekonomov, Phys. Stat. Sol. (a) 12 (1972) 303.
[7] J.F. Dillon, Jr., E. Yi Chen and H.J. Guggenheim, Sol. St. Comm. 16 (1975) 371; AIP Conf. Proc. 29 (1976) 651; and to be published.
[8] M. Blume, L.M. Corliss, J.M. Hastings and E. Schiller, Phys. Rev. Lett. 32 (1974) 544.
[9] N. Giordano and W.P. Wolf, to be published.
N. Giordano, thesis, Yale University.
[10] J.F. Dillon, Jr., E. Yi Chen, N. Giordano and W.P. Wolf, Phys. Rev. Lett. 33 (1974) 98.
J.F. Dillon, Jr., E. Yi Chen and H.J. Guggenheim, AIP Conf. Proc. 24 (1975) 300.
[11] M.E. Fisher, Rept. Prog. Phys. 30 (1967) 615.
D.R. Nelson and M.E. Fisher, Phys. Rev. B11 (1975) 1030.
[12] P. Heller, Rept. Prog. Phys. 30 (1967) 731.
[13] E.K. Riedel and F.J. Wegner, Phys. Rev. Lett. 29 (1972) 349.

THERMODYNAMIC PROPERTIES OF THE METAMAGNET $CsCoCl_3 \cdot 2D_2O$ NEAR THE MULTICRITICAL POINT

A.L.M. BONGAARTS and W.J.M. DE JONGE

Department of Physics, Eindhoven University of Technology, Eindhoven, The Netherlands

The cross-over from critical to multicritical behaviour of $M_{st}(H,T)$ and $M(H,T)$ has been investigated near the multicritical point in the magnetic phase diagram of the metamagnet $CsCoCl_3 \cdot 2D_2O$.

In the present contribution we report the results of neutron-diffraction experiments on the critical behaviour of the staggered magnetization $M_{st}(H,T)$ and the uniform magnetization $M(H,T)$ along the phase boundary $H = H_c(T)$ in the magnetic phase diagram of the metamagnet $CsCoCl_3 \cdot 2D_2O$ with the emphasis on the cross-over near the multicritical point (MCP).

$CsCoCl_3 \cdot 2D_2O$ is a pseudo one-dimensional compound with strongly anisotropic (Ising) interactions [1]. In zero magnetic field long-range magnetic order sets in at $T_N = 3.30$ K [2, 3]. The magnetic structure, space group $P_{2b}cca'$, consists of canted antiferromagnetic chains along the **a** direction, coupled ferromagnetically along **c** and antiferromagnetically along **b**. When the **a** component of an external magnetic field exceeds the critical value $H_c(T)$, the moments in alternate ac-planes reverse, giving rise to a metamagnetic phase transition. In the magnetic (H,T) phase diagram a MCP is located at $T_m = 1.85$ K and $H_m = 2.75$ kOe [2], see fig. 1.

The neutron-diffraction data were obtained from the intensity measurement of the magnetic super-lattice reflections $(1\tfrac{1}{2}0)$ and (100), which are proportional to M_{st}^2 and M^2, respectively. The experiments were performed as function of the magnetic field at various fixed temperatures.

The critical behaviour of $M_{st}(H,T)$ and $M(H,T)$ for $T \geq T_m$ in isothermal scans may be expressed by the powerlaws:

$$M_{st}(H,T) \sim [1 - H/H_c(T)]^\beta, \qquad (1)$$

$$M_c(T) - M(H,T) \sim [1 - H/H_c(T)]^{1-\alpha'}. \qquad (2)$$

In order to determine the values of the critical exponents α and β and to obtain the temperature dependence of $H_c(T)$ and $M_c(T)$ the experimental data of $CsCoCl_3 \cdot 2D_2O$ for $T \geq T_m$ close to the phase boundary, corrected for critical scattering and demagnetizing effects, were analyzed by fitting them to (1) and (2). The main problem was to correct for demagnetization. Due to an imperfect ellipsoidal shape of the sample the internal magnetic field was not entirely homogenious. Therefore the data corresponding to $[1 - H/H_c(T)] \leq 5 \times 10^{-3}$ could not be used in the analyses.

In the temperature region $1.850\,\text{K} \leq T \leq 1.875\,\text{K}$ the data could be analyzed by a full least-squares fit to the powerlaws (1) and (2), from which the multicritical exponents $\beta_m = 0.15 \pm 0.02$ and $\alpha_m = 0.65 \pm 0.05$ were determined. For $1.875\,\text{K} \leq T \leq 2.0\,\text{K}$ cross-over from critical to multicritical behaviour was observed in both the $M_{st} \sim H_{int}$ and M vs. H_{int} curves. Above $T \approx 2.0$ K no cross-over was present and the exponent β was found to be independent of the position on the critical line, with an average value $\beta = 0.295 \pm 0.010$ [3]. The exponent α was much harder to obtain since H becomes more and more an independent direction with increasing temperature. Fixing $H_c(T)$ at the value of the corresponding M_{st} data we found that

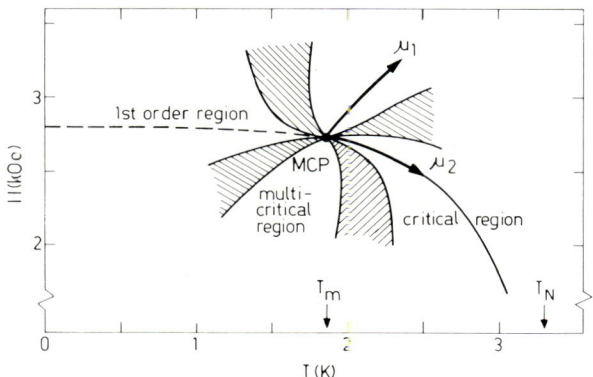

Fig. 1. Magnetic phase diagram of $CsCoCl_3 \cdot 2D_2O$. The scaling fields μ_1 and μ_2 are defined in (5). Shaded regions correspond to the 25%-cross-over regions.

plots of $\ln[M(T) - M(T)]$ vs. $\ln[1 - H/H_c(T)]$ approached a limiting slope corresponding to $\alpha' \approx 0.15$.

The cross-over in the vicinity of the MCP was further analyzed by constructing the scaling functions as introduced by Riedel [4] and Riedel and Wegner [5]. These authors showed that the singular part of any extensive thermodynamic quantity $B(\mu_1, \mu_2)$ with critical exponent a and multicritical exponent a_m, in terms of the scaling fields μ_1 and μ_2 must satisfy:

$$\mu_2^{-a_m\phi} \cdot B(\mu_1, \mu_2) = B^*(\mu_1/\mu_2^\phi). \qquad (3)$$

The scaling fields μ_1 and μ_2 are analytic functions of H and T; μ_2 must be chosen parallel to the phase boundary at the MCP. The exponent ϕ is the cross-over exponent, which describes the shape of the phase boundaries in field space. In (3) the scaling function B^* depends only on μ_1/μ_2^ϕ and is expected to exhibit the following asymptotical behaviour [4]:

$B^* \sim (\mu_1/\mu_2^\phi)^a$, critical region,

$B^* \sim (\mu_1/\mu_1^\phi)^{a_m}$, multicritical region, (4)

$B^* =$ constant, 1st order region.

For the scaling of the experimental data of $CsCoCl_3 \cdot 2D_2O$ we used the values of α_m, β_m, $H_c(T)$ and $M_c(T)$ obtained from the analyses of the isothermal scans. For the scaling fields μ_1 and μ_2 we chose (see fig. 1):

$$\mu_1 = \frac{H - H_c(T)}{H_m} \text{ and}$$
$$\mu_2 = \frac{T - T_m}{T_m} - \frac{H - H_c(T)}{H_m}. \qquad (5)$$

The cross-over exponent ϕ was determined from the data $M = M_m$, which resulted in $\phi = 2.0 \pm 0.2$ in agreement with the value derived from the temperature dependence of the discontinuities of M_{st} and M along the first-order phase boundary for $T < T_m$.

According to (3) the data for $M(\mu_1, \mu_2) - M(0, \mu_2)$ and for $M_{st}(\mu_1, \mu_2)$ were scaled by $\mu_2^{(1-\alpha_m)\phi}$ and $\mu_2^{\beta_m\phi}$, respectively. The scaled data did fall on smooth curves as can be seen for

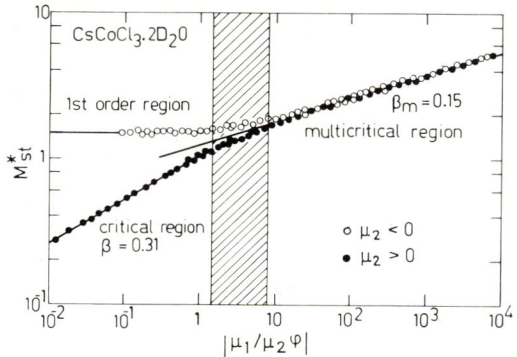

Fig. 2. Double logarithmic plot of $M^*(\mu_1, \mu_2)$ defined in (3) as a function of μ_1/μ_2^ϕ for $\beta_m = 0.15$ and $\phi = 2$. The asymptotic behaviour for $\mu_1/\mu_2^\phi \ll 1$ and $\mu_1/\mu_2 \gg 1$ is represented by the drawn lines. The shaded region corresponds to the 25% cross-over region.

M_{st}^* in fig. 2. The asymptotic behaviour in the critical region, $|\mu_1/\mu_2^\phi| \ll 1$ for $\mu_2 > 0$, in the multicritical region, $|\mu_1/\mu_2^\phi| \gg 1$, and in the first-order region, $|\mu_1/\mu_2^\phi| \ll 1$ for $\mu_2 < 0$, showed the theoretically expected behaviour of (4), with the exponents $\alpha = \alpha' = 0.12 \pm 0.05$, $\beta = 0.31 \pm 0.01$, $\alpha_m = \alpha_m' = 0.65$ and $\beta_m = 0.15$. The intermediate region, $|\mu_1/\mu_2^\phi| \approx 4$, is the cross-over region. In figs. 1 and 2 we have indicated the 25% cross-over regions, i.e. the region in which the slopes of $\ln B^*$ vs. $\ln \mu_1/\mu_2^\phi$ deviate by more than 25% of $|a - a_m|$ from either a or a_m [5].

The results of the present analysis may be summarized by stating that these experiments show that for $T_m < T \leq T_N$ the line of $H = -H_c(T)$ consists of ordinary critical points with the exponents of the $d = 3$ Ising model. At the MCP the exponents have changed to a new set: $\alpha_m = 0.65$, $\beta_m = 0.15$ and $\phi = 2$. The cross-over in the vicinity of the MCP is well described by the cross-over scaling theory of Riedel [4] and Riedel and Wegner [5].

References

[1] A. Herweyer, W.J.M. de Jonge, A.C. Botterman, A.L.M. Bongaarts and J.A. Cowen, Phys. Rev. B5 (1972) 4618.
[2] A.L.M. Bongaarts (Ph.D. thesis), RCN-235, Reactor Centrum Nederland, Petten, The Netherlands.
[3] A.L.M. Bongaarts, W.J.M. de Jonge and P. van der Leeden, Phys. Rev. Letter 37 (1976) 1007.
[4] E.K. Riedel, Phys. Rev. Letter 28 (1972) 675.
[5] E.K. Riedel and F.J. Wegner, Phys. Rev. B9 (1974) 294.

A SPIN-FLOP BICRITICAL LINE IN GdAlO$_3$

H. ROHRER, B. DERIGHETTI and Ch. GERBER

IBM Zurich Research Laboratory, 8803 Rüschlikon, Switzerland

In antiferromagnets with orthorhombic symmetry, the surface of first-order spin-flop transitions is connected to the paramagnetic phase boundary in a bicritical line, provided the field lies in the easy-hard plane. We have experimentally verified this spin-flop bicritical line in GdAlO$_3$.

Multicritical points are ideally suited for quantitative experimental investigations of crossover phenomena. In particular, the antiferromagnetic spin-flop bicritical point has been used to determine the anisotropy crossover exponent ϕ [1, 2] and to verify predictions [3] of renormalization group theory. In uniaxial antiferromagnets like MnF$_2$ a spin-flop bicritical point exists only for perfect alignment of the easy axis with the magnetic field [2]. For orthorhombic anisotropy as in GdAlO$_3$ however, the molecular field approximation (MFA) predicts a bicritical line provided the applied field lies in a plane containing the easy axis of magnetization and the axis of hardest anisotropy perpendicular to it (easy-hard plane).

For uniaxial anisotropy (fig. 1a), a first-order spin-flop (SF) transition exists only within a narrow angle between easy axis and applied field [4]. This angle decreases with increasing temperature giving rise to a triangular-like shelf of first-order SF transitions which touches the phase boundary to the paramagnetic (PM) state at the "umbilical" bicritical point [3]. In a top view (fig. 1b), the SF transition shelf plays the role of an intermediate phase bounded by two second-order phase transitions. All four second-order transitions of fig. 1b should now meet tangentially at the umbilical point which is now a tetracritical point.

The orthorhombic case is more complicated. In the easy-hard plane, spanned by H_e and H_h in fig. 1c, the first-order SF transition line extends all the way to the paramagnetic phase boundary. It separates a symmetric configuration S from an asymmetric configuration AS [5]. In the coordinates (H_e, H_h, T), the surface of SF transitions is, therefore, connected to the PM phase boundary in a bicritical line forming a bicritical "fold". Consequently, the general points B and B' on fig. 1d are bicritical points in contrast to the simple critical points A and A' in fig. 1b. If

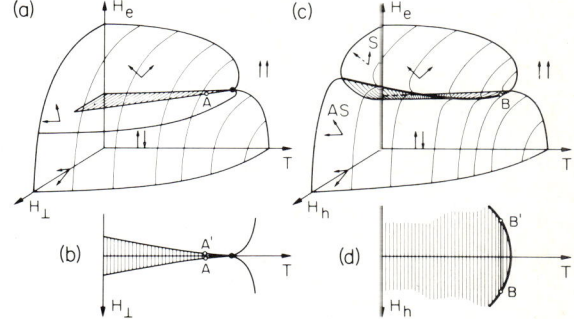

Fig. 1. Comparison of phase diagrams of antiferromagnets with uniaxial (a, b) and orthorhombic anisotropy (c, d). The arrows indicate the direction of the sublattices. H_e and H_h point along the easy and hard anisotropy axes, respectively; $H_\perp$ is an arbitrary axis perpendicular to H_e. The shaded parts are surfaces of first-order spin-flop transitions shown projected in b and d. The heavy line in c and d is the bicritical line.

the field perpendicular to H_e does not point along the hard axis, then the topology of the phase diagram is similar to that of fig. 1a. Thus, the hard axis of anisotropy introduces a fourth variable which spreads the bicritical "umbilicus" into a bicritical "fold".

We have now determined the bicritical line in GdAlO$_3$ by measuring the isothermal and adiabatic susceptibilities and rf absorption. The experiments have been performed both on a GdAlO$_3$ single-crystal sphere of 3 mm diameter and a long bar 12 mm × 0.5 mm × 0.5 mm. The crystals were X-ray oriented and mounted in the cryostat with the easy axis parallel to the magnetic field within an accuracy of 0.2 degrees. Fine adjustment of the applied-field direction was accomplished with the aid of an additional perpendicular field δH^t from Helmholz coils outside the dewar. In order to avoid complications due to demagnetization, we only discuss here the results on the sphere where the bicritical point for fields along the easy axis lies at $H_B = 12.495$ kOe and $T_B = 3.124$ K.

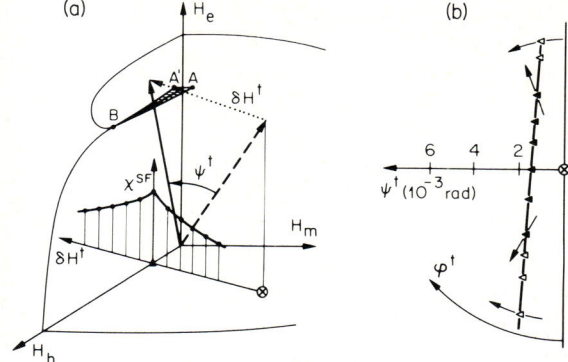

Fig. 2a. Phase diagram at constant temperature (H_m is the axis of medium anisotropy). The points A, A' and B correspond to those of fig. 1. The direction of the applied field (broken line) leads initially past the shaded SF transition shelf and is tilted by δH^t into the easy-hard plane. The height of the susceptibility peak χ^{SF} as a function of δH^t is also shown.

Fig. 2b. Locus of the susceptibility peak maxima as a function of tilting angle ψ^t and tilting direction φ^t. Initial direction of the applied field: $\otimes$. The filled triangles were obtained by tilting at constant azimuth φ^t, the open ones by rotating a constant tilting angle ψ^t.

In order to find the bicritical line, we have located the easy-hard plane of fig. 2a as follows. The isothermal susceptibility versus applied field exhibits a sharp peak at the SF transition, but if the field direction leads past the SF transition shelf, only a small and wide peak is observed at a field corresponding to the SF transition field. Close to the bicritical temperature T_B, the width A–A' of the SF transition shelf is narrower than 10 e or 10^{-4} rad at 10 mK below T_B and cannot be resolved experimentally. Thus, the susceptibility peak χ^{SF} as a function of δH^t is maximal when the applied field lies in the easy-hard plane, as shown in fig. 2a for an arbitrary azimuthal direction φ^t. The polar plot of fig. 2b, centered at the initial direction of the applied field, shows the locus of these susceptibility peak maxima as a function of the size and direction φ^t of the tilting angle ψ^t at 5 mK below T_B. Each point on the resulting line corresponds to a particular applied-field direction in the easy-hard plane. The direction of the line is found to be independent of temperature. All points on the line have identical magnetic properties as seen from isothermal- and adiabatic-susceptibility and rf-absorption measurements. In particular, no changes in the bicritical behaviour of the PM boundaries [1] or in the bicritical temperature could be observed. The latter is in agreement with MFA, which predicts a decrease in the bicritical temperature of less than 0.3 mK within the range of experimentally available field tilting angles ($\psi^t < 0.5°$). The present experiment therefore demonstrates the existence of a bicritical "fold" in $GdAlO_3$ whose temperature dependence is compatible with the MFA prediction.

References

[1] H. Rohrer, Phys. Rev. Letters 34 (1975) 1638.
  N.F. Oliveira, Jr., A. Paduan-Filho and S.R. Salinas, Phys. Letters 55A (1975) 293.
  Y. Shapira and C.C. Becerra, Phys. Letters 57A (1976) 483.
[2] A.R. King and H. Rohrer, in: Magnetism and Magnetic Materials 1975, AIP Conf. Proc. No. 29 (American Institute of Physics, New York, 1976) pp. 420–421.
[3] M.E. Fisher and D. Nelson, Phys. Rev. Letters 32 (1974) 1550.
  M.E. Fisher, Phys. Rev. Letters 34 (1975) 1634.
  M.E. Fisher in: Magnetism and Magnetic Materials 1974, AIP Conf. Proc. No. 24 (American Institute of Physics, New York, 1975).
[4] H. Rohrer and H. Thomas, J. Appl. Phys. 40 (1969) 1025.
[5] C.J. Gorter and J. Haanties, Physica 18 (1952) 285.

OPTICAL DETERMINATION OF THE MAGNETIC PHASE DIAGRAM AND CRITICAL EXPONENTS OF DyPO$_4$

M. REGIS, J. FERRE* and Y. FARGE
Laboratoire de Physique des Solides†, Université Paris-Sud, 91405 Orsay, France

I.R. JAHN
Institut für Kristallographie der Universität, Tübingen, BRD

The magnetization of the 3D metamagnet DyPO$_4$ below T_N is measured from the magnetic circular dichroism. The magnetic phase diagram is plotted and a new value for the tricritical point is reported $T_T = 2.19 \pm 0.02$ K. The critical exponent α is measured near T_T and a cross-over from tricritical to critical behaviour is observed.

1. Introduction

Dysprosium phosphate is a tetragonal crystal with diamond structure. It is a three dimensional Ising antiferromagnet below $T_N = 3.39$ K with a high anisotropy ($g_\parallel = 19.5$; $g_\perp = 0.2$). The magnetic moments of the ions are directed along the c axis.

The zero-field susceptibility, heat capacity and sublattice magnetization were measured [1] and compared with series expansions based on nearest neighbour coupling within the Ising model, with a good agreement. On the contrary, the behaviour with applied magnetic field was found to depend on a non-negligible contribution from the long-range dipole interactions [2].

Below T_N, in applied fields parallel to the c axis, DyPO$_4$ undergoes a metamagnetic phase transition, second order above the tricritical temperature and first order below.

The value of the tricritical temperature $T_T = 0.75$ K was reported by Koonce et al. in 1971. In 1975, Faraday rotation experiments indicated $T_T = 1.95$ K [3]. In 1975 also, neutron scattering data gave $T_T \simeq 2.0$ K [4]. More recently, Becker et al. [5] studied the linear magnetic birefringence and found $T_T = 2.00$ K.

In this paper, we report some measurements of the magnetization, by means of the magnetic circular dichroism (MCD) of some absorption lines of Dy^{3+} (4680 Å $< \lambda <$ 4770 Å).

2. Magnetic phase diagram

The changes of the amplitude of the MCD at the antiferromagnetic-paramagnetic phase transition allow us to plot the magnetic phase diagram of DyPO$_4$. Experimentally, the tricritical point is defined as the temperature and field at which a mixed phase begins to exist. At this point, the transition becomes first order and a jump in the total magnetization appears.

In the particular case of DyPO$_4$, the demagnetizing field is not homogeneous in all the sample (which cannot be cut as an ellipsoid because of its brittleness). To avoid this difficulty, we have put an electron microscope diaphragm ($D = 100\mu$) before our sample, and focused the light beam on the other side of the crystal. So, the illuminated part of the sample diaphragm ($D = 100\mu$) in front of our sample, and focused the light beam on the other side of the crystal. So, the illuminated part of the sample (0.05 × 0.05 × 2.5 mm) can be approximated by an ellipsoïdal shape with $c/a = 50$. The demagnetizing field was about 200 Oe.
stant field by varying the temperature. So, we have plotted the magnetic phase diagram and observed the jump of the magnetization up to $T_T = 2.19$ K ± 0.02 and $H = 4.20$ kOe (fig. 1).

Four experimental facts allow us to locate the TCP at this point: 1) the shape of the $M = f(H)$ curves changes at this temperature; 2) the shape of the $M = f(T)$ curves changes also at this temperature, exhibiting a small constant region, corresponding to the crossing of the metamagnetic phase; 3) we have observed an hysteresis, due to the presence of domains in the mixed phase up to this temperature; 4) as we shall see in the following part, the critical change of the total magnetization exhibits a cross-over behaviour in the temperature region surrounding 2.20 K and no cross-over can be observed at lower temperature (critical experiments were

---

* On leave from the Laboratoire d'Optique Physique, EPCI, Paris.
† Laboratoire associé au CNRS.

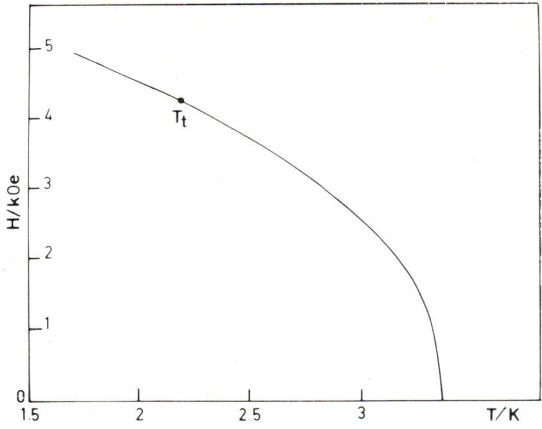

Fig. 1. Magnetic phase diagram of DyPO$_4$ determined optically.

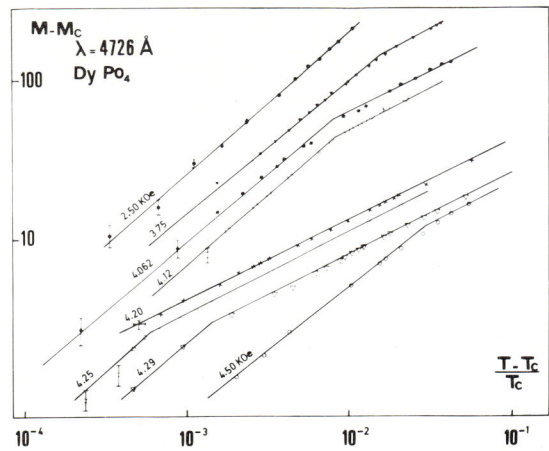

Fig. 2. Change of the MCD versus the reduced temperature. At 2.50 kOe, the slope of the line corresponds to $\alpha = 0.10$. At $H = 4.20$ kOe, the slope corresponds to the tricritical exponent α_T. For the intermediate values of the field, a cross-over between the critical and tricritical behaviour can be seen.

performed at 1.92 and 1.97 K and do not exhibit any cross-over).

3. Critical exponents

The intensity of the MCD signal is correlated with the total magnetization M of the sample. In the neighbourhood of the tricritical point, the singular part of the Gibbs free energy of the system can be written [6]:

$$G \sim \mu_1^{2-\alpha_T} g\left(\frac{h}{\mu_1^s}, 1, 0\right),$$

in which μ_1 is one of the two scaling fields and h is the magnetic field coupled with the order parameter. The scaling field μ_1 is a function of both $\Delta T = T - T_T$ and $\Delta H = H - H_T$ which play the same role and involve similar singularities.

So, near the TCP:

$$M \sim \partial G/\partial H \sim \Delta T^{1-\alpha_T} \sim \Delta H^{1-\alpha_T},$$

where α_T is the tricritical exponent correlated with the specific heat. The singularity of M near the TCP can be studied either as a function of the reduced temperature or of the reduced applied magnetic field.

Fig. 2 shows a log–log plot around 2.20 K of the MCD signal versus the reduced temperature. The slope of the straight lines in the critical region are such that the reduced value of α is:

$$\alpha^+ = 0.10 \pm 0.04 \quad \text{for} \quad T > T_N(H),$$

$$\alpha^- = 0.13 \pm 0.03 \quad \text{for} \quad T < T_N(H).$$

In the limit of the experimental precision, the values of α above and below the transition line are the same, in good agreement with theoretical predictions for a 3-d Ising system: $\alpha = \frac{1}{8}$. The same results are obtained from log–log plots of the MCD in function of the reduced magnetic field $(H - H_C)/H_C$, at constant temperature.

The range of the critical region becomes smaller when the TCP is approached and a sudden change in slope is observed at the critical-tricritical cross-over. The new measured exponents in the tricritical region are:

$$\alpha_T^+ = 0.50 \pm 0.04, \quad \alpha_T^- = 0.46 \pm 0.04.$$

The two values are in good agreement with the calculation of Riedel and Wegner giving:

$$\alpha_T^+ = \alpha_T^- = \tfrac{1}{2}.$$

By measuring the cross-over point for different temperatures and fields, the boundaries of the tricritical region can be plotted and the cross-over lines can be described by the expression: $\mu_1 \sim \mu_2^{(1/\phi_T)}$, where μ_2 is a coordinate along an axis tangent to the λ line at the TCP and μ_2 along an axis perpendicular, ϕ_T being the cross-over exponent. The determination of the cross-over lines is not very precise, however it allows a rough determination of ϕ_T and we find:

$0.50 < \phi_T < 0.75$.

This estimation is consistent with theoretical predictions [7] ($\phi_T = 0.5$) and previous measurements in the case of ^{3}He–^{4}He mixtures [8].

In conclusion, the different values of T_T obtained in the various experiments can certainly be attributed to the severe experimental difficulties in the tricritical region, due to fluctuations and inhomogeneity of the demagnetizing field. We plan to do light scattering experiments on DyPO$_4$ to check the value of T_T found by MCD measurements.

We are pleased to thank P. Pfeuty for his constant interest and his help during this work.

References

[1] J.C. Wright, H.W. Moos, J.H. Colwell, B.W. Mangum and D.D. Thornton, Phys. Rev. B 3 (1971) 843.
[2] C.S. Koonce, B.W. Mangum and D.D. Thornton, Phys. Rev. B 4 (1971) 4054.
[3] J.E. Battison, A. Kasten, M.J.M. Leask and J.B. Lowry, Sol. Stat. Comm. 17 (1975) 1363.
[4] N. Schibuya, K. Knorr, H. Dachs, M. Steiner and B. Wanklyn, Sol. Stat. Comm. 17 (1975) 1305.
[5] P.J. Becker, I.R. Jahn and S.H. Smith, J. Phys. C: Solid State Phys. 9 (1976) L505.
[6] G. Toulouse and P. Pfeuty, Introduction au Groupe de Renormalisation et à ses Applications (Presses Universitaires de Grenoble, 1975).
[7] E.K. Riedel and F.J. Wegner, Phys. Rev. Lett. 29 (1972) 349.
[8] E.K. Riedel, H. Meyer and R.P. Behringer, J. Low. Temp. Phys. 22 (1976) 369.

HYPERSCALING IN THE ISING MODEL*

G.A. BAKER, Jr.

Theoretical Division, University of California, Los Alamos Scientific Laboratory, Los Alamos, New Mexico, 87545

The high temperature series expansions relevant to the hyperscaling relation $\Delta = \frac{1}{2}(d\nu + \gamma)$ are extended and re-examined. Hyperscaling is found to hold in 2 dimensions as expected, but fails in 3 and 4 dimensions. The triviality of hyperstrong-coupling, Euclidean, Boson, ϕ^4 field theory follows.

The modern renormalization group theory of critical phenomena [1], as currently practiced, implicitly assumes that the hyperscaling relationship holds between the critical indices. Specifically, we consider the usual spin-$\frac{1}{2}$, Ising model defined by the partition function

$$Z = \sum_{\{\sigma_i = \pm 1\}} \exp\left[K \sum_{\text{nearest neighbors}} \sigma_i \sigma_j + H \sum_i \sigma_i\right], \quad (1)$$

where the sum in the exponential is over all the nearest neighbor pairs (i, j) of a regular space-lattice in d-dimensions of N sites. In space dimensions d higher than 1, this model has a critical temperature, T_c at which the correlation length ξ diverges to infinity. Letting $\tau = (1 - T_c/T)$ we have, just above the critical temperature, the behavior

$$\chi = N^{-1} \frac{\partial^2 \ln Z}{\partial H^2} \approx A_+ \tau^{-\gamma}, \quad \frac{\partial^2 \chi}{\partial H^2} \approx -B_+ \tau^{-\gamma - 2\Delta}, \quad (2)$$

$$\xi^2 = \sum_r r^2 \langle \sigma_0 \sigma_r \rangle / (2d\chi) \approx D_+^2 \tau^{-2\nu}.$$

The hyperscaling relation is then

$$\Delta = \frac{1}{2}(d\nu + \gamma) \quad (3)$$

This relation has been proven for $d = 2$ by Kadanoff [2] and in general it has been shown that $\Delta \leq \frac{1}{2}(d\nu + \gamma)$ by Baker, Krinsky, and Schrader [3]. The best previously available high temperature series estimates [4, 5] yield $\frac{1}{2}(d\nu + \gamma) - \Delta \approx 0.019 \pm 0.006$ in $d = 3$ which is suggestive of a failure of hyperscaling. The consequences of such a possible failure are significant. For example, the R.G. theory of critical phenomena would need to be revised, and the quantum field theory $\lambda : \phi^4 : d$ would be trivial (i.e., free of real scattering).

We can check on this possibility, and incidently compute a number of interest to field theory by considering the dimensionless, renormalized coupling constant

$$w = gm^{d-4} \propto -v \frac{\partial^2 \chi}{\partial H^2} \bigg/ \chi^2 \xi^d$$

$$\approx [vB_+/A_+^2 (aD_+)^d] \tau^{d\nu + \gamma - 2\Delta} \quad (4)$$

which is finite, if hyperscaling holds, and vanishes if it does not. The quantity, v, is the specific hyper-volume per site on the lattice, and a, the lattice spacing.

The series $\partial^2 \chi / \partial H^2$ for $d > 3$ is not available. We have computed this series on the hyper-simple-cubic lattice by the method of Rushbrooke and Scoins [7] through order $(\tanh K)^9$, where $u = \tanh K$.

The general methods of series analysis employed are those of Padé approximation [8]. In particular, heavy use is made of two ideas. The first is "critical-point renormalization" [5, 8]. If we have

$$f(x) = \sum f_j x^j \approx A(1 - yx)^{-\psi} + B,$$
$$g(x) = \sum g_j x^j \approx C(1 - yx)^{-\psi} + D, \quad (5)$$

for the same value of y, then

$$h(x) = \sum f_j x^j / g_j \propto (1 - x)^{-(\psi - \phi + 1)} \quad (6)$$

for x near 1. The second idea is that one can obtain an estimate of the difference $(\psi - \phi)$ in

---

* Work performed under the auspices of the Energy Research and Development Administration.

(6) by forming Padé approximations to

$$(1-x)(d/dx)\log h(x) \quad (7)$$

and evaluating them at $x = 1$.

One of the problems which can arise in the analysis of series such as these is the occurrence of confluent singularities. By this remark is meant that near the singular point

$$k(x) \approx (1-yx)^{-\phi}\{1 + A(1-yx)^{+\tau} + o[(1-yx)^{\tau}]\}, \quad (8)$$

where $\tau < 1$. For the 3-dimensional, spin-$\frac{1}{2}$ Ising model case we are treating, Camp et al. [9] have found, both for the susceptibility χ, and the second moment series used to define ξ in eq. (2) that such confluent singularities are absent. (These singularities do occur, however, for $s > \frac{1}{2}$.) We have examined the structure of $\partial^2\chi/\partial H^2$. A Padé analysis of d log $\partial^2\chi/\partial H^2/dK$ reveals a clearly-defined, isolated pole corresponding to the critical point. The confluent singularity analysis of Baker and Hunter [10] when applied here is strongly interfered with on open lattices by the presence of non-confluent singularities. On the close-packed lattices there are no apparent confluent singularities.

We have analysed the dimensionless quantity (also independent of spin-length scale) $[(\partial^2\chi/\partial H^2)/\chi^2]$. This quantity leads to stable Padé estimates. By the procedures of eqs. (5)–(7) we conclude that $2\Delta - \gamma$ has the values

S.Q., 1.998 ± 0.002; T., 2.00 ± 0.01;
S.C., 1.885 ± 0.006;
B.C.C., 1.886 ± 0.003; F.C.C., 1.8868 ± 0.001;
H.S.C., 1.918 ± 0.006. (9)

It is to be noted that the results for the 2-dimensional lattices are consistent with the exact value of 2.

We have reanalysed carefully the series for ξ^2. By the methods of eqs. (5)–(7) we conclude for ν

S.Q., 0.97 ± 0.1; T., 1.0 ± 0.05;
S.C., 0.642 ± 0.002;
B.C.C., 0.6384 ± 0.0007; F.C.C., 0.6384 ± 0.0006;

H.S.C., 0.555 ± 0.008. (10)

We remark that $\nu = 1$ is the exact answer for $d = 2$. The results of (9) and (10) agree with hyperscaling for $d = 2$. For $d = 3$, the F.C.C. results are $2\Delta - \gamma - d\nu = -0.028 \pm 0.0028$ which contradicts the R.G. hypothesis of zero, in my opinion, and implies that w of (4) is zero. One concludes that the present R.G. is unfaithful and does not include the Ising fixed point, if any.

For the case of four dimensions we compute $2\Delta - \gamma - d\nu = -0.302 \pm 0.038$. We conclude that w of (4) is zero in four dimensions. Renormalization group theory [1] has predicted that w vanishes like $1/(\log \tau)$ instead of $\tau^{0.3}$, but we have not been successful in attempting to verify this hypothesis. We would have expected $[(\partial^2\chi/\partial H^2)/\chi]$ to be free of log terms, but it has a more erratic Padé analysis than $[(\partial^2\chi/\partial H^2)/\chi^2]$. Also our attempts to treat $\chi\xi^4$ as though it was proportional to $\log \tau$ were not successful either.

Finally, we present in table I the amplitude factors of eq. (4). The values of (B_+/A_+^2) were estimated by the method of Essam and Hunter [4] and D_+ by the method of Fisher and Burford [6]. These amplitudes assume (except for $d = 4$) that hyperscaling holds and compute ν from $2\Delta - \gamma$. The relation to the Callan–Symanzik equation approach of Baker et al. [11] is given by how ν^* compares to $9w/(48\tau)$, or 1.42 compares to 1.44 ± 0.02. The values of D_+ in 3 dimensions are biased high by about 5%. If the best estimate for ν is used the values are closer to those of Fisher and Burford [6].

Table I
Amplitude factors

Lattice	(B_+/A_+^2)	D_+	w
S.Q.	4.72	0.5672	14.68
T.	4.683	0.526	14.66
S.C.	3.173	0.505	24.64
B.C.C.	3.231	0.469	24.11
F.C.C.	3.266	0.457	24.20
H.S.C.	1.221	0.400	0

The author is pleased to acknowledge helpful discussions with B. Nickel, and R. Schrader as well as assistance by D. Meiron with some of the computations reported herein.

References

[1] K.G. Wilson and J.B. Kogut, Phys. Rep. 12C (1974) 75.
E. Brézin, J.C. LeGuillou and J. Zinn-Justin, Phase

Transitions and Critical Phenomena, C. Domb and M.S. Green, eds., Vol. 6, (Academic Press, New York, to be published).
[2] L.P. Kadanoff, Phys. Rev. 188 (1969) 859.
This proof depends on a hypothesis shown by J. Stephenson, J. Math. Phys. 5 (1964) 1009, as explained by B. McCoy and T.T. Wu, The Two-Dimensional Ising Model, (Harvard Univ. Press, Cambridge, 1973) pp. 186–199. I am grateful to C. Tracy for this reference.
[3] G.A. Baker, Jr. and S. Krinsky, J. Math. Phys. (to be published).
R. Schrader, Phys. Rev. B 14 (1976) 172.
[4] J.W. Essam and D.L. Hunter, J. Phys. C 1 (1968) 392 and private communication (1976).
[5] D.S. Gaunt and G.A. Baker, Jr., Phys. Rev. B 1 (1970) 1184.
M.A. Moore, D. Jasnow and M. Wortis, Phys. Rev. Letters 22 (1969) 940.
[6] M.E. Fisher and D.S. Gaunt, Phys. Rev. 133 (1964) A224.
M.E. Fisher and R.J. Burford, Phys. Rev. 156 (1967) 583.
M.A. Moore, Phys. Rev. B 1 (1970) 2238.
G.A. Baker, Jr., Phys. Rev. B 9 (1974) 4908.
C. Domb, Phase Transitions and Critical Phenomena, C. Domb and M.S. Green, eds., Vol. 3, (Academic Press, New York, 1974) p. 357 et seq.
[7] G.S. Rushbrooke and H.I. Scoins, J. Math. Phys. 3 (1962) 176.
[8] See, for example, G.A. Baker, Jr., Essentials of Padé Approximants (Academic Press, New York, 1975).
[9] W.J. Camp and J.P. Van Dyke, Phys. Rev. B 11 (1975) 2579.
W.J. Camp, D.M. Saul, J.P. Van Dyke and M. Wortis, preprint.
[10] G.A. Baker, Jr. and D.L. Hunter, Phys. Rev. B 7 (1973) 3377.
[11] G.A. Baker, Jr., B.G. Nickel, M.S. Green and D.I. Meiron, Phys. Rev. Letters 36 (1976) 1351.

EXACTLY SOLVABLE MODEL WITH A MULTICRITICAL POINT

M. LUBAN*
Department of Physics, Bar-Ilan University, Ramat-Gan, Israel

R.M. HORNREICH and SHTRIKMAN‡
Department of Electonics, The Weizmann Institute of Science, Rehovot, Israel

A hypercubic d-dimensional lattice of spins with multineighbor ferro and antiferromagnetic coupling is studied in the spherical model limit ($n \to \infty$) and is found to exhibit a multicritical point of the uniaxial Lifshitz type. Analytic expressions are given for the shift exponent $\psi(d)$ and its amplitudes $A_\pm(d)$.

We deal here with an n-component, d-dimensional lattice spin model that can be solved exactly in the spherical model limit ($n \to \infty$) and which exhibits a multicritical point on the λ line, of the uniaxial Lifshitz type [1], i.e. a triple point for paramagnetic, ferromagnetic, and helicoidal ferromagnetic phases. Order parameter fluctuations play a primary role for the present model as evidenced by the fact that the λ line exhibits a singularity of the type $T_c(p, d) \sim T_c(d) + A_\pm(d)|p|^{1/\psi(d)}$ as one varies a parameter p. Here $T_L(d) = T_c(0, d)$ is the Lifshitz temperature, ψ is the shift exponent and the amplitudes A_+, A_- apply for $p > 0$, $p < 0$, respectively. We present explicit analytic expressions for ψ and $A_\pm$ as functions of dimensionality d in the range $2 < d \le 3.5$. For $d > 3.5$, ψ "sticks" at the value $\psi = 1$. An unexpected feature is the change of sign of $A_-(d)$ at $d = 3$. As discussed below, there are several rare earth alloy systems that appear to display a multicritical point of the uniaxial Lifshitz type, and the existing data for the shape of the λ line is suggestive of that calculated here for $d = 3$.

We consider a system of classical, $n(\to \infty)$-component spins occupying the sites of a d-dimensional hypercubic lattice with unit lattice spacing. A given spin at lattice site R_l is coupled to each of its $2d$ nearest neighbors via a ferromagnetic interaction J_1 (>0), and also, by an interaction $J_2 = \frac{1}{4}(p-1)J_1$, to the spins at the pair of sites $R_l \pm 2\hat{e}_1$, where $\hat{e}_1$ is a unit vector in the x_1 direction.

* Also at Department of Electronics, The Weizmann Institute of Science, Rehovot, Israel.
‡ Work supported in part by the Commission for Basic Research of the Israel Academy of Sciences and Humanities.

For $p > 0$ the ordered state is a spatially homogeneous ferromagnetic phase, characterized by the wave vectors $\pm k_s (= k_s \hat{e}_1)$, where $\cos k_s = (1-p)^{-1}$. In the immediate vicinity of the Lifshitz point at $p = 0$ we have $k_s \sim (-p)^{\beta_k}$ as $p \to 0^-$, with $\beta_k = 1/2$ for all d.

We have obtained an analytic expression for the leading singular term of T_c for small values of $|p|$ and $2 < d \le 3.5$. The strong dependence on dimensionality can be summarized as follows: For $2 < d < 2.5$, $T_L(d)$ is zero and we have $\psi = |2.5 - d|^{-1}$ with $A_+/A_- = \sqrt{2} \sin \frac{1}{2}\pi(3-d)$. For $d = 2.5$, $T_L(2.5)$ is still zero and $T_c(p \to 0, 2.5)$ decreases to zero as $(\ln|p|)^{-1}$ with $A_+ = A_-$. For $d > 2.5$, the Lifshitz temperature $T_L(d)$ increases from zero monotonically with increasing d, and, in the interval $2.5 < d < 3.5$, $\psi = |2.5 - d|^{-1}$ as before, but now $A_-/A_+ = \sqrt{2} \sin \frac{1}{2}\pi(3-d)$. There is a dramatic change in the shape of the λ line as one passes through $d = 3$ due to the change of sign of A_- at this dimensionality, and T_c increases linearly with $|p|$ for $d = 3$ and $p < 0$. We remark that we have obtained the result $A_- = 0$ for $d = 3$ also for a continuous spin model in the $n \to \infty$ limit. This reflects the more general fact that the ratio A_+/A_- is universal for a system with a uniaxial Lifshitz point, at least in the $n \to \infty$ limit. At $d = 3.5$, T_c displays a very weak singularity of the form $|p \ln|p||$ with $A_- = -A_+$. Finally, for $d > 3.5$, $T_c(p \to 0, d)$ is dominated by a term linear in $|p|$ with $A_- = -A_+$, so that the shift exponent "sticks" at $\psi = 1$ for all $d > 3.5$. By contrast, for a uniaxial Lifshitz point, the thermodynamic critical exponents adopt classical values only when $d > 4.5$ [1]. A more complete description of our results for this model are given elsewhere [2].

As is well known, scaling arguments predict

that ψ is equal to the crossover exponent. In the present context ϕ is the crossover exponent in the scaling form for the free energy $F(T,p) \sim t^{2-\alpha}f(p/t^{\phi})$, where $t = (T - T_c(p,d))/T_c(p,d)$.

Using renormalization group methods we have found [1] $\phi = \nu_{l4}(2 - \eta_{l4})$, and, for the uniaxial Lifshitz point considered in the spherical model limit, we have $\phi = (d - 2.5)^{-1}$ for $2.5 < d < 4.5$. Thus we confirm the equality $\psi = \phi$ for $2.5 < d < 3.5$, although for $d = 3$ one has the additional feature that the amplitude A_- vanishes.

There are several magnetic systems which appear to display a multicritical point of the type described here. Of particular interest are the alloys Gd–Y and Gd–Sc where neutron diffraction data [3] indicate that the helicoidal phase is characterized by a wavevector which decreases continuously to zero, as required at a Lifshitz point, when the parameter p (here alloy composition) is varied. As regards the dependence of T_c on p in the vicinity of the multicritical point, only a limited amount of data is available. However, for both Gd–Y [4] and Gd–Sc [5], the existing data is suggestive of that derived here for the case $d = 3$. The same is true for the alloys Gd–La [4] and Gd–Dy [6], although for these compounds no data exists for the composition dependence of the wavevector characterizing the helicoidal phase. Precise measurements of both the λ line and the critical exponents in the vicinity of the Lifshitz point would be of great value.

References

[1] R.M. Hornreich, M. Luban and S. Shtrikman, Phys. Rev. Lett. 35 (1975) 1678; Phys. Lett. 55A (1975) 269.
[2] R.M. Hornreich, M. Luban and S. Shtrikman, Physica A (in press).
[3] H.R. Child and J.W. Cable, J. Appl. Phys. 40 (1969) 1003.
[4] W.C. Thoburn, S. Legvold and F.H. Spedding, Phys. Rev. 10 (1958) 1298.
[5] H.E. Nigh, S. Legvold, F.H. Spedding and B.J. Beaudry, J. Chem. Phys. 41 (1964) 3799.
[6] F. Milstein and L.B. Robinson, Phys. Rev. 159 (1967) 466.
D.M. Sweger, R. Segnan and J.J. Rhyne, Phys. Rev. B9 (1974) 3864.

ON THE OBSERVATION OF CROSS-OVER EXPONENTS

G.A. GEHRING

Department of Theoretical Physics of University of Oxford, Oxford OX1 3PQ, U.K.

The critical behaviour of the quadrupole ordering parameter is discussed for $n \geq 2$. Phase transitions characterised by an n dimensional order parameter at a unique k value are discussed separately from those in which the soft mode occurs at two or more inequivalent k points.

1. Introduction

A system with an order parameter ϕ^α of dimensionality $n \geq 2$ has an ordered phase characterised by non-vanishing values of both $\langle \phi^\alpha \rangle$ and higher order combinations transforming like spherical Harmonics [1]. The quadrupole terms which are characterised by the Heisenberg–Ising cross over exponent will be discussed here [1, 2]. The higher order spherical Harmonics correspond to irrelevant operators [1] for $n < 4$. Experimental methods which allow for the observation of quadrupole terms are discussed as well as the microscopic mechanisms which will determine the size of the observable effects.

Two types of phase transition of $n \geq 2$ may occur. Those in which an n dimensional soft mode occurs at a unique point in the Brillouin zone and those in which the soft modes occur at inequivalent k points.

2. Ordering characterised by unique k vector

Examples of this type of transition are Heisenberg ferromagnets, simple antiferromagnets (eg $RbMnF_3$) or distortive transitions like that in $SrTiO_3$.

The following quadrupole operator will have a non-vanishing value below the phase transition associated with an ordered state in which

$$\langle \phi^\alpha(R_i) \rangle \neq 0, \quad \langle \phi^\beta(R_i) \rangle = 0, \quad \beta \neq \alpha,$$

$$Q^{\alpha\alpha} = \sum_{ij} f(R_i - R_j) \left[\phi^\alpha(R_i)\phi^\alpha(R_j) - \frac{1}{n} \sum_{\beta=1}^{n} \phi^\beta(R_i)\phi^\beta(R_j) \right]. \quad (1)$$

The renormalization group equations for a Heisenberg system with a perturbation given by $gQ^{\alpha\alpha}$ have been studied extensively [1, 2]. It is found that the quadrupole ordering and susceptibility vary like,

$$\langle Q^{\alpha\alpha} \rangle \propto t^{2-\alpha-\phi}, \quad T < T_c, \quad (2)$$

$$\chi_Q \propto t^{2-\alpha-2\phi}. \quad (3)$$

(Actually $2 - \alpha - \phi$ does not differ much from the molecular field prediction of 2β).

The quantity $\langle Q^{\alpha\alpha} \rangle$ is a component of a second rank tensor. It may be related linearly to other tensors such as the magnetoelastic strain or the dielectric tensor. Microscopically the magnetoelastic strain may occur because of single ion magnetostriction or because of anisotropic exchange-striction. The change in the electric polarisibility may arise from changes in the electronic wave functions [3] or due to bulk strains [4] or due to internal distortions (an optic phonon normal mode displacement) [5].

The effect of the lattice coupling to **Q** on the phase transition should be considered.

If we only allow coupling to optic phonon coordinates (which are likely to be most important) then there are no free surface effects and the magnetoelastic Hamiltonian is given by,

$$\sum_{k\alpha} \sum A(\alpha k) Q^{\alpha\alpha}(k) U^\alpha(-k)$$

$$+ \sum_{k\alpha} \sum (P^\alpha(k) P^\alpha(-k) + \omega^2(k\alpha) U^\alpha(k) U^\alpha(-k). \quad (4)$$

The lattice coordinates may be integrated out and the resulting effective Hamiltonian has an additional negative cubic anisotropy. A similar result was found by Aharony and Bruce [6] for bulk strain coupling. Thus a magnetoelastic interaction of this symmetry does not cause renormalisation of the critical exponents but may lead to a first order phase transition if the total cubic anistropy is sufficiently negative [6].

The critical behaviour of $\langle Q^{\alpha\alpha}\rangle$ and χ_Q may be studied using optical birefringence in which the difference of two refractive indices may be measured very accurately [7, 8]. The experiments are difficult because a single domaining field is required but accurate because the fluctuations are averaged over the light path [5].

The single domaining field may be an external stress or a magnetic field for an antiferromagnet. For a ferromagnet scaling arguments predict the critical exponent for the variation of $\langle Q^{\alpha\alpha}\rangle$ with a magnetic field.

$$F_{\text{sing}}(T, g, h) = t^{2-\alpha} f(gt^{-\phi}, ht^{-\Delta}),$$

$$\left.\frac{\partial \langle Q\rangle}{\partial h}\right|_0 = t^{2-\alpha-\phi-\Delta} \left.\frac{\partial^2 f(x, y)}{\partial x \partial y}\right|_{x=0, y=0}. \quad (5)$$

Such measurements would give the cross over exponent ϕ directly.

3. Ordering occurring at inequivalent q values

Examples of phase transitions of this type are the ferrodistortive phase transitions of cubic NH_4Br and the face centred cubic antiferromagnets. The symmetry of the crystal is lowered at the phase transition because the ordering takes place characterised by one of the possible q values. There may be an additional lowering of symmetry if the direction of the spins is not uniquely determined by the ordering q vector but this may be treated as in section 2.

Let us define $Q^{\alpha\alpha}$ corresponding to ordering in the q^α mode.

$$Q^{\alpha\alpha} = \sum_q f(q)\left[\phi_{q^\alpha+q}\phi_{q^\alpha-q} - \frac{1}{n}\sum_{\beta=1}^n \phi_{q^\beta+q}\phi_{q^\beta-q}\right] \quad (6)$$

$$= \sum_{ij} f(\mathbf{R}_i - \mathbf{R}_j)\left[e^{i(\mathbf{R}_i-\mathbf{R}_j)\cdot \mathbf{q}^\alpha}\right.$$

$$\left.- \frac{1}{n}\sum_{\beta=1}^n e^{i(\mathbf{R}_i-\mathbf{R}_j)\cdot \mathbf{q}^\beta}\right]\phi(\mathbf{R}_i)\phi(\mathbf{R}_j). \quad (7)$$

The strain arising from exchange structure has exactly this form [5] and probably causes the birefringence [7, 9]; but this may also be due to the reduction of the Brillouin zone so that more virtual transitions become allowed at $k=0$.

Renormalisation group calculations [10, 11] show that the phase transitions should be first order in these systems. However, thermodynamic perturbation theory indicates that $\langle Q^{\alpha\alpha}\rangle \propto t^{2\beta}$ for $T < T_N$ until the 'run away' to first order behaviour occurs [5] in agreement with experiment [7, 9]. Above the phase transition an applied stress of appropriate symmetry would force the system to order in one q^α mode so a cross-over should occur to a second order Ising like transition.

References

[1] F.J. Wegner, Phys. Rev. B6 (1972) 1891.
[2] M.E. Fisher, Rev. Mod. Phys. 47 (1974) 597.
[3] T. Moriya, J. Phys. Soc. Japan 23 (1967) 490.
[4] G.A. Smolenskiĭ, R.V. Pisarev and I.G. Siniĭ, Sov. Phys. Usp. 18 (1976) 410.
[5] G.A. Gehring, to be published.
[6] A. Aharony and A.D. Bruce, Phys. Rev. Lett. 33 (1974) 429.
[7] K.H. Germann, K. Maier and E. Strauss, Phys. Stat. Sol. B61 (1974) 449; Solid St. Commun. 14 (1974) 1309.
[8] E. Courtens, Phys. Rev. Lett. 29 (1972) 1380.
[9] G. Egert, I.R. Jahn and D. Renz, Solid St. Commun. 9 (1971) 775.
I.H. Brunskill, I.R. Jahn and H. Dachs, Solid St. Commun. 16 (1975) 835.
[10] D. Mukamel and S. Krinsky, Phys. Rev. B13 (1976) 5065, 5078.
[11] H. Horner, to be published.

PHYSICAL REALISATION OF $n \geq 4$ VECTOR MODELS*

D. MUKAMEL

Baker Laboratory and Materials Science Center, Cornell University, Ithaca, NY 14853, USA

and

S. KRINSKY and P. BAK†

Brookhaven National Laboratory, Upton, NY 11973, USA

We have performed renormalisation group analysis of several $n \geq 4$ systems. The first-order transitions in Cr, Eu, UO_2, MnO and TbP can be explained by noting that the Hamiltonians have no stable fixed-points in $4-\epsilon$ dimensions. The phase transitions in $TbAu_2$, DyC_2, Tb, Ho, Dy, TbD_2, Nd, MnS_2 and K_2IrCl_6 are described by $n \geq 4$ Hamiltonians with a stable fixed point, and we have calculated critical exponents corresponding to this fixed point.

In recent years the importance of symmetry in classifying second-order transitions has been recognized [1], and much progress has been made in grouping second-order transitions into universality classes. According to universality critical behavior should depend only upon a small number of a system's properties, such as the spatial dimensionality d, the number of components of the order-parameter n, and the symmetry of the Hamiltonian. In this paper, we wish to emphasize the importance of symmetry considerations in classifying first-order transitions. The relation between a system's symmetry properties and its phase transitions can be studied using the phenomenological theory developed by Landau. Wilson [1] has extended the Landau theory to include the effects of fluctuations by using an expansion in $\epsilon = 4 - d$. We have used the group theoretical method of Landau and Lifshitz to construct effective Hamiltonians describing certain paramagnetic to antiferromagnetic transitions having order-parameters with $n \geq 4$ components, and we have performed a renormalization group analysis in $4-\epsilon$ dimensions. We have found systems for which the renormalization group equations possess no stable fixed points [2] and suggest that this lack of a stable fixed point is indicative of a first-order transition. We present support for this rule and propose specific experiments to test it.

* Work supported, in part, by Energy Research and Development Administration.
† Present address: NORDITA, Blegdamsvej 17, 2100 Copenhagen Ø, Denmark.

According to the theory of Landau and Lifshitz, the symmetry breaking order-parameter associated with a second-order transition transforms according to an irreducible representation of the symmetry group of the disordered phase. The number of independent components of the order-parameter is equal to the dimensionality n of the representation according to which it transforms. It has been emphasized [3] that when the unit cell is increased in one or more directions the order-parameter transforms according to the space group of the high symmetry phase, which may have representations with dimensionality $n \geq 4$. The renormalization group equations in $4 - \epsilon$ dimensions are dramatically different for systems with $n \leq 3$ from that for those with $n \geq 4$. When $n \leq 3$, the isotropic fixed point is always stable, while for $n \geq 4$ it is always unstable. Hence, $n \geq 4$ systems may have either no stable fixed point or a stable anisotropic fixed point.

We have found five different $n \geq 4$ component Hamiltonians for which there exist no stable fixed points in $4 - \epsilon$ dimensions. We now describe the physical systems corresponding to these Hamiltonians:

Type I antiferromagnets $\boldsymbol{m} \perp \boldsymbol{k}$ (e.g., UO_2). The transition is described by a 6-component Hamiltonian. The transition is known to be first-order.

Type II antiferromagnets $\boldsymbol{m} \| \boldsymbol{k}$ (e.g., TbAs, TbSb, TbP, CeS, TbSe, NdSe, NdTe). The phase transition is described by an $n = 4$ model, where the four components correspond to the four equivalent $(\frac{1}{2}\frac{1}{2}\frac{1}{2})$ vectors of the fcc structure.

TbP has recently been shown to exhibit a first order transition [4]. We suggest it is worthwhile to study the critical behavior of all of these systems.

Type II antiferromagnets $m \perp k$ (e.g., MnO, NiO, ErSb, EuTe). The order parameter has 8 components, and the Hamiltonian is constructed from the six fourth order invariants which can be formed from the order parameter. It has recently been shown that MnO has a first order transition. The nature of the phase transition in the remaining compounds is not yet well established. We urge experimental work be done on these systems.

Cr (sinusoidal magnetic structure), Eu (helical structure): These systems are described by $n = 12$ Hamiltonians, and the phase transitions in both Cr and Eu are known to be first-order.

Type III antiferromagnets $m \perp [100]$, $k = [\frac{1}{2}01]$: The dimensionality of the order parameter is twelve. We are not aware of any fcc systems with this structure.

There has been a very interesting recent experiment [5] using neutron scattering measurements to study the antiferromagnetic order-parameter in MnO as a function of temperature and applied uniaxial stress. It was found that a large [111] stress can change the transition from first-order to second-order. This is consistent with our ideas, since the [111] stress breaks the cubic symmetry and the transition in the system under stress is described by an $n = 2$ order-parameter [6], so the $n = 2$ isotropic fixed point is stable. We have also studied a $2m$-component Hamiltonian which possesses a stable fixed point. For $m = 2$, the Hamiltonian describes the paramagnetic to antiferromagnetic transition in $TbAu_2$, DyC_2, Tb, Ho and Dy, and the structural transition in NbO_2. If these transitions are second order we predict they all belong to the same universality class. If we set $\epsilon = 1$ in the expansion of exponents, we find $\beta = 0.39$, $\nu = 0.70$ and $\gamma = 1.39$. These are the same as for the isotropic $n = 4$ model. For Ho, β has experimentally been found to be 0.39 (+0.04, −0.03) [7]. When $2m = 6$, the Hamiltonian describes the antiferromagnetic transitions in TbD_2, Nd, K_2IrCl_6 and MnS_2. We obtain $\beta = 0.39$, $\nu = 0.59$, $\gamma = 1.38$.

Of course, there is a possibility that the actual Hamiltonians describing these physical systems are outside the regions of convergence to the stable fixed point. In this case, the phase transition is expected to be first order. This may be the case for the first order transition in MnS_2 [8].

The $n \geq 4$ component systems we have been discussing provide an exciting subject for experimental investigation. In particular, it would be of great interest to find new examples of first-order paramagnetic to antiferromagnetic transitions. We hope a comprehensive study of these $n \geq 4$ systems will be undertaken.

References

[1] M.E. Fisher, Rep. Progr. Phys. 30 (1967) 615.
K.G. Wilson and J. Kogut, Phys. Rept. 12C (1974) 75.
[2] D. Mukamel and S. Krinsky, J. Phys. C8 (1975) L1496.
P. Bak, S. Krinsky and D. Mukamel, Phys. Rev. Lett. 36 (1976) 52 (and references therein).
D. Mukamel and S. Krinsky, Phys. Rev. B13 (1976) 5065, 5078.
P. Bak and D. Mukamel, Phys. Rev. B13 (1976) 5086.
[3] D. Mukamel, Phys. Rev. Lett. 34 (1975) 481.
See also R. Alben, C.R. Acad. Sci. B279 (1974) 111.
[4] E. Bucher, J.P. Maita, G.W. Hull, Jr., L.D. Longinotti, B. Lüthi and P.S. Wang, Z. für Physik (to be published).
[5] D. Bloch, D. Hermann-Ronzaud, C. Vettier, W.B. Yelon and R. Alben, Phys. Rev. Lett. 35 (1975) 963.
[6] P. Bak, S. Krinsky and D. Mukamel, Phys. Rev. Lett. 36 (1976) 829.
[7] J. Eckert and G. Shirane, Sol. Stat. Comm. (to be published).
[8] J.M. Hastings and L.M. Corliss, Phys. Rev. B14 (1976) 1995.

UNIVERSAL RELATIONS AMONG CRITICAL AMPLITUDES

A. AHARONY

Department of Physics and Astronomy, Tel-Aviv University, Ramat-Aviv, Israel

and

P.C. HOHENBERG

Bell Laboratories, Murray Hill, N.J. 07974, USA, and Physik Department, Technische Universität, München, 8046 Garching, W. Germany

The hypothesis of universality implies four universal ratios among the six usually defined thermodynamic critical amplitudes. Theoretical information is presented on the values of these ratios for short ranged Ising and Heisenberg systems and for dipolar systems, and experiments are discussed. Additional ratios are also mentioned.

1. Introduction

The main purpose of this note is to emphasize the importance of *critical amplitudes*. Experiments measuring thermodynamic quantities near critical points usually yield both a critical exponent and a critical amplitude. It is well known that the exponents have the same values for materials belonging to the same universality class. We wish to stress the fact that, in addition, there exist several ratios among critical amplitudes which should also be universal in the same way. Experimental results should therefore be analyzed with the aim of classifying materials through both their exponents and their amplitude ratios.

2. Critical amplitudes

We consider here six critical amplitudes. At zero magnetic field, the singular term in the specific heat may be written as

$$C_s = (A/\alpha) t^{-\alpha} \qquad (t > 0, H = 0), \qquad (1)$$

where $t = (T - T_c)/T_c$, and H is the magnetic field, or

$$C_s = (A'/\alpha')(-t)^{-\alpha'} \qquad (t < 0, H = 0). \qquad (2)$$

Similarly, the zero field susceptibility is written as

$$\chi = \Gamma t^{-\gamma} \qquad (t > 0, H = 0) \qquad (3)$$

or

$$\chi = \Gamma'(-t)^{-\gamma'} \qquad (t < 0, H = 0), \qquad (4)$$

the magnetization is

$$M = B(-t)^\beta \qquad (t < 0, H = 0), \qquad (5)$$

and the critical isotherm is

$$H = DM|M|^{\delta-1} \qquad t = 0. \qquad (6)$$

3. Scaling and universality

By the hypothesis of *scaling*, the equation of state may be written in the form

$$H = M|M|^{\delta-1} h(t|M|^{-1/\beta}). \qquad (7)$$

This implies the four exponent relations

$$\alpha = \alpha', \qquad \gamma = \gamma', \qquad \gamma = \beta(\delta - 1),$$
$$\alpha = 2 - 2\beta - \gamma. \qquad (8)$$

The hypothesis of *universality* states that the normalized function

$$\bar{h}(\bar{x}) = h(x/x_0)/h_0 \qquad [h_0 = h(0), h(-x_0) = 0], \qquad (9)$$

is the same for all systems within a given universality class. This leads to the universality of the four amplitude ratios

$$A/A', \qquad \Gamma/\Gamma', \qquad R_\chi = \Gamma D B^{\delta-1} \qquad (10)$$

and

$$R_C = AB^{-2}\Gamma. \qquad (11)$$

Note that the amplitudes are defined in terms of dimensionless variables, in which C is measured in units of the gas constant R.

4. Numerical results

Theoretical information on the equation of state is available from both series and renormalization group studies. The latter involves

Table I
Summary of thermodynamic amplitude ratios

	$d=2, n=1$ exact	$d=3, n=1$ series	$d=3, n=1$ ϵ-expansion	$d=3, n=3$ series	$d=3, n=3$ ϵ-expansion
A/A'	1	0.51	0.55	1.52	1.36
Γ/Γ'	37.69	5.07	4.80	—	—
R_C	6.78	0.059	0.066	0.165	0.17
R_χ	0.319	1.75	1.6	1.23	1.33

expansions in $\epsilon = 4 - d$, where d is the dimensionality of the system. These expansions have to be extrapolated to $d = 3$.

Detailed ϵ-expansions for all these amplitude ratios are given in ref. 1. Appropriate extrapolations then yield the values presented in table I. Similar ϵ-expansions exist for dipolar systems, with $n > 1$ [1].

For the three-dimensional uniaxial dipolar Ising model, the power laws, eqs. (1)–(6), are modified by logarithmic corrections [2], e.g.

$$M = \hat{B}(-t)^{\frac{1}{2}} |\ln(-t)|^{\frac{1}{3}}, \quad \chi = \hat{\Gamma} t^{-1} |\ln t|^{\frac{1}{3}},$$
$$C = \hat{A} |\ln t|^{\frac{1}{3}}. \tag{12}$$

For these amplitudes one finds exactly

$$\hat{A}/\hat{A}' = \tfrac{1}{4}, \quad \hat{\Gamma}/\hat{\Gamma}' = 2, \quad \hat{R}_c = \tfrac{1}{6}, \quad \hat{R}_\chi = \tfrac{1}{2}. \tag{13}$$

Experiments on the Heisenberg ferromagnets Ni and EuO, and on the uniaxial dipolar Ising ferromagnet LiTbF$_4$ give results which are close to those given above [1]. However, the experimental information is rather crude.

5. Two scale factor universality

In addition to the above ratios, the hypothesis of the two scale factor universality [3] predicts a relation between the specific heat amplitude A and that of the correlation length ξ_0,

$$\xi = \xi_0 t^{-\nu} \quad (t > 0, H = 0), \tag{14}$$

i.e.

$$\xi_0 = R_\xi^+ A^{-1/d}, \tag{15}$$

where R_ξ^+ is universal. Similar relations exist between A' and ξ_0' (for $n = 1$) or ξ_0^T (for $n > 1$), where ξ_0^T is the amplitude in $\xi_T = \xi_0^T(-t)^{-\nu}$, which is defined through the transverse response function

$$\chi_T(q, t, m) = M^2 \xi_T^{d-2}/k_B T q^2$$
$$(q \to 0, t < 0, H = 0). \tag{16}$$

Details are given in ref. 4.

In conclusion, more accurate experimental determination of the various amplitude ratios will be very helpful in the classification of materials by their universality properties.

References

[1] A. Aharony and P.C. Hohenberg, Phys. Rev. B13 (1976) 3081.
[2] A.I. Larkin and D.E. Khmel'nitzkii, Zh. Eksp. Theor. Fiz. 56 (1969) 2087. [Sov. Phys. JETP 29 (1969) 1123.
A. Aharony, Phys. Rev. B8 (1973) 3363; Phys. Rev. 9 (1974) 3946 (E).
[3] D. Stauffer, M. Ferer and M. Wortis, Phys. Rev. Lett. 29 (1972) 345.
[4] P.C. Hohenberg, A. Aharony, B.I. Halperin and E.D. Siggia, Phys. Rev. B13 (1976) 2986.
A. Aharony and B.I. Halperin, Phys. Rev. Lett. 35 (1975) 1308.

DOUBLE SPIN ONE-HALF LATTICE GAS MODEL

J. SIVARDIERE

Laboratoire de Chimie Physique Nucléaire, Département de Recherche Fondamentale, Centre d'Etudes Nucleaires de Grenoble, 85X-38041 Grenoble Cedex, France

A double spin one-half lattice gas model is introduced in order to describe a magnetic annealed binary alloy or a nematic mixture. The Ising-like hamiltonian is solved in the molecular field approximation. Various phase diagrams are found, exhibiting unmixing critical points, magnetic critical lines, triple and tricritical points.

Consider first a magnetic alloy. Two fictitious spins $\frac{1}{2}$ are introduced at each site i of the lattice. The first one, $\sigma_i = \pm\frac{1}{2}$, describes the type of atom, A or B, which is found at this site. The second spin, $\tau_i = \pm\frac{1}{2}$, describes the orientation of the magnetic moment. The system is characterized by three order parameters: $M = \langle \sigma_i \rangle$ is the deviation from equiconcentration; $Q = \langle \tau_i \rangle$ is the net magnetization; $R = \langle \sigma_i \tau_i \rangle = \langle \lambda_i \rangle$ is the difference between the magnetizations of A and B atoms.

A binary fluid mixture of elongated molecules is described in the same way: $\sigma_i = \pm\frac{1}{2}$ gives the type of molecule, A or B, which occupies the site i; $\tau_i = \pm\frac{1}{2}$ gives the orientation, "right" or "wrong", of this molecule. $Q = \langle \tau_i \rangle$ is the nematic order parameter.

We introduce the interatomic interactions J_{AA}, J_{BB} and J_{AB}, the positive magnetic interactions j_{AA}, j_{BB} and j_{AB} and the chemical potentials μ_A and μ_B. The hamiltonian is

$$\mathcal{H} = -H\sum_i \sigma_i - \sum_{i,j} [4J\sigma_i\sigma_j + 4k\tau_i\tau_j + 16j\lambda_i\lambda_j + 8l(\lambda_i\tau_j + \lambda_j\tau_i)],$$

with:

$4J = J_{AA} + J_{BB} - 2J_{AB}, \qquad 4k = j_{AA} + j_{BB} + 2j_{AB},$
$4l = j_{AA} - j_{BB}, \qquad 4j = j_{AA} + j_{BB} - 2j_{AB},$
$H = \mu_A - \mu_B + J_{AA} - J_{BB}.$

The three equations of state have been determined in the molecular field approximation. If only one solution of these equations is found at a given temperature T for given values of J, k, l, j and H, it represents a homogeneous mixture. If n solutions are found with the same free energy, they represent n homogeneous phases at equilibrium. Consequently, in order to determine the phase diagram of the mixture, we choose some values of the interactions J, k, l, j and of the field H, and look for a first-order transition as T is varied.

The onset of ferromagnetism gives rise to a critical line; its equation is found from a Landau development of the free energy around $Q = R = 0$:

$$(1 - 2\beta j - 4\beta l M)(1 - 2\beta k - 4\beta l M) = 4\beta^2 (1 - 2jM)(1 + 2kM).$$

Possible tricritical points are determined from the quartic term of the Landau development.

1. Phase diagrams of ideal mixtures ($J = 0$)

We consider first symmetrical mixtures ($l = 0$). If $j_{AB} > j_{AA} = j_{BB}(j < 0)$ the critical line presents a maximum. The solid solution is always homogeneous, homogeneity is favored by the onset of ferromagnetism. If $j_{AA} = j_{BB} > j_{AB}(j > 0)$, the critical line presents a minimum. At lower temperatures phase separation is induced by the magnetic interactions.

If $j \simeq 0$ and $k \gg l$, no phase separation is found in the ferromagnetic region. If k/l is lower and $j \simeq 0$, a tricritical point C' and magnetically induced phase separation are found, the critical line is made of two branches. Phase separation disappears as $T \to 0$. If $k \simeq l \simeq j$, the phase diagram is similar to that of the He3–He4 mixture [1], except that the critical line is made of two branches (fig. 1).

2. Phase diagrams of non-ideal mixtures ($J > 0$)

The case $k = l = j$ has already been considered [1]: phase separation may be induced either by interatomic or magnetic interactions. We consider now only symmetrical mixtures.

If $j < 0$, phase separation induced by interatomic interactions is hindered by magnetic interactions. If $j > 0$, phase separation is en-

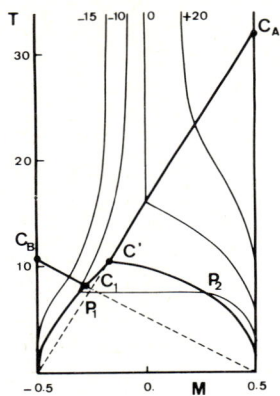

Fig. 1. Phase diagram of an ideal magnetic mixture with: $k = 5.33$, $l = 2.66$, $j = 5.33$. The critical lines $C_A C'$ and $C_B C_1$ are straight lines. Lines of constant H are shown. P_1 and P_2 are two phases at equilibrium.

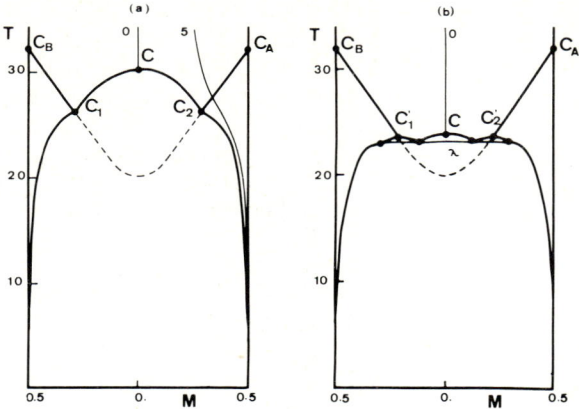

Fig. 2. Phase diagram of a magnetic mixture with: $k = 10$, $l = 0$, $j = 6$. (a) $J = 15$; (b) $J = 12$: C' and C' are tricritical points, λ is a quadruple point. The dashed line represents the critical line if $J = 0$.

hanced by magnetic interactions: if J is large, phase separation is found in the paramagnetic region of the (M, T) plane [fig. 2(a)]. For lower values of J a second-order phase transition in M is followed by a first-order phase transition in Q and R as T is decreased for $H = 0$. This leads to the phase diagram of fig. 2(b). For still lower values of J, the two phase transitions considered above coalesce, so that the paramagnetic binodal curve disappears.

As discussed by Wortis [2], our model of a solid magnetic binary mixture is valid if the atoms are mobile. Our lattice–gas model applies also to the description of mixtures of nematic compounds. The possibility of finding tricritical points in dilute liquid crystals has already been suggested [3].

References

[1] M. Blume, V.J. Emery and R.B. Griffiths, Phys. Rev. A4 (1971) 1071.
[2] M. Wortis, Phys. Lett. 47A (1974) 445.
[3] M. Papoular and J.P. Laheurte, Solid State Commun. 12 (1973) 71.

ON THE CRITICAL DYNAMICS OF DISORDERED SPIN SYSTEMS

U. KREY

Fachbereich Physik der Universität Regensburg, D-8400 Regensburg, Germany

The dynamic critical behaviour of impure or amorphous spin systems is discussed by renormalization group techniques. For systems with critical behaviour of the Ising type in three dimensions the critical slowing down is possibly enhanced by impurities. Near the critical temperature the energy density mode will always decouple from the order parameter.

1. Introduction

Recent renormalization calculations [1–3] have shown that the static critical behaviour of spin systems changes with the introduction of isotropic impurities if the exponent of the specific heat is positive in the pure system: $\alpha_{\text{pure}} > 0$. This condition is fulfilled for Ising systems ($n = 1$) in three dimensions, but not for XY ($n = 2$) or Heisenberg systems ($n = 3$).

Hence, for impure Ising systems in three dimensions a crossover to a "random" critical behaviour is expected, where the static exponents differ from their values in the pure system. Especially, if $\alpha_{\text{pure}} > 0$, then $\alpha_{\text{impure}} < 0$.

In renormalization calculations for "$4 - \epsilon$ dimensions" with $\epsilon \ll 1$, one would obtain $\alpha_{\text{pure}} > 0$ for $n < 4 - 0(\epsilon)$; accordingly, in 4-ϵ dimensions the exponents change for $n = 1, 2, 3$.

Moreover, if besides the isotropic impurities there is also a randomness of local anisotropy directions ("amorphous systems"), then according to [4] also in three dimensions, and for Heisenberg systems, the critical statics should change, but it is unclear whether there is a sharp transition in that case.

In the following, the *dynamic* critical behaviour of these two kinds of disordered systems is discussed.

2. The model

It is well known that systems with identical statics can still differ in their critical dynamics. The simplest model with nonconventional dynamics is the Halperin–Hohenberg–Ma model [5] with the equation of motion

$$\frac{\partial \phi}{\partial t} = -\Gamma\left(\frac{\delta H}{\delta \phi} - h\right) + \eta. \qquad (1)$$

Here $\phi(x, t)$ is the real spin density vector field with n components; H is the classical hamiltonian

$$H = \int dx^d \{\tfrac{1}{2}[(\nabla \phi)^2 + r\phi^2 + V(x)\phi^2] + u\phi^4\}, \qquad (2)$$

where $V(x)$ is the impurity scattering potential; the remaining quantities having their usual meaning (especially $h(x, t) =$ external field; $\eta(x, t) =$ gaussian noise source with the width 2Γ).

Eq. (1) corresponds to systems where the order parameter is nonconserved; for $n = 1$ it represents a continuum analogue of the kinetic Ising model. For $n > 1$, similar to [1–3], we have assumed isotropy of the spin–spin interaction: If there are, additionally, random anisotropy directions present, then a further contribution $-\int dx^D D \cdot (a \cdot \phi)^2$ should be added to H, where $a(x)$ is a randomly distributed unit vector with n components.

3. Formalism

The critical slowing down of the relaxation is calculated from the configuration averaged inverse dynamic susceptibility

$$\langle G_k^{-1}(w) \rangle = -iw/\Gamma_0 + r_0 + k^2 - \sum(w, k), \qquad (3)$$

where w is the complex frequency, k the wave number, Γ_0 the bare relaxation rate, $r_0^{-1/2}$ the bare correlation length, and $\Sigma(w, k)$ the complex self-energy. This quantity can be calculated by diagram techniques.

To a first order in ϵ, only the familiar impurity scattering diagram ⚊ contributes to the dynamics; to a second order there are five additional diagrams. In the integrals corresponding to these diagrams, wave numbers are integrated as usual from a lower cutoff radius b^{-1} to the upper cutoff 1. These and similar points will be treated in a more detailed paper (to be published elsewhere); here we give only the resulting recursion relation (4) for Γ, where we have assumed b large enough such that terms $\sim \epsilon^2 \cdot 1$ can be neglected against terms $\sim \epsilon^2 \ln b$ and $\epsilon^2 (\ln b)^2$. But finally, only the terms linear in $\ln b$ have been written down since the correct exponents can already be deduced from these terms (see e.g. [6]):

$$-i\left(\frac{w}{\Gamma_1}\right) = -i\left(\frac{w}{\Gamma_0}\right) b^{2-\eta-z} \{1 + [\Delta_c^2 \cdot K_d + \tfrac{3}{2}\Delta_c^4 K_d^2 \\ + 48(n+2)u_c^2 \cdot \ln(\tfrac{4}{3}) \cdot K_d^2 \\ - 4(n+2)u_c \cdot \Delta_c^2 \cdot K_d^2] \cdot \ln b\}. \quad (4)$$

Here z is the dynamic critical exponent to be determined; Δ_c^2 and u_c are the fixed point values of the mean squared scattering potential ($\Delta^2 = \langle |V_k|^2 \rangle_{k \to 0}$) and of the four-spin interaction, respectively; η is the well-known spin rescaling exponent, and $K_d \cdot (2\pi)^d$ is the surface of the d-dimensional unit sphere.

4. Results

Now the exponent z is determined such that the l.h.s. and r.h.s. of eq. (4) agree for $\Gamma_1 = \Gamma_0$. The necessary values of Δ_c^2, u_c and η can be found in [1] for $n = 2, 3$ and in [2] for $n = 1$. Especially, if $\alpha_{\text{pure}} < 0$ (that is, for $n \geq 4$ in 4-ϵ dimensions), then $\Delta_c = 0$ for the stable fixed point, and the critical behaviour of the disordered system and of the ordered system are the same, namely $z = 2 + [6 \cdot \ln(\tfrac{4}{3}) - 1] \cdot \eta$. For $n < 4 - 0(\epsilon)$, however, in $4 - \epsilon$ dimensions, $\alpha_{\text{pure}} > 0$ and $\Delta_c \neq 0$; consequently z changes. For $n = 2, 3$ it is

$$z = 2 + \epsilon \frac{(4-n)}{8(n-1)} + \frac{\epsilon^2}{1024(n-1)^3} \\ \times [128 - 640n + 104n^2 - 69n^3 \\ + 192(n+2)(n-1) \cdot \ln(\tfrac{4}{3})], \quad (5)$$

while for $n = 1$, to the same order in powers of u and Δ^2,

$$z = 2 + \left(\frac{6\epsilon}{53}\right)^{1/2} + \epsilon \cdot [\tfrac{24}{53} \cdot \ln(\tfrac{4}{3}) - \tfrac{17}{106}]. \quad (6)$$

To this order, the fixed point corresponding to eq. (6) is marginal, but it may become stable for $d \to 3$; especially, to the next order in powers of u and Δ^2, the coefficient of ϵ in eq. (6) changes, too. Both in eqs. (5) and (6), the third term on the r.h.s. is negative.

For $n = 1$, with $\epsilon = 1$, one would obtain $z_{\text{pure}} = 2.121$ and $z_{\text{impure}} = 2.306$: The critical slowing down is enhanced. But for $n = 2, 3$, an enhancement is obtained only from the terms $\sim \epsilon$, whereas inclusion of the terms $\sim \epsilon^2$ with $\epsilon = 1$ would result in a reduction of z. However, this reduction should not be taken literally, since in fact α_{pure} is negative for $n = 2, 3$ and $d = 3$. On the other hand, the 10% enhancement of z for $n = 1$ might be essentially correct.

To a first order in ϵ, eq. (4) can also be applied if additionally random anisotropies are present: one simply sets in that case

$$\Delta_c^2 = \langle |V_k|^2 \rangle_{k \to 0} + (4D^2/n) \cdot [1 - (1/n)], \quad (7)$$

which corresponds to $-8[u + (m+1)w]$ in Aharony's paper [4].

In [4], the static behaviour has only been calculated to the order ϵ. In this approximation, for the most interesting case $n = 3$, three of the eight fixed points (III, V, VII) are unphysical, i.e. they lead to a $\Delta_c^2 < 0$. Of the remaining fixed points, none is stable apart for the case of vanishing anisotropies (fixed points I, II, IV). This situation might correspond either to a smeared transition or to a transition of first order, but in an exact calculation for $d = 3$ the fixed point VI might turn out as stable [4]. Anyhow, sufficiently close to the "amorphous" fixed points VI and VIII one would observe the enhanced exponents $z_{\text{VI}} = 2 + (\tfrac{5}{4})\epsilon$ and $z_{\text{VIII}} = 2 + 0.0274\epsilon$ for $d = 4 - \epsilon$, $\epsilon \ll 1$.

5. Modifications

Finally, let us conclude with remarks on two important modifications of the model. In the case of a conserved order parameter ϕ, where Γ is replaced by $-\lambda \nabla^2$, one has the conventional behaviour ($z = 4 - \eta$) both in the ordered and in the disordered system. The second case is that of an additional conserved mode $e(x, t)$, representing, for example, the energy density or a strain mode, which is coupled to the nonconserved $\phi(x, t)$ by a term $\int dx^d [\gamma \cdot e \cdot \phi^2 + e^2/2C]$ in the hamiltonian. It is well known from [5] that such modes can lead to important modifications of the critical dynamics in case of ordered crystals. But analogously to the analysis in [5] one can show that γ renormalizes to zero if

$\alpha < 0$, which in impure systems is apparently always the case at the relevant fixed points [1–3].

This means that the energy mode becomes always decoupled from ϕ in the impure case. Furthermore, in case of models with additional dynamical mode–mode coupling as discussed in [7], the asymmetrical models are ruled out. To the first order, the results have already been published [8]. A more detailed version will also treat systems with dynamic mode–mode coupling (planar and isotropic antiferromagnets).

References

[1] T.C. Lubensky, Phys. Rev. B11 (1975) 3573.
[2] G. Grinstein and A. Luther, Phys. Rev. B13 (1976) 1329
[3] U. Krey, Phys. Lett. 51A (1975) 189; Phys. Lett. 54A (1975) 21.
[4] A. Aharony, Phys. Rev. B12 (1975) 1038.
[5] B.I. Halperin, P.C. Hohenberg and Shang-keng Ma, Phys. REv. B10 (1974) 139.
[6] A. Bruce, M. Droz and A. Aharony, J. Phys. C7 (1974) 3673.
[7] B.I. Halperin, P.C. Hohenberg and D.E. Siggia, Phys. Rev. B13 (1976) 1299.
[8] U. Krey, Phys. Letters 57A (1976) 275.

ONSET OF HELICAL ORDER

J.F. NICOLL, G.F. TUTHILL, T.S. CHANG
Physics Department, Massachusetts Institute of Technology, Cambridge, Mass. 02139, USA

and

H.E. STANLEY
Physics Department, Massachusetts Institute of Technology, Cambridge, Mass. and Physics Department, Boston University, Boston, Mass. 02215, USA

Renormalization group methods are used to describe systems which model critical phenomena at the onset of helical order. This onset is marked by a change in the "bare propagator" used in perturbation theory from a k^2-dependence to a more general form. We consider systems which in the non-helical region exhibit $\mathcal{O}$ simultaneously critical phases. Results are given to first order in an ϵ-expansion. For the isotropic case of k^{2L} dependence and $\mathcal{O} = 2$, we give η to first order in $1/n$ for $d_- \leq d \leq d_+$ where $d_\pm$ are upper and lower borderline dimensions.

In 1959 Yoshimori [1], Villain [2], and Kaplan [3] independently proposed that for materials with certain forms of competing exchange interactions there could exist a ground state spin configuration in which the mean value of the order parameter varied periodically in space with a characteristic wavelength that depended on the exchange interactions and in general was not commensurate with the lattice constants of the material. Since their work, many materials have been found to display such helicoidal ordering [4].

The original theoretical work was concerned with ground state spin configurations, and was carried out in the molecular field approximation. Very recently Hornreich et al. [5] have used renormalization group methods to study phenomena associated with the onset of helicoidal ordering. This occurs at a specific point– termed a Lifshitz point [6]–in a phase diagram [fig. 1] in which temperature is plotted against some parameter p which may be conveniently thought of as a ratio of competing exchange interactions.

The onset of helical order can be incorporated into a Landau–Ginsberg model by including higher order derivatives of the magnetization. In particular, when the thermodynamic potential is written in terms of the Fourier transform of the magnetization, powers of the wave-vector k other than k^2 will occur. The usual uniform $k = 0$ phase will be thermodynamically favored as long as the k^2-term dominates for $k \to 0$. If the coefficient of the k^2-term can be made to vanish,

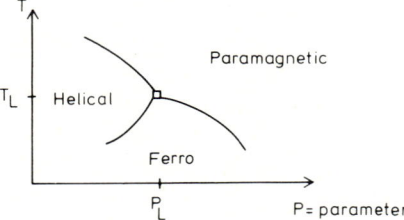

Fig. 1. Schematic phase diagram, indicating the occurrence of a Lifshitz point (T_L, p_L).

then the phases are still uniform but will have drastically altered correlation functions. If the coefficient of k^2 can be made negative, the free energy minimum will be achieved by a phase with non-zero k–a helical phase. Thus, even within the uniform phase region, the onset of helical order can be identified with the change in character of the derivatives in the thermodynamic potential, or equivalently, the bare propagator.

The simplest case is the "isotropic Lifshitz point" for which the bare propagator is given by

$$G^{-1} = (k^2)^L. \tag{1}$$

A more general situation is the anisotropic Lifshitz point with

$$G^{-1} = \sum_i k_i^{2L_i}, \tag{2}$$

with each k_i a d_i-dimensional vector; the propagator exponents L_i need not be integers.

We consider systems which in the uniform region encompass $\mathcal{O}$ simultaneously critical phases of the sort described in [7]. The case $\mathcal{O} = 2$, $L_i = \{1, 2\}$ was considered in ref. 5. We use renormalization group methods [8] to extend this to arbitrary $\mathcal{O}$ and $\{L_i\}$. The upper borderline dimension d_+ (above which mean-field exponents are correct) and the lower borderline dimension d_- (at which infrared divergences commence) depend on $\mathcal{O}$ and $\{d_i, L_i\}$. Thus the universality class is now presumably determined by $\mathcal{O}$, $\{d_i, L_i\}$ and n, where n is the spin symmetry; e.g. $\gamma = \gamma(d_i, n, \mathcal{O}, L_i)$.

In what follows we shall summarize our results, treating first the isotropic Lifshitz point and then the anisotropic Lifshitz point [8].

Case I. Isotropic Lifshitz point

A. ($\mathcal{O} = 2$) critical points
1. Upper and lower borderline dimensionalities $d_{\pm}(L)$ (fig. 2)

$$d_+(L) = 4L, \qquad d_-(L) = 2L. \tag{3a,b}$$

Note that for $L > 1$, the lower borderline dimension is greater than 3. Thus, the isotropic case is principally of academic interest.

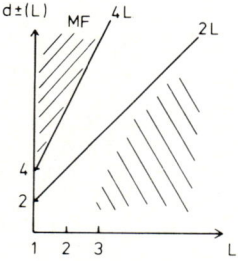

Fig. 2. Dependence of upper and lower critical dimensionalities $d_{\pm}(L)$ upon Lifshitz character L for the case of an isotropic Lifshitz point.

2. Exponents

Critical exponents were calculated to lowest order in

$$\epsilon \equiv d_+ - d = 4L - d, \tag{4}$$

and for η to lowest order in $1/n$. The result for the exponent η describing the decay of correlations at the Lifshitz point,

$$\eta(d, n, \mathcal{O} = 2, L)$$

$$= \frac{(-1)^{L+1}}{n} \left[\frac{(4L - d) \sin \tfrac{1}{2}\pi(4L - d)}{\tfrac{1}{2}\pi} \right] \frac{1}{L}$$

$$\times \frac{\Gamma(d - 2L)\Gamma(2L)}{\Gamma(\tfrac{1}{2}d + L)\Gamma(\tfrac{1}{2}d - L)}, \tag{5}$$

is particularly interesting since it is an oscillatory function of d with $L + 1$ zeros (cf. fig. 3).

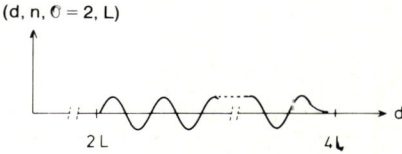

Fig. 3. Dependence upon lattice dimensionality d of the critical exponent η for an $\mathcal{O} = 2$ critical point for the case of an isotropic Lifshitz point of Lifshitz character L.

3. Scaling laws

The various critical exponents are predicted to be related to one another by 2-exponent and 3-exponent scaling laws that are formally analogous to the 2-exponent and 3-exponent scaling laws relating the familiar $L = 1$ exponents of an ordinary critical point. Thus, for example, we find for general L that

$$2 - \alpha_L = d\nu_L, \tag{6a}$$

$$\gamma_L = (2L - \eta_L)\nu_L, \tag{6b}$$

$$\delta_L = \left(\frac{d}{2L - \eta_L} + 1\right) \bigg/ \left(\frac{d}{2L - \eta_L} - 1\right). \tag{6c}$$

Since eq. (6) contains some unfamiliar expressions, it is worthwhile to note that the usual thermodynamic scaling relations are maintained, e.g.

$$\delta_L = (2 - \alpha_L + \gamma_L)/(2 - \alpha_L - \gamma_L). \tag{7}$$

B. Critical point of arbitrary $\mathcal{O}$
1. Upper and lower borderline dimensionalities $d_{\pm}(\mathcal{O}, L)$

The generalization of eqs. (3) for arbitrary $\mathcal{O}$ is

$$d_+(\mathcal{O}, L) = 2L\mathcal{O}/(\mathcal{O} - 1), \qquad d_-(\mathcal{O}, L) = 2L. \tag{8a,b}$$

Thus the lower borderline dimension d_- is independent of $\mathcal{O}$, while the upper borderline dimension decreases with $\mathcal{O}$ and approaches d_- as $\mathcal{O}$ approaches infinity.

2. Exponents

As for the case $\mathcal{O} = 2$, critical exponents were calculated to lowest order in

$$\epsilon \equiv (\mathcal{O} - 1)(d_+ - d) = 2L\mathcal{O} - d(\mathcal{O} - 1). \tag{9}$$

3. Scaling laws

For each value of $\mathcal{O}$, there are a family of scaling laws relating the appropriate exponents.

Case II. Uniaxial Lifshitz anisotropy

Next we treat the physically interesting case in which one of the components of the d-dimensional wave vector $\boldsymbol{k} \equiv (k_1, k_2, \ldots, k_d)$ is raised to the power $2L_1$ so that eq. (1) becomes

$$G^{-1} = + k_1^{2L_1} + k_2^2 + k_3^2 + \ldots + k_d^2. \tag{10}$$

1. Upper and lower borderline dimensionalities $d_\pm(\mathcal{O}, L_1)$

$$d_+(\mathcal{O}, L_1) = (3\mathcal{O} - 1)/(\mathcal{O} - 1) - 1/L_1 \tag{11a}$$
$$d_-(\mathcal{O}, L_1) = 3 - 1/L_1. \tag{11b}$$

Thus, $d_- < 3 \leq d_+$ for $\mathcal{O} \leq 2L_1 + 1$.

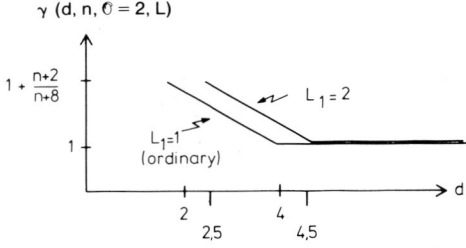

Fig. 4. Dependence upon d of the susceptibility critical exponent γ for the case of uniaxial Lifshitz anisotropy. Curves for anisotropy parameter L_1 greater than two are similarly shifted to the left, with the upper and lower borderline dimensionalities being given by eq. (11) with $\mathcal{O} = 2$.

2. Exponents

Critical exponents were calculated to lowest order in

$$\epsilon \equiv (\mathcal{O} - 1)(d_+ - d) = (3\mathcal{O} - 1) - (\mathcal{O} - 1)(d + 1/L_1).$$

The results bear many resemblances to the results for the case of an ordinary critical point ($L_1 = 1$). For example, fig. 4 shows the dependence on d of the critical exponent $\gamma(d, n, \mathcal{O} = 2, L_1)$ for the susceptibility.

3. Scaling laws

The scaling laws corresponding to eqs. (6) for an isotropic Lifshitz point are

$$2 - \alpha_L = \nu_{L_1} + (d-1)\nu_\perp, \tag{13a}$$

$$\gamma = (2L_1 - \eta_{L_1})\nu_{L_1} = (2 - \eta_\perp)\nu_\perp, \tag{13b}$$

$$\delta = \left(\frac{1}{2L_1 - \eta_{L_1}} + \frac{d-1}{2 - \eta_\perp}\right) \bigg/ \left(\frac{1}{2L_1 - \eta_{L_1}} - \frac{d-1}{2 - \eta_\perp}\right), \tag{13c}$$

where the exponents with a subscript L_1 denote the behavior of correlations between spins joined by a vector whose components lie entirely along the "1" direction, and "⊥" denotes directions perpendicular to "1".

In [8], the general anisotropic case is considered and explicit expressions are given for all the exponents.

References

[1] A. Yoshimori, J. Phys. Soc. Jap. 14 (1959) 807.
[2] J. Villain, J. Chem. Phys. Solids 11 (1959) 303.
[3] T.A. Kaplan, Phys. Rev. 116 (1959) 888; Phys. Rev. 124 (1961) 329.
[4] For a recent review, see D.E. Cox, IEEE Trans. Magn. 8 (1972) 161.
[5] R.M. Hornreich, M. Luban and S. Shtrikman, Phys. Rev. Lett. 35 (1975) 1678; Phys. Lett. 55A (1975) 269; paper 4D7, this conference.
[6] For a discussion of the origin of this terminology, see S. Goshen, D. Mukamel and S. Shtrikman, Int. J. Magn. 6 (1974) 221. An equally appropriate name would be a "YVK point" after the authors of refs. 1–3.
[7] T.S. Chang, G.F. Tuthill, and H.E. Stanley, Phys. Rev. B 9 (1974) 4882; J.F. Nicoll, T.S. Chang, and H.E. Stanley, Phys. Rev. Lett 33 (1974) 540.
[8] The detailed derivation of these results are given elsewhere: J.F. Nicoll, G.F. Tuthill, T.S. Chang and H.E. Stanley, Phys. Lett. 58A (1976) 1. For a discussion of the methods employed see G.F. Tuthill, J.F. Nicoll, and H.E. Stanley, Phys. Rev. B 11 (1975) 4579, and J.F. Nicoll, T.S. Chang, and H.E. Stanley, Phys. Rev A13 (1976) 1251, A 14 (1976) 1921.

REAL SPACE RENORMALIZATION GROUP CALCULATIONS FOR THE 3-DIMENSIONAL ISING MODEL

J. OITMAA and M.N. BARBER

School of Physics and Department of Applied Mathematics, The University of New South Wales, Kensington, N.S.W. 2033, Australia

We report the results of real space renormalization group calculations for the spin $\frac{1}{2}$ Ising model on the simple cubic lattice. Several transformation schemes between site and cell variables are used. The results are only in fair agreement with series estimates.

A large number of real space renormalization group calculations for two-dimensional spin $\frac{1}{2}$ Ising systems have been carried out using the formalism developed by Niemeijer and van Leeuwen [1, 2]. Apart from a recent calculation by Fields and Fogel [3] for a spin 1 model, no applications of this approach in three dimensions have been reported in the literature.

We have used this formalism to carry out a number of renormalization calculations for the spin $\frac{1}{2}$ Ising model on the simple cubic lattice. We choose as the basic cell a cube of 8 spins, shown in fig. 1(a), and associate with each cell a spin $\tilde{s}_i = \pm 1$, denoting the site spins within a cell by s_i^α, $\alpha = 1, 2, \ldots, 8$. A renormalized hamiltonian for the cell spins $\tilde{\mathcal{H}}(\tilde{s})$ is defined, as usual, by

$$\exp[G + \tilde{\mathcal{H}}(\tilde{s})] = \sum_{\{s\}} \prod_i p(\tilde{s}_i, s_i) \exp[\mathcal{H}(s)]. \quad (1)$$

The choice of mapping $p(\tilde{s}_i, s_i)$, between site and cell spin configurations, is in principle (almost) arbitrary subject to $\Sigma_{\{\tilde{s}\}} p(\tilde{s}, s) = 1$ to conserve the partition function. In cases where the basic cell contains an odd number of spins the usual choice has been

$$p(\tilde{s}_i, s_i) = \delta\left[\tilde{s}_i, \operatorname{sgn}\left(\sum_\sigma s_i^\alpha\right)\right]. \quad (2)$$

In the present situation, in which cell configurations with zero total spin occur, the above choice is not possible and we have used the following transformations

(a) $p(\tilde{s}_i, s_i) = \delta\left[\tilde{s}_i, \operatorname{sgn}\left(\sum_\alpha s_i^\alpha\right)\right]$
$\qquad + \frac{1}{2}\delta\left[0, \operatorname{sgn}\left(\sum_\alpha s_i^\alpha\right)\right] \quad (3)$

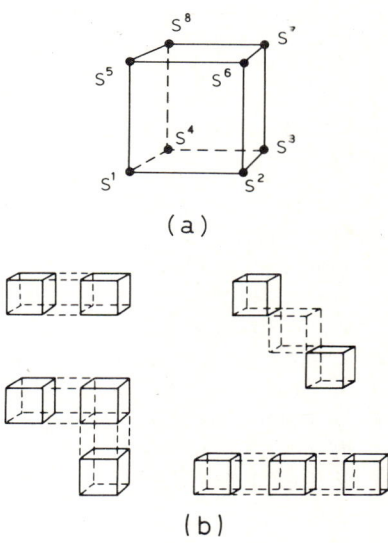

Fig. 1. (a) The basic cell; (b) cell configurations which contribute to the cumulant expansion to second order.

and

(b) $\quad p(\tilde{s}_i, s_i) = \frac{1}{2} + \frac{1}{2}\rho\tilde{s}_i \sum_\alpha s_i^\alpha. \quad (4)$

Case (b) is an example of a "linear weight factor" [2] and consequently the parameter ρ should be constrained to ensure a critical decay of correlations at the fixed point. However, it is difficult to meet this constraint in an approximate calculation (see [4]) and we have used a range of values. The results given in table I correspond to $\rho = 0.18, 0.20$.

In the cumulant expansion method we write the hamiltonian as the sum of an intracell hamiltonian $\mathcal{H}_0(s)$ and an intercell hamiltonian $V(s)$. The renormalized hamiltonian is then

Physica 86–88B (1977) 621–622 © North-Holland

Table I
Numerical results for simple cubic lattice

	K_c	λ_T	λ_H	$2-\alpha$	γ
1st order cumulant (A)	0.298	2.371	6.908	2.41	2.07
2nd order cumulant (A)	0.227	2.339	6.324	2.45	1.86
2nd order cumulant ($\rho = 0.18$)	0.230	3.027	7.055	1.88	1.65
2nd order cumulant ($\rho = 0.2$)	0.237	2.781	5.879	2.03	1.43
2-cell cluster (A)	0.305	2.322	6.942	2.47	2.13
2-cell cluster ($\rho = 0.2$)	0.268	3.05	5.896	1.868	1.32
Series	0.222	3.031	5.657	1.875	1.25

given by

$$\tilde{\mathcal{H}}(\tilde{s}) = -G + \ln Z_0 + \langle V \rangle_0 + \tfrac{1}{2}[\langle V^2 \rangle_0 - \langle V \rangle_0^2] + \ldots . \quad (5)$$

Evaluation of the averages in eq. (5) leads, in the usual way [1,2], to a set of recursion relations among coupling constants. In the present case, to second order in zero field, we need to include nearest neighbour interactions K, second neighbour interactions L, and fourth neighbour interactions M. The interactions L and M are regarded as of order K^2. In fig. 1(b) we show the types of cell configurations which contribute to this order. The recursion relations are

$$\tilde{K} = (4f_1^2)K + (8f_1^2)L + (8f_1^2)M,$$
$$\tilde{L} = 2f_1^2(3f_{12} + 3f_{13} + f_{17} + 1 - 8f_1^2)K^2 + (2f_1^2)L,$$
$$\tilde{M} = 2f_1^2(f_{12} + 2f_{13} + f_{17} - 4f_1^2)K^2, \quad (6)$$

where the f values are single cell spin averages

$$f_1 = \tilde{s}_i \langle s_i^\alpha \rangle_0, \qquad f_{\alpha\beta} = \langle s_i^\alpha s_i^\beta \rangle_0 \quad (7)$$

and the numbering of sites within a cell is as shown in fig. 1(a). The fixed point of the recursion relations (6) is found numerically to be

$$(K^*, L^*, M^*) = (0.2222, 0.0259, -0.00264). \quad (8)$$

The recursion relations for the case of non-zero external field are considerably more complicated and are not given here. They involve three additional coupling constants to second order, a single site field and two kinds of three spin interaction. The relevant eigenvalues and critical parameters obtained from the analysis are shown in table I.

We have also carried out a very simple cluster calculation based on a two cell cluster. Only nearest neighbour interactions occur and in zero field the recursion relation is

$$\tilde{K} = \tanh^{-1} \left\{ \frac{4vf_1^2 + 4v^3 f_{123}^2}{1 + v^2(4f_{12}^2 + 2f_{13}^2) + v^4 f_{1234}^2} \right\}, \quad (9)$$

where $v = \tanh K$ and the f values are spin averages defined analogously to eq. (7). The results of the analysis are shown in table I.

Comparison of these results with the accurate series estimates, also shown in table I, shows that the renormalization results are rather irregular. There is some evidence of convergence in going from first to second order cumulants, particularly with regard to K_c. It is likely that third order results would show a significant improvement since it is in this order that essentially "three-dimensional" cell configurations first arise. It is also likely that improved convergence would be obtained by taking a larger basic cell [5], in this case a cell of 27 spins – this would however require a large amount of computing time. The cluster calculation also does not adequately sample the cubic structure and it is likely that an eight cell cluster would be needed to obtain good results.

References

[1] Th. Niemeijer and J.M.J. van Leeuwen, Physica 71 (1974) 17.
[2] Th. Niemeijer and J.M.J. van Leeuwen, preprint to be published in Phase Transitions and Critical Phenomena, Vol. 6, C. Domb and M.S. Green (Eds.), (Academic Press, New York, 1976).
[3] J.N. Fields and M.B. Fogel, Physica 80A (1975) 411.
[4] K. Subbarao, Phys. Rev. B11 (1975) 1165.
[5] A.S. Sudbø and P.C. Hemmer, Phys. Rev. B13 (1976) 980.

ENTANGLED SCALE TRANSFORMATION ON A LATTICE AND UNIVERSALITY OF THE CRITICAL EXPONENTS OF THE ISING MODELS

P. PERETTO

Centre d'etudes Nucleaires de Grenoble, Départment de Recherche Fondamentale, CPN/Groupe d'Interactions Hyperfines, 85X-38041 Grenoble Cedex, France

In this paper we use the properties of the scale transformations on a lattice [1] to prove the universality of the critical exponents of the Ising models, i.e. their independence with respect to the crystal lattice symmetry. Let us suppose that one can define a transformation F_{21} of a system 1 into a system 2 and a transformation F_{12} of a system 2 into a system 1 which leave the partition function unchanged. For example summing out one in every three spins transforms a triangular lattice into a honeycomb lattice (sites A, J, K, ... in fig. 1). A further decimation (on sites B, D, F, ...) recovers an enlarged triangular lattice.

Then the transformation $F_1 = F_{12} \cdot F_{21}$ is a scale transformation of the system 1 whereas the transformation $F_2 = F_{21} \cdot F_{12}$ is a scale transformation of the system 2. We shall show that the critical exponents deduced from these two entangled transformations are identical.

1. The fixed points of the transformations

Let

$$\mathcal{H}_1 = \beta \sum_i K_1^i S_1^i,$$

$$\mathcal{H}_2 = \beta \sum_j K_2^j S_2^j, \quad (\beta = 1/k_B T)$$

be the hamiltonians of the two systems. S_1^i is the product of the elementary spins $\sigma = \pm 1$ of a given set i of sites of the lattice.

Let $\mathbf{K}_1^*$ and $\mathbf{K}_2^*$ be the fixed points of the transformations $\mathbf{F}_1$ and $\mathbf{F}_2$. Then if

$$\mathbf{K}_2^0 = \mathbf{F}_{21}(\mathbf{K}_1^*)$$

we have $\mathbf{F}_2(\mathbf{K}_2^0) = \mathbf{K}_2^0$ and therefore

$$\mathbf{K}_2^* = \mathbf{F}_{21}(\mathbf{K}_1^*), \qquad \mathbf{K}_1^* = \mathbf{F}_{12}(\mathbf{K}_2^*).$$

2. A mapping between two systems

We build a mapping between two systems which have the same dimensionality but different symmetries, for example the square and the triangular lattices. The transformation must leave the partition function unchanged.

When two systems have the same dimensionality we can always superimpose, say the system 1, over the system 2 by a topological deformation. For example, the shear deformation 1 (see fig. 1) transforms a triangular lattice into a square lattice. The interactions are kept constant in the transformation so that the hamiltonian of the deformed lattice $\mathcal{H}_2^d (\equiv \mathcal{H}_1)$ is the hamiltonian of a system of spins on a new

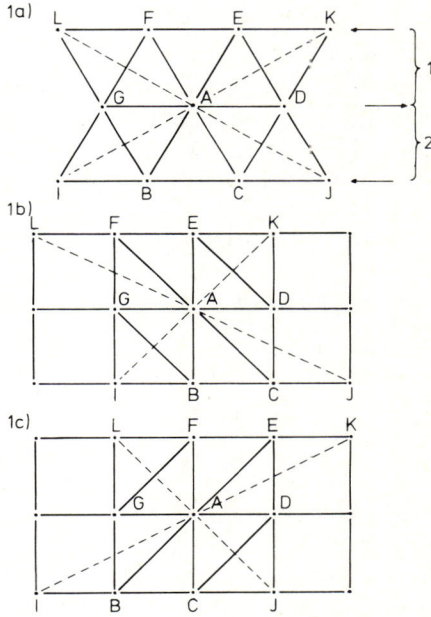

Fig. 1. Mapping of a triangular lattice [fig. 1(a)] into a square lattice. The shear deformation 1 (resp. 2) transforms fig. 1(a) into fig. 1(b) [resp. 1(c)]. Only n.n. and n.n.n. interactions are displayed. Superimposing 1(b) and 1(c) recovers the square symmetry of the interactions on the square lattice.

lattice interacting via the old interaction constants. Let us now consider the various systems $\mathcal{H}_{2,\alpha}^d$ generated from the deformed system by the g symmetry operations of the lattice 2. In the case of fig. 1 only the two deformations 1 and 2 have to be considered to recover the square symmetry. Let us pack all these systems together. On every site of the lattice 2 there are g spins $\sigma_{1,\alpha}$. We perform a partial summation on every site j of the lattice 2:

$$\exp -\mathcal{H}_2(\sigma_2) = \sum_{(\sigma_{1,\alpha}^j ; \sigma_2^j)} \exp -\frac{1}{g} \sum_\alpha \mathcal{H}_{2,\alpha}^d(\sigma_{1,\alpha}), \quad (1)$$

with, as usual [1], $\sigma_2^j = \operatorname{sgn} \sum_\alpha \sigma_{1,\alpha}^j$, the configurations with $\sum_\alpha \sigma_{1,\alpha}^j = 0$ being shared between the two configurations of σ_2^j. As symbolized by $(\sigma_1^j; \boldsymbol{\sigma}_2^j)$, all the configurations σ_1^j which preserve a given configuration σ_2^j are summed out.

The partial summation couples the spins $\sigma_{1,\alpha}^j$ into one spin σ_2^j and the hamiltonian $\mathcal{H}_2$ has the lattice 2 symmetry. Moreover we have

$$\sum_{(\sigma_2)} \exp -\mathcal{H}(\sigma_2) = \sum_{(\sigma_{1,\alpha})} \exp -\frac{1}{g} \sum_\alpha \mathcal{H}_{2,\alpha}^d$$

$$= \sum_{(\sigma_1)} \exp -\mathcal{H}_1(\sigma_1).$$

Therefore the transformation (1) is a mapping of the interactions K_1 of the system 1 into the interactions K_2 of the system 2 which leaves the partition function unchanged. Combining this mapping with classical scale transformations we can build two transformations $\mathbf{F}_{12}$ and $\mathbf{F}_{21}$.

3. Equality of the critical exponents

The critical exponents, ν, are given by the eigenvalues $\lambda = t^\nu$ of the following matrices:

$$T_1 = \frac{\partial \mathbf{F}_1}{\partial \mathbf{K}_1}\bigg]_{\mathbf{K}_1^*}; \qquad T_2 = \frac{\partial \mathbf{F}_2}{\partial \mathbf{K}_2}\bigg]_{\mathbf{K}_2^*},$$

for a couple of connected fixed points $\mathbf{K}_1^*, \mathbf{K}_2^*$. t is the dilation parameters of both transformations. Using the properties of the partial derivatives, we can write $T_1 = UV$; $T_2 = VU$, with

$$U = \frac{\partial \mathbf{F}_{21}}{\partial \mathbf{K}_1}\bigg]_{\mathbf{K}_1^*}; \qquad V = \frac{\partial \mathbf{F}_{12}}{\partial \mathbf{K}_2}\bigg]_{\mathbf{K}_2^*}.$$

Moreover, it can be shown that a linearized scale transformation can be written as the product of two symmetric matrices [2]:

$$T_1 = A_1 B_1; \qquad T_2 = A_2 B_2;$$
$$(A_1 = A_1^+, \ldots, \text{ but } [A_1, B_1] \neq 0, \ldots).$$

Because T_1, T_2, U and V all are scale transformations we have

$$U = CD; \qquad V = EF; \qquad (C = C^+, \ldots).$$

Using the identity

$$\sum_{(\sigma_1; \sigma_3)} \sum_{(\sigma_4; \sigma_1)} = \sum_{(\sigma_4; \sigma_3)} = \sum_{(\sigma_2; \sigma_3)} \sum_{(\sigma_4; \sigma_2)}$$

we also have $A_1 = CE$, $A_2 = EC$. Therefore $A_1 = A_1^+ = CE = EC = A_2$. Similarly $B_1 = B_2$ and $T_1 = T_2$. The critical exponents are the same.

4. Remark

This type of argument is quite general and can be used to prove other universality properties. For example we can try to build a mapping between lattices with different spin magnitudes [3]. The packing procedure introduces naturally new operators such as σ_i^2, $\sigma_i^2 \cdot \sigma_j^2$ and so on.... When the corresponding fields are fixed it is then possible to prove the independence of the critical exponents with respect to the spin magnitude.

References

[1] Th. Niemeijer and J.M.J. Van Leeuwen, Physica 71 (1974) 17.
[2] P. Peretto, Solid State Commun. 235 (1976) 19.
[3] N. Beker, to be published.

RENORMALIZATION GROUP TREATMENT OF A COMPRESSIBLE ISING MODEL

Z. FRIEDMAN

Department of Physics, Duke University, Durham, N.C., USA

and

L. GUNTHER

Department of Physics, Tufts University, Medford, MA, USA

We rigorously transform a compressible Ising model into a continuous spin hamiltonian. We find a constrained Ising tricritical point at a pressure which can be shape-dependent, in which case the second order transition may be "pseudo-critical" due to instabilities recently discussed in the literature.

1. Introduction

We have derived a rigorous method of transforming a compressible Ising model into a field theory. We were motivated by the following factors.

(a) While the Baker–Essam model [1] is exactly solvable and was shown to have a tricritical point [2], it is unphysical in that it lacks shear forces, which when added make the model unsolvable. Various approximation schemes [3–5] on Ising models with shear forces have been used, but they are not reliable near the critical point. The renormalization group method (RG) allows one to derive the critical behavior without obtaining an exact solution.

(b) RG starts with a hamiltonian having continuous fields as its degrees of freedom. This hamiltonian can either be postulated in an ad hoc fashion [6], or can sometimes be derived from a microscopic model. The latter method has the advantage of providing us first with assurance that the field theory accurately reflects the microscopic model and second with a direct relationship between the parameters of the field theory and the parameters of the microscopic model, which are sometimes accessible to experiment [7].

(c) Milošević et al. [8] showed that incorrect results are obtained if one transforms the exactly solvable Baker–Essam model into a field theory by merely replacing the discrete spins by *continuous* spins s_i and adding a weighting factor to the partition function of the form $\exp[-\Sigma_i(rs_i^2 + us_i^4)]$. Reference 8 provided the ultimate impetus for our work.

2. Theory

The method we have developed is applicable to any Ising model interacting linearly with the lattice vibrations. For simplicity, we will discuss the model of Baker and Essam [4] which includes only a nearest neighbor compression force constant ϕ_2 and a nearest neighbor shear force constant Λ in a simple cubic lattice. After integrating over the lattice variables, one obtains an effective spin hamiltonian

$$H_s = \sum_\alpha \left\{ -[J_0 + J_1(a-a_0)]Q_0^\alpha + \frac{kT}{2} g \sum_{k\neq 0} Y_\alpha(k)|Q_k^\alpha|^2 \right\} \quad (1)$$

where k is the wave vector, $\alpha = (x, y, z)$, J_1 is the magneto-elastic coupling constant, a and a_0 are the lattice spacing with $P \neq 0 \neq J_1$ and $P = 0 = J_1$, respectively, k is Boltzmann's constant, T is the temperature and g is a dimensionless parameter given by $J_1^2/kT\phi_2$. The other two functions are:

$$Q_k^\alpha = \sum_i e^{ik\cdot R_i}\sigma_i\sigma_{i_\alpha}, \quad (2)$$

where R_i is the lattice vector of the ith atom, i_α is the neighbor in the αth direction of the ith atom, and

$$Y_x(k) = \frac{\phi_2 \sin^2 k_x a}{\phi_2 \sin^2 k_x a + \Lambda(\sin^2 k_y a + \sin^2 k_z a)}. \quad (3)$$

To obtain the RG hamiltonian, we perform *two* consecutive transformations of the form

$$\exp\left(\tfrac{1}{2}\sum_{ij} A_{ij}\nu_i\nu_j\right) = (2\pi)^{-N/2}\int \delta z \exp\left[-\tfrac{1}{2}\sum_i z_i^2 - \sum_{ij}(A^{1/2})_{ij}\nu_i z_j\right], \quad (4)$$

where δz indicates a multiple integration over the variables $\{z_i\}$. The first transforms the last term of H_s, which is quartic in the spin variables $\{\sigma_i\}$, into an Ising model with coupling constants

K_{ij} which depend upon new field variables $\{z_i\}$. The $\{z_i\}$ represent the fluctuating atomic positions. The second transformation results in a hamiltonian which is linear in the spins $\{\sigma_i\}$ (whose values can now be summed over) and introduces the continuous spin field variables $\{s_i\}$. The resulting partition function has the form:

$$Z \propto \int \delta s \int \delta z \, \exp\left(-\tfrac{1}{2}\sum_i z_i^2\right)$$
$$\times \exp\left[-\sum_{ij} K_{ij}(\{z_i\}) s_i s_j + \log \cosh s_i\right]. \quad (5)$$

We have not been able to integrate out the variables $\{z_i\}$ exactly. What we have done is to carry out an expansion in the parameter g, which is typically $O(kT_c/\phi_2 a^2) = O(10^{-2})$.

Our results are as follows:

(a) To first order in J_1, the dependence of the transition temperature on J_1 agrees with Domb's result [9] and has been confirmed by recent experimental results on EuO [7].

(b) To first order in g, the RG hamiltonian has the same form as that postulated by de Moura et al. [10]. It has a constrained Ising fixed point. When

$$P < P_t \equiv \frac{\phi_2}{2a}\left(\frac{1-\langle Y_0\rangle}{\langle Y_0\rangle}\right), \quad (6)$$

where

$$\langle Y_0\rangle \equiv \lim_{k\to 0} \tfrac{1}{3} \sum_\alpha Y_\alpha(\boldsymbol{k}) \quad (7)$$

the transition is first order. For $P > P_t$, the transition is second order, with renormalized exponents.

One should note that P_t is independent of J_1 to order J_1^2. Furthermore, expression (6) for P_t is identical to that conjectured by Friedman and Gunther [5], with $\langle Y_0\rangle$ replaced by an isotropic average of $\lim_{k\to 0} Y_\alpha(\boldsymbol{k})$.

(c) From eqs. (3), (6), and (7) we see that $\langle Y_0\rangle$ and hence P_t are shape dependent.

(d) We pointed out in ref. 5 that this model has a macroscopic shear instability when $P > P_s \equiv \Lambda/a$. The pressure P_s does not coincide with P_t as is the case in Imry's work [6].

(e) de Moura et al. [10] have shown that the constrained Ising fixed point is unstable with respect to anisotropic elastic perturbations, with a first order transition probably resulting. Bergman and Halperin [11] have studied this problem using an RG technique which retains the lattice variables throughout the renormalization process and have shown that this instability is associated with a microscopic shear instability (mSI) which may lead to a domain structure. It is significant that both this instability and the shape dependence of P_t depend upon the anisotropy of $\langle Y_0\rangle$.

(f) The cross-over from Ising to renormalized Ising behavior occurs at a reduced temperature g^{1/α_I}, where α_I is the Ising specific heat exponent. de Moura et al. [10] have shown that the cross-over between Ising or renormalized Ising behavior to mSI behavior occurs at a reduced temperature on the order of $(fg)^{1/\alpha_I}$ and $(fg)^{1/|\alpha_R|}$, respectively, where $\alpha_R = -\alpha_I/(1-\alpha_I)$ is the renormalized Ising specific heat exponent and where f is a measure of the degree of anisotropy of $\langle Y_0\rangle$ and is assumed to be much less than unity.

Since $g \ll 1$, $f \ll 1$, and $|\alpha_R| > \alpha_I$, we see that as we approach the transition, the system will pass from Ising behavior to renormalized Ising behavior and then to mSI behavior – which is an expected result. If, on the other hand, $f = O(1)$, the system may never exhibit renormalized Ising behavior, passing directly from Ising to mSI behavior.

(g) We expect the higher order terms in our expansion in the parameter g to affect P_t, but not the critical behavior.

References

[1] G. E. Baker, Jr. and J. W. Essam, Phys. Rev. Lett. 24 (1970) 447.

[2] L. Gunther, D. J. Bergman and Y. Imry, Phys. Rev. Lett. 27 (1971) 558.
D. J. Bergman, Y. Imry and L. Gunther, J. Stat. Phys. 7 (1973) 337.

[3] H. Wagner and J. Swift, Z. Phys. 239 (1970) 182.

[4] G. A. Baker, Jr. and J. W. Essam, J. Chem. Phys. 55 (1971) 861.

[5] Z. Friedman and L. Gunther, Phys. Rev. B12 (1975) 5123.

[6] See, for example, Y. Imry, Phys. Rev. Lett. 33 (1974) 1304.

[7] Y. Shapira, R. D. Yacovitch, C. C. Becerra, S. Foner, E. J. McNiff, Jr., D. Nelson and L. Gunther, to be published in Phys. Rev.

[8] S. Milošević, N. Švrakić and B. Tadić, Nuovo Cim. Lett. 14 (1975) 421.

[9] C. Domb, J. Chem. Phys. 25 (1956) 783.

[10] M. A. de Moura, T. C. Lubensky, Y. Imry and A. Aharony, Phys. Rev. B13 (1976) 2176.

[11] D. J. Bergman and B. I. Halperin, Phys. Rev. B13 (1976) 2145.

QUANTUM RENORMALIZATION FOR THE ANISOTROPIC HEISENBERG MODEL*

R. DEKEYSER, M. REYNAERT and M.H. LEE[†]

Instituut voor Teoretische Fysika, Katholieke Universiteit Leuven, Celestijnenlaan 200 D, B-3030 Heverlee, Belgium

It is shown that the quantum renormalization method may be applied to the two-dimensional spin $\frac{1}{2}$ anisotropic Heisenberg model. An appropriate set of transformations is chosen in such a way that the symmetry properties of the hamiltonian are not disturbed by the renormalization. The fixed points and the flow diagram are determined.

Recently, attempts have been made to extend the renormalization group techniques of Niemeijer and van Leeuwen [1] to quantum models [2, 3]. If the original hamiltonian $\mathcal{H}(\{S_j\})$ is expressed in terms of the spin vectors S_j at the lattice sites j, the essential steps in this method are:

(a) The usual choice of cells, numbered by an index J.

(b) A unitary transformation, within each cell, from the base vectors in spin space $|\{S_j^z\}\rangle$ to a new base $|S_J^z, \tau_J\rangle$, where S_J^z will be the z-component of the cell spin and τ_J is a dummy variable.

(c) The evaluation of the renormalized cell spin hamiltonian $\mathcal{H}'(\{S_J\})$ by taking the partial traces over the τ_J:

$$\langle\{S_J^z\}|e^{\mathcal{H}'}|\{S_J^{z'}\}\rangle \equiv \sum_{\{\tau_J\}} \langle\{S_j^z, \tau_J\}|e^{\mathcal{H}}|\{S_j^{z'}, \tau_J\}\rangle. \tag{1}$$

or

$$\langle\{S_J^z\}|e^{\mathcal{H}'}|\{S_J^{z'}\}\rangle \equiv \sum_k p_k \sum_{\{\tau_J\}} \langle\{S_j^z, \tau_J\}; k|e^{\mathcal{H}}|\{S_j^{z'}, \tau_J\}; k\rangle \tag{2}$$

if one wishes to take an average (with normalized weight factors p_k) over several choices of unitary transformations from step b.

One may need to use the more complicated formula (2) in order to preserve, through the renormalization, some symmetry properties of the hamiltonian. The simple formula (1) disturbs the isotropy of the Heisenberg hamiltonian, since it allows the z-axis to play a special role (see ref. 2). One thus needs similar transformations where the z-axis is replaced by the x- and y-axes. The general procedure is as follows.

Suppose the hamiltonian is invariant under a group G, whose elements k are unitarily represented in the space of the jth spin by $U(k, j)$. The renormalized hamiltonian will then preserve the symmetry of the group G if we use formula (2) with all basic sets of the form

$$|S_J^z, \tau_J; k\rangle \equiv U^+(k, J) \prod_{j \in J} U(k, j)|S_J^z, \tau_J\rangle. \tag{3}$$

In approximate calculations it is not necessary in eq. (2) to average over the whole group, but an averaging over an appropriate subgroup can do the same job, as will be illustrated in what follows.

We have applied this method to the anisotropic spin $\frac{1}{2}$ Heisenberg hamiltonian

$$\mathcal{H} = \sum_{\langle ij \rangle} (K_1(S_i^+ S_j^- + S_j^+ S_i^-) + 2K_2 S_i^z S_j^z), \tag{4}$$

where the sum runs over all nearest-neighbour pairs on a triangular lattice. We have chosen the triangular cells as in ref. 1; and we may write

$$\mathcal{H} = \mathcal{H}_0 + V, \tag{5}$$

where $\mathcal{H}_0$ contains all intracell interactions.

Starting with the simple set of new base vectors

$$\begin{aligned}
|+, 1\rangle &= |++-\rangle & |-, 1\rangle &= |--+\rangle \\
|+, 2\rangle &= |+-+\rangle & |-, 2\rangle &= |-+-\rangle \\
|+, 3\rangle &= |-++\rangle & |-, 3\rangle &= |+--\rangle \\
|+, 4\rangle &= |+++\rangle & |-, 4\rangle &= |---\rangle
\end{aligned} \tag{6}$$

we have obtained new sets by the procedure (3), using the subgroup of O_3, generated by the rotations of 90° around the x-, y- and z-axes and

---

* This research has been partially supported by a NATO Research Grant, No. 1024.

† Permanent address: University of Georgia, Athens, USA.

by the inversion. In order to obtain the renormalized hamiltonian, we apply the Baker–Campbell–Hausdorff formula and the cumulant expansion as explained in refs. 2 and 3. Anisotropic Heisenberg interactions between next and third nearest neighbours are generated by the same procedure, and their interaction strengths are labeled K_3–K_6. The resulting renormalization equations,

$$K'_j = f_j(K_1, \ldots, K_6), \qquad (7)$$

have been evaluated up to second order in K_1 and K_2 (this is to first order in K_3–K_6). These equations now preserve the isotropy since, e.g.

$$f_1(a, a, b, b, c, c) = f_2(a, a, b, b, c, c).$$

Eqs. (7) exhibit fixed points at $K_j = 0$ and at infinity, and also at three non-trivial fixed points:

(a) a Heisenberg fixed point at $K_1 = K_2 = 1.427$, $K_3 = K_4 = 0.360$, $K_5 = K_6 = 0.0095$;

(b) an Ising fixed point at $K_1 = K_3 = K_5 = 0$, $K_2 = 0.699$, $K_4 = 0.135$, $K_6 = 0.045$; and

(c) an XY fixed point at $K_1 = 0.739$, $K_2 = 0.032$, $K_3 = 0.137$, $K_4 = -0.00002$, $K_5 = 0.043$, $K_6 = -0.00003$.

The Ising and XY fixed points have the correct orders of magnitude [note in the definition (4) the difference from the usual Ising expansion parameter by a factor 2]. The Heisenberg fixed point is unreliable since K_1 is too large to allow us to truncate the cumulant expansion after the second order. It may well be that in a higher

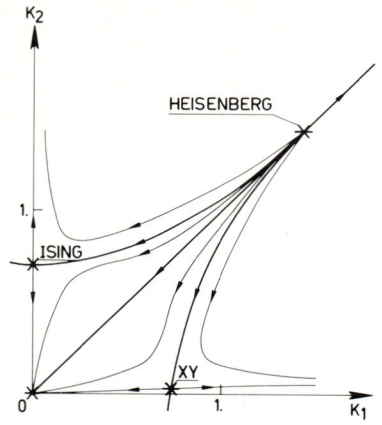

Fig. 1. Fixed points and flow diagram for the renormalization equations of the anisotropic Heisenberg model (projection on the (K_1, K_2)-plane).

order calculation this fixed point disappears, or it shifts to infinity ($T_c = 0$). Linearization around this point yields an anisotropy crossover exponent $\phi \simeq 6.8$ (in $1/n$ expansion [4]) it is infinity at $d = 2$).

Although the numerical results may be further improved, the resulting flow diagram between the fixed points (see fig. 1) clearly shows the qualitative features that one believes will persist for $d = 3$.

References

[1] Th. Niemeijer and J.M.J. van Leeuwen, Phys. Rev. Lett. 31 (1973) 1411, and Physica 71 (1974) 17.
[2] J. Rogiers and R. Dekeyser, Phys. Rev. B13 (1976) 4886.
[3] D.D. Betts and M. Plischke, preprint.
[4] S. Hikami and R. Abe, Progr. Theor. Phys. 52 (1974) 369.

A NEW TYPE OF CRITICAL BEHAVIOR: k-SPACE INSTABILITY ON THE λ LINE

R. M. HORNREICH, S. SHTRIKMAN*
Department of Electronics, The Weizmann Institute of Science, Rehovot, Israel

and

M. LUBAN[†]
Department of Physics, Bar-Ilan University, Ramat-Gan, Israel

Critical exponents at a general Lifshitz point are calculated in the spherical model limit and to first order in a double ϵ expansion. In addition, the exponents at an isotropic Lifshitz point are calculated to second order in ϵ and to first order in $1/n$.

Following Wilson's application of renormalization group methods to critical point phenomena, important strides have been made towards developing successful descriptions of both static and dynamic properties. Only recently, however, have these methods been used to study the critical properties of systems whose ordered phase, characterized by an order parameter $\langle M_k \rangle$, consists of two regions, in one of which the equilibrium wavevector k is equal to a fixed value, k_0, allowed by the Lifshitz condition [1], while in the second, k varies continuously from k_0 when an external parameter P, such as pressure or material composition, is changed. For such a system the T–P plane will exhibit two transition "lines", $T_\lambda(P)$, separating the ordered and disordered phases, and $T_H(P)$, separating the ordered phase into regions with $k = k_0$ and $k \neq k_0$. These transition lines meet at a point (T_L, P_L) which we have termed the Lifshitz point [2]. Note that the wavevector k characterizing the helicoidal structure changes continuously from $k = k_0$ at the Lifshitz point. We summarize here our results for the critical exponents at a Lifshitz point as calculated using renormalization group techniques.

The Lifshitz point is characterized by an instability in k-space associated with the absence of quadratic terms of the form k_i^2 in the free energy functional, for all $i = 1, 2, \ldots, m \leq d$, where d is the dimensionality. This instability gives rise to two distinct inverse correlation lengths $\kappa_{\ell 4}$ and $\kappa_{\ell 2}$, described by critical exponents $\nu_{\ell 4}$ and $\nu_{\ell 2}$, and appropriate to correlations between spins joined by a vector whose components do or do not lie exclusively in the m dimensional subspace. When $d > 4 + \frac{1}{2}m$, the critical exponents assume their "classical" values, i.e. $\nu_{\ell 4} = \frac{1}{4}$, $\nu_{\ell 2} = \frac{1}{2}$, $\eta_{\ell 4} = \eta_{\ell 2} = 0$.

Thus one can calculate critical exponents for a system associated with a point (d, m) in the nonclassical region by expanding in a double power series in $\epsilon_\alpha = m_0 - m$ and $\epsilon_\beta = (d_0 - m_0) - (d - m)$, where (d_0, m_0) is any point on the critical-dimensionality line $d_0(m) = 4 + \frac{1}{2}m$, $m \leq 8$. This constitutes a generalization of the usual Wilson–Fisher ϵ expansion [3].

To first order in ϵ_α and ϵ_β the results of the renormalization group calculation are

$$\nu_{\ell 4} = \frac{\nu_{\ell 2}}{2} = \frac{1}{4}\left(1 + \frac{(n+2)}{2(n+8)} \epsilon_\ell\right) + O(\epsilon_\ell^2);$$

$$\eta_{\ell 2}, \eta_{\ell 4} = 0 + O(\epsilon_\ell^2), \qquad (1)$$

where $\epsilon_\ell = 4 - d + \frac{1}{2}m = \frac{1}{2}\epsilon_\alpha + \epsilon_\beta$. For the completely isotropic ($m = d$) model, the results of a second-order ϵ calculation for the Lifshitz–Heisenberg fixed point are

$$\eta_{\ell 4} = -\frac{3}{20} \frac{(n+2)}{(n+8)^2} \epsilon_\alpha^2 + O(\epsilon_\alpha^3),$$

$$\nu_{\ell 4} = \frac{1}{4} + \frac{(n+2)}{16(n+8)} \epsilon_\alpha$$

$$+ \frac{(n+2)(15n^2 + 89n + 4)}{960(n+8)^3} \epsilon_\alpha^2 + O(\epsilon_\alpha^3). \qquad (2)$$

In the spherical model limit ($n \to \infty$), the criti-

---

* Supported in part by the Commission for Basic Research of the Israel Academy of Sciences and Humanities.
† Also with the Department of Electronics, Weizmann Institute of Science, Rehovot, Israel.

cal exponents were obtained by modifying the renormalization group equations of Ma [4]. The results, for $2+\frac{1}{2}m < d < 4+\frac{1}{2}m$, are

$$\eta_{\ell 2} = \eta_{\ell 4} = 0, \qquad \nu_{\ell 2} = 2\nu_{\ell 4} = (d-2-\tfrac{1}{2}m)^{-1},$$
$$\beta_\ell = \tfrac{1}{2}, \qquad \alpha_\ell = (d-4-\tfrac{1}{2}m)/(d-2-\tfrac{1}{2}m), \qquad (3)$$
$$\gamma_\ell = 2(d-2-\tfrac{1}{2}m)^{-1},$$
$$\delta_\ell = (d+2-\tfrac{1}{2}m)/(d-2-\tfrac{1}{2}m).$$

For $m = d$, we extended Ma's $0(1/n)$ calculations [5] and obtained for $4 < d < 8$

$$\eta_{\ell 4} = \frac{3 \times 2^{d-2}(8-d)\sin(\pi d/2)\Gamma[\tfrac{1}{2}(d-3)]}{\pi^{3/2} d(d+2)\Gamma(\tfrac{1}{2}d)} \frac{1}{n}$$
$$+ 0\left(\frac{1}{n^2}\right),$$

$$\gamma_\ell = (\tfrac{1}{4}d-1)^{-1} - (\tfrac{1}{4}d-1)^{-2} \qquad (4)$$
$$\times \frac{\Gamma(d-4)}{\Gamma(\tfrac{1}{2}d)\Gamma^2(\tfrac{1}{2}d-2)\Gamma(4-\tfrac{1}{2}d)}$$
$$\times [1 + (10-d)(d-5)/3$$
$$+ 3(d-6)(d-8)/4(d+2)]\frac{1}{n} + 0\left(\frac{1}{n^2}\right).$$

Note that $\eta_{\ell 4}$ changes sign at $d = 6$ and approaches zero as d approaches 4 and 8.

The new scaling relations appropriate to a general (nonisotropic) Lifshitz point are [6]

$$m\nu_{\ell 4} + (d-m)\nu_{\ell 2} = 2 - \alpha_\ell, \qquad (5a)$$
$$\gamma_\ell = (4-\eta_{\ell 4})\nu_{\ell 4} = (2-\eta_{\ell 2})\nu_{\ell 2}. \qquad (5b)$$

Using eq. (5b), eqs. (2) and (4) are readily shown to be in exact agreement in the overlap region. For the case $m < d$, similar agreement is obtained.

In the vicinity of the Lifshitz point we expect the free energy to have the generalized scaling form $F(T, P) \sim t^{2-\alpha} f(p/t^\phi)$, where $t = [T - T_\lambda(P)]/T_\lambda(P)$, $p = (P - P_L)/P_L$, and ϕ is the crossover exponent. We find that $\phi = \nu_{\ell 4}(2 - \eta_{\ell 4})$.

Upon approaching the Lifshitz point on the helicoidal segment of the λ line we expect that the wavevector k will be related to the reduced variable p by $k \sim |p|^{\beta_k}$, $p < 0$. A preliminary renormalization-group study indicates that β_k cannot be expressed solely in terms of the Lifshitz point critical exponents and that $\beta_k = \tfrac{1}{2} + 0(\epsilon^2)$.

Further details of our calculations are given elsewhere [2].

References

[1] The Lifshitz condition restricts k to one of a small number of wavevectors in the first Brillouin zone. See L.D. Landau and E.M. Lifshitz, Statistical Physics, 2nd edn. (Pergamon, New York, 1968) ch. XIV.
[2] R.M. Hornreich, M. Luban and S. Shtrikman, Phys. Rev. Lett. 35 (1975) 1678.
R.M. Hornreich, M. Luban and S. Shtrikman, Phys. Lett. 55A (1975) 269.
[3] For a review, see K.G. Wilson and J. Kogut, Phys. Rep. 12C (1974) 75.
[4] S.-K. Ma, Rev. Mod. Phys. 45 (1973) 589.
[5] S.-K. Ma, Phys. Rev. A7 (1973) 2172.
[6] If $T_L = 0$ K, one must replace $2 - \alpha_\ell$ and γ_ℓ by $-\alpha_\ell$ and $\gamma_\ell - 1$, respectively, in eq. (5). Modifications of exponent relations for ordinary critical points at 0 K have been discussed by G.A. Baker, Jr. and J.C. Bonner, Phys. Rev. B12 (1975) 3741.

HELIMAGNETIC–FERROMAGNETIC TRANSITION

J. VILLAIN

DRF/DN, Centre d'Etudes Nucléaires, 85 X, 38041 Grenoble Cedex, France

Spin systems described by an isotropic, temperature independent spin hamiltonian and having a helical spin structure at low temperature may have a ferromagnetic state in an intermediate temperature range. For two-dimensional spins, the ferromagnetic–helimagnetic transition is shown to be of second order for a lattice dimensionality $D \geq 3$, and of first order for $D = 2$.

1. Introduction

It is well known that for certain values of the exchange coupling constants, Heisenberg (or XY) magnets can have at low temperature T a "helimagnetic" structure:

$$\langle S_i \rangle = S(u \cos Q \cdot R_i + v \sin Q \cdot R_i), \quad (1)$$

where u, v are orthogonal unit vectors, and S_i is the spin at site R_i.

In the mean field approximation, the "structure vector" Q does not depend on T. It will be shown here that Q does depend on T (for a temperature-independent hamiltonian), and can even be zero for certain temperatures (the system is then ferromagnetic). The investigation will be restricted to two-dimensional, classical spins $S_i = (\cos \phi_i, \sin \phi_i)$.

2. The model

We start with the standard, bilinear, classical, XY hamiltonian:

$$\mathcal{H} = -\sum J_{ij} S_i \cdot S_j = -\sum J_{ij} \cos(\phi_i - \phi_j). \quad (2)$$

Space isotropy will be assumed for simplicity. For example, one can work on a D-dimensional, simple cubic lattice with interactions between first and $(D+1)$th neighbours: $J_{ij} = J > 0$ if i, j are nearest neighbours, and $J_{ij} = J' < 0$ if i and j have a common neighbour l and are symmetric with respect to l. The ground state is helimagnetic if $\epsilon > 0$, where

$$\epsilon = -(J + 4J'). \quad (3)$$

ϵ will be assumed to be small throughout this work. An elementary calculation transforms (2) into

$$\mathcal{H} = \sum_{ij} U_{ij}(\phi_i - \phi_j)$$
$$+ \sum_{ijl} G_{ijl} \cos(\phi_i - \phi_j) \sin^2(\phi_l - \tfrac{1}{2}\phi_i - \tfrac{1}{2}\phi_j), \quad (4)$$

where $G_{ijl} = -2J' = 2|J'|$ if i and j are neighbours of l and symmetric with respect to l, and $G_{ijl} = 0$ otherwise.

$$U_{ij}(\psi) = -J \cos \psi - J' \cos 2\psi = U(\psi) \quad (5)$$

if i and j are neighbours, and $U_{ij} = 0$ otherwise. If $0 < \epsilon \ll J$, $U(\psi)$ consists of successive minima separated by alternating high barriers (of height about J) and low barriers (of height about ϵ^2/J).

Intuitively, the system is expected to be paramagnetic if $K_B T > J$, helimagnetic if $K_B T \leq \epsilon^2/J$, and ferromagnetic in some intermediate temperature region. The corresponding phase diagram is shown by fig. 1 and exhibits a "Lifshitz point" [1]. A quantitative approach to this problem is developed in the following sections.

3. Self-consistent harmonic approximation (SCHA)

In the ferromagnetic region, the hamiltonian (2) can be approximated by a harmonic hamiltonian [2, 3]

$$\tilde{\mathcal{H}} = \tfrac{1}{2} \sum \tilde{J}_{ij}(\phi_i - \phi_j)^2, \quad (6)$$

where

$$\tilde{J}_{ij} = J_{ij} \langle \cos(\phi_i - \phi_j) \rangle^{\tilde{}} \quad (7)$$

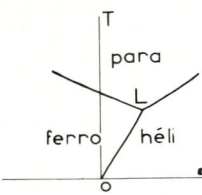

Fig. 1. Phase diagram.

in which

$$\langle \tilde{X} \rangle = (1/\tilde{Z}) \operatorname{Tr} X \, e^{-\beta \tilde{\mathcal{H}}}, \qquad \tilde{Z} = \operatorname{Tr} e^{-\beta \tilde{\mathcal{H}}}$$

and

$$\operatorname{Tr} = \prod_{i=1}^{N} \int_{-\infty}^{\infty} d\phi_i \qquad (8)$$

In the helimagnetic region, it can be shown that the best harmonic hamiltonian is

$$\tilde{\mathcal{H}} = \tfrac{1}{2} \sum_{ij} \tilde{J}_{ij}(\theta_i - \theta_j)^2 \qquad (9)$$

where

$$\theta_i = \phi_i - \mathbf{Q} \cdot \mathbf{R}_i. \qquad (10)$$

$$\tilde{J}_{ij} = J_{ij} \langle \cos(\theta_i - \theta_j) \rangle^{\tilde{}} \cos \mathbf{Q} \cdot (\mathbf{R}_i - \mathbf{R}_j). \qquad (11)$$

Eq. (11) is derived from the variational principle which states that

$$\mathcal{F} = -K_B T \log \tilde{Z} + \langle \mathcal{H} \rangle^{\tilde{}} \qquad (12)$$

should be minimum. Similarly, Q should minimize $\mathcal{F}$, therefore

$$0 = \frac{\partial}{\partial Q_\alpha} \mathcal{F} = \frac{\partial}{\partial Q_\alpha} \langle \mathcal{H} \rangle^{\tilde{}}$$

$$= -\frac{\partial}{\partial Q_\alpha} \sum_{ij} J_{ij} \langle \cos(\theta_i - \theta_j) \rangle^{\tilde{}} \cos \mathbf{Q} \cdot (\mathbf{R}_j - \mathbf{R}_i)$$

$$= \sum_{ij} J_{ij} (R_j^\alpha - R_i^\alpha) \langle \cos(\theta_i - \theta_j) \rangle^{\tilde{}} \sin \mathbf{Q} \cdot (\mathbf{R}_j - \mathbf{R}_i).$$

If a *second order transition* is assumed from ferromagnetism to helimagnetism at some temperature T_{hf}, $Q = 0$ at T_{hf} and the above condition yields, at T_{hf}, after insertion of (11):

$$\sum_{ij} \tilde{J}_{ij}(R_j^\alpha - R_i^\alpha)^2 = 0. \qquad (13)$$

In the formalism of section 2, the SCHA replaces (5) above T_{hf} by

$$\tilde{U}(\psi) = -\tfrac{1}{2} \tilde{\epsilon} \psi^2, \quad \text{where } \tilde{\epsilon} = -(\tilde{J} + 4\tilde{J}') < 0. \qquad (14)$$

Eq. (13) implies

$$\tilde{\epsilon} = 0 \quad (T = T_{hf}). \qquad (15)$$

The interpretation of (15) is that, below T_{hf}, $\psi = 0$, thus $\tilde{\epsilon} > 0$ below T_{hf} and $\tilde{\epsilon} = 0$ at T_{hf}.

4. Three-dimensional case

Eq. (7) yields at low T

$$\tilde{J}_{ij} \simeq J_{ij} - \tfrac{1}{2} J_{ij} \langle (\phi_i - \phi_j)^2 \rangle$$

or, in the special case of section 2:

$$\tilde{J} \simeq J - \tfrac{1}{2} TJ \int \frac{k^4}{\tilde{\epsilon} k^2 + Jk^4} dk,$$

$$\tilde{J}' \simeq J' - 2J' T \int \frac{k^4}{\tilde{\epsilon} k^2 + Jk^4} dk.$$

At T_{hf}, eq. (15) yields

$$\tilde{\epsilon} \simeq \epsilon + \tfrac{1}{2}(J + 16J') T_{hf} \int \frac{k^4}{\tilde{\epsilon} k^2 + Jk^4} dk = 0.$$

Hence, $T_{hf} \approx \epsilon$, higher than the speculation of section 2. A complete treatment can be given, and confirms the existence of a second order transition from ferro- to helimagnetism.

5. Two-dimensional case, $D = 2$

If eq. (15) were satisfied, the SCHA equations would predict strong fluctuations $\langle (\phi_i - \phi_j)^2 \rangle = \infty$ and the SCHA would not be applicable because of eq. (8). What happens, in fact, is that there is, for all $T \leq J$, a consistent solution of the SCHA equations for $Q = 0$, which always implies $\tilde{\epsilon} < 0$: we conclude that this solution corresponds to a relative minimum of eq. (12) with respect to Q, and therefore describes a metastable state at low T. The stable state must of course correspond to $Q \neq 0$, and the transition from ferromagnetism ($Q = 0$) to helimagnetism ($Q \neq 0$) must be of first order.

6. Renormalization group treatment

Consider the model defined by eqs. (4) and (5) on a D-dimensional, simple cubic lattice. We wish to use the so-called "decimation" procedure, i.e. to eliminate a fraction of the sites (say, 3 sites over 4) in order to obtain the free energy as a function of the phases ϕ_r of the remaining sites. At low T, the second term of eq. (4) can be replaced by a harmonic expression:

$$\mathcal{H}_{\mathrm{har}} = \tfrac{1}{4} \sum G_{ijl}(2\phi_l - \phi_i - \phi_j)^2.$$

It is convenient to define scaled variables

$$\phi'_r = 2^{(D/2-2)}\phi_r \tag{16}$$

so that the harmonic part of the free energy takes after a few steps the following form, invariant under scaling:

$$F_{\mathrm{har}} = \sum Jk^4 \phi'_{-k}\phi'_k,$$

where k is scaled in such a way that the Brillouin zone is not modified by scaling.

At low temperature, $T \ll T_{\mathrm{hf}}$, the first term of eq. (4) gives rise, after decimation, to an anharmonic part of the free energy:

$$\begin{aligned}
F_{\mathrm{anh}} &= 2^D \check{\sum} U\left(\frac{\phi_r - \phi_{r'}}{2}\right) \\
&= 2^D \check{\sum} U\left(\frac{\phi'_r - \phi'_{r'}}{2^{(D/2-1)}}\right),
\end{aligned} \tag{17}$$

where $\check{\Sigma}$ denotes a sum over nearest neighbours, the factor 2^D compensates the loss of matter and the denominator 2 appears because in the ordered region, the angle between neighbouring spins is doubled by decimation.

It is seen from eq. (17) that, at low T, the potential barrier g is increased by a factor 2^D by decimation, and the distance φ between the minima is increased for $D > 2$ and is constant for $D = 2$. At higher T, for $D > 2$, g and φ increases less rapidly, and above T_{hf}, both g and φ decrease and vanish above a few steps. At T_{hf}, g and φ can reasonably be expected to remain constant so that one has a fixed point in the space of the hamiltonians. Such a fixed point is excluded for $D = 2$, because the low T approximation (17) predicts φ to be constant: a correct treatment would always predict a decrease of φ, so that at T_{hf}, g has to increase and a fixed point is impossible. Although a second order transformation without fixed point is not strictly forbidden, more detailed argument shows that it is unlikely, so that a first order transition is expected.

References

[1] R.M. Hornreich et al. Phys. Rev. Lett. 55A (1975) 269.
[2] G. Sarma, Solid State Commun. 10 (1972) 1049.
[3] J. Villain, J. Phys. 35 (1974) 27.
[4] L.P. Kadanoff and A. Houghton, Phys. Rev. BII (1975) 377.

MAGNETIC ORDERING IN CuSO$_4$.5H$_2$O AND RELATED LINEAR CHAIN COMPOUNDS IN EXTERNAL FIELD

T.O. KLAASSEN, L.S.J.M. HENKENS and N.J. POULIS

Kamerlingh Onnes Laboratorium der Rijksuniversitiet, Leiden, The Netherlands

(Invited paper)

The magnetic behaviour of a system of weakly coupled $S = \frac{1}{2}$ antiferromagnetic Heisenberg linear chains in external field is discussed. From NMR experiments detailed information is obtained concerning the influence of the low dimensionality of the magnetic system on the properties of the 3D long range ordered state at very low temperatures. The experimentally observed large spin reduction is found to lead to a peculiar temperature and fielddependence of the sub-lattice magnetization.

1. Introduction

The low temperature magnetic behaviour of one-dimensional systems has been given much attention during the last decade. A large number of substances have been found in which the magnetic ions are coupled antiferromagnetically in linear chains. Theory predicts that a system of isolated linear chains does not show long range order at any non-zero temperature. However, due to small interactions between the chains, that will be present in real crystals, a transition to a 3D ordered state will always occur at some finite temperature. Although the magnetic behaviour in the short range order regime is understood quite well, both experimentally and theoretically, little information exists on the properties of a quasi one-dimensional system in the long range ordered state.

We will discuss here the results of a very extensive experimental investigation into the magnetic properties of three isomorphous Heisenberg linear chain salts, CuSO$_4$.5H$_2$O, CuSeO$_4$.5H$_2$O and CuBeF$_4$.5H$_2$O, in the short range ordered state as well as in the long range ordered state. The results show that, in order to obtain detailed information on the 3D ordered state, one has to apply NMR techniques, as macroscopic susceptibility and magnetization measurements appear to yield information on the short range order processes only. Moreover it is found to be of prime importance for a thorough determination of the characteristic features of the long range order in quasi-1D systems to study the properties in external fields up to the critical field H_c at which the chain system saturates paramagnetically for $T = 0$. There appear to be only a few chain salts known, for which the intrachain interaction is small enough as to enable experiments in a static field $H_0 \approx H_c$.

The mentioned copper compounds prove to be ideal for a thorough investigation of both the static and the dynamic properties of a system of weakly coupled magnetic chains. The critical fields are small ($H_c \le 40$ kOe), and due to the presence of hydrogen nuclei, for which the crystallographic positions are known, NMR experiments can be performed, and the results analysed, accurately. The exchange interaction is isotropic and the spinvalue is low ($S = \frac{1}{2}$), so it can be expected that quantum effects like zero point motion, characteristic for a low dimensional system, will be very pronounced. The triclinic unit cell of these compounds contains two unequivalent copper ions. The one type at the $(\frac{1}{2}, \frac{1}{2}, 0)$ position forms a normal paramagnetic system with a Curie–Weiss temperature $\theta \approx 0.05$ K, while the ions at (0, 0, 0) couple in antiferromagnetic linear chains (a.l.c.) along the crystallographic a-axis. The presence of the paramagnetic system is an advantage for the NMR experiments. In low fields the magnetization of the a.l.c. system is very small and consequently in a pure chain salt, because of the frequent overlap of the resonance lines, the NMR spectrum cannot be analysed easily. In the present compounds however the paramagnetic ions provide large additional shifts of the resonance lines by which the resonance spectrum is split up conveniently. Therefore, also in low fields the magnetic behaviour of these chain systems can be studied accurately by NMR experiments. The dynamical behaviour of these a.l.c. systems has been studied as a function of the temperature and the magnitude and direction of the external field by means of nuclear spin–lattice relaxation measurements. Unexpected results for the auto-correlation function of the spins in the chain system are found. These results will be published in a forthcoming paper.

2. Magnetic properties in the short range order region

Miedema [1], Wittekoek [2], Giauque [3], Van Tol [4] and Henkens [5] derived from specific heat, susceptibility and NMR experiments information concerning the magnetic behaviour of these chain systems. From the determination of the NMR frequency of a number of proton resonance lines as a function of H_0 and T, complete and accurate data on the magnetization of both the a.l.c.- and paramagnetic system have been obtained. (For the details of the analysis of the NMR data see for instance ref. 5.) From the temperature dependence of $\langle \mu_{\frac{1}{2}} \rangle$, the time averaged magnetic moment of the copper ions at $(\frac{1}{2}, \frac{1}{2}, 0)$, in external field, the ferromagnetic Curie–Weiss temperatures for the paramagnetic system have been calculated (see table I). The zero field susceptibility $\chi_0(T)$ of the a.l.c. systems has been determined from the temperature dependence of the magnetic moment $\langle \mu_0 \rangle$ of the copper ions at $(0, 0, 0)$ in low fields.

Experiments performed in various crystallographic directions prove that, apart from the anisotropy due to the anisotropic electronic g factor, $\chi_0(T)$ is isotropic within the experimental accuracy. Therefore the experimental susceptibility curves have been compared with the results of the numerical calculations by Bonner and Fisher [6] for the $S = \frac{1}{2}$ isotropic Heisenberg a.l.c. Fitting the experimental data to the theoretical curve the value of the intrachain interaction J/k has been calculated.

The field dependent magnetization of an a.l.c.

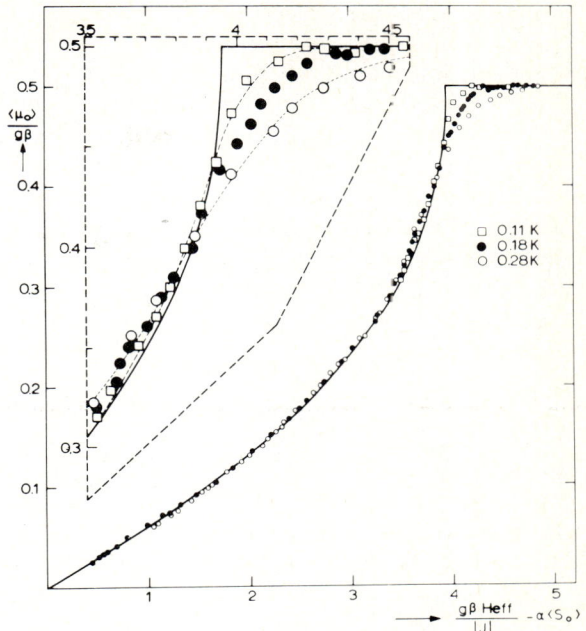

Fig. 1. Magnetization isotherms of the a.l.c. system in $CuBeF_4 \cdot 5H_2O$. The drawn lines represent the theoretical curves, corrected for the interchain interaction $z'J'/k = 0.14$ K resulting in a critical field $H_c = 3.95|J|/g\beta$ ($\alpha = z'J'/|J| = 0.1$). The experimental points have been corrected for the interaction $zJ_{12}/k = 0.15$ K.

system exhibits its specific behaviour only at temperatures $T \ll |J/k|$. Experiments have been performed consequently down to $T = 70$ mK using an adiabatic demagnetization apparatus or a ^{3}He–^{4}He dilution refrigerator. In fig. 1 a few magnetization isotherms for the chain system in $CuBeF_4 \cdot 5H_2O$ are shown. Complete saturation of the magnetization is observed for fields $H_0 > H_c = |4J/g\beta|$. Theoretical $\langle \mu_0(H) \rangle$ curves have been obtained by extrapolation to $N = \infty$ of numerical calculations on finite rings with up to $N = 13$ spins. Fitting the experimental data to these theoretical curves a value for J/k is obtained that deviates from that, calculated from the zero field susceptibility. The discrepancy

Table I
Interaction parameters determined for the three linear chain compounds.

	θ (K)	J/k (K)	zJ_{12}/k (K)	$z'J'/k$ (K)	T_c (K)	H_c (kOe)
$CuSeO_4 \cdot 5H_2O$	0.05	−0.85	0.2		0.125	24
$CuSO_4 \cdot 5H_2O$	0.02	−1.45	0.3	0.14	0.100	40
$CuBeF_4 \cdot 5H_2O$	0.07	−1.45	0.2	0.14	0.091	40

falls outside the experimental errors and is found to be due to interaction between the chains (J'/k) and interaction between the chain- and the paramagnetic system (J_{12}/k).

A very good agreement between experiment and theory for both magnetization and the susceptibility has been obtained by taking into account the influence of these interactions via a molecular field approximation. The so calculated values of the exchange interaction constants are listed in table I.

3. Properties in the long range ordered state

Van Tol [4] was the first to observe the onset of 3D antiferromagnetic long range order in external field in $CuSO_4 \cdot 5H_2O$ from the doubling of the protonresonance lines below $T = T_c \approx 100\,mK$. This second order phase transition is also apparent from a small but distinct λ-anomaly in the specific heat. Since the a.l.c. system looses the major part of its entropy via short range order processes at higher temperatures, the entropy involved in this phase transition is small. From experiments in relatively high magnetic fields, where the λ peak can be distinguished well from the Schottky curve originating from the paramagnetic system, an entropy gain of approximately 4% of $\frac{1}{2}R \ln 2$ is calculated. Whereas the specific heat data only indicate the occurrence of the phase transition, the NMR experiments yield complete information about the properties of the ordered chain system. In all three compounds the protonresonance spectra (and in $CuBeF_4 \cdot 5H_2O$ also the ^{19}F spectrum) have been determined as a function of direction and magnitude of the applied field and as a function of temperature. The magnetic behaviour of the ordered chain system in the three salts is found to be completely analogous, and the compounds will therefore not be discussed separately. The most complete set of experiments and the most detailed analysis of the data has been carried out for the selenate, as this salt exhibits the highest transition temperature T_c.

For $T > T_c$ ten proton resonance lines, corresponding to the ten magnetically unequivalent proton positions in the unit cell are observed. In moderately high fields the resonance frequencies are temperature independent because the paramagnetic system is completely saturated and the chain magnetization does not change with T anymore. At T_c each line splits up into a doublet, with the two components displaced symmetrically with respect to the single line at $T > T_c$. The temperature dependence of the doublet splittings δ_{ν_i} shows the characteristic features of the onset of a spontaneous magnetization. For $T \ll T_c$ the doublet splittings depend on the direction but not on the magnitude of the external field as long as the paramagnetic system is saturated. For $H_0 \to H_c$ the splittings reduce to zero. The average resonance frequency of each doublet for all fields and temperatures proves always to be equal to the frequency the single resonance line should have had if no transition to the long range ordered state had occurred.

A careful analysis of all these data leads to the following conclusions (Henkens [7, 8]).

1) At $T = T_c$ a transition of the a.l.c. system to a 3D antiferromagnetically ordered state, with two sublattices, occurs.

2) The time averaged magnetic moments $\langle \boldsymbol{\mu}_0^\pm \rangle$ on the two sublattices are equal in magnitude but are tilted symmetrically with respect to the direction of the external field.

3) The magnitude of the induced component of the magnetic moment, parallel to H_0, is not influenced by the long range order, and corresponds always to the magnetization expected for an unordered system of weakly coupled linear chains under the given conditions.

4) The long range ordering is characterized by the presence of a spontaneous component of the magnetic moment. These components $\langle \boldsymbol{\mu}_\perp^+ \rangle$ and $\langle \boldsymbol{\mu}_\perp^- \rangle$ on the two sublattices are antiparallel and equal in magnitude. They are directed perpendicular to H_0 and their magnitude at constant reduced temperature $T/T_c(H_0)$ is field independent up to $H_0 \approx 3/4 H_c$ and reduces to zero for $H_0 \to H_c$. In fig. 2 the temperature dependence of $\langle \boldsymbol{\mu}_\perp \rangle$ in $CuSeO_4 \cdot 5H_2O$ as derived from the doublet splittings is shown. Near T_c the spontaneous magnetization can be described by $M(T)/M(0) = B(1 - T/T_c)^\beta$ with $\beta = 0.34 \pm 0.04$ and $B = 1.65 \pm 0.10$. This high value of β is indicative for the three dimensionality of this long range ordering.

Fitting theoretical resonance patterns, obtained from dipolar field calculations, to the experimentally observed spectra for $T > T_c$ and for $T < T_c$, important additional information is

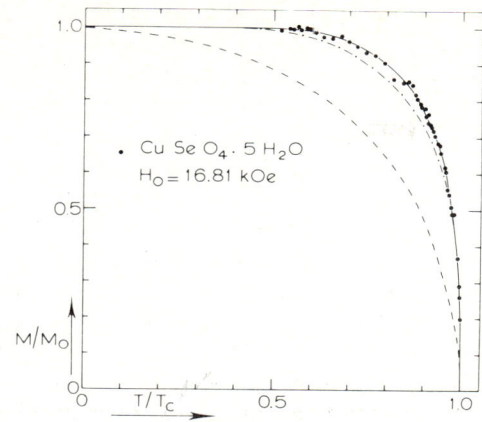

Fig. 2. Spontaneous magnetization of CuSeO$_4$.5H$_2$O for $H_0 = 16.81$ kOe. The dashed and dot–dashed curves represent the theoretical magnetization of respectively a Heisenberg and an Ising 3D s.c. lattice for $S = \frac{1}{2}$.

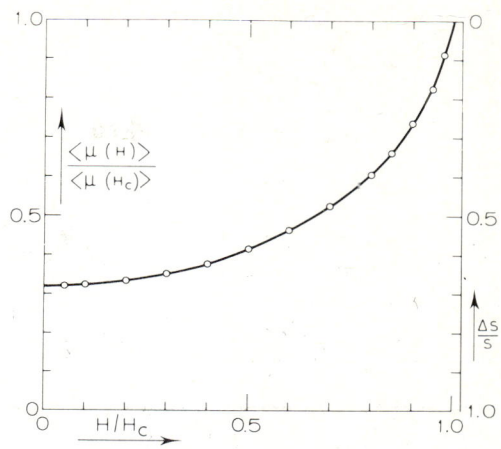

Fig. 3. Field dependence of the total sub-lattice magnetization and of the spin reduction for CuSeO$_4$.5H$_2$O versus field for $T \ll T_c$.

obtained. In CuSeO$_4$.5H$_2$O $\langle \mu_\perp \rangle$, which is always perpendicular to H_0, is found to be confined to a plane with normal vector $(+0.16, -0.03, +0.86)$ (expressed in the triclinic coordinate system). It thus appears that the system is in a "flopped" phase, analogous to that for a slightly anisotropic 3D Heisenberg antiferromagnet above the spin-flop field. Down to $H_0 = 2$ kOe no indication of a spin-flop transition has been observed however.

The spontaneous components of the magnetization are antiparallel for adjacent copper ions along the a-axis (chain axis) and parallel for those along the c-axis. The direction of succeeding moments along the b-axis cannot be determined uniquely as, due to the large distance between these ions, the influence of the relative directions of the moments on the calculated resonance patterns is small. Because of this large distance the interchain interaction j_b along the b-axis will be very small and thus the total interchain interaction J' will be approximately equal to the interaction j_c along the c-axis. The so determined sublattice structure is in agreement with the ferromagnetic sign of the measured interchain interaction J'.

The saturation value of $\langle \mu_- \rangle$ in CuSeO$_4$.5H$_2$O is calculated to be $\langle \mu_\perp(T=0) \rangle = 0.35\, g\beta S$. So in the groundstate ($T = 0$, $H = 0$) where the parallel magnetization is zero, a spin reduction $\Delta S/S = 0.65$ is present. According to the calculations of Ishikawa [9] on the $S = \frac{1}{2}$ Heisenberg antiferromagnet, a spin reduction of this magnitude will exist for an interchain interaction $|J'/J| \approx 10^{-3}$. In fig. 3 the field dependence of the magnitude of the total sublattice magnetization and of the spin reduction in CuSeO$_4$.5H$_2$O, as deduced from the experiments at $T \ll T_c$, is given. The behaviour of the sublattice magnetization deviates strongly from that of a normal 3D Heisenberg antiferromagnet where the magnitude of $\langle \mu^\pm \rangle$ depends mainly on the temperature and nearly not on the applied field. Moreover the component of $\langle \mu^\pm \rangle \perp H_0$ is strongly fielddependent in a 3D antiferromagnet.

In fig. 4 the phase boundary curves of all three compounds as measured both by NMR and specific heat experiments are depicted. For $H_0 < 2$ kOe it has been impossible until now to determine T_c because of a strong broadening of the resonance lines and the unobservability of the λ-peak due to the large Schottky anomaly of the $(\frac{1}{2}, \frac{1}{2}, 0)$ ions. Characteristic is the large region for which T_c does nearly not depend on H_0. The increase of T_c towards low field is due to the influence of the paramagnetic system on the effective interchain interaction. Apart from the coupling j_b between adjacent spins along the b-axis, also a coupling between the chains in the ab plane via an intermediate paramagnetic ion will exist. The canted magnetic moments of the chain system induce a canting in the paramagnetic system via the coupling J_{12}. It can be shown in a molecular field approximation that this leads to an additional interchain interaction of about $-J_{12}^2 \langle \mu_{1/2} \rangle / g^2 \beta^2 H_0$. The total interaction

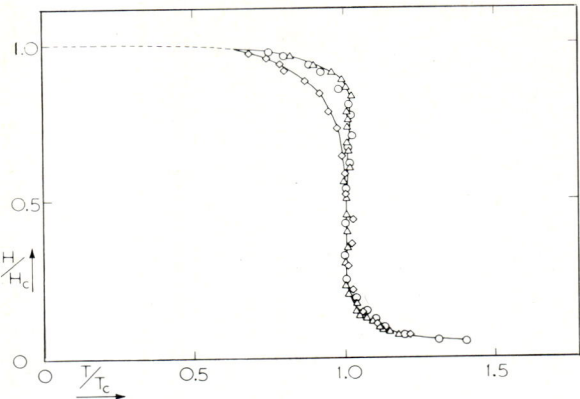

Fig. 4. Phase boundary curves for the three chain compounds on a reduced field- and temperature scale. ○ $CuSO_4.5H_2O$, △ $CuBeF_4.5H_2O$, ◇ $CuSeO_4.5H_2O$.

between chains in the ab plane becomes therefore $J_b = j_b - J_{12}^2 \langle \mu_{1/2} \rangle / g^2 \beta^2 H_0$ and will increase, if $j_b < 0$, continuously towards lower fields because of the increase of $\langle \mu_{1/2} \rangle / H_0$ for $H_0 \to 0$. The induced canting of the paramagnetic moments has been observed actually from the increase of the doublet splittings with decreasing field for $H < 5$ kOe. The field dependence of the doublet splittings agrees well with the predictions based on the molecular field approximation.

In $CuSO_4.5H_2O$ for $10 < H_0 < 30$ kOe a slight increase of T_c with increasing field has been observed. Probably this effect can be explained by the increase of T_c with increasing H_0 predicted by Imry [10] to occur in ordered quasi 1-D antiferromagnetic systems in which quantum effects play an important role.

The phase boundary curve of $CuSO_4.5H_2O$ has been estimated from the formalisms given by Hennessy [11] (T_c as a function of two unequal interchain coupling constants) and by Imry [9]. In the calculation it has been assumed that the "quantum correction" introduced by Imry is equal to the experimentally determined spin reduction $\Delta S/S$. Up to $H_0 \approx 0.7 H_c$ a reasonable fit can be obtained if it is assumed that $j_c/k = 4 \times 10^{-2}$ K and $j_b/k = -10^{-3}$ K, $J/k = -1.45$ K and $J_{12}/k = 8 \times 10^{-2}$ K. The small qualitative differences between the three phase boundary curves are mainly due to the difference in the ratio J_{12}/J for the three compounds.

In the field region where T_c is nearly field independent, T_c has been determined as a function of the direction of H_0. The observed large anisotropy in T_c (10%) is probably due to the fact that T_c strongly depends on the small interaction j_b that may be quite anisotropic because of dipolar contributions. The anisotropic 3D character is also reflected in the temperature dependence of the spontaneous magnetization. The theoretical curve for the 3D Ising model fits rather well the experimental points (see fig. 2).

In conclusion we can say that the properties of a system of weakly coupled antiferromagnetic linear chains in the 3D long range ordered state are strongly determined by the 1D character of the interactions. Especially the occurrence of substantial spinreduction leads to a peculiar behaviour of the sublattice magnetization. The increase of T_c with H_0, that may be characteristic for ordered 1D systems, is largely masked by the influence of the paramagnetic system. However, the field dependence of the interchain interaction J' provides a unique opportunity to study in very low fields the relation between J', T_c and the spin reduction. This investigation has been planned for the near future.

References

[1] A.R. Miedema, H. van Kempen, T. Haseda and W.J. Huiskamp, Physica 28 (1962) 119 (Commun. Kamerlingh Onnes Lab., Leiden No. 331a).
[2] S. Wittekoek, N.J. Poulis and A.R. Miedema, Physica 30 (1964) 1051 (Commun. Kamerlingh Onnes Lab., Leiden No. 338b).
[3] W.F. Giauque, R.A. Fisher, E.W. Horning and G.A. Brodale, J. Chem. Phys. 53 (1970) 3733.
[4] M.W. van Tol and N.J. Poulis, Physica 69 (1973) 341 (Commun. Kamerlingh Onnes Lab., Leiden No. 403e).
[5] L.S.J.M. Henkens, K.M. Diederix, T.O. Klaassen and N.J. Poulis, Physica 81B (1976) 259 (Commun. Kamerlingh Onnes Lab., Leiden No. 418c).
[6] J.C. Bonner and M.E. Fisher, Phys. Rev. 135 (1964) A640.
[7] L.S.J.M. Henkens, M.W. van Tol, K.M. Diederix, T.O. Klaassen and N.J. Poulis, Phys. Rev. Lett. 36 (1976) 1252.
[8] L.S.J.M. Henkens, K.M. Diederix, T.O. Klaassen and N.J. Poulis, Physica 83B (1976) 147 (Commun. Kamerlingh Onnes Lab. No. 421a).
[9] T. Ishikawa and T. Oguchi, Prog. Theor. Phys. 54 (1975) 1282.
[10] Y. Imry, P. Pincus and D. Scalapino, Phys. Rev. B12 (1975) 1978.
[11] M.J. Hennessy, C.D. McElwee and P.M. Richards, Phys. Rev. B7 (1973) 930.

NEUTRON SCATTERING STUDIES OF LOW DIMENSIONAL MAGNETIC SYSTEMS

G. SHIRANE

Brookhaven National Laboratory*, Upton, New York 11973, U.S.A.

and

R.J. BIRGENEAU

Department of Physics, Massachusetts Institute of Technology†, Cambridge, Massachusetts 02139, USA

(Invited paper)

Within the last few years, neutron scattering studies have been carried out on several new prototypes of low dimensional (2D or 1D) magnetic systems. We will discuss in some detail a 2D Ising antiferromagnet K_2CoF_4 (Ikeda and Hirakawa), as well as our current results on K_2NiF_4, $Rb_2Mn_{0.5}Ni_{0.5}F_4$ and $Rb_2Mn_{0.5}Mg_{0.5}F_4$. We will also report our new results of high resolution studies of TMMC. Recent neutron experiments at Brookhaven have been carried out in collaboration with J. Als-Nielsen (Risø), R.A. Cowley (Edinburgh), and H.J. Guggenheim (Bell Laboratories).

1. Introduction

Since the last Intern. Magnetism Conf. in Moscow 1973, we have witnessed significant progress in the neutron scattering studies of low dimensional magnetic systems. In two-dimensional (2D) magnetism, Ikeda and Hirakawa [1] in 1974 reported their remarkable results on the 2D Ising antiferromagnet K_2CoF_4; the experiments show complete agreement with the exact solutions of Onsager [2]. Previously, a Heisenberg antiferromagnet K_2NiF_4 had been thoroughly investigated and its 2D characteristics were demonstrated in its static as well as dynamic magnetic properties [3]. It appears now that "pure" 2D systems are reasonably well understood. Very recently, a systematic investigation was initiated for random 2D antiferromagnets such as $Rb_2Mn_{0.5}Ni_{0.5}F_4$ and $Rb_2Mn_{0.5}Mg_{0.5}F_4$.

In one-dimensional (1D) magnetic systems, definitive experiments were carried out on two model systems. Endoh et al. [4] studied the spin dynamics in $CuCl_2 \cdot 2NC_5D_5$, a 1D $S = \frac{1}{2}$ Heisenberg antiferromagnet. Well-defined excitations are observed over the whole zone with energies in agreement with the des Cloizeaux–Peason exact solution. The second 1D sample is $CsNiF_3$, investigated by Steiner et al. [5] This represents an easy plane ferromagnetic chain and it shows different responses in "in-plane" and "out-of-plane" correlations.

In this brief review, we will discuss these neutron scattering results in some detail. Two comprehensive and excellent review articles have recently treated some of these model systems. In 1974, de Jongh and Miedema [6] gave detailed accounts of both 1 and 2D systems. The K_2CoF_4 study appeared after this review. On the other hand, both $CuCl_2 \cdot 2NC_5D_5$ and $CsNiF_3$ were covered extensively in a very recent review article on 1D systems by Steiner et al. [7]. Thus we will not discuss these two compounds. Finally, we will present some new results of high resolution studies of $(CD_3)_4N\,MnCl_3(TMMC)$, a 1D $S = \frac{5}{2}$ Heisenberg antiferromagnet [8]. These reveal interesting features in the spin dynamics at longer wavelengths than previously investigated.

2. 2D Ising system K_2CoF_4

Neutron scattering studies on 2D magnetic systems started, in 1969, with a near Heisenberg antiferromagnet K_2NiF_4 [3]. Then they were extended to Rb_2MnF_4 [3] and K_2MnF_4 [4]. These measurements demonstrated unambiguously almost ideal 2D characteristics in all aspects of the scattering function $\mathscr{S}(Q, \omega)$. Observed critical quantities β, γ, ν could not be compared directly with Onsager's exact solutions, since the latter apply to the Ising system. It was then highly desirable to find a compound which is planar and Ising-like in its interaction. As it

---

* Work at Brookhaven performed under the auspices of the U.S. Energy Research and Development Administration.
† Work at Massachusetts Institute of Technology supported by National Science Foundation-MRL Grant no. DMR12-03027-ADS.

turned out, K_2CoF_4 and Rb_2CoF_4 exhibit these characteristics.

The spin Hamiltonian for these compounds may be written

$$\mathcal{H} = \sum_{i>j} J_\| S_i^z S_j^z + J_\perp (S_i^x S_j^x + S_i^y S_j^y)$$

with fictitious spin $S = \frac{1}{2}$. Breed et al. [10] have shown that $J_\|/J_\perp \sim 4$ for these 2D Co compounds. Also the neutron experiments demonstrate these compounds to be almost ideal Ising systems. Samuelson [11] reported neutron scattering measurements on Rb_2CoF_4 and, independently, Hirakawa and collaborators have carried out extensive studies on K_2CoF_4 not only by neutron techniques but also by specific heat measurement [12].

Fig. 1 shows the sublattice magnetization M_s of K_2CoF_4 measured by Ikeda and Hirakawa. It is truly remarkable that, for the *entire* temperature range, M_s follows Onsager's analytic formula

$$\frac{M_s(T)}{M_0} = \left(1 - \sinh^{-4} \frac{2J}{kT}\right)^{1/8}.$$

Near $T_N = 108$ K, the magnetization can be approximated by

$$\frac{M_s(T)}{M_0} = B\left(\frac{T_N - T}{T_N}\right)^\beta$$

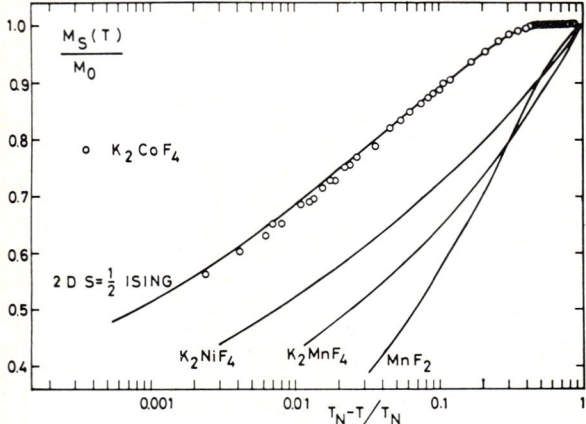

Fig. 1. Sublattice magnetization for K_2CoF_4, after Ikeda and Hirakawa [1]. The solid line marked as 2D Ising is Onsager's exact solution. Observed curves for other antiferromagnets are shown schematically.

with $\beta = 0.123 \pm 0.008$, in excellent agreement with the exact solution of 0.125. It is quite instructive to compare this result with those for Heisenberg antiferromagnets K_2NiF_4 and K_2MnF_4, which are also shown schematically in fig. 1 (see also table I). The exponents β are quite similar; 0.14 for K_2NiF_4 [3] and 0.15 for K_2MnF_4 [13]. However, the low temperature behavior of M_s is entirely different, reflecting the differences in spin-value and in the amount of spin-anisotropy.

Other critical exponents are also in good agreement with theoretical predictions. The cross sections above T_N were analyzed by the formula of Fisher and Burford [14]

$$\chi(q) = \frac{A(q)}{(\kappa^{\|2} + q^2)^{1-\frac{1}{2}\eta}}$$

where $A(q)$ is a slowly changing function with temperature. Here, T_N, $\kappa^\|$ and $\chi(0)$ may be expressed as

$$\kappa^\| \propto \left(\frac{T - T_N}{T_N}\right)^\nu, \quad \chi(0) \propto \left(\frac{T - T_N}{T_N}\right)^\gamma.$$

Ikeda and Hirakawa obtain $\nu = 0.97$; $\gamma = 1.71$ and $\eta = 0.2$, in excellent agreement with Onsager's exact solutions (see table I). In addition, the specific heat of K_2CoF_4 and Rb_2CoF_4 was investigated by Ikeda et al. [12]. Again they observed a symmetric logarithmic singularity as predicted by the theory.

3. 2D Heisenberg antiferromagnet

In the previous subsection we have discussed the experiments on the highly anisotropic 2D antiferromagnets K_2CoF_4 and Rb_2CoF_4. As we have seen, there is good absolute agreement with the exact results of Onsager for the 2D Ising model. These experiments employed techniques developed earlier in an extensive study of the 2D near-Heisenberg antiferromag-

Table I
Critical exponents for 2D magnetic systems

	β	ν	γ	Reference
2D Ising	0.125	1.0	1.75	Onsager [2]
K_2CoF_4	0.123	0.97	1.71	Ikeda & Hirakawa [1]
K_2NiF_4	0.14	0.9	1.65	Als-Nielsen et al. [16]
$Rb_2Mn_{0.5}Ni_{0.5}F_4$	0.16	0.9	1.65	Als-Nielsen et al. [16]

net K_2NiF_4 by Birgeneau, Skalyo, and Shirane [15]. Somewhat ironically, it has turned out that the analysis of the 2D Heisenberg systems is much more complex, and it is only recently that the critical behavior of these systems has been properly elucidated. In particular, the apparent mean-field-like exponents $\gamma \sim 1$, $\nu \sim \frac{1}{2}$ originally reported by Birgeneau et al. [15] for K_2NiF_4 are incorrect due to an improper treatment of the transverse susceptibility. In this subsection, we discuss recent experiments and analyses by Als-Nielsen, Birgeneau, Guggenheim, and Shirane [16] on K_2NiF_4, K_2MnF_4 and $Rb_2Mn_{0.5}Ni_{0.5}F_4$.

The spin Hamiltonian for these systems may be written

$$\mathcal{H} = \sum_{i>j} J_{ij} S_i S_j + \sum g_i \mu_B H_i^A S_i^z$$

The exchange is predominantly between nearest neighbors only. The anisotropy field H_i^A originates from a combination of dipolar spin–spin interactions and single-ion crystal-field anisotropy. The strength of this Ising-type spin anisotropy relative to the isotropic exchange field varies from 0.2 to 0.7% for the three compounds.

The general behavior of the fluctuations in these systems has been extensively reviewed previously [15], so we shall not repeat that here. We simply remind the reader that above T_N the crystal exhibits a ridge of critical scattering where the ridge is uniform in the direction perpendicular to the sheets. In the critical region, only the zz component of the wave-vector dependent susceptibility $\chi^{\alpha\alpha}(q)$ is divergent. At and below T_N a 3D Bragg peak appears, but the critical scattering remains 2D. As noted in the previous section, in K_2NiF_4 and K_2MnF_4 the sublattice magnetization follows a simple power law with $\beta = 0.138 \pm 0.004$ and 0.15 ± 0.01, respectively. In both cases $B \sim 1.0$. The random system is complicated by some smearing of T_N. Empirically, it is found that the amount of smearing scales approximately with the size of the crystal. In their smallest sample Als-Nielsen et al. found that the Bragg intensity is accurately described by the square of the above power law convoluted with a Gaussian distribution of Néel temperatures. Least squares fits yield $\beta = 0.16 \pm 0.01$, $B = 1.0$, $\langle T_N \rangle = 68.72 \pm 0.02$ K and a relative Gaussian smearing of $\sigma/\langle T_N \rangle = 0.6\%$. It is evident that both B and β are extremely close, if not identical, to those in the pure systems. Furthermore, β is close to the asymptotic 2D Ising value of 0.125.

We now consider the critical scattering measurements in the three systems for $T > T_N$; these were performed in the $(h, 0, l)$ plane, where l is the coordinate perpendicular to the sheets. For purely two-dimensional spin correlations the critical scattering should take the form of ridges along l at $h = 1, 3, \ldots$. At all temperatures, aside from known geometrical effects, the critical scattering is indeed independent of l. For each system the detailed measurements across the ridge were carried out around $(1, 0, \frac{1}{2})$, where, for an incident neutron energy of 13.6 meV, the quasielastic approximation becomes nearly exact. The cross section at this position is given by

$$\sigma(q) = T[0.95 \chi^{\parallel}(q) + 1.05 \chi^{\perp}(q)].$$

Here $\parallel$ and $\perp$ are with respect to the Ising axis and $q = (q_x^2 + q_y^2)^{1/2}$. We now consider the appropriate functional form for $\chi(q)$. Tracy and McCoy [17] have shown that for the 2D Ising model the Lorentzian approximation $\chi(q) = A(t)/(q^2 + \kappa^2)$ is essentially exact for $q \lesssim 10 \kappa$. In order to keep the analysis of the results at the simplest level, Als-Nielsen et al. discard data with $q > 10 \kappa$ and assume a cross section $\sigma(q)$ of the form

$$\sigma(q) = T[A^{\parallel}/(q^2 + \kappa^{\parallel 2}) + A^{\perp}/(q^2 + \kappa^{\perp 2})].$$

The most complex aspect of the data analysis originates in the difficulty of separating out $\chi^{\parallel}$ and $\chi^{\perp}$. In their original study Birgeneau et al. assumed that $\chi^{\perp}$ could be absorbed into the background near T_N; this assumption was later proved to be unjustified. As discussed in detail in recent papers we believe that the above technique remedied this difficulty with $\kappa^{\perp}$. The results for the correlation length $\kappa^{\parallel}$ are shown in fig. 2 together with the corresponding data for the pure materials K_2NiF_4 and K_2MnF_4. From the figures it is evident that over the reduced temperature range studied, the critical behavior is identical in all three systems. Detailed fits to simple power laws for the correlation length and staggered susceptibility yield $\nu = 0.9 \pm 0.1$ and $\gamma = 1.65 \pm 0.15$, respectively, for all three

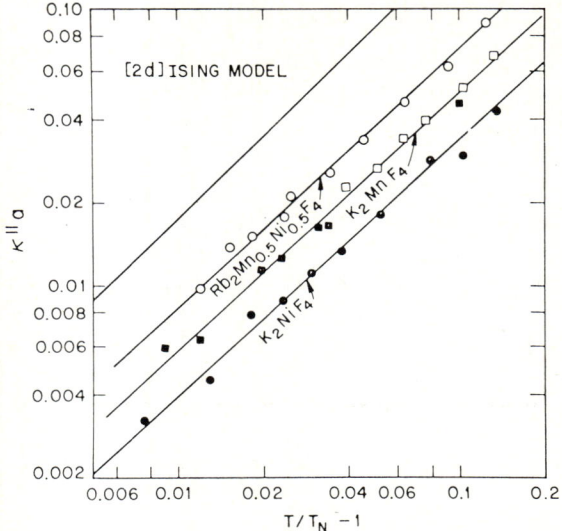

Fig. 2. Inverse correlation lengths for three 2D antiferromagnets, together with the exact result for 2D Ising model, after Als-Nielsen et al. [16].

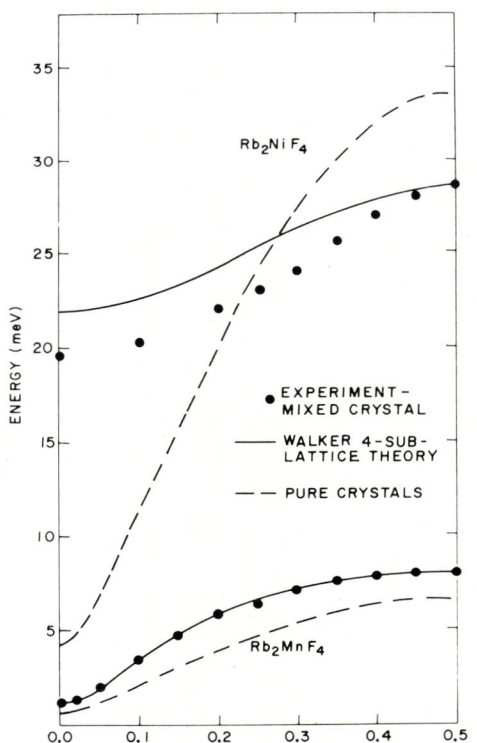

Fig. 3. Dispersion relations in the [ζ00] direction for $Rb_2Mn_{0.5}Ni_{0.5}F_4$, together with those for the pure Ni and Mn salts. After Als-Nielsen et al. [16].

systems. We do not regard the differences from the ideal 2D Ising values of 1.0 and 1.75 as significant, given both the approximations in the data analysis and the possible presence of higher-order corrections to the asymptotic power laws.

In summary, then, the near-Heisenberg 2D antiferromagnets of the K_2NiF_4 class all exhibit 2D Ising-like phase transitions driven by the Ising anisotropy, whether single-ion or dipolar in origin. Randomness appears to have no measurable effect on the observed critical behavior.

On the other hand, the dynamic properties [18] exhibit rather striking effects for $Rb_2Mn_{0.5}Ni_{0.5}F_4$ due to the random nature of the system. Fig. 3 shows the results of inelastic measurements for $Rb_2Mn_{0.5}Ni_{0.5}F_4$. They are characterized by two well-defined bands of excitations. The width and energies of the zone boundary modes are accurately predicted using a simple Ising cluster model while overall dispersion is correctly given by the Walker mean-crystal model (see ref. 18). An extensive algebraic and "computer experiment" analysis has been carried out by Kirkpatrick and Harris [19], in good overall agreement with the experiment.

4. High resolution study of TMMC

In the last five years extensive work has been done on the dynamics of one-dimensional systems. Experiments have been concentrated on three model systems:

(1) $(CD_3)_4NMnCl_3$ (TMMC), a 1D $S = 5/2$ Heisenberg antiferromagnet
(2) $CuCl_2 \cdot 2NC_5D_5$, a 1D $S = 1/2$ Heisenberg antiferromagnet
(3) $CsNiF_3$, a 1D easy plane $S = 1$ Heisenberg ferromagnet.

Experiments and theory appropriate to these three systems have recently been reviewed in detail by Steiner, Villain, and Windsor [7]. We shall therefore limit our discussion in this section to a report of some new high-resolution measurements in TMMC.

The exchange Hamiltonian for TMMC may be written [8]

$$\mathcal{H} = -2J_{nn} \sum_i S_i \cdot S_{i+1}$$

with $J_{nn} = 6.5$ K. In the classical spin approximation [20] the static inverse correlation length is given simply by

$$a\kappa_s = (1+u)/(-u)^{1/2},$$

where $u = \coth K - 1/K$, $K = 2J_{nn}S(S+1)/k_B T$, where a is the nearest neighbor Mn^{++} spin separation.

In their original measurements Hutchings et al. [8] showed that at $T = 1.9$ K ($T_N = 0.845$ K) TMMC exhibits well-defined spin-wave excitations throughout most of the second half of the 1D Brillouin zone, that is for $\frac{1}{2}(\pi/a) \leq q \leq 0.95 \ (\pi/a)$. Further, the q-dependence of the intensities is well described by simple spin-wave theory. At 1.9 K the inverse correlation length is approximately $\kappa_s = 0.005 \ (\pi/a)$ so that the existence of well-defined propagating modes in the above q-range need not be considered as surprising.

For completeness we have extended the measurements of Hutchings et al. to cover the range

$$0.1 \ (\pi/a) \leq q \leq 0.985 \ (\pi/a)$$

at $T = 1.45$ K. Again sharp excitations are observed over this whole range. The resulting dispersion relation is shown in fig. 4. As expected a double-sine curve is obtained with the periodicity of the antiferromagnetic Brillouin zone, that is, with a double periodicity in the true nuclear Brillouin zone appropriate to the paramagnetic phase. Once more, the excitation intensities are accurately predicted by simple classical spin-wave theory.

Hutchings et al. also reported some measurements of the temperature dependence of the spin excitations between 1.9 K and 40 K at $q = 0.95 \ (\pi/a)$ and $q = 0.9 \ (\pi/a)$. At both wave vectors the excitation becomes progressively broader as the temperature is increased ultimately becoming overdamped at the higher temperatures. As discussed by Steiner et al. [7], these measurements stimulated an enormous amount of theoretical activity on 1D paramagnetic spin dynamics. By now an impressive array of theories have appeared and, not surprisingly, all are relatively successful in accounting for the above-mentioned results. However, since the experiments of Hutchings et al. are at rather large wave vectors, they do not really provide a very critical test for the theories. In order to resolve a number of basic physical questions as well as to provide a firm test of current and future theories, we have performed high-resolution measurements at low temperatures and long wavelengths in TMMC. We have chosen the scans so as to probe two basic questions:

(1) Is there any evidence for a strong dynamic central peak arising from "correlated spin cluster" diffusion?

(2) Is there a dynamic critical wave vector κ_0 characterizing the evolution from underdamped to overdamped behavior as a function of q and how is it related to the static inverse correlation length κ_s.

Measurements at $q^* = \pi/a - q = 0.015$, 0.03, and 0.04 (π/a) were performed using 5.5 meV incident neutrons. Background, including the incoherent scattering around $E = 0$, was determined by several measurements near $q \sim (\pi/2a)$ where no low-energy magnetic scattering is expected. This background was subtracted point-by-point from the small q^* data. The experimental results after this subtraction are shown for $q^* = 0.015$, 0.03 (π/a) in figs. 5, 6. These data have *not* been corrected for any instrumental factors. In order to transform the data to the form $\mathcal{S}(q, \omega)$ for detailed comparison with theory the data should be multiplied by the factor $[5.5/(5.5 - \hbar\omega)]^2$. The instrumental resolution is 0.13 meV, full width at half maximum (FWHM). Measurements were also per-

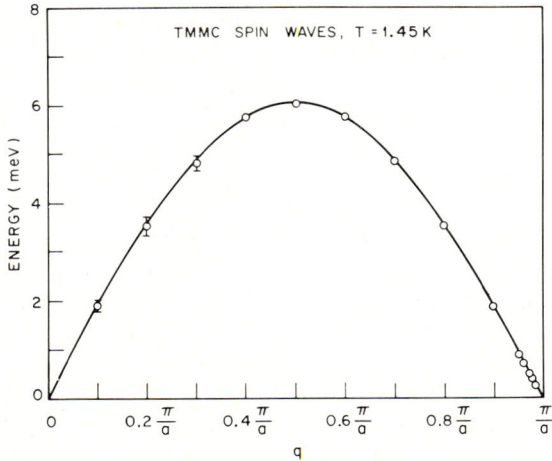

Fig. 4. Entire dispersion curve for TMMC spin waves at $T = 1.45$ K.

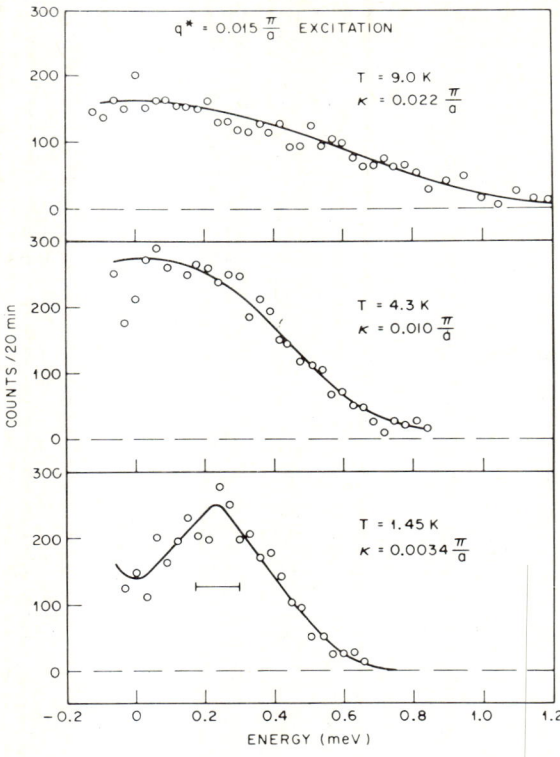

Fig. 5. High resolution data for TMMC at $q^* = 0.015\ (\pi/a)$.

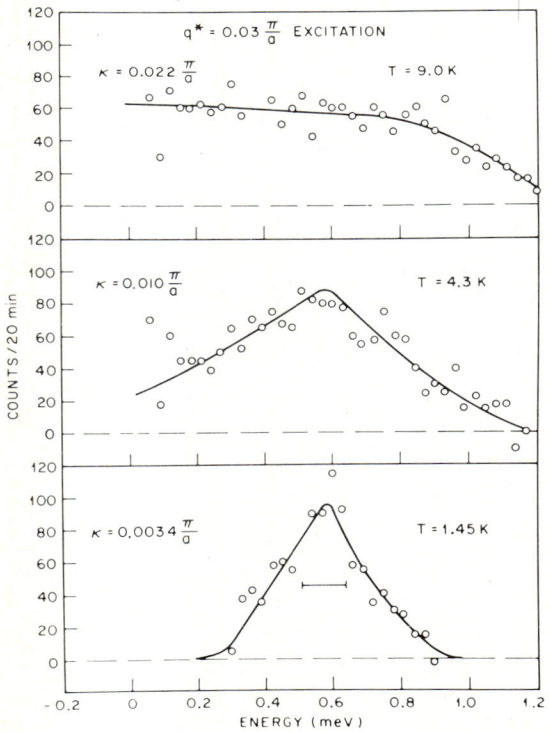

Fig. 6. High resolution data for TMMC at $q^* = 0.03\ (\pi/a)$.

formed at $q^* = 0.05\ (\pi/a)$ using 13.7 meV incident neutrons. A similar background subtraction was carried out but in this case the incoherent scattering around $E = 0$ was too large to allow meaningful results to be obtained near $E = 0$. These data have been corrected for instrumental intensity factors so that they represent the true Van Hove scattering function $\mathcal{S}(q, \omega)$ convoluted with an instrumental resolution function with FWHM 0.60 meV. The $q^* = 0.05\ (\pi/a)$ results are shown in fig. 7.

It is evident that these experiments provide direct answers to both questions (1) and (2). Firstly, somewhat to our surprise, there is no evidence for a distinct central peak at $q^* = 0.015\ (\pi/a)$ or $q^* = 0.03\ (\pi/a)$ at any temperature. This would appear to place severe restrictions on theories based on "pseudo-static" clusters. This absence of a central peak appears to be unique to the one-dimensional case since in both two- and three-dimensional Heisenberg systems triple-peaked functions are generally observed above T_N. Secondly, at each wave vector it is apparent that as q^* goes from $\sim 2\kappa_s$ to κ_s the dynamic line shape changes from being peaked at finite energy to being overdamped. Thus there is indeed a dynamic critical wave vector κ_c and, not surprisingly, it is simply related to the static inverse correlation length κ_s. From figs. 5–7, one would estimate $\kappa_c \sim 1.5\ \kappa_s$.

This contradicts explicitly the Lovesey–Meserve theory [21], which predicts κ_c to vary with the square root of κ_s. Correspondingly, the Lovesey–Meserve theory predicts diffusive behavior through the q, T range covered by figs. 5 and 6, in disagreement with the experiment.

There are also a number of theories [22] which assume $\kappa_c = \kappa_s$ as a starting point and further have built into them the assumption that the system will exhibit propagating excitations for $q > \kappa_s$. These are all somewhat unsatisfactory in that they assume much of the empirical answer. Nevertheless, they do provide an important first step towards a genuine first-principles theory. Such calculations have been performed by McLean and Blume [22] for $q^* = 0.015\ (\pi/a)$. The agreement is only partially satisfactory. It would undoubtedly be improved considerably by choosing $\kappa_c \simeq 1.5\ \kappa_s$. Clearly, much further work remains to be done on 1D spin dynamics. We hope that these new results

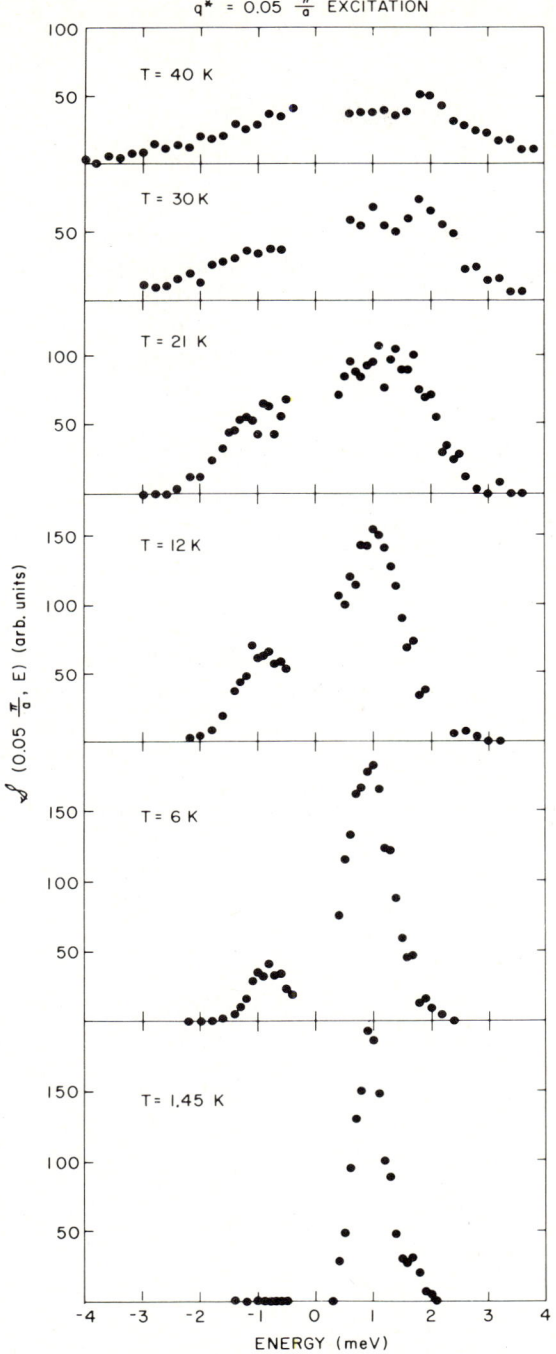

Fig. 7. Neutron energy-gain and loss spectrum for TMMC at $q^* = 0.05 \, (\pi/a)$.

5. Concluding remarks

We have discussed several recent studies of low-dimensional magnetic systems. We have now representative prototypes of 1D and 2D model systems well characterized by neutron scattering techniques. Recently, attention has been directed towards the percolation transition in magnetic model systems. Here again low-dimensional systems may play a vital role because of their simplicity. Experimental work is now underway for the $Rb_2Mn_cMg_{1-c}F_4$ system [23].

For the square lattice with nn interactions alone, the percolation limit is $C_p = 0.59$. The experiments have been carried out on two specimens with $C = 0.54 \pm 0.02$ and 0.57 ± 0.02. The critical scattering displays fascinating characteristics and a preliminary report will be given by Cowley et al. [23] at this conference. We limit ourselves to a brief mentioning of the dynamical properties. The spectrum of the $C = 0.54$ samples shows well-defined peaks as shown in fig. 8. They are also quite different from those observed in $Rb_2Mn_{0.5}Ni_{0.5}F_4$ (fig. 3). The observed cross sections are excellently described by computer simulation by Alben and Thorpe [24] based on the linear spin-wave approximation. Further experiments along these lines are now in progress.

will provide valuable intuitive guidance to future theories as well as a substantial testing ground for quantitative comparisons.

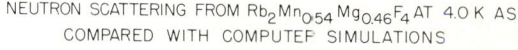
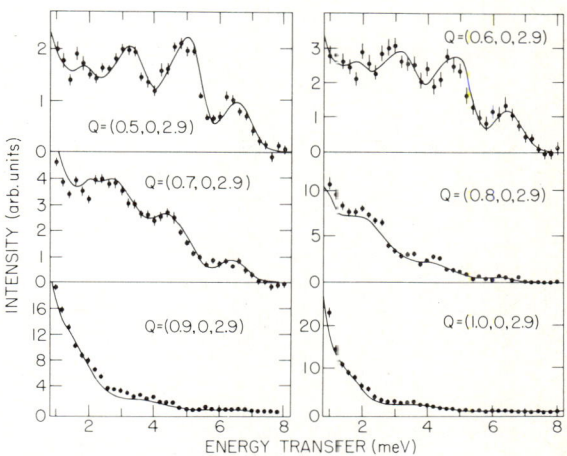

Fig. 8. Cross sections for $Rb_2Mn_{0.54}Mg_{0.46}F_4$. Solid lines are computer simulation by Alben and Thorpe [24] for this composition.

We would like to thank J. Als-Nielsen, M. Blume, and R.A. Cowley for many stimulating discussions.

References

[1] H. Ikeda and K. Hirakawa, Solid State Commun. 14 (1974) 529.
[2] L. Onsager, Phys. Rev. 65 (1944) 117.
B. Kaufman and L. Onsager, Phys. Rev. 76 (1949) 1244.
[3] R.J. Birgeneau, H.J. Guggenheim and G. Shirane, Phys. Rev. Lett. 22 (1969) 720; Phys. Rev. B1 (1970) 2211.
[4] Y. Endoh, G. Shirane, R.J. Birgeneau, P.M. Richards and S.L. Holt, Phys. Rev. Lett. 32 (1974) 170.
[5] M. Steiner and B. Dorner, Solid State Commun. 12 (1973) 537.
M. Steiner, B. Dorner and J. Villain, J. Phys. C8 (1975) 165.
[6] L.J. de Jongh and A.R. Miedema, Adv. Phys. 23 (1974) 1.
[7] M. Steiner, J. Villain and C.G. Windsor, Adv. Phys. 25 (1976) 87.
[8] R.J. Birgeneau, R. Dingle, M.T. Hutchings, G. Shirane and S.L. Holt, Phys. Rev. Lett. 26 (1971) 718.
M.T. Hutchings, G. Shirane, R.J. Birgeneau and S.L. Holt, Phys. Rev. B5 (1972) 1999.
R.J. Birgeneau, G. Shirane and T.A. Kitchens, Proc. 13th Conf. Low Temp. 2 (1973) 371.
[9] H. Ikeda and K. Hirakawa, J. Phys. Soc. Japan 35 (1973) 617.
[10] D.J. Breed, K. Gilijamse and A.R. Miedema, Physica 45 (1969) 205.
[11] E.J. Samuelsen, Phys. Rev. Lett. 31 (1973) 936.
[12] H. Ikeda, I. Hatta, A. Ikushima and K. Hirakawa, J. Phys. Soc. Japan 39 (1975) 827.
H. Ikeda, I. Hatta and M. Tanaka, J. Phys. Soc. Japan 40 (1976) 334.
[13] R.J. Birgeneau, H.J. Guggenheim and G. Shirane, Phys. Rev. 8 (1973) 304.
[14] M.E. Fisher and R.J. Burford, Phys. Rev. 156 (1967) 583.
[15] R.J. Birgeneau, J. Skalyo, Jr. and G. Shirane, Phys. Rev. 133 (1971) 1736.
[16] J. Als-Nielsen, R.J. Birgeneau, H.J. Guggenheim and G. Shirane, J. Phys. C9 (1976) L121: (and to be published).
[17] C.A. Tracy and B.M. M'Coy, Phys. Rev. Lett. 31 (1973) 1500.
[18] J. Als-Nielsen, R.J. Birgeneau, H.J. Guggenheim and G. Shirane, J. Phys. C (1976) L121; Phys. Rev. B12 (1975) 4963.
[19] S. Kirkpatrick and A.B. Harris, Phys. Rev. B12 (1975) 4980.
[20] M.E. Fisher, Am. J. Phys. 32 (1964) 343.
[21] S.W. Lovesey and R.A. Meserve, Phys. Rev. Lett. 28 (1972) 614.
S.W. Lovesey, J. Phys. C7 (1974) 2008.
K. Tomita and H. Mashiyama, Prog. Theor. Phys. 48 (1972) 1133.
[22] F.B. McLean and M. Blume, Phys. Rev. B7 (1973) 1149.
S.H. Liu, Phys. Rev. B13 (1976) 2979.
[23] R.A. Cowley, G. Shirane, R.J. Birgeneau and H.J. Guggenheim, this Conference 3 F2 (and to be published).
[24] R. Alben and M.F. Thorpe, AIP Conf. Proc. 29 (1976) 250.

SUBLATTICE MAGNETIZATION OF $K_3Mn_2F_7$, A QUADRATIC DOUBLE-LAYER HEISENBERG ANTIFERROMAGNET

A.F.M. ARTS and H.W. DE WIJN
Fysisch Laboratorium, Rijksuniversiteit, Utrecht, The Netherlands

The temperature dependence of the sublattice magnetization of the double-layer antiferromagnet $K_3Mn_2F_7$ has been measured by NMR techniques. A least-squares adjustment to spin-wave theory including higher-order Oguchi corrections yields for the exchange constant $J/k_B = -7.59 \pm 0.03$ K and for the zero-temperature spin-wave gap $T_G(0) = 5.99 \pm 0.06$ K.

An analysis is presented of the temperature dependence of the sublattice magnetization in the antiferromagnetic double layer $K_3Mn_2F_7$, as measured by NMR techniques. The crystallographic structure of $K_3Mn_2F_7$ is closely related to the single-layer structure K_2MnF_4, in the sense that Mn–F layers are substituted by sheets of unit cells of the perovskite $KMnF_3$. Magnetically, the lattice consists of sheets of two adjoining quadratic layers of antiferromagnetically ordered Mn^{2+} spins, with each spin surrounded by five nearest Mn^{2+} neighbours, one of which is located in the adjoining layer. There is no net interaction between two adjacent double layers, while the interaction between next-nearest double layers is smaller by orders of magnitude relative to the intralayer exchange, because of the large distance.

The temperature dependence of the sublattice magnetization has been measured by tracking the NMR frequency $f(T)$ of ^{19}F nuclei adjacent to Mn^{2+} ions in the double layer (fig. 1). These nuclei resonate in a transferred hyperfine field, scaling with the sublattice magnetization, as expressed by $f(T) = A_{19}\langle S^z \rangle / h$, where A_{19} is the transferred hyperfine coupling constant. The NMR frequency was measured by beating the free induction decay, following the excitation of the nuclei with a high power r.f. pulse, with a standard oscillator.

The data have been analyzed, with least-squares adjustment, in terms of a two-dimensional four-sublattice spin-wave theory with inclusion of temperature dependent and independent renormalization as formulated by Oguchi [1] as well as proper renormalization of the $k = 0$ magnon energy. In the adjustment three parameters were used, namely, the exchange constant J, the zero-temperature energy gap $T_g(0)$ and the zero-temperature resonance

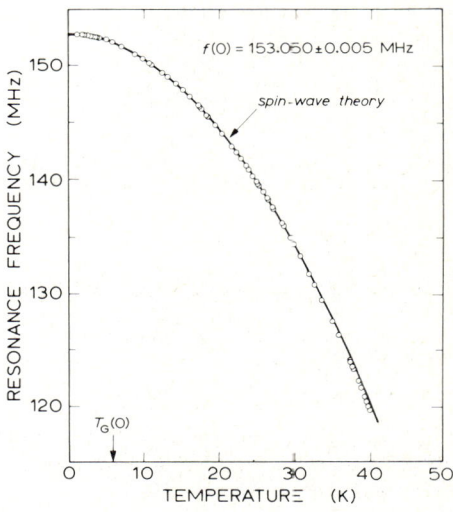

Fig. 1. NMR frequency $f(T)$ vs. temperature T. The solid curve is calculated from renormalized spin-wave theory, as described in the text.

frequency $f(0)$. The latter is however determined with high precision and turns out to be only weakly correlated with the other two parameters. To find the largest temperature region of concurrence between theory and experiment, within the experimental errors, we have extended the temperature range of the least-squares adjustment from the data-points at the lowest temperatures (1.2 K) to a series of different upper temperature limits. Above 27.5 K the data could not be fitted within the experimental errors and it is believed that this upper limit represents the onset of the breakdown from spin-wave theory including Oguchi corrections. From the fit up to 27.5 K we find for the exchange constant $J/k_B = -7.59 \pm 0.03$ K and for the zero-temperature gap $T_G(0) = 5.99 \pm 0.06$ K. With these values we calculated the solid line in fig. 1. To demonstrate the breakdown of spin-wave theory more clearly the deviation between experimental and calculated NMR frequencies $f_{\text{exp}}(T) - f_{\text{theor}}(T)$ is plotted vs. the temperature in fig. 2 as the open circles. To show the improvement that could be obtained by including the temperature-dependent Oguchi corrections also the results are presented, as the closed circles, of a calculation without these corrections. While below 17 K there is no significant difference between the two calculations, the onset of the breakdown is reduced by, say, 8 K.

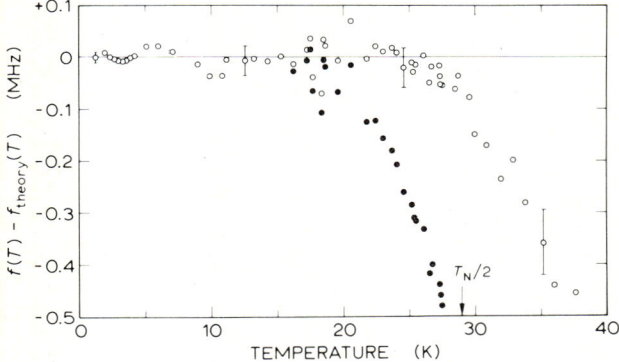

Fig. 2. The deviation of the experimental resonance frequency $f(T)$ from the frequency computed with renormalized spin-wave theory (open circles). Substantial improvement is obtained over spin-wave theory without temperature-dependent renormalization (closed circles).

At the point of breakdown the magnetization has dropped by 10% relative to $T = 0$ K. In three and two dimensions these values are respectively 50% for MnF_2 [2] and only 7% for K_2MnF_4 [3]. Our value obtained for J is in excellent agreement with recent Raman scattering experiments [4], which yielded $J/k_B = -7.54 \pm 0.12$ K. However, disaccord is found with the value $J/k_B = -8.08 \pm 0.32$ K obtained from susceptibility measurements just above T_N [5] by use of a high temperature series for the susceptibility of n-layer Heisenberg films [6]. The zero-temperature gap corresponds to an anisotropy field of 1.40 kG, close to the dipolar field of 1.20 kG, and to a spin-flop field of 44.6 ± 0.5 kG.

From the difference in zero-temperature NMR frequencies in $K_3Mn_2F_7$ and K_2MnF_4, we finally deduced the zero-point spin reduction. Assuming the transferred hyperfine coupling A_{19} to be the same in the two compounds, but incorporating a slight change in the dipolar contribution, we find $\Delta_0 = 0.12 \pm 0.03$, in good agreement with the spin-wave value $\Delta_0 = 0.124$.

This work has been supported by the Foundations F.O.M. and Z.W.O.

References

[1] T. Oguchi, Phys. Rev. 117 (1960) 117.
[2] G.G. Low, Proc. Phys. Soc. 82 (1963) 992.
[3] H.W. de Wijn, L.R. Walker and R.W. Walstedt, Phys. Rev. B8 (1973) 285.
[4] A. van der Pol, M.P.H. Thurlings and H.W. de Wijn, private communication.
[5] R. Navarro, J.J. Smit, L.J. de Jongh, W.J. Crama and D.J.W. Ydo, Physica 83B (1976) 97.
[6] D.S. Ritchie and M.E. Fisher, Phys. Rev. B7 (1973) 480.

PARALLEL PUMPING OF TWO-DIMENSIONAL SPIN-WAVES IN $(C_4H_9NH_3)_2CuCl_4$

H. YAMAZAKI

Department of Physics, Faculty of Science, Okayama University, Tsushima, Okayama 700, Japan

Parallel pumping and ferromagnetic resonance experiments on $(C_4H_9NH_3)_2CuCl_4$ show that the width of the spin-wave band is narrower than that for a three-dimensional ferromagnet, and the spin-wave frequency of each layer is different from those of the adjacent layers. These results agree with a two-dimensional spin-wave theory.

In one- or two-dimensional magnetic materials the temperature dependence of the shift or width of electron spin resonance above the critical temperature has been an important tool to study the effect of short-range magnetic order. On the other hand at a temperature well below the critical point ferro- or antiferromagnetic resonance are observed because of the existence of long range order induced by three-dimensional coupling although the latter may be weak. In systems in which chains or layers are magnetically inequivalent and the coupling between them is very weak, ferro- or antiferromagnetic resonance line may split into two lines [1].

The spins within a layer in $(C_4H_9NH_3)_2CuCl_4$ are tightly coupled with each other by relatively strong ferromagnetic exchange interaction ($H_E = 5 \times 10^5$ Oe), while the exchange interaction between adjacent layers is very weak ($H'_E = -17 \pm 3$ Oe) and is smaller than the anisotropy within the ab-plane ($H_{A1} = 234 \pm 15$ Oe) [2, 3]. Since H'_E is very small, a phase transition from the spin-flop phase to the paramagnetic phase occurs at a very low magnetic field applied along the ab-plane. Above this transition the configuration of the spin system is identical to that of a ferromagnet.

Ferromagnetic resonance experiments were made in the temperature range from 1.4 to 4.2 K and at 8.94 GHz. As shown in fig. 1 two separate resonance lines were observed when the direction of the external field was tilted away from the a- and b-axes. The minima of the resonance field occur in the field $\pm 17°$ away from the a-axis. From this result it is seen that there are two inequivalent magnetic sites of which the directions of easy magnetization are $\pm 17°$ away from the a-axis. Accordingly it appears that the resonance frequency of one ab-layer is different from those of the ab-layers immediately above and below except along the a and b directions, and amalgamation of the two lines does not occur because of the very small H'_E.

Parallel pumping absorption above a threshold for spin-wave instability was measured applying the magnetic field parallel to the ab-plane at 1.4 K and at a pumping frequency of 8.91 GHz. The absorption spectrum typically has two peaks; one is due to absorption by the $k \simeq 0$ magnons propagating perpendicular to the ab-plane and the other by the $k \simeq 0$ magnons propagating parallel to the magnetic field. The field difference between these peaks corresponds to the width of the spin-wave band. As shown in fig. 2 this variation of the width cannot be explained by the spin-wave theory for a three-dimensional ferromagnet.

The width of the spin-wave band arises from the dipolar interaction between the transverse components of the spin precessions. The frequency of the $k \perp H$ magnon is higher than that of the $k \| H$ magnon. The difference between them is evaluated by the term $4\pi M$ of the spin-wave frequency formula [4, 5]. If the spin-wave frequency for one layer is different enough from that for the adjacent layers, the

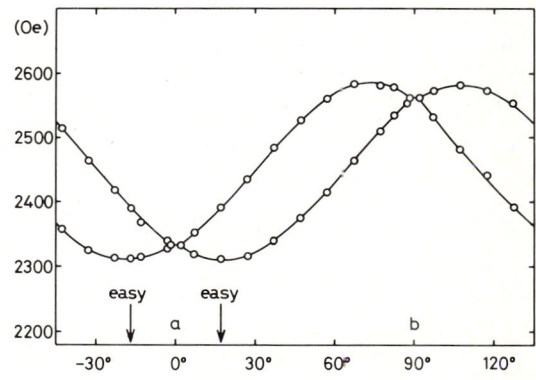

Fig. 1. Angular dependence of ferromagnetic resonance lines in the ab-plane at 4.2 K and at 8.94 GHz.

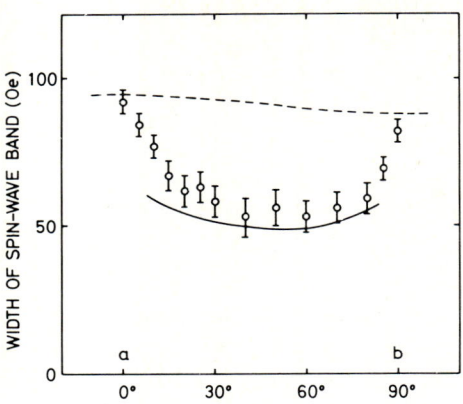

Fig. 2. Angular dependence of the width of the spin-wave band. The dashed and solid lines are the predictions of three- and two-dimensional spin-wave theories, respectively.

dipolar interaction of the transverse components of the spin precessions between adjacent layers vanishes. In this case the width is determined from the dipolar interaction among the spins within a given sublattice only. Therefore the width is reduced because the term $4\pi M$ must be replaced by $2\pi M$. As the direction of the external field approaches the a- or b-axis the dipolar interaction between adjacent layers comes to revive. The width with the field $\mathbf{H} \| a$- and b-axes is in good agreement with the theory for a three-dimensional ferromagnet.

In order to calculate quantitatively the width of the spin-wave band we consider two magnetically inequivalent ferromagnetic layers. Two spin-wave branches have been obtained from the hamiltonian consisting of the intra- and interlayer exchange interactions, intra- and interlayer dipolar interactions, the Zeeman and the anisotropy terms. Details of the theory will be published elsewhere [5]. The numerically calculated width of the spin-wave band is shown in fig. 2 as the solid line. The agreement between theory and experiment appears to be good, except in the vicinity of the a- and b-axes where two spin-wave bands overlap each other. The theory may be invalid in this region because the exchange narrowing effect between two spin-wave bands has not been considered.

As verified by the reduction of the width of the spin-wave band, the exchange and dipolar interactions of the transverse components of the spin precessions between adjacent layers vanish. Therefore it can be concluded that the spin-wave spectrum is indeed two-dimensional, except along the a and b directions and their vicinities.

References

[1] M. Date, A. Nakanishi and K. Oshima, Proc. Int. Conf. Magnetism, Vol. 4 (NAUKA, Moscow, 1974) p. 98.
[2] H. Yamazaki, J. Phys. Soc. Jap. to be published.
[3] L.J. de Jongh and A.R. Miedema, Advan. Phys. 23 (1974) 1.
[4] R.M. White, Quantum Theory of Magnetism (McGraw-Hill, New York, 1970) p. 158.
[5] H. Yamazaki, J. Phys. Soc. Jap. to be published.

ULTRASONIC ATTENTUATION IN THE ONE-DIMENSIONAL ANTIFERROMAGNET CsNiCl₃*

D.P. ALMOND

School of Materials Science, University of Bath, Bath, England

Attenuation measurements in CsNiCl₃ show a broad peak at 30 K and a small critical anomaly at 4.36 K. The critical behaviour is consistent with mode-mode coupling theory and the 30 K peak is a measure of the spin energy density. The Néel temperature is found to increase with magnetic field.

CsNiCl₃ is a compound which has been shown [1] to behave as a one-dimensional antiferromagnet. It has a hexagonal crystal structure with Ni^{2+} ions coupled in chains along the c-axis. The interchain coupling is ~1% of the intrachain coupling but is sufficient to achieve three-dimensional antiferromagnetic ordering at ~4.3 K.

Measurements were made of the ultrasonic attenuation of longitudinal waves propagated along the c-axis of a 1.4 cm long single crystal of CsNiCl₃. The attenuation of 150 MHz ultrasound in the temperature range 0–60 K is shown in fig. 1. The data shows a Schottky-like anomaly with a peak at ~30 K. Similar measurements, covering a range of frequencies [2], showed the attenuation to vary quadratically with frequency over the whole temperature range. A small critical anomaly was found at the Néel temperature. This was studied in detail at the higher frequency of 330 MHz. The critical attenuation was found to vary with a critical exponent of 1.35 ± 0.03 [2].

Longitudinal waves interact with magnetic systems by the modulation of exchange fields between spins and by the modulation of crystalline fields at spin sites. In CsNiCl₃ the single-ion anisotropy is small (~10%) in comparison with the exchange coupling, and for a first approximation we shall assume that magnetostrictive coupling is the dominant interaction. Ultrasonic attenuation is generally considered to be caused by an exchange of energy between the ultrasonic phonons and long wavelength components of the fluctuations in the order in a spin system. Usually this only occurs in the critical region where long

* Work completed with J.A. Rayne at Carnegie-Mellon University, Pittsburgh USA and supported by the National Science Foundation.

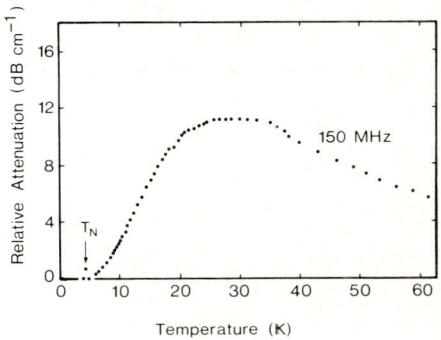

Fig. 1. Temperature dependence of the relative attenuation of 150 MHz longitudinal waves propagating along [0001] in CsNiCl₃.

wavelength components of spin order appear. The main attenuation observed here, away from the Néel temperature, is not thought to be caused by the above effect, as long range order of 20 μm – the ultrasonic wavelength – does not occur in one-dimensional magnets much above 0 K. It is suggested, however, that a longitudinal wave, propagated along the strong coupling directing of a one-dimensional spin system, couples directly to the energy density of the system. The condition for this type of interaction is that the magnetostrictive coupling hamiltonian and the total exchange hamiltonian are proportional to each other [3]. Here both hamiltonians would be calculated using the same number of neighbouring spin interactions – two for nearest neighbours – and the Hamiltonians would be proportional to each other. In this case the attenuation takes the form

$$\alpha = q^2 \left(\frac{a}{J} \frac{dJ}{da} \right)^2 \frac{TCD}{V^3 M}.$$

The attenuation α is proportional to the product of the magnetic heat capacity C, the spin energy

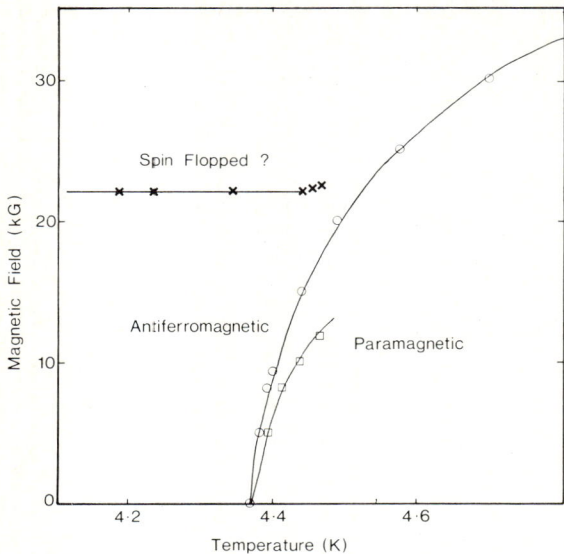

Fig. 2. Phase diagram in CsNiCl$_3$; ⊙ and × magnetic field applied parallel to [0001], ⊡ magnetic field applied perpendicular to [0001].

diffusion constant D and the absolute temperature T. All other symbols are the same as used in [3]. This relationship between attenuation and heat capacity goes some way to explain the resemblance of the attenuation data, fig. 1, to the familiar behaviour of the heat capacity of one-dimensional spin systems. It is hoped that further measurements of changes in sound velocity will reveal the full temperature dependence of D.

The critical exponent of 1.35 is in excellent agreement with the prediction of $\frac{4}{3}$ from mode–mode coupling theory [4]; in contrast with other insulating magnets which exhibit much smaller critical exponents. The latter is attributed to a preponderance of the same energy density coupling which does not exhibit critical behaviour [3]. In our experiments, however, the sound wave does not interact directly with the small interchain exchange coupling, assumed responsible for three-dimensional ordering. Our magnetostrictive coupling hamiltonian is therefore not proportional to this small addition to the exchange hamiltonian. The attenuation in the critical region is thought to be caused by interactions with the long wavelength components of spin fluctuations alone; which are described by the mode–mode coupling theory. Part of the phase diagram in CsNiCl$_3$, fig. 2, was plotted by monitoring the position of the critical anomaly in magnetic fields up to 3.5 T. Unlike other antiferromagnets, the Néel temperature increases monotonically with magnetic field. This has been observed in one other one-dimensional antiferromagnet [5] and has been explained by Imry et al. [6]. Our results show the predicted parabolic increase in Néel temperature with magnetic field but show increases which are about an order of magnitude greater than the calculations suggest [7]. A line of large attenuation anomalies were observed at 2.2 T which we suspect separates the antiferromagnetic from a spin flopped state. The critical anomalies were not sharp enough to establish the nature of the intersection of the phase boundaries at the bicritical point.

References

[1] N. Achiwa, J. Phys. Soc. Jap. 27 (1969) 561.
[2] D.P. Almond and J.A. Rayne, Phys. Lett. 54A (1975) 295.
[3] B. Luthi and R.J. Pollina, Phys. Rev. Lett. 22 (1969) 717.
[4] G.E. Laramore and L.P. Kadanoff, Phys. Rev. 187 (1969) 619.
[5] W.G. Clark et al Proc 14th Int. Conf. Low Temp. Phys. (Helsinki 1975).
[6] Y. Imry, P. Pincus and D. Scalapino, Phys. Rev. B12 (1975) 1978.
[7] D. Scalapino, private communication.

EXACT THEORY OF MAGNETIC COOLING FOR A SYSTEM WHICH OBEYS THE THIRD LAW*

J.C. BONNER†

University of Rhode Island, Kingston, RI 02881, USA

and

J.D. JOHNSON

Los Alamos Scientific Laboratory, Los Alamos, NM 87545, USA

An exact solution has been developed for a reasonably realistic magnetic model known to obey the third law. Results indicate that the 1-D systems are actually more efficient coolers than 3-D systems and that adiabatic demagnetization of an *antiferromagnet* down to the (upper) critical field is a very powerful cooling process.

The study of magnetic cooling and the search for new magnetic thermometers and refrigerators has not seen much progress in recent years. The extant theories of magnetic cooling are either unrealistic at low temperatures because they fail to take account of spin–spin interactions (classical and quantum paramagnets) or actually disobey the third law of thermodynamics (classical paramagnet and antiferromagnetic Ising models and mean field theories).

The time seems ripe for a more sophisticated, statistical mechanical approach and, accordingly, a solution has now been worked out for a well-known and reasonably realistic model (the spin-$\frac{1}{2}$, nearest-neighbour, one-dimensional (1-D) Heisenberg–Ising model). This solution is exact in the sense that the problem is analytically reduced to the solution of an infinite set of coupled, non-linear integral equations which are then treated numerically to an arbitrary degree of accuracy [1, 2]. This numerical approach is supplemented and supported by exact analytic results in cases of special interest. The model is known to obey the third law of thermodynamics [2, 3].

The results obtained so far include an extensive determination of the entropy isotherms, isentropes (cooling curves) and specific heat as a function of temperature and applied magnetic field. From these quantitative results, computations can then be made [2] of the relative cooling efficiency over different ranges of applied magnetic field (a) for varying degrees of anisotropy and (b) for the antiferromagnet in comparison with the corresponding ferromagnet. A plot of the antiferromagnetic isentropes for intermediate uniaxial anisotropy is shown in fig. 1. The anisotropic system is characterized by *two* critical fields, H_{c1} and H_{c2}. It is amusing that four different types of cooling are possible: (1) cooling by adiabatic magnetization to the lower critical field H_{c1}; (2) cooling by adiabatic demagnetization from a field H intermediate between the two critical fields, i.e. $H_{c2} > H > H_{c1}$, down to H_{c1}; (3) cooling by adiabatic magnetization from an intermediate field to the up-

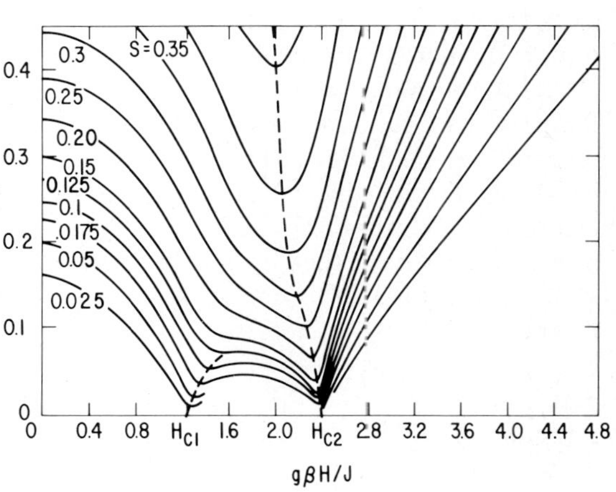

Fig. 1. Cooling isentropes for a 1-D antiferromagnet with intermediate anisotropy, plotted in terms of reduced temperature KT/J and magnetic field $g\beta H/J$. J is the 1-D nearest-neighbour exchange constant.

---

* Work performed under the auspices of the US Energy Research and Development Administration.
† Work performed while at Brookhaven National Laboratory, Upton, New York, 11973 USA.

per critical field H_{c2} and; (4) cooling by adiabatic demagnetization down to the upper critical field H_{c2}. It might be noted that this multiple cooling phenomenon is not only a property of systems with uniaxial anisotropy: very similar cooling behaviour is seen in the case of the weakly-interacting magnetic cluster (dimer) system $Cu(NO_3)_2 \cdot 2.5H_2O$ [4], where the spin hamiltonian is effectively isotropic (Heisenberg) in type. However, in the Heisenberg–Ising system described here, H_{c1} tends to zero as the isotropic, Heisenberg limit is approached, leaving a single critical field $H_{c2} = H_c$.

Of the four cooling processes observable in fig. 1, clearly the most efficient is cooling by adiabatic demagnetization down to the upper critical field. (This is actually the analogue of cooling an ideal paramagnet down to *its* critical field $H_c = 0$). The cooling efficiency [of process (4)] for the antiferromagnet improves as the anisotropy increases. The reverse is true in the case of the ferromagnet. When a comparison is made of the antiferromagnet at its upper critical field ($H_c = H_{c2}$) with the ferromagnet of *corresponding* anisotropy ($H_c = 0$, always), the antiferromagnet turns out to be a strikingly better cooler. For the antiferromagnet, the entropy decreases to zero rather slowly as the temperature T decreases, as a power law ($T^{\frac{1}{2}}$). The entropy of the ferromagnet, by contrast, vanishes exponentially fast. The reason is, of course, that the uniaxial ferromagnet at $H = 0$ always has a gap between the ground state and the first excited states (which decreases as the anisotropy decreases, vanishing in the isotropic limit [5]) whereas the antiferromagnet at $H = H_c$ (H_{c2}) has no gap. To be useful in attaining low temperatures, a system should have as large an entropy as possible down to as low a temperature as possible. The antiferromagnet is clearly much superior in this respect. It is interesting to note that by similar arguments, a 1-D antiferromagnet is also superior to a higher-D antiferromagnet, since the critical entropy is expected to behave as $T^{D/2}$, where D is the dimensionality. Of course, ultimately weak interchain interactions cause a quasi-1-D real system to cross over to 3-D behaviour at low enough temperatures, but with relatively minor effect on the argument above. The existence of a phase transition is not important in connection with magnetic cooling [6].

The practical consequences of the foregoing discussion are as follows. In a search for an efficient new magnetic cooler, it appears preferable to examine first the ranks of the 1-D rather than the 3-D magnets, and it appears also that antiferromagnets are preferable to ferro- (para-) magnets. Further, the cooling salts in common use, such as cerium magnesium nitrate, are magnetically dilute to remain paramagnetic down to very low temperatures. This reduces the magnetic specific heat available for refrigeration. A (quasi-) 1-D system may actually be less magnetically dilute than the common paramagnets. Accordingly, cooling experiments are in progress on quasi-1-D, spin-$\frac{1}{2}$ copper salts [7]. The one disadvantage of antiferromagnets is that their (upper) critical field may be rather large in comparison with the maximum field obtainable with the current available superconducting magnets (100–120 kOe). The (upper) critical field is related to the intra-chain exchange integral J, which should be as small as possible.

Unfortunately, a 1-D *nuclear* spin antiferromagnet is not available, but magnetic cooling by both adiabatic magnetization and demagnetization should be possible for the BCC nuclear spin-$\frac{1}{2}$ antiferromagnet, solid ^{3}He.

References

[1] J.D. Johnson, Phys. Rev. A9 (1974) 1743.
J.D. Johnson and B.M. McCoy, Phys. Rev. A6 (1972) 1613.
[2] J.C. Bonner and J.D. Johnson, in preparation. A much more detailed mathematical presentation of which the qualitative highlights are reported here.
[3] J.C. Bonner and M.E. Fisher, Proc. Phys. Soc. 80 (1962) 508.
[4] M.W. van Tol, K.M. Diederix and N.J. Poulis, Physica 64 (1973) 363.
J.C. Bonner and S.A. Friedberg, in: Phase Transitions 1973 (Proc. Conf. on Phase Transitions and their Applications in Materials Science), H.K. Henisch, R. Roy and L.E. Cross (Eds.) (Pergamon Press, 1973) p. 429. See also J.C. Bonner and S.A. Friedberg, Proc. 19th Conf. on Magnetism and Magnetic Materials, C.D. Graham, Jr. and J.J. Rhyne (Eds.) (1973) p. 1311.
[5] J.D. Johnson and J.C. Bonner, unpublished work.
[6] J.C. Bonner and J.F. Nagle, Phys. Rev. A5 (1972) 2293.
[7] S.A. Friedberg, private communication.

ONE-DIMENSIONAL MAGNETIC VARIETY IN A FAMILY OF TTF-BIS-DITHIOLENE METAL COMPLEX COMPOUNDS*

I.S. JACOBS,† H.R. HART, JR.,† L.V. INTERRANTE,† J.W. BRAY†
J.S. KASPER,† G.D. WATKINS,‡ D.E. PROBER,§ W.P. WOLF§ and
J.C. BONNER**

The donor–acceptor compounds TTF·$MS_4C_4(CF_3)_4$ where TTF = tetrathiafulvalene and M = Cu, Au, Pt, Ni are mixed-stack quasi-one-dimensional magnetic insulators of remarkable variety. For M = Cu, Au, a novel progressive spin-lattice dimerization occurs (spin-Peierls transition). For M = Pt, Ni, ferro/ferrimagnetic chains develop with moderate/weak antiferromagnetic coupling.

Quasi-one-dimensional (1-D) systems have attracted much interest for over a decade for their unique physical properties [1]. We have prepared [2] and examined a family of π-donor–acceptor compounds of the type TTF·$MS_4C_4(CF_3)_4$, where TTF = tetrathiafulvalene and $MS_4C_4(CF_3)_4$ is a series of bis-dithiolene metal complexes where M = Cu, Au, Pt and Ni. The compounds crystallize to form pseudo 1-D stacks in which the planar TTF$^+$ and $MS_4C_4(CF_3)_4^-$ ions alternate. They constitute a unique class of magnetic insulators in which a subtle change in molecular composition produces profound changes in magnetic properties. We report here the results of magnetization and susceptibility measurements considered in the context of structural information [2] and other physical property data [3–7].

The M = Pt, Cu, and Au compounds are isostructural with triclinic symmetry and alternate stacking of ions along the c-axis of the structure. The ionic alternation also occurs along the a and b axes. The structure of the M = Ni derivative is similar in the a–c plane; however, along b the repeat unit is three chains. The ions of this compound are tilted in such a way to suggest more interchain interaction.

In fig. 1 is shown a composite of the susceptibilities for the TTF·$MS_4C_4(CF_3)_4$ family. The asymptotic high-temperature values reflect the fact that the TTF$^+$ ions are spin-$\frac{1}{2}$ paramagnetic ($g \approx 2$) while the $MS_4C_4(CF_3)_4^-$ ions are diamag-

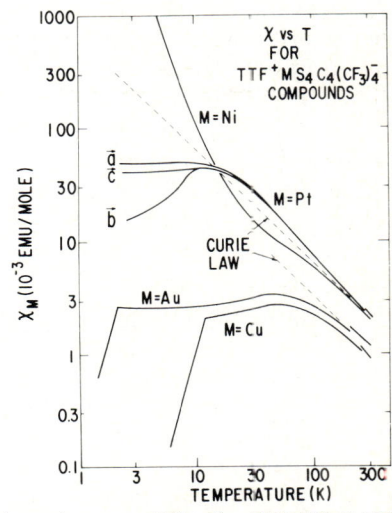

Fig. 1. Magnetic susceptibility for TTF·$MS_4C_4(CF_3)_4$ family.

netic for M = Au, Cu, but paramagnetic for M = Ni, Pt ($S = \frac{1}{2}$, $g \approx 2$).

M = Cu, Au: The susceptibility is isotropic at all temperatures. For the Cu compound, between 12 and 250 K the susceptibility is modeled very accurately by a system of antiferromagnetic (AF) Heisenberg chains of spins with $J/k_B = 77$ K where $\mathcal{H}_{ij} = J S_i \cdot S_j$. Below 12 K, the susceptibility decreases sharply toward zero. The isotropy of this decrease is incompatible with 3-D ordering of the spin system. From similar data for the Au compound we have $J/k_B = 68$ K and $T_c = 2.1$ K. (The breaks occurring at 250 and 200 K, respectively, are associated with a lattice distortion tied to rotation of the CF_3 groups.)

These data display with classic simplicity the behavior expected for a "spin–Peierls" transition, i.e. a progressive spin-lattice dimerization occurring below a transition temperature in a system of AF Heisenberg chains. The transition is second-order and is driven by the spin subsystem coupled to a 3-D phonon field. At the

---

* Supported in part by AFOSR under Contract F-44620-71-C-0129; for WPW by AROD; and for JCB by ERDA for work performed at Brookhaven National Laboratory.
† General Electric Corp. Res. and Dev., Schenectady, NY 12301, USA.
‡ Physics Dept., Lehigh Univ. Bethlehem, PA 18015 USA.
§ Dept. of Engineering and Applied Science, Yale Univ., New Haven, CT, 06520, USA.
** Physics Dept., Univ. of Rhode Island, Kingston, RI, 02881, USA.

lowest temperature the system is in a singlet ground state with a magnetic gap. Details are published elsewhere [4, 5].

M = Pt: The positive deviation of χ from a Curie-law between 270 and 40 K indicates dominant ferromagnetic coupling, presumably in c-axis chains. At lower temperatures antiferromagnetic coupling is manifested as a marked drop of the b-axis χ. While this resembles a 3-D AF ordering near 8 K, preliminary specific heat results [9] suggest a broad anomaly between 5 and 9 K instead of the sharp anomaly associated with the onset of long-range order. Magnetization isotherms to 40 kOe ($\boldsymbol{H}\|b$) between 2.8 and 10 K show a temperature-dependent upward concavity not characteristic of long range ordered behavior [1] but consistent with the destruction of intermediate range AF order. These properties may be intrinsic to the remarkable anisotropy of the structure [2, 4]. Alternatively, they suggest "incomplete 3-D order" [8] (as in a spin-glass) perhaps due to defects which interrupt intrachain order. The deliberate addition of up to 20% Au to interrupt the FM chains, shown in fig. 2, depresses the temperature of the maximum in χ.

M = Ni: This compound shows (fig. 1) dominant antiferromagnetic behavior above 40 K and ferromagnetic behavior below. At low temperatures and high fields the magnetization approaches a saturation of $2\mu_B/3$ per formula unit [3]. The $\frac{1}{3}$-factor recalls the b-axis structure repeat and suggests linear AF trimers as units of a ferrimagnetic structure. The detailed model includes additional weak coupling of the spins ferromagnetically along the c-axis ("trimer ladder"). For some samples (type A of fig. 3), no

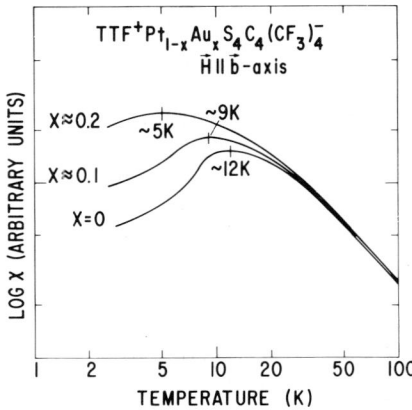

Fig. 2. Magnetic susceptibility for TTF·Pt$_{1-x}$Au$_x$S$_4$C$_4$(CF$_3$)$_4$.

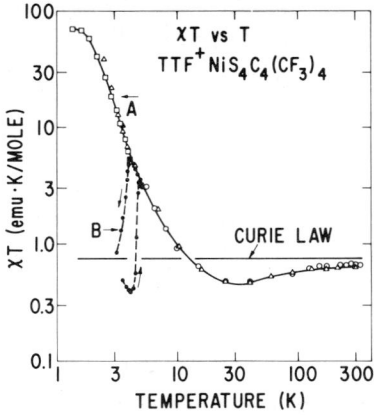

Fig. 3. Comparative behavior of samples of TTF·NiS$_4$C$_4$(CF$_3$)$_4$.

long-range order occurs down to 1.5 K, while for other samples (type B) an abrupt AF transition (first-order with thermal hysteresis) occurs near 4 K. Below the transition the magnetization isotherms are metamagnetic, with considerable field hysteresis. The properties of this transition suggest magneto-elastic behavior [10]. That samples of higher purity favor type B behavior may again signal the important role of defects in quasi-1-D systems.

References

[1] L.J. de Jongh and A.R. Miedema, Advan. Phys. 23 (1974) 1.
H.J. Keller (Ed.), Low-Dimensional Cooperative Phenomena, NATO Adv. Study Inst. Vol. B7 (Plenum Press, New York, 1975).
[2] L.V. Interrante, K.W. Browall, H.R. Hart, Jr., I.S. Jacobs, G.D. Watkins and S.H. Wee, J. Amer. Chem. Soc. 97 (1975) 889.
J.S. Kasper and L.V. Interrante, Acta Cryst. B32 (1976) 2914.
[3] I.S. Jacobs, L.V. Interrante and H.R. Hart, Jr., AIP Conf. Proc. 24 (1975) 355.
[4] J.W. Bray, H.R. Hart, Jr., L.V. Interrante, I.S. Jacobs, J.S. Kasper, G.D. Watkins, S.H. Wee and J.C. Bonner, Phys. Rev. Lett. 35 (1975) 744.
[5] I.S. Jacobs, J.W. Bray, H.R. Hart, Jr., L.V. Interrante, J.S. Kasper, G.D. Watkins, D.E. Prober and J.C. Bonner, Phys. Rev. B14 (1976) 3036.
[6] L.S. Smith, E. Ehrenfreund, A.J. Heeger, L.V. Interrante, J.W. Bray, H.R. Hart, Jr. and I.S. Jacobs, Solid State Commun. 19 (1976) 377.
[7] T. Wei, A.J. Heeger, M.B. Salamon and G.E. Delker, Solid State Commun. 21 (1977) in press.
[8] D. Hone, P.A. Montano, T. Tonegawa and Y. Imry, Phys. Rev. B12 (1975) 5141.
[9] T. Wei and A.J. Heeger, private communication.
[10] D.S. Rodbell and C.P. Bean, J. Appl. Phys. 33 (1962) 1037.
C.P. Bean and D.S. Rodbell, Phys. Rev. 126 (1962) 104.

NEUTRON SCATTERING STUDY OF THE MAGNETISM OF Rb_2CrCl_4, A TWO-DIMENSIONAL EASY-PLANE FERROMAGNET

M.J. FAIR, A.K. GREGSON, P. DAY
Inorganic Chemistry Laboratory, South Parks Road, Oxford, England

and

M.T. HUTCHINGS
Materials Physics Division, Harwell, Didcot, OX11 ORA, England

The crystal and magnetic structure, and the temperature variation of the moment, of Rb_2CrCl_4 have been investigated using neutron diffraction. The Cr^{2+}, $S = 2$, spins order ferromagnetically at $T_c = 57 \pm 2$ K. At 4.2 K they lie perpendicular to the tetragonal axis with a moment of $3.7 \pm 0.4\ \mu_B$/ion. As the temperature is raised the moment falls rapidly in a manner characteristic of a two-dimensional easy-plane ferromagnet.

1. Introduction

Recent susceptibility [1] and optical experiments [2, 3] have suggested that the chromous compounds A_2CrCl_4, where A is K^+, Rb^+ or Cs^+, order ferromagnetically in three dimensions with relatively high Curie temperatures T_c of the order of 60 K. Single crystals of the K^+ and Rb^+ compounds may be grown [4] and as these are transparent in the visible region they have possible technological interest as modulation devices. They also form a unique system for the study of the fundamental properties of ferromagnetism in insulators of which there are only a few known examples [5]. The ground state of the chromous ion in Rb_2CrCl_4 is $3d^4$, $^5A_1$, with $S = 2$. There are therefore similarities to, and important differences from – including the possibility of single-ion anisotropy effects – the extensively studied K_2CuF_4 [6, 7], where the Cu^{2+} ion has $S = \frac{1}{2}$ and $T_c = 6.25$ K. The spin wave excitations in Rb_2CrCl_4 have recently been investigated, and shown to follow closely those of a two-dimensional [2d] easy-plane ferromagnet [8].

The crystal structure of Rb_2CrCl_4 was first examined by Seifert and Klatyk [9], who determined the lattice constants of the K_2NiF_4-type, D_{4h}^{17}, unit cell at room temperature using powder X-ray diffraction. The related compound K_2CuF_4 was recently shown [10] to have the distorted structure D_{2d}^{10}, rather than D_{4h}^{17} as originally thought.

2. Crystal and magnetic structure

In order to determine the crystal and magnetic structure of Rb_2CrCl_4 both powder and single crystal diffraction measurements were carried out using the PANDA and MARK VI diffractometers, respectively, at Harwell. The samples were prepared by P.J. Walker of the Clarendon Laboratory, Oxford [4].

The powder diffraction data could be indexed on a D_{4h}^{17} unit cell and showed no superlattice peaks from the D_{2d}^{10} structure of K_2CuF_4. A careful search was also made for such peaks at 293 and 4.2 K using the more sensitive single crystal diffraction, but none were found, and we conclude from our data that the structure is not D_{2d}^{10}. The structure parameters were derived by fitting to the powder intensity profiles [11]. The values deduced at 293 and 4.2 K are listed in table I. Single crystal data at 85 and 4.5 K yielded the same positional parameters as the powder data, but gave slightly larger thermal parameters.

Correlations between the magnitude of the moment and its angle to the c-axis, ϕ_c, made it difficult to obtain both simultaneously from the diffraction intensities. A separate experiment was therefore carried out on a small single crystal in order to determine ϕ_c. A magnetic field **H** was applied perpendicular to the scattering plane, which coincided with the (010) plane of the crystal. The intensities of peaks (h01) were measured with $H = 0$ and 1.8 T. Typically we found that the intensities of the

Table I
Results of powder neutron diffraction from Rb$_2$CrCl$_4$, $\lambda = 1.54$ Å

Temp. (K)	a_0(Å)	c_0(Å)	z_{Rb}	z_{Cl}	B_{Cr}	B_{Rb}	B_{Cl_1} (z-axis)	B_{Cl_2} (In plane)	$\mu(\mu_B)$	Weighted R (Expected R)
293	5.142 ($\pm$0.001)	15.769 ($\pm$0.005)	0.362 ($\pm$0.001)	0.1544 ($\pm$0.0005)	0.84 ($\pm$0.50)	3.3 ($\pm$0.5)	1.06 ($\pm$0.30)	1.66 ($\pm$0.35)		17.16 (11.69)
4.2	5.086 ($\pm$0.001)	15.715 ($\pm$0.005)	0.360 ($\pm$0.001)	0.1552 ($\pm$0.0005)	0.172 ($\pm$0.138)	-0.163 ($\pm$0.289)	0.088 ($\pm$0.113)	1.05 ($\pm$0.93)	3.94 ($\pm$0.16)	17.32 (10.07)

The temperature factors B are defined as $\exp(-W_j) = \exp(-B_j \sin^2\theta/\lambda^2)$ multiplying the structure factor contribution from ion j. Weighted R factors are defined for the whole profile [11].

(101) and (004) changed from 3150 ± 56 and 1643 ± 40 to 4469 ± 66 and 1706 ± 41, respectively. As the intensity is proportional to $\sin^2\alpha$, where α is the angle between the moment and scattering vector, we deduce α (004) is constant within our errors at 90°, and therefore $\phi_c = 90°$, i.e. the field only rotates the moments in the basal plane.

The magnitude of the moment found by fitting the powder intensities using the free-ion Cr^{2+} calculated form factor is given in table I. A more detailed investigation of the form factor has been carried out using polarized neutrons and will be reported elsewhere. The moment deduced from the single crystal data, $3.3 \pm 0.5\,\mu_B$, was somewhat lower than that from the powder data but had a relatively large uncertainty. In view of this difference we cannot quote the moment more accurately than $3.7 \pm 0.4\,\mu_B$.

3. Temperature variation of magnetization

The temperature variation of the magnetization was investigated using two crystals of different size on the MARK VI diffractometer at Harwell. The results from the smaller crystal, ~ 12 mm^3, have been briefly reported [12], and we here present more accurate data on a larger crystal $\sim 9 \times 5 \times 4$ mm^3, with broad mosaic. Integrated intensities of three peaks (002), (101), and (004), whose intensities decrease progressively, were measured at a number of temperatures between 4.45 and 91 K. T_c was found to be 57 ± 2 K from the peak of the [2d] rod-like intensity, and from the [3d] magnetic scattering. Variation in intensity due to thermal vibrations was shown to be negligible, and the intensity above T_c was subtracted from each peak to give the magnetic intensity, I_M. The intensity at 0 K was extrapolated from that at 4.45 K, the earlier data down to 1.7 K having indicated that such an extrapolation was justified. The variation of the three peaks coincided exactly, indicating that extinction effects did not affect their intensity. The moment, proportional to $\sqrt{I_M}$, found from the (101) reflection is shown normalized to unity at 0 K in fig. 1. Fitting $(1 - M_T/M_0) = CT^a$ to the data up to 30 K we find $a = 1.27 \pm 0.08$ and $C = 0.0025 \pm 0.00025$.

A number of theoretical treatments of the problem of the temperature variation of the

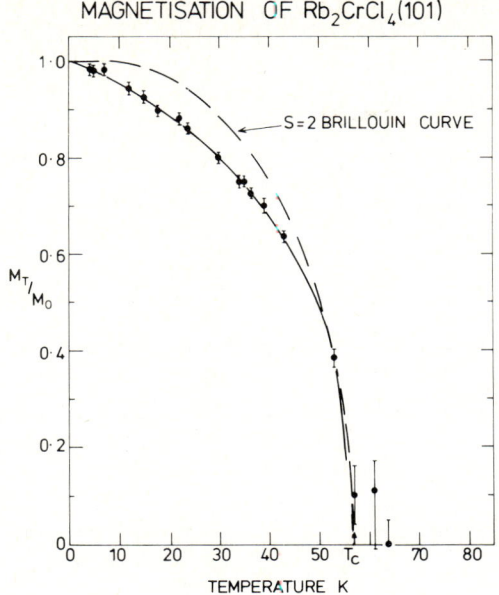

Fig. 1. Variation of moment with temperature deduced from the integrated intensity of the (101) reflection. Solid line is best fit of power law for $T < 30$ K, and is guide to eye only for $T > 30$ K.

magnetization of [2d] ferromagnets have now been given [13], and these show a strong dependence on the form and size of any anisotropy and small interplanar interactions. However, the case of single-ion anisotropy has not yet been considered explicitly. We may note that if spin wave theory and the measured spin wave energy dispersion at 5 K are used to calculate M_T vs. T, assuming an extremely small gap at $q = 0$ by omitting the $q = 0$ point from the summations, we find $\alpha \sim 1.1$. Measurements of the renormalization of the spin wave energies are currently in progress.

References

[1] See A.K. Gregson, P. Day, D.H. Leech, M.J. Fair and W.E. Gardner, J. Chem. Soc. (Dalton) (1975) 1306, and references therein.
[2] P. Day, A.K. Gregson and D.H. Leech, Phys. Rev. Lett. 30 (1973) 19.
[3] M.J. Fair, Chemistry, Part II, Thesis, Oxford University (1973).
[4] G. Garton and P.J. Walker, J. Cryst. Growth 33 (1976) 61.
[5] L.J. de Jongh and A.R. Miedema, Advan. Phys. 23 (1974) 1.
[6] See H. Kubo, J. Phys. Soc. Jap. 36 (1974) 675, and references therein.
[7] K. Hirakawa and H. Ikeda, J. Phys. Soc. Jap. 35 (1973) 1328.
[8] M.T. Hutchings, M.J. Fair, P. Day and P.J. Walker, J. Phys. C9 (1976) L.55.
[9] H-J. Seifert and K. Klatyk, Z. Anorg. Allg. Chem. 334 (1964) 113.
[10] R. Haegele and D. Babel, Z. Anorg. Allg. Chem. 409 (1974) 11.
[11] H.M. Rietveld, J. Appl. Cryst. 2 (1969) 65.
[12] M.T. Hutchings, P. Day, M. Fair and A.K. Gregson, AIP Conf. Proc. 29 (1976) 263.
[13] See, for example, K. Yamaji and J. Kondo, J. Phys. Soc. Jap 39 (1975) 1239.
R. Shanker and R.A. Singh, J. Phys. Soc. Jap. 40 (1976) 657.

MAGNETIC ORDERING IN A PLANAR X–Y MODEL: $BaCo_2(AsO_4)_2$

L.P. REGNAULT, P. BURLET and J. ROSSAT-MIGNOD

DRF/DN, Centre d'Etudes Nucléaires, 85X, 38041 Grenoble Cedex, France

Magnetic and neutron diffraction experiments have been performed on a single crystal of the layered compound $BaCo_2(AsO_4)_2$. The cobalt moments lie in a plane (X–Y model) and order abruptly at $T_c = 5.4$ K with a helical structure. An in-plane applied field induces a quite original 2D "ferrimagnetic" structure. At lower temperature the transition becomes irreversible and this structure remains in zero field.

1. Introduction

During the last years much attention has been devoted to two-dimensional magnetism. A lot of experimental and theoretical work has been done on Ising and Heisenberg systems; however, planar systems (X-Y model) have been less studied. $BaCo_2(AsO_4)_2$ is a good candidate for this model. This compound has a rhombohedral structure (space group $R\bar{3}$) [1]; the cobalt atoms are located at honeycomb lattice points in (001) planes spaced by 7.8 Å. The in-plane Co–Co distance is 2.9 Å. The planar anisotropy has been evidenced by magnetic measurements on single crystal which are reported in this paper together with neutron diffraction studies of the 3D order and magnetization processes [2].

2. Experimental

Experiments have been performed on single-crystals which have been grown by cooling slowly the melted compound [2]. Magnetic susceptibilities in the temperature range 1.5–300 K and magnetization curves in fields up to 140 kOe have been measured [3]. Neutron experiments have been carried out using a double-axis spectrometer operating at the reactor Siloe. The crystal was placed inside a cryostat-electromagnet assembly which produces a field up to 20 kOe and allows a temperature range 1.5–300 K.

3. Experimental results

3.1. Magnetic susceptibility

The susceptibility parallel to the c-axis ($\chi_\parallel$) is much lower than the perpendicular susceptibility ($\chi_\perp$) in the whole temperature range. At high temperature ($T > 150$ K) $\chi_\parallel$ and $\chi_\perp$ follow approximatively a Curie–Weiss law ($C_\parallel = 6.9$ uem/mole, $\theta_\parallel = -90$ K: $C_\perp = 5.9$ uem/mole, $\theta_\perp = 35$ K). Below $T \approx 50$ K a large enhancement of $\chi_\perp$ is observed which cannot be explained by crystal-field effects but by strong in-plane correlations. $\chi_\perp$ shows a maximum at $T = 6.3$ K and an inflexion point at $T_c = 5.4$ K which corresponds to the 3D ordering (fig. 1).

3.2. 3D magnetic order

By neutron diffraction experiments, below T_c, magnetic superlattice peaks have been observed by scanning along reciprocal directions. They are associated to a propagation vector $\mathbf{k} = (0.261, 0, -\tfrac{4}{3})$. It corresponds to a regular staking of helimagnetic (001) planes in which the moments are confined. From superlattice peak intensities the phase angle between the moments of the two Bravais lattices has been deduced ($\phi = 83 \pm 10°$). The in-plane magnetic

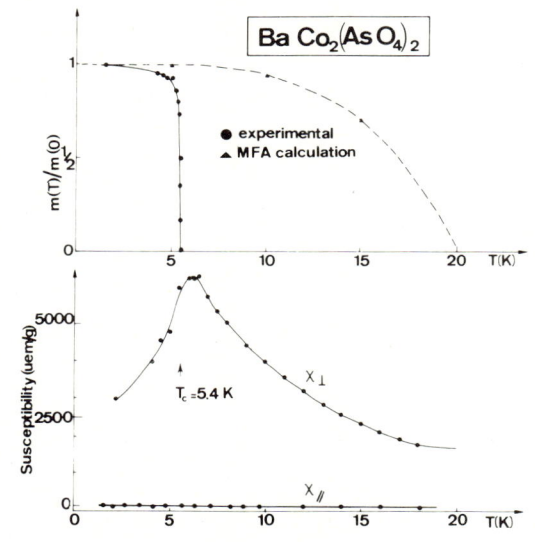

Fig. 1. Low temperature susceptibilities $\chi_\parallel$ and $\chi_\perp$ and staggered magnetization of $BaCo_2(AsO_4)_2$.

structure is reported in fig. 2. By adding the three domains contribution a magnetic moment value $m = 2.9 \pm 0.2\, \mu_B$ is obtained at $T = 4.2$ K. This value is practically temperature independent except very near to T_c where it decreases very abruptly (fig. 1). This transition seems to be of first order but no hysteresis and cell parameter discontinuity have been detected.

3.3. Magnetization process

High-field measurements of $\sigma_\parallel$ and $\sigma_\perp$ reflect the planar anisotropy (for $H = 100$ kOe: $\sigma_\parallel = 0.48\, \mu_B$ and $\sigma_\perp = 2.95\, \mu_B$). Below T_c an in-plane applied field induces a metamagnetic behaviour (fig. 3). In low field the magnetization $\sigma_\perp$ increases accordingly with a large initial susceptibility up to a plateau, between H_{c1} ($\simeq 2$ kOe) and H_{c2}, ($\simeq 4.5$ kOe) with a value of one-third the saturated moment value. Neutron measurements of the variation of lattice peak intensities (fig. 3) show a similar behaviour. In addition to the nuclear intensity, a ferrimagnetic contribution appears between H_{c1} and H_{c2} which corresponds to one-ninth the saturated ferromagnetic intensity. This means that the intermediate "ferrimagnetic" structure has a net magnetization of $\sigma_S/3$. Moreover the magnetic field induces a modification of the superlattice peaks. At 4.2 K (fig. 4) the k_x component of the k-vector increases up to $\frac{1}{3}$ ($k = \frac{1}{3}, 0, \frac{4}{3}$) and remains constant until the saturated paramagnetic state. This transition produces a large broadening of the superlattice peaks; in fact a scan along the $(\frac{1}{3}, 0, L)$ direction shows a diffusion ridge with an intensity maximum at $k_z = -\frac{4}{3}$. At lower temperature ($T = 1.7$ K) the transitions are more abrupt and when the field is switched off this "ferrimagnetic" state remains. This irreversible behaviour disappears progressively above 2 K.

4. Discussion

These results have evidenced the X–Y anisotropy of the Co^{2+} ions in the $BaCo_2(AsO_4)_2$ layered structure. This compound exhibits a para-helimagnetic transition which has a surprising nature. The 3D order consists in a periodic stacking of helimagnetic layers giving a quite original structure. It is strange that an applied field seems to destroy the 3D ordering. From the experimental results the "ferrimagnetic" state would be described by a 2D collinear ferrimagnetic order in (001) planes. It

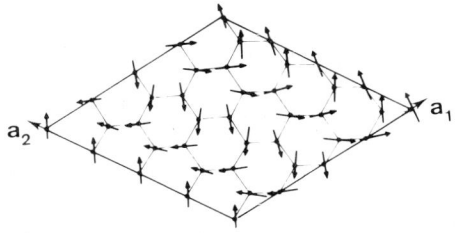

Fig. 2. In-plane magnetic order of Co^{2+} moments in $BaCo_2(AsO_4)_2$.

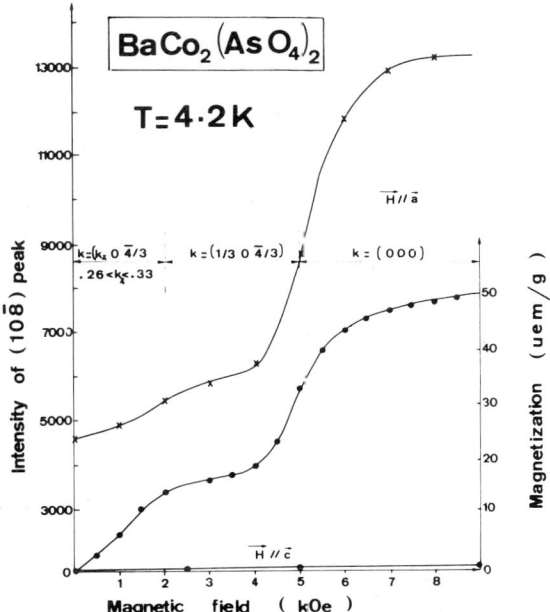

Fig. 3. Variation vs. applied field of (1) parallel and perpendicular magnetization, (2) $(10\bar{8})$ peak intensity, and (3) k_x component of the wave vector.

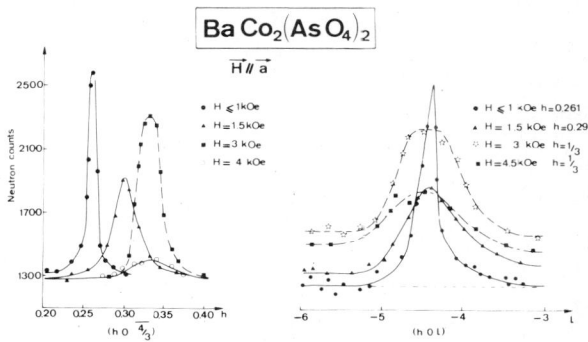

Fig. 4. Typical scans showing the large broadening of the superlattice peaks when a field is applied.

would consist in ferromagnetic chains which progate along the a-axis according to a $++-$ sequence. Between the planes only a short-range order exists. This structure can explain the value $k_x = \frac{1}{3}$, the net magnetization of the intermediate state ($\sigma_S/3$) and the modulation of the Bragg ridges.

The stability of the zero-field structure requires at least three in-plane exchange integrals. They have been determined by a classical analysis. These values ($J_1 \simeq 36$ K, $J_2 \simeq 4$ K, $J_3 \simeq -18$ K for a $S = \frac{1}{2}$ spin hamiltonian) allow the existence of the ferrimagnetic structure described above for an intermediate field but lead, using MFA, to an ordering temperature (~ 20 K) much larger than the experimental one ($T_c = 5.4$ K). These results will be published in more detail. They are very promising for more extensive experimental studies and for theoretical developments.

References

[1] S. Eymond and A. Durif, Mater. Res. Bull. 4 (1969) 595.
[2] L.P. Regnault, 3e Cycle Thesis, University of Grenoble, France (1976).
[3] The experiments have been carried out at the CNRS and SNCI Grenoble, France.

FIELD DEPENDENCE OF THE MAGNETIZATION OF THE TWO-DIMENSIONAL ANTIFERROMAGNET K_2MnF_4

A.F.M. ARTS, C.M.J. VAN UIJEN and H.W. DE WIJN

Fysisch Laboratorium, Rijksuniversiteit, Utrecht, The Netherlands

Both the net magnetization and the sublattice magnetization of the 2D antiferromagnet K_2MnF_4 have been measured as a function of temperature in external magnetic fields up to the spin-flop value. Renormalization is found to become increasingly important with field.

In this paper we present, together with an interpretation in terms of spin-wave theory, data on both the sublattice and net magnetization of the two-dimensional Heisenberg antiferromagnet K_2MnF_4 ($T_N = 42.2$ K) in magnetic fields up to the spin-flop value. Experimentally these properties are reflected in the NMR frequencies of the out-of- and in-layer ^{19}F nuclei, $^{19}F^I$ and $^{19}F^{II}$, respectively, resonating in the hyperfine field of their magnetic neighbours. It is noted that the $^{19}F^{II}$ NMR can only be observed in a non-zero external field, which induces an unbalance between the sublattices.

The interpretation is based on a hamiltonian including Heisenberg exchange, a uniaxial staggered anisotropy field H_A, and an external magnetic field H_0, both fields being directed along the tetragonal c-axis [1]. The magnetization of the down and up sublattices is then given by

$$\langle S^z \rangle_T^{(1,2)} = S - \Delta_0 - \Delta S_1(H_0, T) \mp \Delta S_2(H_0, T), \quad (1)$$

where $S = \frac{5}{2}$ for Mn^{2+}, Δ_0 is the zero-point spin reduction, and $\Delta S_1(H_0, T)$, only weakly dependent on H_0, can be written

$$\Delta S_1(H_0, T) = \frac{1}{2N_0} \sum_k \frac{1+\alpha}{[(1+\alpha)^2 - \gamma_k^2]^{1/2}} (n_k^{(1)} + n_k^{(2)}). \quad (2)$$

Here $\alpha = g\mu_B H_A/4|J|S$, γ_k is a geometrical factor, and $n_k^{(1,2)} = [\exp(\epsilon_k^{(1,2)}/k_B T) - 1]^{-1}$ are the Bose occupation numbers of the two magnon modes. In the calculation we use the spin-wave dispersion $\epsilon_k^{(1,2)}$ with inclusion of higher-order corrections according to Oguchi [3] (see eq. (3) of [1]). The term $\Delta S_2(H_0, T)$, half the net magnetization, is given by

$$\Delta S_2(H_0, T) = \frac{1}{2N_0} \sum_k (n_k^{(1)} - n_k^{(2)}). \quad (3)$$

At zero field, Oguchi-type renormalization extends the range of concurrence between theory and the experimental sublattice magnetization by 3 K [2]. Since in a finite external field one branch of the spin-wave spectrum is lowered, and thus becomes more heavily populated at finite temperatures, spin-wave interactions are expected to be more pronounced than in the fieldless case.

In fig. 1 the decrement of the magnetization of the down sublattice in external fields of 10.1 kG and 20.0 kG is shown, as reflected in the shift of the NMR frequency relative to the frequency at zero field, with the latter calculated from [2]. The solid curve has been computed with renormalized spin-wave theory by use of the

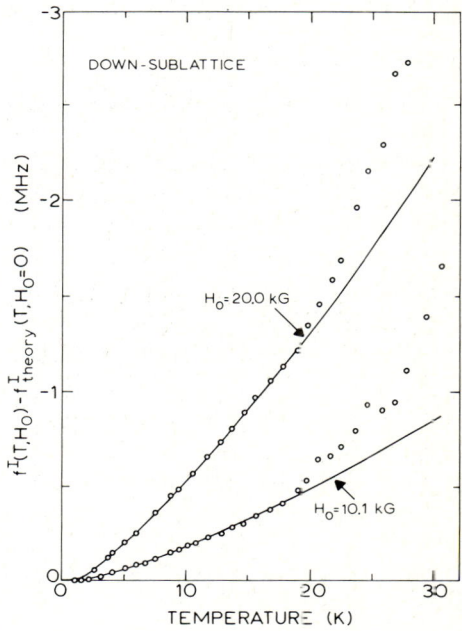

Fig. 1. Resonance frequencies vs. temperature of the $^{19}F^I$ nuclei adjacent to the down sublattice with reference to the frequency at zero-field. The solid curves are calculated from renormalized spin-wave theory.

spin-wave parameters of [2], $J/k_B = -8.41$ K and $T_G(0) = 7.54$ K. As in the zero-field case, experiment is in excellent agreement with theory up to, say, 18 K. With unrenormalized theory, however, the sublattice magnetization at 20 kG can only be described up to 9 K, in contrast with an upper limit of 15 K for zero field [2]. In fig. 2 the experimental results have been plotted for the net magnetization at $H_0 = 20.0$ kG and 50.0 kG, the latter being close to the spin-flop field. At 20 kG, spin-wave theory with the parameters of [2] again agrees with experiment up to 18 K, with an improvement of 9 K due to inclusion of renormalization. The data taken at 50.0 kG concur with theory to a substantially lower temperature of 10 K, whereas unrenormalized theory fits up to only 4 K.

The general conclusion may be summarized by saying that in a field, renormalization indeed becomes increasingly important. Good agreement between experiment and renormalized theory is obtained up to temperatures and fields, where the thermal decrement of $\langle S^z \rangle$ is only 7% in addition to 7% due to zero point motion, i.e. at a point where the number of magnons excited is still very small. Finally, from the inflection point in the net magnetization versus temperature one may deduce the development of the antiferromagnetic-paramagnetic transition temperature with field. This yields a decrement of T_N by 3 K in an external field of 50 kG.

This work has been supported by the Foundations F.O.M. and Z.W.O.

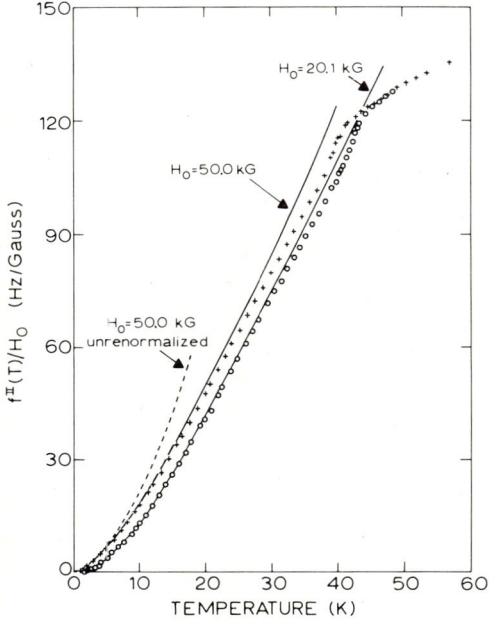

Fig. 2. Resonance frequencies of the $^{19}F^{II}$ nuclei, reflecting the net magnetization, vs. temperature at two external fields H_0. For comparison the results are divided by H_0. The solid curves are calculated from renormalized spin-wave theory. For $H_0 = 50.0$ kG, the result of a calculation with unrenormalized theory is also shown.

References

[1] H.W. de Wijn, L.R. Walker, S. Geschwind and H.J. Guggenheim, Phys. Rev. B8 (1973) 299.
[2] H.W. de Wijn, L.R. Walker and R.E. Walstedt, Phys. Rev. B8 (1973) 285.
[3] T. Oguchi, Phys. Rev. 117 (1960) 117.

LINEWIDTH OF TWO-MAGNON RAMAN SCATTERING IN TWO-DIMENSIONAL ANTIFERROMAGNETS AT FINITE TEMPERATURE

A. VAN DER POL, G. DE KORTE, G. BOSMAN, A.J. VAN DER WAL and H.W. DE WIJN

Fysisch Laboratorium, Rijksuniversiteit, Utrecht, The Netherlands

The temperature dependence of two-magnon Raman scattering in the quadratic-layer antiferromagnets K_2NiF_4 and K_2MnF_4 is found to agree with a higher-order Green function theory.

Raman scattering in antiferromagnetics by two-magnon processes provides information on spin waves near the Brillouin-zone edge. In the quadratic-layer Heisenberg antiferromagnets, experiments have to date been reported on K_2NiF_4 ($S = 1$) [1–3] and its isomorph K_2MnF_4 ($S = \frac{5}{2}$) [4]. Here we present, in addition to new experimental data for K_2MnF_4, the results of a higher-order Green function calculation in the two-dimensional (2D) antiferromagnetic systems. The hamiltonian, with inclusion of a staggered anisotropy field, reads

$$\mathcal{H} = J \sum_{1,m} S_1 \cdot S_m - g\mu_B H_A \left(\sum_1 S_1^z - \sum_m S_m^z \right), \quad (1)$$

where 1 and m run over the up and down sublattices, respectively, and the first summation is restricted to nearest neighbours on opposite sublattices. This leads to the renormalized one-magnon dispersion relation [5]

$$\bar{\Omega}_k(T) = \alpha(T)\Omega_k + \frac{(zJS)^2}{\Omega_k} \Delta(1+\Delta)[1-\alpha(T)], \quad (2)$$

in which Ω_k is the spin-wave energy without Oguchi corrections [6], $\Delta = g\mu_B H_A/zJS$, and the coefficient $\alpha(T)$ describes the renormalization of the spin-wave energy at temperature T. The Stokes scattering cross-section K as a function of the shifted energy ω is now obtained as the imaginary part of the Green function of the Raman transition operator of the two-magnon Raman scattering process, which may be reduced to [7]

$$K(\omega) = \frac{c}{1 - e^{-\omega/k_B T}} \alpha^2(T) S^2 \left(-\frac{1}{2\pi}\right) \mathrm{Im} \frac{L_0(\omega)}{1 - JL_0(\omega)}. \quad (3)$$

In a second-order theory we have for L_0 [7]

$$L_0(\omega) = -\frac{1}{N} \sum_k (\cos k_x a - \cos k_y a)^2 \frac{2n_k + 1}{\omega - 2\bar{\Omega}_k + 2i\Gamma_k}, \quad (4)$$

where the summation is over the 2D first Brillouin zone. The one-magnon damping constant Γ_k, which is absent in the first-order theory, is written into a mathematically manageable form by approximating the unrenormalized density of states by two rectangles. Over the interval $zJS(2\Delta)^{1/2} < \Omega_k < Y$, covering the larger part of the Brillouin zone, the density is equated to a constant ρ_1, while near the Brillouin zone edge, $Y < \Omega_k < zJS(1+\Delta)$, the density is ρ_2, with ρ_2 substantially larger than ρ_1. The parameters ρ_1, ρ_2, and Y are determined by a 2D version of the

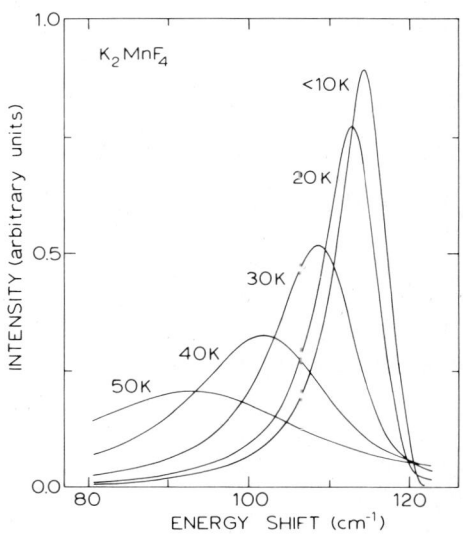

Fig. 1. Computed two-magnon Raman spectra in K_2MnF_4 at various temperatures. Below 10 K the curves coincide on the scale of the drawing.

rules given by Balucani and Tognetti [7].

Representative two-magnon spectra in K_2MnF_4, computed with the above scheme, are displayed in fig. 1. At this point it should be noted that there are no adjustable parameters in the calculations. The exchange constant J is taken from ref. 4. For Δ at zero temperature the values from antiferromagnetic resonance [9, 10] are adopted while the temperature dependence of Δ is assumed to scale with the square of the sublattice magnetization [9].

Comparison of second-order theory with experimental data on the width of the Raman peak is presented in fig. 2 for K_2MnF_4 and K_2NiF_4.

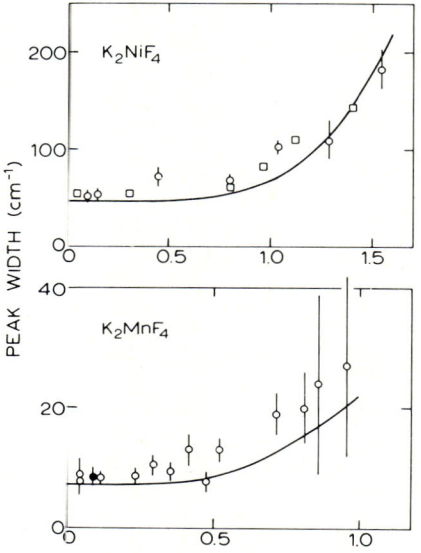

Fig. 2. Computed full width at half height of the two-magnon Raman-scattering line in K_2MnF_4 ($S = \frac{5}{2}$, $J = 8.41$ K, $T_N = 42.1$ K, while $\Delta = 0.0038$ at $T = 0$ K), and K_2NiF_4 ($S = 1$, $J = 102.1$ K, $T_N = 97.1$ K, while $\Delta = 0.0021$ at $T = 0$ K). The data on K_2MnF_4 are from the present paper, except the closed circle, which is taken from ref. 4. The data on K_2NiF_4 are from ref. 1 (○) and ref. 3 (□).

Apart from a single point at 4.2 K [4], the data on K_2MnF_4 were obtained by us with a conventional Raman scattering set-up, using a 488 nm 500 mW argon laser, a double monochromator and photon counting techniques, while the data on K_2NiF_4 are taken from refs. 1 and 3. The computed peak positions fit experiment equally well as in first-order theory. However, while first-order theory predicts at best a peak width constant with temperature, the one-magnon damping provided by second-order theory is seen to contribute substantially to the peak width at high temperature. There remains a slight tendency to underestimate the experimental values, which at low temperatures could possibly be explained by the finite resolving power of the spectrometer. For our data on K_2MnF_4 the instrumental profile is ~ 2.5 cm^{-1} wide; for K_2NiF_4 [1, 3] the instrumental width has unfortunately not been reported. In summary, the temperature dependence of the two-magnon Raman spectrum in 2D antiferromagnets appears to be in concurrence with second-order Green function theory up to temperatures above T_N.

This work has been supported by the Foundations F.O.M. and Z.W.O.

References

[1] P.A. Fleury and H.J. Guggenheim, Phys. Rev. Lett. 24 (1970) 1346.
[2] S.R. Chinn, H.J. Zeiger and J.R. O'Connor, Phys. Rev. B3 (1971) 1709.
[3] D.J. Toms, W.J. O'Sullivan and H.J. Guggenheim, Solid State Commun. 14 (1974) 715.
[4] W. Lehmann and R. Weber, Phys. Lett. A45 (1973) 33.
[5] S.R. Chinn, R.W. Davies and H.J. Zeiger, Phys. Rev. B4 (1971) 4017.
[6] T. Oguchi, Phys. Rev. 117 (1960) 117.
[7] U. Balucani and V. Tognetti, Phys. Rev. B8 (1973) 4247.
[8] H.W. de Wijn, L.R. Walker and R.E. Walstedt, Phys. Rev. B8 (1973) 285.
[9] H.W. de Wijn, L.R. Walker, S. Geschwind and H.J. Guggenheim, Phys. Rev. B8 (1973) 299.
[10] R.J. Birgeneau, F. DeRosa and H.J. Guggenheim, Solid State Commun. 8 (1970) 13.

CRITICAL BEHAVIOUR OF Mn(CH$_3$COO)$_2\cdot$4H$_2$O

P. BEAUVILLAIN and J.P. RENARD

Institut d'Electronique Fondamentale, Laboratorie associé au CNRS, Bâtiment 220, Université Paris-Sud, 91405 Orsay-Cedex, France

We have measured the parallel and perpendicular magnetic susceptibility from 1.2 to 300 K and the spontaneous magnetization down to 0.4 K of Mn(CH$_3$COO)$_2\cdot$4H$_2$O. The critical exponents $\gamma = 1.59$ and $\beta = 0.16$ are near to the theoretical values for a two-dimensional Ising ferromagnet; crossovers in the spin and lattice-dimensionality are also observed.

The crystal structure of Mn(CH$_3$COO)$_2\cdot$4H$_2$O consists of Mn acetate layers (ab planes) well separated from each other by H$_2$O layers. The unit cell parameters are $a = 9.099$ Å, $b = 17.60$ Å, $c = 19.56$ Å and $\beta = 94.52°$ [1–3]. Flippen and Friedberg [4] first studied the three-dimensional magnetic order ($T_c = 3.18$ K). They found that the saturation magnetization along the three orthogonal axes a, b, c^* is equal to $5/3\,\mu_B$ instead of the value of $5\,\mu_B$ expected for Mn^{2+}. The neutron diffraction study [3] has confirmed the existence of antiferromagnetic Mn$_2$–O–Mn$_1$–O–Mn$_2$ groups (triplets). The Mn triplets belonging to the same ab plane are mutually coupled by ferromagnetic exchange while the weak interaction between adjacent Mn layers is antiferromagnetic.

We measured the magnetic susceptibilities of single crystals in the temperature range 1.2–300 K by means of a mutual inductance bridge operating at 70 Hz. The a.c. field was kept as low as 2 Oe. Between 1.2 and 4.2 K, the sample was immersed in a pumped ^{4}He bath. Temperature was measured with a 100 Ω Allen Bradley resistor anchored to the sample by a bundle of 50 copper wires. This resistor was first calibrated against ^{4}He vapour pressure and CrK alum susceptibility. It was carefully recalibrated during each experimental run against (superfluid) ^{4}He vapour pressure. Above 4.2 K, the sample was set in a small glass cryostat and thermally linked to the ^{4}He bath by 1 torr ^{3}He gas and was heated by a Ag–Au resistor up to 30 K. Temperature was measured from a carbon resistor calibrated against Cr–K alum susceptibility. Above 30 K, we used a cold ^{4}He gas flow cryostat and measured temperature by a copper constantan thermocouple. The susceptibility measurements were performed along the easy axis a and along the b and c^* axes. Below 4.2 K, χ_a exhibits a net maximum and decreases abruptly below $T_c = 3.18 \pm 0.01$ K.

This behavior was previously observed by Friedberg et al. [4]. The measured susceptibilities along a, b and c^* corrected for shape effect are plotted versus $\epsilon = T/T_c + 1$ on logarithmic scales in fig. 1. Experimental errors on $T - T_c$ are estimated to be 1 mK near T_c. The non-divergence of χ_b and χ_{c^*} is clearly due to the magnetic anisotropy of the salt. In the reduced temperature range $\epsilon = T/T_c - 1 < 2 \times 10^{-1}$, the Ising-type anisotropy is predominant. In the range $6 \times 10^{-2} < \epsilon < 3 \times 10^{-1}$, the experimental results are well described by the law $\chi_a \sim \epsilon^{-\gamma}$ with $\gamma = 1.59$. This effective value of γ is rather near the theoretical γ value of the two-dimensional Ising model equal to 1.75. Below $\epsilon = 6 \times 10^{-2}$ three-dimensional antiferromagnetic interaction becomes effective and prevents χ_a from diverging.

In order to estimate the lattice dimensionality

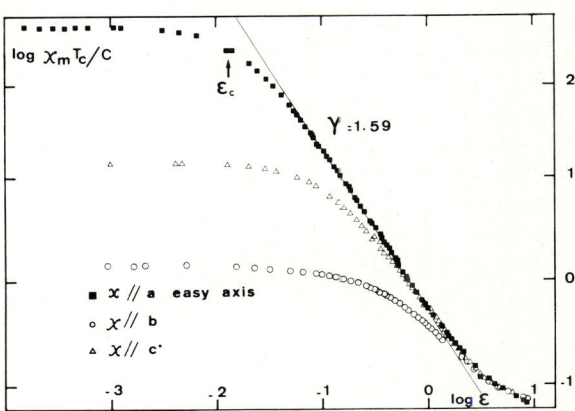

Fig. 1. Log $\chi_m T_c/C$ vs. log ϵ. T_c is the Curie temperature, C is the Curie–Weiss constant for "triplets", χ_m is the magnetic susceptibility per gram corrected from shape effect, ϵ is the reduced temperature $T/T_c - 1$. ϵ_c is the reduced temperature of the crossover.

crossover temperature ϵ_c we estimated the intralayer exchange field H_{ex} from the critical temperature [5,6] and the interlayer exchange field H'_{ex} from the field induced transition at low temperature. Using $\epsilon_c = R^{1/\varphi}$ with $R = H'_{ex}/H_{ex} = 8 \times 10^{-4}$ and $\varphi = \gamma I_{2d} = 1.75$, gives $\epsilon_c = 1.6 \times 10^{-2}$ in good agreement with the experimental value.

We have performed NMR experiments on single crystals with the radio-frequency field along b. The NMR signals were detected by using a frequency swept marginal oscillator. We have observed the dependence of the proton NMR frequency with temperature in the range 1.2–3.18 K from 2.5 to 5.2 MHz and from 10 to 13.5 MHz. The seven observed NMR lines have not exactly the same variation with temperature. This might be due to small displacements of water and acetate molecules leading to a slight temperature dependence of proton-manganese interactions. Nevertheless this may be neglected because of the restricted range of the critical region and we have a good estimate of the critical exponent β from the NMR frequency vs. temperature. The measurements were performed on the narrow structureless CH_3 line of frequency 4.75 MHz at 1.2 K previously observed by Spence [7]. In fig. 2 we have plotted the frequency vs. temperature on logarithmic scale; in the range $3 \times 10^{-2} < |\epsilon| < 6 \times 10^{-1}$ the spontaneous magnetization deduced from proton NMR is proportional to $|\epsilon|^\beta$ with $\beta = 0.16 \pm 0.01$.

By magnetization measurements with a continuously recording pickup-coil-flux integration device we have also [7,8] studied the antiferroferrimagnetic transition of $Mn(CH_3COO)_2 \cdot 4H_2O$. We did not observe the irreversible behaviour of the magnetization isotherms previously reported by Schmidt and Friedberg. We have measured the transition field vs. temperature down to 0.4 K. The extrapolated value of this field to 0 K is $H'_{ex} = 6.10 \pm 0.02$ Oe. The critical behaviour of the metamagnetic transition field is

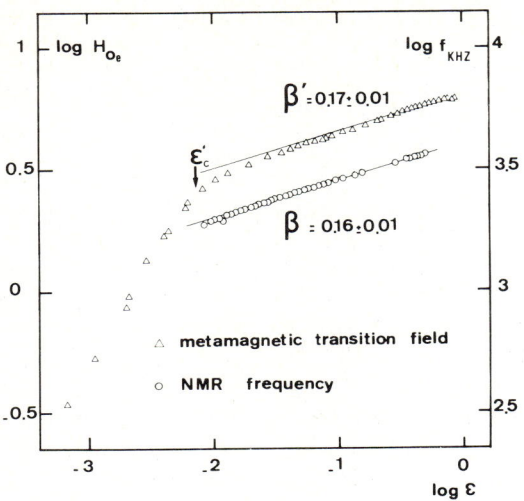

Fig. 2. $\log H$ and $\log f$ vs. $\log \epsilon$. H is the metamagnetic transition field, f the NMR frequency and ϵ is the reduced temperature $T/T_c - 1$. ϵ'_c is the reduced temperature of the crossover.

also given in fig. 2. A power law dependence $|\epsilon|^{\beta'}$ in the range of temperature: $3 \times 10^{-2} < |\epsilon| < 6 \times 10^{-1}$ is observed with $\beta' = 0.17 \pm 0.01$. The metamagnetic transition field appears to be proportional to the spontaneous magnetization and we could show the existence of a crossover with lattice dimensionality in the order state for the spontaneous magnetization at $\epsilon = 6 \times 10^{-3}$.

References

[1] H. Abe and H. Morigaki, Proc. 1st Int. Conf. on Paramagnetic Resonance, Jérusalem (1962).
[2] P. Burlet and E.F. Bertaut, Solid State Commun. 14 (1974) 665.
[3] P. Burlet, Thèse de Doctorat d'Etat, Grenoble (1973).
[4] P.B. Flippen and S.A. Friedberg, Phys. Rev. 121 (6) (1961) 1591.
[5] H.E. Stanley and T.A. Kaplan, Phys. Rev. Lett. 17 (1966) 913.
[6] M.E. Lines, Phys. Rev. B3 (1971) 1749.
[7] R.D. Spence, J. of Chem. Phys. 62 (9) (1975) 3659.
[8] V.A. Schmidt and S.A. Friedberg, Phys. Rev. 188 (2) (1969) 809.

ANTIFERROMAGNETIC, SPIN FLOP AND PARAMAGNETIC RESONANCES OF HEXAHYDRATED COBALT CHLORIDE AT VERY LOW TEMPERATURES

W. GHIDALIA

Laboratoire de Physique XII, Université Pierre et Marie Curie, Tour 23, 4 place Jussieu, 75230 Paris Cedex 05, France

J. TUCHENDLER and J. MAGARIÑO

Laboratoire de Physique des Solides de l'Ecole Normale Supérieure 24, rue Lhomond, 75231 Paris Cedex 05, France

From the EPR diagram at $T = 4.2$ K (80 GHz $< \omega <$ 310 GHz) we deduce $g_c = 4.04$ and $g_b = 4.25$. The difference with the known values is attributed to exchange. At $T = 1.5$ K and for each axis, we observe two similar sets of resonances that we explain by a four-sublattices magnetic structure model.

The hexahydrated cobalt chloride is well known as an *XY* model [1]. This salt has been studied by antiferromagnetic resonance [2], susceptibility [3–5], heat capacity measurements [6] and neutron diffraction experiments [7]. However, persistent discrepancies exist between these experimental results. This situation arises probably from the crystallographic and/or the magnetic structure.

$CoCl_2 \cdot 6H_2O$ is monoclinic with the space group C2/m. The crystallographic *ac* plane is the mirror and the *b* axis is the axis of symmetry, $\beta = 122°19'$ [8]. At temperatures lower than $T_N = 2.29$ K, this salt has an antiferromagnetic structure with magnetizations in the *ac* plane probably close to the *c* axis.

We report the results of magnetic resonance experiments performed in a large range of frequencies (80 GHz $\leq \omega \leq$ 310 GHz), in d.c. magnetic fields up to 57.5 kOe.

In the paramagnetic state, at $T = 4.2$ K, this gives an accurate determination of the *g* factors. At $T = 1.5$ K, the complete diagram of antiferromagnetic, spin–flop and paramagnetic resonances, depending on the d.c. magnetic field, are observed. The experimental set up is described elsewhere [9].

Fig. 1 shows the frequency dependence of the observed paramagnetic resonances as a function of the magnetic field *H* along the *c* and *b* axes. Figs. 2 and 3 show similarly the frequency dependence of the resonances below the Néel temperature. The resonances (1, 3) were already known [2].

The resonances (2) and (4) are observed for the first time. They occur in the spin–flop region at about 138 GHz and remain constant in frequency for a large range of magnetic field.

We note on the same figures the presence of a new set of resonances (1', 3') already seen in previous experiments [10]. For high values of ω and H, several resonances which most likely are magnetostatic modes are observed.

We interpret these results by means of Nagamiya and Yosida's phenomenological

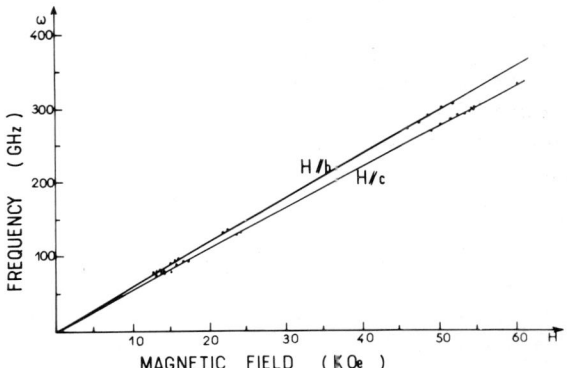

Fig. 1. Paramagnetic resonance diagram $\omega = f(H)$ for *c* and *b* axes, with $T = 4.2$ K $> T_N$.

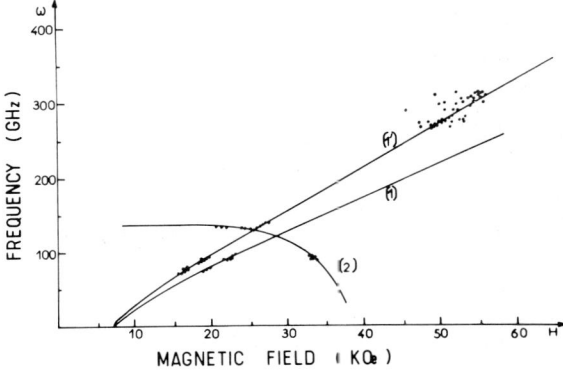

Fig. 2. Resonance diagram $\omega = f(H)$ for *c* axis with $T = 1.5$ K $< T_N$.

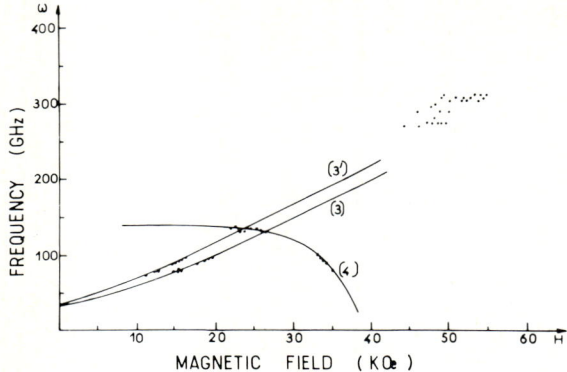

Fig. 3. Resonance diagram $\omega = f(H)$ for b axis with $T = 1.5\,\text{K} < T_N$.

theory including anisotropic exchange. With our experimental g-factors, $g_c = 4.04$ and $g_b = 4.25$ and by fitting the antiferromagnetic resonances (1, 2, 3, 4) we obtain the exchange parameters. From these we get the Curie–Weiss temperature $\theta = 5.4\,\text{K}$, which is not in agreement with the value $\theta = 8.9\,\text{K}$ obtained from the susceptibility measurements [5].

In fact, the g-values measured by paramagnetic resonance at 4.2 K are g-factors modified by exchange near the Néel temperature:

$$g_{b\,\text{meas}} = g_b\{1 + \tfrac{1}{2}(P_x - P_y)\chi\}\{1 + \tfrac{1}{2}(P_z - P_y)\chi\}, \quad (1)$$

$$g_{c\,\text{meas}} = g_c\{1 + \tfrac{1}{2}(P_y - P_x)\chi\}\{1 + \tfrac{1}{2}(P_z - P_x)\chi\}, \quad (2)$$

where the high temperature g-factors are $g_b = 4.9$ and $g_c = 4.93$ [3].

In this case the best fit of the antiferromagnetic diagram (1, 2, 3, 4) is obtained for the following values of the exchange parameters:

$A = 870 \quad A'_x = 270 \quad A'_y = 225 \quad A'_z = -495$

$\Gamma = 495 \quad \Gamma'_x = 165 \quad \Gamma'_y = 155 \quad \Gamma'_z = -320$

in e.m.u. per gram.

From these values we deduce $\theta = 8.6\,\text{K}$ and for the intralayer interactions ratios: $j_x/j_y = 1.04$, $j_z/j_x = 0.32$ and $j_x/j_y = 1.02$, $j_z/j_x = 0.26$, respectively, calculated from the A, A' and Γ, Γ' factors, in good agreement with previous experiments [1].

The value of $\chi = 5.4 \times 10^{-4}$ e.m.u., obtained by substitution of these parameters in eqs. (1) and (2), is in good agreement with the value $\chi = 4.9 \times 10^{-4}$ e.m.u. obtained at $T = 1.5\,\text{K}$ in the paramagnetic phase at high magnetic field.

To explain the existence of the second set of resonances (1') and (3') (also observed in preliminary experiments on $CoBr_2 \cdot 6H_2O$), we suggest the following hypothesis. There are two Co^{2+} ions magnetically inequivalent according to the relation to the next-nearest water molecules [3], for which the superexchange interactions may be different and this has no influence on the g-factors in accordance with our EPR results. This difference in the superexchange interactions generates two directions of easy magnetization (four crossed sublattices) near each other and lying in the ac plane near the c axis.

References

[1] J.W. Metselaar, D. de Klerk and C.J. de Jongh, Physica 63 (1973) 191.
[2] M. Date, J. Phys. Soc. Jap. 16 (1961) 1337.
[3] T. Haseda, J. Phys. Soc. Jap. 15 (1960) 483.
[4] J.E. Rives and S.N. Bhatia, Phys. Rev. B12 (1975) 1920.
[5] J.P. Renard and B. Lecuyer, unpublished results (private communication).
[6] J. Skalyo Jr. and S.A. Friedberg, Phys. Rev. Lett. 13 (1964) 133.
[7] R. Kleinberg, J. Chem. Phys. 53 (1970) 2660.
[8] J. Mizuno, K. Ukei and T. Sugawara, J. Phys. Soc. Japan 14 (1959) 383.
[9] J. Magariño, J. Tuchendler, J.P. D'Haenens and A. Linz, Phys. Rev. B13 (1976) 2805.
[10] H. Benoi and W. Ghidalia, Physica 78 (1974) 233.

SOME MAGNETIC PROPERTIES OF THE PSEUDO ONE-DIMENSIONAL ISING-LIKE MAGNETIC SYSTEMS CsCoCl$_3$·2H$_2$O AND RbFeCl$_3$·2H$_2$O

K. KOPINGA, Q.A.G. VAN VLIMMEREN, A.L.M. BONGAARTS and W.J.M. DE JONGE

Department of Physics, Eindhoven University of Technology, Eindhoven, The Netherlands

The heat capacity of CsCoCl$_3$·2H$_2$O and RbFeCl$_3$·2H$_2$O has been analyzed using a $d=1$ and $d=2$ Ising model, respectively, yielding $|J/k| = 38.6$ K for CsCOCl$_3$·2H$_2$O and $|J/k| = 39$ K, $|J'/k| = 0.7$ K for RbFeCl$_3$·2H$_2$O. RbFeCl$_3$·2H$_2$O shows two metamagnetic transitions in applied fields along the c axis.

CsCoCl$_3$·2H$_2$O and RbFeCl$_3$·2H$_2$O belong to a series of isomorphic transition metal halides. All members of this series, including the well-known CsMnCl$_3$·2H$_2$O, show pronounced linear chain ($d=1$) characteristics with ratios of interchain to intrachain exchange interactions of 10^{-2}–10^{-3}. In this paper we will report the results of specific-heat measurements on both compounds and some preliminary results of magnetization and susceptibility measurements on the iron isomorph.

In the past few years, CsCoCl$_3$·2H$_2$O has been the subject of a number of investigations [1–4]. The canted magnetic structure in the ordered state and the small interchain interactions give rise to an interesting low-field metamagnetic transition including a multicritical point, which has been the subject of detailed neutron-diffraction investigations. The susceptibility results and neutron-scattering results indicate a strongly 1d behaviour above T_N. So far the interpretation of the specific-heat results was hampered by the fact that no diamagnetic isomorph was available.

We investigated the overall magnetic heat capacity of CsCoCl$_3$·2H$_2$O by subtracting a scaled lattice heat capacity of CsMnCl$_3$·2H$_2$O [5] from the experimental data, using a simple temperature-independent scaling factor. A least-squares fit of the Ising model and the lattice contribution to the experimental data above 9 K, by varying both the exchange parameter J_a and the scaling factor, yielded $|J_a/k| = 38.6$ K. The result is plotted in fig. 1. The magnetic heat capacity below 9 K is rather high compared to the theoretical prediction, which is also reflected in the value of the evaluated magnetic entropy increase, which is too high by $\sim 10\%$. This might indicate the existence of non-zero transverse components of the exchange interaction. The

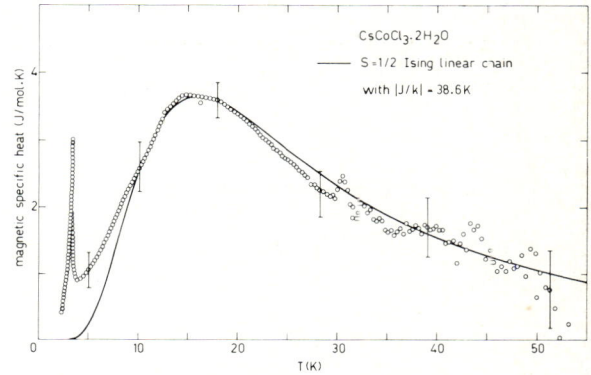

Fig. 1. Magnetic specific heat of CsCoCl$_3$·2H$_2$O. The open circles are the experimental data points corrected for the lattice contribution. The drawn curve denotes the theoretical prediction. Errorbars reflect the uncertainty in the evaluation of the lattice heat capacity of CsMnCl$_3$·2H$_2$O, which has been used in the scaling procedure.

magnitude of the intrachain interaction is consistent with measurements of the correlation length in the a direction by means of neutron-scattering experiments, which give $|J_a/k| > 25$ K [3]. From the metamagnetic transition in applied fields along the a axis an interchain coupling $J_b/k = -0.1$ K was deduced [2]. The magnitude of the ferromagnetic coupling in the c direction J_c can be estimated from the location of the multicritical point in the magnetic phase diagram [3]; $J_c/k = 0.2 \pm 0.2$ K. The ratio of inter- to intrachain interactions $(|J_b|+|J_c|)/|J_a| \sim 8 \times 10^{-3}$ indicates that this substance has the same degree of one-dimensionality as the isomorphic CsMnCl$_3$·2H$_2$O.

The magnetic properties of RbFeCl$_3$·2H$_2$O have not yet been reported. X-ray experiments show that RbFeCl$_3$·2H$_2$O is isomorphic with CsCoCl$_3$·2H$_2$O. NMR and magnetization experiments suggest a canted array with the magnetic moments situated in the ac plane, like

$CsCoCl_3 \cdot 2H_2O$, but individual chains having a net moment along c instead of a. Measurements of the magnetization in the ordered state reveal two metamagnetic transitions in applied fields along the c axis at 8.1 and 12.4 kOe, respectively. Fig. 2 shows that the magnitude of the discontinuities in the magnetization at both metamagnetic transitions is equal. From these measurements and from preliminary results of *spin cluster resonance*, an Ising-like anisotropy at low temperatures can be concluded.

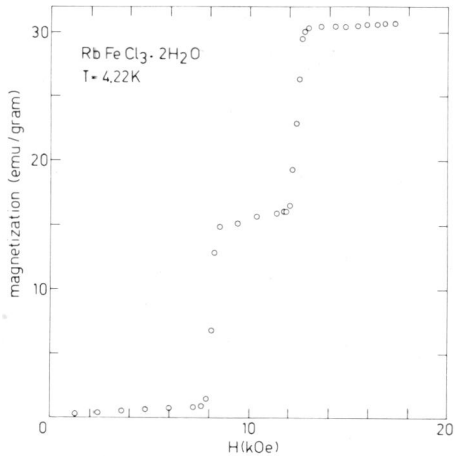

Fig. 2. Magnetization of $RbFeCl_3 \cdot 2H_2O$ versus applied field along the c axis at $T = 4.22$ K.

The overall magnetic heat capacity was investigated by subtracting a scaled lattice heat capacity of the isomorphic $\alpha RbMnCl_3 \cdot 2H_2O$. The resulting magnetic contribution could not be represented satisfactorily by a purely one-dimensional model system. However, as is shown in fig. 3, the rectangular $S = \frac{1}{2}$ Ising model with $|J'/J| = 1.8 \times 10^{-2}$ and $|J/k| = 39$ K was found to give a remarkably good description both above and below the ordering temperature. The apparent small inconsistency with this model close to the three-dimensional ordering temperature is probably caused by the remaining small interchain interactions, which have not been taken into account by the theoretical model. These interactions tend to shift the actual ordering towards slightly higher temperatures. One should bear in mind that the treatment given above is somewhat tentative and further experimental evidence is needed in order to determine the ratio of the various interchain interactions.

A more detailed analysis of the heat capacity in the paramagnetic region suggests that one or more excited levels of the Fe^{2+} single-ion spectrum may be present at ~ 80 K above the ground-state doublet. This is consistent with preliminary measurements of the susceptibility at room temperature, which can be described with $g \sim 2$ and $S = 2$. Since it is likely that in this compound the orbital degeneracy of the Fe^{2+} ion has been removed completely, this indicates that all levels of the Fe^{2+} spin-quintet are populated at room temperature.

One of the authors (Q.A.G. v. V) gratefully acknowledges financial support from "Stichting voor Fundamenteel Onderzoek der Materie" (F.O.M.) and the "Nederlandse Organisatie voor Zuiver Wetenschappelijk Onderzoek" (Z.W.O.).

References

[1] W.J.M. de Jonge, K.V.S. Rama Rao, C.H.W. Swüste and A.C. Botterman, Physica 51 (1971) 620.
[2] A. Herweijer, W.J.M. de Jonge, A.C. Botterman, A.L.M. Bongaarts and J.A. Cowen, Phys. Rev. B5 (1972) 4618.
[3] A.L.M. Bongaarts, W.J.M. de Jonge and P. van der Leeden, Phys. Rev. Lett. 37 (1976) 1007.
A.L.M. Bongaarts and W.J.M. de Jonge, to be published in Phys. Rev. B15.
A.L.M. Bongaarts, Thesis, Eindhoven (1975).
[4] N. Thorup and H. Soling, Acta Chem. Scand. 23 (1969) 2933.
[5] K. Kopinga, T. de Neef and W.J.M. de Jonge, Phys. Rev. B11 (1975) 2364.

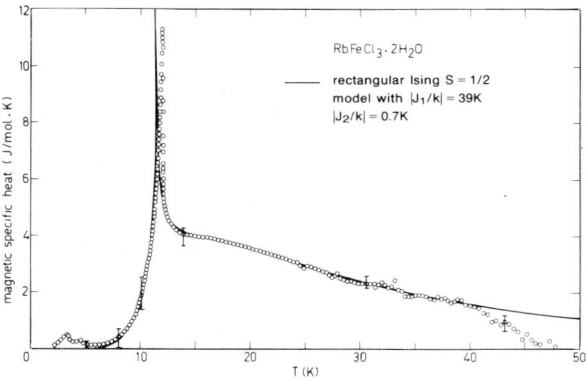

Fig. 3. Magnetic specific heat of $RbFeCl_3 \cdot 2H_2O$. The open circles are the experimental data points corrected for the lattice contribution. The drawn curve denotes the theoretical prediction. Errorbars reflect the uncertainty in the evaluation of the lattice heat capacity of $\alpha RbMnCl_3 \cdot 2H_2O$, which has been used in the scaling procedure. The small anomalies at 3.3 and 4.6 K are due to sample impurities.

FIR-INVESTIGATIONS ON THE ONE-DIMENSIONAL FERROMAGNET CsNiF$_3$

R.J. GRILL, U. DÜRR and R. WEBER*

Physikal. Institut der Universität Stuttgart, Teilinstitut 2, Pfaffenwaldring 57, 7000 Stuttgart 80, W. Germany

The magnetic field and temperature dependence of the ferromagnetic resonance of the 1d easy plane type magnet CsNiF$_3$ was determined by far infrared techniques. The single ion anisotropy of the Ni^{2+} ions turned out to be $D = 3.3$ cm^{-1}. The g-factor for the magnetic field within the easy plane was found to be 2.28. The temperature dependence of the position of the resonance can be satisfactory explained by spin wave renormalization.

1. Introduction

CsNiF$_3$ crystallizes in the hexagonal BaNiO$_3$ structure, corresponding to the space group P6$_3$/mmc [1]. The Ni-ions are surrounded by trigonally distorted fluorine octahedra which, sharing opposite faces, form chains along the hexagonal c-axis. The single ion anisotropy favours an orientation of the spins within a plane perpendicular to the c-axis. As the in-plane anisotropy is negligible the system has an easy magnetization plane.

Neutron diffraction and magnetization measurements [2] showed that above the three-dimensional antiferromagnetic ordering temperature ($T_N = 2.61$ K) there exists a one-dimensional (1d) ferromagnetic short range order along the chains. The magnon dispersion relation for a 1d easy plane ferromagnet with a magnetic field H applied perpendicular to the c-axis has the following form [3, 4]:

$$\hbar\omega = \left[\left\{4JS\left(1 - \cos k\frac{a}{2}\right) + 2DS + g_\perp\mu_B H_\perp\right\}\left\{4JS\left(1 - \cos k\frac{a}{2}\right) + g_\perp\mu_B H_\perp\right\}\right]^{1/2}. \quad (1)$$

From inelastic neutron scattering experiments Steiner [2] has proved the existence of spin waves in the region $2.5 < T < 20$ K. He determined the dispersion-curve except for small values of wave vector k.

Here we report on the first far infrared investigation on the ferromagnetic resonance (FMR) (corresponding to the excitation of a spin wave with $k \cong 0$) of a one-dimensional easy plane magnet.

* Present address: Fachbereich Physik, Universität Konstanz.

2. The magnetic field dependence of the FMR

For the measurements we used a Fourier spectrometer of the lamellar grating type, in connection with a ^{3}He cooled bolometer. The low-frequency-limit of the instrument was about 3 cm^{-1}.

The dependence of the resonance frequency on the magnetic field at 4.2 K is shown in fig. 1. The values below 3 cm^{-1} are taken from microwave measurements by Rosinski et al. [5]. The solid line represents a least square fit of the theoretical field dependence of the FMR, given by eq. (1) to our experimental data. For $k = 0$ eq. (1) reads

$$\hbar\omega = \sqrt{g_\perp\mu_B H_\perp(2DS + g_\perp\mu_B H_\perp)}. \quad (2)$$

The parameters D and $g_\perp$ were determined to be

$D = (3.3 \pm 0.1)$ cm^{-1} $\{D/k_B = (4.7 \pm 0.2)$ K$\}$,
$g_\perp = 2.28 \pm 0.02$.

The value of the anisotropy parameter, D, is in good agreement with results obtained by inelas-

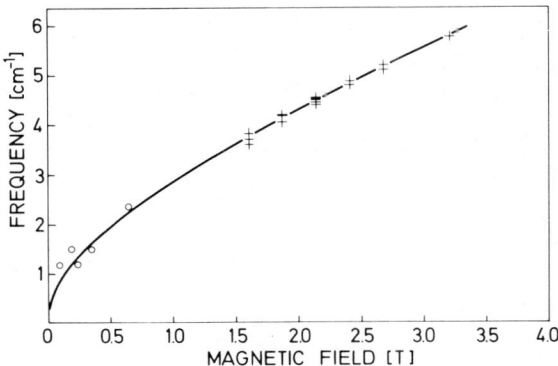

Fig. 1. Magnetic field dependence of the FMR at 4.2 K. The solid line represents the theoretical curve for $D = 3.3$ cm^{-1} and $g_\perp = 2.28$. ○ microwave results [5]. + this work.

tic neutron scattering, reported by Steiner [2], $D/k_B = (4.9 \pm 1)$ K.

For the external field $H \| c$ no magnetic field dependent absorption was found up to 3 T. Calculations show that the resonance line should lie at zero frequency for fields smaller than $H = 2D/g_\| \mu_B$.

3. The temperature dependence of the FMR

Investigations have been performed in the temperature range $1.5 \text{ K} < T < 14 \text{ K}$. The temperature of the exchange gas coupled crystal was kept constant within 0.2 K.

Within the accuracy of our measurements no influence of the three-dimensional phase transition could be observed. This can be understood from the magnetic phase diagram of $CsNiF_3$, reported by Steiner et al. [6]. If the external field exceeds a critical field H_c, which is of the order of 0.3 T at 1.5 K, the antiferromagnetic system undergoes a metamagnetic phase transition to a ferromagnetic phase with the spins parallel to the external field. The temperature-dependence measurements were performed at an external field of 2.675 T, which means that the system was in the ferromagnetic phase even at the lowest temperature.

Fig. 2 shows the temperature-dependence of the resonance frequency. The decrease of the magnon energy $E_{(T)}$ with increasing temperature can be described by means of the Loudon–Keffer renormalization of the spin waves [7].

We have calculated $E_{(T)}$ by an iterative method by means of the dispersion relation, eq. (1), in connection with the above reported values of D and $g_\perp$. For J we used the value of $J/k_B = 11.8$ K, determined by Steiner [2]. The result is shown as solid line in fig. 2. The agreement with the experimental data is quite good.

The integrated intensity of the resonance line increases with increasing temperature. A possible explanation is given by the decreasing correlation length of the ferromagnetically coupled spins in the chain when the temperature is raised. The total spin component with respect

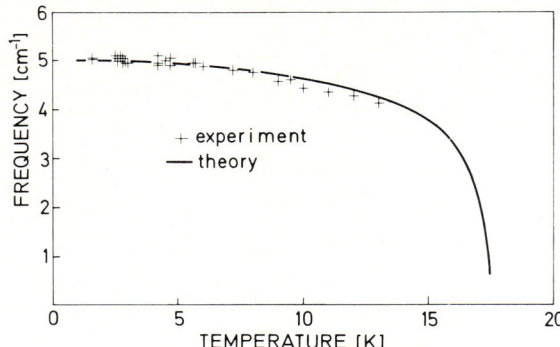

Fig. 2. Temperature dependence of the FMR at constant external magnetic field ($H_\perp = 2.675$ T). The solid line shows the result of a spin wave renormalization calculation.

to the quantisation axis can only be changed by integral numbers. This means in a semiclassical picture, that the precession amplitude (u) of the excited correlated spins depends on the number of correlated spins:

$$u = (1/\sqrt{N}) \text{ const.}$$

Hence, the intregrated intensity depends inversely on the number of correlated spins. This is in good agreement with our measurements.

The authors thank Prof. Pick for his interest and support of this work. Financial support by the Sonderforschungsbereich 67 (Defektstrukturen in festen Stoffen) is gratefully acknowledged.

References

[1] L. Babel, Z. Anorg. Allg. Chemie 369 (1969) 117.
[2] See, for example, M. Steiner, Int. J. Magn. 5 (1973) 95.
[3] J.V. Lebesque, J. Snel and J.J. Smit, Solid State Commun 13 (1973) 371.
[4] J. Villain, J. Phys. C6 (1973) L97 and J. de Physique 35 (1974) 27.
[5] Ch. Rosinski, B. Elschner, to be published in J. Magn. Magn. Mater.
[6] M. Steiner and H. Dachs, Solid State Commun. 14 (1974) 841.
[7] F. Keffer, in: Handbuch der Physik, Vol. XVIII/2 (Springer Verlag, Berlin/Heidelberg/New York, 1966).
[8] M. Steiner, B. Dorner and J. Villain, J. Phys. C8 (1975) 165.

ANTIFERROMAGNETIC STRUCTURE IN ADSORBED O_2 MONOLAYERS

M. NIELSEN and J.P. McTAGUE*

Research Establishment Risø, Roskilde, Denmark

Neutron diffraction from monolayers of O_2 adsorbed on graphite shows structural arrangements similar to the dense planes of bulk O_2. At monolayer completion and above, a magnetic superlattice reflection shows well-developed antiferromagnetic order for $T \leq 10$ K. The submonolayer phase also shows signs of antiferromagnetic correlations at low temperature.

It has been known for some time that the low temperature phase of bulk O_2 (α-O_2) is antiferromagnetic [1], and the magnetic structure has been directly observed by neutron diffraction. O_2 is particularly interesting as it is the only known direct overlap antiferromagnetic insulator. Since thermodynamic and neutron scattering studies [2] show that monolayers adsorbed on Grafoil, a very uniform high surface area graphite, behave as essentially two-dimensional (2-d) systems we have made neutron diffraction investigations of the crystal structure and magnetic order in adsorbed O_2 films as functions of temperature and coverage.

Fig. 1 shows neutron groups from structural peaks around $Q = 2.1$–2.3 Å^{-1} at $T = 4.2$ K. The lower coverage, $\rho = 1.57 \rho_0$, is 90% of a monolayer for the observed lattice spacing, while the $\rho = 1.96 \rho_0$ one has 105% coverage. The film density ρ_0 which is used as a unit of the coverage is defined as one O_2 adsorbed molecule per three surface graphite hexagons. At $\rho = 1.96 \rho_0$ there is a well-developed splitting in this peak, indicating two distinct sets of near neighbors arranged like those in the closest packed $(a-b)$ plane of bulk $\alpha - O_2$ [3]. The weak peak at half the momentum transfer of the high Q structural peak confirms that the monolayer is antiferromagnetic, with four nearest neighbors ($d_0 = 3.20$ Å) antiferromagnetically coupled, and two ferromagnetic near neighbors ($d_1 = 3.40$ Å). The sublattice magnetization points along the direction of the ferromagnetic neighbors (see fig. 2). Both the structural and magnetic peak widths indicate a correlation length $L \sim 125$ Å. This is known to be the upper limit imposed by the substrate.

As the temperature increases both nuclear and magnetic features evolve continuously in the range 9–11 K (see fig. 3) to a structure (β-O_2) having all six neighbors equidistant at $d_0 = 3.27$ Å, and no long range magnetic order. For comparison, a 2D $S = 1$ Heisenberg system with four nearest neighbors and with $J = 5.75$ K [4], the bulk exchange constant for O_2, has a diver-

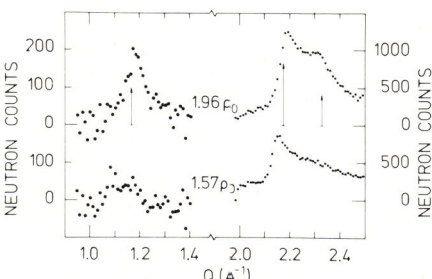

Fig. 1. Neutron diffraction from O_2 adsorbed on Grafoil at $T = 4.2$ K. The numbers in the figure give the coverages in units of ρ_0, which corresponds to one O_2 molecule adsorbed per 3 carbon hexagons on the graphite surface. The groups to the right show the difference spectrum $I(Q)$ (Grafoil + O_2) − $I(Q)$ (Grafoil) while the groups to the left are $I(Q, T = 4.2$ K$) - I(Q, T = 20$ K$)$ (Grafoil + O_2). Note the five-fold difference in scales for the two regions. The three arrows indicate the positions and the relative intensities of Bragg peaks from monolayers of O_2 identical to the $(a-b)$ planes of bulk α-O_2.

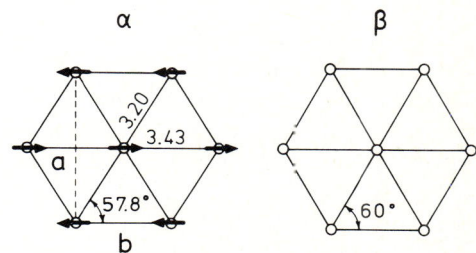

Fig. 2. The structures of the closest packed planes in α- and β-O_2. The molecular axes are perpendicular to the plane. Arrows indicate the directions of the magnetic moments. The β-phase has no long range magnetic order.

* John Simon Guggenheim Fellow. Permanent address: University of California, Los Angeles, California 90024, USA.

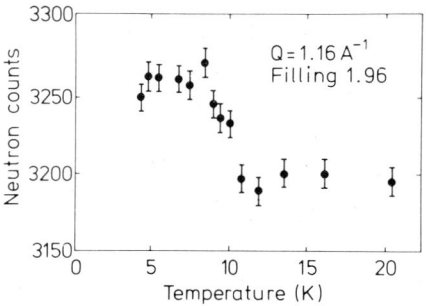

Fig. 3. The scattering intensity of the magnetic sublattice reflection at $Q = 1.16$ Å^{-1} as function of temperature. The coverage is 1.96 ρ_0. The count rate 3200 corresponds to the background scattering from the Grafoil in this region of Q.

gent susceptibility at $T \sim 10.3$ K [5]. Bulk O_2 has a transition between similar structures at $T = 23.8$ K.

The nature of the low coverage ($\rho < 1.7 \rho_0$) δ phase is less clear. Antiferromagnetic correlations are implied by the low Q magnetic scattering, while the shape of the structural group, with excess intensity in the high Q wing, suggests the existence of magnetostrictive fluctuations, but lack of long range structural order. Furthermore, the density is somewhat less than that of the α and β phases, but in all three packing restrictions imply that the molecules stand essentially normal to the graphite (001) plane. Unlike the α–β transition, the coexistence of δ and β peaks over a range of T and coverage indicates that this transition is of first order.

None of the observed structures is commensurate with the underlying graphite plane, and the 2-D α and β phases are almost identical to the a–b planes in bulk α- and β-O_2, indicating that the substrate plays no major role in the properties of these essentially 2-D phases. Indeed the structures observed correspond closely to those predicted for a 2-D sheet of O_2 molecules [6].

References

[1] M.F. Collins, Proc. Phys. Soc. 89 (1966) 415.
[2] J.G. Dash, Films on Solid Surfaces (Academic Press, New York, 1975).
    J.K. Kjems, L. Passell, H. Taub, J.G. Dash and A.D. Novaco, Phys. Rev. B13 (1976) 1446.
[3] C.S. Barrett, L. Meyer and J. Wasserman, J. Chem. Phys. 47 (1967) 592.
[4] E.J. Wachtel and R.G. Wheeler, Phys. Rev. Lett. 24 (1970) 233.
[5] H.E. Stanley and T.A. Kaplan, Phys. Rev. Lett. 17 (1966) 913.
    H.E. Stanley, J. Appl. Phys. 40 (1969) 1546.
[6] C.A. English and J.A. Venables, Proc. Roy. Soc. Lond. A340 (1974) 81.

MAGNETIC ORDER IN CoBr$_2$·6{xD$_2$O(1 − x)H$_2$O}

J.A.J. BASTEN
Netherlands Energy Research Foundation (E.C.N.), Petten (N.H.), The Netherlands

and

A.L.M. BONGAARTS
Department of Physics, Eindhoven University of Technology, Eindhoven, The Netherlands

The transition to long-range order in the 2d XY compound CoBr$_2$·6{xD$_2$O(1 − x)H$_2$O} with $x \approx 0.48$ has been investigated by neutron diffraction. A change in β is observed, which is not caused by a spread in T_N. Close to T_N the critical exponents β, γ, γ' and ν approach 3d values.

1. Introduction

The monoclinic compounds CoBr$_2$·6H$_2$O and CoCl$_2$·6H$_2$O are good examples of the two-dimensional (2d) XY model [1]. The exchange interaction between successive ab-planes is small compared with the exchange interactions within these layers. The bc-plane is the XY-plane. The transition to 3d antiferromagnetic order at 3.2 K and 2.3 K, respectively, is due to small deviations from the 2d XY model. In the ordered phase the direction of the magnetic moments is near the c-axis. In order to establish whether the transition to long-range order (LRO) is triggered by the small interactions between the ab-layers or by the anisotropy in the XY-plane, we decided to investigate the critical region of these compounds by means of neutron scattering.

Because of the large incoherent neutron-scattering cross section of hydrogen, in such experiments a deuterated sample is preferred. However, in a previous study [2] we pointed out that CoBr$_2$·6D$_2$O and CoCl$_2$·6D$_2$O are subject to a crystallographic phase transition from monoclinic (C 2/m) into triclinic (P $\bar{1}$), accompanied by the occurrence of crystallographic domains. In order to avoid this effect, which in the bromide occurs for deuterium fractions $x > 0.55$ and in the chloride for $x > 0.035$, the compound CoBr$_2$·6{xD$_2$O(1 − x)H$_2$O} with $x = 0.48$ was selected for the present work. This material will further be denoted as CB48.

2. Results

A single crystal of about 1.5 cm^3 has been used for the determination of critical scattering near the (1 0 $\tfrac{1}{2}$) reflection, both above and below the Néel temperature T_N. For the measurements of the temperature dependence of the sublattice magnetization $M_s(T)$, also a single crystal of 0.1 cm^3 has been used in order to check on absorption and extinction effects. Experiments were confined to the [0 1 0] zone. The critical scattering data were analyzed, taking into account the three-dimensional effects of the finite resolution of the diffractometer. A modified version of the Fisher approximant

$$\chi(q)/\chi(0) = \{1 + (q_x/k_x)^2 + (q_y/k_y)^2 + (q_z/k_z)^2\}^{-1+\eta/2} \quad (1)$$

has been used, which applies to low-dimensional systems. In eq. (1) $q = Q - \tau$; Q is the momentum loss of the neutron and τ is the nearest reciprocal lattice point; η is the exponent describing the deviation from a simple lorentzian; $\chi(0)$ is the staggered susceptibility; k_x, k_y, and k_z are the inverse ranges of correlations in the a, b, and c^* directions, respectively. As the exchange interactions along the a and b axis are equally strong [1], we assume $k_x = k_y$.

The most important result of the critical scattering analysis in CB48 was the possibility to obtain both the correlation range within the ab-planes ($1/k_x$) and the correlation range between ab-layers ($1/k_z$) over almost the entire accessible temperature range. Fig. 1(a) shows that the temperature dependence of k_x and k_z is the same, which means that the transition to LRO is essentially 3d in character. Only the ratio $k_z/k_x \approx 4$ reflects the expected 2d characteristics.

The critical scattering data and the sublattice

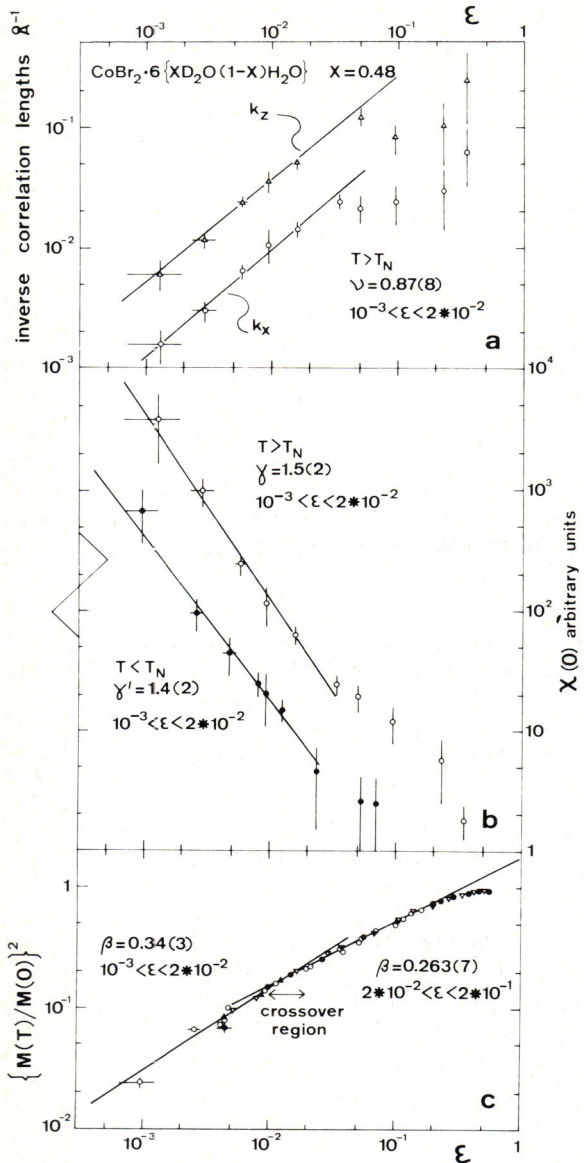

Fig. 1. Single power laws [eqs. (2)–(4)] fitted to the data with $T_N = 3.117(1)$ K. Best fits are represented by drawn lines; the corresponding exponents are given in the figures. Error bars indicate one standard deviation. (a) Inverse correlations within ab-layers (k_x) and between ab-layers (k_z). (b) Susceptibility $\chi(0)$ above and below T_N; $C_+/C_- = 8(2)$. (c) Reduced intensity of several magnetic Bragg reflections $I(T)/I(0) = [M_s(T)/M_s(0)]^2$. In the region $10^{-2} < \epsilon < 2*10^{-2}$ a change in β is observed.

magnetization were described by the single power laws

$$M_s(T)/M_s(0) = B\epsilon^\beta, \qquad (2)$$

$$\chi(T) = C\epsilon^{-\gamma} \quad \text{and} \quad k_\alpha(T) = K_\alpha \epsilon^\nu, \qquad (3,4)$$

where $\epsilon = |T/T_N - 1|$. Best fits in the region $10^{-3} < \epsilon < 2*10^{-2}$ yield an average Néel temperature $T_N = 3.117(1)$ K. (Throughout this paper standard deviations are given between parentheses in units of the least significant digit.) With this choice of T_N the following values were obtained: $\nu = 0.87(8)$, $K_z/K_x \approx 4$ [fig. 1(a)], $\gamma = 1.5(2)$ for $T > T_N$ and $\gamma' = 1.4(2)$ for $T < T_N$. The ratio of the susceptibility amplitudes above and below T_N is $C_+/C_- = 8(2)$ [fig. 1(b)]. The sublattice magnetization shows a change from $\beta = 0.263(7)$ with $B = 1.31(2)$ in the region $2*10^{-2} < \epsilon < 2*10^{-1}$, to $\beta = 0.34(3)$ with $B = 1.8(2)$ in the region $10^{-3} < \epsilon < 10^{-2}$ [fig. 1(c)]. Best fits to the scattering profiles were achieved with $\eta = 0.1(1)$.

3. Discussion

An interesting aspect of the critical behaviour in CB48 is the change in β. In ref. 3 several examples have been mentioned, in which an apparently increased β at small ϵ is simply due to a distribution in T_N. In CB48 this can be caused by an inhomogeneous distribution of hydrogen and deuterium over the crystal. Ref. 3 presents explicit calculations on this effect in 2d Ising systems. The main reason for the seeming 3d values of β close to T_N is the erroneous identification of T_N with the maximum in $\chi(T)$. By the spread σ in T_N, this maximum shifts to higher T values, by about 1.5 σ, due to the large ratio $C_+/C_- = 37$ in 2d Ising systems.

There are several reasons why the above mentioned argument is not valid for CB48:

(i) Whereas the 2d Ising systems discussed in [3] show an apparent 2d behaviour of the critical scattering, the 3d character in CB48 is clear from the identical temperature variation of k_x and k_z.

(2) In $CoBr_2 \cdot 6\{xD_2O(1-x)H_2O\}$ T_N shows a minimum at $x = 0.5$ [4]. Even a large spread of 0.05 in x will result in a spread σ in T_N of only 6 mK for CB48.

(3) The ratio $C_+/C_- = 8(2)$ in CB48 is close to the expected value of 5 for 3d systems. This is much smaller than the 2d Ising value of 37. Therefore the shift in the maximum of $\chi(T)$ will be much smaller, at most 0.4 σ. If one fits formula (3) to the susceptibility data for $T > T_N$ in the region $10^{-3} < \epsilon < 2*10^{-2}$, a value of $T_N = 3.119(1)$ K results. So, our choice of an averaged

T_N of 3.117(1) K includes already a shift of the maximum of χ over 2(1) mK.

From these arguments we conclude that the observed β in the region $\epsilon < 10^{-2}$ has to be considered as a real 3d-value. Possibly the calculated exponents would become lower and approach 3d-values better, if a spread of a few mK in T_N is taken into account. However, the large uncertainties in the exponents, due to the small temperature region in which the values had to be determined, impede the distinction between 3d XY-behaviour ($\beta = 0.33$, $\gamma = 1.33$, $\nu = 0.69$) and 3d Ising-behaviour ($\beta = 0.31$, $\gamma = 1.25$, $\nu = 0.64$).

It should be pointed out that the change in β near $\epsilon = 2 \ast 10^{-2}$ cannot be interpreted as a "crossover" in the sense of a transition from a specific "ideal model" into another one (e.g. 2d $XY \rightarrow$ 3d XY or 2d Ising $\rightarrow$ 3d Ising). It seems possible that the observed change in β for CB48 indicates a transition from a quasi-2d state, which does not correspond to an "ideal model", into 3d behaviour. A possible change in γ, which is expected at still higher ϵ-values, cannot be observed because of the large uncertainties in $\chi(0)$ near $\epsilon \approx 0.1$. A more extensive report and analysis of the present experiment will be presented elsewhere.

References

[1] J.P.A.M. Hijmans, Q.A.G. van Vlimmeren and W.J.M. de Jonge, Phys. Rev. B12 (1975) 3859, and references therein.
[2] J.A.J. Basten and A.L.M. Bongaarts, Phys. Rev. B14 (1976) 2119.
[3] R.J. Birgenau, H.J. Guggenheim and G. Shirane, Phys. Rev. B8 (1973) 304.
[4] J.P. Legrand and J.P. Renard, in: XVII Congress Ampere, V. Hovi (Ed.) (North-Holland, Amsterdam, 1973) p. 452.

PHASE TRANSITION OF TWO DIMENSIONAL HEISENBERG ANTIFERROMAGNET OF $S=\frac{1}{2}$

M. MATSUURA, Y. YAMAMOTO, H. YAMAKAWA, T. HASEDA
Faculty of Engineering Science, Osaka University, Toyonaka 560, Japan

and

Y. AJIRO
Department of Chemistry, Faculty of Science, Kyoto University, Kyoto 606, Japan

From the analysis of the susceptibility of $Cu(HCOO) \cdot 4H_2O$ in which the Cu^{2+} ions are on two inequivalent sites, the staggered susceptibility of a two-dimensional Heisenberg antiferromagnet is examined. From the results, and from proton NMR and heat capacity on the related salts, characteristic features of the phase transition of a two-dimensional Heisenberg antiferromagnet are inferred.

Ordering of a two-dimensional Heisenberg (2dH) spin system is a most interesting unsolved problem. A rigorous proof was put forward by Mermin and Wagner that no spontaneous magnetization appears at a finite temperature in a 2dH ferromagnet [1]. Stanley and Kaplan have conjectured that 2dH ferromagnets of $s \geq 1$ may have a new type of transition, involving a divergence of the susceptibility at T_c, but no spontaneous magnetization below T_C [2]. Many experimental studies have been done on these attractive problems, mostly in ferromagnets [3, 4].

In the present work, we investigate 2dH antiferromagnet of $s = \frac{1}{2}$, which may be an even more interesting case than the ferromagnet. Since the staggered magnetization does not commute with the total hamiltonian, the antiferromagnet may be even more difficult to order than the ferromagnet.

$Cu(HCOO)_2 \cdot 4H_2O$ (CuF4H) has a layer-structure along the c axis and approximates a 2dH antiferromagnet. From the symmetry, we know the g-tensors of Cu^{2+} ions at $(0, 0, 0)$ and $(\frac{1}{2}, \frac{1}{2}, 0)$ sites, $\mathbf{g}_1$ and $\mathbf{g}_2$ are not equivalent. In such a crystallographic two-sublattice (C_2) system, not only a uniform magnetization but also a staggered magnetization is induced by applying a uniform field [5]. The reason is that under a field H each spin of Cu^{2+} at a $(0, 0, 0)$ site has a response to $\mathbf{g}_1 H$ while each at $(\frac{1}{2}, \frac{1}{2}, 0)$ site responds to $\mathbf{g}_2 H$. This circumstance is quite equivalent to the following: the Cu^{2+} ion system with a g-tensor $\mathbf{g}[=\frac{1}{2}(\mathbf{g}_1 + \mathbf{g}_2)]$ (hereafter C_1 system) feels a uniform field H and a staggered field $\mathbf{H}_s(=\mathbf{g}^{-1}\mathbf{d}H)$ where $\mathbf{d}$ is defined by $\frac{1}{2}(\mathbf{g}_1 - \mathbf{g}_2)$. The corresponding uniform and intermode susceptibilities of the C_2 system $\mathbf{X}_u^{(2)}$ and $X_{su}^{(2)}$ can be expressed by [6]

$$\mathbf{X}_u^{(2)} = \mathbf{X}_u^{(1)} + \mathbf{d}\mathbf{g}^{-1}\mathbf{X}_s^{(1)}\mathbf{g}^{-1}\mathbf{d}, \tag{1}$$

$$\mathbf{X}_{su}^{(2)} = \mathbf{X}_s^{(1)}\mathbf{g}^{-1}\mathbf{d}, \tag{2}$$

where $\mathbf{X}_u^{(1)}$ and $\mathbf{X}_s^{(1)}$ are the uniform and staggered susceptibilities of the C_1 system.

The temperature dependence of $\mathbf{X}_u^{(2)}$ is shown in fig. 1 [7]. The principal values along $L_1(\sim c)$ and $L_2(b)$ axes increases rapidly down to T_N. The characteristic feature is certainly coming from $\mathbf{X}_s^{(1)}$ because $\mathbf{X}_u^{(1)}$ does not show any such anomaly across T_N as seen in $\mathbf{X}_{L_3(\sim a)}$.

$\mathbf{X}_s^{(1)}$ is the linear response coefficient for $\mathbf{H}_s$ which is staggered in the ab plane and uniform along the c axis. So we sign the field as $\mathbf{H}_s(\pi, 0)$ and the corresponding $\mathbf{X}_s^{(1)}$ as $\mathbf{X}_s^{(1)}(\pi, 0)$. We can

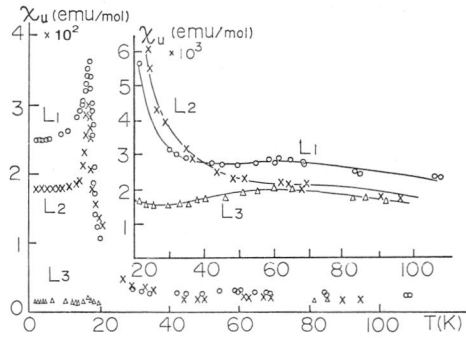

Fig. 1. Susceptibility of CuF4H along the principal directions [7].

define another staggered field $H_s(\pi, \pi)$, which is staggered along the c axis too, and the corresponding susceptibility $X_s^{(1)}(\pi, \pi)$. If the interplanar interaction along the c axis J' is zero, these two quantities are equal to $nX_s^{(1)}(\pi)$ where n is the number of ab planes and $X_s^{(1)}(\pi)$ is the staggered susceptibility of one ab plane. If J' is finite and ferromagnetic, $H_s(\pi, 0)$ enhances the cooperative action of the system while $H_s(\pi, \pi)$ suppresses it. So $X_s^{(1)}(\pi, 0) \geqslant nX_s^{(1)}(\pi) \geqslant X_s^{(1)}(\pi, \pi)$. If J' is finite and antiferromagnetic, then $X_s^{(1)}(\pi, 0) \leqslant nX_s^{(1)}(\pi) \leqslant X_s^{(1)}(\pi, \pi)$.

As the increase of the measured $X_u^{(2)}$ near T_N is coming from $X_s^{(1)}(\pi, 0)$ and J' is antiferromagnetic from proton NMR below T_N [8] the following two cases will be considered with reference to the experiments. (1) If we could make J' smaller, $X_u^{(2)}$ should increase more significantly. (2) If we apply an external field H corresponding to H_s of the order of J', a nonlinear effect of staggered magnetization may be observed.

The temperature dependence of $X_u^{(2)}$ of $Cu(HCOO)_2 \cdot 2CO(NH_2)_2 \cdot 2H_2O$ (CuFUH) which can be seen as derived from CuF4H by partial substitution of H_2O by urea molecules [9], is compared with that of CuF4H in fig. 2(a). The intraplanar interactions in both salts are about the same [9]. However, the interplanar interaction of CuFUH is estimated to be much weaker than that of CuF4H from the larger interplanar separation [10]. The more significant increase of $X_u^{(2)}$ in CuFUH supports the first prediction. The temperature dependence of proton NMR frequency shift in CuF4D, a substituted form of CuF4H by D_2O, is shown in fig. 2(b). The shift depends on the applied field intensity and is different from $X_u^{(2)}$ even at the field corresponding to an H_s of 10^{-5} times the exchange field. This remarkable field effect is evidence for the second prediction.

Recent heat capacity measurement showed that the anomaly at T_N was extremely small in CuF4H [11] and CuFUH [9]. The entropy change associated with the anomaly was 0.01% of the total magnetic entropy ($R \ln 2$) for CuF4H and much smaller for CuFUH (fig. 3).

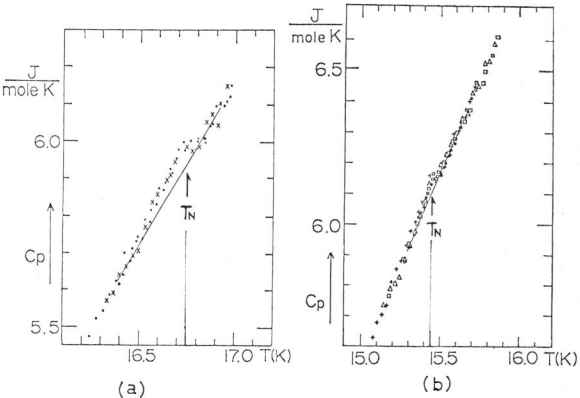

Fig. 3. Heat capacity of (a) CuF4H [11] and (b) CuFUH [9] near T_N.

From these results, we conclude that there is a phase transition at a finite temperature T_N far above 0 K at which the staggered susceptibility shows a remarkable peak, but the heat capacity shows little anomaly, in such a quasi-two-dimensional nearly Heisenberg antiferromagnet.

References

[1] N.D. Mermin and H. Wagner, Phys. Rev. Lett. 17 (1966) 1133.
[2] H.E. Stanley and T.A. Kaplan, Phys. Rev. Lett. 17 (1966). 913.
[3] A.R. Miedema, P. Bloembergen, J.H.P. Colpa, F.W. Gorter, L.J. de Jongh and L. Noordermeer, Proc. 19th M.M.M. Conf. Boston (1973).
[4] K. Hirakawa and H. Ikeda, J. Phys. Soc. Jap. 35 (1973) 1328.
[5] M. Matsuura and Y. Ajiro, J. Phys. Soc, Jap. 41 (1976) 44.

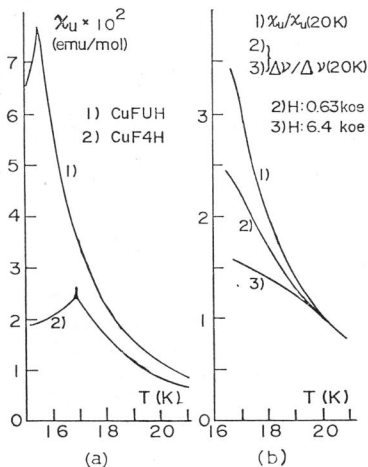

Fig. 2. (a) Susceptibility of CuF4H and CuFUH in powdered form [9]. (b) NMR frequency shift normalized to 1.0 at 20.4 K at which the shift changes linearly with external field intensity up to 8 kOe.

[6] Generally, these include also intermode susceptibilities $\mathbf{X}_{su}^{(1)}$ and $\mathbf{X}_{us}^{(1)}$ due to Dzyaloshinsky–Moriya type interaction. From the analysis of angular dependence of proton NMR frequency, these contribution is found much smaller than that of $X_s^{(1)}$ (Ajiro et al., to be published).

[7] H. Kobayashi and T. Haseda, J. Phys. Soc. Jap. 19 (1963) 541.

[8] A. Dupas and J.P. Renard, Phys. Lett. 33A (1970) 470.

[9] Y. Yamamoto, M. Matsuura and T. Haseda, J. Phys. Soc. Jap. 38 (1975) 1776.

[10] H. Kiriyama and K. Kitahama, Acta. Cryst. B32 (1976) 330.

[11] Y. Yamamoto, M. Matsuura and T. Haseda, J. Phys. Soc. Jap. 40 (1976) 1300.

MÖSSBAUER STUDY OF THE TWO-DIMENSIONAL ANTIFERROMAGNET $(CH_3NH_3)_2FeCl_4$*

H. KELLER, W. KÜNDIG
Physics Institute, University of Zurich, Switzerland

and

H. AREND
Laboratory of Solid State Physics, ETH-Zurich, Switzerland

Mössbauer-effect measurements of the hyperfine interaction of ^{57}Fe nuclei in the layered antiferromagnet $(CH_3NH_3)_2FeCl_4$ have been used to study the temperature dependence of the sublattice magnetization. Near the Néel temperature $(2 \times 10^{-4} < 1 - T/T_N < 10^{-2})$ a fit to the power law $(1 - T/T_N)^\beta$ yields $\beta = 0.146 \pm 0.005$ and $T_N = 94.46 \pm 0.02$ K. This value for β is close to those found for other two-dimensional magnetic systems.

In the face-centred tetragonal structures of $(C_nH_{2n+1}NH_3)_2MCl_4$ (M = Mn, Cu, Fe; $n = 1, 2, 3\ldots$) the magnetic Fe^{2+} ions are arranged in perovskite-type layers separated by long organic chains of variable length. This class of compounds are ideal candidates in which to study the magnetic behaviour of quasi two-dimensional magnetic systems.

^{57}Fe Mössbauer spectra of powdered and single crystal samples were measured as a function of temperature in the range 6–300 K [1–3]. Extrapolation to 0 K yields the following parameters: the magnetic hyperfine field $H(0) = 280 \pm 3$ kOe, the quadrupole interaction $\frac{1}{2}eQV_{zz} = 2.58 \pm 0.04$ mm/s (the asymmetry parameter η of the electric field gradient in this tetragonal structure is assumed to be zero), the angle $\theta = \sphericalangle(V_{zz}, H) = 85 \pm 1°$.

The value of θ indicates that the spins are almost parallel to the layers. Below the Néel temperature T_N, V_{zz} and θ were found to be nearly temperature independent. T_N was determined by two self-consistent methods: by an extrapolation of the measured hyperfine field $H(T)$ to zero (fig. 1) and by the characteristic peak in the linewidth (fig. 2). The two methods give a value of $T_N = 94.5 \pm 0.1$ K in agreement with susceptibility measurements [4].

If one assumes that the hyperfine field $H(T)$ at the Fe^{2+} nucleus is proportional to the sublattice magnetization $M(T)$, one may write near T_N,

$$M(T)/M(0) = H(T)/H(0) = D(1 - T/T_N)^\beta, \quad (1)$$

* Supported by the Schweizerische Nationalfonds.

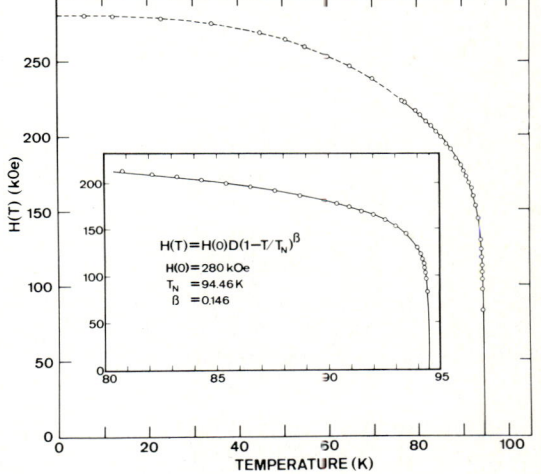

Fig. 1. Hyperfine field as a function of temperature. The dashed line corresponds to the measured points and the solid line corresponds to the calculated power-law magnetization curve.

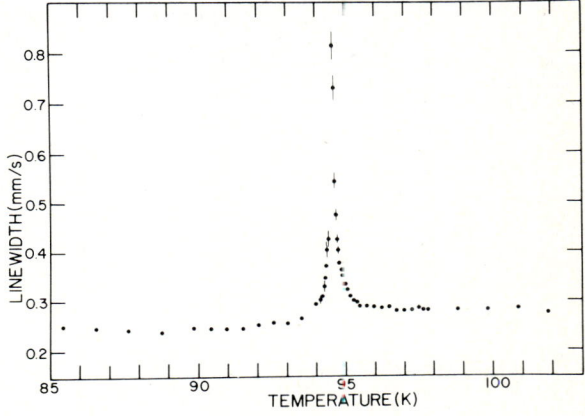

Fig. 2. Temperature dependence of the linewidth near T_N.

where $H(0)$ is the hyperfine field at $T = 0$ K, D is a reduction factor matching this high-temperature formula to $M(0)$, and β is the critical exponent. For most magnetic model systems, this power-law behaviour is limited to the temperature region $1 - T/T_N < 4 \times 10^{-2}$ [5]. A least square fit to the data in the range $2 \times 10^{-4} < 1 - T/T_N < 10^{-2}$ yields (figs. 1 and 3.), $D = 1.01 \pm 0.02$, $T_N = 94.46 \pm 0.02$, $\beta = 0.146 \pm 0.005$.

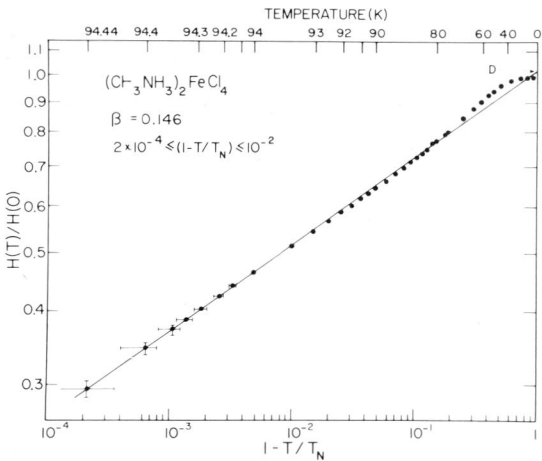

Fig. 3. Log–log plot of the reduced hyperfine field as a function of the reduced temperature. The slope of the straight line is the critical exponent β.

A comparison of the experimentally determined value β and the theoretically predicted values given in table I indicates that $(CH_3NH_3)_2FeCl_4$ is a good example for a quasi-two-dimensional magnetic system with a dominant magnetic exchange within the layers. Since the spins are almost parallel to the layers, the 2d-XY-model is most applicable. The 2d

Table I
Critical exponent β as predicted by various models

	2d	3d
Ising	0.125	0.31
XY	0.13 ± 0.03[a]	0.33
Heisenberg	–	0.36

[a] 2d-XY-model with quartic anisotropy (ref. [7]).

character is further supported by susceptibility measurements of Willet and Gerstein [4]. They found for the critical exponents, $\gamma = 1.67$ and $\gamma' = 1.60$ (2d-Ising: $\gamma = \gamma' = 1.75$).

The present value for β is smaller than those given earlier ($\beta = 0.23$ [2] and $\beta = 0.22$ [6]). This deviation may be explained by the fact that the earlier results were obtained for $1 - T/T_N > 2 \times 10^{-2}$, a temperature region not close enough to T_N to justify the use of the power law.

References

[1] M.F. Mostafa and R.D. Willet, Phys. Rev. B4 (1971) 2213.
[2] J.L. Schurter, R.G. Barnes and R.D. Willet, Proc. 20th Conf. on Magnetism and Magnetic Materials, San Francisco (1974).
[3] H. Keller, W. Kündig and H. Arend, Proc. of the Mössbauer Conf., Vol. 1, Cracow (1975).
[4] R.D. Willet and B.G. Gerstein, Phys. Lett. 44A (1973) 153.
[5] R.F. Wielinga, Progr. Low Temp. Phys., Vol VI (North-Holland, 1970) p. 333.
[6] H. Keller, W. Kündig and H. Arend, Helv. Phys. Acta 49 (1976) 148.
[7] T. Schneider and E. Stoll, Phys. Rev. Lett. 36 (1976) 1501.

MAGNETIZATION MEASUREMENTS ON QUASI-TWO-DIMENSIONAL SPIN SYSTEMS $(CH_2)_n(NH_3)_2MnCl_4$

K. BABERSCHKE
Institut für Atom- u. Festkörperphysik, Freie Universität Berlin, Boltzmannstr. 20, D-1000 Berlin 33, W. Germany

F. RYS
Institut für Theoretische Festkörperphysik, Freie Universität Berlin, Arnimallee 3, D-1000 Berlin 33, W. Germany

and

H. AREND
Institut für Festkörperphysik ETH-Z, Hönggerberg, CH-8049 Zürich, Switzerland

The biammonium compounds $(CH_2)_n(NH_3)_2MnCl_4$ form quasi-two-dimensional antiferromagnetic layer compounds, which are similar to the monoammonium compounds. Especially for $n = 3$ sharp spin–flop transitions are observed. The magnetization as a function of temperature and field was measured. The spin–flop field H_{sf} varies with temperature from 24.5 to 28.5 kG. The critical temperature $T_c^{\|}(H\|c$-axis) from the ordered to the disordered phase (Néel point) is approx. $T_c \approx 43$ K. From both results the bicritical point T_b was determined to $T_b \approx 40 \pm 2$ K.

1. Introduction

Diammonium layer-compounds of the type $(CH_2)_n(NH_3)_2MnCl_4$ order antiferromagnetically. The easy axis (c-axis) is perpendicular to the layer. They undergo a spin–flop transition at approx. 25 kG [1]. In this paper we report magnetization measurements for the compound $n = 3$, which shows the sharpest transition.

The study of the magnetization of this quasi-two-dimensional Heisenberg-antiferromagnet [2] is of interest because it yields part of the phase diagram. Following Fisher and Nelson [3] (fig. 1) the discontinuity in the magnetization at the spin flop transition scales with reduced temperature:

$$\Delta M_\|(T) \propto t^{\tilde{\beta}},$$

where $t = (T - T_b)/T_b$, T_b being bicritical temperature at which the transition vanishes. The critical exponent is given for a two-(three-)dimensional system as $\tilde{\beta} = 0.88(0.85)$ [2].

2. Experimental results

The magnetization was measured with a commercial sample vibrating magnetometer. Crystals of about 10–20 mgr were placed with the c-axis parallel resp. perpendicular to the vibrating axis (equivalent to the magnetic field axis). Because we focused only on the change of the magnetization as a function of the applied field H and the temperature T the experimental data are not correct for diamagnetism and demagnetizating effects.

Fig. 2 shows $M_\|$ as a function of temperature.

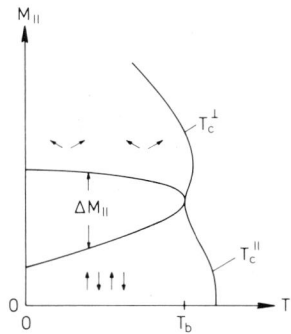

Fig. 1. Schematic phase diagram of an anisotropic antiferromagnet with the field H parallel to the easy (c)-axis [3].

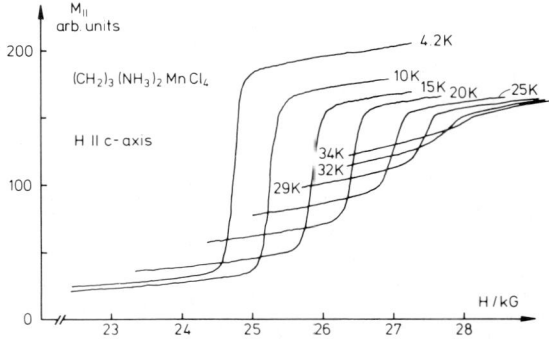

Fig. 2. Magnetization $M_\|$ as a function of the applied field H. The highest temperature at which a spin–flop transition could be observed was 38 K.

It can be seen from the graph that the width of the transition increases with temperature [4]. This may be caused by spin fluctuations at temperatures near the critical temperature T_b. Because of this broadening a precise determination of T_b is difficult: $T_b = (40 \pm 2)$ K. The spin–flop field H_{sf} changes from $H_{sf}(T=0) = 24.5$ kG to $H_{sf}(T_b) = 28.5$ kG. At fields $H > H_{sf}$ the magnetization $M_\parallel$ agrees within the error bar with $M_\perp$. Thus, experimentally no intermediate single phase [3], which would occur near a tetracritical point, could be observed.

In order to determine the Néel temperature T_c (fig. 1) the temperature at which the antiferromagnetic ordering changes into the disordered state we follow the definition as it was used for the two-dimensional antiferromagnets K_2MnF_4 and $RbMnF_4$ [5]. Here the ordering temperature is considered to be the temperature at which the difference between the magnetizations (susceptibilities) parallel and perpendicular to the easy axis sharply starts to increase. In both systems the manganese fluorides and the manganese biammonium compounds exists already a small difference between $M_\parallel$ and $M_\perp$ at $T > T_c$. We extracted $T_c^\parallel$ $(H \to 0) \approx 43$ K. It decreases very little with the applied field (fig. 3).

The experiments also show a quite abnormal behaviour at T_c (1). Decreasing the temperature below T_c M sharply increases ($H = 1$ kG) and (2) $M_\parallel$ shows a small bump (20 kG). Both effects seem to indicate a small ferromagnetic component in the layer plane.

3. Discussion

The magnetic properties of the diammonium compounds [1] are very similar to the corresponding manoammoniums $(C_nH_{2n+1})_2(NH_3)_2MnCl_4$ [6]. In this report we show the magnetization measurements $M(H, T)$ for $n = 3$. A sharp spin–flop transition enables us to determine part of the M versus T phase diagram [3,6]. Scaling M to the reduced temperature t yields a critical exponent $\tilde\beta = 0.8 \pm 0.2$ which is in rough agreement with the theory [2]. Unfortunately the absence of specific heat data (to determine T_b) and the increasing width of the transition at $t \ll 1$ make a precise determination of $\tilde\beta$ impossible.

The transition from the antiferromagnetic phase ($H\|c$-axis) to the paramagnetic one seems to be roughly field independent. $T_c^\parallel$ $(H=0) \approx 43$ K. For increasing field T_c moves slowly toward $T_b = 40$ K. De Jongh and Miedema discussed a different T_c^+ at which $\partial\chi/\partial T$ reaches infinitely in a Ising model resp. a maximum value for a Heisenberg model. From our experiment T_c^+ is very difficult to extract because M increases roughly linearly with temperature (2 K $< T <$ 30 K).

W. Wisny is acknowledged for assistance in the experiments.

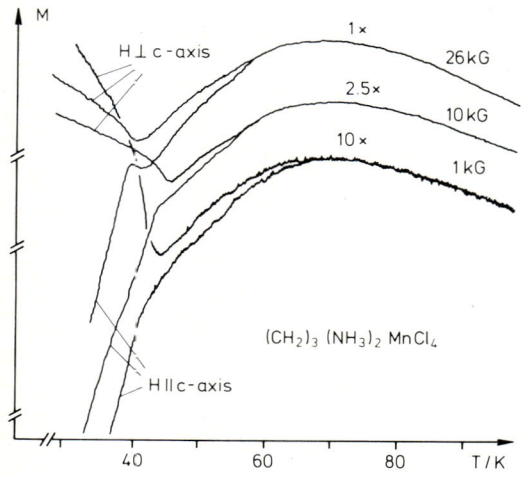

Fig. 3. Magnetization as a function of the temperature with the field parallel and perpendicular to the c-axis. The different gain is given in the graph.

References

[1] H. Arend, K. Tichy, K. Baberschke and F. Rys, Solid State Commun. 18 (1976) 999.
F. Rys and K. Baberschke, Helv. Phys. Acta 48 (1975) 438.
[2] F. Rys, Proc. 21st Conf. on Magnetism and Magn. Materials, Philadelphia (1975) p. 458.
[3] M.E. Fisher and D.R. Nelson, Phys. Rev. Lett. 32 (1974) 1350.
[4] All samples which were measured show the same broadening effect.
[5] D.J. Breed, Thesis, University of Amsterdam (1969).
[6] L.J. De Jongh and A.R. Miedema, Advan. Phys. 23 (1974) 1.

MAGNETIC PROPERTIES OF QUASI-2d HEISENBERG FERROMAGNETS $[C_6H_5(CH_2)_nNH_3]_2$ $CuCl_4$ WITH $n = 1, 2, 3$

A. DUPAS, LE DANG KHOÏ, J.P. RENARD and P. VEILLET

Institut d'Electronique Fondamentale, laboratoire associé au C.N.R.S., Bâtiment 220, Université Paris-Sud, 91405 Orsay-Cedex, France

We have measured the magnetic susceptibility and the NMR of copper and chlorine nuclei in monocrystals of a new family of quasi-2d Heisenberg ferromagnets $[C_6H_5(CH_2)_nNH_3]_2$ $CuCl_4$ with $n = 1, 2, 3$. From these experiments we have deduced the Curie temperatures, the intra-layer ferromagnetic exchange integrals and the anisotropy parameters. We have also shown that the spontaneous magnetization below $\frac{1}{2}T_c$ is in good agreement with a simple spin-wave treatment.

Quasi-2d Heisenberg ferromagnets (HF) deserve a wide attention owing to the peculiar low-temperature and critical properties of the ideal model [1]. The study of a new family of such compounds $[C_6H_5(CH_2)_nNH_3]_2$ $CuCl_4$ ($\Phi C_n Cl$) is thus of a great interest. These compounds are isomorphous to the previously well known family $[(C_nH_{2n+1}NH_3]_2CuCl_4$ (C_nCl), with two-dimensional layers of square planar $CuCl_4^{2-}$ ions, widely separated by $C_6H_5(CH_2)_nNH_3$ groups. Their physical properties have been investigated by Daoud et al. [2] using X-ray diffraction, Raman and EPR spectroscopy up to $n = 4$.

At low temperatures the crystals of $\Phi C_n Cl$, with $n = 1, 2, 3$ exhibit a 3d ferromagnetic order which shows that the inter-layer exchange is ferromagnetic. This property is very interesting since previously only two examples of quasi-2d HF with ferromagnetic inter-layer exchange were known that could be grown in large single-crystals: K_2CuF_4 and C_1Cl. The Curie temperatures of $\Phi C_n Cl$ with $n = 1, 2, 3$, as found from results of magnetic susceptibility measurements by a mutual inductance bridge operating at 70 H_z, are given in table I.

The susceptibility perpendicular to the $CuCl_2$ layers (along the x-axis) was measured on small, nearly cubic, samples. It is rather low-valued and the demagnetizing correction is less than 10%. The susceptibility parallel to the $CuCl_2$ layers (bc-plane) was measured on thin square crystals $6 \times 6 \times 1$ mm to minimize the demagnetizing factor. A static field of up to 340 Oe could be applied. It appears that a is the hard axis, with the easy axis and the next-preferred axis lying in bc-plane. The out-of-plane anisotropy field H_{Aout} was deduced from the extrapolation of $\chi_a(T)$ to 0 K. The in-plane aniso-

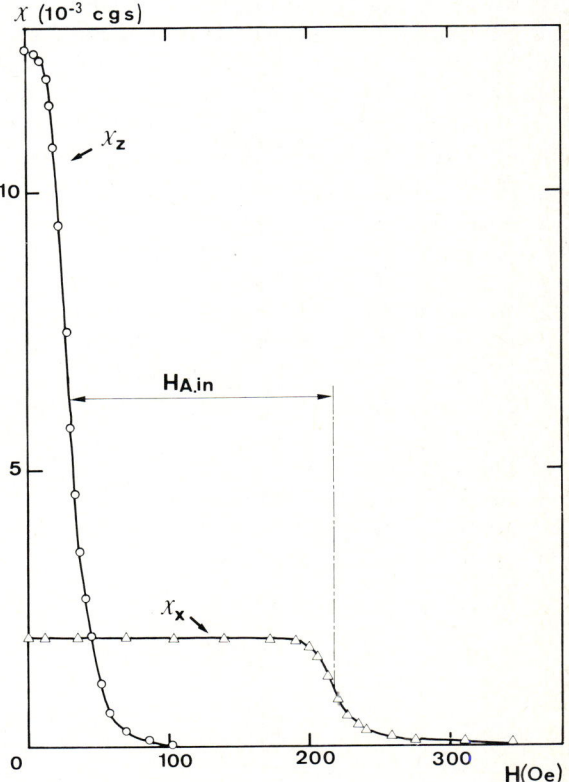

Fig. 1. Principal magnetic susceptibilities χ_x (along in plane hard axis) and χ_z (along easy axis b) vs. applied field H at $T = 1.23$ K in $\Phi C_2 Cl$.

Table I
Curie temperatures intra-layer exchange and anisotropy fields of $\Phi C_n Cl$ with $n = 1, 2, 3$

	T_c(K)	J/k(K)	H_{Ain}(Oe)	H_{Aout}(Oe)
$\Phi C_1 Cl$	7.7	16.5	24.5	1280
$\Phi C_2 Cl$	9.2	19	200	1500
$\Phi C_3 Cl$	7.8	16.5	250	1370

tropy field H_{Ain} was determined from the field dependence of the susceptibilities measured along the two principal magnetic axis z and x of the bc-plane (fig. 1). The values of H_{Aout} and H_{Ain} are given in table I.

The NMR of ^{35}Cl, ^{37}Cl, ^{63}Cu and ^{65}Cu was observed by a spin–echo method between 1.4 and 4.2 K in single crystals of $\Phi C_n Cl$, with $n =$ 1, 2, 3. Measurements were performed in zero applied field as well as in applied field up to 10 kOe. These NMR experiments will be discussed elsewhere [3]. The point here is that the resonance frequency of the Cu^{2+} central line in an applied field $H > H_A$ is simply related to the reduced magnetization $\langle S_z \rangle / S$

$$\langle S_z \rangle / S = C^{te}(\nu + H),$$

where γ is the nuclear gyromagnetic ratio. The field dependence of $\langle S_z \rangle / S$ was compared to theoretical curves, computed with a simple spin-wave model [4] for different values of the intra-layer exchange integral J, using the experimental values of the anisotropy fields. The inter-layer exchange was neglected since according to preliminary measurements by Yamazaki [5] the inter-layer exchange field is much smaller than $\sqrt{H_{Ain} \times H_{Acut}}$. The values of J were found by fitting theory and experiments as shown in fig. 2 for $\Phi C_1 Cl$ at two temperatures.

The spontaneous magnetization below 4.2 K was also determined from chlorine NMR. Here again a good fit was observed between experiments and the results of a simple spin-wave model with experimental values for H_{Ain}, H_{Aout} and J.

The values of anisotropy fields given in table I show that the compounds $\Phi C_n Cl$ have a dominant XY-like anisotropy, and a small Ising-like anisotropy which takes values differing by an order of magnitude in the family. These

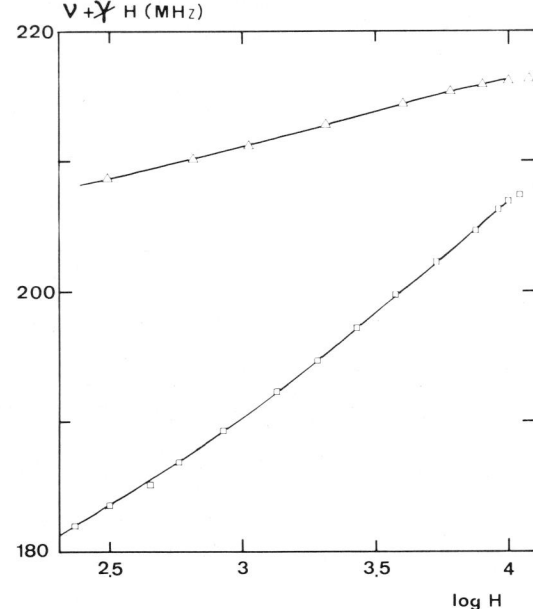

Fig. 2. Magnetization of $\Phi C_1 Cl$ vs. applied field at two temperatures [□: $T = 4.3$ K; △: $T = 1.6$ K. We have plotted $\nu + H$ on the vertical scale, where ν is the central NMR frequency and γ the gyromagnetic ratio of ^{63}Cu. This quantity is proportional to the magnetization. The curves are the theoretical ones for $J/k = 16.5$ K, adjusted to fit the experimental data at $H = 10$ kOe.

compounds are thus very promising for the study of the cross-over phenomena in the critical region. Further work is in progress.

References

[1] L.J. de Jongh and A.R. Miédema, Advan. Phys. 23 (1974) 1.
[2] A. Daoud, A. Thierr-Sorel, R. Perret, B. Chaillot and J.E. Guerchais, Bul. Soc. Chimique de France (1975) 535.
A. Daoud, thèse de Doctorat, Dijon (1976).
[3] Accepted for publication in J. Chem. Phys.
[4] A. Dupas and J.P. Renard, C.R. Acad. Sci., Paris 275 B (1972) 919.
[5] To be published in J. Phys. Soc, Jap.

RENORMALIZED SPIN–WAVE THEORY OF NEARLY TWO-DIMENSIONAL HEISENBERG FERROMAGNET K_2CuF_4

N. URYÛ and K. TSURU
Department of Applied Science, Faculty of Engineering, Kyushu University, Fukuoka 812, Japan

The temperature dependence of the magnetization of the ferromagnetic layer-type compound K_2CuF_4 is analyzed by the renormalized spin–wave theory. The spin–wave interaction plays rather an important role in the nearly two-dimensional magnetic system and the apparent $T^{3/2}$ temperature dependence of the magnetization can reasonably be explained.

1. Introduction

Low temperature magnetic properties of K_2CuF_4, a typical example of the nearly two-dimensional Heisenberg ferromagnet, have been widely investigated [1]. The spin system of K_2CuF_4 consists of weakly coupled ferromagnetic layers whose intra- and inter-layer exchange coupling constants J and J' have been estimated as $J = 11.2\,k_B$ [2], $J'/J = 8.5 \times 10^{-4}$ [3]. Below the Curie temperature $T_c = 6.25$ K, the spins lie in the c plane and have very small XY-like anisotropy.

Recently it has been reported that the magnetization $M(T)$ of K_2CuF_4 shows a temperature dependence $M(T) \propto (1 - CT^{3/2})$ ($C \simeq 0.030$) in the temperature range of $(0.25$–$0.86)T_c$ [4], which is in contradiction to the two-dimensionality of this system. Furthermore, a strong magnetic field dependence of the magnetization has been reported [5].

In the present paper, the magnetization of K_2CuF_4 is calculated with the use of the renormalized spin wave theory. For sufficiently weak magnetic anisotropy and interplane exchange interaction as compared with the predominant intraplane one, the analytical expressions of magnetization and renormalization constants are obtained and the behaviors of the magnetization can reasonably be explained. The dynamical interaction in the two-dimensional case is shown to be much more important than in the case of the three-dimensional magnetic system.

2. Theory

Let the easy axis of the spin be the z axis and the x axis be normal to it in the c plane. Then the hamiltonian of the present system can be written as follows:

$$\mathcal{H} = -2J \sum_{\substack{\text{intra-}\\\text{plane}}} [(1-\xi)S_i^x S_j^x + (1-\eta)S_i^y S_j^y + S_i^z S_j^z]$$
$$-2J' \sum_{\substack{\text{inter-}\\\text{plane}}} \mathbf{S}_i \cdot \mathbf{S}_l - g\mu_B H \sum S_i^z, \quad (1)$$

where $\xi(=J_a^x/J)$ and $\eta(=J_a^y/J)$ denote the in-plane and the out-of-plane anisotropies, respectively.

Following the spin wave theory [6], eq. (1) is rewritten in terms of magnon operators. Neglecting the higher order terms than sixth degree, the hamiltonian can be diagonalized by the Bogoliubov transformation, provided that the Hartree approximation is applied for the fourth order terms.

With the dimensionless quantities $t = k_B T/\pi \mathcal{J}$, $\lambda = g\mu_B H/\mathcal{J}$, and $\nu = 2z_0'J'S/\mathcal{J}$, reduced by $\mathcal{J} = 2z_0 JS$, the reduced magnetization $\sigma(t, \lambda)$ $[= M(t, \lambda)/Ng\mu_B]$ in the free spin wave approximation is obtained as

$$\sigma_{FSW}(t, \lambda) = S - t \log t' - \frac{\pi^3}{12} t^2 - \frac{\lambda}{2\pi} - \cdots, \quad (2a)$$

$$t' \equiv 2\pi t/\kappa, \quad (2b)$$

$$\kappa = \tfrac{1}{2}(\sqrt{\lambda + \xi} + \sqrt{\lambda + \xi + 2\nu}) \\ \times (\sqrt{\lambda + \eta} + \sqrt{\lambda + \eta + 2\nu}), \quad (2c)$$

neglecting the terms of linear in ξ, η, ν and of quadratic in λ.

The effects of the dynamical interaction can be comprised in renormalization constants ξ_i ($i = 1, 2, 3$) which modify the exchange integrals included implicitly in eqs. (2a–c) as functions of t and λ and are defined by

$$J_{\text{eff}}(t, \lambda) = J\xi_1(t, \lambda),$$

$$J'_{\text{eff}}(t, \lambda) = J'\xi_2(t, \lambda), \quad (3)$$

$$J^{x,y}_{a\,\text{eff}}(t, \lambda) = J^{x,y}_a \xi_3(t, \lambda).$$

These renormalization constants can be obtained as follows:

$$\xi_1(t, \lambda) = 1 - \frac{t^2}{S}\left[\frac{\pi^3}{6} - \frac{\lambda}{t}\left(\log\frac{\pi t}{\lambda} + 1\right)\right] - \cdots,$$

$$\xi_2(t, \lambda) \simeq \frac{1}{S}[\sigma_{\text{FSW}}(t, \lambda) + tQ(\lambda)], \quad (4)$$

$$\xi_3(t, \lambda) = \frac{1}{S}[2\sigma_{\text{FSW}}(t, \lambda) - S],$$

where

$$Q(\lambda) = \frac{1}{2}\left[\left(\sqrt{1 + \frac{\xi(xi) + \lambda}{2\nu}} - \sqrt{\frac{\xi(xi) + \lambda}{2\nu}}\right)^2 + \left(\sqrt{1 + \frac{\eta + \lambda}{2\nu}} - \sqrt{\frac{\eta + \lambda}{2\nu}}\right)^2\right].$$

3. Results and discussion

The results of calculation of the magnetization of K_2CuF_4 are shown in figs. 1 and 2. The values of parameters used in the calculation are those of $\nu = H'_E/H_E = 1.7 \times 10^{-3}$ [3], $J^y_a = 0.05\,k_B$ [5] (equivalent to $\eta = 4.5 \times 10^{-3}$) and $\xi \sim 10^{-5}$ [1]. In fig. 1, the best fit with the experiment can be obtained with a slightly modified value of $J = 11.0\,k_B$. The errors included in eq. (2a) are estimated as $\sqrt{\nu(\nu + \eta)}/2\pi \sim 5 \times 10^{-4}$.

For simplicity, consider the case of $\xi = \eta = 0$. In the absence of an external field, the spontaneous magnetization $\sigma(t)$ can be obtained from eqs. (2a–c) and eq. (3) as

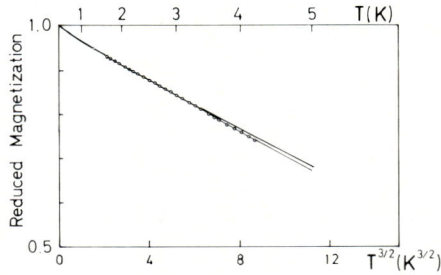

Fig. 1. Temperature dependence of the spontaneous magnetization of K_2CuF_4. The thick line has been calculated with the use of $J = 11.2\,k_B$ and the fine line with $J = 11.0\,k_B$ and normalized at $T = 1.66$ K. The open circles denote the experiment by Kubo.

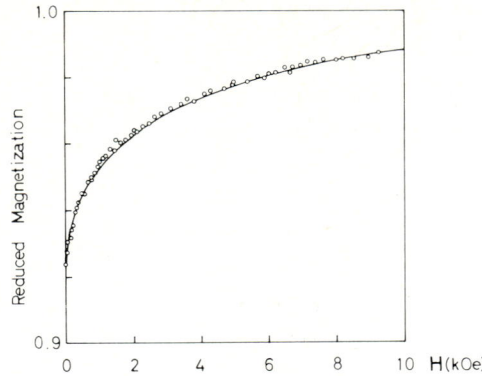

Fig. 2. Magnetic field dependence of the magnetization of K_2CuF_4 at $T = 1.77$ K. The full line shows the present calculation normalized at $H = 0$ and the open circles denote the experiment by Kubo and Kaneshima.

$$\sigma(t) = S - t\log t' - \frac{t^2}{S}(\log t' - 1) - \frac{\pi^3}{12}t^2 - \cdots. \quad (5)$$

The third term comes from the dynamical interaction. The apparent $T^{3/2}$ temperature dependence is caused by competing effects of both the free spin wave and the spin wave interaction in a certain temperature range. The increment of magnetization due to an external field is given by

$$\sigma(t, \lambda) - \sigma(t, 0) = 2t\,\text{arcsinh}\left(\frac{g\mu_B H}{4z'_0 J'S}\right)^{1/2} - \frac{\lambda}{2\pi} + \cdots. \quad (6)$$

The strong magnetic field dependence of the magnetization as compared with the three-dimensional case is attributable to the very small value of J'.

The present results are also applicable to the layer-type ferromagnet having an antiferromagnetic interlayer coupling ($J' < 0$) as far as an external magnetic field is absent.

References

[1] L.J. de Jongh and A.R. Miedema, Advan. Phys. 23 (1974) 1.
[2] I. Yamada, J. Phys. Soc. Jap. 33 (1972) 979.
[3] H. Yamazaki, J. Phys. Soc. Jap. 34 (1973) 270.
[4] H. Kubo, J. Phys. Soc. Jap. 36 (1974) 675.
[5] H. Kubo and N. Kaneshima, private communication.
[6] F. Keffer, Encyclopedia of Physics, Vol. 18/2 (Springer-Verlag, 1966) p. 1.

HIGH-TEMPERATURE SUSCEPTIBILITY OF COPPER COMPOUNDS CONTAINING $CuCl_4^{2-}$ OR $Cu_2Cl_6^{2-}$ IONS

M.F. MOSTAFA, M. SEMARY and M.A. AHMED

Physics Department, Faculty of Science, Cairo University, Giza, Egypt

We have measured the magnetic susceptibility of $[(C_2H_5)_2NH_2]_2CuCl_4$, $(iso-C_3H_7NH_3)_2CuCl_4$, and $(iso-C_3H_7NH_3)CuCl_3$ in the temperature range $80 < T < 400$ K. These compounds show thermochromic first order phase transitions at 315.5, 329.5 and 324.5 K, respectively, which is related to the change in geometry of the chlorine coordination of Cu^{2+}. From the susceptibility data, estimates of the super-exchange interaction can be inferred. For $[(C_2H_5)_2NH_2]_2CuCl_4$ and $(iso-C_3H_7NH_3)_2CuCl_4$ our results evidenced the presence of ferromagnetic interactions between the Cu ions within the network formed by $CuCl_4^{2-}$ ions. For the $(iso-C_3H_7NH_3)CuCl_3$ the interaction between two Cu^{2+} ions forming the dimer can be estimated. The interaction between the dimers is found to be much weaker.

The compounds $(iso-C_3H_7NH_3)_2CuCl_4$, $[(C_2H_5)_2NH_2]_2CuCl_4$ and $(iso-C_3H_7NH_3)CuCl_3$, henceforth, $(IPA)_2CuCl_4$, $(DEA)_2CuCl_4$ and $(IPA)CuCl_3$, respectively, have been investigated. The first two compounds contain $CuCl_4^{2-}$ ions while the third is made up of $Cu_2Cl_6^{2-}$ ions. The compounds show thermochromic transitions at 315.5, 329.5 and 324.5 K, respectively, caused by a change in the structural symmetry of the transition metal ion from tetrahedral into square planar [1, 2]. These changes will be reflected in the magnetic behaviour of these compounds.

The magnetic susceptibility as a function of temperature for the three compounds, in the range 300–400 K, is shown in fig. 1. The midpoints between susceptibility maxima and minima occur at the corresponding thermochromic temperature (T_{thermo}). The thermal hysteresis observed in the neighbourhood of T_{thermo} is an indication that the transition is of first order.

The near infra-red spectra obtained for $(IPA)CuCl_3$ below and above T_{thermo} indicates that the Cu(II) ions have a square planar geometry and tetrahedral distortion respectively. This is similar to the case found for the other two compounds. Willett et al. [1] have reported that both $(IPA)_2CuCl_4$ and $(DEA)_2CuCl_4$ consist of $CuCl_4^{2-}$ ions of tetrahedrally distorted geometry at $T > T_{thermo}$. However, the building units in $(IPA)CuCl_3$ are $Cu_2Cl_6^{2-}$ dimers as indicated by its pleochromic property and its similarity to the other compounds of known crystal structures.

At temperatures above T_{thermo}, the molar susceptibility could be fitted to the Curie–Weiss law with $X_M = 0.55/(T + 18)$, $X_M = 0.54/(T + 58)$, and $X_M = 1.17/(T - 30)$ for $(IPA)_2CuCl_4$, $(DEA)_2CuCl_4$ and $(IPA)_2Cu_2Cl_6$, respectively. The corresponding μ_{eff} are 2.09, 2.09 and 3.06 B.M., respectively. At temperatures below T_{thermo}, the data can be approximated by $X_M = 0.47/(T - 20)$, $X_M = 0.46/(T - 9)$ and $X_M = 0.96/(T + 4)$ for the three compounds, respectively. The corresponding values of μ_{eff} are 1.95, 1.92 and 2.77 B.M. The results are summarized in table I. The values of μ_{eff} obtained above and below T_{thermo} are in agreement with what would be expected for $CuCl_4^{2-}$ and $Cu_2Cl_6^{2-}$ ions of tetrahedral and square planar geometries.

The plot of the reciprocal molar magnetic susceptibility vs. temperature showed a small

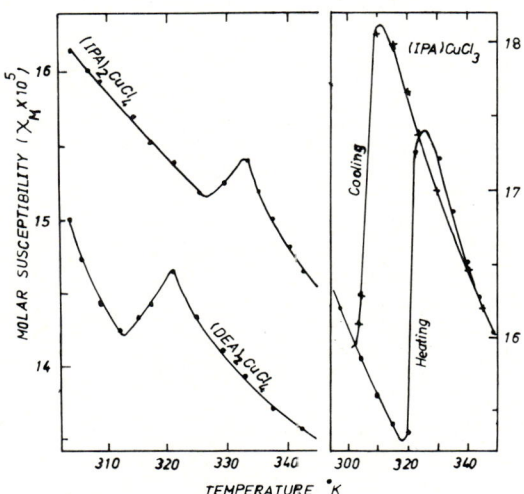

Fig. 1. Molar magnetic susceptibility vs. temperature in the range of the thermochromic transitions.

Table I

Compound	$T > T_{thermo}$			$T < T_{thermo}$		
	C	θ	μ_{eff}B.M.	C	θ	μ_{eff}B.M.
$(IPA)_2CuCl_4$	0.55	−18	2.09	0.47	20	1.95
$(DEA)_2CuCl_4$	0.54	−58	2.09	0.46	9	1.92
$(IPA)_2CuCl_6$	1.17	30	3.06	0.96	−4	2.77

but significant deviation from the Curie–Weiss behaviour below 200 K for the first two compounds, and below 95 K for the (IPA)CuCl$_3$. For the former this is due to a ferromagnetic interaction in the system, as is evidenced when plotting $X_M T$ vs. T. As the temperature decreases below 200 K, the value of $X_M T$, for the first two compounds, increases implying that the magnetic moment/particle is increasing. This proposed ferromagnetic interaction, is supported by the results of De Jongh et al. [3] on the magnetic interaction of Cu(II) complexes of the general formula $(C_nH_{2n+1}NH_3)_2CuCl_4$. These complexes where the $CuCl_4^{2-}$ ions are almost of square planar geometry were reported to behave as two-dimensional Heisenberg ferromagnets at low temperatures.

The results obtained for $(IPA)_2CuCl_4$ and $(DEA)_2CuCl_4$ were compared to the series expansion of $X_M T/C$ in powers of J/kT for both the linear chain and the quadratic Heisenberg ferromagnets [4, 5]. A good fit is obtained with the simple quadratic model giving $|J/k| = 12.5$ and 4.5, respectively (see fig. 2).

The $(IPA)_2Cu_2Cl_6$ compound shows a quite interesting behaviour. At $T > T_{thermo}$, the obtained value of C corresponds to a $g = 2.27$ for a system of N spin 1 moments. It is thus possible, from structural information and magnetic behaviour of similar compounds, to predict that the compound consists of magnetically isolated $Cu_2Cl_6^{2-}$ dimers which are tetrahedrally distorted at $T > T_{thermo}$, but are of square planar geometry at $T < T_{thermo}$. The high temperature expansion of the Bower–Bleany equation [6] yields $\theta = 2J/3k$. This gives $|J/k| = 45$ K in reasonable agreement with the value obtained for similar compounds [7]. At $T < T_{thermo}$. The value of C corresponds to $g = 2.26$ for a system of $2N$ spin $\frac{1}{2}$ ions with $|J/k| = 6$ K. The deviation from the

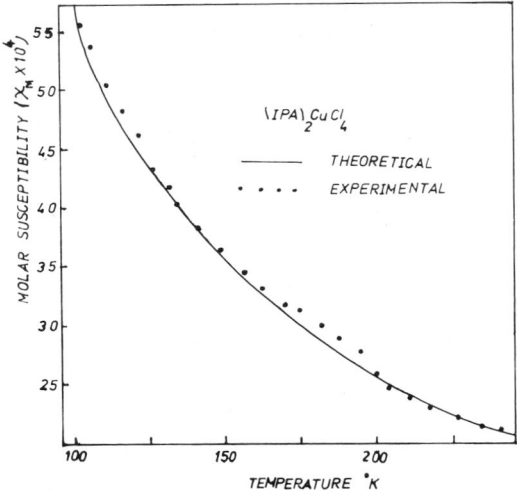

Fig. 2. Theoretical and experimental molar susceptibility as function of temperature for $(IPA)_2CuCl_4$.

Curie–Weiss behaviour observed below 95 K could be a result of a dimer–dimer interaction and the system becomes a linear chain (possibly antiferromagnet) at lower temperatures. This argument is supported by the EPR results obtained by Willett et al. [7]. At 78 K, they observed a single exchange narrowed anisotropic line, rather an $S = 1$ spectrum. Detailed magnetic behaviour of this compound has to await crystal structure analysis and low temperature magnetic work.

The authors wish to thank Prof. R.D. Willett for the enlightening discussions.

References

[1] R.D. Willett, J.A. Haugen, J. Lebsak and J. Morry, Inorg. Chem. 13 (1974) 2670.
[2] M.A. Ahmed, M.Sc. Thesis, University of Cairo, Egypt (1976).
[3] L.J. De Jongh and W.D. van Amstel, J. de Physique 32 C1 (1971) 880.
[4] G.S. Rushbrook and P.J. Wood, Proc. Phys. Soc. 68A (1955) 116.
[5] G.A. Baker, H.E. Gilbert, J. Eve and G.S. Rushbrook, Phys. Lett. 25A (1967) 207.
[6] B.Bleany and K.D. Bowers, Proc. Roy. Soc. (London) A214 (1952) 451.
[7] R.D. Willett, private communication, Washington State University, Pullman, Washington.

CALCULATED SPECIFIC HEAT AND SUSCEPTIBILITY FOR THE QUADRATIC $S = \frac{1}{2}$ HEISENBERG ANTIFERROMAGNET, COMPARED WITH EXPERIMENTAL DATA ON $Cu(C_5H_5NO)_6(BF_4)_2$

R. NAVARRO*, H.A. ALGRA, L.J. DE JONGH
Kamerlingh Onnes Laboratory, University of Leiden, The Netherlands

R.L. CARLIN and C.J. O'CONNOR
Dept. of Chemistry, U.I.C.C., Chicago, Illinois 60680, USA

Predictions for the thermodynamic behavior of the quadratic $S = \frac{1}{2}$ Heisenberg antiferromagnet are obtained from high-temperature series expansions and spin-wave theory. Comparison is made with data on $Cu(C_5H_5NO)_6(BF_4)_2$. Large zero-point effects are observed.

The high-temperature series [1] for the magnetic specific heat (C_m) and susceptibility (χ) of the quadratic-layer Heisenberg antiferromagnet with spinvalue $S = \frac{1}{2}$ has been analysed using the Padé-approximant technique. The results are shown in figs. 1 and 2, respectively, as the curves labelled HTS. Vertical bars denote the ranges in which the direct approximants are found. For $t \equiv kT/|J| \le 1.2$ the uncertainty is very large. However, C_m appears to have a maximum at $t \simeq 1.2$ (with $C_m/R \simeq 0.45$). Here J denotes the exchange constant. The antiferromagnetic χ has a maximum at $t = 1.89 \pm 0.04$ (with $\bar{\chi} = 0.0469 \pm 0.0001$).

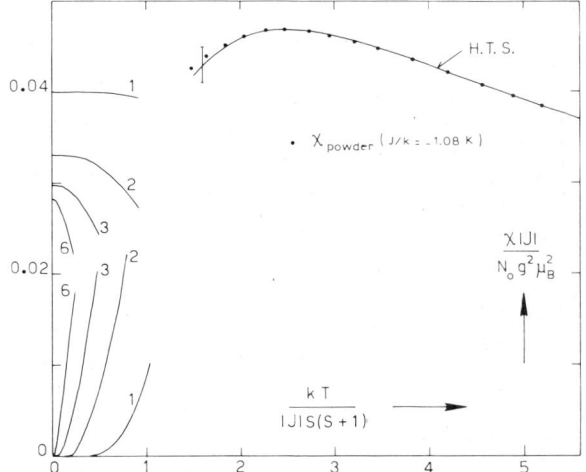

Fig. 2. Calculated results for the susceptibility $\bar{\chi} = \chi |J|/N_0 g^2 \mu_B^2$.

Low-temperature predictions are obtained from spin-wave theory, adding a small uniaxial anisotropy term to the Heisenberg hamiltonian

$$\mathcal{H} = -2J \sum_{\langle i,j \rangle} \mathbf{S}_i \cdot \mathbf{S}_j.$$

The calculational procedures have been previously outlined [2]. Dynamical spin-wave interactions [3] are accounted for. The numbers 1–6 labelling the curves in figs. 1 and 2 correspond to values for $-\log \alpha$, where α is the anisotropy parameter $\alpha \equiv H_A/H_E$. The Kubo prediction [4], $C_m/R = 0.287\, t^2$, for $\alpha = 0$ is included in fig. 1. The difference with our curves for $\alpha = 10^{-4} - 10^{-6}$ is mainly due to the correction for dynamical interactions. Although for finite α

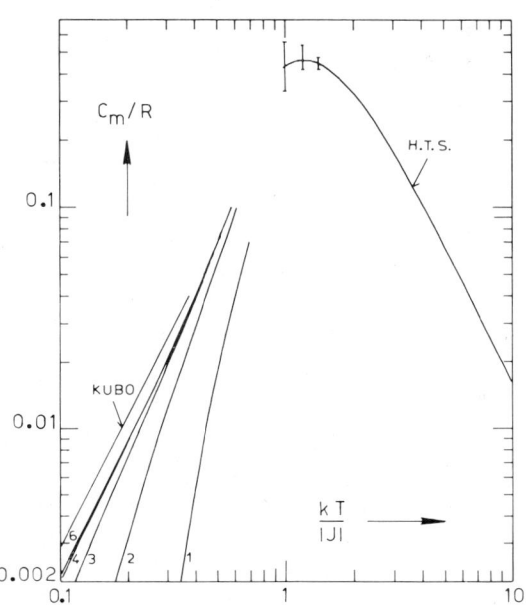

Fig. 1. Calculated results for the specific heat.

* On leave from University of Zaragoza, Spain.

the behavior of C_m for $t \to 0$ is exponential, the curves for $\alpha \leq 10^{-4}$ display an apparent t^2 dependence in the region shown in fig. 1, up to $t \simeq 0.25$.

As regards χ the upper and lower curves in fig. 2 refer, respectively, to the perpendicular ($\chi_\perp$) and the parallel ($\chi_\parallel$) part of the susceptibility. For $\alpha = 0$ the behavior of both $\chi_\parallel$ and $\chi_\perp$ becomes anomalous. However, in the formalism applied no account is taken of the kinematical spin-wave interactions [3] (KSI), which will be very important for the quadratic $S = \frac{1}{2}$ antiferromagnet [5, 6]. We have applied (and partially extended) Herbert's treatment [5] to calculate the zero-point spin reduction $\Delta S(\alpha)$ and the ground-state energy shift $e(\alpha)$, including the KSI. For $\alpha = 10^{-3}$, for example, we find [7] at $T = 0$: $\Delta S(\alpha) = 0.134$ and $e(\alpha) = 0.72$, to be compared with the values $\Delta S'(0) = 0.183$ and $e'(0) = 0.632$ uncorrected for KSI. Accordingly, the value of $\bar{\chi}_\perp (T = 0)$ would change from 0.02975 (fig. 2, curve 3) to 0.0345. We have not yet succeeded in obtaining $e(\alpha)$ as a function of T, corrected for KSI, so that the effect of KSI upon the T-dependence of χ and C_m is still to be established.

We now compare these results with data on $CuL_6(BF_4)_2$, where $L = C_5H_5NO$, which compound apparently mimics a quadratic antiferromagnet. This contrasts with the isomorphous Co^{2+} salts which have a simple cubic magnetic structure [8]. We attribute this to a Jahn–Teller distortion of the octahedral oxygen environment of the Cu^{2+} ions, leading to a strong lattice-anisotropy in the superexchange. This assumption has yet to be verified, however, by e.g. crystallographic studies at low T.

The specific heat was measured for $0.1 < T < 20$ K, and representative data are shown in fig. 3 (up to 9 K). For $T > 1.5$ K the data are well described by curve b, which is the sum of a Debije lattice term aT^3, with $a = (2.6 \pm 0.2) \times 10^{-3}$ K^{-3} (curve a) and the series prediction of fig. 1 (curve HTS), the scaling upon the experiment yielding $J/k = -1.10 \pm 0.05$ K. Below the short-range order maximum ($C_m/R = 0.44 - 0.45$ at $T_{max} \simeq 1.4$ K) a minor discontinuity is observed at $T_c = 0.63 \pm 0.02$ K, which is attributed to the onset of 3-d long-range order. For $0.1 < T < 0.3$ K the data are described by curve e, which is the sum of a cT^2 term (curve c) with $c = 0.35 \pm 0.02$ K^{-2},

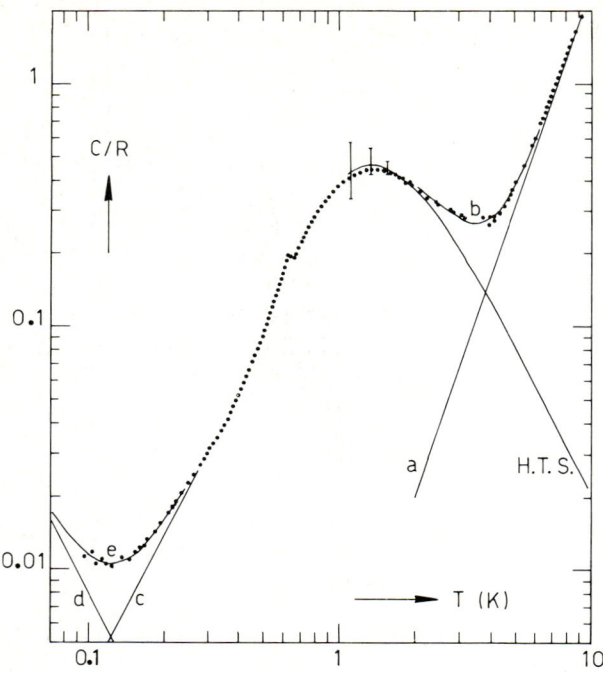

Fig. 3. Experimental specific heat data (●) on $Cu(C_5H_5NO)_6(BF_4)_2$.

and a hyperfine contribution dT^{-2} (curve d) with $d = (8 \pm 0.5) \times 10^{-5}$ K^2. The total electronic entropy equals $R \ln 2$ within 2%, whereas the total magnetic energy is obtained as $U_{tot}/R = 1.5 \pm 0.1$. Theoretically, $U_{tot}/N_0 J = zS^2 (1 + e(\alpha)/zS)$. With $z = 4$, $S = \frac{1}{2}$ and the above J/k, we thus find $e(\alpha) = 0.7 \pm 0.2$, in accord with theory.

The hyperfine contribution can be calculated from the expression $CT^2/R = (A_\parallel^2 + 2A_\perp^2)S(S+1)I(I+1)/9k^2$, using the values $A_\parallel = (131 \pm 2) \times 10^{-4}$ cm^{-1} and $A_\perp = (30 \pm 3) \times 10^{-4}$ cm^{-1} as found [9] from ESR on Cu^{2+} doped in $ZnL_6(BF_4)_2$. With $I = \frac{3}{2}$, $S = \frac{1}{2}$, one has $CT^2/R = (12.4 \pm 0.4) \times 10^{-5}$ K^2. Comparison with the experimental value $d = (8 \pm 0.5) \times 10^{-5}$ K^2 points to a spin reduction $\Delta S = 0.14 \pm 0.03$, again in accord with theory.

The observed CT^2 spin-wave term would imply a low value for α (cf. fig. 1). However the value for c is unexpectedly large. The Kubo prediction leads, with $J/k = -1.10$ K, to $c = 0.24$ K^{-2}, and application of the dynamical corrections leads to an even lower (apparent) value: $c \simeq 0.17$ (cf. fig. 1). Although other factors may play a role, we suspect that the discrepancy will be due at least in part to the neglect of kinematical interactions in the theory.

As regards χ, we have thus far only available [10] the T-dependence of χ_{powder}, for $1.2 < T < 4.2$ K. These data are scaled on the HTS-curve in fig. 2, giving $J/k = -1.08 \pm 0.03$ K, in agreement with the value found from C_m. Since χ was not measured in absolute units the evidence is still unconclusive. However, the χ data clearly show the magnetic interaction to be antiferromagnetic.

The investigations are supported in part by Stichting "F.O.M.". One of us (R.N.) thanks the "Juan March" Foundation for financial support.

References

[1] G.A. Baker Jr., H.E. Gilbert, J. Eve and G.S. Rushbrooke, Phys. Lett. 25A (1967) 207.
[2] R. Navarro, J.J. Smit, L.J. de Jongh, W.J. Crama and D.J.W. IJdo, Physica 83B (1976) 97.
[3] F.J. Dyson, Phys. Rev. 102 (1956) 1217.
[4] R. Kubo, Phys. Rev. 87 (1952) 568.
[5] D.C. Herbert, J. Math. Phys. 10 (1969) 2255
[6] T. Ishikawa and T. Oguchi, Progr. Theor. Phys. 54 (1975) 1282.
[7] R. Navarro, Thesis, Univ. of Zaragoza (1976).
[8] H.A. Algra, L.J. de Jongh, W.J. Huiskamp and R.L. Carlin, Physica 83B (1976) 71.
[9] J. Reedijk, Delft Univ. of Technology, private communication.
[10] A.J. van Duyneveldt, private communication.

SPIN WAVES IN THE QUASI TWO-DIMENSIONAL HEISENBERG FERROMAGNET K_2CuF_4: NEUTRON EXPERIMENTS AND THEORY

F. MOUSSA
Laboratoire Léon Brillouin, CEN Saclay, BP no. 2, 91190 Gif-sur-Yvette, France

and

J. VILLAIN
DRF-DN, CEN Grenoble, 85X 38041 Grenoble, Cédex, France

We interpret, by a theory giving the scattering law $S(Q, \omega)$, the spin wave spectrum of K_2CuF_4 and its line shape, from 2 to 14 K (the Curie temperature of K_2CuF_4 is 6.25 K). The agreement between theory and experiments is good up to 9 K and only qualitative above this temperature.

1. Introduction

K_2CuF_4 is a good example of a quasi two-dimensional (2-D) Heisenberg ferromagnet. Its crystallographic and thermodynamic properties have been described in refs. 1 and 2. K_2CuF_4 exhibits long-range magnetic order (LRO) below the Curie temperature $T_c = 6.25$ K [1, 2]. However, short-range order persists well above T_c; on the other hand, LRO is only partial between T_c and, say, $\frac{1}{2}T_c$, and in this region, short-range order is also an important feature. A characteristic line shape arises in neutron scattering experiments, and the purpose of the present report is the analysis of this lineshape.

2. Neutron inelastic scattering experiments

The experimental results have been reported elsewhere [3] and will be briefly recalled here. There is no dispersion along the c axis, as is expected for a quasi-2-D magnet. Zone boundary magnons are observed up to 14 K, showing the existence of an intense short-range order up to $2 T_c$; magnons with lower wave vector q disappear at lower temperature, as is expected; the broadening is important.

3. Theory

K_2CuF_4 will be treated as a Heisenberg, $s = \frac{1}{2}$, ferromagnet on a square lattice, with nearest neighbour interaction J (energy difference between singlet and triplet state for one pair: $2J$).

A kind of local magnetization M_i can be defined on each site i by [4]:

$$M_i = \frac{1}{\sqrt{N}} \sum_{k<k_0} S_k e^{-ik \cdot R_i},$$

where

$$S_k = \frac{1}{\sqrt{N}} \sum_i S_i e^{ik \cdot R_i}.$$

Here, N is the number of sites and k_0 is a cut-off, much smaller than wave vectors accessible to neutrons in the first Brillouin zone, but large enough to satisfy

$$s^2 - \langle M_i^2 \rangle \ll s^2, \quad \text{with} \quad s = \tfrac{1}{2}.$$

For each site i, one can define a set of orthogonal unit vectors u_i, v_i, w_i, with

$$w_i = M_i/|M_i|.$$

The definition of u_i and v_i is not complete yet: we shall come back to this point. We define operators S_i^z, S_i^x, S_i^y by

$$S_i = S_i^z w_i + S_i^x u_i + S_i^y v_i.$$

Bose operators a_i^+ and a_i can now be defined by the usual Holstein–Primakoff transformation:

$$S_i^z = S - a_i^+ a_i,$$

$$S_i^+ = S_i^x + iS_i^y = \sqrt{2S}\left(1 - \frac{a_i^+ a_i}{2S}\right)^{1/2} a_i,$$

$$S_i^- = S_i^x - iS_i^y = \sqrt{2S} a_i^+ \left(1 - \frac{a_i^+ a_i}{2S}\right)^{1/2},$$

The Heisenberg hamiltonian can now be rewritten as

$$\mathcal{H} = -\tfrac{1}{2}J \sum_{\langle i,j \rangle} w_i, w_j + \mathcal{H}_{sw} + \mathcal{H}_{int},$$

where

$$\mathcal{H}_{sw} = J \sum_{\langle i,j \rangle} (a_i^+ a_i + a_j^+ a_j - a_i^+ a_j - a_j^+ a_i),$$

and $\mathcal{H}_{int}$ is a complicated function of the w_i, u_i, v_i, a_i^+ and a_i values. It can be shown [5] that $\mathcal{H}_{int}$ can be neglected for a convenient choice of the u_i values. Then a straightforward calculation yields the spin pair correlation function:

$$\langle S_i \cdot S_j(t) \rangle = \langle w_i \cdot w_j(t) \rangle$$

$$\times \left(S^2 - \frac{2S}{N} \sum_{k > k_0} \frac{1}{\exp[\beta \hbar \omega_k] - 1} \right)$$

$$+ \langle u_i \cdot u_j(t) + v_i \cdot v_j(t) \rangle$$

$$\times \frac{S}{2N} \sum_{k > k_0} \exp[i k \cdot (R_j - R_i) + i \omega_k t]$$

$$+ \frac{2 \cos[k \cdot (R_j - R_i) + \omega_k t]}{\exp[\beta \hbar \omega_k] - 1},$$

where $\hbar \omega_q = 4J(1 - \cos \pi q)$.

It can be shown [5] that the choice of the u_i values implies, for Heisenberg ferromagnets,

$$\langle u_i \cdot u_j(t) + v_i \cdot v_j(t) \rangle \simeq \langle u_i \cdot u_j + v_i \cdot v_j \rangle$$

$$\simeq \frac{1}{S} \langle M_i \cdot M_j \rangle^{1/2}$$

$$\simeq \frac{1}{S} \langle S_i \cdot S_j \rangle^{1/2}.$$

The above equations yield the time-dependent spin-pair correlation function (and, therefore, the neutron inelastic scattering cross section) in terms of the equal time correlation function; for an ideal, 2-D Heisenberg magnet, the latter is approximately given by:

$$\langle S_i \cdot S_j \rangle \simeq s^2 r_{ij}^{-\eta} \exp[-r_{ij}/\xi],$$

where the exponent η can be determined at low temperature [6] as well as the correlation length ξ [7, 8]

4. Comparison between theory and experiment

The experimental dispersion law is correctly reproduced if $J = 11.4 \, \text{K} \pm 0.1 \, \text{K}$. The experimental lineshape is well reproduced below T_c (fig. 1) and above T_c (fig. 2) except at the

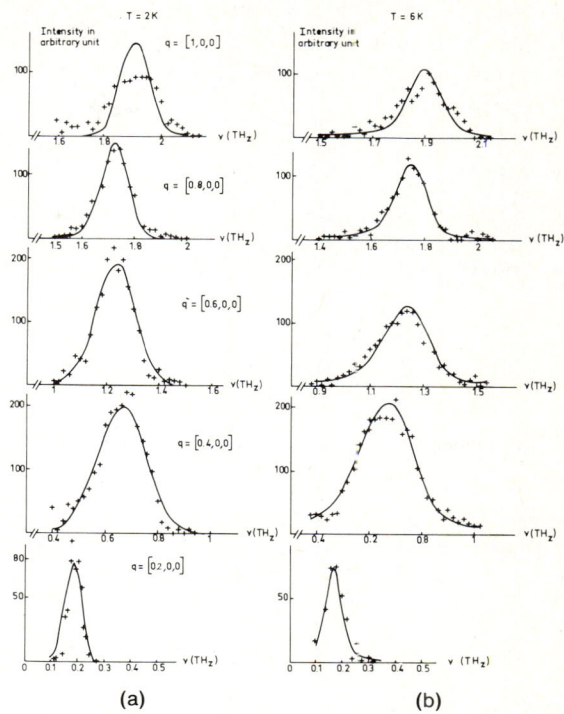

Fig. 1. (a) Fit of magnons at $T = 2$ K; (b) fit of magnons at $T = 6$ K.

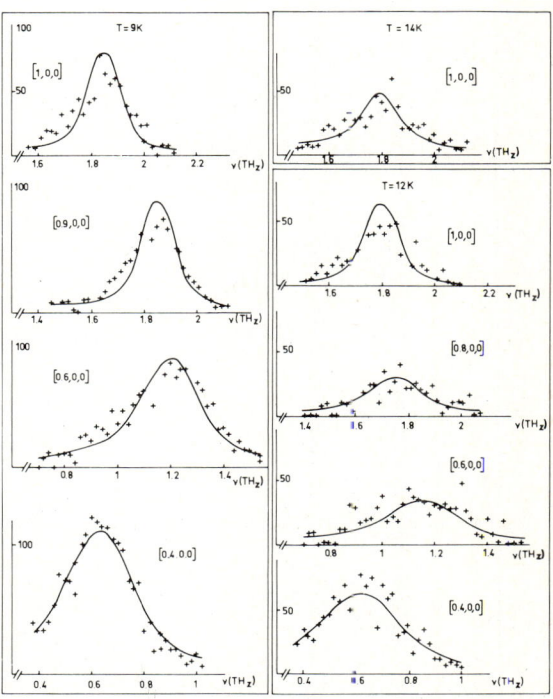

Fig. 2. Fit of magnons at $T = 9$, 12 and 14 K.

zone boundary. The total intensity is well reproduced, except at 12 K: the fit in fig. 2 was only obtained after multiplication by an arbitrary factor $\frac{1}{2}$.

References

[1] I. Yamada, J. Phys. Soc. Jap. 33 (1972) 979.
[2] K. Hirakawa and H. Ikeda, J. Phys. Soc. Jap. 35 (1973) 1328 and 1608.
[3] S. Funahashi, F. Moussa and M. Steiner, Solid State Commun. 18 (1976) 433.
[4] Y. Kuramoto, Progr. Theor. Phys. 40 (1968) 36. We disagree with Kuramoto's choice of the u_i values, and therefore with his results.
[5] F. Moussa and J. Villain, to be published in J. Phys. C.
[6] V.L. Berezinskii and Y. Blank, Sov. Phys. JETP 37 (1973) 369.
[7] A.M. Polyakov, Phys. Lett. 59B (1975) 79.
[8] E. Brézin and J. Zinn-Justin, Phys. Rev. Lett. 36 (1976) 691.

ONE- AND TWO-DIMENSIONAL ANTIFERROMAGNETISM IN SOME NEW IRON III FLUORIDES

J.M. DANCE, F. MENIL, D. HANŽEL*, R. SABATIER, A. TRESSAUD, G. LE FLEM and P. HAGENMULLER

Laboratoire de Chimie du Solide du CNRS, Université de Bordeaux-I, 351, cours de la Libération 33405 Talence, France

Magnetic and Mössbauer resonance studies have allowed for the characterization of three new one- and two-dimensional Fe^{3+} fluorides: $N_2H_6FeF_5$, $NaFeF_4$ and NH_4FeF_4. The intra-chains and intra-layers exchange integrals are calculated from theoretical results. In the case of the two-dimensional compounds the critical exponent β was also determined.

The influence of lattice-dimensionality on the magnetic properties of transition metal compounds has been widely studied these last years [1]. We report here three new examples of one- and two-dimensional magnetic systems with $S = \frac{5}{2}$.

1. One-dimensional antiferromagnetism in $N_2H_6FeF_5$

Chemically chained compounds behave generally as one-dimensional systems and are particularly amenable to theoretical calculations. $N_2H_6FeF_5$, obtained by reaction of hydrazine difluoride with iron fluoride, can be structurally considered as derived from the $A^{II}M^{III}F_5$ and $A_2^IM^{III}F_5$ compounds (A^I = monovalent element, A^{II} = divalent element, M^{III} = 3d element, Al, Ga) which are generally constituted of corner shared (MF_6) octahedra chains isolated by A units.

Powder magnetic measurements on $N_2H_6FeF_5$ show for χ^{-1} vs. T a broad minimum around 100 K and a small dip at 10 K (fig. 1). Weng's calculations [2] for a one-dimensional Heisenberg model have been used to determine the intra-chain exchange integral J/k. In this case, the calculation for an infinite chain with $S = \frac{5}{2}$ leads to $J/k = -10.2$ K.

Mössbauer effect experiments performed down to liquid helium give the value of the Néel temperature at which the chains are three-dimensionally coupled ($T_N = 9$ K). This temperature corresponds roughly to the appearance of a weak ferromagnetic component (fig. 1). The ratio $\eta = |J'/J|$ of the inter-J' and intra-chain J

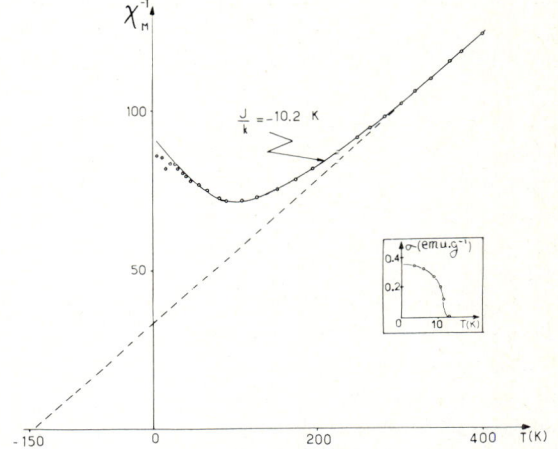

Fig. 1. Magnetic susceptibility for $N_2H_6FeF_5$ (points: experimental values; full curve: theoretical values after Weng's calculations for $S = \frac{5}{2}$).

exchange integrals can be evaluated by Oguchi's formula [3] for an Heisenberg system, giving $\eta \simeq 2 \times 10^{-3}$.

2. Two bidimensional antiferromagnets: $NaFeF_4$ and NH_4FeF_4

2.1. $NaFeF_4$

$NaFeF_4$ is isostructural with $NaNbO_2F_2$ [4]; a view of the structure along $0z$ is shown in fig. 2. The four shared corners are not situated in the same plane leading to "zig-zag layers" in opposition with the $TlAlF_4$ type where planar layers are found [$A^IM^{III}F_4$ phases (A^I = K, Rb, Cs, Tl, NH_4; M = 3d el., Al, Ga)].

Neutron diffraction experiments at 4.2 K on $NaFeF_4$ show a magnetic cell doubled along $0x$ due to the three-dimensional ordering and a lowering of symmetry, the crystallographic unit

---

* Permanent address: Institute Josef Stefan, University of Ljubljana, Ljubljana, Yugoslavia.

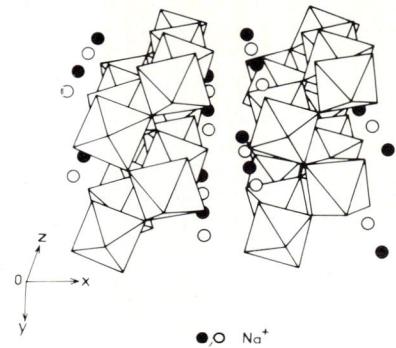

Fig. 2. View of the NaFeF$_4$ structure along the 0z axis.

cell becoming triclinic. The comparison of the calculated and observed intensities of the diagram are in good agreement for a G-mode with the magnetic moments aligned along the **b** axis. Therefore, at 4.2 K, each Fe^{3+} spin is coupled antiparallel with its nearest neighbours of a same layer and with the Fe^{3+} ions of each neighbouring layer deduced by the translation along 0x [5].

Magnetic susceptibility measurements are shown in fig. 3. Above 450 K χ_M^{-1} follows a Curie–Weiss law with $\theta = -325$ K and $C = 4.40$. Below 450 K a flattening of the curve is observed down to 120 K. There is a sharp minimum at about 105 K which corresponds to the three-dimensional ordering temperature. We have used the Rushbrooke and Wood [6] high temperature series expansion to calculate the magnetic susceptibility for an $S = \frac{5}{2}$ quadratic Heisenberg antiferromagnet. The theoretical curves give by interpolation the value of the exchange integral $J/k = -23$ K.

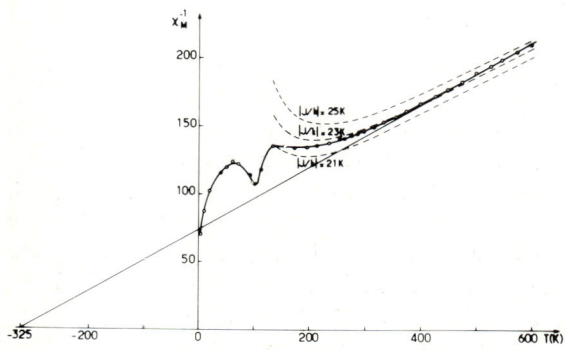

Fig. 3. Magnetic susceptibility for NaFeF$_4$ (full curve: experimental data; dotted lines: theoretical curves calculated from the H.T.S.).

The Mössbauer spectrum at 293 K gives the values of the isomer shift (0.46 mm/s) and the quadrupole splitting (0.69 mm/s). At 111.5 K the Zeeman splitting appears, thus giving the value of the Néel temperature of NaFeF$_4$. The temperature dependence of the hyperfine field (supposed proportional to the magnetization of the sublattice) leads to the independent parameters $D = 1.02$; $\beta = 0.25$.

2.2. NH$_4$FeF$_4$

A similar magnetic study was performed on NH$_4$FeF$_4$ which is isostructural with the TlAlF$_4$ type fluorides (KFeF$_4$, RbFeF$_4$, CsFeF$_4$). The magnetic susceptibility is shown on fig. 4. Using the high temperature series expansion technique we obtain the value of $J/k = -26$ K. From the Mössbauer effect data the three-dimensional ordering temperature was found to be $T_N = 135$ K. The values of D and β were also obtained from the temperature dependence of the hyperfine field: $D = 1.08$; $\beta = 0.27$.

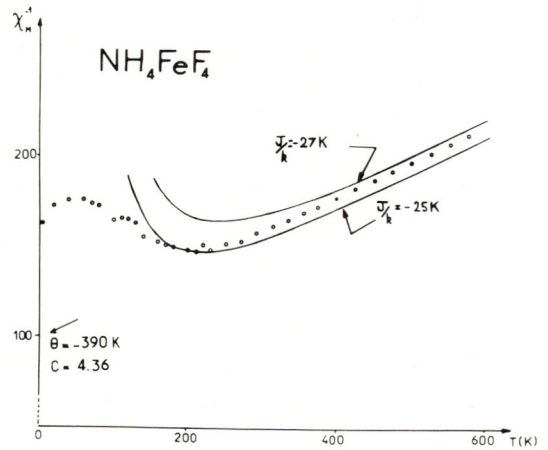

Fig. 4. Magnetic susceptibility for NH$_4$FeF$_4$ (points: experimental; full curves: theoretical curves calculated from H.T.S.).

The exchange integrals of NaFeF$_4$ and NH$_4$FeF$_4$ are similar to the values already found for bidimensional Fe^{3+} fluorides [7]. The discrepancies of the experimental β-values with the theoretical value 0.125 (two-dimensional Ising model), suggests the presence of simultaneous two- and three-dimensional correlations above the Néel temperature.

References

[1] A review is given by L.J. de Jongh and A.R. Miedema in Advan. Phys. 23 (1974) 1.
[2] Chi-Yuan Weng, Ph.D. Thesis, Carnegie-Mellon University, USA (1968).
[3] T. Oguchi, Phys. Rev. 133 (1964) 1098.
[4] S. Andersson and J. Galy, Acta Cryst B25 (1969) 847.
[5] J.M. Dance, R. Sabatier, F. Meril, M. Wintenberger, J.C. Cousseins, G. Le Flem and A. Tressaud, Solid State Commun. 19 (1976) 1059.
[6] G.S. Rushbrooke and J.P. Wood, Mol. Phys. 6 (1963) 409.
[7] M. Eibschutz, G.R. Davidson and H.J. Guggenheim, Phys. Rev. B9 (1974) 3885.

MAGNETIC BEHAVIOUR OF VANADIUM +II IN ONE- AND TWO-DIMENSIONAL SYSTEMS

M. NIEL, C. CROS, G. LE FLEM, M. POUCHARD and P. HAGENMULLER

Laboratoire de Chimie du Solide du C.N.R.S., Université de Bordeaux I, 351 cours de la Libération, 33405 Talence, France

The magnetic behavior of vanadium +II studied in $CsVX_3$ and VX_2 halogenides (X = Cl, Br) can be explained respectively by a 1D and 2D Heisenberg model. J/k values are discussed on the basis of Goodenough–Kanamori superexchange rules.

The synthesis of vanadium +II halogenides has been carried out by solid state reactions. Single crystals have been grown by both Bridgman and sublimation methods for the $CsVX_3$ phases and only by sublimation for VCl_2 and VI_2.

The structure of $CsVCl_3$ was entirely determined by single crystal X-rays techniques [1]. All the $CsVX_3$ phases crystallize with the 1D $CsNiCl_3$ type structure (space group $P6_3/mmc$). The parameters of the cell and the intrachain vanadium–vanadium distances are given in table I.

Table I

$CsVX_3$	a (Å)	c (Å)	c/a	$(V-V)_{intra}$ (Å)
$CsVCl_3$	7.228	6.030	0.834	3.015
$CsVBr_3$	7.56	6.32	0.836	3.16
$CsVI_3$	8.10	6.77	0.836	3.39

The VX_2 phases crystallize in the 2D CdI_2 type structure (space group $C\bar{3}m$). The parameter of the cell and interatomic distances are given in table II.

Magnetic susceptibilities of all these compounds have been measured from 4.2 to 800 K. According to the model of Figgis the experimental susceptibility χ_M^{mes} is related to the spin only value χ_M^S of the $^4A_{2g}$ term by the expression

$$\chi_M^S = \frac{\chi_M^{mes} - 8 k_l^2 N\mu_B^2/\Delta}{1 - 8 k_l^2 \lambda_0/\Delta},$$

where N represents Avogadro's number, μ_B the Bohr magneton, g the Landé factor (measured by E.P.R.), Δ the octahedral crystal field splitting, determined by optical measurements (table

Table II

X	a (Å)	c (Å)	$(V-V)_{intra}$ (Å)	$(V-V)_{inter}$ (Å)	V–X (Å)	$r(V^{2+})+r(X^-)$ (Å)
Cl	3.60	5.84	3.60	5.84	2.54	2.55
Br	3.78	6.23	3.78	6.23	2.67	2.70
I	4.06	6.75	4.06	6.75	2.87	2.94

Table III

	g	k_l	Δ (cm^{-1})
VCl_2	1.97	0.78	9200
VBr_2	1.99	0.43	8600
VI_2	2.02	0	7900
$CsVCl_3$	1.98	0.67	10260
$CsVBr_3$	2.00	0	9060
$CsVI_3$	2.06	0	7850

III), and λ_0 the spin–orbit coupling parameter ($\lambda_0 - 57$ cm^{-1} for the free ion V^{2+}) [2].

Both types of structures, were investigated using a Heisenberg hamiltonian. The thermal variation of the corrected susceptibilities of $CsVCl_3$, $CsVBr_3$ and $CsVI_3$ is given in figs. 1–3. Intra-chain interactions have been obtained by fitting to the Fisher's chain model with $S = \frac{3}{2}$ [3].

For $CsVCl_3$ $J/k = -115$ K, but in the case of $CsVBr_3$ and $CsVI_3$ J/k increases with temperature respectively from -91 to -80 K and from -67 to -54 K. This fact seems to be due to the thermal expansion of the cell, which is more important for the bromide and iodide than for the chloride.

The thermal variation of the susceptibility of VCl_2, VBr_2 and VI_2 are given in fig. 4. The magnetic constants are given in table IV. Intralayer interactions have been calculated using the Rushbrooke and Wood high temperature series expansion with a plane triangular cell [4].

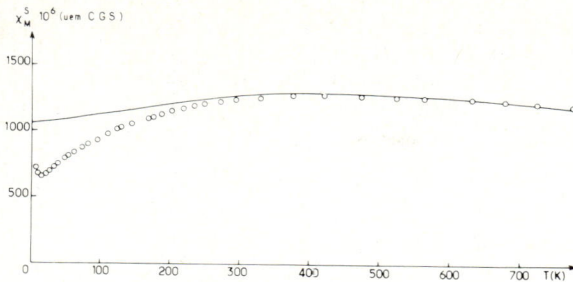

Fig. 1. Thermal variation of the corrected magnetic susceptibility of CsVCl$_3$.

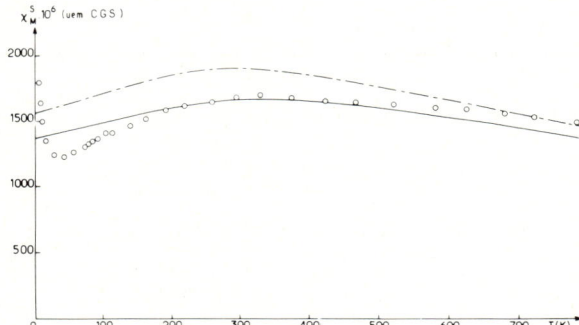

Fig. 2. Thermal variation of the corrected magnetic susceptibility of CsVBr$_3$.

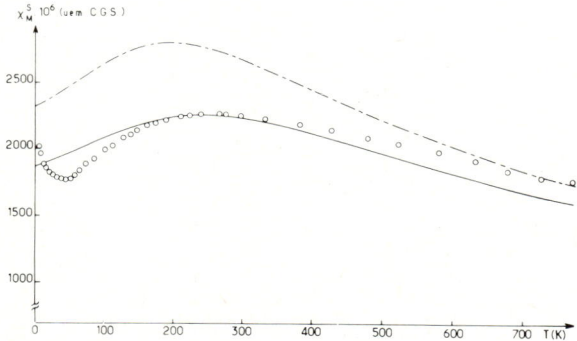

Fig. 3. Thermal variation of the corrected magnetic susceptibility of CsVI$_3$.

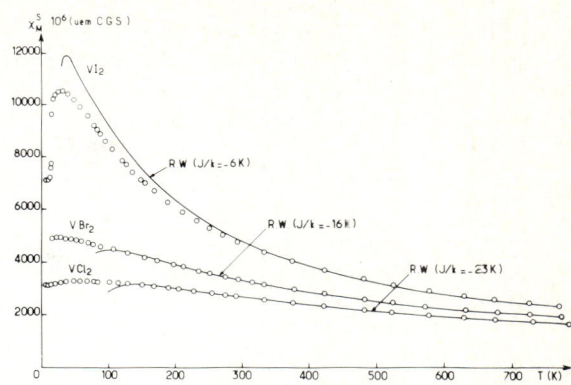

Fig. 4. Thermal variation of the magnetic susceptibility of VX$_2$ halogenides.

Table IV

	C_M	μ_{eff} (μ_B)	$d\sqrt{S(S+1)}$ (μ_B)	θ_P (K)
VCl$_2$	1.96	3.96	3.815	-437
VBr$_2$	2.07	4.07	3.854	-335
VI$_2$	2.07	4.07	3.912	-143

Table V

Direct interactions	t_{2g}–t_{2g}	↑↓
Superexchange interactions (delocalization)	t_{2g}–t_{2g}–t_{2g}	↑↓
	t_{2g}–p–e_g	↑↑
	e_g–s–e_g	↑↓
Superexchange interactions (correlations)	t_{2g}–p–e_g	↑↑
	e_g–p_σ–p_σ–e_g	↑↑
	t_{2g}–p_π–p_π–e_g	↑↑
Result for VX$_2$ and CsVX$_3$		↑↓

Fig. 4 shows that experimental curves are well described when VCl$_2$, VBr$_2$ and VI$_2$ have a J/k value of -23, -16 and -6 K, respectively. In addition a 3D ordering temperature appears at 15 K for VI$_2$.

The signs of direct cation–cation interactions and cation–anion–cation superexchange interactions are reported in table V according to the Goodenough–Kanamori rules for a d^3 ion.

The strongly antiferromagnetic exchange in CsVX$_3$ is clearly the result of direct t_{2g}–t_{2g} interactions and superexchange e_g–s–e_g and t_{2g}–t_{2g}–t_{2g} interactions (this last implying the contribution of the empty t_{2g} halogen orbitals).

The intralayer exchange integrals of VX$_2$ halogenides increase from VCl$_2$ ($J/k = -23$ K) to VBr$_2$ ($J/k = -16$ K) and VI$_2$. ($J/k = -6$ K). These values can be compared with those of the 2D CrX$_3$ compounds [5]. For CrCl$_3$ J/k is positive but the interlayer J'/k is negative; for CrBr$_3$ and CrI$_3$ both intralayer and interlayer exchange integrals involve ferromagnetic couplings ($J/k = 8.25$ and 13.5 K, respectively).

The distinct behaviour between VX$_2$ and CrX$_3$ compounds can be explained by the difference in oxidation states of V^{2+} and Cr^{3+}. For this last ion, the covalency plays a greater role and enhances the delocalization ferromagnetic t_{2g}–p–e_g coupling. The ferromagnetic exchange increases with the sequence: VCl_2–VBr_2–VI_2–$CrCl_3$–$CrBr_3$–CrI_3.

References

[1] M. Niel, C. Cros, M. Vlasse, M. Pouchard and P. Hagenmuller, Mater. Res. Bull. 11(7) (1976).
[2] B.N. Figgis, Introduction to Ligand Fields (Intersciences Publ., New York, 1961).
[3] M.E. Fisher, Amer. J. Phys. 32 (1964) 343.
[4] G.S. Rushbrooke and P.J. Wood, Mol. Phys. 6 (1965) 409.
[5] E.J. Samuelson, R. Silberglitt and G. Shirane, Phys. Rev. B3(1) (1971) 157.

EFFECT OF MAGNETIC AND NON-MAGNETIC IMPURITIES ON THE SUSCEPTIBILITY AND NÉEL TEMPERATURE OF THE LINEAR ANTIFERROMAGNET TMMC

C. DUPAS and J.P. RENARD

Institut d'Electronique Fondamentale, Laboratoire Associé au C.N.R.S., bâtiment 220, Université Paris-Sud, 91405 Orsay-Cedex, France

The paramagnetic susceptibility in 1d TMMC diluted with magnetic (Cu, Ni) and diamagnetic (Cd) impurities increases at low temperature, in agreement with theory. The Néel temperature is strongly depressed by the impurities and this effect increases when the impurity–host coupling decreases. For 0.4 at .% Cd, $T_N(x)/T_N(0) = 0.67$.

1. Introduction

The study of the magnetic properties of one-dimensional (1d) compounds diluted with magnetic and non-magnetic impurities has recently received much attention. Owing to the restricted dimensionality, the effect of impurities is much more important in 1d compounds than in 3d ones. We report here the results of susceptibility and Néel temperature (T_N) measurements in the quasi 1d antiferromagnet (AFM) TMMC with minute concentrations of magnetic and non-magnetic impurities.

The compound $(CH_3)_4 NMnCl_3$ (TMMC) is actually known as the best physical realization of the 1d Heisenberg antiferromagnet. The chains of Mn^{2+} ions are well separated from each other by large tetramethylammonium ions, leading to a small ratio of interchain to intrachain interaction $J'/J \simeq 10^{-4}$. A 3d antiferromagnetic ordered state sets up at $T_N = (0.853 \pm 0.005)$ K, which corresponds to $kT_N/|J|S(S+1) = 0.014$ with $S = \frac{5}{2}$ and $J/k = -7$ K.

We succeeded in growing several monocrystalline samples of TMMC with two different magnetic impurities: Cu^{2+} ($S' = \frac{1}{2}$) and Ni^{2+} ($S' = 1$). In order to realize a more effective breaking of the magnetic chains, we tried to prepare TMMC with diamagnetic impurities. Only Cd^{2+} is found to replace Mn^{2+}, but homogeneous large crystals are difficult to grow. The magnetic susceptibilities of these samples were studied in the He^4 and He^3 temperature ranges by means of a mutual inductance bridge operating at 70 Hz.

2. Paramagnetic susceptibilities

When neglecting the impurity–host interaction J_1, the susceptibility of a 1d isotropic AFM with an atomic concentration of impurity x is, to the first order in x:

$$\chi = \frac{x}{2} \frac{N g^2 \mu_B^2 S(S+1)}{3 k_B T} + (1-x)\chi_0 + x \frac{N g'^2 \mu_B^2 S'(S'+1)}{3 k_B T}.$$

In addition to the pure chain susceptibility χ_0 there are two terms proportional to T^{-1} which correspond to the paramagnetic susceptibility of the $\frac{1}{2}Nx$ finite chain units with an odd number of Mn spins and to the susceptibility of the Nx impurities with spin S'.

We measured the susceptibility of two powder samples of TMMC: Cd for which the chemical analysis gave $x = 0.027$ and 0.09. By plotting $1/\Delta\chi$ versus T, where $\Delta\chi = \chi - \chi_0$, the expected Curie law is well observed (fig. 1). To our knowledge, it is the first observation of the T^{-1} behaviour of AFM chain units without interaction. We did not observe this T^{-1} behaviour in our TMMC samples with magnetic impurities because $J_1 \neq 0$, as already seen in the organic 1d TCNQ compounds [1].

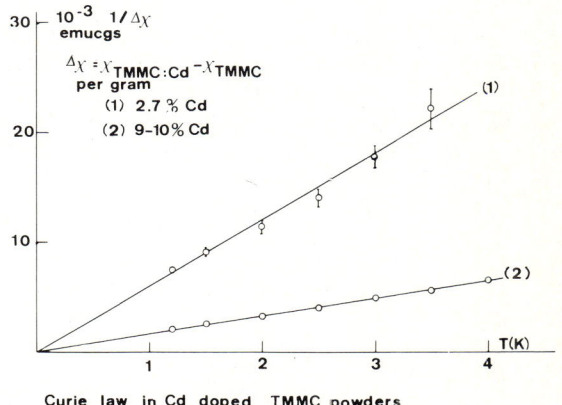

Fig. 1. Curie law in powdered TMMC samples diluted with diamagnetic Cd^{2+} impurities.

The effect of classical spin magnetic impurities has been theoretically studied by Tonegawa et al. [2] and this work was recently extended by Richards [3] to impurities with $S = \frac{1}{2}$. In agreement with Richards we observed an anisotropic behaviour of the susceptibilities. Both $\chi_\parallel$ and $\chi_\perp$ (where $\parallel$ and $\perp$ refer to the chain axis) are larger than those of pure TMMC and increase at low temperature. The effect is more important for the sample with 4% Cu than for that with 4% Ni.

The plot of $3/(\chi_\parallel + 2\chi_\perp)$ versus T suggests that J_1 is antiferromagnetic for TMMC:Ni and ferromagnetic for TMMC:Cu. In this last case, we found the best fit with theory for $J_1 = 1.6$ K in agreement with Richards.

3. Néel temperature

The effect of impurities on the Néel temperature of quasi 1d systems is quite drastic, since they severely limit the 1d correlation length at low temperature. This leads to a strong T_N reduction which was first suggested by Imry et al. [4] A theoretical treatment of this effect for magnetic and non-magnetic impurities was performed by Hone et al. [5] for classical spins. We first reported the experimental observation of this effect by preliminary measurements on TMMC:Cu [6]. As to extend this work, we also measured T_N in TMMC:Ni and TMMC:Cd by the same technique, i.e. by determining the temperature at which a sharp peak occurs in the susceptibility parallel to c.

In all cases, T_N is drastically depressed by the presence of impurities. For TMMC:Cu, the variation of $T_N(x)/T_N(0)$ is in good agreement with the theory for $J'/|J| = 10^{-4}$ and $J_1 = 1.6$ K, which corresponds to $j_1 = J_1 S'/J(S+1) \cong 3 \times 10^{-2}$ (j_1 is a reasonable estimation of the classical impurity–host interaction to host-host interaction). For TMMC:Ni, the effect is less important and corresponds to $|j_1| \simeq 7 \times 10^{-2}$.

As expected, the most drastic effect is ob-

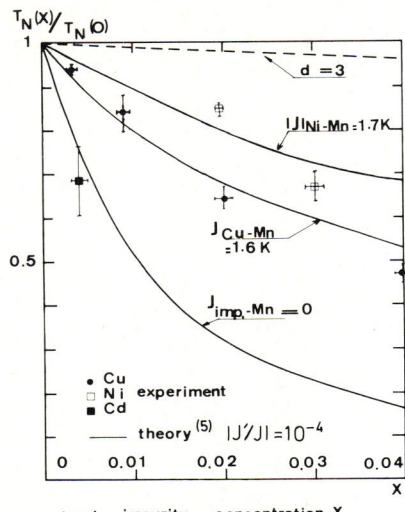

Fig. 2. Variation of the Néel temperature T_N versus atomic impurity concentration x of Ni, Cu and Cd in TMMC.

tained with the diamagnetic Cd impurity since $J_1 = 0$. Indeed for a monocrystalline sample with $x = 0.004$, we found $T_N = 0.57$ K, i.e. $T_N(x)/T_N(0) = 0.68$. This value is close to the theoretical one for $J_1 = 0$ (fig. 2). This is to our knowledge the larger T_N reduction observed as yet in a diluted magnetic compound.

References

[1] L.N. Bulaevskii, A.V. Zvarykina, Yu.S. Karimov, R.B. Lyubovskii and I.F. Shchegolev, Zh. Eksp. Teor. Fiz, 62 (1972) 725 [Sov. Phys. JETP 35 (1972) 384].
[2] T. Tonegawa, H. Shiba and P. Pincus, Phys. Rev. B11 (1975) 4683.
[3] P.M. Richards, Phys. Rev. B14 (1976) 1239.
[4] Y. Imry, P.A. Montano and D. Hone, Phys. Rev. B12 (1975) 253.
[5] D. Hone, P.A. Montano, T. Tonegawa and Y. Imry, Phys. Rev. B12 (1975) 5141.
[6] C. Dupas and J.P. Renard, Phys. Lett. 55 A (1975) 181.

MAGNETIC SPECIFIC HEAT OF $Cs_2Co_{1-x}Zn_xCl_4$. EFFECT OF DILUTION OF A LINEAR CHAIN XY ANTIFERROMAGNET

F.G. BARTOLOME*, H.A. ALGRA, L.J. DE JONGH and W.J. HUISKAMP.

Kamerlingh Onnes Laboratory, University of Leiden, The Netherlands

Specific heat data on $Cs_2Co_{1-x}Zn_xCl_4$ are presented and compared with calculations for a system of isolated diluted XY chains. The contributions from interchain interactions are also studied, in particular the effect of dilution upon the λ-anomaly.

It has recently been shown [1] that the magnetic behavior of Cs_2CoCl_4 approximates that of the linear chain XY antiferromagnet with $S = \frac{1}{2}$. We report here on heat capacity data obtained below 1 K on Cs_2CoCl_4 diluted with Zn.

Samples of $Cs_2Co_{1-x}Zn_xCl_4$ were prepared by melting a mixture of Cs_2ZnCl_4 and Cs_2CoCl_4 powder in the desired proportion in a vacuum-sealed quartz tube. To approximate quenched-site conditions, the mixtures were rapidly cooled from about 800°C to room temperature (in $\simeq 100$ s) and then powdered. Purity and composition of the materials were checked by chemical (and X-ray) analyses.

Representative data for different x are shown in fig. 1, with data [1] for $x = 0$ included. The pure Co salt has a broad maximum near 1 K, resulting from the intrachain exchange (J), and a small λ-peak at $T_c = 0.222 \pm 0.005$ K, reflecting long-range ordering due to the weaker interchain interactions (J'). We note that the specific heat contributions from the lattice and from the higher-lying energy levels (the $|\pm \frac{3}{2}\rangle$ level, which is $\simeq 14$ K above the $|\pm \frac{1}{2}\rangle$ ground-state doublet) have been subtracted from the x

* On leave from University of Zaragoza, Spain.

$= 0$ data. These only become important for $T > 2$ K; for the present data corrections were necessary only for largest values of x.

We first note the drastic reduction of T_c upon dilution: for $x = 0.02$ we find $T_c(x)/T_c(0) \simeq 0.83$ already. This is exemplified by the enlarged plot in fig. 2. It is further seen that the amplitude of the λ-peak is also greatly reduced, such that for $x \geq 0.04$ this anomaly can no longer be resolved. These features confirm the expected [2] critical influence of impurities upon the long-range order in systems of weakly coupled chains. Our experiments also indicate that the small sizes of the λ-peaks, as observed in undiluted quasi one- or two-dimensional magnetic systems [3], may be due in part to the presence of minute amounts of impurity (or other lattice defects) in the experimental samples.

Secondly, we discuss the effect of dilution upon the short-range order maximum arising from the intrachain interaction. As seen in fig. 1

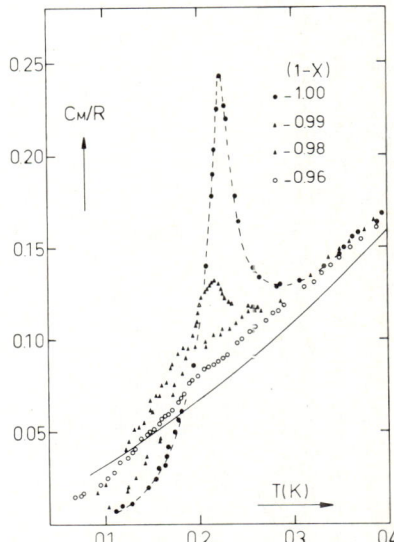

Fig. 2. Enlarged plot of three-dimensional ordering anomalies, —— C_M of a linear chain with $J_\parallel/J_\perp = \frac{1}{4}$, $J_\perp = -1.35$ K. ----- Guide to the eye.

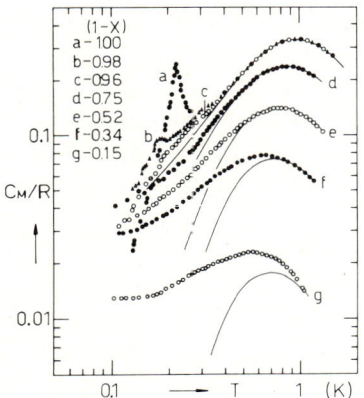

Fig. 1. Magnetic specific heat per mole of $Cs_2Co_{1-x}Zn_xCl_4$.

this contribution is gradually reduced upon dilution. For temperatures near to and above the maxima, the experimental curves agree reasonably well with calculations of Blöte [4] of our laboratory (solid curves) for a system of randomly diluted isolated chains ($J' = 0$). For given x, the number of chains containing n spins is given by $N_0 x^2 (1-x)^n$ per mole, where N_0 is Avogadro's number. The specific heat is thus calculated from the weighed contributions of chains of finite lengths n. Blöte has recently [5] obtained results for $S = \frac{1}{2}$ chains with $n \leq 11$, using the uniaxial hamiltonian.

$$\mathcal{H} = -2 \sum_{\langle i,j \rangle} \{ J_\perp (S_{ix}S_{jx} + S_{iy}S_{jy}) + J_\parallel S_{iz}S_{jz} \},$$

and varying the ratio $J_\parallel / J_\perp$. The behavior for chains of length $n > 11$ can be estimated by extrapolation procedures, that can be checked for the exactly solvable cases $J_\perp = 0$ and $J_\parallel = 0$ (Ising and XY model, respectively). For the present materials, the ratio $J_\parallel / J_\perp = \frac{1}{4}$ was adopted, as has been argued [1] to apply to Cs_2CoCl_4. Also, the exchange constant $J_\perp / k = -1.35$ K found [1] for Cs_2CoCl_4 has been used in all the calculations. Since the lattice constants of Cs_2ZnCl_4 and Cs_2CoCl_4 differ by 0.2% only, we do not expect the possible variation of $J_\perp / k$ with x to exceed more than a few per cent.

The agreement with the so-calculated chain contribution is particularly good for small values of x (fig. 1). For larger x, the short-range order maxima predicted by the model still agree with experiment as far as height and position are concerned. So, the condition of random dilution seems to be achieved, with negligible variation of the intrachain exchange. Yet an increasing amount of excess specific heat above the chain contributions is observed in the temperature region below the maxima.

We attribute these to the presence of the weaker interchain interactions (J'), leading to the formation of clusters of Co^{2+} ions belonging to different chains. These will give rise to an additional Schottky-type (short-range order) contribution [6]. Since the easy-planes of magnetization of the Co^{2+} ions belonging to adjacent chains have been shown [1] to be nearly perpendicular to each other, one would expect the interaction J' to be of the Ising rather than the XY type.

In fig. 3 we show that such an additional

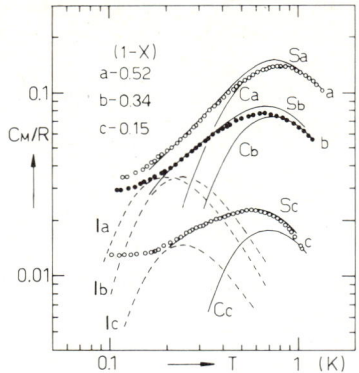

Fig. 3. Plot of the data for largest x compared with calculated curves. C_i: chain contributions. I_i: Schottky-anomalies from interchain interactions. S_i: sum of C_i and I_i ($i = a, b, c$).

Ising–Schottky term for clusters of two ions may account for a large part of the excess specific heat observed for $1 - x = 0.52$, 0.34 and 0.15. In all three cases the Schottky maxima are near $T \simeq 0.25$ K, from which it would follow that $|J'/k| \simeq 0.5$ K. Compared to $|J_\perp/k| = 1.35$ K, this value of J' appears to be rather large, in view of the strongly one-dimensional character of the pure compound. However, from the crystal structure a variety of possible interchain interaction pathways can be distinguished. It is thus conceivable that different interchain interactions exist which partly cancel each other in the ordered state of the pure compound, leading to a net interchain coupling much smaller than the individual interactions between ions belonging to different chains. In that case the above derived J'/k would comprise only part of the different interchain interactions, and still other interchain interactions (presumably smaller) would exist, giving additional contributions at lower temperature. This may explain the upward trends in the observed specific heats for $T \lesssim 0.15$ in fig. 3. We remark in this connection that the hyperfine interactions in Cs_2CoCl_4 are expected [7] to be too small to account for these upward deviations.

These investigations are supported in part by Stichting "F.O.M.". We are much indebted to Dr. H.W.J. Blöte for performing the chain specific heat calculations as well as to W.J. Crama for advice in the sample preparation and Dr. J. Reedijk for the chemical analyses. One of us (J.B.) thanks the Spanish M.E.C. for a post-doctoral grant.

References

[1] H.A. Algra, L.J. de Jongh, H.W.J. Blöte, W.J. Huiskamp and R.L. Carlin, Physica 82B (1976) 239. J.N. McElearney, S. Merchant, G.E. Shankle and R.L. Carlin, to be published in J. Chem. Phys., October 1976.

[2] Y. Imry, P.A. Montano and D. Hone, Phys. Rev. B12 (1975) 253.

[3] For a review see L.J. de Jongh and A.R. Miedema, Advan. Phys. 23 (1974) 1.

[4] H.W.J. Blöte, private communication.

[5] H.W.J. Blöte, Physica 79B (1975) 427.

[6] See, e.g., H.A. Algra et al., this conference.

[7] G.F. Shankle, J.N. McElearney, R.W. Schwartz, A.R. Kampf and R.L. Carlin, J. Chem. Phys. 56 (8) (1972) 3750.

TWO-MAGNON RAMAN SCATTERING IN $KNi_xMn_{1-x}F_3$

G.J. COOMBS and D.J. LOCKWOOD

Physics Department, Edinburgh University, Edinburgh EH9 3JZ, Scotland

The two-magnon excitations in two single crystals of the substitutionally disordered antiferromagnet $KNi_xMn_{1-x}F_3$ with nominal concentrations $x = 0.50$ and 0.75 have been studied as a function of temperature by Raman spectroscopy. Two well-defined bands that broaden and shift to lower energy with increasing temperature were observed. A cluster model calculation is in reasonable agreement with the low-temperature results.

The advantages of antiferromagnets as examples of substitutionally disordered systems for the study of collective excitations are well known [1]. The material $KNi_xMn_{1-x}F_3$ is a particularly simple example of such a system as both constituents are antiferromagnets, with the cubic perovskite structure at high temperatures, whose magnetic-excitation spectra may be characterised by standard spin-wave theory based on the Heisenberg model with nearest-neighbour interactions. As might be expected from the large differences in the exchange constants of the two constituents, the magnon dispersion curves, as measured by neutron scattering for $x = 0.75$ [2], show two branches lying in well-separated bands both having appreciable dispersion.

We have measured the Raman spectra over a wide range of temperatures of two crystals of $KNi_xMn_{1-x}F_3$ having nominal concentrations of nickel of 0.75 and 0.50. The results at the lowest temperatures have been compared with theoretical calculations based on a simple Ising cluster model.

The crystals of $KNi_xMn_{1-x}F_3$, provided by the late D.A. Jones, were cut into cubes of approximately 5 mm edge with [100] faces. These samples were mounted in a flow cryostat and the Raman spectrum was excited with 300 mW of argon laser light at 514.5 nm with the scattered light analysed using equipment described previously [3]. The spectra reported here were recorded with $X(YY)Z$ polarization, where X, Y and Z refer to the crystal cubic axes. No magnon features were seen in the off-diagonal polarizations. The spectral resolution was $4.0\ cm^{-1}$ at 514.5 nm.

As can be seen from the results shown in Figs. 1 and 2, there are three features in the spectra at low temperatures. The temperature dependences of the peak positions are shown in

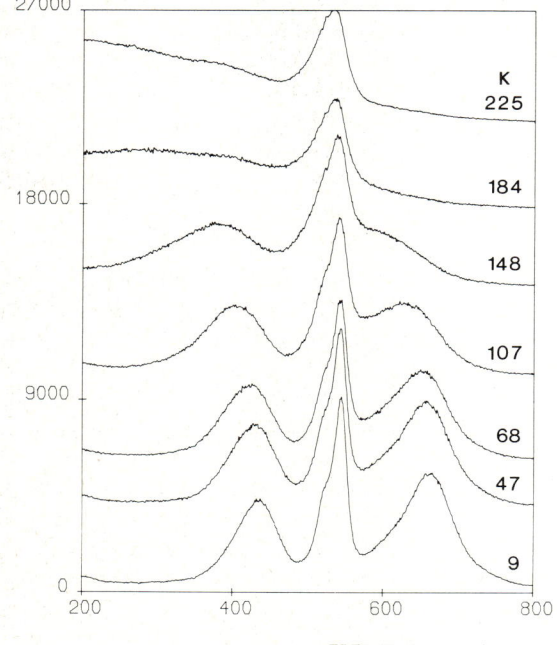

Fig. 1. Raman spectrum of $KNi_{0.75}Mn_{0.25}F_3$ recorded at various temperatures with diagonal polarization.

Fig. 3. On the basis of their marked temperature dependences and theoretical considerations given below, the highest frequency ($\sim 640\ cm^{-1}$) and lowest frequency ($\sim 420\ cm^{-1}$) peaks are identified as being due to two-magnon excitations on nickel–nickel pairs and manganese–nickel pairs, respectively. The weakly temperature-dependant peak in the middle, at approximately $540\ cm^{-1}$, is believed to be due to two-phonon scattering. A similar peak is observed in the YY spectrum of cubic $KMnF_3$ [4]. Neither the manganese–manganese two-magnon peak nor the two zone-centre one-magnon peaks were observed.

The low-temperature linewidths of the lower and upper two-magnon bands are approximately

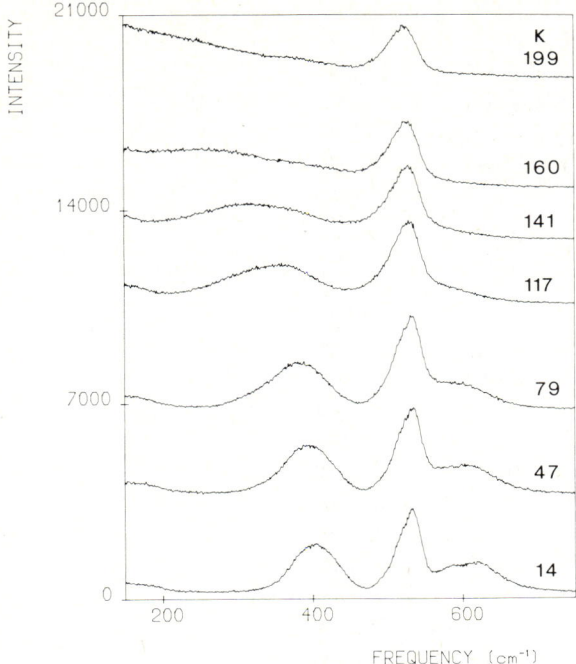

Fig. 2. Raman spectrum of $KNi_{0.5}Mn_{0.5}F_3$ recorded at various temperatures with diagonal polarization.

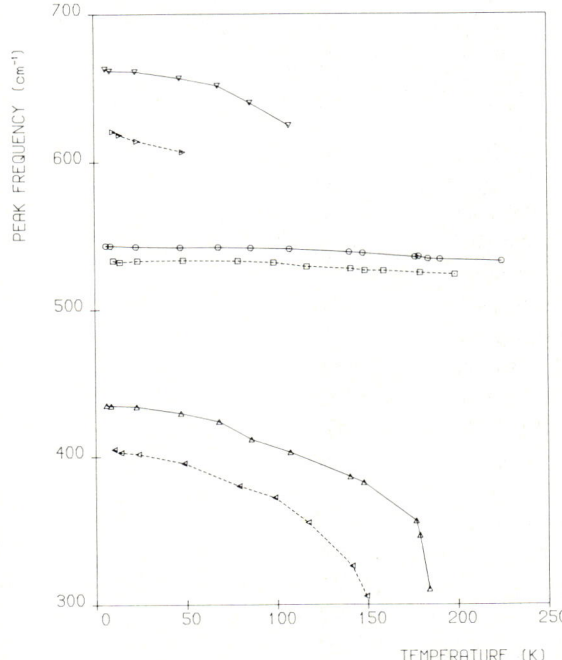

Fig. 3. Temperature dependence of the peak frequencies of the two magnon bands ($\triangledown, \triangle$ for $x = 0.75$ and $\triangleright, \triangleleft$ for $x = 0.50$) and two-phonon band (0 for $x = 0.75$ and $\square$ for $x = 0.50$) in the Raman spectrum of $KNi_xMn_{1-x}F_3$.

70 and 85 cm^{-1} for both concentrations. As the temperature increases, both peaks increase in width and shift to lower frequencies until they are lost in the background. The excitations exist at temperatures well above the Néel temperatures (T_N) estimated from the nominal concentrations to be 167 and 207 K.

No structural phase transition, as found in $KMnF_3$ [4], was observed in these mixed crystals for temperatures down to about 10 K.

In an attempt to describe the results for the two-magnon scattering, we have used the Ising cluster model. This model is derived, and its limitations discussed, by Buchanan et al. [5] who apply it with some success to $Mn_xZn_{1-x}F_2$. The scattering function is given by a sum of delta functions positioned at the Ising energies of the various possible configurations of the cluster and each weighted by the probability of the occurrence of the particular configuration. We use the expressions given by Fleury and Guggenheim [6].

To facilitate the comparison of theory and experiment, the delta functions have been replaced by Lorentzians with a width arbitrarily chosen to be 15 cm^{-1}. The values of the model parameters used are taken from Cowley and Buyers [7]. The theoretical two-magnon spectrum consists of three quite distinct peaks, which may be regarded as due to excitations on nickel–nickel, nickel–manganese and manganese–manganese nearest-neighbour pairs, in order of decreasing energy. The scattering function thus obtained has been fitted to the lowest temperature spectrum for each concentration with the scattering strengths of the manganese and nickel ions taken as separately variable parameters. To take account of the two-phonon band, a harmonic oscillator response function has been used. This does not give a particularly good representation of a clearly complex peak but it should give a reasonable description of the lower part of the peak where it overlaps the magnetic scattering. A fixed, flat background has been added to the fitting function. In the fitting procedures it was found that for both crystals the agreement could be improved by reducing the concentration from the nominal value. Ultimately the values used were 0.42 and 0.57 instead of the nominal 0.50 and 0.75. Whether the fault lies with the model or the values of the nominal concentrations is

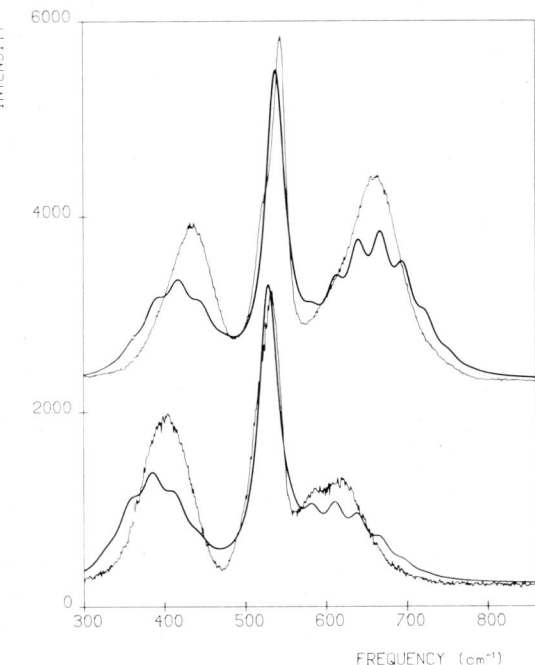

Fig. 4. Comparison of Ising cluster theory with the Raman spectra of $KNi_xMn_{1-x}F_3$. The upper spectrum is for $x = 0.75$ at a temperature of 9 K and has been divided by a factor of 5. The lower is for $x = 0.50$ at 14 K. In each case the heavy line is the theory.

unknown but the latter is suspected and it is intended to obtain accurate values of both the Néel temperatures and concentrations at some future time. The results of the fitting are shown in Fig. 4. The theoretical widths for all the two-magnon peaks may be seen to be greater than the experimental widths, which is surprising as the model includes only the effects of the disorder broadening and ignores the finite bandwidths.

References

[1] R.A. Cowley, Proc. 21st Conf. on Magnetism, Philadelphia (1975).
[2] G.J. Coombs, R.A. Cowley, D.A. Jones, G. Parisot and D. Tocchetti, Proc. 21st Conf. on Magnetism, Philadelphia (1975).
[3] D.J. Lockwood and G.J. Coombs, J. Phys. C: Solid State Phys. 8 (1975) 4062.
[4] D.J. Lockwood and B.H. Torrie, J. Phys. C: Solid State Phys. 7 (1974) 2729.
[5] M. Buchanan, W.J.L. Buyers, R.J. Elliot, R.T. Harley, W. Hayes, A.M. Perry and I.D. Saville, J. Phys. C: Solid State Phys. 5 (1972) 2011.
[6] P.A. Fleury and H.J. Guggenheim, Phys. Rev. B12 (1975) 985.
[7] R.A. Cowley and W.J.L. Buyers, Rev. Mod. Phys. 44 (1972) 406.

EXCITATION IN TWO DIMENSIONAL DILUTED FERROMAGNET

T. FUJIWARA

Department of Theoretical Physics, University of Oxford, England, and Department of Applied Physics, University of Tokyo, Japan

Excitation in two-dimensional diluted ferromagnet with a single-ion anisotropy and an anisotropic exchange interaction is studied theoretically by using the moment expansion method. The spectral function is presented and the impurity band is discussed at $k = 0$.

One of the successful approximations treating the spin wave spectrum in diluted or mixed crystals is an extended CPA [1]. However, this is not completely satisfactory since it cannot predict the fine structure in the spectrum and always gives incorrectly a sharp cut-off at the band edge [2]. This is a serious shortcoming in the comparison with the response at $k = 0$, such as is observed in optical experiments. In this situation the moment expansion method [3], which is an "exact" calculation in the sense that one can calculate moments exactly up to an arbitrary order in principle, is more satisfactorally. Experimentally, the $k = 0$ response has been observed in $FeCl_2$: $MgCl_2$ [4] and the line shape is found to be symmetric, which may be the result of the impurity band caused by the density fluctuation of non-magnetic ions. In this paper, the moment expansion method will be applied to a two-dimensional ferromagnetic system with a single-ion anisotropy and an anisotropic exchange interaction between the nearest-neighbour pairs.

The hamiltonian is given by

$$\mathcal{H} = \sum_i p_i D \{(S_i^z)^2 - S(S+1)/3\} - \sum_{i>j} p_i p_j (2K_{ij}S_i^z S_j^z + 2J_{ij}\mathbf{S}_i \cdot \mathbf{S}_j), \quad (1)$$

where p_i are random variables taking values 1 and 0 with the probability C and $1 - C$, respectively. Restricting ourselves to the one-magnon excitation region, the hamiltonian in units $2JS$ can be rewritten using boson operators as,

$$\mathcal{H} = \sum_i p_i r a_i^+ a_i + \sum_{i,j} p_i p_j (q a_i^+ a_i - a_i^+ a_j), \quad (2)$$

where $r = -D(2S-1)/(2JS)$ and $q = 1 + K/J$. The first term in eq. (2) causes only the energy shift r. The single particle response is given by the Green function

$$\mathcal{G}_k(\omega) = \sum_{\mathbf{R}_{ij}} e^{i\mathbf{k}\cdot\mathbf{R}_{ij}} \langle p_i a_i (\omega - \mathcal{H})^{-1} p_j a_j^+ \rangle, \quad (3)$$

which can be expanded into a power series as

$$\mathcal{G}_k(\omega) = \sum_{n=0}^{\infty} M_n(c, \mathbf{k})(\omega - r)^{-n-1}. \quad (4)$$

We used as a model a system of triangular (two-dimensional) lattice, bearing $FeCl_2$: $MgCl_2$ in mind, and used clusters containing seven or fewer spins together with the formulae for the single impurity problem in order to get moments up to eighth order. The table for the moments with parameters k and q and calculations at arbitrary k values will be published in the near future.

The region of interest is outside the circle of convergence of the moment series eq. (4), and therefore we use two subsequent transformations, i.e.

$$\omega = \omega' + 2a + r \quad (5)$$

and

$$\omega' = az + a/z. \quad (6)$$

where $a = (6q + 3)/4$. As a result, the physical sheet in the complex ω-plane is mapped on to the region exterior to the circle $|z| = 1$. To obtain the response function, we write the resulting series as a Padé-approximant.

The numerical result at $k = 0$ and $q = 1.3$, which corresponds to the value in $FeCl_2$: $MgCl_2$ [4], is shown in fig. 1. The response from finite clusters containing four or less spins was calculated separately and the corresponding moments were subtracted from the results (4). The remaining part is a series characterising a continuous response and the calculated results are shown by heavy lines in the figure. The results for concentrations C equal to or larger than 0.7 are based on a [4, 4], the other results on a [3, 3] Padé-approximant. Though a [4, 4] Padé does not give unphysical poles for $C \gtrsim 0.5$, it shows a

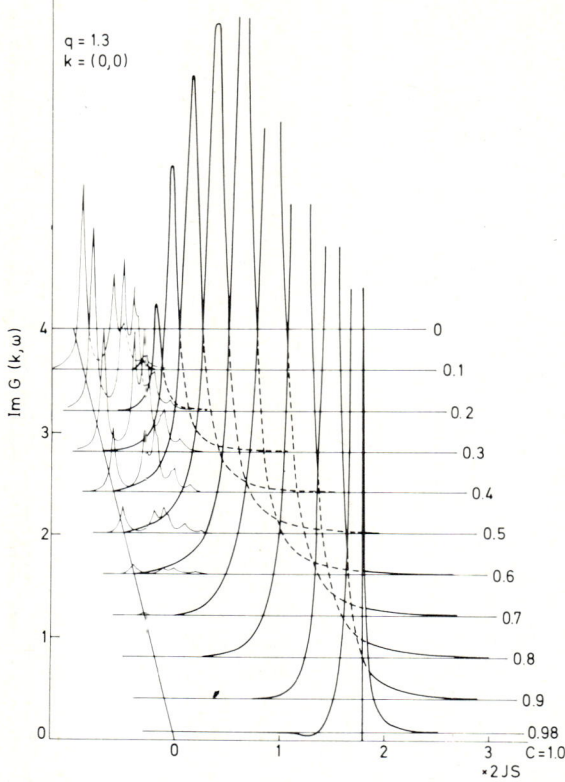

Fig. 1. Spectral function at $k = 0$, $q = 1.3$. The response of the finite clusters containing four or less spins is shown separately by the thin solid line with the imaginary frequency $\delta = 0.03$. The origin of the horizontal axis is $\omega = r$.

two-peaked structure around $C \sim 0.55$ as does a [3, 4] Padé. These are discarded since the trajectory of poles in the two cases is quite different and therefore the two-peaked structure in the region $C \sim 0.55$ is probably not real. This is, of course, due to a poor convergence of the series expansion eq. (4) around the percolation concentration. The resulting spectral function is surprisingly symmetric and broad even at low concentration of non-magnetic ions. The first, second and third moments at $k = 0$ are given as

$$\langle \omega \rangle = r + 6C(q-1),$$
$$\langle (\omega - \langle \omega \rangle)^2 \rangle = 6C(1-C)(q-1)^2,$$
$$\langle (\omega - \langle \omega \rangle)^3 \rangle = 6C(1-C)(q-1)^2\{(1-2C)(q-1) + 3C\}. \quad (7)$$

Though the spectral function extends over a wide range of $\omega > \langle \omega \rangle$, we see from the eqs. (7) and fig. 1 that the spectral weight is concentrated at $\omega < \langle \omega \rangle$. At low concentrations of non-magnetic ions a local mode of A_1-symmetry is predicted to occur when $G_{00} - qG_{01} = [G_{00}\{\omega - r - 12(q-1)\} - q]/6 = 0$, where G_{00} and G_{01} are the Green's functions of spins in the pure crystal for the central site and between the nearest-neighbour sites, respectively. It is easily shown that the local mode in the region of the energy gap exists for $q > 1$, because of the logarithmic singularity of the real part of G_{00}. In the limit $C \to 0$, we calculate Im $G_{00}(\omega = \langle \omega \rangle) \approx -0.14$ and Re $G_{00}(\omega - \langle \omega \rangle = -1) \approx -0.18$, and we take

$$\text{Re } G_{00}(\omega) = \text{Re } G_{00}(\omega - \langle \omega \rangle = -1)$$
$$- \pi^{-1} \text{Im } G_{00}(\omega = \langle \omega \rangle) \cdot \text{Ln } |\omega - \langle \omega \rangle|,$$

the separation of the local mode is 10^{-5}–10^{-6} in units $2JS$. On the other hand, the quasi-particle pole for $C = 0.98$ is at $1.759 + 0.012\,i$ and the resulting width of the spectrum covers the separation of the local mode completely. Since no evidence, apart from an anomalous line-width, is seen in the moment calculation, we believe that the impurity level must broaden rapidly with C and merge into the host band.

The author thanks Professor R.J. Elliott for many comments and fruitful discussions also to Dr. D.W. Taylor for suggesting the existence of local mode of A_1-symmetry.

References

[1] A.B. Harris, P.L. Leath, B.G. Nickel and R.J. Elliott, J. Phys. C7 (1974) 1693.
[2] I.M. Lifshitz, Advan. Phys. 13 (1964) 483.
[3] B.G. Nickel, J. Phys. C7(1974) 1719.
[4] M.C.K. Wiltshire, Private communication.

NEUTRON SCATTERING AND MAGNETIZATION MEASUREMENTS IN $CsMnFeF_6$

W. KURTZ
Institut für Kristallographie der Universität Tübigen, D-74 Tübigen, W. Germany

and

S. ROTH
Max-Planck-Institut, HML, F-38042, Grenoble 166 X, France

Neutron scattering indicates the absence of magnetic long range order and the importance of antiferromagnetic short range order in the modified pyrochlore $CsMnFeF_6$. The magnetization is consistent with a magnetic cluster model.

1. Introduction

Increasing efforts are made to investigate materials showing such phenomena as spin-glass behaviour, mictomagnetism and superparamagnetism, which are intermediate to well established magnetic long range order and paramagnetism. $CsMnFeF_6$ is an interesting member of this class of substances, being an insulator of modified pyrochlore structure with relatively close packed Fe^{3+} and Mn^{2+} ions occupying identical crystallographic sites with equal probability [1]. Mössbauer line splitting occurs below $T_0 = 26$ K suggesting a phase transition to an ordered state at this temperature [2]. From magnetic susceptibility data at $T > 150$ K Banks et al. [3] deduce $\theta = -287$ K in agreement with the assumption of strong antiferromagnetic nearest neighbour interaction predicted by Goodenough's superexchange rules. The appearance of remanent magnetization below T_0 leads Banks et al. [3] to suggest a canted antiferromagnetic structure for $CsMnFeF_6$.

2. Neutron diffraction

Neutron diffraction experiments have been carried out on powdered samples of 1 g of $CsMnFeF_6$ in the temperature range from 3 to 600 K. The high temperature diffraction pattern at 600 K shows the Bragg peaks of the nuclear structure superposed upon a smooth paramagnetic background. The analysis of the nuclear structure verifies the randomness of the Mn and Fe ions on their common sites. On lowering the temperature a broad diffuse peak arises around the (1 1 1) nuclear peak, which is interpreted as indication of antiferromagnetic short range order. This diffuse peak is already well developed at 500 K and increases in intensity down to a temperature of some 50 K, below this value the peak is almost independent of temperature. Notably, scans above and below 25 K do not show any significant change in intensity at any scattering angle within the experimental error of 4%.

In fig. 1 an attempt is made to compare the shape of the diffuse peak at 3 K with the intensity distribution expected from a simple magnetic cluster model considering magnetic correlation only within the first two coordination shells [4]. The solid line is the result of a least square fit and corresponds to antiferromagnetic coupling of the first and ferromagnetic coupling of the second shell.

3. Magnetization in high fields

Magnetization has been measured on powder samples at temperatures between 4.2 K and 200 K. We verify the spontaneous magnetization below 25 K reported by Banks et al. [3]. At 4.2 K we find a spontaneous moment of 14.7 emu/g (not corrected for demagnetization),

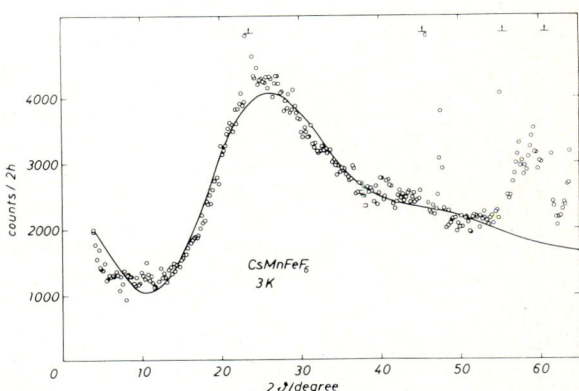

Fig. 1. Fit of magnetic cluster model to neutron diffraction data.

which is about 25% less than the value of ref. 3. Within our experimental limits we cannot resolve any hysteresis.

Fig. 2 shows high field magnetization curves for various temperatures. The magnetization does not saturate up to 13 T even at the lowest temperatures. In fig. 3 we have plotted susceptibility versus temperature at several fields. At low fields these curves show a cusp-like peak close to 26 K. In increasing fields this peak becomes broader and moves to higher temperatures.

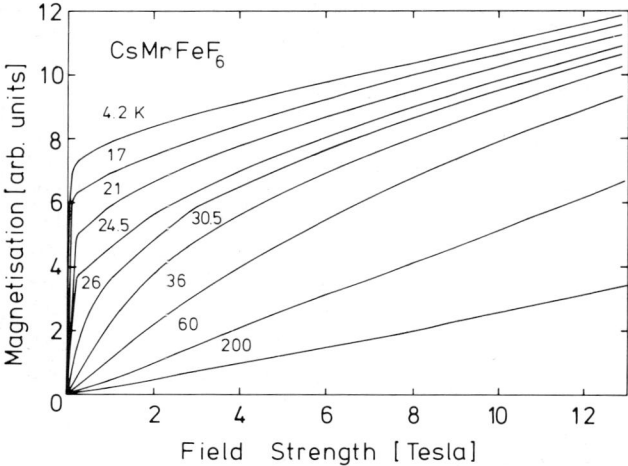

Fig. 2. High field magnetization of $CsMnFeF_6$ at various temperatures.

4. Discussion

The neutron diffraction experiments have shown that there exists no magnetic long range order in $CsMnFeF_6$ within the experimental accuracy. The very pronounced magnetic short range order can be explained by a simple cluster model with antiferromagnetic nearest neighbour coupling and without canted or screw structures. This is consistent with the magnetization data which are very similar to those of some disordered alloys such as the Fe–Cr system [5], where remanence is the result of non-complete spin compensation in antiferromagnetic clusters and susceptibility peaks arise from the competing effect of dissolving clusters (thus liberating paramagnetic ions) and thermal disorder. The sharp transition temperature in the Mössbauer results is not contradictory to the absence of long range order since a Mössbauer experiment

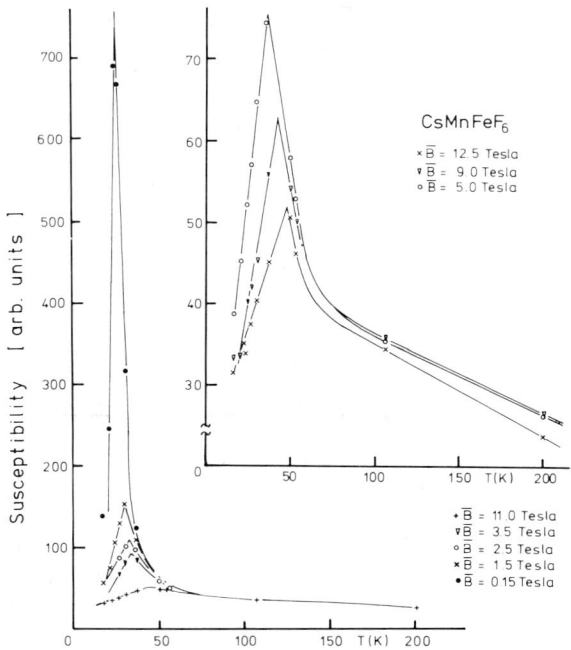

Fig. 3. Temperature dependence of magnetic susceptibility.

senses only local properties. The discrepancy of $T_0 \sim 26$ K from Mössbauer and low field magnetization measurements and neutron scattering intensity saturating near about 50 K should be explained by the different sampling times of these experimental techniques.

The present neutron experiments have been performed at the FR2-Reactor, Karlsruhe, and on the D1b multicounter instrument of the ILL reactor, Grenoble. The help of Drs. M. Steiner and P. Convert is gratefully acknowledged. Also we thank Prof. H. Dachs for valuable contributions. Part of this work has been supported by the Deutsche Forschungsgemeinschaft and by the Stiftung Volkswagenwerk.

References

[1] D. Babel, Z. Anorg. allg. Chem. 387 (1972) 161.
[2] R. Geller and R. Müller, Proc. 3rd Int. Conf. Magn. Moscow, vol. I/2 (1973) 326.
[3] E. Banks, J.A. De Luca and O. Berkooz, J. Solid State Chem. 6 (1973) 569.
[4] W. Kurtz, R. Geller, H. Dachs and P. Convert, Solid State Commun. 18 (1976) 1479.
    H. Dachs, W. Kurtz, J. Magn. Magn. Mater. (in print).
[5] B. Loegel, J. Phys. F: Metal Phys. 5 (1975) 497.
    Y. Ishikawa, R. Tournier and J. Filippi, J. Phys. Chem. Solids 26 (1965) 1727.

SOME NEW RESULTS ON SOLUBLE ISING MODELS FOR THE CRITICAL BEHAVIOR OF IMPERFECT MAGNETIC CRYSTALS

I. DECKER, K. RANDER and H. HAHN

Institut A für Theoretische Physik der Technischen Universität Braunschweig, Mendelssohnstrasse 1A, D-3300 Braunschweig, W. Germany

We report on some more results on the behaviour of two-dimensional Ising models with spatial inhomogeneities of the nearest neighbour coupling. The use of a continuum approximation first suggested by Ferrell [1] was especially helpful in simplifying the otherwise exact calculations.

1. Motivation

The theory of critical phenomena in spatially homogeneous systems is by now reasonably well developed. On the other hand, however, almost any actual experiment is done on a sample which deviates more or less strongly from the ideal behaviour. For example, practically every ferromagnetic crystal contains structural imperfections (as dislocations, grain bounderies, etc.), i.e. deviations from the ideal, translationally invariant crystal (and coupling) structure, resulting in "rounding" and other effects on the magnetic phase transition. A systematic theoretical study of the influence of such spatial inhomogeneities on the cooperative behaviour is therefore in appropriate order.

2. The model

To steer clear of any theoretical artefacts that might be produced by uncontrolled approximations, we have chosen to consider an exactly solvable model, namely a two-dimensional Ising model with a layer-type inhomogeneity of the nearest neighbour exchange coupling ("layered Ising model"). We have studied (1) the dependence of the zero-field specific heat on the parameters characterizing the spatial inhomogeneity, and
(2) the coordinate dependence of the spontaneous magnetization.

3. Specific heat at zero field

3.1. Starting point

We use the transfer matrix method (in the usual fermion formulation, with "transfer" direction along the layers) to calculate the zero-field partition function. For the special case of a weak and slowly varying spatial inhomogeneity, the critical behaviour of the system is solely determined by the long-wavelength fermions, so that an expansion of the transfer matrix for small wavenumbers is permitted. This leads to a formal correspondence with the time-development operator of a quantized Dirac field in *one* spatial dimension without interaction.

In this picture, the spatial inhomogeneity enters via the coordinate dependence of an effective "rest mass" of the Dirac particles. This rest mass is including the local exchange parameters and the temperature of the system in such a way that it acquires the physical meaning of a "*local reciprocal correlation length*" [2].

The partition function (i.e. the largest eigenvalue of the transfer matrix) is proportional to an exponential of the sum of the one-fermion energies; in the thermodynamic limit, this sum can be replaced by an integral. The calculation of the thermodynamic functions is therefore reduced to the determination of a one-particle density of states as a function of the one-fermion energy.

3.2. Application

In the case of a periodic coordinate-dependence of the exchange coupling, the problem is solved by standard quantum-mechanical methods. This results in a band structure of the one-particle spectrum which, in turn, is directly related to the temperature dependence of the specific heat, and thus giving direct insight into the phase transition of an inhomogeneous system.

Apart from the parameters characterizing the coupling inhomogeneity (amplitude, length of period), the lowest energy band (corresponding to excitations with wavelength exceeding the period of the inhomogeneity!) depends only on the *spatial average* of the local reciprocal correlation length that has the meaning of a "*global*" reciprocal correlation length of the system as a whole. This band alone is responsible for the logarithmic singularity always occurring in a *periodic* Ising model. The "global" critical temperature at which this divergence takes place, and at which long range order throughout the whole crystal sets in, is given by the above mentioned global correlation length diverging.

The higher-energy bands, on the other hand, are best visualized for the special case of extended homogeneous subdomains – each with a different coupling – within one spatial period. Here it turns out to be a good approximation to produce these bands by just superposing the energy-level densities (level δ-functions) which would result for the single subdomains (finite in one direction) if these were isolated. It follows that the higher bands mainly depend on the *local* correlation lengths. Consequently, the specific heat shows an (analytically "rounded"!) maximum wherever one of the local correlation lengths reaches the width of the corresponding homogeneous subdomain. (This corresponds to the onset of an alignment of all the spins within that subdomain!)

The formulation of the specific heat in terms of a density of one-fermion states *remains valid* for the (more realistic) case of a *statistical* spatial inhomogeneity in the coupling. As with a periodic inhomogeneity, a singularity in the specific heat (necessarily a global effect, if it exists!) will be determined by the density of states in a narrow vicinity of energy zero. First calculations in this "critical" regime corroborate the result of McCoy and Wu [3] who proved (albeit for a rather special case) that – unless it is confined to a part of the crystal finite in at least one dimension – even the slightest deviation of the spatial inhomogeneity from strict periodicity results in a specific heat loosing its logarithmic divergence, and showing a smooth temperature dependence throughout (although possessing an essential singularity – but no divergence – at the global critical temperature).

Our calculations do not depend on the special form of the inhomogeneity statistics. Rather, the continuum approximation described in section 3.1. is just constructed in such a way that the one-particle density of states in the critical (low energy) regime is determined by the first and second moments of the inhomogeneity probability distribution.

By analyzing the wave functions belonging to the lowest one-particle energies, we hope to get more detailed information of the disappearance of the logarithmic singularity, and in the origin of the non-analyticity in the specific heat which indicates the existence of a critical temperature and of some kind of phase transition even in an Ising model with statistically fluctuating (quenched) inhomogeneities of the coupling constant.

4. Coordinate dependence of spontaneous magnetization for periodic layering

Without any approximation we have given a relatively simple exact expression for the *spontaneous magnetization on the mirror-symmetry lines* of an Ising model with two types of (internally homogeneous) sublayers alternating periodically. For a large width $2d$ of a sublayer, the corresponding magnetization value has now been shown rigorously – for temperatures below the critical one – to decay asymptotically as $d^{-\frac{3}{2}}e^{-d\xi}$ to its $d \to \infty$ limit, with ξ the local correlation length of that layer below its "critical" temperature. Finally, the *asymptotic spatial decay* of the magnetization away from a sublayer interface is shown to follow the same law, if d is the distance from the interface [4].

References

[1] R.A. Ferrell, J. Stat. Phys. 8 (1973) 265.
[2] I. Decker and H. Hahn, Physica 83A (1976) 143 and to be published.
[3] B.M. McCoy and T.T. Wu, Phys. Rev. 176 (1968) 631.
[4] K. Rander and H. Hahn, Phys. Lett. A 53 (1975) 287.

HEISENBERG FERROMAGNET CONTAINING A NON-MAGNETIC IMPURITY. LOCAL MEAN SPIN DEVIATIONS*

P. MODRAK

Department of Physics, The University of British Columbia, Vancouver, B.C., Canada†

The effect of a single non-magnetic impurity on the local mean spin values has been calculated. The independent random walk method (which is equivalent to the non-interacting spin wave method) has been used and no further approximations have been introduced. The low temperature expansion of the local mean spin deviations in the vicinity of the impurity for five lattice sites of a simple cubic lattice has been obtained.

1. Introduction

In the previous papers [1, 2] the effect of non-magnetic impurities on macroscopic magnetic properties of a Heisenberg ferromagnet has been calculated. The random walk method [3] has been used and the results obtained show that the first term in the low temperature expansion (proportional to $T^{3/2}$, where T is the absolute temperature) of the spontaneous magnetization as well as of the specific heat is affected by the presence of non-magnetic impurities. The same conclusion results from the similar calculations performed by the Green function method [4] (the coefficients of the $T^{3/2}$-term in the low temperature expansion of the specific heat obtained in refs. 2 and 4 have the same value within numerical accuracy of ref. 2 if a misprint contained in ref. 4 is corrected [5]). On the other hand the calculations [6–8] of the local effects, that is, the local mean spin deviations $\Delta S(j)$ in the vicinity of a single non-magnetic impurity show that the $T^{3/2}$-term in the low temperature expansion of $\Delta S(j)$ vanishes. The question arises whether this incompatibility of the results for the local effects and macroscopic effects is real or simply a consequence of inadequate approximations? The calculations of the local effects in refs. 7 and 8 are based on the assumption that the only magnetic atoms affected by a non-magnetic impurity are those occupying the nearest neighbour positions of the impurity. This simplifying assumption has not been used in the calculations given in refs. 1 and 2. That is why it is instructive to calculate also the local effects without using that assumption. What is more, the random walk method applied to a hamiltonian which neglects the interactions of spin waves is exact, while the calculation in ref. 6 is performed to the first order of perturbation theory.

2. Method and results

The calculations of the local effects have been performed for a Heisenberg ferromagnet containing a single non-magnetic impurity and described by the same hamiltonian as in ref. 1. In order to calculate the local mean spin value at the position j, $S_z(j) = \{Z^{-1}\mathrm{Tr}[\exp(-\beta\epsilon')\hat{S}_z(j)]\}_{a \to 0}$, where $\hat{S}_z(j)$ is the z component of the spin operator of the atom at the position j and the remaining symbols have the same meaning as in ref. 1, we have followed the method of calculation of the partition function used in ref. 1, and obtained, without any approximations except those equivalent to non-interacting spin wave approximation, that

$$S_z^i(j) = S - \sum_{q=1}^{\infty} e^{-4fS\beta q} \sum_{t=0}^{\infty} T_S^t(j, q), \qquad (1)$$

$$T_S^t(j, q) = \sum_{l=0}^{\infty} [(2f)^l (2Sq\beta)^{l+t}/(l+t)!] \\ \times \sum_{l_1^+ \ldots l_{t+1}^+ = l} \sum_{j_{n_1} \ldots j_{n_t}} T_{l_1}(j, j_{n_1}, \mathbf{Q}) \ldots T_{l_{t+1}}(j_{n_t}, j, \mathbf{Q}). \qquad (2)$$

where again the meaning of the symbols is the same as in ref. 1. The low temperature expan-

---

* Supported in part by a grant from the National Research Council of Canada.
† On leave from Institute of Physical Chemistry of the Polish Academy of Sciences, Kaspraka 44, Warsaw, Poland.

sion of the local mean spin deviation $\Delta S(j) = S_z^i(j) - S_z^p(j)$ where S_z^i and S_z^p are the local mean spin values for impure and pure ferromagnet, respectively, has been obtained for s.c. lattice using the same methods as in ref. 1 and has the following form:

$$\Delta S_z(j) = -(2\pi)^{-3/2}\left\{a_2(j)\zeta\left(\frac{5}{2}\right)\gamma^{-5/2} + a_3(j)\zeta\left(\frac{7}{2}\right)\gamma^{-7/2}\right\}, \quad (3)$$

here $\gamma = 4SJ/kT$, $\zeta(m)$ is the Riemann zeta function, $a_2(j) = 0.085, -0.069, -0.082, 0.044, -0.029$ and $a_3(j) = 3.6, 3.8, 4.4, 5.3, 8.4$ for $j = (1,0,0), (1,1,0), (1,1,1), (2,0,0)$ and $(2,2,2)$, respectively (the impurity is located at $j = (0,0,0)$). The $T^{3/2}$-term in the low temperature expansion of $S_z(j)$ in the absence of impurity is thus not modified by its presence. Before discussing this point we shall first consider the fact that the values of $a_2(j)$ are of different signs for various positions j, because this fact is incompatible with the conclusion obtained by considering the spins in the vicinity of the impurity. The spin of a nearest neighbour of the impurity can assume the direction opposite to the direction of the external field easier (less energy being supplied) than a spin far from the impurity. The turning of the spin at the nearest neighbour position of the impurity makes the turning of spins on the adjacent sites easier. As a result, the coefficient $a_2(j)$ should be positive for all j. However, the mechanism just described is not included in the non-interacting spin wave or independent random walk method. In this method the energy needed for turning a spin is calculated as if the spins occupying all adjacent lattice sites were always parallel to the direction of the external magnetic field, the effect of the impurity thus being reduced to exclusively two effects. The first one is connected with the passing of the spin in its random walk through the nearest neighbour position of the impurity and the second one consists in eliminating all trajectories leading through the lattice site occupied by the impurity. It is not surprising that for certain positions in the lattice the competition of only these two effects can give the negative values of $a_2(j)$.

However, the inaccuracy in the calculation of $\Delta S_z(j)$ caused by the approximation does not affect the conclusion that the first term in the low temperature expansion of $S_z^i(j)$ is the same as that for $S_z^p(j)$. We can always choose the temperature to be low enough to make the probability that two spins oriented in the direction opposite to the direction of the magnetic field occupy adjacent lattice sites to be very small even in the vicinity of the impurity. In this case all the effects neglected in the approximation do not affect the result. Thus we obtain two conclusions which are independent of the introduced approximation: (i) a single impurity does not affect the first term in the low temperature expansion of the local mean spin values in its vicinity, and (ii) even a small concentration of the impurities affects the first term in the expansions of the spontaneous magnetization and the specific heat. These two conclusions are not incompatible. They only imply that the summation over j of the expression (1) and the expanding in asymptotic series are not interchangeable. However, this also means that the effect of the impurity extends over many lattice sites.

The approximation which has turned out to be somewhat doubtful in the case of the calculations of the local effects should not affect the results of the calculations of the macroscopic effects to the same extent. The calculations of the mean spin deviations for the sites not too close to the impurity are less affected by the approximation, and these spin deviations significantly contribute to the macroscopic effects.

The exact solution of the impurity problem would probably also give a non-monotonic behaviour of $\Delta S_z(j)$ as a function of j at least for some temperatures because such a behaviour is not likely to disappear entirely if the effects neglected in the present calculations were taken into account.

The author wishes to express his gratitude to Professor W. Opechowski for many valuable discussions and critical reading of the manuscript. Part of the work was done at the Institute of Physical Chemistry of the Polish Academy of Sciences. My thanks are also due to Mr. T. Kwiatkowski from this Institute for his assistance in the numerical computations.

References

[1] P. Modrak, Physica 72 (1974) 43.
[2] P. Modrak, Physica 76 (1974) 186.
[3] W. Opechowski, Physica 25 (1959) 476.
[4] A. B. Harris, P. L. Leath, B.G. Nickel and R.J. Elliot, J. Phys. C: Solid State Phys. 7 (1974) 1693.
[5] Prof. A.B. Harris' private communication to Prof. W. Opechowski.
[6] H.P. Van de Braak and W.J. Caspers, Z. Phys. 200 (1967) 270.
[7] H.P. Van de Braak and W.J. Caspers, Phys. Status Solidi 24 (1967) 733.
[8] R.H. Swendsen, Phys. Rev. B6 (1972) 1903.

RAMAN SCATTERING FROM ZEEMAN SPLIT LEVELS IN PARAMAGNETIC $Co_cZn_{1-c}F_2$

J.P. GOSSO, J. QUAZZA, P. MOCH and J. LABBE

Laboratoire PMTM, Université Paris XIII, Av. J.B. Clément, 93430 Villetaneuse, France

We report Raman measurements of the splitting by an applied magnetic field of the lowest Kramers doublets of the paramagnetic Co^{2+} ion, diluted in the diamagnetic host ZnF_2. We derive the **g** tensor components for both doublets, and discuss the Raman intensities and polarization selection rules of the transitions.

1. Introduction

The study of substitutionally disordered magnetic compounds has known a considerable interest in the past years. One distinguishes two cases: the first involves two magnetic ions, such as Mn^{2+} and Co^{2+} and has been extensively studied, using neutron scattering [1], and optical techniques [2, 3]. In the second case, one of the two ions is non-magnetic (Zn, Mg) [4, 5]. We study here the $Co_cZn_{1-c}F_2$ system, with c low enough to prevent magnetic ordering even at $T = 0$ K ($c = 0.01$ and 0.05).

2. Theoretical predictions

$Co_cZn_{1-c}F_2$ has the rutile quadratic structure; the Co^{2+} site group is D_{2h} (fig. 1). The cubic crystal field, acting on the ^4F free ion ground state, gives rise to a $^4\Gamma_4^+$ ground state which is subsequently split into six Kramers' doublets by the lower symmetry crystal field terms and spin-orbit coupling. The nth doublet is well described by (6):

$$|n\pm\rangle = a_n|0,\pm\tfrac{1}{2}\rangle + b_n|\pm 1,\mp\tfrac{1}{2}\rangle + c_n|\mp 1,\pm\tfrac{3}{2}\rangle + d_n|0,\mp\tfrac{3}{2}\rangle + e_n|\mp 1,\mp\tfrac{1}{2}\rangle + f_n|\pm 1,\pm\tfrac{3}{2}\rangle,$$

where the first number refers to an "effective" $L = 1$ orbital momentum and the second one to a $S = \tfrac{3}{2}$ spin. (The matrix elements of the true orbital momentum are proportional to those of L within a ^4P state.) A magnetic field along the z axis splits symmetrically each Kramers' doublet, the splitting being $E_n = g^n_{zz}\beta H$, where $g^n_{zz} = 2|\langle n+|L_z + 2S_z|n+\rangle|$. (For the used fields, up to 50 kG, the quadratic terms are smaller than 5%.) The two kinds of available Co^{2+} sites (a and b on fig. 1) feel different perturbations when the field, inclined of θ with X is in the (x, y) plane. The splitting is

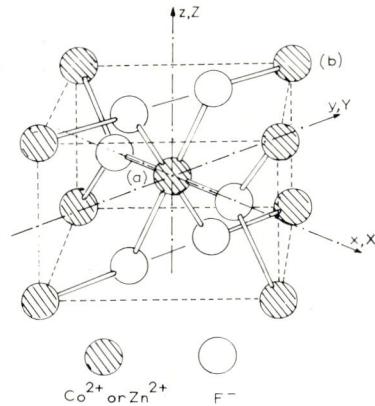

Fig. 1. $Co_cZn_{1-c}F_2$ unit cell.

$\beta H(g^{n^2}_{xx}\cos^2\theta + g^{n^2}_{yy}\sin^2\theta)^{1/2}$, for the a sites,

$\beta H(g^{n^2}_{xx}\sin^2\theta + g^{n^2}_{yy}\cos^2\theta)^{1/2}$, for the b sites,

where $g^n_{xx} = 2|\langle n^+|L_x + 2S_x|n^-\rangle|;$
$g^n_{yy} = 2|\langle n^+|L_y + 2S_y|n^-\rangle|.$

Hence for the two lowest Kramers' doublets, three distinct Raman lines are predicted for $H//z$, and six lines for $H \perp z$ (fig. 2), Polarization selection rules follow the group theoretical analysis summarized in fig. 2. The D_{2h} symmetry reduces to $C_{2h}^{x,y,z}$ respectively for **H** parallel to x, y or z, and to C_i for other directions (no selection rules.)

For a simple theoretical analysis of the Raman intensities, we calculate an explicit expression of the Co^{2+} wave functions by perturbation theory. The quadratic and orthorhombic crystal field terms split the $^4\Gamma_4^+$ level into three sublevels associated respectively with the $|u_+, m_s\rangle, |u_0, m_s\rangle, |u_-, m_s\rangle$ wave functions, where $u_\pm = (2)^{-1/2}(u_{+1}\pm u_{-1}); u_0, u_{\pm 1}$ are the eigenfunctions of an effective L_z with eigenvalues 0, ± 1, and $m_s = \pm\tfrac{1}{2}, \pm\tfrac{3}{2}$ is a spin quantum number.

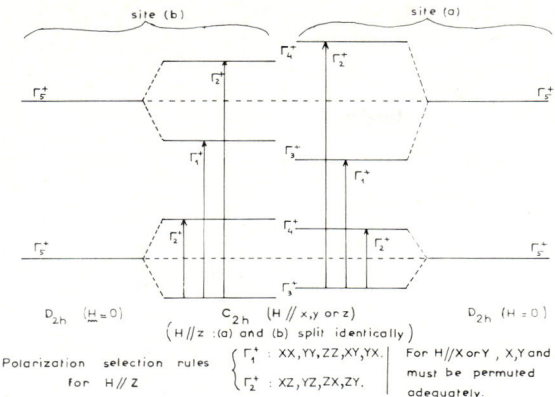

Fig. 2. Raman polarization selection rules under an applied magnetic field.

The $|u_+, m_s\rangle$ ground state is separated from the two other levels by energy intervals D_0 and D_-. For $H//z$, the residual perturbation is $\Lambda L \cdot S + \beta H(L_z + 2S_z)$. Λ is much larger than βH but only the βH term has non-vanishing matrix elements within $|u_+, m_s\rangle$. The corresponding 4×4 matrix involves then terms in Λ^2/D_0, Λ^2/D_- and βH. We find a non-vanishing Raman transition only by taking into account the perturbation of the $|u_+, m_i\rangle$ by the spin–orbit coupling. The appropriate wave functions are

$$|\phi_i\rangle = |u_+, m_i\rangle + \sum_{m_s, \alpha \neq +} (D_\alpha)^{-1} |u_\alpha, m_s\rangle\langle u_\alpha, m_s|$$
$$\Lambda L \cdot S |u_+, m_i\rangle.$$

The Raman transition probability $|\phi_i\rangle \to |\phi_j\rangle$ is proportional to [7]

$$|M_{ij}|^2 = \left| \sum_{N'L'M'S'M'_s} \left(\frac{\langle \phi_j | A_2 \cdot p | N'L'M'S'M'_s\rangle\langle N'L'M'S'M'_s | A_1^* \cdot p^* | \phi_i\rangle}{E_{N'L'} - E_0 - \hbar\omega_1} \right.\right.$$
$$\left.\left. + \frac{\langle \phi_j | A_1^* \cdot p^* | N'L'M'S'M'_s\rangle\langle N'L'M'S'M'_s | A_2 \cdot p | \phi_i\rangle}{E_{N'L'} - E_0 + \hbar\omega_2} \right) \right|^2.$$

This expression involves an infinite number of intermediate levels. However, using the Wigner–Eckart theorem, the result can be expressed in terms of only the following four expressions:

$$\sum_{N'} \frac{(L\|P\|L')(L'\|P\|L)}{E_{N'L'} - E_0 - \hbar\omega_1}; \quad \sum_{N'} \frac{(L\|P\|L')(L'\|P\|L)}{E_{N'L'} - E_0 + \hbar\omega_2},$$

where $L' = L \pm 1$.

Expanding in powers of $(E_{N'L'} - E_0)^{-1}$, the resulting Raman tensor has, in the lowest approximation, a frequency independent symmetrical part proportional to $\Lambda^2/D_0 D_-$ and a frequency dependent antisymmetrical part proportional to $(\Lambda/D_0)(\hbar\omega/\epsilon_{N'L'})$ where $\epsilon_{N'L'} = E_{N'L'} - E_0$.

3. Experiments

In order to observe the low frequency line coming from the splitting of the lowest Kramers' doublet, we used a triple monochromator Raman spectrometer. The exciting frequencies were the 4880 and 4765 Å A$^+$ ion laser lines. Notice that in the vicinity of 4880 Å line, an A$^+$ plasma line at 9.5 cm^{-1} implies some precautions in experimental conditions and analysis of the spectra. The magnetic field was obtained with two superconducting Helmholtz plates. Both coil and sample were immersed in superfluid He ($T = 2$ K). The magnetic field was parallel to the scattered direction and the sample cut in a way to allow $H//z$ and $H \perp z$ measurements. The $\theta = 30°$ value was chosen to provide low frequency lines far enough from the laser line to be observed.

4. Results and discussion

The Raman spectrum agrees with the scheme of fig. 2 (where X, Y and Z refer to the laboratory axes) for $H \neq 0$ as well as for $H = 0$ (single line centered at 155.5 cm^{-1}). The experimental and calculated g values are given in table I. Notice that for $H//z$, g_{zz}^1 can be evaluated in two ways, using

Table I
Measured and calculated g values

Kramers doublet →	g_{xx}		g_{yy}		g_{zz}	
	1	2	1	2	1	2
calculated	5.84	0.78	2.14	5.38	4.12	0.78
exp. Raman	6.2	0.6	2.5	5.0	4.2	0.7
exp. R.P.E.[6]	6.027		2.296		4.240	

the lowest doublet splitting or the excited one. Experimental polarization selection rules well agree with theoretical predictions when $H//z$. An exhaustive discussion of the observed Raman intensities compared with the calculated ones, will be reported later.

References

[1] E.C. Svensson, W.J.L. Buyers, T.M. Holden, R.A. Cowley and R.W.H. Stevenson, Can. J. Phys. 47 (1969) 1983.
[2] B. Enders, P.L. Richards, W.E. Tennant and E. Catalano, AIP Conf. Proc. 10 (1973) 179.
[3] J.P. Gosso and P. Moch, Proc. 3rd Int. Conf. Light Scat. Sol. (Flammarion, Paris, 1975) p. 214.
[4] M. Buchanan, W.J.L. Buyers, R.J. Elliott, R.T. Harley, W. Hayes, A.M. Perry and I.D. Saville, J. Phys. C5 (1975) 2011.
[5] R.A. Cowley, O.W. Dietrich and D.A. Jones, J. Phys. C8 (1975) 3023.
[6] H.M. Gladney, Phys. Rev. 146 (1966) 253.
[7] J.E. Kardontchik, E. Cohen and J. Makovsky, Phys. Rev. 13 (1975) 2955.

PERCOLATION PROCESSES IN THREE AND MORE DIMENSIONS

D.S. GAUNT

Physics Department, King's College, Strand, London WC2R 2LS, England

Series expansions have been used to estimate the critical concentration p_c for random site or bond mixtures on the usual three-dimensional lattices. Estimates have been obtained for the critical exponents γ, β and δ, and these satisfy the scaling relation $\delta - 1 = \gamma/\beta$. Series estimates of γ for simple hypercubical lattices of dimension $d = 4, 5$ and 6 support Toulouse's conjecture regarding the critical dimensionality for percolation processes.

Series expansions required for a study of random mixtures of sites or bonds on a three-dimensional lattice have been derived and analysed [1, 2]. We have described the theoretical background and introduced the series method in earlier papers [3–7] devoted to two-dimensional lattices. More specifically, the mean size of finite clusters $S(p)$ has been expanded in powers of the concentration p [1], while for p close to unity expansions in powers of $q = 1 - p$ have been derived [2] for the percolation probability $P(p)$. Up to 15 coefficients have been derived for $S(p)$ and up to 46 coefficients for $P(p)$. First we used the series for $S(p)$ to estimate the critical concentration p_c and then investigated the following hypotheses for the critical behaviour:

$$S(p) \simeq C(p_c - p)^{-\gamma} \quad (p \to p_c -), \tag{1}$$

$$P(p) \simeq B(p - p_c)^\beta \quad (p \to p_c +), \tag{2}$$

where the exponents γ and β are dimensional invariants. Using the ratio and Padé approximant techniques [8] we estimate [1] for the bond problem on the usual three-dimensional lattices

$$p_c(\text{f.c.c.}) = 0.119 \pm 0.001, \quad p_c(\text{b.c.c.}) = 0.1785 \pm 0.002,$$
$$p_c(\text{s.c.}) = 0.247 \pm 0.003, \quad p_c(\text{d}) = 0.388 \pm 0.005, \tag{3}$$

and for the site problem

$$p_c(\text{f.c.c.}) = 0.198 \pm 0.003, \quad p_c(\text{b.c.c.}) = 0.245 \pm 0.004,$$
$$p_c(\text{s.c.}) = 0.310 \pm 0.004, \quad p_c(\text{d}) = 0.428 \pm 0.004. \tag{4}$$

Although we have found it very difficult to draw precise conclusions, all the available series appear to be reasonably consistent with the hypothesis that γ and β are dimensional invariants for both bond and site mixtures in three dimensions. Our best estimates

$$\gamma = 1.66 \pm 0.07, \quad \beta = 0.42 \pm 0.06 \tag{5}$$

are close to the simple fractions

$$\gamma = 1\tfrac{2}{3}, \quad \beta = \tfrac{5}{12}, \tag{6}$$

which we adopt as convenient mnemonics [1, 2]. In two dimensions, the analogous results are [4, 6]

$$\gamma = 2.43 \pm 0.03, \quad \beta = 0.138 \pm 0.007 \tag{7}$$

which are close to

$$\gamma = 2\tfrac{3}{7}, \quad \beta = \tfrac{1}{7}. \tag{8}$$

By introducing a notional field variable λ into the percolation problem, a function $P_c(\lambda)$ can be defined whose Ising analogue is the magnetic field variation of the magnetization along the critical isotherm. Using expansions up to λ^{14}, we have studied the expected critical behaviour:

$$P_c(\lambda) \simeq E(1 - \lambda)^{1/\delta}, \quad (p = p_c, \lambda \to 1-), \tag{9}$$

with δ a dimensional invariant. A similar analysis for two-dimensional lattices was undertaken by Gaunt and Sykes [7] with the result

$$\delta = 18.0 \pm 0.75 \tag{10}$$

suggesting the conjecture

$$\delta = 18. \tag{11}$$

In three dimensions, a preliminary analysis of

the series data gives

$$\delta = 5.0 \pm 0.8 \qquad (12)$$

suggesting the simple mnemonic

$$\delta = 5. \qquad (13)$$

A detailed analysis will be published in due course.

The estimate (12) is in good agreement with that predicted from the scaling law [9]

$$\delta - 1 = \gamma/\beta \qquad (14)$$

using the series estimates (5) for γ and β, namely

$$\delta = 4.95 \pm \frac{0.86}{0.64}. \qquad (15)$$

In addition, (6) and (13) satisfy (14) exactly. Thus, the scaling law appears to hold in three dimensions and incidentally in two dimensions [7], as may be checked explicitly using (7) and (10) or alternatively (8) and (11).

It has been conjectured [10] that the critical dimension for percolation is $d_c = 6$, instead of $d_c = 4$ found for second-order phase transitions with short-range interactions. This means that for $d \geq d_c$ the critical exponents should be classical ($\gamma = 1$, $\beta = 1$, $\delta = 2$) and dimension independent. We [11] have derived series for the mean cluster size of site mixtures on a d-dimensional simple hypercubical lattice with the aim of testing this conjecture. Our results are broadly in agreement with those of Kirkpatrick [12] who has studied this problem using Monte Carlo methods. We find the following values for $\gamma(d)$:

$$\gamma(4) = 1.41 \pm 0.25, \qquad \gamma(5) = 1.25 \pm 0.15,$$
$$\gamma(6) = 1.06 \pm 0.20. \qquad (16)$$

Within the accuracy attainable they support Toulouse's hypothesis that $d_c = 6$ for percolation processes. We have also derived series for the closely related cluster growth problem and analysis of these indicates that $d_c = 6$ in this case also [11].

References

[1] M.F. Sykes, D.S. Gaunt and Maureen Glen, J. Phys. A: Math. Gen. 9 (1976) 1705.
[2] M.F. Sykes, D.S. Gaunt and J.W. Essam, J. Phys. A: Math. Gen. 9 (1976) L43.
[3] M.F. Sykes and Maureen Glen, J. Phys. A: Math. Gen. 9 (1976) 87.
[4] M.F. Sykes, D.S. Gaunt and Maureen Glen, J. Phys. A: Math. Gen. 9 (1976) 97.
[5] M.F. Sykes, D.S. Gaunt and Maureen Glen, J. Phys. A: Math. Gen. 9 (1976) 715.
[6] M.F. Sykes, D.S. Gaunt and Maureen Glen, J. Phys. A: Math. Gen. 9 (1976) 725.
[7] D.S. Gaunt and M.F. Sykes, J. Phys. A: Math. Gen. 9 (1976) 1109.
[8] D.S. Gaunt and A.J. Guttmann, in: Phase Transitions and Critical Phenomena, Vol 3, C. Domb and M.S. Green (Eds.) (Academic Press, New York, 1974) p. 181.
[9] J.W. Essam and K.M. Gwilym, J. Phys. C: Solid State Phys. 4 (1971) L228.
[10] G. Toulouse, Nuovo Cim. 23B (1974) 234.
[11] D.S. Gaunt, M.F. Sykes and Heather Ruskin, J. Phys. A: Math. Gen. 9 (1976) 1899.
[12] S. Kirkpatrick, Phys. Rev. Lett. 36 (1976) 69.

MAGNETIC CORRELATIONS IN $Rb_2Mn_{0.5}Mg_{0.5}F_4$

R.A. COWLEY*, G. SHIRANE

Brookhaven National Laboratory†, Upton, New York 11973, U.S.A.

R.J. BIRGENEAU

Department of Physics, MIT§, Cambridge, Massachusetts 02139, U.S.A.

and

H.J. GUGGENHEIM

Bell Laboratories, Murray Hill, New Jersey 07974, U.S.A.

Neutron scattering measurements in the dilute antiferromagnets $Rb_2Mn_cMg_{1-c}F_4$ with $c = 0.54 \pm 0.02$, 0.57 ± 0.02 have been performed. The critical scattering intensity diverges for $T \to 0$ until the correlation length exceeds the size of the clusters. Classical random-walk calculations are presented which elucidate the important physical effects near the percolation point.

Despite considerable theoretical interest [1–4] in percolation, there has as yet been little microscopic experimental information available on real systems. In this paper we report on diffuse neutron scattering measurements [5] designed to elucidate the development of the magnetic correlations close to a percolation point. Two specimens of the dilute [2d] antiferromagnets $Rb_2Mn_cMg_{1-c}F_4$ with $c = 0.54 \pm 0.02$ and 0.57 ± 0.02 have been examined. In these materials the exchange interactions are largely between nearest neighbors only so that the percolation concentration is $c_p = 0.59$ [1].

The critical scattering measurements were carried out on a two-axis spectrometer at the Brookhaven High Flux Reactor with a configuration similar to that described in ref. 6. The data were analyzed by assuming a [2d] lorentzian form for the intensity, $I(Q) = A/(K^2 + Q^2)$. The parameters A and K were determined by convoluting this expression with the experimental resolution function and fitting it in a least-squares sense to the experimental data. A good fit resulted showing that the correlations decrease as $R^{-\frac{1}{2}}e^{-KR}$ in real space; the resulting K values are shown in fig. 1. Not unexpectedly the results show that the effect of dilution below the percolation limit is to sup-

---

* Permanent address: Department of Physics, Edinburgh University, Scotland.
† Work at Brookhaven supported by U.S. Energy Research and Development Administration.
§ Work at MIT supported by the National Science Foundation, MRL Grant No. DMR72-03027-ADS.

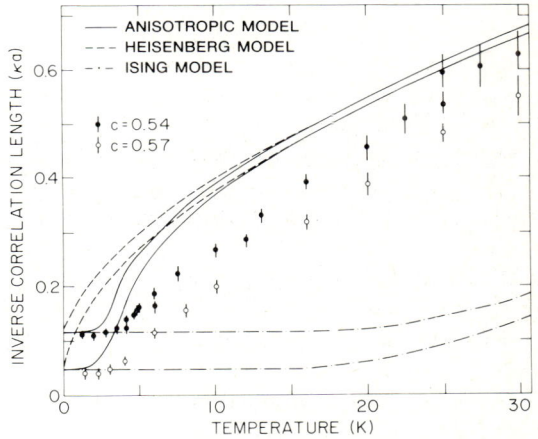

Fig. 1. The temperature dependence of the inverse correlation length in $Rb_2Mn_cMg_{1-c}F_4$ compared with calculations based on chains of N spins placed randomly on the lattice. The probability of a chain of N spins depends on the concentration as given for clusters on a square lattice by Leath [2].

press the development of the spin correlations. The difference between the K values for $c = 0.54$ and 0.57 suggests the approximate relation, $K = K_T + K_G$, where K_T is the inverse correlation length due to thermal fluctuations and K_G that due to the geometrical disorder. As the temperature is lowered, K_T decreases and the correlation length increases until at low temperatures it reaches the size of the largest clusters when it of necessity saturates; the pair-connectivity correlation lengths obtained from the measurements at low temperatures are 9 and

Fig. 2. A [2d] square lattice showing the clusters present when there are nearest neighbor interactions and $c = 0.50$.

25 lattice constants for $c = 0.54$ and 0.57, respectively.

A more quantitative understanding of these results may be achieved by considering the shape of the clusters as illustrated in fig. 2. Although the larger clusters are very complex they all have many [1d] "weak links" and it is to be expected that the development of the magnetic correlations over large distances will be controlled especially at low temperatures by these links.

A simple model of the clusters is then to approximate them as linear chains. In another publication [5] we have shown that treating the clusters as self-avoiding walks on a square lattice gives a good account of the temperature dependence of K. In this paper we use a model in which it is easily possible to take account of the finite size of the clusters. The details of the model will be published elsewhere; in essence, it consists of placing a linear chain of N spins randomly on the lattice. The probability of a chain of N spins is taken from Leath's calculations [2] of the cluster sizes on a square lattice. The results shown in fig. 1 were calculated using a classical nearest neighbor model with $JS(S + 1) = 6.4$ meV, and assuming the interactions to be (a) Heisenberg-like, (b) Ising-like and (c) Heisenberg-like with a small single ion anisotropy chosen to be the same size as in pure Rb_2MnF_4. Only the calculations with model (c) show qualitatively the same behavior as the experimental results showing that the inclusion of the small anisotropy is necessary to understand our measurements. The temperature dependence with this model is not given as satisfactorily as with the self-avoiding chains described earlier [5].

In conclusion, we have measured the development of the magnetic correlations near to a percolation transition. The results show that the temperature dependence is dominated by the behavior of the [1d] weak links in the clusters but that a complete theory will need to take account of the size of the clusters and also of the effect of the small anisotropy in the magnetic interactions.

We have benefited from discussions with P.W. Anderson, H.E. Stanley, and T.C. Lubensky and also from the provision of a computer programme to calculate correlations functions of an anisotropic [1d] linear chain by M. Blume.

References

[1] J.W. Essam, in: Phase Transitions and Critical Phenomena II, C. Domb and M.S. Green (Eds) (Academic Press, New York, 1972) p. 197.
[2] P.L. Leath, Phys. Rev. Lett. 36 (1976) 921.
[3] A.B. Harris, T.C. Lubensky, W.K. Holcomb and C. Dasgupta, Phys. Rev. Lett. 35 (1975) 327.
[4] D. Stauffer, Z. Phys. B22 (1975) 161.
[5] R.J. Birgeneau, R.A. Cowley, G. Shirane and H.J. Guggenheim, to be published.
[6] J. Als-Nielsen, G. Shirane, R.J. Birgeneau and H.J. Guggenheim, Phys. Rev. B12 (1975) 4963.

INFLUENCE OF ANISOTROPY ON $d = 2$ AND $d = 3$ MAGNETIC ORDER IN LAYER-TYPE ANTIFERROMAGNETS $K_2Mn_{1-x}Fe_xF_4$

L. BEVAART, J.V. LEBESQUE*, E. FRIKKEE and L.J. DE JONGH†,
Netherlands Energy Research Foundation ECN, Petten (N.H.), The Netherlands

Single-crystals of $K_2Mn_{1-x}Fe_xF_4$ with $x = 2.3$, 2.8, 6.1 and 12 at.% have been studied by neutron diffraction and susceptibility experiments. The type and magnitude of the spin anisotropy change with x, which has a profound effect on the magnetic ordering phenomena.

An almost ideal example of the two-dimensional ($d = 2$) Heisenberg model is K_2MnF_4, the principal deviation being a small axial dipolar anisotropy, hence the system behaves $d = 2$ Ising-like. A transition to antiferromagnetic long-range order (LRO) is observed at $T_c = 42.2$ K [1]. For the Fe^{2+}-ions in Rb_2FeF_4 ($T_c = 56.3$ K) [2] the dipolar anisotropy is very weak in comparison with the planar single-ion anisotropy. Thus, it may be expected that substituting Mn^{2+} by Fe^{2+} in K_2MnF_4 will initially lead to a decrease of the (overall) axial anisotropy to some minimum value, thereafter to increasing planar anisotropy.

From susceptibility measurements on mixed compounds $K_2Mn_{1-x}Fe_xF_4$ one may conclude that this minimum in the average anisotropy occurs for $0.02 < x < 0.06$. In order to investigate the influence of these small anisotropy perturbations on the magnetic ordering in more detail, elastic neutron scattering experiments were carried out on four single crystals with $x = 0.023$, 0.028, 0.061, and 0.12.

The susceptibility of the four samples has been measured between $4 K < T < 300 K$ parallel and perpendicular to the c-axis. For $x = 0.023$ the susceptibility shows the same overall features as for $x = 0$; the differences are a lower T_c (corresponding to a maximum in $\partial \chi_\parallel / \partial T$) and a discontinuity in both $\chi_\parallel$ and $\chi_\perp$ at $T \simeq 15$ K. For $x = 0.028$, 0.061 and 0.12, however, the susceptibility below $T = 60$ K differs widely from the $x = 0$ case, in that $\chi_\parallel$ is lower than $\chi_\perp$. This suggests that the preferred spin-direction is changed from the c-axis to the ab-plane with increasing Fe^{2+}-concentration. Also a discontinuity in both $\chi_\parallel$ and $\chi_\perp$ is observed at $T \simeq 19$ K for $x = 0.028$.

* Natuurkundig Laboratorium van de Universiteit van Amsterdam, The Netherlands.
† Kamerlingh Onnes Laboratorium, Leiden, The Netherlands.

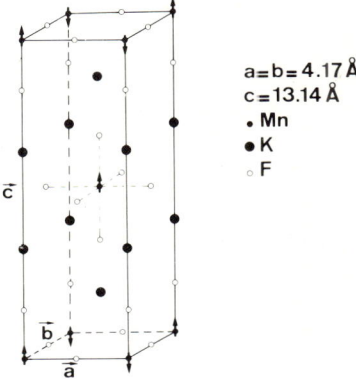

Fig. 1. Chemical unit cell of K_2MnF_4. The indicated spin structure corresponds to domain type 1.

Neutron scattering experiments have been done by measuring the reflections corresponding to points in the $(\bar{1}10)$ reciprocal lattice plane. The magnetic structure (fig. 1) allows the formation of two domain types. Domains 1 with parallel spins at positions $(0, 0, 0)$ and $(\frac{1}{2}, \frac{1}{2}, \frac{1}{2})$ give rise to superlattice reflections $(\frac{1}{2}, \frac{1}{2}, 1)$, $(\frac{1}{2}, \frac{1}{2}, 3)$, etc. whereas the reflections $(\frac{1}{2}, \frac{1}{2}, 0)$, $(\frac{1}{2}, \frac{1}{2}, 2)$, etc. are due to domains 2, for which the spins in the plane $z = \frac{1}{2}$ are reversed. The line $(\frac{1}{2}, \frac{1}{2}, \zeta)$ in reciprocal space corresponds to $d = 2$ LRO in the ab-plane. For each sample the intensities of the four magnetic reflections $(\frac{1}{2}, \frac{1}{2}, 0)$, $(\frac{1}{2}, \frac{1}{2}, 1)$, $(\frac{1}{2}, \frac{1}{2}, 2)$, and $(\frac{1}{2}, \frac{1}{2}, 3)$ have been measured for $4 K < T < 60 K$. The results can be used to determine the spin direction. Also, measurements were made on the $d = 2$ spin–spin correlations by scanning along and across the ridge $(\frac{1}{2}, \frac{1}{2}, \zeta)$. In all samples $d = 3$ LRO starts to develop at a well defined T_c. For each sample the intensities of the reflections satisfy the relation $I(T)/I(0) = B^2(1 - T/T_c)^{2\beta}$ in a temperature range $19 K < T < T_c$. Besides, for $x = 0.023$ and 0.028 the spins are found to reorien-

Table I
Ordering characteristics for $d = 3$ LRO in $K_2Mn_{1-x}Fe_xF_4$

x	0.023	0.028	0.061	0.12
T_R(K)	14(2)	19(2)	–	–
T_c(K)	40.20(15)	36.91(5)	42.30(5)	44.0(1)
β	0.19(1)	0.27(2)	0.26(2)	0.21(2)
ϕ	$T > T_R$: $\phi = 0°$	$T > T_R$: $\phi = 90°$	$\phi = 90°$	$\phi = 90°$
	$T = 4.4$ K: $\phi = 35(5)°$	$T = 4.4$ K: $\phi = 59(5)°$		

Uncertainties are given in brackets in units of the last decimal.
ϕ indicates the angle between the spins and the [001] direction.

tate at a temperature $T_R < T_c$. A summary of the $d = 3$ LRO characteristics is given in table I.

For $x = 0.061$ and 0.12 the intensity of the ridge $(\frac{1}{2}, \frac{1}{2}, \zeta)$ shows a decrease below T_c which is proportional to the increase of the intensity of the magnetic Bragg reflections. However, for $x = 0.023$ and 0.028 the ridge intensity remains constant after the initial decrease just below T_c, until it tends to zero below T_R (see fig. 2).

For an explanation of the experimental results two effects seem to be most significant. (1) The competition between the dipolar anisotropy, which increases with decreasing T, and the nearly temperature-independent single-ion anisotropy of the Fe^{2+}-ions. (2) Spatial variations of the local effective anisotropy as a result of the statistical distribution of the Fe^{2+}-ions.

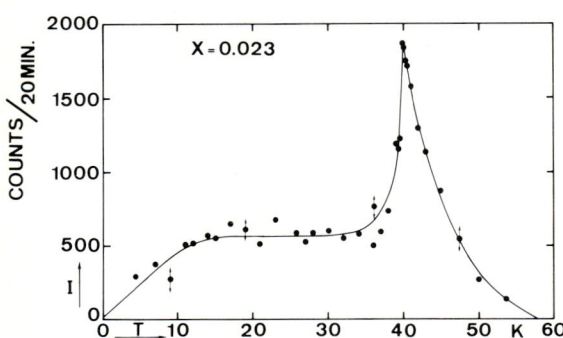

Fig. 2. Intensity observed at a fixed point ($\zeta = 0.36$) on the magnetic ridge $(\frac{1}{2}\frac{1}{2}\zeta)$. Similar results are obtained for the $x = 0.028$ sample.

These fluctuations will be particularly important if the average anisotropy is small. The coexistence of the $d = 2$ and $d = 3$ LRO between T_c and T_R is probably a result of the local fluctuations in the anisotropy. In the samples with $x = 0.023$ and 0.028 the average anisotropy is smallest [3], and hence the establishment of LRO and the preferred spin direction will be affected by local variations in the anisotropy. This could lead to a mismatching of ordering patterns and may explain why in different parts of the same crystal $d = 2$ and $d = 3$ LRO are coexisting. Below T_R full $d = 3$ LRO is gradually established, probably because the magnetic dipolar anisotropy becomes larger than the local crystal-field anisotropy, forcing the spins in the remaining $d = 2$ regions into a $d = 3$ LRO pattern. For $x = 0.061$ and 0.12 the magnetic structure is the same as in Rb_2FeF_4, because the single-ion anisotropy is too large to be compensated by the dipolar anisotropy.

A full report will appear elsewhere [3].

References

[1] R.J. Birgeneau, H.J. Guggenheim and G. Shirane, Phys. Rev. B8 (1973) 304.
[2] R.J. Birgeneau, H.J. Guggenheim and G. Shirane, Phys. Rev. B1 (1970) 2211.
[3] L. Bevaart, J.V. Lebesque, E. Frikkee and L.J. de Jongh, to be published.

CRITICAL BEHAVIOR OF THE SITE IMPURE SIMPLE CUBIC ISING MODEL*

D.P. LANDAU

Department of Physics and Astronomy, University of Georgia, Athens, Georgia 30602, USA

A Monte-Carlo method has been used to study the simple cubic Ising Model containing up to 30% of random, quenched, non-magnetic impurities. The critical temperature is found to decrease linearly with increasing impurity content but no evidence was found for any significant changes in critical exponents.

1. Introduction

The critical behavior of impure magnetic models is of great theoretical interest as well as relevant to real magnetic systems. It is known that the behavior for mobile (annealed) impurities is related to the pure lattice properties through a set of renormalized critical exponents [1, 2]. Our understanding of fixed (quenched) impurities is less complete although considerable progress has been made recently. McCoy and Wu [3] provided the exact result for random infinitely long row defects in two dimensions and Fisher and Au-Yang [4] have studied regularly spaced point defects. No exact results are available, however, for the more interesting problem of random, point impurities. Renormalization group studies [5] suggest that the phase transition remains "sharp" in the quenched case and that the critical exponents do not depend continuously on impurity concentration. If the specific heat exponent α of the corresponding pure system is negative, then the exponents are unaffected by impurities; but if $\alpha > 0$ a new random fixed point should become stable leading to new critical exponents. The 2-d case which is "borderline" since $\alpha = 0$ has already been studied by Monte-Carlo calculations [6]. In this work we shall report results for three dimensions where $\alpha > 0$ and the theoretical predictions can be tested.

2. Method of study

We have studied $N \times N \times N$ simple cubic Ising lattices using an importance sampling Monte-Carlo method which has already been described elsewhere [7, 8]. Systems with ferromagnetic nearest-neighbour coupling J and periodic boundary conditions were studied for $6 \leq N \leq 20$. Up to $x = 0.3$ of non-interacting impurities were distributed in the lattices randomly. Each data point was taken from at least two different starting configurations and two different impurity distributions. Between 1000 and 5000 M.C. steps were kept for averages for each data point.

3. Results and discussion

The data obtained were qualitatively similar to those for the pure lattice [8]. Specific heat results for several x values are shown in Fig. 1. The data show clearly that the specific heat peak is strongly depressed by the addition of the impurities. The susceptibility data, however, were not affected as strongly. Qualitatively similar behavior had been observed in two dimensions [6] and had in fact been predicted by Stauffer [9]. The ordering temperature $T_c(x)$ was determined for each lattice size and impurity concentration from the peak in the specific heat. Each infinite lattice T_c at fixed x was estimated by extrapolating T_c vs. $N^{-1/\nu}$ in ac-

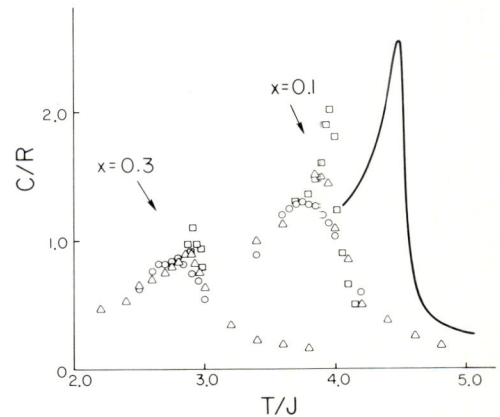

Fig. 1. Specific heat data for : $N = 6$, ○; $N = 10$, △; $N = 20$, □. The solid line is a smooth curve through the $N = 20$ data for $x = 0$ (see ref. 8).

---

* Research supported in part by the National Science Foundation.

cord with finite size scaling (see [7] and [8] and references therein). A certain amount of care must be exercised since a value of ν must be assumed and the infinite lattice T_c obtained will later be used to determine other critical exponents. The total extrapolation from $N = 20$ to $N = \infty$ was about 1% in all cases so any error due to an erroneous choice of ν should be quite small. If, however, our choice of ν was very innaccurate, other critical exponents should be significantly altered from their pure system value and this should be reflected in the exponent analysis.

If the critical behavior is unchanged by the addition of impurities, we can use the generalized form of the scaling hypothesis proposed by Fisher and Au-Yang [4] which predicts

$$\delta T_c = [T_c(0) - T_c(x)]/T_c(0) = a_1 x + a_2 x^{2-\alpha}, \ldots \quad (1)$$

However, if the exponents change we can use the crossover scaling hypothesis [10]

$$\delta T_c = A x^{1/\phi}. \quad (2)$$

The variation of T_c with x is consistent with either Eq. (1) with $a_1 = 1.09 \pm 0.03$ and $a_2 = 0.12 \pm 0.04$ or Eq. (2) with $\phi = 0.98 \pm 0.04$ and $A = 1.17 \pm 0.10$.

The critical behavior of the spontaneous magnetization and high temperature susceptibility was analyzed in terms of the usual power law forms using the values of $T_c(x)$ determined above. In all cases the exponents were within experimental error ($\sim 10\%$) of the pure lattice values. If, however, exponents do change slightly for $x > 0$, then the critical amplitudes should vary as [10]:

$$B(x) \propto x^{(\beta_x - \beta)/\phi} \quad \text{and} \quad C(x) \propto x^{(\gamma - \gamma_x)/\phi}, \quad (3a,b)$$

where β_x, γ_x, etc. are the exponents for the impure system. The observed variation of critical amplitudes with x is consistent with small differences ($\sim 5\%$) between the pure and impure exponent values. A definitive treatment, however, must include proper scaling axes [4] (which by itself introduces a small variation of critical amplitudes).

Clearly more extensive studies on large lattices are needed. Any small change in critical exponents for $x > 0$ will be observed only within a narrow region near $T_c(x)$ and for larger values of $|T - T_c(x)|$ crossover to pure system behavior will occur. Harris [11] has estimated the width of the "impure" critical region to vary as $\sim x^{1/\alpha}$; if this estimate is correct then very large values of x and N will be needed to detect crossover to impure lattice behavior.

We wish to thank the Institut fur Festkörperforschung der KFA Jülich for its hospitality during a portion of the time this work was carried out.

References

[1] I. Syozi, Progr. Theor. Phys. 34 (1965) 189.
[2] H. Garelick and J.W. Essam, Proc. Phys. Soc. 92 (1967) 136.
[3] B.M. McCoy and T.T. Wu, Phys. Rev. Lett. 21 (1968) 549; Phys. Rev. 176 (1968) 631.
[4] M.E. Fisher and H. Au-Yang, J. Phys. C8 (1975) L418.
[5] See G. Grinstein and A. Luther, Phys. Rev. B13 (1976) 1329 and references therein.
[6] W.Y. Ching and D.L. Huber, Phys. Rev. B13 (1976) 2962. E. Stoll and T. Schneider, AIP Conf. Proc. 29 (1976) 490 and to be published. R. Fisch and A.B. Harris, AIP Conf. Proc. 29 (1976) 488.
[7] D.P. Landau, Phys. Rev. B13 (1976) 2997.
[8] D.P. Landau, Phys. Rev. B14 (1976) 255.
[9] D. Stauffer, Z. Physik B22 (1975) 161.
[10] E.K. Riedel and F.J. Wegner, Z. Phys. 225 (1969) 195.
[11] A.B. Harris, J. Phys. C7 (1974) 1671.

LOW FREQUENCY MAGNONS IN $Ni_{1-x}Co_xO$

V. WAGNER*

Institut Laue-Langevin, 156X Centre de Tri, 38042 Grenoble Cedex, France

C.R. BECKER and R. GEICK

Physik Institut der Universität, Röntgenring 8, 87 Würzburg, W. Germany

The magnon dispersion in $Ni_{1-x}Co_xO$ has been measured by inelastic n-scattering techniques along the symmetry direction [111] for $x = 0.07$, 0.10 and 0.30 and magnon energies ≤6.5 THz. As in pure NiO the magnon branch is split due to anisotropy and domain effects. The results are discussed within the terms of the virtual crystal approximation.

1. Introduction

NiO and CoO are f.c.c. antiferromagnets with a collinear spin arrangement of type II with very different anisotropy. In NiO it arises mainly from dipole interaction confining the spins in (111) planes. In CoO there is a strong single ion anisotropy with preferred axis out of the (111) plane. The study of the low frequency magnons in the mixed systems is of interest, because those modes at $q \approx 0$ depend strongly on the exchange interaction and anisotropy. Therefore the dependence of the anisotropy on the composition can be investigated, if the exchange interactions are known. In this context an investigation of the magnon dispersion curves in the mixed crystal for small q by inelastic n-scattering techniques appears very valuable, as thereby the exchange field in the mixed crystal is probed directly. Furthermore, such measurements complete AFMR experiments on the dilute ends [1, 2]. Dispersion curves in the whole range of magnon energies were recently reported for $x = 0.30$ [3].

2. Experimental details and results

The measurements were done at the triple axis crystal spectrometer IN3 at a thermal neutron guide of the HFBR at Grenoble. Most of the results (cf. fig. 1) were obtained from two samples with Co concentrations $x = 0.07$ and 0.30 and a mosaicity of 30'. Some additional scans were made on a sample with $x = 0.10$ of poor mosaicity (~1°30'). The measurements were made along [ζζζ] in different magnetic zones at 4.2 K. As the samples were mul-

tidomain samples with respect to the spin alignment in the different (111) planes, magnons at the M-point on the zone boundary in [11$\bar{1}$] direction were observed at the same time.

The bars in fig. 1 represent the observed linewidth of the scattered neutron groups rather than estimated errors. There is some evidence from the dependence of the scattering intensity on the magnetic zone that the lowest branch has to be assigned to the in-plane magnon mode and the next branch to the out-of-plane magnon. For low Co-concentration ($x = 0.07$ and 0.10) the two zone-boundary modes could be distinguished, while for $Ni_{0.70}Co_{0.30}O$ one AFMR and one zone boundary mode merge into the same branch.

The dispersion relation

$$\nu_j^2(\zeta) = \nu_j^2(0) + C^2\zeta^2, \qquad (1)$$

where $\nu_j(0)$ is obtained from the experiment and

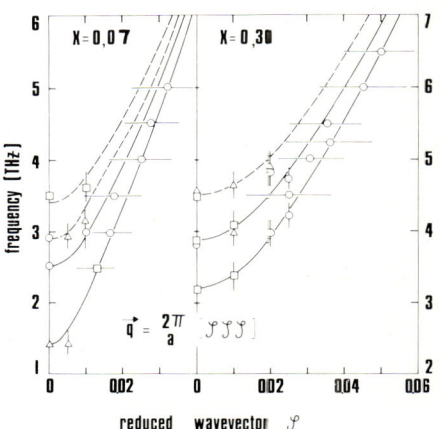

Fig. 1. Magnon dispersion in $Ni_{1-x}Co_xO$ at 4.2 K. The experimental points indicate different magnetic zones: $\bigcirc(\frac{1}{2}\,\frac{1}{2}\,\frac{3}{2})$, $\triangle(\frac{3}{2}\,\frac{3}{2}\,\frac{1}{2})$, $\square(\frac{3}{2}\,\frac{3}{2}\,\frac{3}{2})$. The lines result from VCA as explained in the text.

---

* Guest scientist from Physik Institut der Universität, Würzburg, W. Germany.

Table I
Parameters for effective spin hamiltonian. All interactions are given in THz

	x	$\langle S \rangle$	J_1	J_2	$6\langle J_1S_jS_{j+1}+J_2S_jS_{j+2}\rangle$	$A_{1,2}$	
NiO[a]	0	1	−0.33	4.60	25.62	0	0.023
Ni$_{1-x}$Co$_x$O	0.07	1.03			23.96 ⎫	0.04	0.10 ⎫
	0.10	1.05			22.92 ⎬[c]	0.05	0.17 ⎬[d]
	0.30	1.15			18.34 ⎭	0.28	0.43 ⎭
CoO[b]	1	$\tfrac{3}{2}$	−0.14	1.04	8.10		1.16

[a] From ref. 4. [b] From dispersion curves along [111] ref. 5. [c] From eq. (2) with $J' = (JJ'')^{1/2}$. [d] This work.

C is predicted from the virtual crystal approximation, is a satisfactory description for both zone center and zone boundary magnons (full and dashed lines in fig. 1).

3. Discussion

In the virtual crystal approximation the disordered two-component system is replaced by a perfect one-component system, the properties of which are defined as averages of the properties of the original components. In terms of the parameters of an effective spin hamiltonian this means that a single ion property is replaced by its mean value, e.g. $\langle S \rangle = (1-x)S + xS'$ and an exchange interaction is replaced by its average over pairs

$$\langle J_\delta S_j S_{j+\delta}\rangle = (1-x)^2 J_\delta S_j S_{j+\delta} \\ + 2x(1-x)J'_\delta S_j S'_{j+\delta} + x^2 J''_\delta S'_j S'_{j+\delta}, \quad (2)$$

where the prime refers to the impurity spin.

Then we find for magnons propagating along $[\zeta\zeta\zeta]$

$$C = \{6 \langle J_1 S_j S_{j+1} + J_2 S_j S_{j+2}\rangle/\langle S\rangle\},$$

where we assumed that all spins were collinear in the mixed crystal and where J_1 and J_2 is the isotropic exchange between first and second neighbours, respectively, in a hamiltonian with an effective anisotropy AS_j^z (table I). The agreement with the experiment is satisfactory, as would be expected for long wavelength magnons, which average the interaction over many unit cells. Moreover as the phase relation between n.n.n. spins is the same at the zone center and at the M point, one might expect the predictions of the VCA to be useful even at this specific point on the zone boundary if $J_2 \gg |J_1|$ (dashed lines in fig. 1).

When we calculate the effective anisotropy from the observed frequencies and $\nu^2(0) = \langle A\rangle^2 + 2\langle A\rangle C$, we find that $\langle A\rangle$ does not depend linearly on x as would be expected for a single ion anisotropy. This is not surprising, as the model does not account either for a change of the preferred spin direction with composition or for ferrimagnetic fluctuations arising in the mixed system.

A detailed paper will be published elsewhere. One of us (V.W.) acknowledges financial support from the "Bundesministerium Forschung und Technologie" Bonn.

References

[1] C.R. Becker, P. Lau, R. Geick and V. Wagner, Phys. Status Solidi (b) 67 (1975) 653.
[2] G. Geis, R. Geick, C.R. Becker and V. Wagner, this conference, paper 1C1.
[3] V. Wagner, D. Tocchetti and B. Hennion, in: Magnetism and Magnetic Materials – 1975, J.J. Becker, G.H. Lander, J.J. Rhyne (Eds.) (AIP, New York, 1976) p. 255.
[4] M. Hutchings and E.J. Samuelsen, Phys. Rev. B6 (1972) 3447.
[5] V. Wagner and D. Ronzaud, unpublished.

MAGNETIC PROPERTIES OF $Mn_xMg_{1-x}S$ $(0 < x \leq 1)$

H.H. HEIKENS and C.F. VAN BRUGGEN
Laboratory of Inorganic Chemistry, Materials Science Center of the University, Groningen, The Netherlands

Magnetic properties of polycrystalline solid solutions $Mn_xMg_{1-x}S$ are reported. From these data a critical concentration $x_c = 0.13 \pm 0.01$ was obtained, in accordance with calculations from series expansion of the mean cluster size for a f.c.c. lattice with exchange interactions up to 12 n.n. and 6 n.n.n. Magnetic measurements on single crystals of α – MnS show anomalies which are not observed in powder samples.

In previous papers [1, 2] the magnetic properties of polycrystalline solid solutions $Mn_xMg_{1-x}Y_2S_4$ were discussed. From this study it appeared worthwhile to know corresponding data of a crystallographically more simple system. For this purpose the solid solution series $Mn_xMg_{1-x}S$ was chosen. The component materials MnS and MgS both crystallize in the rocksalt structure, with about the same cell edges at room temperature ($a = 5.21$ and 5.19 Å, respectively).

Fig. 1 shows the reciprocal molar susceptibilities (χ^{-1}) from 2 to 300 K for the series $Mn_xMg_{1-x}S$. At high temperatures χ^{-1} vs. T obeys the Curie–Weiss law $\chi^{-1}_{mol\ Mn} = x(T - \theta)/C$, with asymptotic Curie tem-

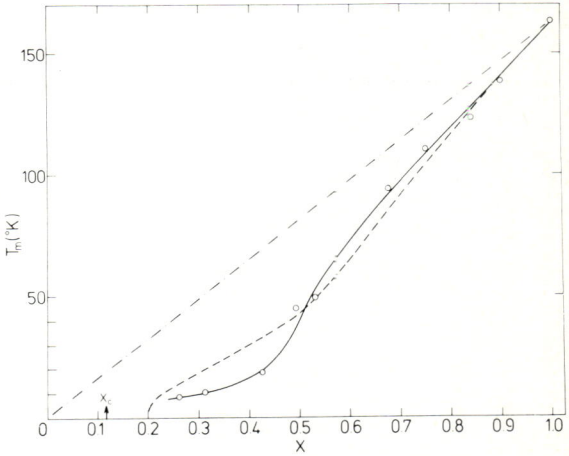

Fig. 2. Composition dependence of experimentally observed T_m (–O–); exact series expansion for f.c.c. with 12 n.n. and $S = \frac{1}{2}$ (– – –); molecular field theory (– · – ·); see text.

peratures θ and Curie constants C. θ decreases linearly with composition x. For $x \geq 0.25$ a minimum or change in slope of χ^{-1} vs. T (onset of short-range antiferromagnetic order) can be observed at temperatures T_m which increase with x (fig 2). The observed x dependence of T_m agrees qualitatively with the results of a series expansion calculation of χ for the f.c.c.-type structure with 12 exchange coupled nearest neighbours (n.n.) [3]. For low concentrations $x \leq 0.25$ no ordering phenomena in χ^{-1} vs. T could be observed; instead a continuous downturn of the χ^{-1} vs. T curve to the origin is observed which is caused by the presence of isolated paramagnetic Mn ions. This behaviour is also known from amorphous magnetic materials [2]. Although ordering phenomena may be obscured by magnetic contributions of finite clusters, no antiferromagnetic ordering will occur below a critical site concentration x_c [4]. For $x \geq x_c$ the mean cluster size is infinite and long-range ordering may occur. The experimental value of x_c

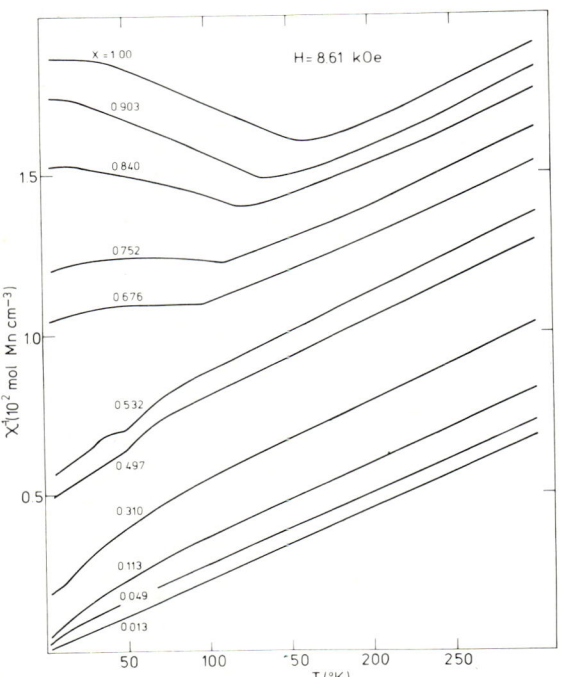

Fig. 1. Reciprocal magnetic susceptibility for $Mn_xMg_{1-x}S$ from 2 to 300 K for an applied magnetic field $H = 8.61$ kOe.

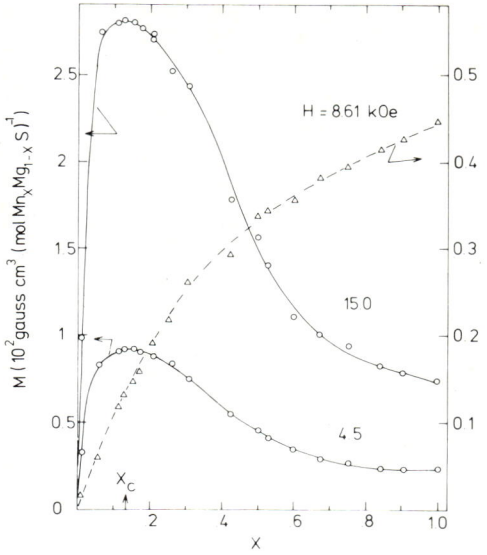

Fig. 3. Magnetization versus composition x at 4.5 K (solid lines) resp. 300 K (broken line); see text.

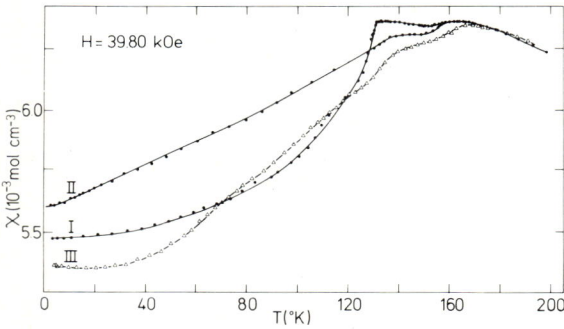

Fig. 4. Susceptibility of α − MnS from 4 to 200 K for an applied magnetic field $H = 39.80$ kOe (I: unstressed crystal, II: [111] stressed crystal, III: randomly oriented powder); see text.

was obtained from magnetization M vs. x curves at 4.5 K (fig. 3; solid lines). These data were obtained from M vs. H measurements and give a maximum at $x = 0.13 \pm 0.01$. The corresponding data at 300 K are represented in the same figure by a broken line; this paramagnetic behaviour is described by $M = Nxg^2\beta^2S(S+1)H/\{3k(T-\theta)\}$. The low-temperature behaviour (solid lines) for $x < x_c$ can formally be described by the same law, when substituting paramagnetism by superparamagnetism by virtue of the presence of finite clusters with a net superparamagnetic moment. However for $x \geqslant x_c$ an infinite mean cluster size implies an increasing antiferromagnetic compensation of magnetic moment with increasing x. Therefore x_c is characterized by a maximum in the M vs. x curve. Calculations based on a series expansion of the mean cluster size and applied to a f.c.c. lattice result in the following critical probabilities: $x_c = 0.195$ (12 n.n.), 0.136 (12 n.n. + 6 n.n.n.), 0.061 (12 n.n. + 6 n.n.n. + 26 n.n.n.n.) (n.n.n. = next-nearest neighbours etc.) [5]. From the close correspondence between experiment and theory, it may be concluded that an interaction sphere up to 6 n.n.n. is adequate to describe the exchange properties of the system $Mn_xMg_{1-x}S$.

Finally, some results are reported for pure α − MnS. Single crystals were grown by iodine vapour transport. Accurate H- and T-dependent measurements on (unstressed) oriented single crystals revealed that its magnetic properties are quite different both from those of the [111]-stressed crystals as from those of randomly oriented powdered samples (fig. 4). For $H = 39.80$ kOe one observes a strong anomalous decrease at $T = 131$ K for the unstressed crystal (curve I); the stressed crystal (curve II) and the powdered sample (curve III) indicate only a weak anomaly at a somewhat higher temperature.

Differential thermal analysis reveals only a Néel temperature $T_N = 147.6$ K, in accordance with results from NMR measurements [6].

M vs. H was measured with H up to 40 kOe at several temperatures (not shown). At $T < 131$ K M vs. H is concave up and shows hysteresis effects. At $131 < T < 148$ K, however, the curves are linear and no hysteresis is observed.

Preliminary X-ray diffraction measurements indicate that the intermediate temperature range from 131 to 148 K is associated with a drop in the intensity of the reflections.

References

[1] H.H. Heikens and C.F. van Bruggen, IV Int. Conf. on Solid Compounds of Transition Elements, Geneva, April (1973).
[2] H.H. Heikens and C.F. van Bruggen, Proc. Int. Conf. on Magnetism, ICM-74, Moscow, Vol. I(2) (1974) p. 83-88.
[3] G.S. Rushbrooke, in: Critical Phenomena in Alloys, Magnets and Superconductors, R.E. Mills, E. Asher and R.I. Jaffee (eds) (Mc-Graw-Hill, 1971) p. 155.
[4] R.J. Elliot and B.R. Heap, Proc. Roy. Soc. (London) A 265 (1962) 264.
[5] C. Domb and N.W. Dalton, Proc. Phys. Soc. 89 (1966) 856.
[6] E.D. Jones, Phys. Rev. 151 (1966) 315.

EXPERIMENTAL STUDY OF THE DILUTED S.C. $S = \frac{1}{2}$ XY ANTIFERROMAGNET: SPECIFIC HEAT OF $Co_{1-x}Zn_x(C_5H_5NO)_6(ClO_4)_2$

H.A. ALGRA, L.J. DE JONGH, W.J. HUISKAMP

Kamerlingh Onnes Laboratory, University of Leiden, The Netherlands

and

J. REEDIJK*

Chemistry Department, Delft University of Technology, The Netherlands

Specific heat data on $Co_{1-x}Zn_x(C_5H_5NO)_6(ClO_4)_2$ are compared with predictions for the quenched-site diluted XY magnet. The variation of $T_c(x)$ is studied. For $x \geq 0.3$ a Schottky-type contribution is resolved.

Compounds of the series ML_6X_2, where M^{2+} is a 3d metal ion, $L = C_5H_5NO$, $X^- = BF_4$ or ClO_4, are isomorphous [1] with lattice constants equal within 1%. Recently [2] we have shown that the Co^{2+} salts approximate the simple cubic (s.c.) XY model of an antiferromagnet. The ordering temperatures are below 1 K, the Co^{2+} ions having effective spins $S = \frac{1}{2}$ with strong XY-type spin-anisotropy.

We report here specific heat data on $CoL_6(ClO_4)_2$ diluted with Zn, comparing the results with predictions [3] for the quenched-site diluted XY model. As the cell-constants of the pure Co and Zn compounds are equal within 0.25%, the variation of the superexchange (J/k) between Co^{2+} ions upon dilution is expected to be minor (a few %).

Diluted (powder) samples were prepared by adding a mixture of $Co(H_2O)_6(ClO_4)_2$ and $Zn(H_2O)_6(ClO_4)_2$, dissolved in the desired ratio in ethanol and triethylorthoformate, to a second solution of C_5H_5NO in ethanol. The desired material then precipitates in the form of microscopic crystallites. After stirring the solution for several days the crystallites are filtered, washed in dry diethylether, and dried in vacuo for several hours at 20–40°C. The percentages of Co and Zn were determined by chemical analyses (estimated error of a few %). Since the metal ions are "encaged" in a surrounding of six C_5H_5NO groups, diffusion processes near room temperature seem unlikely. On the other hand, the speed of precipitation, for example, will depend on the ratio of the various solvents used, so that we still have some uncertainty about the extent in which quenched-site conditions are fulfilled.

Representative specific heat curves obtained below 1 K for $Co_{1-x}Zn_xL_6(ClO_4)_2$ samples for varying x are shown in fig. 1, with data [2] for $x = 0$ included. For $x > 0.3$ a Schottky-type anomaly is resolved; for larger x the specific heat appears to be composed of this short-range order contribution plus a long-range order anomaly (λ-type peak), $T_c(x)$ decreasing gradually upon dilution. Although $T_c(x)$ remains reasonably sharp, the amplitude of the λ-peak is reduced considerably, such that for $x > 0.4$ the peak is hardly resolvable. Furthermore, with increasing x the Schottky maximum broadens

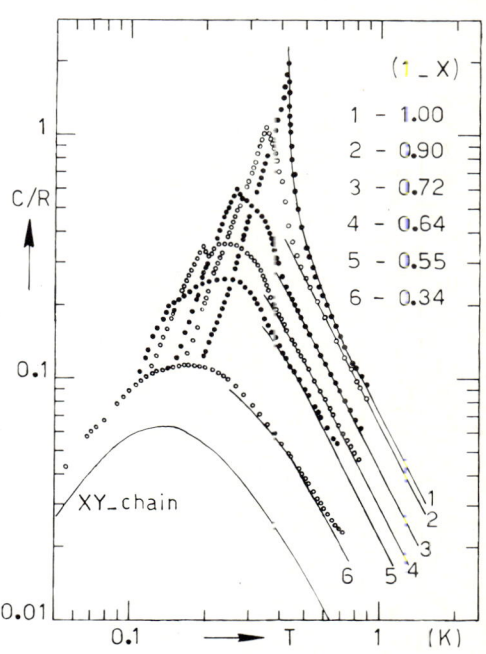

Fig. 1. Logarithmic plot of the specific heat for $Co_{1-x}Zn_xL_6(ClO_4)_2$.

and shifts to lower T. Interestingly, it appears to approach the shape and position (on the T-axis) of the specific heat curve for the linear chain XY model. Katsura's prediction [4] for this model is included in fig. 1, with the specific heat divided by a factor of 5, and using $J/k = -0.212$ K, as found [2] for $CoL_6(ClO_4)_2$. All the above qualitative features are strikingly similar to those of the Syozi model [5] for a diluted decorated square Ising lattice, which represents the only exact solution obtained thusfar for a diluted magnetic system. In this model C_m/R has a logarithmic singularity for $x = 1$ (as predicted also for s.c. XY), but a finite cusp for $x < 1$, the amplitude decreasing with x. For $x \geqslant 0.25$ a Schottky anomaly is resolved which apparently likewise approaches the (shape and position of the) Ising chain specific heat. One is thus tempted to conclude that these characteristics are common to a variety of diluted magnetic systems.

Quantitatively, we may compare our data with the high-temperature series given by Reeve and Betts [3] for the specific heat of the quenched-site diluted s.c. XY magnet. Predictions obtained by direct Padé-approximants to this series [6] are included in fig. 1 (solid curves). Since only three terms are available, relatively little improvement is obtained upon the prediction given by the (asymptotic) first term: $C_m/R = \frac{3}{2}(1 - x)^2(J/kT)^2$ only. For small x, however, the approximants appear to indicate curvatures similar to the observed ones. For $x = 0$ a longer series is available [7], and a logarithmic divergence has been predicted [8]. As reported earlier [2], theory and experiment for $x = 0$ are in accord for $T/T_c \geqslant 1.03$ (cf. fig. 1). For $x \geqslant 0.3$ a small lattice contribution is observable in fig. 1 for $T > 0.6$ K, becoming the more important the larger is x.

In fitting the series predictions to the data we have as a first choice taken J/k as variable, using the experimental x values. This then results in the spread of J/k values observed in table I, where the properties of the samples are summarized. The spread remains within 5% only (except for $1 - x = 0.344$), and we suspect that in reality errors in the x values are largely responsible, J/k remaining nearly constant. The experimentally obtained values for the entropy generally agree within the uncertainty (3%) to the theoretical figures (table I). In calculating the latter, correction is made for the presence of isolated magnetic ions by using $(1 - x)(1 - x^6)$ instead of $(1 - x)$ for the concentration of magnetic ions.

Lastly we compare in fig. 2 the observed $T_c(x)$ variation with theory (solid curve [3], extrapolated to the critical percolation probability [9] $P_c \simeq 0.31$). Slight deviations are found. These may be due, for example, to the fact that the full

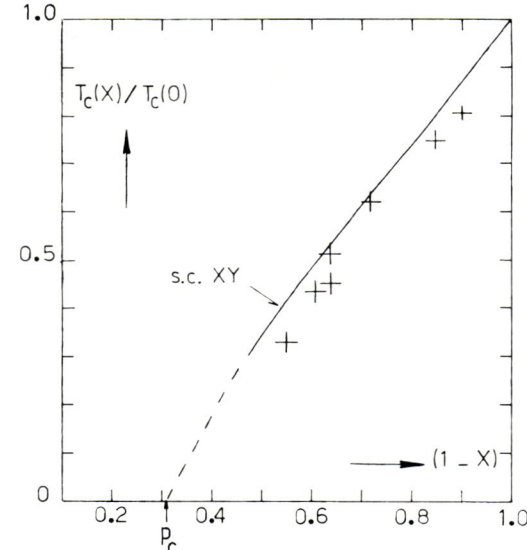

Fig. 2. Dependence of $T_c(x)$ on dilution.

Table I
Properties of the materials

$1 - x$	1.00	0.90	0.845	0.716	0.639	0.635	0.608	0.545	0.344
T_c (K)	0.428(3)	0.345(5)	0.318(5)	0.265(5)	0.194(5)	0.220(5)	0.185(5)	0.140(5)	?
J/k (K)	0.212(1)	0.220(5)	0.208(5)	0.225(5)	0.211(5)	0.227(5)	0.220(5)	0.225(10)	0.24(2)
S_{th}/R	0.693	0.624	0.589	0.496	0.444	0.444	0.416	0.374	0.219
S_{ex}/R	0.71	0.69	0.60	0.54	0.45	0.43	0.43	0.37	0.24

S_{exp} and S_{th} are the experimental and theoretical magnetic entropies, respectively.

XY anisotropy is not reached in the experiment. [2]. On the other hand, we still have to check on the possible influence of the method of sample preparation.

Further experiments are in progress. The investigations are supported in part by Stichting "F.O.M.".

References

[1] A.D. van Ingen Schenau, G.C. Verschoor and C. Romers, Act. Cryst. B30 (1974) 1686.
[2] H.A. Algra, L.J. de Jongh, W.J. Huiskamp and R.L. Carlin, Physica 83B (1976) 71.
[3] J.S. Reeve and D.D. Betts, J. Phys. C8 (1975) 2642.
[4] S. Katsura, Phys. Rev. 127 (1962) 1508.
[5] I. Syozi and S. Miyazima, Progr. Theor. Phys. 36 (1966) 1083; see also K. Takeda et al., J. Phys. Soc. Jap. 28 (1970) 34 (fig. 8).
[6] R. Navarro, private communication.
[7] J.T. Tsai and C.J. Elliott, Phys. Lett. 45A (1973) 295.
[8] D.D. Betts and J.R. Lothian, Can. J. Phys. 51 (1973) 2249.
D.D. Betts, this conference.
[9] M.F. Sykes and J.W. Essam, Phys. Rev. 133A (1964) 310.

FERROMAGNETIC ORDERING AND PHOTOLUMINESCENCE OF $Eu_xSr_{1-x}S$ AND $Eu_xSr_{1-x}O$

K. WESTERHOLT, B. GHOSH*, K. SIRATORI†, S. METHFESSEL

Institut für Experimentalphysik IV, Ruhr-Universität Bochum, 463 Bochum, W. Germany

and

T. PETZEL

Institut für Chemie, Universität Freiburg, 78 Freiburg, W. Germany

The ferromagnetic Curie temperatures of the solid solution systems $Eu_xSr_{1-x}S$ and $Eu_xSr_{1-x}O$ show rather different concentration dependence. This result is analysed by two spin cluster calculations. We get additional information about short range magnetic ordering by studying the photoluminescence of the $4f^7$–$4f^65d$ optical transition, which is strongly influenced by the spin ordering.

The rocksalt type compound EuS is a semiconductor with a ferromagnetic Curie temperature of 16 K and an anomalous dependence of the optical $4f^7$–$4f^65d$ transition on the spin ordering [1]. By dilution of the Eu-sublattice with nonmagnetic Sr, we have examined these properties as functions of the decreasing exchange interactions.

The samples were prepared by precipitating a mixed (Eu–Sr)-oxalate from a mixture of a $Eu(NO_3)_3$-and a $SrCl_2$-solution, followed by sulfuration in an H_2S–H_2-stream at 1100°C for 10 h. The lattice parameters were measured by X-ray powder patterns. The paramagnetic Curie temperatures were obtained from susceptibility measurements using the Faraday method between 300 and 4 K. They both show linearity with concentration as expected for good solid solutions (fig. 1). The ferromagnetic Curie temperatures T_c were determined from the sharp change in the initial susceptibility measured in an a.c. magnetic field of 1 G. In fig. 2 the reduced ferromagnetic Curie temperatures are given for $Eu_xSr_{1-x}S$ and $Eu_xSr_{1-x}O$ as functions of concentration x and compared with theory.

The theoretical curves for the Curie points were obtained from the two spin cluster calculations of Callen and Callen [2] for a Heisenberg magnet with nearest and next nearest neighbour exchange. We adapted their formula for concentrated systems for our diluted system by extending the sum over magnetic neighbours

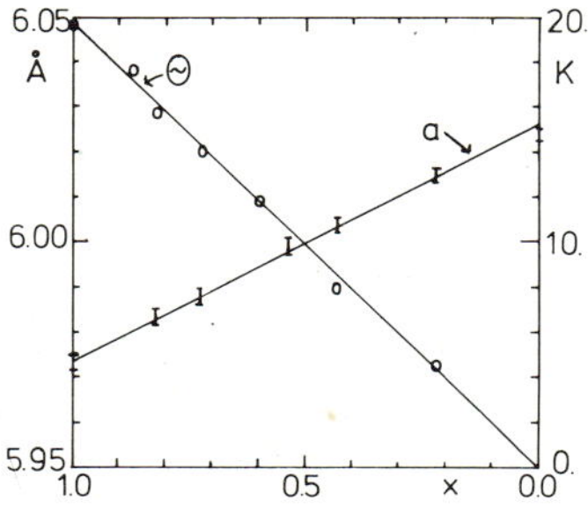

* Now at University of Calcutta, India.
† Now at Osaka University, Toyanaka, Osaka, Japan.

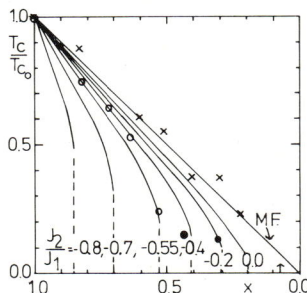

Fig. 1. Lattice parameters a and paramagnetic Curie temperatures Θ as function of x in $Eu_xSr_{1-x}S$.

Fig. 2. Reduced ferromagnetic Curie temperatures as function of x in $Eu_xSr_{1-x}S$ (○○○) with $T_{co} = 16.5$ K and $Eu_xSr_{1-x}O$ (×××) with $T_{co} = 71$ K. The solid lines result from two spin cluster calculations for different J_2/J_1-ratios or molecular field theory (MF).

only. These equations were then solved numerically for different ratios of ferromagnetic nearest neighbour exchange J_1 and antiferromagnetic next nearest neighbour exchange J_2 with the results shown in fig. 2. The broken lines indicate the lower limits in concentration for ferromagnetic ordering. We get paramagnetism as stable solutions for Eu-concentrations smaller than this lower limit.

The experimental points for $Eu_xSr_{1-x}S$ are between the theoretical curves for $J_2/J_1 = -0.4$ and -0.55. This ratio of exchange contants is in reasonable agreement with other measurements [3]. Our experiments give a temperature independent initial susceptibility at low temperatures also when the Eu-concentration becomes smaller than the theoretical lower limit of $x \approx 0.5$. The black points in fig. 2 indicate the temperature where the susceptibility becomes constant.

Probably the magnetization of the samples then breaks up into superparamagnetic clusters in agreement with our optical experiments. It is interesting to note that the concentration dependence of the Curie temperatures follows the molecular field theory in the case of $Eu_xSr_{1-x}O$. This could indicate that in EuO J_2 is also ferromagnetic. Of our samples only the sample with the composition $Eu_{0.07}Sr_{0.93}O$, marked by an arrow in fig. 2, showed no ferromagnetic order down to 4 K.

The optical transition $4f^7 - 4f^65d$ of the Eu-ions can be used as an additional probe for spin ordering. We measured the dependence of the luminescence of this transition on temperature and composition, using the radiation of a high pressure Xe-lamp for excitation. In the whole solid solution range of $Eu_xSr_{1-x}S$ only one type of luminescence with nearly lorentzian intensity distribution and a half width of about 0.25 eV is seen. The peak shifts to lower energies with increasing Eu-concentration and decreasing temperature as shown in fig. 3. The red shift of the luminescence peak with decreasing temperature can be related to the magnetic ordering [1]. The wave-function of the optically excited 5d electron overlaps with the 5d orbitals of the neighbouring Eu-ions and by f–d interatomic exchange its energy depends on the relative orientation of the f-spins. Since the $4f^65d$ exciton is rather localized [4], its energy depends stronger on short range spin order than on the net magnetization. Therefore the red shift is sensitive to spin correlations far above the Curie point and can be influenced strongly by applied magnetic fields (fig. 3). The samples with Eu-concentrations below the limit for long range ordering also show the red shift, in agreement

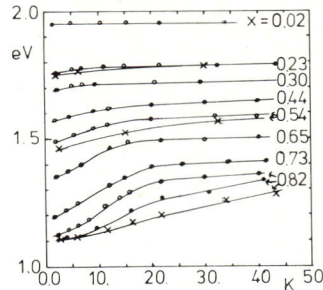

Fig. 3. Energy of the luminescence maximum for different Eu-concentrations in $Eu_xSr_{1-x}S$ as a function of temperature. Samples with the concentrations $x = 0.23$, 0.54 and 0.82 have been measured in magnetic fields: (○○○: $H = 0$, ●●●: $H = 10$ kG, ×××: $H = 40$ kG).

with the assumption of superparamagnetic spin clusters.

Another very interesting but more complex effect of the magnetic ordering on the luminescence is shown in fig. 4. The intensity of the luminescence is quenched at low temperatures. This magnetic quenching is probably caused by the interaction of the $4f^65d$ exciton with impurity centres which provide radiationless transitions. In the paramagnetic regime at not too high temperatures two mechanisms can localize the excited d electron. The first, and larger one, is due to the Coulomb attraction by the $4f^6$ Eu-centre. The second one is a spin polarization effect which traps the excited electron in a cloud of aligned 4f-spins produced by its exchange interactions with the nearest neighbour ions. The spin polaron trapping is quenched by long range ferromagnetic ordering. Then the exciton will spread out into the lattice and increase the probability of interaction with impurity-centres for radiationless transitions.

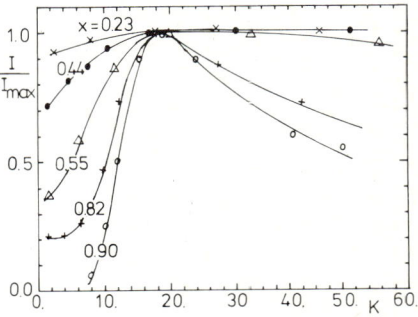

Fig. 4. Relative intensity of the emission peak as a function of temperature for different Eu concentrations x in $Eu_xSr_{1-x}S$.

These centres must be very effective because the luminescence in the paramagnetic state has a quantum efficiency of less than 1% and already small changes in exciton radius cause a strong quenching as shown in fig. 4.

The authors thank the Deutsche Forschungsgemeinschaft for the support of this work.

References

[1] G. Busch, P. Sreit and P. Wachter, Solid State Commun. 8 (1970) 1759.
[2] H.B. Callen and E. Callen, Phys. Rev. A136 and (1964) 1675.
[3] S. Methfessel and D.C. Mattis Handb. der Phys. 18 (1969) 391.
[4] T. Kasuya and A. Yanase Rev. Mod. Phys. 40/4 (1968) 684.

CHAPTER 5

DISORDERED AND AMORPHOUS SYSTEMS

Theory; Disorder (*Session 2A*)	745
Amorphous alloys I, rare earths-transition metals (*Session 6D*)	755
Amorphous alloys II (*Session 7X*)	775
Amorphous alloys III (*Session 8V*)	796
Spin glasses (*Invited paper 7BI2*)	813
Spin glasses I (*Session 4U*)	820
Spin glasses II (*Session 5C*)	842
Spin glasses III (*Session 6U*)	856

MAGNETIC MOMENT DISTRIBUTION OF FERROMAGNETIC Ni-Rh ALLOYS

J.W. CABLE and E.O. WOLLAN

Solid State Division, Oak Ridge National Laboratory*, Oak Ridge, Tennessee 37830, USA

The diffuse scattering of polarized and unpolarized neutrons was used to determine the spatial distribution of the magnetic moment for ferromagnetic Ni-Rh alloys. The average Ni moment remains near 0.6 μ_B to 12 at.% Rh and then decreases toward zero at the critical concentration of 37 at.% Rh. There is an initial rapid decrease in the Rh moment that follows a P_{12} dependence and corresponds to a moment of 2 μ_B for isolated Rh atoms. The data indicate moment fluctuations at both the Ni and the Rh sites that are associated with local environment.

The magnetic behavior of ferromagnetic Ni-Rh alloys is anomalous and not well understood. It has been shown [1] that the increase in magnetization at low Rh content [2] is due to a Rh moment of about 2 μ_B. However, with increasing Rh, the magnetization passes through a maximum near 4 at.% Rh and then decreases to zero near 37 at.% Rh. It is essential to know the magnetic moment distribution in order to determine if this rapid loss of moment is due to competing interactions, as in the Ni-Mn system, or to some other local-environment effect that destroys the Rh moment. We have determined these moment distributions for Ni-Rh alloys by neutron diffuse scattering methods.

Both polarized and unpolarized-neutron cross sections were measured but, in this paper, we concentrate on the polarized data. The polarized-neutron diffuse-scattering cross section for these random alloys is

$$\Delta\frac{d\sigma}{d\Omega} = 1.08c(1-c)(b_{Rh} - b_{Ni})M(K). \quad (1)$$

in which c is the Rh content, $b_{Rh} = 0.584$ and $b_{Ni} = 1.03$. In the linear-superposition model of Marshall [3], $M(K)$ is given by

$$M(K) = \bar{\mu}_{Rh}f_{Rh} - \bar{\mu}_{Ni}f_{Ni} + (1-c)G(K)f_{Ni} + cH(K)f_{Rh}, \quad (2)$$

where $\bar{\mu}_{Rh}$ and $\bar{\mu}_{Ni}$ are the average Rh and Ni moments with associated form factors f_{Rh} and f_{Ni}. $G(K)$ and $H(K)$ are Fourier transforms of the Rh-induced moment-disturbance at Ni and Rh sites, respectively. The observed $M(K)$ are shown in fig. 1 along with the corresponding functions, $T(K)^{1/2}$, from unpolarized

* Operated by Union Carbide Corporation for the USERDA.

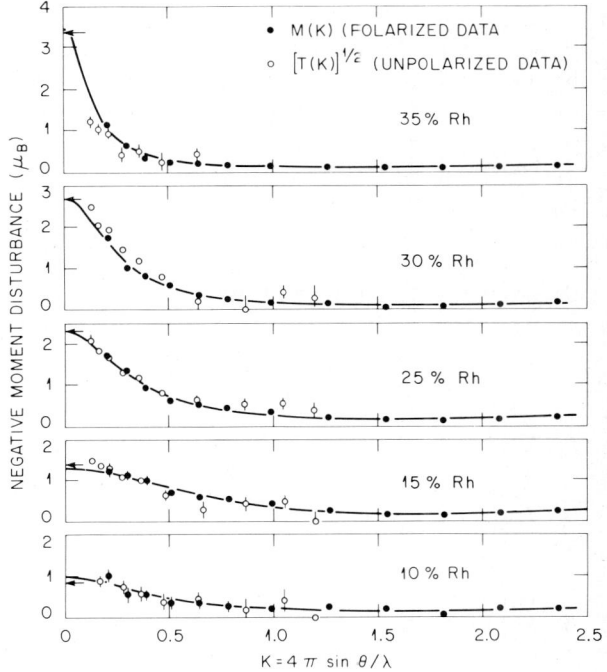

Fig. 1. The K-dependent moment-disturbances for Ni-Rh alloys. The solid curves are fitted to the $M(K)$ data and the arrows indicate magnetization results for $d\bar{\mu}/dc$.

measurements. The data were obtained for annealed and quenched, polycrystalline samples at 4.2 K in a 20 kOe field.

Since $G(K)$ and $H(K)$ have the same form, the Ni and Rh moment disturbance parameters cannot be determined independently. We have therefore fitted the data to the expression

$$M(K)/f_{Ni} = (f_{Rh}/f_{Ni})\mu_{Rh} - \mu_{Ni} + \sum_{R_i} Z(R_i)\phi(R) \sin KR_i/KR_i, \quad (3)$$

where $Z(R_i)$ is the number of atoms in the ith

shell and $\phi(R_i)$ is the total Rh-induced moment-disturbance. We assume a Yukawa form for $\phi(R_i)$, i.e.

$$\phi(R_i) = \phi(R_1) R_1 R_i^{-1} \exp[-\kappa(R_i - R_1)]. \qquad (4)$$

and least squares fit with the parameters $(f_{Rh}/f_{Ni})\bar{\mu}_{Rh} - \mu_{Ni}$, $\phi(R_1)$ and κ. The fitted curves are shown in fig. 1 where it can be seen that this lorentzian form for $\phi(K)$ adequately represents the data.

The parameter $(f_{Rh}/f_{Ni})\bar{\mu}_{Rh} - \mu_{Ni}$, along with magnetization data and form factor assumptions, yields directly the average Rh and Ni moments. We use the published magnetization data [2, 4] and the Pd and Ni form factors and obtain the moment values shown in fig. 2. The 2 at.% Rh data point is from unpolarized-neutron measurements and agrees with the previous result [1] in this concentration region. The curve labelled 2P$_{12}$ is twice the probability that an atom has 12 Ni nearest neighbors. This describes the initial Rh moment behavior reasonably well and suggests a moment of 2 μ_B for isolated Rh atoms. Clearly, this is not simply an on-off situation since a small $\bar{\mu}_{Rh}$ persists right up to the critical concentration.

Although the $\phi(R_1)$ and κ parameters give the range and magnitude of the total moment-disturbance, they are weighted averages and are not unambiguously resolvable into parameters describing the disturbances at Ni and Rh sites. Some insight can, however, be gained by taking the $K = 0$ limit of $M(K)$. Marshall [3] shows that $M(0) = d\bar{\mu}/dc$, $G(0) = d\bar{\mu}_{Ni}/dc$ and $H(0) = d\bar{\mu}_{Rh}/dc$ if the moment disturbances are associated with local environment effects. That $M(0) \simeq d\bar{\mu}/dc$ can be seen in fig. 1 where the arrows indicate $d\bar{\mu}/dc$ values from the magnetization data [2, 4]. This indicates that the moment disturbances are indeed due to local-environment effects and that $G(0)$ and $H(0)$ can be taken from the concentration derivatives of $\bar{\mu}_{Rh}$ and $\bar{\mu}_{Ni}$. From fig. 2 and eq. (2) it can be seen that the Rh moment fluctuations dominate $M(K)$ below 10 at.% Rh while the Ni moment fluctuations become the most important above 15 at.% Rh. In the latter region, the sharper K dependence of $M(K)$ with increasing Rh content should therefore be associated mostly with an increase in the range of the Rh-induced moment disturbance at Ni sites. This also occurs for Ni–Cu alloys [5, 6] and has been interpreted in terms of a cooperative magnetic environment effect. We expect that magnetic environment is also an important factor in the moment distribution of these alloys.

References

[1] J.B. Comly, T.M. Holden and G.G. Low, J. Phys. C1 (1968) 458.
[2] J. Crangle and D. Parsons, Proc. Roy. Soc. (London) A255 (1960) 509.
[3] W. Marshall, J. Phys. C1 (1968) 88.
[4] W.C. Muellner and J.S. Kouvel, Phys. Rev. B11 (1975) 4552.
[5] A.T. Aldred, B.D. Rainford, T.J. Hicks and J.S. Kouvel, Phys. Rev. B7 (1973) 218.
[6] R.A. Medina and J.W. Cable, to be published.

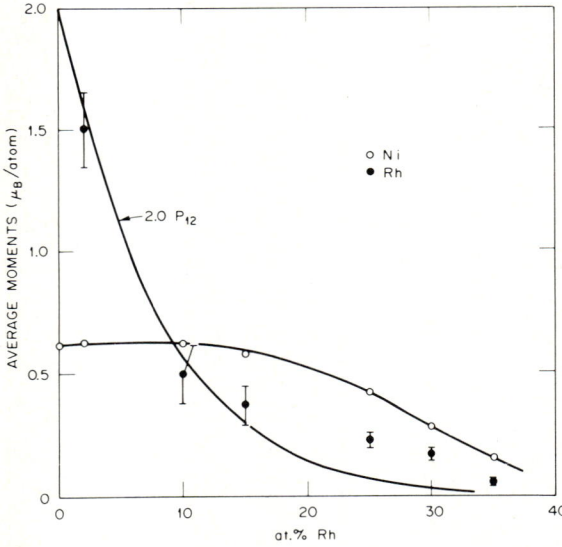

Fig. 2. The average Rh and Ni moments vs. Rh content. The curve labelled 2P$_{12}$ is twice the probability that an atom has 12 Ni nearest neighbors.

MAGNETIC STATE OF Mn ATOM IN FERROMAGNETIC Ni–Fe–Mn AND Ni–Mn ALLOYS: LOCAL ENVIRONMENT EFFECT

T. JO

Department of Physics, Faculty of Science, Osaka University, Toyonaka, Osaka 560, Japan

The effect of the local atomic environment on the magnetic state of Mn atom in ferromagnetic Ni–Fe–Mn and Ni–Mn alloys is investigated theoretically. The calculations well explain the frequency distribution of Mn NMR spin echo signal for these alloys.

Recently, Kitaoka and Asayama [1] presented direct evidence for the coexistence of Mn atoms with moments parallel and antiparallel to the bulk magnetization in ferromagnetic Ni–Fe–Mn and Ni–Mn alloys from the shift of the NMR spectrum due to external field. Previously, we investigated the existence of two magnetic states generally and discussed the effect of the local atomic environment on the magnetic state of Mn impurity in Ni–Co alloys [2]. The purpose of the present work is to discuss the magnetic state of Mn in Ni–Fe–Mn and Ni–Mn alloys by extending the previous approach and to compare the calculated results with the above-mentioned NMR experiment.

We first discuss the magnetic state of a Mn impurity in $Ni_{1-x}Fe_x$ alloys. Our model is the tight binding one with intra-atomic Coulomb interaction, which is treated by the Hartree–Fock (HF) approximation. The hamiltonian of σ spin electrons is given by

$$H^\sigma = t \sum_{i,j} a^+_{i\sigma} a_{j\sigma} + \sum_{i\sigma} E_{i\sigma} a^+_{i\sigma} a_{i\sigma},$$

where i and j denote the lattice sites. The first term represents the electron transfer between nearest neighbors and we adopt the same density of states of pure metal (f.c.c.) corresponding to this term as that used in ref. 2. The second term is the spin dependent atomic energy level given by the HF condition, $E_{i\sigma} = E_i + U_i n_{i-\sigma}$, where E_i and U_i are the spin independent atomic energy level and the intraatomic coulomb integral of atom occupying i site; $n_{i-\sigma}$ is the number of $-\sigma$ spin electrons.

When the concentration of Fe, x is given, we can carry out the coherent potential approximation (CPA) combined with the HF approximation done by Hasegawa and Kanamori [3] to determine the magnetic states of Ni and Fe atoms and the effective medium described by the spin dependent coherent potential. We place, in the effective medium, the atomic cluster composed of a central Mn atom and its nearest neighboring Ni and Fe atoms. We adopt the spin dependent atomic energy levels of the Ni and Fe atoms $E_{k\sigma}$ ($k =$ Ni and Fe) determined by the CPA calculation for Ni–Fe. The energy distribution of electrons at Mn site is calculated for each nearest neighbor configuration specified by the number of Ni or Fe atom (N_{Ni} or N_{Fe}) by the same procedure as in ref. 2. The number of electrons of Mn for each spin state is obtained by integrating the energy distribution up to the Fermi level. The magnetic state of Mn is determined by the local HF equation, $E_{Mn\sigma} = E_{Mn} + U_{Mn} n_{Mn-\sigma}$.

As shown in the case of Mn impurity in Ni–Co alloys, there also exist, in the present system, two HF solutions for the magnetic state of Mn with moments parallel and antiparallel to the bulk magnetization. By comparing the energies of the two solutions, the magnetic moment of Mn is found to be antiparallel when N_{Fe} is equal to or more than a critical number, N_c, while it is parallel when N_{Fe} is less than N_c. Furthermore, N_c decreases with increasing Fe concentration, x (see fig. 1). This result is consistent with the Fe concentration dependence of the relative NMR spin echo intensity corresponding to the state with parallel moment (see fig. 2).

The magnitude of the parallel moment decreases slightly with increasing N_{Fe}. On the other hand, the magnitude of antiparallel moment decreases to a considerable extent with increasing N_{Fe}. According to the NMR data [1], the width of the hyperfine field distribution corresponding to the antiparallel moment is wider than that for parallel moment, which is consistent with our result.

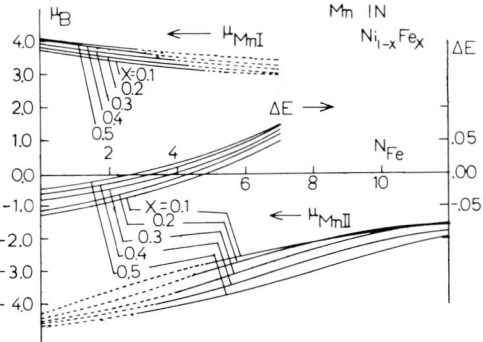

Fig. 1. The magnetic moment of a Mn impurity with moment parallel (I) or antiparallel (II) to the magnetization vs. the number of Fe atoms among the nearest neighbors. ΔE is the energy of the parallel state measured from the antiparallel one; $U_{Ni} = 3.50$, $U_{Fe} = 1.25$, $U_{Mn} = 1.10$, $E_{Fe} - E_{Ni} = 2.68$ and $E_{Mn} - E_{Ni} = 3.09$ in the unit of a half the band width of pure metal.

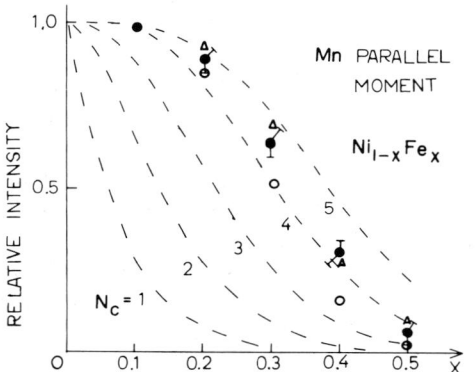

Fig. 2. Dependence of the NMR intensity of the parallel Mn on the concentration of Fe in Ni–Fe–Mn (solid circles) [1]; open circles are calculated results with parameters shown above; triangles are the results with $U_{Mn} = 1.40$. Dashed curves are calculated with the assumption that N_c (see text) is independent of x.

The NMR result shows also in Ni–Mn alloys the coexistence of Mn atoms with parallel and antiparallel moments. The spin–echo signal corresponding to parallel Mn has a satellite in a lower frequency range than the main peak. We focus our attention on this satellite. We can calculate the magnetic state of a Mn atom by a similar procedure as for Ni–Fe–Mn alloy with the assumption that Mn atoms with parallel (I) and antiparallel (II) moments and Ni atoms occupy the nearest neighbor sites. The states of these nearest neighbor atoms are determined by the CPA ternary alloy approach [4]. The farther neighbors are described by the coherent potential in the ordinary HF-CPA.

The calculation shows that the parallel Mn moment decreases by $\sim 0.5\,\mu_B$ when a MnII atom occupies one of the nearest neighbor sites, whereas the effect of the occupation of MnI is very small. According to the NMR data, the satellite intensity grows with increasing Mn concentration in parallel with the intensity of the signal which is ascribed to the antiparallel Mn atom (MnII). We therefore conclude that the satellite arises from the parallel Mn moment having one nearest neighboring MnII.

The author thanks Professor K. Asayama and Mr. Y. Kitaoka for informing him of their experimental data prior to publication.

References

[1] Y. Kitaoka and K. Asayama, J. Phys. Soc. Jap. 40 (1976) 1521, and unpublished work.
[2] T. Jo and H. Miwa, J. Phys. Soc. Jap. 40 (1976) 706.
[3] H. Hasegawa and J. Kanamori, J. Phys. Soc. Jap. 31 (1971) 382.
[4] T. Jo, J. Phys. Soc. Jap. 40 (1976) 715.

MAGNETIC PROPERTIES OF DISORDERED ALLOYS

J. MATHON
Department of Mathematics, The City University, London EC1 4PB, England

The Landau model [1] is applied to PdFe, PdNi, and PtNi alloys at finite temperatures. The generalized Landau equation is linearized and then solved within CPA. The temperature dependence is incorporated via the Landau parameter $A(T)$. The model explains well the temperature dependence of the susceptibility and the concentration dependences of the magnetization and T_c.

1. Introduction

The generalized Landau model [1] is extended to finite temperatures and applied to PdFe, PdNi and PtNi. The application to PdFe alloys is especially interesting since the existing itinerant theories [2, 3] cannot explain all their properties satisfactorily. The Landau model treats the spatial inhomogeneity in the same approximation as in [2], describes correctly the interaction of giant moments, and incorporates the essential nonlinearity of the problem [3].

2. Landau equation at finite temperature

The Landau equation for the magnetization of a disordered alloy takes the form [1]

$$-C\nabla^2 M(r) + \left[A_0(T) + \sum_{\{i\}} V_i(r)\right] M(r) + BM^3(r) = H(r), \quad (1)$$

where A_0 is equal to the reciprocal of the matrix susceptibility, B and C are assumed constant, H is the applied field, V_i is the effective impurity potential, and $\{i\}$ denotes the summation over all impurities. The only temperature-dependent parameter in this equation is the matrix susceptibility $\chi_0(T)$ given by

$$\chi_0(T) = \frac{1}{A_0(T)} = \frac{\chi_0(0)}{1 + aT^2}, \quad (2)$$

where a depends on the matrix density of states and on the matrix exchange interaction I [4].

Eq. (1) should be now solved for every impurity configuration and the resulting magnetization averaged over all impurity configurations. This is feasible only if the cubic term is linearized as follows [5]:

$$M^3(r) = [M_f + m(r)]^3 \approx M_f^3 + 3M_f^2 m(r), \quad (3)$$

where $M_f = (1/\Omega)\int M(r)\,dr$ is the average magnetization for a given configuration. The linearized equation assumes the form of an inhomogeneous Schrödinger equation

$$\left[A_0(T) + 3BM_f^2 - C\nabla^2 + \sum_{\{i\}} V_i(r)\right] M(r) = H(r) + 2BM_f^3 \quad (4)$$

For PdFe alloys, the linearization fails in the dilute limit and a comparison of the linearized Landau equation with the numerical solution of the full cubic equation for a lattice model [5] shows that the linearization is valid for concentrations of Fe atoms not lower than $\approx 0.5\%$. For PdNi and PtNi this problem largely does not arise since the dilute limit is of no special interest.

To solve eq. (4), we shall approximate the impurity potential by a delta-function potential of strength $-V_0$. With this approximation, the equation for the Green's function of eq. (4) is formally equivalent to the Schrödinger equation for the wave function of an ordinary (non-magnetic) disordered binary alloy and can be readily solved. The configuration average of the magnetization $M = \langle M_f \rangle$ satisfies the following equation [5]:

$$\left[A_0(T) + \Sigma - 3B\langle \Delta M^2\rangle\right]M + BM^3 = H + 2B\langle \Delta M^3\rangle, \quad (5)$$

where Σ is the "self-energy" of the equivalent binary alloy and $\langle \Delta M^2\rangle$, $\langle \Delta M^3\rangle$ (arising from the configuration averaging) can be determined from the two- and three-particle Green's functions [5]. Since Σ can be obtained explicitly in CPA, the concentration, temperature, and field dependences of the magnetization can be determined from eq. (5).

3. PdFe, PdNi and PtNi Alloys

The linearized equation (4) contains the parameters A_0, a, B and C characterizing the matrix and the impurity potential V_0. For Pd matrix, $1/A_0 = \chi(0) = 7 \times 10^{-6}$ emu/g, B was determined in [5] to be $\sim 0.01 A_0$, C is directly related to the correlation length of the matrix [1] and V_0 will be treated as an adjustable parameter. To obtain V_0 for PdFe, the calculated magnetization was fitted to the observed value [6] for $c = 1\%$ Fe, which yields $V_{0Fe} = 23.22 A_0$.

At finite temperatures, the value of a is required. In principle, it could be determined from the Pd band structure, but in practice this is not feasible because of the anomalous temperature dependence of χ_{Pd}. To avoid this problem, we shall treat a as an adjustable parameter. We believe that this is justified since a small amount of impurities is known to remove the anomaly and the temperature dependence eq. (2), is then appropriate to an itinerant paramagnet. The value of a obtained by fitting calculated T_c for PdFe to the observed values [6] is $a_{Pd} = 1.32 \times 10^{-4}$ deg^{-2}. The calculated curves $\chi(T)$ and $M(c)$ are shown in figs. 1 and 2 and the calculated values of T_c are compared with the observed values [6] in table I [T_c is defined by $M(T_c) = 0$]. The temperature dependence of χ for PdNi and PtNi alloys is also shown in fig. 1. The values of V_0 for PdNi and

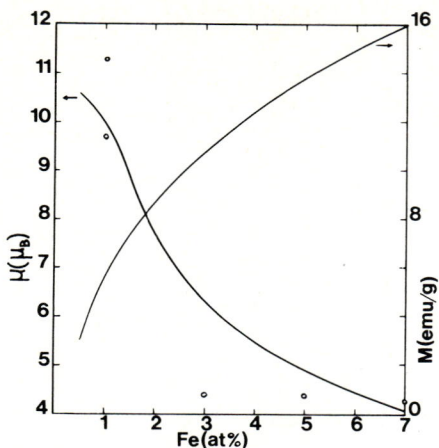

Fig. 2. Concentration dependences of the giant moment μ and of the magnetization M. Experimental points were taken from [6].

Table I
Concentration dependence of T_c

Fe (at.%)	1	3	5	10
T_c^{cal}	82.6	139.0	168.5	212.7
T_c^{exp}	66.0	122.0	162.0	236.0

PtNi were determined from the maximum in the susceptibility at the critical concentration for ferromagnetism ($V_{0PdNi} = 15.1 A_0$, $V_{0PtNi} = 2.24 A_0$) and $B_{Pt} = 0.00271 A_0$, $a_{Pt} = 1.25 \times 10^{-5}$ deg^{-2} were obtained by fitting the calculated values of $M(0)$ and T_c to the observed values [7].

4. Discussion

The present results can be summarized as follows. The phase transition for homogeneous PtNi alloys is very sharp, i.e. χ is "nearly infinite" at T_c. For PdFe the susceptibility is far from infinite, χ has only a small broad peak at T_c and the transition is spread over a wide temperature range. PdNi lies between these two extremes.

The Landau model explains the concentration dependence of the giant moment (fig. 2) although the calculated reduction of the giant moment is not as rapid as the observed one.

The concentration dependence of T_c is in good qualitative agreement with experiment (the experimental points correspond to the magnetization measurements quoted in [6]).

Despite its success, the simple Landau theory fails in the dilute limit when the coupling bet-

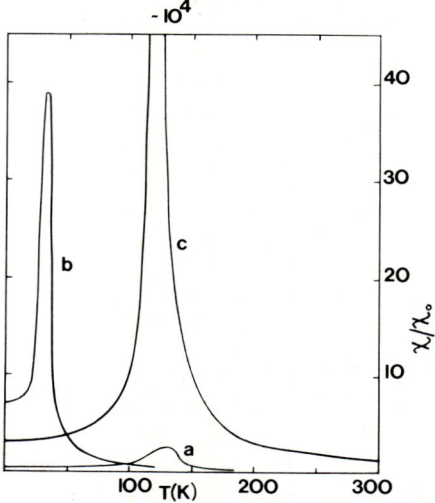

Fig. 1. Temperature dependence of the susceptibility. (a) 3% PdFe ($T_c = 139$ K); (b) 3% PdNi ($T_c = 35$ K); (c) 50.2% PtNi ($T_c = 120$ K).

ween giant moments may be antiferromagnetic. The linearization itself also fails in this limit [5]. To rectify this fault, it is necessary to include higher-order terms in the gradients and develop a new method of solution of the nonlinear Landau equation. Calculations along these lines are now in progress.

References

[1] D.M. Edwards, J. Mathon and E.P. Wohlfarth, J. Phys. F3 (1973) 161.
[2] S. Doniach and E.P. Wohlfarth, Proc. Roy. Soc. (London) A296 (1967) 442.
[3] T. Takahashi and M. Shimizu, J. Phys. Soc. Jap. 20 (1965) 26.
[4] D.M. Edwards and E.P. Wohlfarth, Proc. Roy. Soc. (London) A303 (1968) 127.
[5] T. Kato and J. Mathon, J. Phys. F6 (1976) 1341.
[6] G.J. Nieuwenhuys, Ph.D. Thesis, Leiden University, The Netherlands (1974).
[7] J. Beille, D. Bloch and M.J. Besnus, J. Phys. F4 (1974) 1275.

AN ELECTRONIC MODEL FOR AMORPHOUS SYSTEMS

J. SCHREIBER
Joint institute for Nuclear Research, Head Post office, P.O. Box 79, 101000 Moscow, USSR

Linearly coupled diagonal and off-diagonal randomness are considered using a "cluster-effective-lattice" method where the potential parameters in the cluster are assumed to fluctuate according to a Lorentzian distribution. The density of states, the magnetic properties within a Stoner-like theory, and the localization of electrons are studied.

In amorphous or liquid systems diagonal disorder and, first of all, off-diagonal randomness (ODR) appear [1, 2] whereas in general both kinds of disorder are coupled. Hence we use the following tight binding model for a system of N identical atoms forming an amorphous structure. Instead of the total Hamiltonian we only consider the projection onto the subspace spanned by the atomic orbitals $|\rho_i\rangle$. Using a second quantization representation the approximated hamiltonian may be written as [2]

$$H = \sum_{ij\sigma} V_{ij} \tilde{a}^+_{i\sigma} \tilde{a}_{j\sigma} + \tfrac{1}{2} \sum_{i\sigma} U_i \tilde{a}^+_{i\sigma} \tilde{a}_{i\sigma} \tilde{a}^+_{i-\sigma} \tilde{a}_{i-\sigma}. \qquad (1)$$

Because of the non-orthogonality of $|\rho_i\rangle$ we have $[\tilde{a}_{i\sigma}, \tilde{a}^+_{i\sigma_+}] = (S^{-1})_{ij}$, where S is the matrix of overlap integrals. The second term in eq. (1) represents only the Hubbard-like part of electron correlation. We restrict our discussion to cases where (i) the atomic orbitals are well enough localized and (ii) the amorphous structure has a well established short range order. Therefore we can suppose that V_{ij} and S_{ij} are different from zero only for $i = j$ and for nearest-neighbours (NN). Considering s-like states $|\rho_i\rangle$, $\epsilon_i = V_{ii}$, V_{ij}, and S_{ij} are functions of the atomic distances. Therefore the structure fluctuations of these quantities are coupled so that really only one random variable exists. In a first approximation, expanding ϵ_i, V_{ij}, and S_{ij} linearly in the variations of the distances $|i - j|$ and neglecting three centre integrals, the following simple relations can be obtained:

$$\epsilon_i = A \sum_{i \neq j} (V_{ij} - V_0) + B,$$

$$S_{ij} = C(V_{ij} - V_0) + D, \qquad (2)$$

where V_0 belongs to the averaged atomic positions.

The density of states is given by the relation (cf. [2])

$$\rho(E) = -1/\pi N \sum_{i,j} S_{ij}\, \mathrm{Im}\, G_{ij}(E + i0^+)$$
$$= 1/N \sum_i \rho_i(E), \qquad (3)$$

where G_{ij} is the Zubarev-Green function. We find that G_{ij} obeys the equation

$$\sum_l (ES_{il} - V_{il}) G_{lj} = \delta_{ij}. \qquad (4)$$

The structure averaging for $\rho(E)$ is performed by a self-consistent "cluster-effective-lattice" method. Thereby it is required that the averaged local density of states $\rho_i(E)$ for an atom surrounded by its Z NN and embedded in an effective medium – characterized by an effective lattice (the Be lattice or regular one with equivalent short range order) and a coherent hopping integral V_c – is equal to the corresponding quantity for the effective medium. Numerical computations become easy if we assume that in the cluster the V_{ij} fluctuate statistically independent according to a Lorentzian distribution with mean value V_0 and width Γ (cf. [2]). Results are shown for a Ni-like effective structure (fig. 1) [3, 4]. Switching on ODR, an asymmetrical change of $\rho(E)$ is obtained, where the sign of the coupling between ϵ_i and V_{ij} influences the results qualitatively. We note that the case $AV_0 < 0$ is the physically realistic one since the ϵ_i-level will be lowered if $|V_{ij}|$ becomes greater. The consideration of random overlap integrals ($C \neq 0$) leads also to an asymmetrical change of $\rho(E)$.

Now we investigate the amorphous magnetism within the Hartree-Fock approximation (HF) for the model (1) (cf. [3, 4]). Assuming that U_i is only determined by the sort of atoms, and incorporating the results for $\rho(E)$, the Stoner-

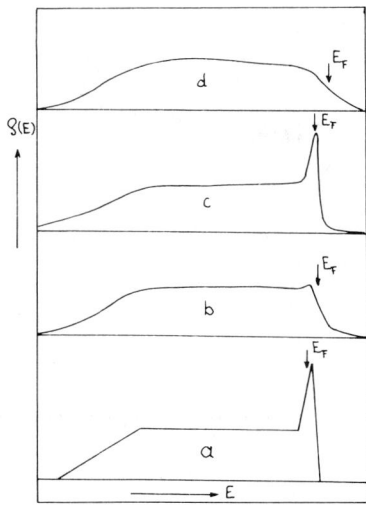

Fig. 1. Ni-like density of states $\rho(E)$ for different kind of disorder. (a) $\Gamma = 0$ (corresponding to crystal); (b) $\Gamma \to 0$, $|A|\Gamma = 13/12 \times 0.06$; (c) $\Gamma = 0.06$, $A = 13/12$; (d) $\Gamma = 0.06$, $A = -13/12$. $V_0 < 0$, $B = C = D = 0$. E_F = Fermi level for electron number per atom $n = 1.88$.

like theory gives:

(i) The Stoner criterion $U\rho_p(E_F) > 1$. The index p denotes the paramagnetic phase. For diagonal disorder and for $AV_0 > 0$ ferromagnetism (FM) is monotonously weakened increasing Γ. In contrast to that for the "physical" case $(AV_0 < 0)$ FM is stabilized in a certain range of Γ. The reason is a shift of the Fermi level to the top of the peak (fig. 1). (ii) The Curie temperature T_c and the magnetization $m(T = 0)$ (fig. 2). For sufficiently great values of Γ both quantities are monotonously weakened by fluctuations of V_{ij} in all cases. Important is the relation between the lowering of T_c and $m(T = 0)$. Experimental data show that in amorphous Ni-samples $m(T = 0)$ is decreased more than T_c compared with the crystalline phase [1]. This tendency is reproduced in the case $AV_0 < 0$. Other kinds of coupling and only diagonal disorder yield the inverse tendency. (iii) The characteristic flattening of the $m(T/T_c)/m(0)$ curve is not found within our model, since the Stoner theory is not appropriate for describing the temperature dependence of $m(T)$.

Recently we have derived some expressions for the localization function $L(E)$ within the Economou and Cohen theory for the model of amorphous systems described above (cf. [2]). Here we discuss only the most successful approximation which was proposed by Licardello and Economou for the Anderson model. Performing the structure averages in the same way as for $\rho(E)$ and choosing a Bethe lattice for the effective structure, we have found the following results ($U_i = 0$). Due to the coupling of ϵ_i and V_{ij} an asymmetrical shift of the mobility edges E_c is caused by increasing potential fluctuations. This is expected from the results for $\rho(E)$. For larger Γ it is observable that the fluctuations of V_{ij} lead to a delocalization effect. Fluctuating overlap integrals yield also asymmetrical delocalization. Consequently, because of the essential effect of ODR in amorphous systems, ODR has to be included when properties of electronic conductivity in these systems are discussed. Regarding amorphous transition metals the electron correlation may have an effect in producing localized states. Within the alloy analogy for the model (1), which yields an effective cellular disorder, the above mentioned localization function can also be calculated [2]. As the density of states for $\Gamma \neq 0$ is smeared out in the gap a conductivity transition is now related to the mobility bands only (fig. 3). Let us assume that

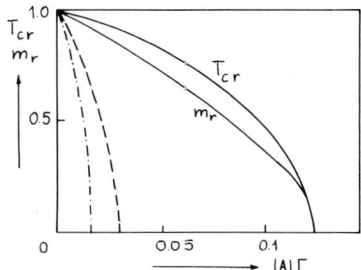

Fig. 2. Dependence of the Curie temperature $T_{cr} = T_c(\Gamma)/T_c(0)$ and magnetization $m_r = m(\Gamma)/m(0)$ at $T = 0$ upon $|A|\Gamma$. —— $A = 13/12$; --- $\Gamma \to 0$, $|A| \to \infty$; -·-·- $A = -13/12$. $V_0 < 0$, $U/|V_0| = 15.0$, $B = C = D = 0$.

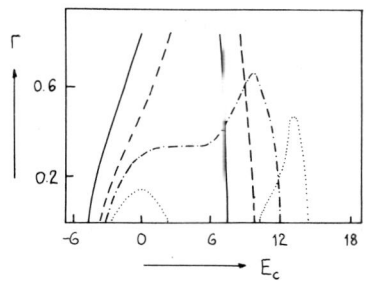

Fig. 3. Mobility bands for the Hubbard model in the paramagnetic case ($n\uparrow = n\downarrow = \frac{1}{2}$). —— $U = 3$; --- $U = 6$; -·-· $U = 9$; ···· $U = 12$. $V_0 > 0$, $A = -1$, $Z = 6$, $B = C = D = 0$.

U/V_0 and Γ will be lowered if pressure is applied to amorphous non-metallic systems. Then we meet the possibility of realization of a non-metal–metal transition. This transition is a mixed one of the Mott and Anderson type where the dominant mechanism is due to electron correlation.

The author is very grateful to Dr. W. John and Mr. J. Richter for valuable discussions.

References

[1] G.W. Wright, Amorphous Transition Metal Films, 7th Int. Colloquium on Magnetic Thin Films, Regensburg (1975).
[2] J. Schreiber, in: Proc. of VII-th Autumn School on Magnetism, Gaussig (1975).
[3] J. Kanemori, J. de Physique C4 (Suppl.) (1974) 131.
[4] J. Richter, J. Schreiber and K. Handrich, Phys. Status Solidi (b) 74 (1976) K125.

MAGNETIC ANISOTROPY IN AMORPHOUS METALS*

R. HARRIS, B.G. MULIMANI† and D. ZOBIN‡

Eaton Electronics Laboratory, McGill University Montreal, Quebec, Canada H3C 3G1

A model hamiltonian incorporating random local crystal fields, and appropriate to the description of the magnetic properties of amorphous rare-earth transition metal alloys [5], has recently been extended to describe a possible "spin-glass-like" phase in these materials [6]. We propose that a model hamiltonian of similar form can be derived for amorphous metals without localized spins, and for which the magnetic anisotropy is due to the spin–orbit interaction. We suggest that the magnitude of such anisotropy may be large enough to explain, for example [1] the anomalous properties of amorphous YFe$_2$ in terms of a "spin-glass-like" ordering. Discussion of this alloy was specifically outside the scope of the older model [5].

Amorphous rare-earth transition-metal (RE–TM) alloys have attracted considerable attention during the past few years. They have been found to display unusual magnetic properties which often differ significantly from those of crystalline alloys of the same composition. For example, alloys of Tb, Dy, and Ho with Fe show values of the spontaneous magnetization and Curie temperature greatly reduced compared with their crystalline counterparts [1, 2], and also show anomalous low-temperature "coercive-like" fields with values around 30 kOe at 4.2 K [1, 3]. In contrast, the amorphous alloy YFe$_2$ appears to show no magnetic ordering down to 3.5 K [1], although the existing spin resonance data [4] is anomalous and although the crystalline alloy is ferro-magnetic with a Curie temperature around 535 K.

One possible explanation of the reduced magnetization and Curie temperature is afforded by the random-anisotropy model of Harris et al. [5]. As originally proposed, this model depends on the response of the localized rare-earth f-electrons to strong electrostatic "crystal" fields, and thus is not applicable to alloys of Gd or Y with either Fe or Co. In addition, the original molecular field formulation of the model was incapable of explaining the anomalous low-temperature behaviour of Tb-, Dy- and HoFe$_2$. However, a modified version [6] which takes some account of local fluctuations in the molecular field, and which can lead to a kind of "spin-glass-like" ordering, when the anisotropy energy is comparable to or larger than the exchange, has made some progress in this direction.

The behaviour of amorphous YFe$_2$, however, remains to be explained. This alloy contains no localized f-electrons; indeed, the ferromagnetism of the crystalline compound is best described by the itinerant (Stoner) model. Thus, even though the magnetization curves of the amorphous alloy resemble those discussed by Harris and Zobin [6], the random anisotropy model would appear to be inappropriate. However, in this paper we propose that the spin–orbit interaction, responsible for the magnetic anisotropy energy of itinerant ferromagnetic materials [7], may give rise to local anisotropy sufficient in magnitude to account for the properties of amorphous YFe$_2$. Our discussion is necessarily only qualitative, since available data is insufficient for precise calculations, but, we believe, it points the way for future investigations.

In crystalline itinerant ferromagnets, the anisotropy energy is calculated by introducing the spin–orbit interaction as a perturbation to the electronic energy-band structure. It is well known that in cubic materials the perturbation expansion must be carried out to fourth order in the coupling constant, whereas in hexagonal materials second order is sufficient. The perturbation is conveniently expressed in the tight-binding model, where the spatial part of the wave function is written as a linear-combination of atomic orbitals

$$\Psi(\mathbf{k}, \mathbf{r}) = \frac{1}{\sqrt{N}} \sum_m \alpha_m(\mathbf{k}) \sum_\mathbf{R} e^{i\mathbf{k}\cdot\mathbf{R}} \phi_m(\mathbf{r} - \mathbf{R}), \quad (1)$$

---

* Work supported by the National Research Council of Canada.
† Permanent address: Dept. of Physics, Karnatak University, Dharwar-580003 (Karnatak), India.
‡ Present address: Dept. of Physics, Simon Fraser University, Burnaby, B.C., Canada.

with usual notation. The spin–orbit interaction then has matrix elements

$$H_{k\sigma,k'\sigma'} = \langle \Psi_\sigma(k) | \tfrac{1}{2}\lambda \sum_R l_R \cdot \sigma | \Psi_{\sigma'}(k') \rangle,$$

where l_R is the orbital angular momentum operator which acts on the atomic states $\phi_m(r - R)$, and the electron spin σ introduced by the appropriate spinor operators is quantized in a direction either parallel or antiparallel to the overall bulk magnetization vector M. For convenience we define M to be in a reference direction z so that the index m takes on its usual significance as $\langle l_z \rangle$, and, for example,

$$H^z_{k\sigma,k'\sigma} = \pm \frac{\lambda}{2N} \sum_R e^{i(k'-k)\cdot R} \left(\sum_m \alpha_m^*(k)\alpha_m(k')m \right). \quad (2)$$

It is then easy to see that contributions to the anisotropy come only from states related by $k' - k = G$, when the sum over sites gives a non-vanishing result $H_{k\sigma,k'\sigma} \approx \lambda$. Macroscopic anisotropy arises when such pairs of states k and k' have perturbed energies close to but on opposite sides of the Fermi level.

In an amorphous yet itinerant ferromagnet it is possible, in principle, to carry through very much the same kind of calculation, with the appropriate linear combinations of atomic orbitals having the form

$$\Psi(j, r) = \frac{1}{\sqrt{N}} \sum_{mR} C_{mR}(j) \phi_m(r - R), \quad (3)$$

where the coefficients $C_{mR}(j)$ are now random, although still of order unity. With the spin quantized in a unique z-direction, as before, typical matrix elements of the spin–orbit interaction are then given by

$$H_{j\sigma,j'\sigma} = \pm \frac{1}{2}\frac{\lambda}{N} \sum_R \left(\sum_m C^*_{mR}(j) C_{mR}(j')m \right)$$

$$\simeq 0\left(\frac{\lambda}{\sqrt{N}}\right) \simeq 0. \quad (4)$$

The crucial difference from the crystalline case arises from the randomness of the coefficients $C_{mR}(j)$ which will cause the value of each matrix element to be very small, leading, at first sight, to the ineffectiveness of the spin–orbit interaction.

However, by allowing the spin of an electron in state $\Psi(j, r)$ to be quantized in a direction, other than $\pm z$, determined by the particular coefficients $C_{mR}(j)$, it is possible to see how the overall energy of the electron sea can be lowered. The optimum choice of the directions of quantization will be that which leads to a maximum value of $H_{j\sigma,j'\sigma}$ for pairs of states j and j' near the Fermi energy. This is a sufficiently weak requirement – placing no constraints on electrons far from the Fermi energy – that there should be no difficulty in satisfying it. The effect of this choice will be to cause the magnetization vector $M(r)$ to fluctuate randomly in direction and magnitude throughout the alloy, yielding a "spin-glass-like" ordering similar to that discussed by Harris and Zobin [6]. On this same macroscopic scale, the net anisotropy energy referred to any unique z-axis will average to zero, indicating the absence of any macroscopic axis of easy magnetization.

The stability of such ordering requires that the spin–orbit "anisotropy" energy per electron be at least comparable with the exchange energy per electron. For electrons near the Fermi energy and satisfying the above criterion, the spin–orbit energy per electron will be at least as large as that in the crystal, which, taking the value for crystalline Fe [8], is around 0.07 eV. However, in crystalline (cubic) YFe$_2$, the existing data [9] indicate a value of around 10^{-4} eV/atom for the fourth order contribution to the magnetic anisotropy energy. This value is much higher than that in crystalline Fe (5×10^{-6} eV/atom) and may be related to narrower energy bands characteristic of Laves phase compounds [10] and possibly to a somewhat larger value for the effective spin–orbit coupling constant. In view of this we might expect the spin–orbit energy in amorphous YFe$_2$ to be of the order of 0.1 eV. For the exchange energy per electron, the appropriate value can be derived from the magnetic properties of crystalline YFe$_2$ in the phenomenological manner suggested by Wohlfarth [11], yielding a value around 0.1–0.2 eV.

Thus the first order spin–orbit energy can indeed be of the same order as the exchange energy, and the "spin-glass-like" ordering is at least possible as a ground state for the electron gas. The possibility arises because the exchange interaction and the anisotropic spin–orbit interaction are in a sense in opposition: in contrast, in a crystal, they are quite compatible. The situation is similar in concept to the original idea of local anisotropy put forward by Harris et al. [5], although the mechanism is quite different, and applies to alloys without f-electrons where the original model is quite inapplicable. Although only speculative, the present mechanism is readily susceptible to experimental verification: in particular, Mössbauer spectroscopy would be an excellent probe of the local hyperfine fields consistent with the model.

References

[1] J.J. Rhyne, J.H. Schelleng and N.C. Koon, Phys. Rev. B10 (1974) 4672.

[2] N. Heiman and K. Lee, Phys. Rev. Lett. 33 (1974) 778.

[3] A. Clark, Appl. Phys. Lett. 23 (1973) 642.

[4] S.M. Bhagat and D.K. Paul, Phys. Rev. Lett. 35 (1975) 1458.

[5] R. Harris, M. Plischke and M.J. Zuckermann, Phys. Rev. Lett. 31 (1973) 160.

[6] R. Harris and D. Zobin, AIP Conf. Proc. 29 (1976) 156, and to be published.

[7] G.C. Fletcher, Proc. Phys. Soc. 67A (1954) 505.
N. Mori, Y. Fukuda and T. Ukai, J. Phys. Soc. Jap. 37 (1974) 1263.

[8] R. Maglic and F.M. Mueller, Int. J. Magn. 1 (1971) 289.

[9] M.P. Dariel, U. Atzmony and D. Lebenbaum, Phys. Status Solidi (b)59 (1973) 615.
A.M. van Diepen, H.W. de Wijn and K.H.J. Buschow, Phys. Rev. B8 (1973) 1125.

[10] A.C. Switendick, private communication.
R. Haydock, private communication.

[11] E.P. Wohlfarth, in: Elements of Theoretical Magnetism, S. Krupička and J. Šternbeck (Eds.) (CRC Press, Cleveland, 1968) p. 109.

RADIATION DAMAGE IN AMORPHOUS RARE EARTH-TRANSITION METAL FILMS

P.J. GRUNDY, A. ALI and S.S. NANDRA
Department of Pure and Applied Physics, University of Salford, Salford M5 4WT, England

The effect of ion and electron irradiation damage on the stripe domain structure in amorphous rare earth-transition metal films has been investigated. Electron microscope observations show the result of this damage. Implications as to the source of the anisotropy giving this domain structure are discussed.

1. Introduction

Several theories on the association of a perpendicular, uniaxial magnetic anisotropy with forms of atomic arrangements in non-crystalline rare earth-transition metal (RE-TM) films have been proposed, e.g. [1, 2]. This paper describes some transmission (TEM) and scanning (SEM) electron microscope observations extending previous work on the effect of ion irradiation damage on this anisotropy and the associated domain structure [1, 3]. It also introduces some preliminary results of the effect of electron irradiation damage as observed in-situ in a 1 MeV high voltage electron microscope (HVEM).

2. Specimen preparation

Amorphous RE-TM thin films were prepared by d.c. bias sputtering or by co-evaporation onto liquid nitrogen cooled rock salt substrates and were then transferred to electron microscope support grids. Ion irradiations were performed in a 2 MeV Van de Graaff accelerator using Ar^+ ions and similar specimens were also irradiated and observed concurrently in the HVEM. The films remained amorphous during both forms of irradiation.

3. Results and discussion

Sputtered GdCo and GdFe films have been irradiated using 1.5 MeV Ar^+ ion fluxes of 10^{12} ions $cm^{-2} s^{-1}$ and total doses between about 10^{14} and 10^{16} ions cm^{-2}. Doses below 10^{14}–10^{15} ions cm^{-2} had no observable effect on the stripe domain structure in these films and presumably caused no significant reduction in the anisotropy. However, as shown in fig. 1(a), a dose of about 5×10^{15} ions cm^{-2} reached some threshold in sputtered $Gd_{19}Co_{81}$ films about 150 nm thick. The region marked A (a sector of a 400 μm diameter circle of the film exposed to the ion beam through an aperture) has received such a dose. In this area the domains are now magnetised in the plane of the film and form closure structures at the boundary of the shielded area, B, which exhibits typical stripe domains. The gradual decay of the stripes into the closure domains suggests that the edge of the defining aperture had a shadowing effect on the number of ions received.

Examination of these irradiated films revealed two effects of possible significance. In the first, forward sputtering of the films by the ion beam resulted in a thinning of the irradiated area. Calculation of this reduction in thickness from measured values of transmitted intensities gave about 10 nm in 150 nm. This reduction in thickness is not responsible for the loss of the stripes as the anisotropy is well developed in these films at a thickness of 80 nm.

The second observation is that the irradiated parts of these free standing films increase in volume and become extremely contorted, as shown in fig. 1(b). Also, films supported on substrates show "blistering" effects, fig. 1(c), typical of high dose ion implantations and bubble formation in bulk specimens and thin metal foils [4]. Although these observations should be taken further, calculations of the projected ion range (>0.5 μm) suggest that the majority of ions pass through the films. It seems probable, therefore, that the anisotropy is not destroyed by any stress or change in atomic separation due to implantation but by a rearrangement of the atoms in the film.

The HVEM provides a very convenient and controllable means of producing and observing radiation damage at the same time. Fig. 2 shows the effect of irradiating a 250 nm thick evaporated $HoCo_4$ film with a 1 MeV, 10^{-6} A, 10 μm diameter beam to give a total close, D, of about 6×10^{21} e cm^{-2}. It can be seen that where the beam spot was focused the magnetization is in-plane, i.e. the domain structure shows an

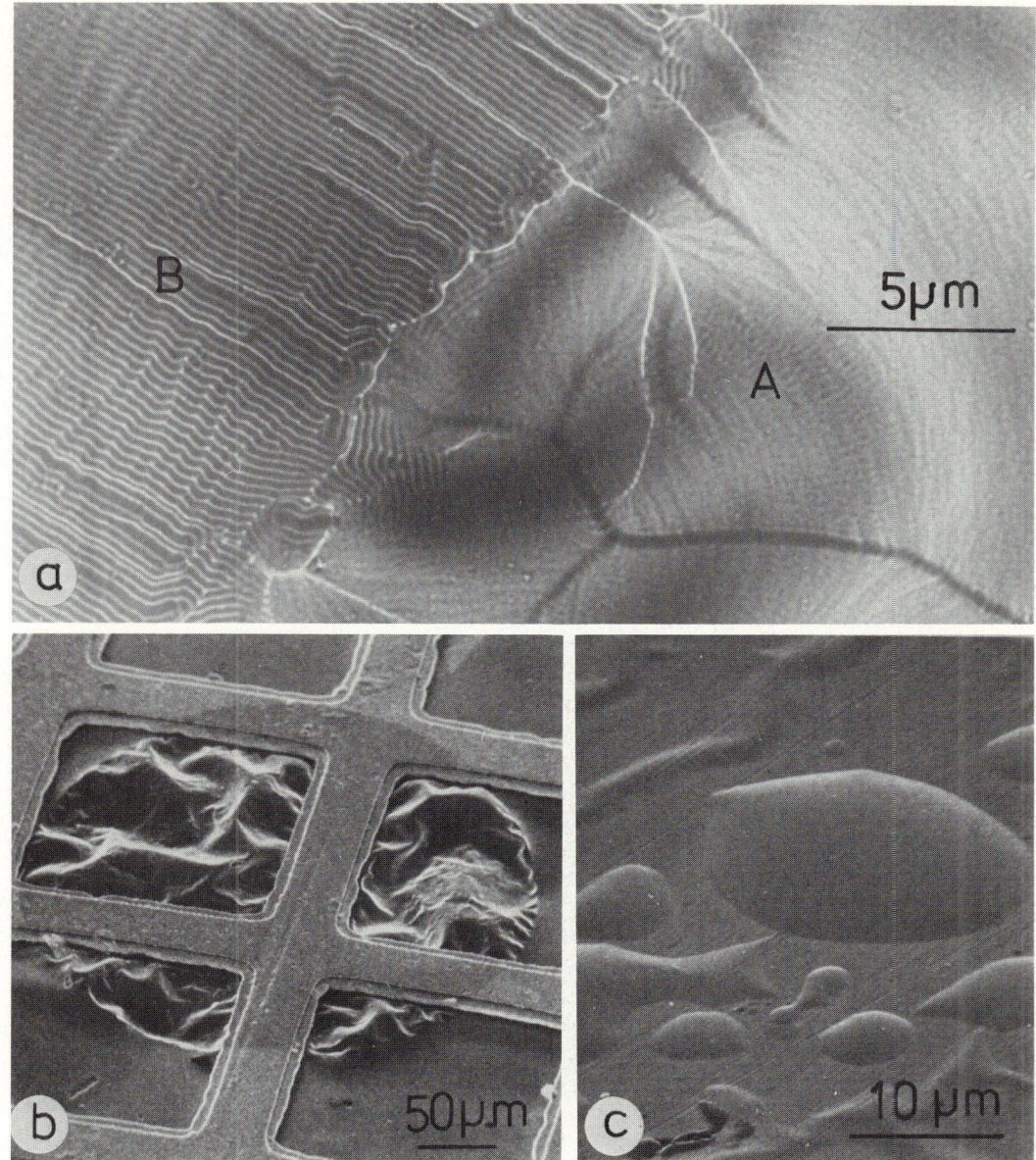

Fig. 1. (a) Lorentz TEM micrograph of the domains in a $Gd_{19}Co_{81}$ film; (b) SEM micrograph of this film showing the irradiated circle and the bars of the supporting grid; and (c) SEM micrograph of blisters in a $GdCo_4$ film deposited on a copper substrate.

in-plane domain wall in contrast to the surrounding stripes.

If the perpendicular anisotropy in Re–Co films is assumed to originate in cobalt atom pair-ordering [1] it can be calculated that each pair contributes about 10^{-15}–10^{-16} erg and therefore about 5×10^{20} pairs cm^{-3} are required to give a typical value of 10^5 erg cm^{-3} for this anisotropy. Assuming each incident ion displaces about 5×10^2–10^3 atoms (a reasonable estimate in these films) a dose of $\sim 10^{14}$ ions cm^{-2} should be sufficient to destroy this anisotropy in a 150 nm thick film. Our threshold dose of $\sim 10^{15}$ ions cm^{-2} is in rough agreement with this.

Assuming an average value of 25 eV for the displacement energy of a cobalt atom in these amorphous films (taken from the results of Howe [5] for crystalline cobalt) we calculate the

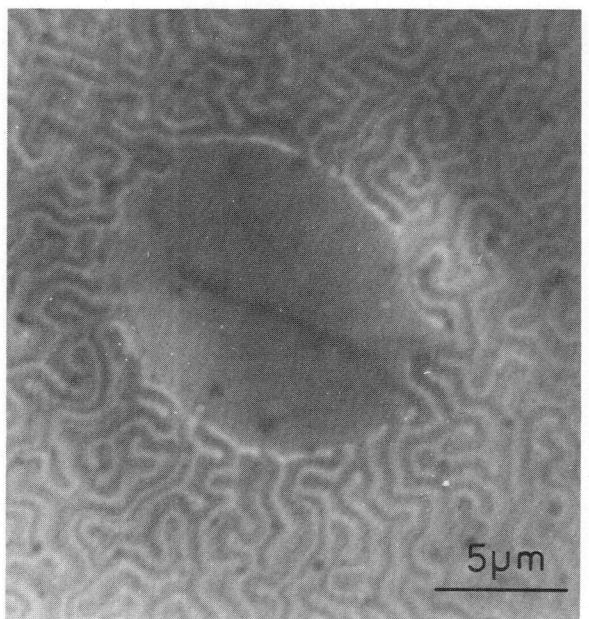

Fig. 2. Lorentz TEM micrograph (at 1 MeV) of the electron irradiated spot in a HoCo$_4$ film.

primary displacement cross-section σ_p as $65b$ for 1 MeV electrons. From this the fractional number of atoms displaced in the HVEM experiments is obtained as $N \simeq \sigma_p D \simeq 0.4$ or 40%. Here we have neglected the presence of the holmium atoms and at 1 MeV it is possible that the holmium atoms have not been displaced. Of interest would be the threshold beam energy for the destruction of the anisotropy – if this were due to cobalt pair-ordering it should be between 400 and 500 keV.

The results of the two types of irradiation together suggest that the anisotropy in these films arises from some particular arrangement of the atoms, be it pair-ordering or some other correlation, and not from any shape or physical growth features in the films.

The authors thank Dr. G.A. Stephens and Mr. P. Cardwell for assistance with the ion irradiations at Salford and the Department of Metallurgy and Science of Materials, University of Oxford for the use of the EM7 HVEM. The support of the Science Research Council in this work is acknowledged.

References

[1] R.J. Gambino, J. Ziegler and J.J. Cuomo, Appl. Phys. Lett. **24** (1974) 99.
[2] N. Heiman, A. Onton, D.F. Kyser, K. Lee and C.R. Guarnieri, AIP Conf. Proc. No. 24 (American Institute of Physics, New York, 1976) p. 573.
[3] A. Ali, P.J. Grundy and G.A. Stephens, J. Phys. D: Appl. Phys. 9 (1976) L69.
[4] J. Roth, in: Applications of Ion Beams to Materials, 1975, (The Institute of Physics, London and Bristol, 1976) p. 281.
[5] L. M. Howe, Phil. Mag. 22 (1970) 965.

MAGNETIC COMPENSATION TEMPERATURES OF AMORPHOUS RARE EARTH–COBALT ALLOYS

A.G. DIRKS, J.W.M BIESTERBOS and K.H.J. BUSCHOW

Philips Research Laboratories, Eindhoven, The Netherlands

The dependence on composition of the compensation temperature (T_{comp}) of various amorphous alloys of Co and rare earths (R = Gd, Tb, Ho, Er) has been systematically investigated, and the results have been analyzed in terms of a molecular field model. Compared with estimates for T_{comp} based on free rare-earth ion moments the observed values of T_{comp} for R ≠ Gd are much too low. This points to a reduction of the R sublattice magnetization on account of local crystal field effects.

The magnetic properties of alloys of rare-earth elements and cobalt, in either a crystalline compound or an amorphous alloy, can be understood as being due to an antiparallel coupling of the rare-earth spins S_R and the Co spins S_{Co}. Since the Co–Co interaction is much stronger than the R–Co interaction, compounds and amorphous alloys over a range of Co concentrations will give rise to a compensation point in the temperature dependence of their magnetization, characterized by a mutual cancellation of the Co and R sublattice magnetization. In this paper we report on a systematic study for amorphous alloys of the dependence of T_{comp} on the relative Co-concentration and the type of R component.

The magnetization of the R–Co films was measured by means of a Faraday balance. The magnetic field strength applied was 10 kOe. The films were prepared by simultaneous electron beam evaporation as described earlier [1]. Some results of these measurements for amorphous Ho–Co alloys are shown in fig. 1. The values of T_{comp}, which we identify with the minima in the $\sigma(T)$ curves, are seen to decrease markedly with Co concentration. The concentration dependence of T_{comp} is shown in fig. 2, together with results obtained with several other R–Co amorphous systems. It can be seen from fig. 2 that changing the Co concentration affects the values of T_{comp} in the Gd–Co alloys to a greater extent than those in the R–Co alloys at the end of the series. In the case of Gd–Co it was extrapolated that T_{comp} tended towards 0 K at a composition of 83 at.% Co. With $\mu_{Gd} = 7\ \mu_B/Gd$ this result can be used to determine the Co moment as $\mu_{Co} = 1.44\ \mu_B/Co$.

In order to investigate the origin of the differences observed in T_{comp} behaviour (fig. 2)

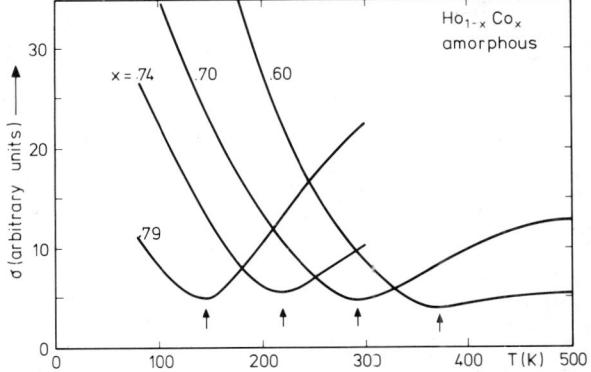

Fig. 1. Temperature dependence of the magnetization (in arbitrary units) of four Ho–Co amorphous alloys as measured with the Faraday balance method. The arrows indicate the magnetic compensation.

we will analyse our data in terms of a molecular field model. The magnetic ordering temperature in the amorphous alloys $R_{1-x}Co_x$ is determined by the strong Co–Co interaction (Curie temperatures up to 1000 K [2]). Since we are interested in the temperature range well below the crystallization temperatures ($T_{cryst} \approx 600$ K) we can safely assume that the Co sublattice magnetization is temperature independent. For μ_{Co} we took the value $1.5\ \mu_B/Co$, which is the average of the value derived above and that derived from direct magnetization measurements on $Gd_{1-x}Co_x$ ($1.6\ \mu_B/Co$ for $x = 0.8$). The R sublattice magnetization is given by

$$M_R = (1-x)N\mu_B gJB_J[gJ\mu_B H_1/kT], \qquad (1)$$

where B_J represents the Brillouin function appropriate to the total angular momentum J of the R component and H_1 the field acting on the R spins. In deriving an expression for H_1 the

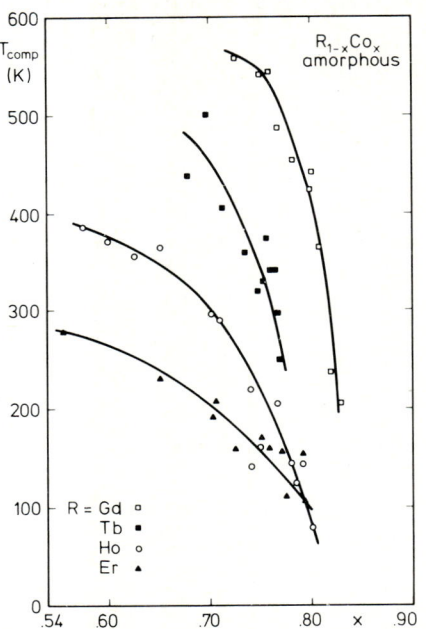

Fig. 2. Concentration dependence of the compensation temperatures (T_{comp}) of several amorphous R–Co evaporated alloys.

molecular field due to the R–R interaction, and also the applied field, can be neglected compared to the field produced by the $12x$ nearest neighbour Co atoms. This leads to

$$H_1 = 12xA(g-1)\mu_{Co}/2g\mu_B^2. \qquad (2)$$

In this expression the constant A reflects the strength of the R–Co exchange interaction. For Co we used a g value equal to 2.

For $Gd_{0.2}Co_{0.8}$, $Ho_{0.2}Co_{0.8}$ and $Er_{0.2}Co_{0.8}$ results are given in fig. 3. The upper curves represent M_R in μ_B per formula unit obtained with eq. (1). The partly broken line is the behaviour assumed for M_{Co} per formula unit. The lower curves represent the total magnetization $M = |M_R - M_{Co}|$. From the calculated curves it becomes clear that (for a given A value) changes in M_{Co} relative to M_R will have a small effect on T_{comp} for R = Ho or Er but a large effect for R = Gd. This agrees with the results shown in fig. 2. Comparison with experiment shows further that for $Gd_{0.2}Co_{0.8}$ the position of T_{comp} is properly predicted by taking a value for A close to -8×10^{-15} erg. In order to obtain T_{comp} values in agreement with experiment for $Ho_{0.2}Co_{0.8}$ and $Er_{0.2}Co_{0.8}$ much lower values of A would be required ($A \approx -2 \times 10^{-15}$ erg and $A \approx -3 \times 10^{-15}$ erg, respectively).

Owing to the fact that the metallic radii of the rare-earth elements are not exactly equal one cannot exclude a slight variation of A within the R–Co alloys. It is unlikely, however, that this variation would be larger than 20%, say. A possible explanation of the apparently too low A values may be the presence of internal electric fields. In principle there is no reason why crystalline fields present in the crystalline R–Co

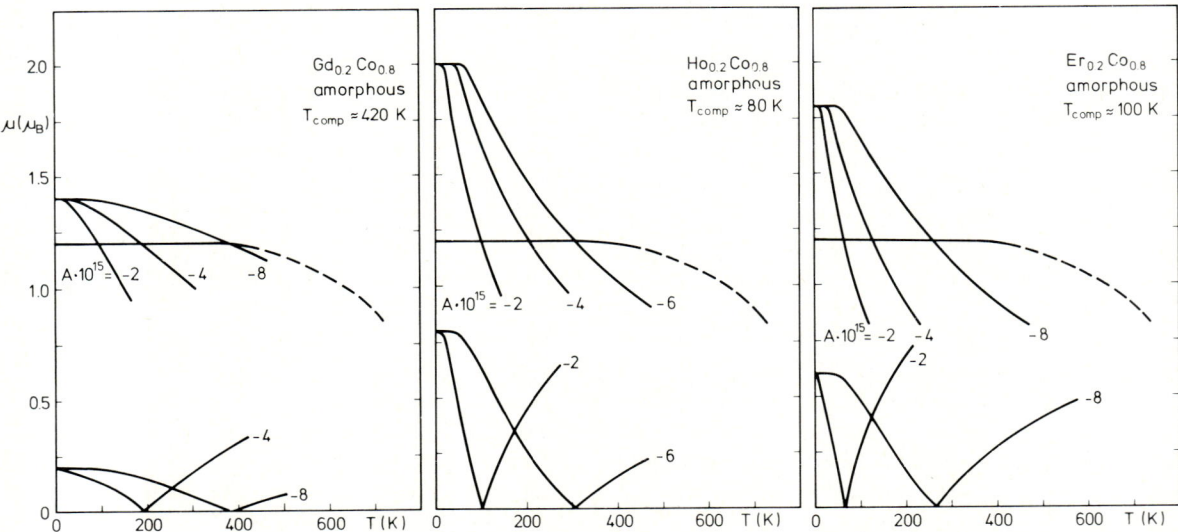

Fig. 3. Results of molecular field calculations for three amorphous alloys of the composition $R_{0.2}Co_{0.8}$ with R = Gd, Ho and Er, respectively. The quantity A (in erg) represents the R–Co exchange interaction.

compounds should be absent in the amorphous materials, although they are of low symmetry and variable from one R site to the other. In case of large crystal field splittings eq. (1) is not justified; a lowering of M_R relative to M_{Co} may provide a fairly good qualitative explanation of the low T_{comp} values for $R_{1-x}Co_x$ with $R \neq Gd$.

In conclusion, we have determined the concentration dependence of T_{comp} in amorphous alloys $R_{1-x}Co_x$ over a range of concentrations for R = Gd, Tb, Ho and Er. From the results of $Gd_{1-x}Co_x$ we have determined the exchange constant for the R–Co interaction. Our value $A = -8 \times 10^{-15}$ erg is of the same order of magnitude than that derived by Hasegawa for ternary Gd–Co–Mo alloys ($A = 2J_{GdCo} = -4.6 \times 10^{-15}$ erg) [3]. For $R \neq Gd$ the compensation temperature in $R_{1-x}Co_x$ is considerably reduced in comparison with estimates involving free rare-earth ion moments, which suggests that the effect of local crystal fields is strong.

References

[1] J.W.M. Biesterbos, M. Brouha and A.G. Dirks, in: AIP Conf. Proc. 29 (1976) 184.
[2] K. Lee and N. Heiman, in: AIP Conf. Proc. 24 (1975) 108.
[3] R. Hasegawa, J. Appl. Phys. 46 (1975) 5263.

MAGNETIC PROPERTIES OF EVAPORATED Gd–Co AMORPHOUS FILMS

I. NAGY, T. TARNÓCZI and Z. FRAIT*

Central Research Institute for Physics, Budapest, Hungary

Magnetization and FMR measurements were performed on Gd–Co amorphous films prepared by vacuum evaporation and bias sputtering. An easy plane type of uniaxial anisotropy was established in evaporated and an easy axis type one in bias sputtered films. These results are interpreted as stress induced- and structural anisotropy respectively.

1. Introduction

In the last few years many investigations have been performed on evaporated and sputtered Gd–Co amorphous films [1]. It is well known from the experiments that a uniaxial magnetic anisotropy with an easy axis perpendicular to the plane may be achieved only in the films prepared by bias field sputtering. But it is not yet clear what kind of structure is responsible for the uniaxial anisotropy in these films and what kind of anisotropy is set up in the evaporated Gd–Co amorphous films. We have investigated both evaporated and sputtered $Gd_{1-x}Co_x$ films in the composition range $0.6 < x < 1$, using ferromagnetic resonance (FMR) and magnetic measurements.

2. Experiments

The evaporated films were prepared by a single source evaporator using premelted alloy and water cooled glass substrate. The sputtered films were deposited in an RF cathod sputtering device using an arc melted $GdCo_2$ target. To illustrate the influence of the bias field on the properties of the films, double layers were prepared: the lower portion of the films were sputtered without bias, the upper one with different bias fields.

The chemical composition of the films was measured by α-backscattering and the amorphous state was established by means of electron diffraction. The thickness of the films was measured by a Talystep machine.

The FMR measurements were made at around 36 and 9 GHz frequencies at room temperature. From the resonance field in parallel (H_1) and perpendicular (H_2) configuration and from the linewidth (ΔH) the effective magnetization $4\pi M_{eff}$, effective field of anisotropy H_u, g-factor, Landau–Lifshitz relaxation constant λ, and Gibert's relaxation parameter α were computed using the following equations:

$$\omega/\gamma = [H_1(H_1 + 4\pi M_{eff})]^{1/2} = (H_2 - 4\pi M_{eff}),$$
$$\gamma = g\mu_B h^{-1}, \qquad H_u = 4\pi(M_s - M_{eff}),$$
$$\Delta H = 2\alpha\omega\gamma^{-1} = 2\lambda\omega\gamma^{-2}(2\pi M_s)^{-1},$$

where ω represents the microwave frequency: μ_B the Bohr magneton: h Plank's constant and M_s the saturation magnetization. The saturation magnetization $4\pi M_s$, was obtained from the in-plane magnetization curve. We have also determined an effective field of uniaxial anisotropy (H_u^M) from the magnetic measurements using the approximate relation: $H_u^M = 4\pi M_s - H_{\perp s}$, where $H_{\perp s}$ represents the external field needed for saturating the film along the normal axis. The character of the anisotropy was determined by the sign of H_u: for $H_u > 0$ we have the case of easy axis anisotropy normal to the film plane and $H_u < 0$ means easy plane anisotropy.

The data of the FMR and magnetic measurements are summarized in table I. The last two rows contain the FMR data of one of the double layers. Considering that both the composition and magnetic anisotropy of the two layers of these films are essentially different, the resonance lines of the layers are separated well in both the parallel and perpendicular configurations.

3. Conclusions

It can be seen in table I that the anisotropy of the evaporated films has a concentration dependence. H_u is negative for all films except the sample with maximum Co concentration ($x = 0.96$) and a change in the sign of H_u takes place around $x = 0.94$.

* Inst. Phys. Acad. Sci., Prague, Czechoslovakia.

Table I
Magnetic and FMR data of evaporated and sputtered Gd–Co amorphous films

Preparation	x	$4\pi M_s$ (kG)	$4\pi M_{eff}$ (kG)	H_s (kG)	H_u (kG)	H_u^M (kG)	g	λ (10^{-8} rad/s)	α (10^2)
Evaporated	0.96	10.20	9.14	9.18	+1.06	+1.02	2.19	9.7	1.1
	0.93	7.89	9.15	9.25	−1.25	−1.36	2.20	1.1	0.92
	0.88	4.90	6.75	6.70	−1.85	−1.85	2.39	1.4	1.8
	0.85	2.43	5.72	4.80	−3.32	−3.22	2.45	0.8	2.1
	0.81	0.46	4.33	3.89	−3.87	−3.43	2.44	0.4	4.8
	0.64	4.58	6.78	6.35	−2.2	−1.77			
Sputtered	0.84	3.0	1.9	–	+1.1	–	2.4	2.4	4.0
	0.74	3.6	3.6	–	0	–	1.62	0.8	1.9

In contrast the anisotropy of the sputtered Gd–Co films is influenced mainly by the bias field. A plausible explanation of the origin and of the compositional dependence of the anisotropy in evaporated Gd–Co films may be given supposing mechanical stress between the substrate and the film. This causes a stress-induced anisotropy, the effective field of which is: $H_u = 3\sigma\lambda'/M_s$, where σ and λ' are the stress and the saturation magnetostriction, respectively. Supposing the continuous dependence of magnetostriction on the composition, the change of its sign around $x = 0.94$ is sufficient to explain the observed behaviour of H_u in evaporated Gd–Co films (with the tacit assumption of equal σ values). This model of the anisotropy is also proved by the FMR linewidth which reaches its minimum value near to zero magnetostriction composition. Such an assumption is supported by an analogous dependence of the magnetostriction and stress-induced anisotropy on the composition in some metallic glasses [2, 3]. It should be noted that in these amorphous magnets λ' changes sign at the same Co concentration (at about 94 at.%) as in the case of Gd–Co amorphous films.

In contrast to this, in sputtered amorphous films the origin of the uniaxial anisotropy may be connected with the superstructure of the films. Our backscattering measurements on the double layers [4] have shown that the bias field during the sputtering did not change the number of Co atoms per unit volume but a remarkable decrease was observed in the number of Gd atoms, and about 0.1 at.% Ar was incorporated in the film. This means not only a change of the chemical composition but a decrease in the density of the film too. The layer sputtered with −200 V bias field had a density calculated from the backscattering spectrum about 3.5% lower than the layer sputtered without bias. Meanwhile the magnetic anisotropy constant changed from $K_u = 1.65 \times 10^3$ erg/cm^3 to $K_u = 0$. Taking into account the decrease of the Gd content and of the density and that the Co content did not increase, we can conclude that first of all Gd atoms were resputtered, and they were only partly replaced by Ar atoms remaining in a lot of "vacancies" in the bias sputtered films. Bearing in mind the rodlike superstructure of amorphous materials which was observed in the sputtered Gd–Co films too [5], one can assume that the most intensive resputtering process takes place between the high density rods. This process causes a fluctuation both in the density and in the composition thereby increasing the shape anisotropy effect which may well provide an important contribution to the uniaxial anisotropy with an easy axis perpendicular to the surface.

The authors wish to thank M. Hossó for her technical assistance and G. Petö for the film preparation.

References

[1] P. Chaudhari, R.J. Gambino and J.J. Cuomo, J. Appl. Phys. Lett. 22 (1973) 337.
Z. Frait, I. Nagy and T. Tarnóczi, Phys. Lett. A55 (1976) 429.
S. Esho and S. Fujiwara, Joint MMM-Intermagn. Conf., Pittsburgh (1976).
[2] R.C. Sherwood, E.M. György, H.S. Chen, S.D. Ferris, G. Norman and H.J. Leamy, in: AIP Conf. Proc. No. 24,

C.D. Graham et al. (Eds.) (American Inst. of Phys., New York, 1975) p. 745.
[3] H. Fujimori, K.I. Arai, H. Shira, H. Saito, T. Masumoto and N. Tsuyay to be published in Jap. J. Appl. Phys. (1976).
[4] T. Tarnóczi, I. Nagy, G. Mezey, T. Nagy, E. Kótai and G. Petö, 2 ème Colloque Int. sur la Pulverisation Cathodique et ses Applications, Nice Comptes rendus (1976) p. 141.
[5] Á. Barna, P. Barna, G. Radnóczi, H. Sugawara and P. Thomas, Int. Conf. on Structure and Excitations of Amorphous Solids, Williamsburgh (1976).

AMORPHOUS MAGNETISM IN BULK SAMPLES OF TERBIUM IRON ALLOYS

H.A. ALPERIN
Naval Surface Weapons Center, White Oak, Maryland 20910, USA and National Bureau of Standards, Washington, D.C., USA

J.R. CULLEN, A.E. CLARK
Naval Surface Weapons Center, White Oak, Maryland 20910, USA

and

E. CALLEN
American University, Washington, D.C. 20016, USA and Naval Surface Weapons Center, White Oak, Maryland 20910, USA

Magnetization measurements made on compositions $x = 0.45$ and 0.75 in the series Tb_xFe_{1-x} show a decreasing average terbium moment with increasing x. Spontaneous moments, coercive forces and Curie temperatures for samples in the range $0.018 \leq x \leq 0.75$ are described in terms of a cluster model.

1. Introduction

Rare earth-transition metal alloys prepared in bulk form by d.c. rapid sputtering [1] are amorphous both atomically and magnetically [2]. They form an interesting class of materials since their compositions may be varied continuously. We have reported previously [3] on the series Tb_xFe_{1-x} with $x = 0.018, 0.117, 0.167, 0.25$ and 0.33 [4]. Measurements are extended here to the terbium-rich compositions $x = 0.45$ and 0.75 where we find evidence for a reduction of the average terbium moment.

2. Experimental

Samples of composition Tb_xFe_{1-x} with $x = 0.45$ and 0.75 and masses 77.4 and 35.7 mg, respectively, were prepared by rapid dc sputtering [1]. Neutron diffraction measurements were performed on larger pieces (volume ~ 50 mm^3) from which these samples were cut and yield amorphous patterns both for the atomic and the magnetic "structure". The atomic arrangement can be understood as a dense random packing of unequal-size spheres. Magnetization measurements to a maximum field of 1.9T as a function of temperature were made using a vibrating sample magnetometer.

3. Results

Magnetization data (σ vs. H) were taken at discrete temperatures from 4.2 to 358 K. No recrystallization effects were observed at these temperatures. The magnetization curves for the $x = 0.45$ sample are similar to the $x < 0.45$ samples [3] and a well-defined Curie temperature (T_c) of 298 K is obtained from an "Arrott plot" of σ^2 vs. H/σ. The hysteresis loops for the $x = 0.75$ sample, however, show severe curvature above 130 K so we have taken the Curie temperature ($T_c = 190$ K) to be the temperature at which the coercive force $H_c = 0$. The spontaneous moment (σ_0) is obtained consistently for all samples as the extrapolation of σ from the highest field back to the demagnetization field. In the middle of the composition range ($x = 0.25$ and 0.33) σ vs. H is quite linear at large H and T_c is quite sharply defined. Towards the ends of the range this is not the case; $x = 0.75$ exhibits the most deviant behavior. σ_0 at $T = 0$ and T_c are given in table I. The coercive force data H_c vs. T for the $x = 0.75$ composition lie fairly close to the $x = 0.018$ curve while the 0.45 curve lies between the 0.018 and 0.1175 curves (see ref. 3).

4. Discussion

We assume oppositely directed terbium and iron moments so that

$$\mu_{tot} = x\mu_{Tb} - (1-x)\mu_{Fe}. \qquad (1)$$

The maximum possible iron moment is obtained when the full terbium moment ($9\,\mu_B$) is used, as given in the iron-rich (left) side of table I. Similarly, the maximum possible terbium mo-

Table I
Spontaneous moment, calculated maximum Fe and Tb moments and Curie temperature for Tb_xFe_{1-x}

x	0.018	0.118	0.167	0.25	0.33 [4]	0.45	0.75	x
σ_0(emu/gm)	74	35	~0	40	85	110	144	σ_0
$(\mu_{Fe})_{max}(\mu_B)$ ($\mu_{Tb} = 9\,\mu_B$)	0.95	1.68	1.81 ± 0.07	2.14	8.2	7.0	5.2	$(\mu_{Tb})_{max}$ ($\mu_{Fe} = 2$)
T_c(K)	245	365	380	404	383	298	190	T_c

ment is obtained by using the maximum iron moment ($2\,\mu_B$) as shown for the terbium-rich compositions. The resulting very low values of μ_{Fe} for small x and correspondingly low values of μ_{Tb} for large x suggest that an "excess" of either iron or terbium atoms results in a misalignment of their moments with one another. This in turn should give rise to magnetization curves with curvature at high fields and relatively ill-defined Curie temperatures, more pronounced at the terbium-rich end, as observed experimentally. In the middle of the range ($x = 0.25$, 0.33), however, all iron and terbium moments are aligned anti-parallel to the maximum possible extent resulting in normal-looking hysteresis loops and well defined T_c values of high value, again in accord with our measurements. In this composition range one should take the iron moment to be $1.6\,\mu_B$, as obtained from amorphous $GdFe_2$ [4], making the average terbium moment $7.4\,\mu_B$.

We therefore assume that for very small x the magnetization can be represented by clusters consisting of a terbium atom (with moment μ_{Tb}^c) surrounded by n_c iron atoms (of moment μ_{Fe}^c each, opposite to the terbium); the clusters in turn, embedded in a matrix of the remaining misaligned iron atoms with average moment μ_F. The magnetic moment per atom is then

$$\mu = -\mu_F[1 - x(n_c + 1)] + (\mu_{Tb}^c - \mu_{Fe}^c n_c)x, \quad (2)$$

and the cluster moment per atom

$$\mu_c = (\mu_{Tb}^c - \mu_{Fe}^c n_c)/(n_c + 1). \quad (3)$$

Since all parameters in eq. (2) are functions of x (that can be described by polynomials) the most one can do is determine $\mu_F(0)$ and $n_c(0)$ provided we extrapolate the experimental values of μ to $x = 0$ and make the very reasonable assumption that $(d\mu_F/dx)_{x \to 0} = 0$. The result will not be very sensitive to the values taken for μ_{Tb}^c and μ_{Fe}^c so for concreteness we use the values $9\,\mu_B$ and $2\,\mu_B$, respectively. Experiment yields $\mu_F(0) = -\mu(0) \sim 0.8\,\mu_B$ and $(d\mu/dx)_{x \to 0} = 2.4$ so from eq. (2) and its derivative we find $n_c(0) = 6.2$ and $\mu_c(0) = -0.47\,\mu_B$.

As the terbium concentration increases and clusters come together and interact it is reasonable to expect n_c to decrease. As mentioned previously, experimental evidence indicates that when x reaches 0.25 and 0.33 all iron atoms are clustered with terbium making $n_c = 3$ and 2, respectively, at these concentrations, i.e. there are no longer any iron atoms left in the matrix.

We apply molecular field theory with two kinds of exchange, single iron–single iron (λ_{F-F}) within the iron matrix and single iron-cluster (λ_{F-c}) between matrix and cluster to derive the following expression:

$$\frac{1}{T_c}\left(\frac{dT_c}{dx}\right)_{x \to 0} = \left(\frac{\mu_{Tb}^c - n_c\mu_{Fe}^c}{\mu_F}\right)^2 \left(\frac{\lambda_{F-c}}{\lambda_{F-F}}\right)^2 - (n_c + w), \quad (4)$$

where w = volume of a Tb atom/volume of an Fe atom. One thus finds $\lambda_{F-c}/\lambda_{F-F} \sim 0.9$ for $x \to 0$.

The magnetization reversal energy $H_c\sigma_0$ extrapolated to $T = 0$ for these alloys falls in the range 0.6×10^7 to 1.8×10^7 erg/cm^3. We also are interested in the reversal energy per terbium atom, K_{Tb} at $T = 0$ as a function of x. This was calculated by subtracting from the measured values of $H_c\sigma_0$ the contribution from the iron matrix. The latter was obtained by multiplying $(H_c)_{x \to 0}$ by the total moment of the iron matrix (assuming $0.8\,\mu_B$/atom). $n_c(x)$ was taken to be a linear function having the values $n_c(0) = 6.2$ and $n_c(0.25) = 3$. There are uncertainties due to the various extrapolations used but qualitatively K_{Tb} shows a minimum at $x = 0.1$, rises to a maximum at 0.45 and then decreases to $x = 0.75$.

This can be understood as follows. The minimum occurs at just the composition where the cluster moment passes through zero thus requiring less energy to reverse the magnetization there. As x increases to 0.25 the cluster moment increases and for greater x, K_{Tb} increases as more terbium atoms with their large values of single-ion anisotropy are added. Eventually however, as x increases further, K_{Tb} must decrease as the average terbium moment/atom decreases.

We are grateful for the invaluable assistance of W. Gillmor. This work was supported by the Naval Surface Weapons Center Independent Research Fund and the Office of Naval Research.

References

[1] Samples prepared at Battelle Northwest Laboratories, Richland, Washington and furnished by R. Allen.
[2] J.J. Rhyne, S.J. Pickart and H.A. Alperin, AIP Conf. Proc. 18 (1974) 563.
[3] H.A. Alperin, J.R. Cullen and A.E. Clark, AIP Conf. Proc. 29 (1976) 186.
[4] J.J. Rhyne, J.H. Schelleng, N.C. Koon, Phys. Rev. B10 (1974) 4672.

THERMOMAGNETIC HISTORY EFFECTS IN AMORPHOUS Y–Fe EVAPORATED ALLOYS

J.W.M. BIESTERBOS, M. BROUHA and A.G. DIRKS

Philips Research Laboratories, Eindhoven, The Netherlands

The temperature and field dependence of the magnetization of amorphous Y–Fe evaporated alloys have been determined. Ferromagnetism is found in case of Fe-rich alloys. At low temperature the films show up to the highest investigated Fe concentration (87 at.%) thermomagnetic history effects corresponding to those observed in spin-glasses or mictomagnets but corresponding also to the behaviour of materials showing a strong temperature dependence of the coercive force, e.g. as a result of a network of small domains. Up to now we could not discriminate between these two possible explanations.

In the class of the frequently investigated amorphous rare earth-transition metal alloys the system Y–Fe occupies a special place because of its complicated magnetic behaviour. In the past few years a number of investigations on the magnetic properties of this system have been published [1]. The results, however, are not in conformity with each other as is clear from, e.g., the T_c-values listed in table I. Preparation techniques of the samples as well as interpretation of the measurements are held responsible for these discrepancies, while also magnetic clusters are presumed to be present in the concentrated Fe alloys.

The $Y_{100-x}Fe_x$ alloys have been prepared by simultaneous electron beam evaporation of the elements (impurities in Y ~ 1%; in Fe ~ 10 ppm) on to amorphous SiO_2 substrates in a vacuum better than 10^{-8} torr. Prior to the evaporation process the pressure is lower than 5×10^{-10} torr. The films have a thickness of 1 μm and are covered with 0.1 μm Al to prevent corrosion. The samples were examined and found to be amorphous for $31 \leq x \leq 87$ at.% by X-ray and electron diffraction experiments. The composition of the films was analyzed chemically and agreed with the desired value within 2%.

When a sample with $59 \leq x \leq 87$ at.% is cooled from room temperature to 4.2 K in zero field and subsequently measured at increasing T and constant H (< 3 kOe) the T-dependence of the magnetization (σ) shows a steep rise, followed by a normal ferromagnetic behaviour. When the film is measured, however, after field cooling ferromagnetic behaviour is also observed at low T. Figs. 1 and 2 show these thermomagnetic history effects of the amorphous Y–Fe films. T_f, the lowest T at which field- and zero-field cooled magnetization behaviour are equal, increases with x (fig. 1) and decreases with H (fig. 2). At a certain field (H_m) the difference in the T-dependence of σ between the field cooled and the zero-field-cooled state disappears. From Arrott-plots it was concluded that samples with $x \geq 64$ at.% are ferromagnetic (cf. table I). Films with $x < 59$ at.% are certainly not ferromagnetic; moreover, no maximum was observed in χ–T plots, even at low external H.

The peculiar T-dependence of σ described above is also reflected in its H-dependence (fig. 3). When the maximum applied H exceeds H_m (~2.5 kOe) the initial σ-curve is situated outside the symmetrical hysteresis loop. However, starting with a field-cooled film, but keeping H below H_m a displaced minor loop is found. The displacement along the H-axis increases roughly proportionally to the field in which the

Table I

x	T_c(K)	Method	Ref.
67	n.f.	Arrott-plot	(1a)
67	400 ⎫		
75	> 450 ⎬	magneto-optical	(1b)
83	> 600 ⎭		
55	280 ⎫		
80	250 ⎬	initial suscept.	(1c)
87	215 ⎭		
64	153 ⎫	Arrott-plot	this work
80	180 ⎭		

T_c-values of amorphous $Y_{100-x}Fe_x$ alloys; x in at.perc.; n.f. = non-ferromagnetic.

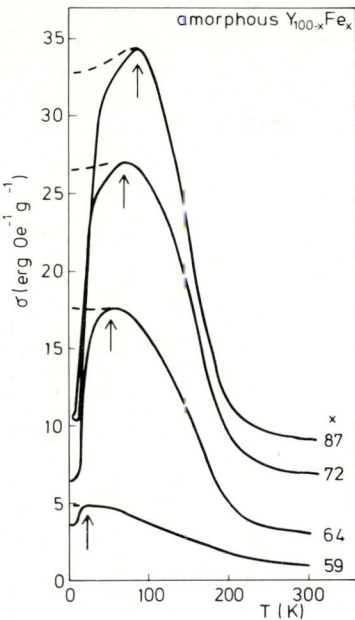

Fig. 1. Magnetization vs. temperature curves for various amorphous Y–Fe alloys. Solid lines represent measurements after zero-field cooling, whereas dashed lines indicate data from field-cooled samples. The measuring as well as the cooling field was 500 Oe. T_f is indicated by an arrow.

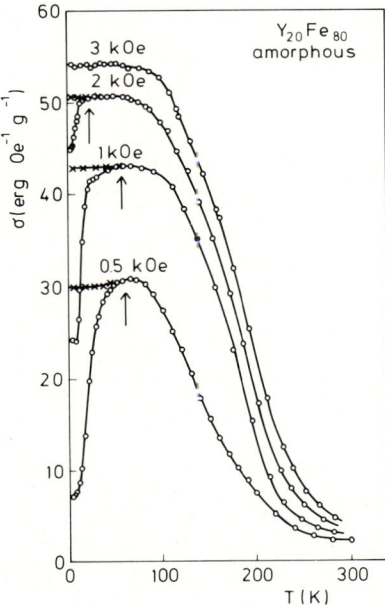

Fig. 2. Temperature dependence of the magnetization in different fields of amorphous $Y_{20}Fe_{80}$ starting from the zero-field-cooled (-0-) and the field-cooled (-x-) state, respectively. In the case of field cooling this field equals the measuring field. T_f is indicated by an arrow.

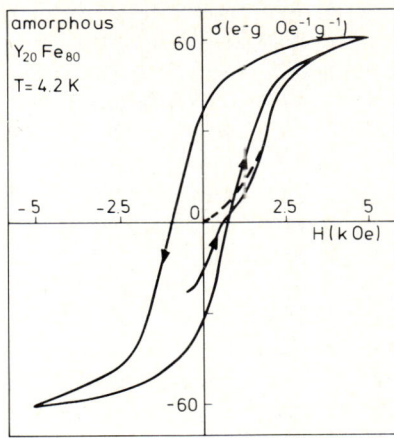

Fig. 3. Hysteresis loop for amorphous $Y_{20}Fe_{80}$. Prior to the measurement the sample has been field cooled at -500 Oe, at which field the magnetization curve also starts. The dashed line indicates the initial magnetization curve of the same sample cooled in zero field.

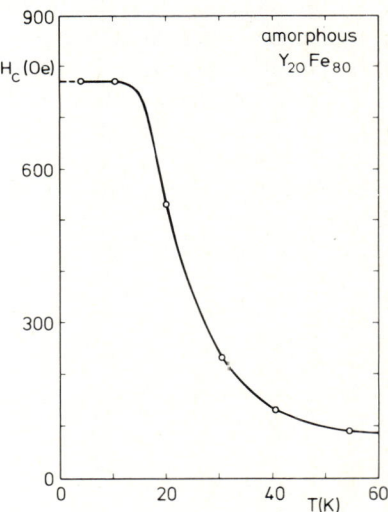

Fig. 4. Coercive field vs. temperature of a zero-field-cooled sample. The maximum applied field is 5 kOe.

sample was cooled. The T-dependence of the coercive force, determined from full hysteresis loops is shown in fig. 4.

In disordered magnetic alloys a behaviour similar to that described above is frequently observed [3]. A material showing these phenomena is called "spin glass" (dilute alloy) or "mictomagnet" (more concentrated alloy containing ferromagnetic clusters) [2]. The striking feature in both types is that at small external H a cusp in χ–T curves occurs. This is ascribed to a cooperative freezing of the spins (or cluster moments) without long-range order, at a well-defined T. Therefore it is impossible to align the spins at low T in a small external H. This results in a small σ at low T, which increases towards the spin freezing T. A spin freezing behaviour can occur in systems with exchange interactions randomly distributed about zero [4] or about a positive mean [5], as well as in systems with a local anisotropy field of random direction [6]. Such conditions can be met in disordered alloys and of course also in amorphous materials. The large discrepancy in our determination of T_c from the initial χ-curves and from Arrott-plots for the various amorphous Y–Fe films (table I) suggest a fair fluctuation in exchange interactions. In combination with the mentioned $\sigma-T$ behaviour this points to an explanation in terms of a spin glass or mictomagnet.

However, the magnetic properties described above can also be understood as a result of the T-dependence of the relatively high intrinsic coercive force as measured in these materials (fig. 4). At present the origin of the coercive force is not clear. It may be caused by the occurence of narrow Bloch walls or a network of small domains.

Of the three possible effects the existence of narrow Bloch walls is very unlikely because the required anisotropy fields are too high by an order of magnitude compared with the measured values of some kOe. Up to now no discrimination can be made between a network of small domains and "random freezing of spins" without supplementary information concerning possible magnetic domains and knowledge of the structure on atomic scale at low T.

The authors enjoyed fruitful discussions with R.P.v. Stapele and H. Zijlstra.

References

[1] J.J. Rhyne, J.H. Schelleng and N.C. Koon, Phys. Rev. B10 (1974) 4672.
K. Lee and N. Heiman, in: AIP Conf. Proc. No. 24 (1975) 108.
J.W.M. Biesterbos, M. Brouha and A.G. Dirks, in: AIP Conf. Proc. No. 29 (1976) 184.
[2] J.A. Mydosh, in: AIP Conf. Proc. No. 24 (1975) 131.
[3] P.A. Beck, Met. Trans. 2 (1971) 2015.
[4] S.F. Edwards and P.W. Anderson, J. Phys. F: Metal Phys. 5 (1975) 965.
[5] D. Sherrington and B.W. Soutern, J. Phys. F. Metal Phys. 5 (1975) L49.
[6] R. Harris and D. Zobin, in: AIP Conf. Proc. No. 29 (1976) 156.

MAGNETIC STRUCTURES AND PROPERTIES OF THE AMORPHOUS ALLOYS DyT_3; T = Fe, Co, Ni

J.P. REBOUILLAT, A. LIENARD, J.M.D. COEY
Centre National de la Recherche Scientifique, 166X, 38042-Grenoble-Cedex, France

R. ARRESE-BOGGIANO and J. CHAPPERT
Centre d'Etudes Nucléaires, 85X, 38041-Grenoble-Cedex, France

Dy carries a well-defined free ion moment in amorphous DyT_3, T = Fe, Co, Ni whereas the Fe and Co moments are 20% greater than in the corresponding crystals; Ni has a very weak moment. The magnetic structures are dominated by the randomly directed local anisotropy of the rare earth, which distributes the Dy moments over all directions within a broad cone.

The rare-earth transition-metal alloys are ideal materials for studying magnetism in amorphous metals. Exchange and single-ion anisotropy, the interactions mainly responsible for the magnetic properties, may be varied at will by changing the constituent atoms or their concentration ratio. Extensive comparisons are possible with corresponding crystalline alloys such as RT_2, RT_3, R_2T_7, RT_5.

In this paper we present the magnetic properties of RT_3 alloys, where R = Dy or Y and T = Fe, Co or Ni. Films prepared by d.c. argon–ion sputtering were all amorphous except $DyNi_3$ which was partially crystalline. We refer to them as RT_3 although their actual compositions lay between $R_{21}T_{79}$ and $R_{28}T_{72}$. Three points to be considered are:

(i) the magnitudes of the moments of the rare-earth and transition metal,

(ii) their magnetic structures at zero temperature, and

(iii) the temperature dependence of the spontaneous magnetization.

Magnetization curves for the six alloys at 4.2 K are shown in fig. 1. YNi_3 is a weak ferromagnet with a moment of 0.04 μ_B/Ni. YCo_3 is an easily saturated ferromagnet with $M_s = 1.2$ μ_B/Co and negligible high field susceptibility, $dM/dH < 10^{-5}$ emu/g in fields greater than 1 kOe. In contrast the magnetization of YFe_3 is strongly field-dependent with $dM/dH = 12 \times 10^{-5}$ emu/g at 100 kOe. The magnetization obtained by extrapolating the high-field data linearly to zero field is only $M_s = 0.8$ μ_B/Fe. Since the moment on the ion derived from the ^{57}Fe hyperfine field at 4.2 K, is 1.8 μ_B, the magnetic structure cannot be collinear.

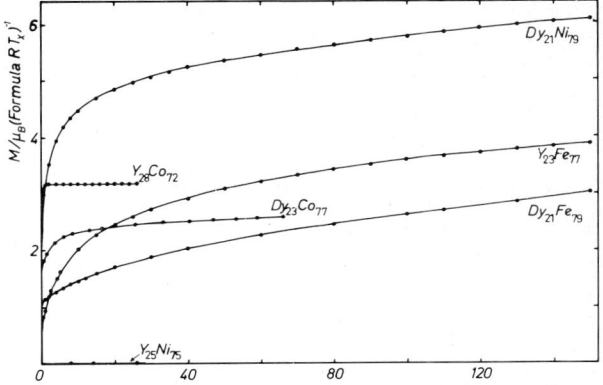

Fig. 1. Magnetization curves for amorphous RT_3 alloys at 4.2 K.

In the alloys with dysprosium, the ^{161}Dy hyperfine field is found to be just 1% less than in crystalline DyT_3, with a relative distribution of only 1% [1]. Its free-ion-like moment of 10 μ_B is barely influenced by the amorphous structure. An average iron moment of 1.9 μ_B in $DyFe_3$ was derived from the ^{57}Fe hyperfine field, but it has a broad relative distribution of 16%, reflecting the variation in nearest neighbour environments in the amorphous structure. Hyperfine fields of ^{57}Fe in RT_3 were found to be between 15 and 35% larger than in the corresponding crystalline alloys. Further details of the Mössbauer results are included in [1]. In the two alloys with cobalt, the cobalt moment is also found to be considerably greater than in the corresponding crystals [2, 3]. However, the nickel moment seems to be essentially zero in both amorphous and crystalline YNi_3 and $DyNi_3$ [1, 4].

Of the five alloys that order magnetically,

only YCo$_3$ has a collinear magnetic structure. The discrepancy between the moment per iron in YFe$_3$ deduced from the high field magnetization data and from the hyperfine field may be resolved if the magnetic structure is *asperomagnetic*, as indicated in fig. 2(a). In the amorphous Y–Fe system, the structure probably evolves continuously from a speromagnetic one (spins randomly distributed over all directions) [5] for YFe$_2$, to an asperomagnetic one [spins distributed randomly over some directions with a probability P(Ψ)] with an increasing net moment as the Fe/Y ratio increases. The non-collinear structures in these alloys is unlikely to have much to do with random single-ion anisotropy, but results from a broad distribution of exchange whose average value is displaced from near zero for YFe$_2$ towards positive values as the Fe/Y ratio increases.

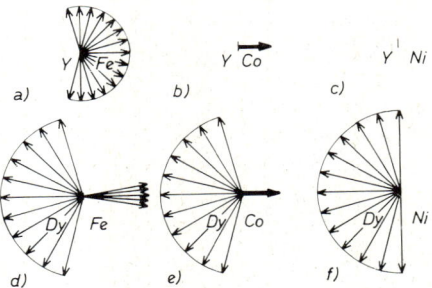

Fig. 2. Magnetic structures proposed for RT$_3$ alloys at 4.2 K (a) and (f) are asperomagnetic, (b) is ferromagnetic and (d) and (e) are sperimagnetic.

The values of M_s at 4.2 K for the alloys with Dy, 1.8, 2.4 and 5.1 μ_B/formula for T = Fe, Co and Ni, respectively, are irreconcilable with a collinear structure for the Dy sublattice. For DyCo$_3$, we showed that the Co sublattice is ferromagnetic, but the Dy moments are strongly correlated with the local crystal field axes, which are distributed, randomly in direction [3], as originally proposed theoretically in [6]. The resulting sperimagnetic structure is indicated in fig. 2(e). The structure of DyFe$_3$ must be similar to give a global moment of 1.8 μ_B with 1.9 μ_B/Fe and 10 μ_B/Dy. There might be a small spread in the iron directions. DyNi$_3$ is an asperomagnet as there is no moment on the Ni, and the Dy–Dy exchange is positive. In each case the net sublattice moment is $\sim 6 \mu_B$/Dy as the Dy moments are distributed over all directions within a broad cone, which would become a hemisphere as the ratio of intrasublattice exchange to single-ion anisotropy ($K \sin^2 \theta$) leads to zero. In this limit, $dM/dH = M_0^2/3 K$, giving "anisotropy fields" of 0.84 and 0.53 MOe for DyCo$_3$ and DyNi$_3$, which are comparable to or bigger than the exchange fields on the Dy. Similar values of the anisotropy field are obtained simply by extrapolating the magnetization curves to a moment corresponding to complete alinement of the Dy. The applied field tends to close the Dy cone in DyCo$_3$ and DyNi$_3$, but opens it in DyFe$_3$.

In fig. 3 we show the temperature dependence of M_s. Only DyCo$_3$ has a compensation point, as expected from the magnetic structures proposed. These magnetization curves have been successfully fitted by Bhattacharjee et al. [8] within the framework of the Harris et al. model [6].

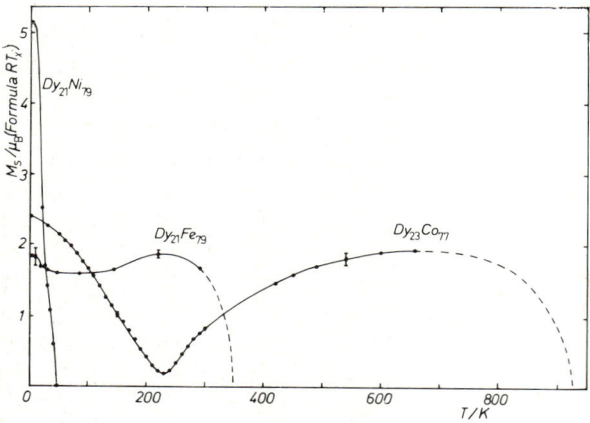

Fig. 3. Temperature dependence of the spontaneous magnetization of DyM$_3$.

References

[1] R. Arrese-Boggiano, J. Chappert, J.M.D. Coey, A. Lienard and J.P. Rebouillat. Conf. on Applications of the Mössbauer effect, Corfu (1976); J. de Physique 37C (in press).
[2] R. Lemaire, Cobalt 33 (1966) 201.
[3] J.M.D. Coey, J. Chappert, J.P. Rebouillat and T.S. Wang, Phys. Rev. Lett. 36 (1976) 1061.
[4] J. Yakinthos and D. Paccard, Solid State Commun. 10 (1972) 989.
[5] J.M.D. Coey and D.W. Schindler, Physica (these Proceedings).
[6] R. Harris, M. Plischke and M.J. Zuckermann, Phys. Rev. Lett. 31 (1973) 160.
[7] H. Jouve, J.P. Rebouillat and R. Meyer, AIP Conf. Proc. 29 (1976) 97.
[8] A.K. Bhattacharjee, R. Jullien and M.J. Zuckermann, submitted for publication.

ROLE OF MAGNETOSTRICTION ON THE HIGH FREQUENCY MAGNETIC CHARACTERISTICS IN AMORPHOUS FERROMAGNETIC RIBBONS

N. TSUYA, K.I. ARAI and M. YAMADA

Research Institute of Electrical Communication, Tohoku University, Sendai, Japan

The initial permeability and the loss factor of amorphous high magnetostrictive $Fe_{80}P_{13}C_7$ and non-magnetostrictive $Fe_5Co_{70}Si_{15}B_{10}$ ribbons were measured. In a magnetostrictive ribbon, the giant ΔE effect of 1.9, the giant magnetomechanical coupling over 0.65 and the change of the sound velocity about 50% were observed.

1. Introduction

The magnetic properties of amorphous iron base materials such as the magnetization [1], the Curie temperature [1] and the magnetostriction [2, 3] were reported. We reported [4, 5] the preliminary results on the magnetostriction and the giant magnetomechanical coupling of the material. In this investigation we extended our experiments to include the loss factors as well as the velocity change in the sound wave under an applied field in the high frequency region.

2. Experimental procedures

The amorphous specimens used in this experiment were made by a roller quenching technique in the form of ribbons. The magnetostriction was measured by a three-terminal capacitance method [2]. The initial permeability μ_i, the loss factor $\tan \delta$ from 3 to 500 kHz could be measured by using a Maxwell Bridge (Ando Electric Co. Ltd.). To measure the change in the sound velocity, a input pulse signal which was modulated by sinsoidal waves from 100 kHz to 6 MHz was applied to a transmitter part, containing a part of the ribbon itself as an element, and was converted into a sound wave pocket in the ribbon. The output signal was detected by a receiver part with a similar structure as above. A solenoidal coil in between them was used in applying a magnetic field to change the sound wave velocity.

3. Results and discussion

The magnetostriction constant of the $Fe_{80}P_{13}C_7$ ribbon was as large as 31×10^{-6}, and a pronounced change from positive to negative in the sign of the magnetostriction constant was observed at the composition $Fe_5Co_{70}Si_{15}B_{10}$. The initial permeability of former was about 400,

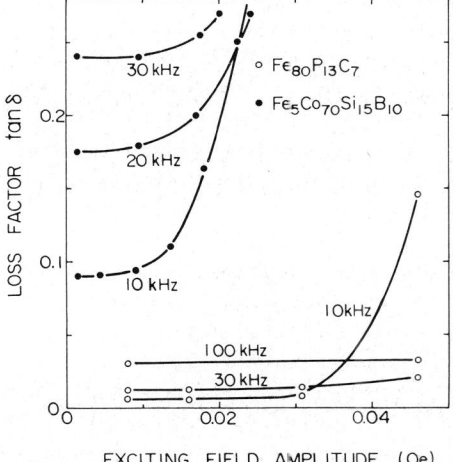

Fig. 1. Exciting field amplitude dependence of the loss factor $\tan \delta$.

and was almost independent of the frequency from 3 to 500 kHz. The behaviour of $\tan \delta$ is shown in fig. 1. In the initial permeability region at high frequency of $Fe_5Co_{70}Si_{15}B_{10}$, the residual loss coefficient and the hysteresis loss coefficient were extremely low, namely, 5×10^{-3} and 150 (cm/A), respectively. Eddy current loss was essential.

Giant values of $\Delta E/E$ 190% and magnetomechanical coupling factor k more than 68% were observed in $Fe_{78}Si_{10}B_{12}$. The sound velocity in this ribbon without applied field was about $2 \mu s/cm$, and the attenuation of the sound wave was very low. When the magnetic field was applied to the ribbon, the delay time of the sound was decreased remarkably. In fig. 2 is shown the output signal and the change in delay time in percentage as a function of the sound frequency, when the applied field is varied. In this figure, the output signal was almost independent of the applied magnetic field and the

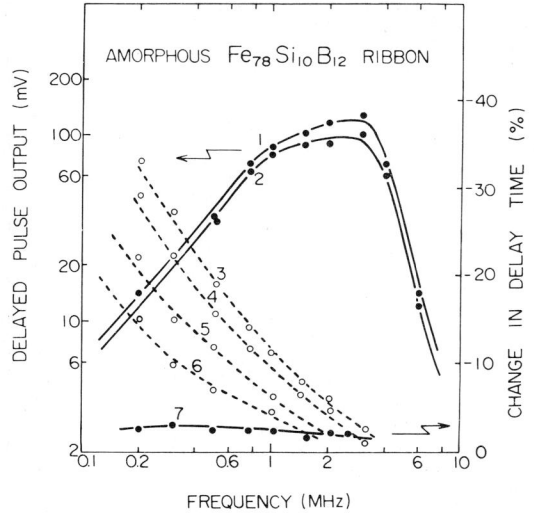

Fig. 2. Sound frequency dependence of the output signal and the change in delay time. ●: as-prepared sample; ○: field-annealed sample; 1: bias field 57 Oe; 2: 9.60 Oe; 3: 9.6 Oe; 4: 3.8 Oe; 5: 1.5 Oe; 6: 1 Oe; 7: 57 Oe.

frequency dependence of the signal was quite similar to that in magnetic recording processes.

In the field annealed ribbon, about 50% change in delay time and in an as prepared ribbon an almost frequency independent change up to 6 MHz were observed.

The authors thank Professor T. Masumoto for preparing the amorphous ribbons and Mr. S. Suda for technical assistance. This work has been partially supported by the Grant in Aid from the Ministry of Education.

References

[1] N. Kazama, T. Masumoto and H. Watanabe, J. Phys. Soc. Jap. 37 (1974) 1171.
[2] N. Tsuya, K. I. Arai, Y. Shiraga and T. Masumoto, Phys. Lett. 51A (1975) 121.
[3] T. Egami, D.J. Flanders and C.D. Graham, Jr., 20 Ann. Conf. on M^3, San Francisco, Dec. (1974).
[4] K.I. Arai, N. Tsuya, M. Yamada and T. Masumoto, Joint M^3 – Intermagn. Conf., Pittsburgh, June (1976) 6D-6.
[5] K.I. Arai, N. Tsuya, M. Yamada and T. Masumoto, Joint M^3 – Intermagn. Conf., Pittsburgh, June (1976) 6D-7.

MAGNETIC PROPERTIES OF AMORPHOUS Gd–Pd–Si ALLOYS

A. ZENTKO, L. POTOCKÝ, S. ULIČIANSKY

Institute of Experimental Physics, Slovak Academy of Sciences, Košice, Czechoslovakia

and

P. DUHAJ

Institute of Physics, Slovak Academy of Sciences, Bratislava, Czechoslovakia

The magnetic susceptibility of amorphous $Gd_xPd_{82-x}Si_{18}$ ($0.5 \leq x \leq 3$ at.%) alloys was measured in the temperature range from 4.2 to 300 K in magnetic fields up to 1.3×10^7 A/m. It was found that this material does not show ferromagnetic properties, but shows a large value of the susceptibility which is practically independent of temperature.

It is known that amorphous Pd–Si alloys containing magnetically active atoms of transition metals (Co, Fe, Ni) show ferromagnetic properties above a certain concentration of these atoms [1–3]. It was pointed out [4–7] that the ferromagnetism of these alloys is connected with the presence of some micro-regions rich in transition element and that the magnetic and other physical properties of the alloys can be explained using the concept of the superposition of the properties of the amorphous basal matrix with that of the above mentioned microregions. The subject of the present paper is the study of the magnetic properties of amorphous Pd–Si alloys containing magnetically active atoms of gadolinium. Alloys of this kind were prepared in the amorphous state with up to 3 at.% Gd. The samples were prepared first as polycrystalline alloys in a plasma jet flame furnace in high purity argon. From these alloys the amorphous samples were prepared using the method of the rapid rotating mill device [8]. The samples obtained had the form of irregular thin foils with a thickness of $20 \pm 2\,\mu$m. The amorphous structure of the samples was verified by electron diffraction. For alloys containing up to 3 at.% Gd the electron diffraction showed diffusion rings, typical for an amorphous structure. It was not possible to prepare amorphous samples with a higher content of Gd.

Measurements of the susceptibility were performed by means of an induction method using an a.c. mutual inductance bridge of the Hartshorn type. The alternating field used was 64 A/m. In combination with this low a.c. field, stationary fields up to 2.4×10^5 A/m and quasi-stationary pulsed fields up to 1.3×10^7 A/m were applied.

Fig. 1 shows the temperature dependence of the gram susceptibility $\chi(T)$ as measured in the low a.c. field. It is evident from this figure, that the susceptibility of all samples attains relatively large values and is practically independent of temperature over a wide range of temperatures. The absolute value of the susceptibility decreases with increasing concentration of the gadolinium in the alloy (fig. 2).

The dependence of $\chi(H)$, resp. gram magnetization $\sigma(H)$ was measured using stationary fields up to 2.4×10^5 A/m resp. pulse field of 1.3×10^7 A/m. The dependence of χ on the field is shown in fig. 3 and that of σ in fig. 4.

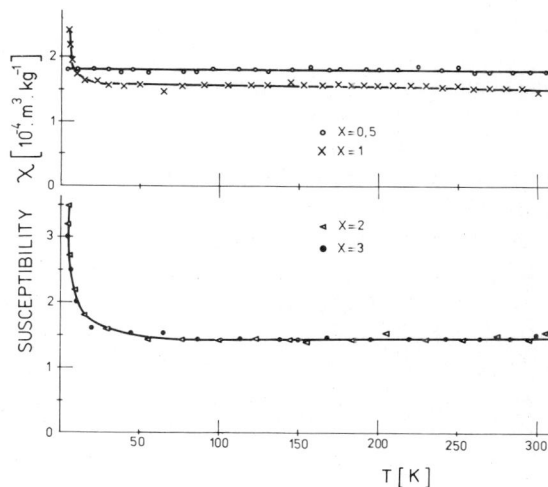

Fig. 1. Temperature dependence of the gram susceptibility of the amorphous alloys $Gd_xPd_{82-x}Si_{18}$ ($0.5 \leq x \leq 3$ at.%).

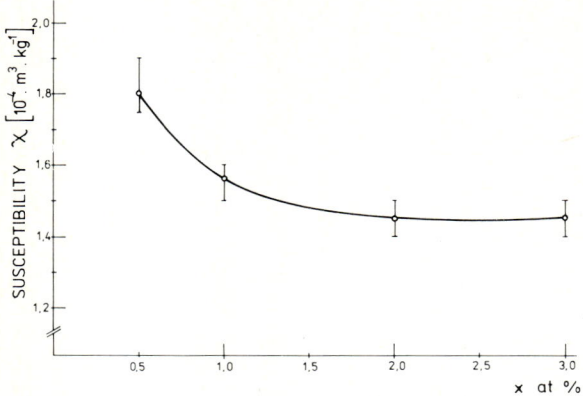

Fig. 2. Dependence of the gram susceptibility of the amorphous alloy $Gd_xPd_{82-x}Si_{18}$ upon the content of gadolinium x in at.%

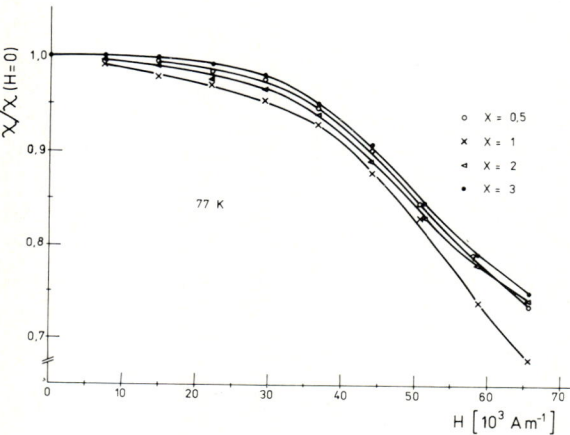

Fig. 3. Dependence of the relative gram susceptibility of the amorphous alloy $Gd_xPd_{82-x}Si_{18}$ ($0.5 \leq x \leq 3$ at.%) upon the intensity of the magnetic field H for $T = 77$ K.

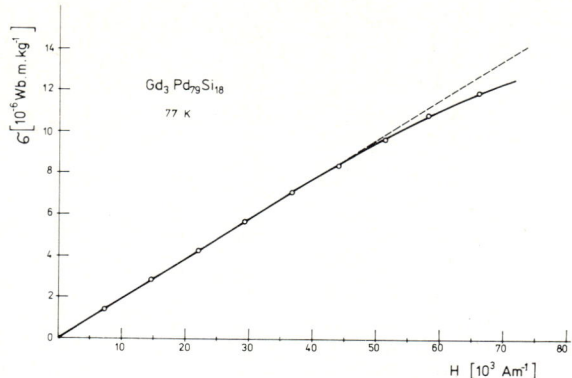

Fig. 4. Dependence of the gram magnetization σ of the amorphous alloy $Gd_3Pd_{79}Si_{18}$ upon the intensity of the magnetic field H for $T = 77$ K.

Measurements of $\chi(H)$ and $\sigma(H)$ showed that these dependences are reversible within the accuracy of our measurements. No hysteresis was observed. Both the $\sigma(H)$ and $\chi(H)$ curves are practically independent of temperature in the temperature range from 40 K to room temperature.

It can be concluded from the results obtained that amorphous alloys $Gd_xPd_{82-x}Si_{18}$ ($0.5 \leq x \leq 3$ at.%) does not show ferromagnetic properties. However, while Pd–Si alloys containing magnetically active atoms of transition metals show the properties of spin glasses up to the certain concentration of these atoms (verified by experiments on the temperature dependence of susceptibility [7]), in amorphous $Gd_xPd_{82-x}Si_{18}$ ($0.5 \leq x \leq 3$ at.%) systems, the susceptibility is practically independent of temperature. Since the susceptibility is independent of temperature already from the low temperature range and there is no possibility for Gd to lose its atomic resp. ionic magnetic moment, it can be assumed that the low temperature contribution may be due to a type of local antiferromagnetic order of Gd moments.

The authors are indebted to Dr. A. Košturiak for chemical analysis of the samples.

References

[1] C.C. Tsuei and P. Duwez, J. Appl. Phys. 37 (1966) 435.
[2] M.E. Weiner, J. Metals 20 (1968) 99A.
[3] R. Hasegawa and C.C. Tsuei, Phys. Rev. B3 (1971) 214.
[4] A. Zentko, L. Potocký, T. Tima, P. Marko and P. Duhaj, in: ICM-73, Vol. II (Nauka, Moscow, 1974) p. 50.
[5] A. Zentko, P. Duhaj, P. Marko, L. Potocký and T. Tima, Phys. Status Solidi (a) 24 (1974) K99.
[6] A. Zentko, P. Duhaj, L. Potocký, T. Tima and P. Marko, Phys. Status Solidi (a) 24 (1974) K103.
[7] A. Zentko, P. Duhaj, L. Potocký, T. Tima, J. Bánsky, Phys. Status Solidi (a) 31 (1975) K41.
[8] P. Duhaj, V. Sládek and P. Mrafko, Čs. Čas. Fyz. A23 (1973) 617.

RESISTANCE AND MAGNETORESISTANCE OF AMORPHOUS FERROMAGNETS*

R.W. COCHRANE and J.O. STRØM-OLSEN

Eaton Electronics Laboratory, McGill University, P.O. Box 6070, Montreal, Quebec, Canada

The resistivity of several amorphous ferromagnets all exhibit the characteristic $-\ln T$ temperature dependence in the liquid helium range. Fields up to 45 kOe have no effect on the negative $\partial\rho/\partial T$ but do contribute a magnetoresistance which saturates with the magnetization. Annealing has negligible effect until the onset of crystallization.

The distinguishing feature of amorphous metallic alloys has been their common structural characteristics described by the model of the dense random packing of hard spheres [1]. The most common physical property of these alloys is the minimum in the resistivity-temperature curve [2, 3] which occurs anywhere from the liquid helium range to well above room temperature. The specific temperature dependence at low temperature is logarithmic [3, 4] over a reasonable temperature range. Moreover, this dependence is unaltered by a magnetic field [4] which rules out magnetic Kondo scattering as the origin of this behaviour. The model which we proposed [4] involves electron scattering from atoms with several equivalent positions and leads to a resistivity dependence of

$$\rho_3 = -A \ln [k^2 T^2 + \Delta^2], \qquad (1)$$

where Δ is a characteristic tunnelling splitting. In this way the resistance anomalies are a direct consequence of the underlying structural disorder of these alloys.

We report here further studies of the resistivity of a number of amorphous ferromagnets (sputtered CoSm, electrodeposited CoP, and rapidly quenched metglas 2826, 2826A and 2605 [5]) from 1.1 to 300 K in magnetic fields up to 45 kOe. The strong magnetic interaction present in these ferromagnets preclude the usual Kondo effect, yet all these alloys exhibit similar $-\ln T$ variations at low temperatures. A typical curve is shown in fig. 1(a) for metglas 2826A ($Fe_{32}Ni_{36}Cr_{14}P_{12}B_6$) where

$$\Delta\rho = \rho(T) - \rho(4.2) \qquad (2)$$

is plotted on a logarithmic temperature scale.

* Research supported by National Research Council of Canada.

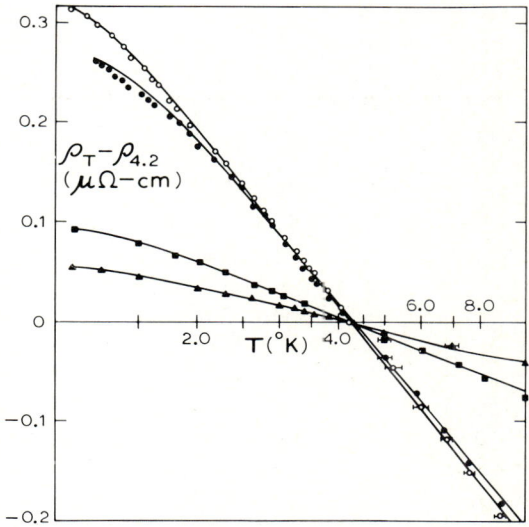

Fig. 1. Resistivity of metglas 2826A plotted on a logarithmic temperature scale. Open circles unannealed; closed symbols annealed at 400°C ●, 600°C ■ and 800°C ▲.

Below 1.5 K the resistivity shows signs of saturation although it is still increasing down to 1.1 K. A fit to eq. (1) can be made with $\Delta/k = 0.6$ K in agreement with our earlier statement that Δ should be less than 1 K. Magnetic fields up to 45 kOe do not alter the temperature dependence, but contribute only a small temperature independent magnetoresistance.

This same metglas 2826A sample has been subjected to a series of 2-h annealing treatments under vacuum from 200 to 600°C, the latter temperature being well above the onset of crystallization at 380°C [6]. A second sample was held at 800°C for 24 h. The anneals at 200, 300 and 350°C produce no noticeable change in either the low temperature resistivity shown in fig. 1(a) or the resistivity at higher temperatures. On the other hand there is a marked change in the mechanical properties in that the sample becomes very brittle even after the 200°C treatment.

The first detectible changes in the resistivity occur after the 400°C anneal: the room temperature value falls from 180 to 140 $\mu\Omega$-cm and the resistivity minimum temperature goes from 270 to 47 K. Concurrently, the Curie temperature has increased to beyond 300 K since the sample showed a ferromagnetic magnetization curve at room temperature. Quantitatively, there is only a small change in the low temperature resistivity as shown in fig. 1(b). Qualitatively, this curve cannot be fitted to eq. (1) as well as the original one; a single value of Δ no longer completely describes the data. As found for the unannealed sample, the application of a magnetic field does not alter $\Delta\rho$.

Annealing at 600°C produces significant quantitative and qualitative changes in the resistivity over the entire temperature range. The room temperature value decreases to 90 $\mu\Omega$-cm and the total change to helium temperatures is almost 40% of this value. At low temperatures a minimum is still evident at 8 K although below this temperature the increase in resistivity is much less than previously observed. The data shown in fig. 1(c) cannot be fitted to eq. (1). Moreover, there is a change in $\Delta\rho(1.1)/\rho(4.2)$ of nearly 5% on application of a 45 kOe field resulting from a temperature dependent magnetoresistance or possibly a Kondo contribution from one of the crystalline phases. Further annealing at 800°C lowers the room temperature resistivity close to 80 $\mu\Omega$-cm and reduces the depth of the low temperature minimum as illustrated in fig. 1(d). The continued presence of a shallow resistivity minimum would seem to attest to some degree of atomic disorder in such a multicomponent system.

Although the magnetic field does not affect the temperature dependence of ρ at low temperatures there is a distinctive longitudinal magnetoresistivity in this regime. Fig. 2(a) shows the variation of ρ at 4.2 K for the unannealed alloy as the field is swept through zero. On annealing at 400 and 800°C [fig. 2(b), (c)] this low field behaviour is still evident. A similar low field magnetoresistance has been reported by Bennett and Wright [7] for amorphous cobalt films. Indeed, identical results are found in the other soft amorphous ferromagnets we have

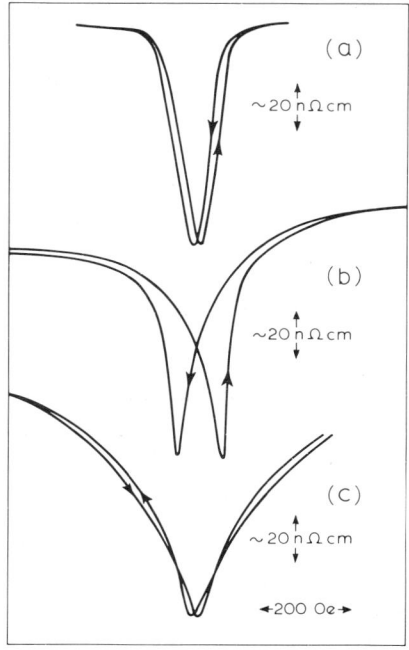

Fig. 2. Longitudinal magnetoresistivity of metglas 2826A at 4.2 K: (a) unannealed, (b) annealed at 400°C, (c) annealed at 800°C.

examined. Independent measurements of the magnetization confirm that this low field magnetoresistivity saturates with the magnetization; the field dependence of ρ appears to be a direct image of $\partial M/\partial H$. The close correlation between the magnetoresistivity and the magnetization suggests that this effect is classical [8] rather than quantum mechanical in contrast to the ln T temperature dependence.

References

[1] G.S. Cargill III, in: Solid State Physics, Vol. 30, H. Ehrenreich et al. (Ed.) (Academic Press, New York, 1975) p. 277.
[2] S.C.H. Lin, J. Appl. Phys. 40 (1969) 2173.
[3] R. Hasegawa and J.A. Dermon, Phys. Lett. 42A (1973) 407.
[4] R.W. Cochrane, R. Harris, J.O. Strøm-Olsen and M.J. Zuckermann, Phys. Rev. Lett. 35 (1975) 676.
[5] The metglas alloys are manufactured by Allied Chemical Corp.
[6] M. Fischer, H.-J. Güntherodt, E. Hauser, H.U. Künzi, M. Liard and R. Müller, Phys. Lett. 55A (1976) 423.
[7] M.R. Bennett and J.G. Wright, Phys. Lett. 38A (1972) 419.
[8] See for example paper 7X7 by E.W. Lee in these proceedings.

RESISTANCE ANISOTROPY IN AMORPHOUS FERROMAGNETIC COBALT FILMS

E.W. LEE

Department of Physics, The University, Southampton, England

An analysis of the resistance anisotropy of amorphous cobalt films shows that such films behave like a two-dimensional assembly of single-domain particles. From this it is concluded that the films, though amorphous, possess a mosaic structure. The linear dimensions of the mosaic blocks are estimated to be about 200 Å.

Structurally amorphous thin films of pure cobalt have been shown by measurements of the resistance anisotropy to be ferromagnetic at low temperatures [1]. We shall demonstrate from the form of the field-dependence of the resistance anisotropy that the films behave as a two-dimensional array of single-domain particles and examine some of the consequences.

Here resistance anisotropy is meant the dependence of the electrical resistance on the angle ϕ between the magnetic moment and the current. For a two-dimensional, isotropic system it can be written

$$\Delta R/R = B(\cos^2 \phi - \tfrac{1}{2}).$$

If there is more than one direction of magnetization within a sample, then since B is a small quantity, the resistance anisotropy of the sample is obtained by replacing $\cos^2 \phi$ by its volume average. This quantity was calculated numerically for a two-dimensional array of single-domain particles using the tables given by Stoner and Wohlfarth [2]. Since these are given in terms of a reduced field h (defined below) the resulting curve, shown in fig. 1, was fitted to the experimental data at $h = 0$ and $h = 2.0$. Better fits could perhaps be obtained but would not alter the crucial feature: $\Delta R/R$ at remanence is zero; between remanence and the coercive field $\Delta R/R$ is negative with respect to its value at saturation. (I am indebted to Drs. Bennett and Wright for confirmation of this.) Such behaviour is not observed in multi-domain systems because of the ease with which they can reverse their magnetization by the formation of 180° reverse domains which leave $\Delta R/R$ unchanged.

If it is accepted that the similarities between the calculated and experimental results are too great to be coincidental then one may proceed to make some tentative deductions concerning the structure of the films. For a particle to be a

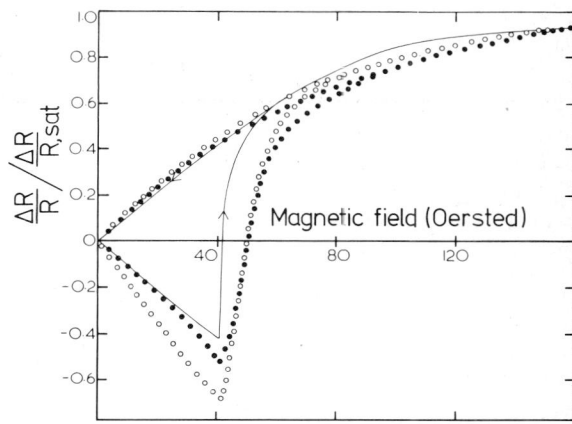

Fig. 1. Relative resistance anisotropy $(\Delta R/R)/(\Delta R/R_{\text{sat}})$ – comparison of measured and calculated field dependence in an amorphous cobalt film at 4.2 K. The continuous line represents the calculated variation; the open circles are the measured values with the field parallel to the current: the full circles are the values with the field perpendicular to the current [1].

single-domain the self energy of its uniformly magnetized state E_d must be less than that of any non-uniformly magnetized state. It is assumed that the particles may be approximated by elliptical discs whose thickness c is equal to that of the film and whose semi-major and semi-minor axes are $R + \delta$ and $R - \delta$, respectively. The principal demagnetizing factors in the plane of the disc are, approximately

$$N_a = \frac{\pi^2 c}{R}\left(1 - \frac{1}{4}\frac{\delta}{R}\right),$$

$$N_b = \frac{\pi^2 c}{R}\left(1 + \frac{5}{4}\frac{\delta}{R}\right),$$

so that $N_b - N_a = 3\pi^2 c\delta/2R^2$. The reduced field

$$h = \frac{H}{(N_b - N_a)M_s} = \frac{2R^2}{3\pi^2 c\delta M_s}H. \qquad (1)$$

An ideally amorphous material has no crystal anisotropy and the non-uniformly magnetized state is taken to be one in which the magnetization rotates in a series of concentric circles in the plane of the disc. The exchange energy in this configuration is readily found, by modification of a standard calculation [3], to be

$$E_{ex} = (\pi Ac/a) \log(R/a),$$

where A is the exchange stiffness constant and a is the mean interatomic spacing. In order to be a single domain $E_d < E_{ex}$ and this means that the radius of the disc must be less than R_c where R_c is given by

$$(\pi a/a) \log(R_c/a) = \tfrac{1}{2}\pi^2 cR_c M_s^2.$$

Setting $A = 1.5 \times 10^{-6}$ erg cm^{-1}, $a = 2.49$ Å, $c = 30$ Å and $M_s = 1420$ G it is found that $R_c \simeq 1.6 \times 10^{-6}$ cm.

The stability of the direction of magnetization of a small ferromagnetic particle of volume V is governed by a relaxation time τ for which

$$\frac{1}{\tau} = \frac{\gamma C}{2\pi M_s} e^{-CV/2kT},$$

where C is the anisotropy of the particle and γ is the magnetomechanical ratio. For the flat particles under consideration this becomes

$$\frac{1}{\tau} = \frac{3\pi\gamma c\delta M_s}{8R^2} e^{-3\pi^2 c^2 \delta M_s/8kT}.$$

This was solved for δ by trial and error, adopting $T = 40$ K, the highest temperature at which measurements were made and setting γ somewhat arbitrarily at 10^3 s, giving $\delta \simeq 2 \times 10^{-7}$ cm. This was then substituted in eq. (1) taking $h = 0.5$ as equivalent to $H_c = 40$ Oe to obtain $R_c \simeq 1.3 \times 10^{-6}$ cm. Thus, the particles must have radii lying between 130 and 160 Å if they are to behave as single-domain particles and not superparamagnetically. Too much significance should not be attached to the actual numerical values since they depend on a number of simplifying assumptions. Nevertheless the values of R_c are rather insensitive to the magnitudes of quantities used and, as revealed by further calculations, to the precise details of the model. Thus, unless the basic contention concerning the single domain character of the films is quite wrong the conclusion that the films have something akin to a kind of mosaic structure seems inevitable. However, a conventional multi-domain structure will have the utmost difficulty in accounting for the variation of magnetoresistance with field. Moreover, since the films are undoubtedly amorphous and can therefore have no overall magnetic anisotropy, even in a multi-domain model it would be necessary to invoke some inhomogeneity in order to account for a coercive force which by multi-domain standards is unusually high. The suggestion that structurally disordered films might have a granular structure is not new. It forms the basis of a structural model put forward by Leung and Wright [4] to account for the low packing fraction of amorphous cobalt films compared with theoretical models which correctly predict the short-range structure factor. They suggested that the films consist of fairly dense clusters of diameter approximately 30 Å separated by thin regions of somewhat lower density. The figure of 30 Å was put forward only because the films are known to form by a process of nucleation and growth and become electrically continuous when the thickness reaches this value. The present analysis suggests that considerable coalescence of nucleated regions takes place. This process cannot go on indefinitely and stops when the films consist of grains approximately 200 Å in diameter separated by highly dislocated but nevertheless amorphous material.

References

[1] M.R. Bennett and J.G. Wright, Phys. Lett. 38A (1972) 419 and private communication.
[2] E.C. Stoner and E.P. Wohlfarth, Phil. Trans. Roy. Soc. (Lond) A240 (1948) 599.
[3] A.H. Morrish, The Physical Principles of Magnetism (Wiley, New York, 1965).
[4] P.K. Leung and J.G. Wright, Phil. Mag. 30 (1974) 185.

MAGNETIC PROPERTIES OF AMORPHOUS Gd–Al AND Gd–Cu

T. MIZOGUCHI*, T.R. McGUIRE, R.J. GAMBINO and S. KIRKPATRICK

IBM Thomas J. Watson Research Center, Yorktown Heights, New York 10598, USA

Sputtered amorphous films of $Gd_{0.41}Cu_{0.59}$ and those with higher concentration of Gd are found to be ferromagnetic with Curie temperatures 75 K and higher. In contrast $Gd_{0.37}Al_{0.63}$ is found to take a spin glass state with a transition temperature of 15.8 K as defined by a sharp asymmetric cusp in the static susceptibility when measured in low magnetic fields. Below 15.8 K relaxation and thermal hysteresis effects of the susceptibility are found, characteristic of a spin glass.

1. Introduction

Amorphous films, prepared by sputtering, with several compositions of Gd with Cu and one composition with Al, $Gd_{0.37}Al_{0.63}$, have been studied by static magnetic susceptibility techniques. The two amorphous alloys $Gd_{0.41}Cu_{0.59}$ and $Gd_{0.37}Al_{0.63}$ are not far in composition from the ordered crystalline intermetallic compounds $GdAl_2$ and $GdCu_2$, These crystalline compounds are respectively ferromagnetic ($GdAl_2: T_c = 170$ K) and antiferromagnetic ($GdCu_2: T_N = 41$ K). In the amorphous form, however, we find that $Gd_{0.41}Cu_{0.59}$ is ferromagnetic with $T_c \approx 75$ K (magnetic properties of amorphous Gd–Cu have also been reported by Heiman and Lee [1]) and $Gd_{0.37}Al_{0.63}$ appears to take a spin glass state below a transition temperature $T_{SG} \approx 15.8$ K. Thus, the magnetic properties of the amorphous samples differ markedly from the crystalline form. In addition the spin glass state which represents a metastable magnetic configuration of frozen-in disorder of the spin system adds another degree of disorder to a system which is already a disordered atomic array.

2. Experimental details and results

The samples were prepared by sputtering as described in detail by Cuomo and Gambino [2]. Static susceptibility measurements were made on film samples of about 3×10^{-4} cm³ by the Faraday method at high fields ($H > 1000$ Oe) and using a "Superconducting Technology SQUID Susceptometer" [3] for measurements at low-fields ($H < 100$ Oe). The SQUID instrument is sensitive to changes in magnetic moment of $\pm 10^{-7}$ emu and this high sensitivity enables us to study the spin glass state in low fields.

We show first the results for $Gd_{0.37}Al_{0.63}$ in fig. 1 where the volume magnetic susceptibility (χ) vs. temperature is plotted. At low fields (0.12, and 9.0 Oe) a sharp asymmetric cusp occurs at $T_{SG} = 15.8$ K which we define as the spin glass transition. When the field is raised the cusp disappears and a broad maximum is found as illustrated for 3000 Oe applied field in fig. 1. At still higher fields (6000 Oe) the maximum in χ is replaced by a level plateau. Below T_{SG} in low magnetic fields thermal hysteresis relaxation effects which are non-exponential are observed.

Fig. 2 shows the θ value from the Curie–Weiss law as a function of at. % Cu for the ferromagnetic Gd–Cu system. The value for $4\pi M_s$ (Gauss) measured in an applied field of 18 kOe parallel to the plane of the film is also shown. With decreasing Gd concentration the ferromagnetic transition temperature is decreasing. The lowest concentration Gd sample, $Gd_{0.29}Cu_{0.71}$, does not show ferromagnetic saturation at 4.2 K.

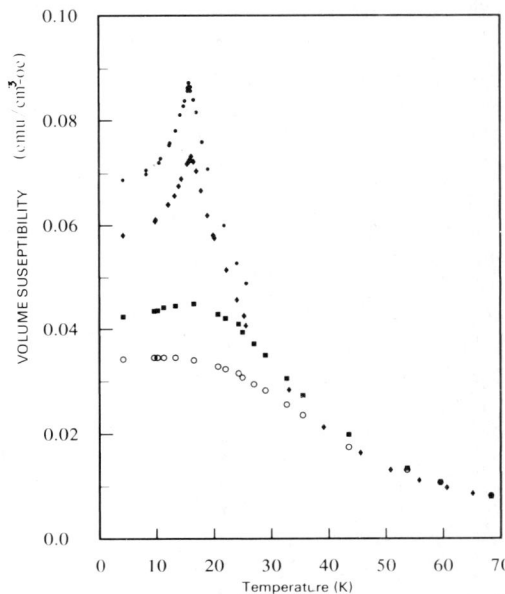

Fig. 1. Magnetic susceptibility of $Gd_{0.37}Al_{0.63}$ vs. temperature in fields of 0.12(●), 9.0(♦), 3000(■) and 6000(○) Oe.

*Permanent address: Gakushuin University, Mejiro, Tokyo, Japan.

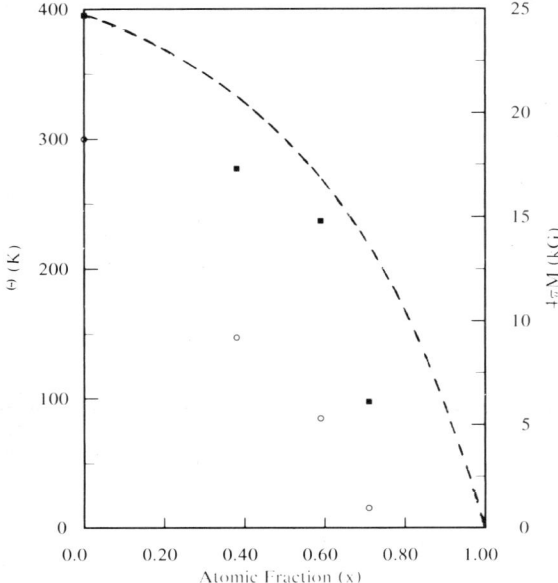

Fig. 2. Magnetization, $4\pi M$ (■) at 18 kOe and 4.2 K and paramagnetic θ values (0) for three compositions of $Gd_{1-x}Cu_x$. Dashed curve is calculated, see text.

3. Discussion

The magnetic properties of $Gd_{0.37}Al_{0.63}$, including the susceptibility cusp at 15.8 K, are characteristic of spin glasses. Some of these effects [4, 5] have been reported for crystalline spin glass materials such as dilute Fe in Au, and Mn in Cu. However, the measurements reported here on $Gd_{0.37}Al_{0.63}$ represent the first time a cusp in χ at low fields has been seen using static magnetic measurements. It must be emphasized that $Gd_{0.37}Al_{0.63}$ represents a new class of spin glass materials because it is not dilute and it is amorphous.

The behavior of the susceptibility which goes from a flat plateau in χ at high fields to a cusp at low fields is consistent with the Edward–Anderson [6] picture of the spin glass transition as one into a random, but rigid state caused by a mixed exchange field of positive and negative sign. It is possible to analyze [7] the shape of the cusp using the Sherrington and Kirkpatrick [8] theory. Using this theory we extract a spin glass order parameter $q(T) \propto (T_{SG} - T)^\beta$ where $0.8 < \beta < 0.9$ fits the data.

$Gd_{1-x}Cu_x$ alloys for $x = 0.38$ and 0.59 show ferromagnetic saturation at 4.2 K with the values given in fig. 2. The measured values of $4\pi M_s$ are lower than those calculated (dotted line in fig. 2) using 7 μ_B for Gd and an average density based on crystalline densities. For $Gd_{0.29}Cu_{0.71}$, the reduction of magnetization may be due to occasional antiferromagnetic interactions which signal the nearby spin glass phase. The susceptibility at 1 kOe shows a broad bump around 8 K similar to that for $Gd_{0.37}Al_{0.63}$. The observed Curie constant is about 20% larger than the calculated value.

The $Gd_{1-x}Cu_x$ data (fig. 2) shows a decrease in θ faster than simple dilution would predict going to $\theta \approx 0$ K at $x \approx 0.8$. Since θ is a measure of the average exchange interaction, negative exchange becomes more important with increasing Cu concentration. This behavior is consistent with RKKY exchange where interatomic distance and conduction electron concentration are both important parameters in determining the strength and sign of the exchange interaction and both can vary in Gd–Cu. The decrease of θ in Gd–Cu with decrease in Gd concentration is similar to that found by Hauser [9] for Gd–Ag.

Observations of spin glass phenomena in our $Gd_{0.37}Al_{0.63}$ sample but not in the $Gd_{0.41}Cu_{0.59}$ does not imply any fundamental difference between the two systems. On the scale of RKKY oscillations $(2k_F)^{-1}$ the Gd atoms in $Gd_{0.37}Al_{0.63}$ may be much further apart than in the $Gd_{0.41}Cu_{0.59}$. Further study of Cu rich Gd–Cu and Gd rich Gd–Al is necessary before valid comparison can be made.

The authors wish to thank D.S. Yu and P.A. Maurer for help with film preparation; F. Cardone and S.O. Ellmann for concentration analysis, and H.R. Lilienthal for magnetic measurements.

References

[1] N. Heiman and K. Lee, Joint MMM-Intermag Conf. Proc. 34 (1976) 319.
[2] J.J. Cuomo and R.J. Gambino, J. Vac. Sci. Tech. 12 (1975) 79.
[3] W.L. Goodmann, V.W. Hesterman, L.H. Rorden and W.S.Goree, Proc. IEEE 61 (1973) 20.
[4] J.A. Mydosh, AIP Conf. Proc. 24 (1974) 131.
[5] J.L. Tholence, thesis, Grenoble (1973); J.L. Tholence and R. Tournier, J. de Physique 35 (1974) C4-229.
[6] S.F. Edwards and P.W. Anderson, J. Phys. F5 (1975) 965.
[7] T. Mizoguchi, T.R. McGuire, S. Kirkpatrick and R.J. Gambino, Phys. Rev. Lett., submitted Aug. 1976.
[8] D. Sherrington and S. Kirkpatrick, Phys. Rev. Lett. 35 (1975) 1792.
[9] J.J. Hauser, Phys. Rev. B 12 (1975) 5160.

EXCITATIONS OF AMORPHOUS FERROMAGNETS

RAYMUND C. JONES and GARY J. YATES

Department of Mathematical Physics, University of Birmingham, Birmingham B15 2TT, England

We present two model calculations of the excitations of an amorphous ferromagnet; in both, the magnetic moments are carried by hard sphere atoms with spin $\frac{1}{2}$. A Monte Carlo calculation is first used to set up an amorphous liquid-like hard sphere structure for which the density of spin wave states and the magnetization are calculated. The second calculation compares these results with those given by an analytical treatment which uses the ideas of liquid theory.

1. Introduction

X-ray scattering data on alloys such as Ni–P and Co–P shows that the structure of these systems is well represented by dense randomly packed hard spheres with both magnetic and non magnetic components having almost identical hard sphere radii [1, 2]. The packing fraction ξ for these systems is in the region 0.56 to 0.63, and about 80% of the atoms are magnetic. The low temperature magnetization varies with temperature, T, as $T^{3/2}$, and the corresponding states curve is always flattened. Such a low temperature behaviour of the magnetization indicates the importance of spin waves in the system.

Consider a set of localized spins $\{S_i\}$ which are at points $\{r_i\}$. We assume that they interact via a Heisenberg hamiltonian. $H = -\Sigma_{ij} J_{ij} S_i \cdot S_j$; for convenience we have chosen $J_{ij} \equiv J(r) \equiv J_0 r^2 e^{-\beta r^4}$. The spins have quantum number one-half. The Spin Green's function $G_{ij} \equiv -i\theta(t)\langle[S_i^+(t), S_j^-(0)]\rangle$ has a linearized equation of motion $(\omega\delta_{ik} - \Phi_{ik})G_{kj}(\omega) = 2\langle S\rangle \delta_{ij}$, where $\langle S \rangle$ is the thermodynamically averaged value of a spin, and the dynamical matrix Φ is given by $\Phi_{ij} = 2\langle S\rangle[\delta_{ij}\Sigma_k J_{ik} - J_{ij}]$. The frequencies of the elementary excitations are given by the solution of $\det \|\mathbf{I}\omega - \mathbf{\Phi}\| = 0$.

2. Monte Carlo studies

We employ the standard computer simulation method described by Metropolis et al. [3], using 128 hard spheres at fixed ξ and cyclic boundary conditions in order to set up a random configuration of hard spheres.

The positions of all the particles thus are now known; using the assumed form of $J(r)$, the dynamical matrix can be calculated for any configuration and hence the set of eigenfrequencies $\omega_1, \omega_2, \ldots$ obtained. From these a histogram of the density of spin wave states is obtained, and the fluctuation dissipation theorem in the form, $(2\langle S\rangle)^{-1} = \Sigma_i \coth(\omega_i/2k_B T)$, is used to obtain self-consistently the thermodynamically averaged spin per site. The whole process was then repeated in order to generate a set of results for each of about 40 different configurations and in this way an averaged density of states histogram and an averaged magnetisation curve (and hence corresponding states plot) were obtained for a given value of ξ. The procedure was repeated for several different values of ξ varying from 0.05 to 0.5. A typical histogram is shown in fig. 1 for $\xi = 0.2$.

We may regard decreasing ξ as a means of increasing the disorder in the system: it was observed that decreasing ξ increased the band width, with a marked high energy tail developing for very small ξ. This may be explained by noting that the peak of the $J(r)$ curve lay in a region near to the closest approach distance of the atoms: thus in systems with small ξ and large disorder there will be a region with closely spaced atoms and a high energy tail is obtained in the density of states. The corresponding states plot was also obtained for several

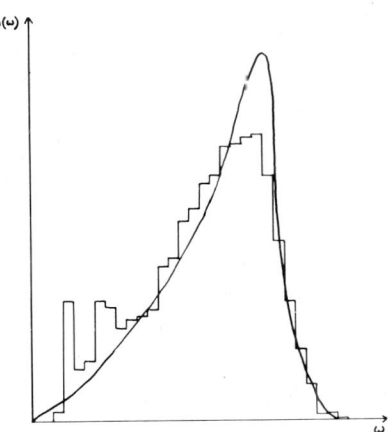

Fig. 1. Averaged density of states for $\xi = 0.2$ (35 configurations in the Monte-Carlo calculation).

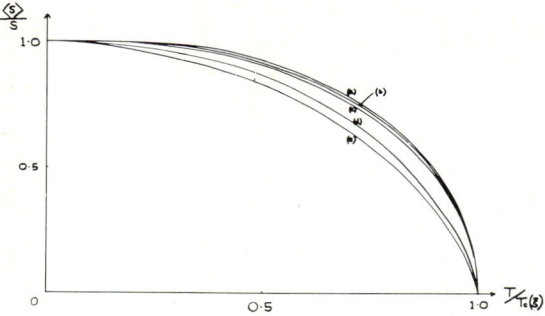

Fig. 2. Corresponding states curves: (a) Crystalline (b.c.c.); (b) $\xi = 0.4$; (c) $\xi = 0.2$ (Monte-Carlo calculations); (d) $\xi = 0.4$; (e) $\xi = 0.2$ (analytical calculations).

different ξ. Two such curves are shown in fig. 2. In all cases the curves were markedly flattened compared with that obtaining for a crystal (also shown): the flattening increased with decreasing ξ with a reduction below the crystalline curve of about 7% for $\xi = 0.4$ and $T/T_c = 0.5$. It should be noted, however, that the low temperature behaviour is exponential rather than following a power law behaviour, but this must be expected when discrete low frequency eigenfrequencies are used.

3. Analytical method

We study the same hard sphere system of N spins in a volume V but use methods based on those used in site or bond disordered problems [4]. The method is briefly outlined below. We wish to calculate the averaged propagator $\langle G(\mathbf{k}, \omega) \rangle$ and use a perturbation expansion about some suitably averaged reference system described by a propagator

$$G^0(\mathbf{k}, \omega) = 2\langle S \rangle \left(\omega - 2\langle S \rangle \frac{N}{V} \int J(r) g(r) (1 - e^{-i\mathbf{k} \cdot \mathbf{r}}) \, d^3r \right)^{-1}$$

and average term by term; $g(r)$ is the pair correlation function for a system of hard spheres. The Kirkwood superposition approximation was used and the assumption is then made that $h(r) [\equiv g(r) - 1]$ is small everywhere. The result is that

$$\langle G(\mathbf{k}, \omega) \rangle = 2\langle S \rangle \left(\omega - \omega_k^0 + 2\langle S \rangle \frac{N}{V} \right.$$
$$\left. \times \int \frac{J(r) h(r) (1 - e^{-i\mathbf{k} \cdot \mathbf{r}}) \, d^3r}{1 + 1/[2\Gamma(r, \omega)]} \right)^{-1}, \quad (1)$$

where

$$\omega_k^0 = 2\langle S \rangle \frac{N}{V} \int J(r) g(r) (1 - e^{-i\mathbf{k} \cdot \mathbf{r}}) \, d^3r$$

and

$$\Gamma(r, \omega) \equiv N^{-1} \sum_j G^0(j, \omega) J(r) (1 - e^{-i\mathbf{j} \cdot \mathbf{r}}),$$

ω has the usual small imaginary part.

The excitation energies of the system (and their associated damping) are obtained from the poles of $\langle G(\mathbf{k}, \omega) \rangle$, and the density of states and magnetisation obtained from the usual relations.

We use the $g(r)$ obtained as a solution of the Percus-Yevick integral equation for hard spheres, for which an analytical closed form exists [5]. In fig. 1 the continuous line shows the density of states for the same packing fraction ξ as used earlier. In the high energy regions, the peak and upper band tails show remarkably close agreement with the Monte Carlo calculations. However, the analytical treatment although giving a parabolic band shape at low frequencies appears to under-estimate the number of low energy modes. A possible explanation is the artificial depletion of states of lowest energy due to the minimum wavelength constraint of the "box" in which the Monte-Carlo simulations are performed. The corresponding states plots obtained from the analytical calculation are shown in fig. 2 and the curves are flatter than those produced by the Monte-Carlo simulations; however, the low temperature behaviour definitely follows a $T^{3/2}$ law characteristic of spin waves and since $\langle S \rangle$ is governed by the low frequency behaviour of the system [inaccurately described by a Monte-Carlo calculation of $n(\omega)$]; this result should not be too surprising. Further calculations, not given here, show that the inverse spin wave lifetimes become comparable with their energies at wavelengths of the order of the average interparticle spacing. This is a physically appealing result.

References

[1] G.S. Cargill, III, J. Appl. Phys. 41 (1970) 2249.
[2] G.S. Cargill, III and R.W. Cochrane, Amorphous Magnetism, H.O. Hooper and A.M. de Graaf (eds.), (Plenum, New York-London, 1973) p. 313.
[3] N. Metropolis, A.W. Rosenbluth, M.N. Rosenbluth and A.H. Teller, J. Chem. Phys. 2 (1953) 1087.
[4] R.C. Jones and G.J. Yates, J. Phys. C. 8 (1975) 1705.
[5] M.S. Wertheim, Phys. Rev. Lett. 10 (1963) 321.

LOW FIELD MAGNETIC SUSCEPTIBILITY OF AMORPHOUS $Co_xPd_{80-x}Si_{20}$ AND $Fe_xPd_{80-x}Si_{20}$ ALLOYS IN THE VICINITY OF THE MAGNETIC TRANSITION TEMPERATURE

A. ZENTKOVÁ
Department of Theoretical Physics and Geophysics, P.J. Šafárik University Košice, Czechoslovakia

P. DUHAJ
Institute of Physics, Slavak Academy of Sciences, Bratislava, Czechoslovakia

A. ZENTKO and T. TIMA
Institute of Experimental Physics, Slovak Academy of Sciences, Košice, Czechoslovakia

The magnetic susceptibility of the amorphous $Co_xPd_{80-x}Si_{20}$ and $Fe_xPd_{80-x}Si_{20}$ alloys near the critical temperature was measured in external magnetic fields up to 350 Oe. It was shown that the observed dependences $\chi(T)$ can be explained using the model of a heterogenous amorphous ferromagnet.

It was shown in a previous paper [1] that a sharp peak appears in the temperature dependence of the low field magnetic susceptibility $\chi(T)$ of amorphous $Co_xPd_{80-x}Si_{20}$ and $Fe_xPd_{80-x}Si_{20}$ alloys. The occurrence of such a maximum was connected with the loss of short range ordering of the amorphous basal matrix. On the other hand, the apparent ferromagnetic behaviour was observed only in the case of samples containing clusters of a composition rich in transition elements [2]. Based on this observation the idea was proposed that the ferromagnetism of these alloys is bound to the presence of such clusters. To verify these assumptions, we investigated the influence of a magnetic field on the dependence $\chi(T)$ in the vicinity of the critical temperature.

The amorphous alloys of the Co–Pd–Si and Fe–Pd–Si systems were prepared using the method of rapid rotating mills. The amorphous structure was verified by electron diffraction. The magnetic susceptibility was measured by an induction method using a Hartshorn-type bridge. In addition to the weak (~ 2 Oe) alternating field a constant magnetic field up to about 1 kOe also could be applied.

Typical experimental results for the low field $\chi(T)$ near the critical temperature for the $Co_xPd_{80-x}Si_{20}$ alloys with concentrations $x = 8$ and $x = 10$ at.% are shown in figs. 1 and 2. It is seen from these figures that already a relatively small constant magnetic field rounds and reduces the peak in $\chi(T)$. In the case of $Fe_xPd_{80-x}Si_{20}$ alloys in addition to rounding and reduction an apparent shift of the peak also occurs (figs. 3 and 4).

Qualitatively the same effect of the magnetic field on the dependence $\chi(T)$ in the region of critical temperature has been observed by Canella [3] in some spin glasses. We therefore assume that the occurrence of a peak in $\chi(T)$ is in our case also connected with the loss of short-range order in the amorphous matrix PdSiCo, which leads to the properties of a spin glass. The anomalous behaviour of $\chi(T)$ for $Co_{10}Pd_{70}Si_{20}$ for temperatures $T > T_0$ (the Curie law is not valid) may be connected with the fact that the samples with cobalt contents $x \geqslant 9$ at.% are ferromagnetic up to room temperatures. The total susceptibility of these samples is $\chi_{tot} = \chi_{sg} + \chi_{fer}$, where χ_{sg} is the susceptibility of the amorphous matrix and χ_{fer} is the susceptibility of the ferromagnetic clusters (by composition variation) emulgated in the matrix. (In samples with $x < 9$ no such composition clusters were observed, so that either they are absent or their linear dimensions are less than 20 Å.)

The situation is more complicated in the Fe–Pd–Si system. First, by the temperature T_0 [at which a peak in $\chi(T)$ appears for the investigated samples] is also the Curie temperature. Transition of the samples into the paramagnetic state occurs in the region $T > T_0$. Therefore it can be assumed that $\chi_{tot} \approx \chi_{fer}$ and that the strongly magnetic clusters play a decisive role in the magnetic properties of these alloys. By applying Landau's theory of the second order phase transition to a system of clusters it can be shown (see [4]) that the apparent transition temperature of such a system will decrease upon increasing the intensity of the external magnetic field. More detailed

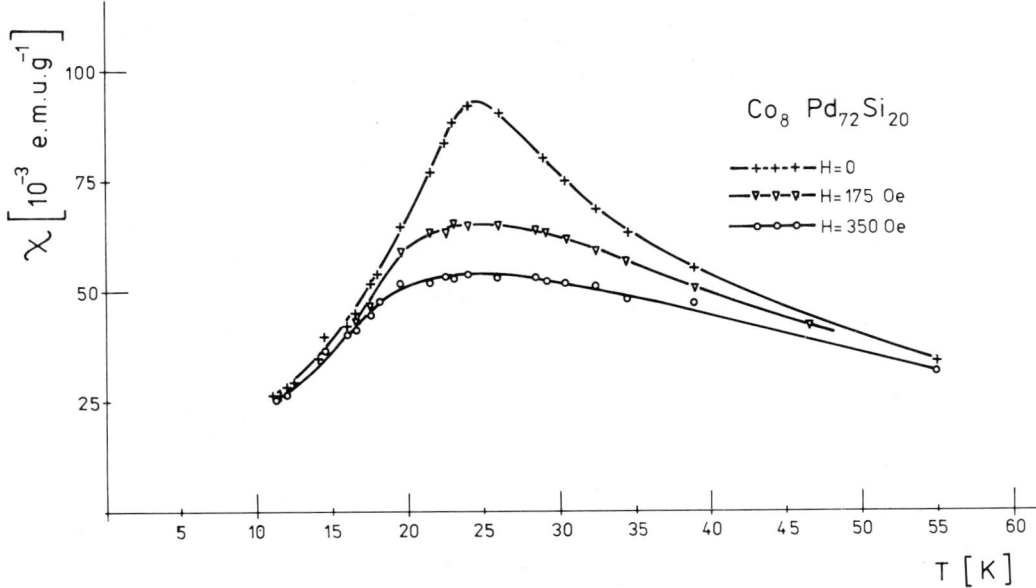

Fig. 1. The low field ac susceptibility of $Co_8Pd_{72}Si_{20}$ amorphous alloy.

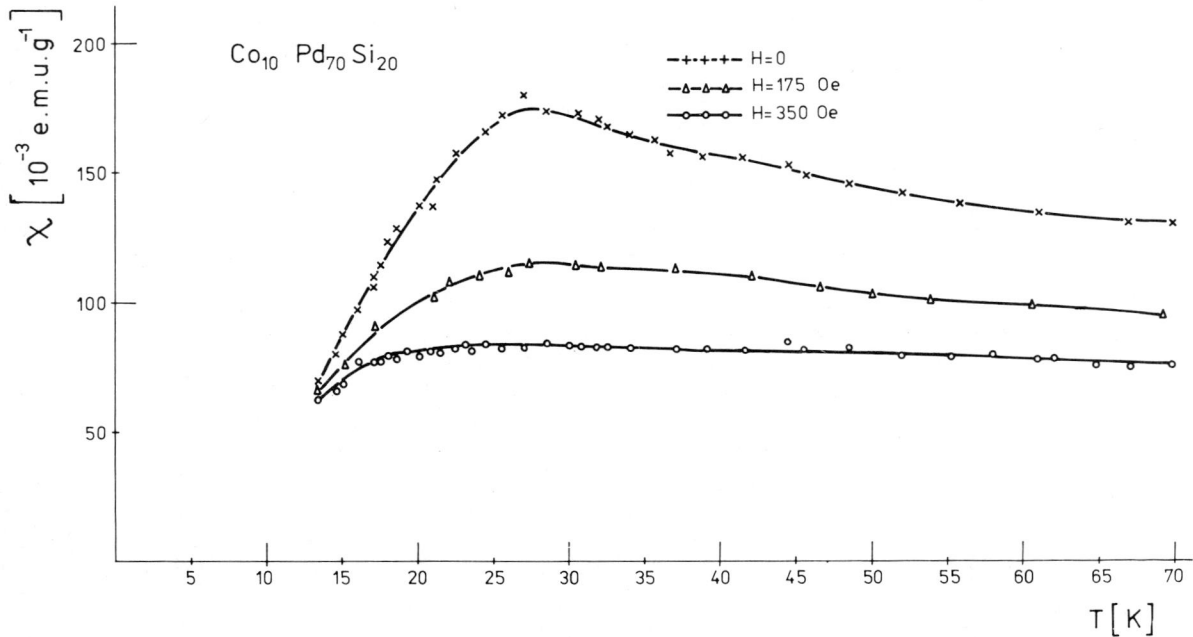

Fig. 2. The low field ac susceptibility of $Co_{10}Pd_{70}Si_{20}$ amorphous alloy.

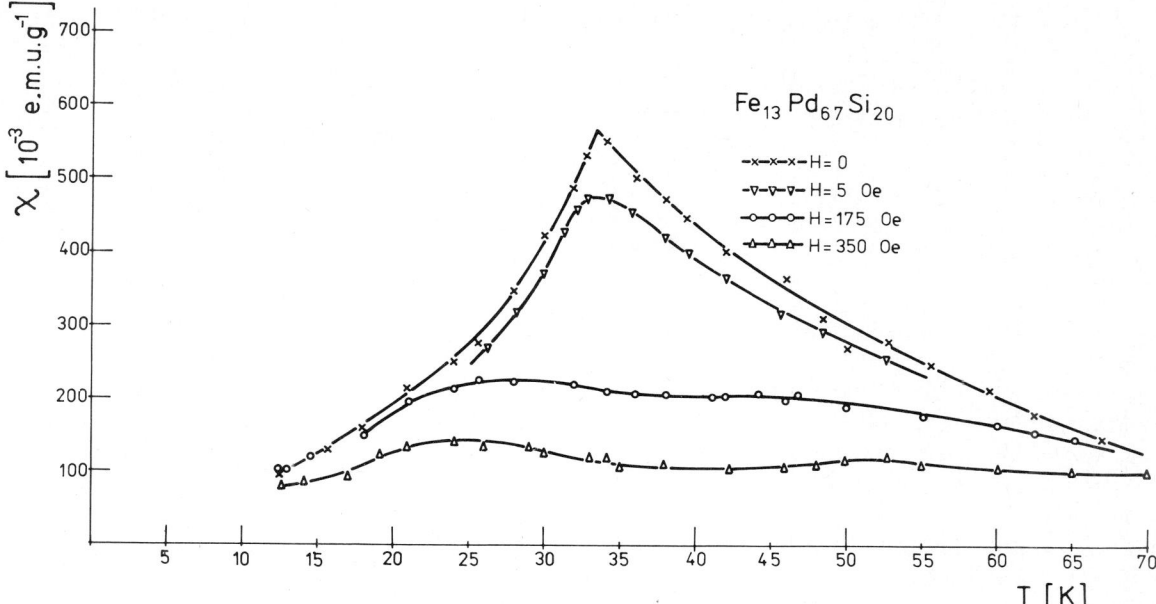

Fig. 3. $\chi(T)$ vs. T for $Fe_{13}Pd_{67}Si_{20}$ alloy.

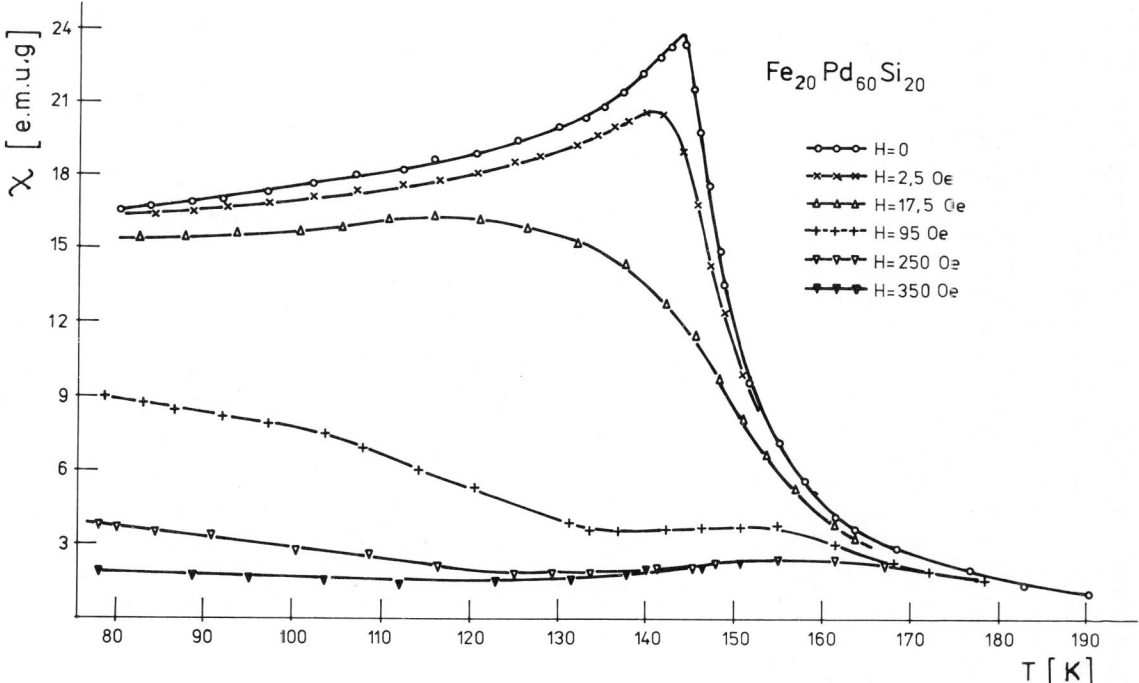

Fig. 4. $\chi(T)$ vs. T for $Fe_{20}Pd_{60}Si_{20}$ alloy.

theoretical considerations on this problem will be published elsewhere.

References

[1] A. Zentko et al., Phys. Status Solidi (a) 31 (1975) K41.
[2] A. Zentko et al., Phys. Status Solidi (a) 24 (1974) K99.
[3] V. Cannella, in: Amorphous Magnetism, H.O. Hooper and A.M. de Graaf (Eds.) (Plenum Press, New York–London, 1973) p. 195.
[4] J. Klamut, J. Sznajd, in: Proc. Int. Conf. Mag. 73, Vol. V (Nauka, Moscow, 1974) p. 300.

MAGNETIC PROPERTIES OF AMORPHOUS NiP ALLOYS

A. BERRADA, M.F. LAPIERRE, B. LOEGEL, P. PANISSOD, C. ROBERT and J. BEILLE*

Laboratoire de Structure Electronique des Solides, 4, rue Blaise Pascal, 67000 Strasbourg, France

An experimental study on amorphous electrodeposited $Ni_{1-c}P_c$ alloys was performed using measurements of magnetic properties, coercive force, NMR and resistivity at low temperatures. The results show that weak ferromagnetism takes place for $0.15 < c < 0.18$ and inhomogeneities appear for $0.18 < c < 0.25$. Furthermore, the anomalies in the transport properties are of structural and not a magnetic origin.

We studied the binary amorphous nickel based $Ni_{1-c}P_c$ system where the phosphorus concentration ranges between $c = 0.15$ and 0.25. The amorphous samples were electrodeposited on a copper cathode and their macroscopic homogeneity was checked by magnetic measurements. The actual phosphorus compositions were determined chemically and spectrophotometrically.

Our experimental investigations include low and high field (the latter were done at the SNCI at Grenoble) magnetization measurements, coercive force, NMR, resistivity and magnetoresistivity measurements. It appears from all these experimental results that the magnetism is not of the same nature below and above a "critical concentration" c_0 ($c_0 \simeq 18$ at.% P) which separates two concentration ranges A ($0.15 < c < c_0$) and B ($c_0 < c < 0.24$) whose characteristics will be discussed below.

In the range A, the saturation magnetization decreases rapidly with increasing phosphorus concentration [fig. 1(a)] and the Arrott plots are linear over large temperature and magnetic field ranges [fig. 2(a)]. These results are similar to those observed for weak homogeneous ferromagnets like NiPt [1]. The low field deviations in the Arrott plots suggest the existence of small local inhomogeneities. The results that we obtained by NMR measurements, show furthermore that there exist, in the range A, only one type of moment; we have no place to give in this paper the detailed results that we obtained (a more complete paper will be published). Briefly, the local magnetization on the phosphorus atom behaves similarly to the mean bulk magnetization [fig. 2(b)], consequently, the fluctuations of the moments on the nickel atoms are very small.

* Laboratoire de Magnetisme, CNRS, BP 166, Grenoble, France.

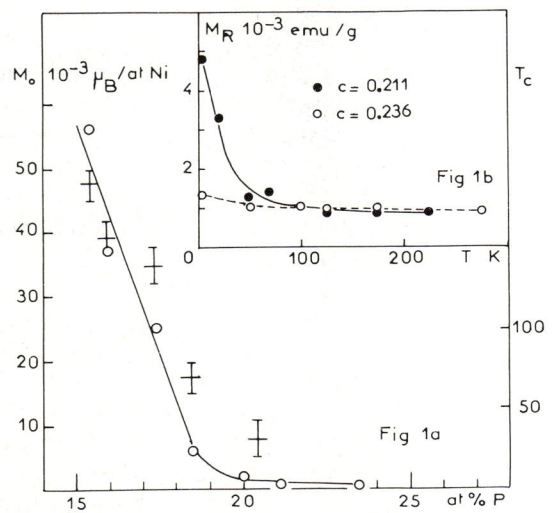

Fig. 1. (a) Saturation magnetization (○) and Curie temperatures (+) vs. concentration for amorphous $Ni_{1-c}P_c$ alloys. (b) Remanent magnetization vs. temperature for two $Ni_{1-c}P_c$ alloys in the B zone.

In the range B, the magnetization can be divided into two contributions: (i) one which varies linearly with magnetic field and is rather temperature independent, and (ii) one which saturates for fields ranging between 5 and 12 kG. Furthermore, the linewidths follow a Curie–Weiss behaviour (i.e. the phosphorus atoms are polarized by a very small magnetic environment induced by the indirect interaction with the neighbouring nickel magnetic clusters). These features lead to the conclusion that the magnetism which arises in the range B must be attributed to the occurrence of inhomogeneities whose size increase with increasing phosphorus content. The existence of these inhomogeneities is confirmed by the observation of low temperature hysteresis phenomena, which appear only in the B range: a remanent magnetization takes place varying with both temperature and phosphorus concentration [fig. 1(b)]. The coer-

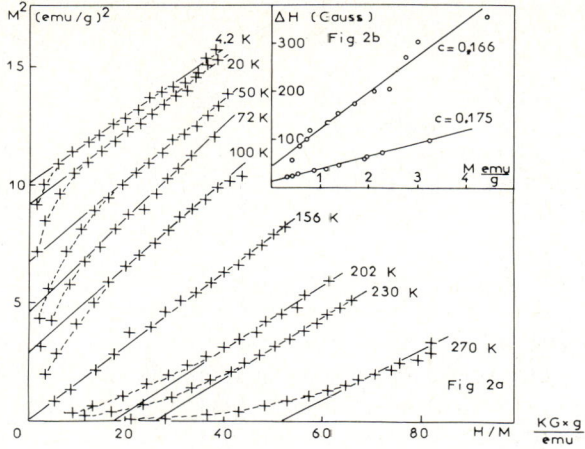

Fig. 2. (a) Arrott plots for $Ni_{0.840}P_{0.160}$. (b) Linewidth ΔH vs. magnetization for two $Ni_{1-c}P_c$ alloys.

cive field is also concentration dependent and exhibits a maximal value of 125 G at 4.2 K for $c = 0.236$.

The origin of these inhomogeneities may be either intrinsic (as proposed in a compact polytetrahedric model or PTC model [2]) or extrinsic (precipitation). They prevent the observation of the appearance of ferromagnetism ("giant moments") as was observed in crystallized iron or nickel based alloys ([3] and references therein).

The transport properties show a similar behaviour in the A and B range around the concentration c_0: in both cases a minimum appears in the temperature dependent part of the resistivity $\rho(T) - \rho(0)$. In fact, $\rho(T) - \rho(0)$ is the sum of two contributions: in the whole temperature range studied and for all samples a good fit is obtained with the phenomenological law $\rho(T) - \rho(0) = AT^2 - K \log(T/\theta)$. The various constants resulting from such an analysis are given in table I. It appears that neither the quadratic nor the logarithmic terms show much dependence on phosphorus concentration. Thus c_0 does not correspond to some critical concentration for the appearance of a new magnetic state, as in crystallized nickel or iron based alloys, where the appearance of ferromagnetism corresponds to a sharp maximum for the coefficient A [3]. In the case of $Ni_{1-c}P_c$ alloys we obtain, to the contrary, the lowest value for the coefficient A at $c = c_0$. It must furthermore be pointed out that a magnetic field has only little effect ($H < 70$ kG), on both the quadratic and – in a

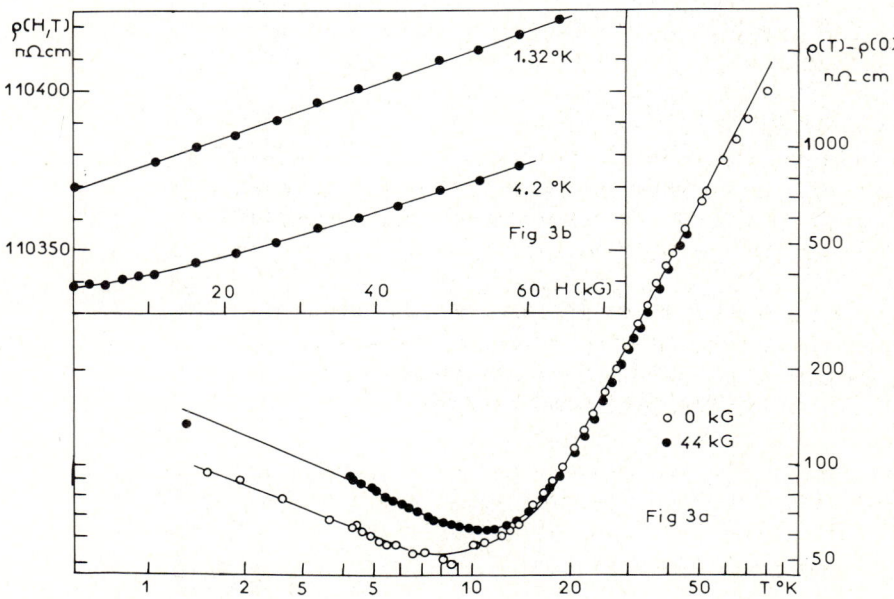

Fig. 3. (a) Temperature dependent part of the resistivity $\rho(T) - \rho(0)$ for $c = 0.201$ in zero magnetic field and for $H = 44$ kG. The solid lines correspond to a fit with $\rho(T) - \rho(0) = AT^2 - K \log(T/\theta)$ (see table I for numerical values). (b) Low temperature resistivity for the same alloys vs. magnetic field.

Table I
Numerical values of the constants, resulting from the analysis of the temperature dependent part of the resistivity, using the law $\rho(T) - \rho(0) = AT^2 - K \log(T/\theta)$. The number in parentheses are given for an applied field of 44 kG

c at.% P	15.50	16.58	17.50	18.16	20.10
$\rho(0)$ ($\mu\Omega$ cm)	88.96 (88.96)	81.73	106.68	105.11	110.28 (110.27)
A (nΩ cm (°K)$^{-2}$)	0.309 (0.309)	0.346	0.180	0.239	0.265 (0.259)
θ (K)	20.3 (21.4)	21.5	16.6	20.0	20.9 (18.7)
K (nΩ cm)	−42.5 (−57.8)	−43.1	−29.6	−38.9	−36.6 (−57.5)

first approximation – the logarithmic terms (fig. 3 and table I) as was already pointed out [4]. These facts – together with the fact that c_0 corresponds roughly to the eutectic composition – show that, as for the magnetic properties, the transport properties are dominated by non-magnetic contributions; thus we conclude that the anomalies observed in the transport properties have mainly a structural origin and not a magnetic one. Similar results for the transport properties were obtained for amorphous FeP alloys [5].

In conclusion, the amorphous binary $Ni_{1-c}P_c$ system remains magnetic in the whole concentration range studied $(0.15 < c < 0.25)$, which can be separated into two regions around the concentration $c_0 \simeq 0.18$ (corresponding roughly to the eutectic composition). For $0.15 < c < c_0$ the alloys are weak homogeneous ferromagnets, while for $c_0 < c < 0.25$ the magnetism arises from chemical inhomogeneities whose size increase with increasing phosphorus concentration. A first report on this work is to be published in Solid State Communications.

References

[1] H.L. Alberts, J. Beille, D. Bloch and E.P. Wohlfarth, Phys. Rev. B, 9 (1975) 2233.
[2] J.F. Sadoc, Thesis, Paris (1976).
[3] A. Amamou, F. Gautier and B. Loegel, J. Phys. F, 5 (1975) 1342.
B. Loegel, J. Phys. F, 5 (1975) 497.
Amamou et al., J. Phys. F, 6 (1976) to be published.
[4] R.W. Cochran, R. Harris, J.O. Strom-Olson and M.J. Zuckermann, Phys. Rev. Lett. 35 (1975) 676.
[5] J. Logan and M. Yung, J. Non-Cryst. Solids, to be published.

SPIN WAVE SPECTRUM OF AMORPHOUS FERROMAGNETS

M.A. CONTINENTINO* and N. RIVIER
Department of Physics, Imperial College, London SW7, England

The dynamic susceptibility of an amorphous ferromagnet is obtained by reducing the spin dynamics to a local frequency modulation. The dispersion law of the spin wave frequencies is the Fourier transform of the radial distribution function times the exchange integral, and represents an invertible relation between magnetic properties and structural information.

Amorphous materials, some of them ferromagnets, like Fe, Co and the alloys FeP and CoP can nowadays be prepared and studied experimentally. However, it has not been possible theoretically to define an equivalence relation between the structures of different amorphous materials, and thus to state on precise grounds whether solid A is more or less amorphous than B. Research on amorphous magnetism has certainly been motivated by the hope that magnetic properties might provide a measure of the degree of amorphousness or at least some direct information on the structure of the material.

We aim to provide a relationship between the readily measurable magnetic properties (like the spin wave spectrum and its width) and the more elusive structural information (pair and triplet correlation functions). This attitude is similar to that of Kaneyoshi [1] who related average magnetization and spin stiffness to the pair correlation function in a mean field approximation. The simplicity of the magnetic elementary excitations – the spin waves being, as is well known, derivable equally well by classical as by quantum mechanics – gives us a handy tool for the theoretical investigation of amorphous matter.

The amorphous magnet is described by the conventional Heisenberg hamiltonian

$$\mathcal{H} = -\tfrac{1}{2}\Sigma J(\mathbf{R}_{ij})\mathbf{S}_i \cdot \mathbf{S}_j, \qquad (1)$$

where the positions $\mathbf{R}_i$ of the spins are random variables. $\mathbf{R}_{ij} = \mathbf{R}_i - \mathbf{R}_j$ and $J(0) = 0$. The linearized equations of motion are obtained directly for the ferromagnet, and yield a formal solution for the dynamic transverse susceptibility,

$$\chi_{ij}^{+-}(t) = c[\exp(-i\hat{\Omega}t)]_{ij}. \qquad (2)$$

* Supported by a grant from the CNPq of Brazil.

This solution is most clearly written in terms of the Fourier transform. We have $(t \geq 0)$

$$\chi_{kk'}^{+-}(t) = c\{\langle k|\exp(-it\Sigma\Omega_{k_1 k_2}|k_1\rangle\langle k_2|)|k'\rangle + (\rho(\mathbf{k}-\mathbf{k}') - \delta_{kk'})\} \qquad (3)$$

where

$$\Omega_{k_1 k_2} = (S/N)\Sigma_i\, e^{-i(k_1-k_2)\cdot R_i}\Sigma_j J_{ij}(1 - e^{ik_2\cdot(R_j-R_i)})$$

and

$$\rho(\mathbf{k}-\mathbf{k}') = N^{-1}\Sigma_i \exp[-i(\mathbf{k}-\mathbf{k}')\mathbf{R}_i].$$

The ensemble averaged susceptibility will be diagonal in this representation which does therefore impose itself naturally at this stage. The susceptibility in eq. (3) obeys the standard equation of motion [1, 2].

However, $\langle \chi_{kk'}^{+-}(t) \rangle_{AV}$ can be evaluated directly. For this purpose, the evolution operator in eq. (3) can be written in the interaction representation

$$\exp(-i\hat{\Omega}t) = \exp(-i\hat{\Omega}_0 t)T \\ \times \left\{\exp - i\int_0^t d\tau[\hat{\Omega}(\tau) - \hat{\Omega}_0(\tau)]\right\}, \qquad (4)$$

with $\hat{\Omega}_0 = \Sigma\langle\Omega_k\rangle_{AV}|k\rangle\langle k|$. The average of eq. (4) will be done considering the simplest random sequence of operators which includes correlation, namely a gaussian distribution. Since a gaussian average corresponds to taking terms up to second order in the cumulant expansion [3], it yields

$$\langle\exp(-i\hat{\Omega}t)\rangle_{AV} = \exp(-i\hat{\Omega}_0 t)T \\ \times \left\{\exp\left[-\tfrac{1}{2}\int_0^t d\tau_1 \int_0^t d\tau_2 \langle T(\hat{\Omega}(\tau_1) - \hat{\Omega}_0(\tau_1))\right.\right. \\ \left.\left. \times (\hat{\Omega}(\tau_2) - \hat{\Omega}_0(\tau_2)))_c\right]\right\},$$

where the cumulant average $\langle\ \rangle_c$, an operator still, can be expressed in terms of the moments as

$$\langle T[\hat{\Omega}(\tau_1) - \hat{\Omega}_0(\tau_1)][\hat{\Omega}(\tau_2) - \hat{\Omega}_0(\tau_2)]\rangle_c$$
$$= \langle T[\hat{\Omega}(\tau_1) - \hat{\Omega}_0(\tau_1)][\hat{\Omega}(\tau_2) - \hat{\Omega}_0(\tau_2)]\rangle_{AV}$$
$$- T\langle\hat{\Omega}(\tau_1) - \hat{\Omega}_0(\tau_1)\rangle_{AV}\langle\hat{\Omega}(\tau_2) - \hat{\Omega}_0(\tau_2)\rangle_{AV}$$
$$= \langle T[\hat{\Omega}(\tau_1) - \hat{\Omega}_0(\tau_1)][\hat{\Omega}(\tau_2) - \hat{\Omega}_0(\tau_2)]\rangle_{AV},$$

where in the last step we have used the definition of $\hat{\Omega}_0$ and $\langle\hat{\Omega}(\tau_1)\rangle_{AV} = \hat{\Omega}_0(\tau_1)$.

If we ignore all time dependence, explicit and implicit (in the time ordering), of the operators, a procedure often called "static approximation", we obtain

$$\langle\chi_k^{+-}(t)\rangle_{AV} = c \exp[-i\Omega_0(k)t - \tfrac{1}{2}t^2\Delta^2(k)], \tag{6}$$

a gaussian propagator and a gaussian spectral density. But the damping rate $\Delta(k) \propto |k|$ in the long wavelength limit and the spin waves are overdamped in this approximation, an unphysical result. Clearly, the fluctuations represented by the cumulants are time-dependent. For long time intervals $|\tau_1 - \tau_2| > \tau_c$ the fluctuations are uncorrelated and the cumulant vanishes. The correlation time τ_c depends on the particular cumulant matrix element, but it certainly is shortest on the hydrodynamic limit ($k \to 0$). If $t \gg \tau_c$ the probability of different fluctuations overlapping is small, the time ordering operator in front of the exponential in eq. (5) becomes redundant, the damping is exponential and the spectral density lorentzian [4]. Thus

$$\langle\chi_k^{+-}(t)\rangle_{AV} = C \exp[-i\Omega_0(k)t - |t|\Delta'(k)], \tag{7}$$

where

$$\Delta'(k) = \tau_c\Delta^2(k)$$
$$= \tau_c[\Sigma_{k'}\langle\Omega_{kk'}\Omega_{k'k}\rangle_{AV} - \langle\Omega_k\rangle_{AV}^2] \propto \tau_c(k)k^2.$$

The spin waves are therefore well defined at long wavelengths for sufficiently fast modulations $J\tau_c \ll 1$. The energy $\Omega_0(k)$ of the spin waves are calculated in the "liquid model" as

$$\Omega_0(k) = (N_0 - 1)\langle J(\mathbf{R})S(1 - \cos\mathbf{k}\cdot\mathbf{R})\rangle$$
$$= 4\pi\rho_0 S\int dR\, R^2 g_2(R)J(R)$$
$$\times [1 - \sin kR/(kR)], \tag{8}$$

$\Omega_0(k)$ involves the radial distribution function $g_2(r)$ of the magnetic atoms (density ρ_0) in the amorphous magnet. We have therefore in eq. (8) the expected relation between magnetic properties and structural information: the spin wave dispersion relation is the Fourier transform of the product of the radial distribution function and the exchange integral. Since $\sin\chi/\chi = j_1(\chi)$ is a spherical Bessel function, eq. (8) is a Bessel transform which can be inverted.

$$g_2(R) = [2\pi^2\rho_0 SJ(R)]^{-1}\int dk\, k^2(\Omega_0(\infty)$$
$$- \Omega_0(k))\sin kR/(kR), \tag{9}$$

$\Omega_0(k)$ increases initially as Dk^2 to reach with oscillations the mean field limit $\Omega_0(\infty) = 4\pi\rho_0 S\int dR\, R^2 g_2(R)J(R)$. $\Omega_0(\infty) = cZ_0 JS$ for a short ranged interaction averaging to J where Z_0 is the coordination number of the amorphous structure ($Z_0 \cong 14$ for a random close packed structure) and c is the concentration of magnetic atoms.

In an insulator, or, less obviously, an amorphous metal with a very short electronic mean free path, the range of the exchange interaction can be limited to nearest neighbours. In this case, the spin wave frequencies are proportional to the Fourier transform of the RDF cut off beyond the first peak. Hence a very good approximation imposes itself: $4\pi\rho_0 R^2 g_2(R)J(R) \cong cZ_0 J\delta(R - a)$ and

$$\Omega_0(k) = cZ_0 JS[1 - \sin ka/(ka)].$$

The spin wave has stiffness $D = cZ_0 JSa^2/6$ and there is a dip $[\Omega_0(\text{dip}) = 0.88\Omega_0(\infty)]$ in the dispersion relation at $k \cong \tfrac{1}{2}5\pi/a$ where the structure factor (Fourier transform of g_2) has itself a dip [5]. The dip is important (when compared to the frequency expected by extrapolation of the low k energies (Dk^2 dip/Ω^0 dip $\cong 10$) and may be the "low lying mode" observed by Mook et al. [6], in amorphous Co_4P although these authors relate it to a peak – rather than a dip – in the structure factor by analogy to the roton dip in He. (The information made available to us at present is insufficient to decide the matter.) The spin wave dispersion law Dk^2 leads to a $T^{3/2}$ decrease of the magnetization at low temperatures (regardless of the damping as long as $\Delta'\alpha k^{2+\epsilon}$ with $\epsilon \geq 0$) as observed [7].

Similarly, the damping rate $\Delta'(k)$ involves the triplet correlation function $g_3(\mathbf{0}, \mathbf{R}_1, \mathbf{R}_2)$. This is as high in the hierarchy of correlation functions as one needs to describe completely the amorphous magnet, since a gaussian distribution correlates its elements $\Omega_{k_1 k_2}$ pairwise only.

References

[1] T. Kaneyoshi, J. Phys. C, 5 (1972) L107.
    T. Kaneyoshi, J. Phys. C, 6 (1973) 3130.
    T. Kaneyoshi and R. Honmura, J. Phys. C, 5 (1972) L65.
[2] S.F. Edwards and R.C. Jones, J. Phys. C, 4 (1974) 2109.
[3] R. Kubo, J. Phys. Soc. of Jap. 17 (1962) 1100.
[4] R. Kubo, in: Fluctuation, Relaxation and Resonance in Magnetic Systems, D. ter Haar (Ed.) (Oliver and Boyd, 1968).
[5] M.W. Johnson, N.H. March, D.I. Page, M. Parinello and M.P. Tosi, J. Phys. C, 8 (1975) 751.
[6] H.A. Mook, N. Wakabayashi and D. Pan, Phys. Rev. Lett. 34 (1975) 1029.
[7] R.W. Cochrane and G.S. Cargill III, Phys. Rev. Lett. 39 (1974) 476.

HALL EFFECT IN METGLAS 2826A AND 2826-DEPENDENCE ON TEMPERATURE AND PRESSURE

R. MALMHÄLL, K.V. RAO*, G. BÄCKSTRÖM
University of Umeå, Umeå, Sweden

and

S.M. BHAGAT†
University of Uppsala, Uppsala, Sweden

The Hall resistivity, ρ_H, of Metglas 2826A is measured for $0.6 \leq T/T_c \leq 1.9$. Below $T_c (\sim 255\text{ K})$, the extraordinary Hall coefficient R_1 is constant at $+4.24 \times 10^{-8}\text{ m}^3/\text{As}$. The magnetic susceptibility, derived from our data, exhibits a Curie–Weiss behaviour above 350 K and varies as $(T - T_c)^{-1.75}$ for $(T - T_c) < 90\text{ K}$. In 2826 at 300 K ρ_H drops linearly with increasing pressure.

Recently we reported measurements on the field dependence of the Hall resistivity, ρ_H, in Metglas 2826 ($Fe_{40}Ni_{40}P_{14}B_6$) and 2826B ($Fe_{29}Ni_{49}P_{14}B_6Si_2$) at 300 K [1]. These materials are ferromagnetic at 300 K ($T_c \geq 400\text{ K}$) and behave rather similarly to that of a crystalline Permalloy. Metglas 2826A ($Fe_{32}Ni_{36}Cr_{14}P_{12}B_6$), also an amorphous material prepared in a similar way, has a T_c which is conveniently below 300 K. In this note we describe measurements on ρ_H, and the electrical resistivity, ρ, in 2826A for $150\text{ K} < T < 470\text{ K}$, i.e. from temperatures well below to well above $T_c (\sim 255\text{ K})$. We find that the initial slope, $R_H = (\partial \rho_H/\partial H)_{H \to 0}$, is roughly a constant below T_c. This then drops rapidly with increasing temperatures as for a crystalline ferromagnet. From these data we calculate and study the behaviour of the magnetic susceptibility, χ, above T_c.

As before, ρ_H was measured by the double-frequency method [1]. In studying the variation with temperature, T, and the applied magnetic field, H_a, ρ_H was measured by keeping T constant and varying H_a up to 1 T and also by keeping H_a constant and letting T drift at a rate of ~ 0.5 K/min. No significant differences were found.

The dependence of ρ_H on H_a and the magnetization, M, is given by

$$\rho_H = R_0(H_a - NM) + R_1 M. \qquad (1)$$

* Permanent address: Clarkson College of Technology, Potsdam, N.Y. 13676, USA.
† Permanent address: University of Maryland, College Park, Md, USA.

From our low field data for $T < 240\text{ K}$ we obtain [2] $R_1 = +4.24 \times 10^{-8}\text{ m}^3/\text{As}$. With $\rho = 1.72\ \mu\Omega\text{m}$ this gives $R_1/\rho^2 = 1.4 \times 10^4$, in good agreement with the data of ref. 1. Since even at high fields ($H_a > 0.6$ T) the ρ_H vs. H_a graph shows no saturation we can only put an upper limit of $1 \times 10^{-9}\text{ m}^3/\text{As}$ for the ordinary Hall coefficient R_0.

The temperature dependence of ρ_H is best represented by the quantity R_H which is shown as a function of T in fig. 1. For $T < 250\text{ K}$ R_H is nearly independent of T. The sharp drop in R_H as T is increased above 260 K is similar in character to that found in crystalline ferromagnets above T_c. That is, immediately above T_c the material is strongly paramagnetic and the value

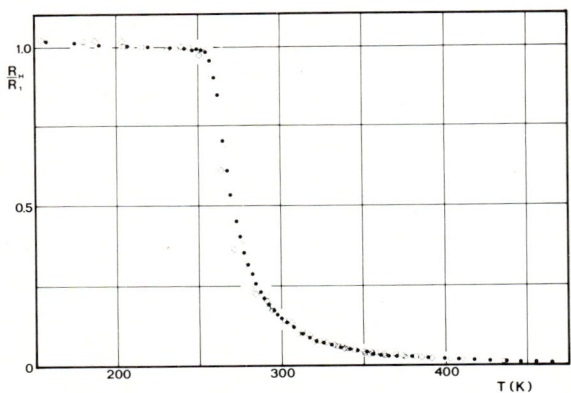

Fig. 1. The initial slope R_H of the Hall resistivity vs. applied magnetic field, normalized with $R_1 = +4.24 \times 10^{-8}\text{ m}^3/\text{As}$, measured as a function of the temperature, T, for Metglas 2826A. The symbols refer to measurements on different samples from the same source.

of R_H is still mainly determined by the magnetization. At high enough temperatures R_H should tend towards R_0. However, we find $R_H = +4.4 \times 10^{-10}$ m^3/As at 465 K, and it is still decreasing with increasing T.

It should be noted that heating the sample to about 450 K has small irreversible effects on ρ_H. However, this only has the effect of shifting the whole curve in fig. 1 to higher T as if T_c had increased by a few degrees [3]. For a sample annealed at 650 K for about 1 h in vacuum ($\sim 10^{-4}$ torr) the R_H vs. T curve changes completely. Instead of fig. 1, we find that R_H is increasing linearly for the same temperature range giving $dR_H/dT \simeq 2 \times 10^{-11}$ m^3/As. Also, ρ is found to increase linearly with T ($d\rho/dT \simeq 3 \times 10^{-4}$ $\mu\Omega$m/K). The room temperature values of R_H and ρ are 2.28×10^{-8} m^3/As and 1.58 $\mu\Omega$m, respectively. Now, R_1/ρ^2 is no more a constant as was seen for the amorphous state.

For the paramagnetic state, since $M = \chi H_a$, we can write eq. (1) as

$$R_H = R_0 + (R_1 - NR_0)\chi(1 + \chi)^{-1}. \quad (2)$$

Assuming constant values for R_1 and R_0 we use eq. (2) to evaluate χ as a function of T. Fig. 2 shows a plot of $1/\chi$ using $N = 1$, $R_1 = +4.24 \times 10^{-8}$ m^3/As and $R_0 = +3 \times 10^{-10}$ m^3/As. This asymptotic value of R_0 is well below the smallest R_H measured by us but rather close to that reported by Fischer et al. [4]. However, it should be noted that for $T < 300$ K, $1/\chi$ is insensitive to the choice of R_0. On the other hand, we cannot expect to get meaningful values for $1/\chi$ for $T > 400$ K as R_H becomes rather close to R_0. Our data thus evaluated agree well with those obtained from low-field studies using a SQUID magnetometer [5] which is also included in fig. 2. It is convenient to discuss the T dependence of χ in two regimes: (i) for $345 < T < 400$ K we observe a Curie–Weiss behaviour, $\chi = C(T - \theta)^{-1}$, with $C = 1.22$ and $\theta = 325$ K. If we treat Metglas as $T_{0.8}G_{0.2}$ (T = transition metal, and G = glass former) we can calculate the paramagnetic moment per "formula" unit to be about 2.3 μ_B. This is surprisingly high. (ii) for $260 < T < 350$ K, χ is found to fit the expression $\chi = \beta(T - T_c)^{-\gamma}$. We plot in fig. 3 log $(1/\chi)$ vs. log $(T - T_c)$ for three choices of T_c. With $T_c = 255$ K giving the best "fit" over the entire range of temperature, we obtain $\gamma = 1.75$ for $(T - T_c) < 90$ K. It is interesting to note that

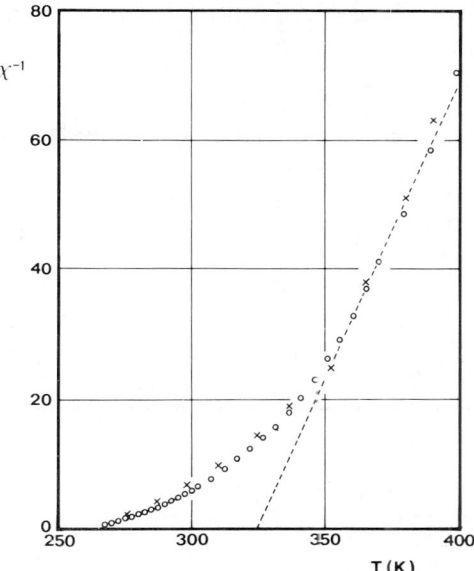

Fig. 2. The inverse magnetic susceptibility $1/\chi$ as a function of temperature [○ = our data using eq. (2); × = data from ref. (5)]. The broken line is a fit to $\chi = C/(T - \theta)$.

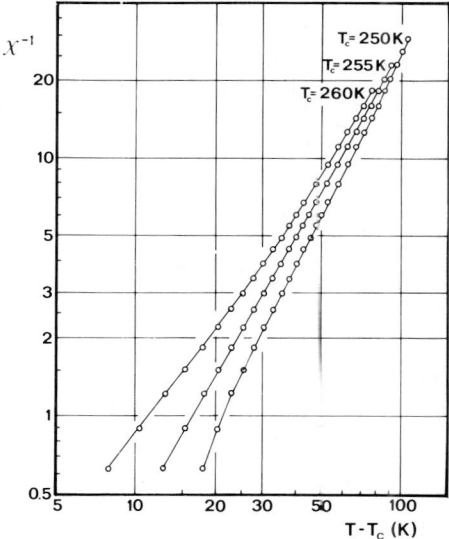

Fig. 3. The inverse magnetic susceptibility, $1/\chi$, as a function of $(T - T_c)$ for three choices of the Curie temperature T_c plotted on a log–log scale.

the data of ref. 5 give $\gamma = 1.65$ and $T_c = 254.5$ K. This value for γ is rather different from any scaling law prediction and in sharp contrast to the behaviour of crystalline soft ferromagnets.

Preliminary measurements of the pressure dependence of ρ, and ρ_H in Metglas 2826 at

300 K indicate that ρ_H drops linearly with increasing pressure. This drop by about 20% up to 10 kbar is significantly larger than that observed in a crystalline ferromagnet [6]. On the other hand ρ, which is almost temperature independent, is found to be invariant with pressure. Further studies on these systems will be reported elsewhere.

S.M.B. thanks Prof. O. Beckman and his colleagues at Uppsala for their kind hospitality during his Sabbatical. This research is supported by the Swedish Scientific Research Council.

References

[1] K.V. Rao, R. Malmhäll, G. Bäckström and S.M. Bhagat, Solid State Commun. 18 (1976) 193.

[2] The present values of R_1 are significantly different from those given in the abstract at this conference. Although we have failed to discover any systematic errors in our earlier data (apart from accidental heating of sample during soldering) we believe the present data to be more reliable as they have been reproduced on several samples.

[3] H.J. Leamy, E.M. Gyorgy, R.C. Sherwood, T. Wokiyama and H.S. Chen, Proc. 21st Magn. & Magn. Mater. Conf. Phil. Pa, Dec. (1975).
H.S. Chen, R.C. Sherwood, H.J. Leamy and E.M. Gyorgy, Proc. 22nd Magn. & Magn. Mater. Conf. Pittsburgh, Pa (1976).

[4] M. Fischer, H.J. Güntherrodt, E. Hauser, H.U. Künzi, M. Liard and R. Müller, Phys. Lett. 55A (1976) 423.

[5] E. Figueroa, L. Lundgren, O. Beckman and S.M. Bhagat, to be published.

[6] T. Hiraoka, J. Sci. Hiroshima Univ., Ser. A-II, 32 (2) (1968) 153.

MAGNETIC PROPERTIES OF AMORPHOUS AND LIQUID Pd–Si ALLOYS

M. MÜLLER, R. MÜLLER, M. LIARD, H.U. KÜNZI, H.-J. GÜNTHERODT
Institut fur Physik, Universitat, Basel, Switzerland

J.D. RILEY, L. LEY, G. GÜNTHERODT
Max-Planck-Institut für Festkörperforschung, Stuttgart, Germany

and

C.C. TSUEI
IBM Research Center, Yorktown Heights, New York, USA

The magnetic susceptibility, the electrical resistivity and the Hall coefficient of $Pd_{81}Si_{19}$ and $Pd_{77.5}Cu_6Si_{16.5}$ in the glassy, crystalline and liquid state have been measured. The small magnetic susceptibility in the glassy and liquid state has been explained by the reduction of the density of states at the Fermi energy as observed in XPS measurements.

In the last few years metallic glasses obtained by rapid quenching of the melt have become technologically as well as scientifically interesting [1]. In view of the amorphous structure of the metallic glasses a comparison with the liquid state is thought to provide a better understanding for the electronic and magnetic properties of glassy metals. We have tried to compare the glassy, crystalline and liquid state of the binary $Pd_{81}Si_{19}$ alloy and of the ternary $Pd_{77.5}Cu_6Si_{16.5}$ alloy. Both alloys are remarkable in the difference of glass forming ability. For the binary alloy a cooling rate of 10^6°C/s and for the ternary alloy only a cooling rate of 10^2°C/s is required.

The binary alloy was rapidly quenched by a splat cooling technique. The ternary alloy was rapidly quenched into water. In view of the limited length of the paper we cannot present the results in detail and therefore we will only discuss the results.

The electrical resistivity of glassy $Pd_{81}Si_{19}$ is slightly larger than for $Pd_{77.5}Cu_6Si_{16.5}$ and corresponds to a value of 80 $\mu\Omega$cm. Upon crystallization, the room temperature resistivity decreases by a factor of three and becomes more temperature dependent. At the melting point the electrical resistivity changes in both alloys to a value of 95 $\mu\Omega$cm. The temperature coefficient of the electrical resistivity is slightly larger than in the glassy state. These liquid state resistivity data follow by extrapolating the resistivity curve of the glassy state. The Hall coefficients are nearly temperature independent. The Hall coefficient of the ternary alloy shows at room temperature a value of -10.7×10^{-11} m³/As which is a larger negative value than observed for the binary alloy. The latter corresponds to -9.6×10^{-11} m³/As. In the liquid state the Hall coefficients of both alloys are smaller by about 15%. This difference does not seem to be significant and it might be difficult to draw final conclusions in the lack of density data. The observed Hall coefficients in the liquid state are comparable to similar liquid transition metal alloys (e.g. Ni–Ge [2]) and indicate that Si, as was observed for liquid Ge, provides four and Pd less than one conduction electron.

The magnetic susceptibilities of the liquid binary and ternary alloys are very small and paramagnetic at higher temperatures, but tend to diamagnetic values at lower temperatures. The magnetic susceptibility of $Pd_{77.5}Cu_6Si_{16.5}$ [3] in the glassy state is diamagnetic corresponding very closely to the susceptibility of liquid noble metals [2] such as Ag, Au and Cu. With increasing temperature the susceptibility becomes more paramagnetic. This analogy suggests that also the susceptibilities in the glassy and liquid state show some correspondence.

The strong paramagnetism of pure Pd is drastically reduced in the glassy and liquid state by alloying with Si. This small susceptibility can be attributed to a smaller density of states at the Fermi energy E_F in the alloys compared to pure Pd.

X-ray and UV photoemission data have been published [4] for $Pd_{77.5}Cu_6Si_{16.5}$, which are in agreement with the smaller density of states. In fact the valence band looks very like a density of states in Cu. In this work X-ray photoemission (XPS) measurements of glassy $Pd_{81}Si_{19}$ have been performed. Preliminary results from XPS measurements of the valence band show a reduction of the density of states at the Fermi energy and what appears to be a narrowing of the Pd d-band width compared to polycrystalline Pd. The reduction in the density of states at the Fermi energy is further supported by the absence of asymmetry of the Pd core levels in glassy $Pd_{81}Si_{19}$.

The main conclusions of our measurements are that the glassy state appears as a continuous extension of the liquid state and that the reported experimental results for the binary and ternary alloy do not differ very much from each other.

The first observation greatly helps to get a better understanding for both the liquid and glassy state. The strong similarities between these two states might justify an extension of the theories developed for liquid transition metals to amorphous metals. On the other hand a number of physical properties which are difficult to be investigated in the liquid state can now be studied in glassy materials to get reasonable suggestions for the situation in the liquid state. As an example measurements of the magnetic susceptibility in liquid transition metal alloys have indicated some models for the density of states in the liquid state [2]. But due to experimental difficulties of photoemission works in liquid transition metals such experiments have never been carried out. However, on glassy materials photoemission work is relatively easy and results of such measurements can be used to test the suggested models for the density of states in the liquid state.

References

[1] J.J. Gilman, Phys. Today 28 (1975) 46.
[2] G. Busch and H.-J. Güntherodt, Solid State Phys. Vol. 29 (Academic Press, New York, 1974) p. 235.
[3] B.G. Bagley and F.J. DiSalvo, in: Amorphous Magnetism, H.O. Hooper and A.M. de Graaf (Eds) (Plenum Press, New York, 1973) p. 143.
[4] S.R. Nagel, B.G. Fischer, J. Tauc and B.G. Bagley, Phys. Rev. B13 (1976) 3284.

DIFFERENCE BETWEEN THE MAGNETIC PROPERTIES OF AMORPHOUS AND CRYSTALLINE Co–B–Si

H. WATANABE, T. MASUMOTO, M. KAMEDA, N. KAZAMA and H. YAMAUCHI

The Research Institute for Iron, Steel and Other Metals, Tohoku University, Sendai, Japan

The temperature dependence of the magnetization and the magnetization curves of amorphous and crystalline Co–B–Si have been measured at low temperatures. It has been found that the difference in the magnetization at 0 K and the Curie temperature of amorphous and crystalline states is not so much as that between the demagnetization characteristics arising from the excitation of spin waves.

1. Introduction

In the studies of amorphous ferromagnetic alloys of 3d-transition metal with elements of Group IIIA, IVA or VA, containing approximately 20 at.% of the latter, it has so far not been clarified to what extent the magnetic properties are affected by the structural disorder. Amorphous materials usually contain glass former atoms such as P, Si, C and B and if we want to get rid of the effect of these atoms on the magnetic properties of the pure metal, it is desirable to have a single phase crystalline state after the crystallization process of an amorphous material. Co–B–Si presents an example of such a system.

In the present study, we try to distinguish the effect of structural disorder from the effect of glass former atoms by comparing the magnetic properties of amorphous $Co_{75}B_{10}Si_{15}$, in which the subscripts denote the atomic concentration of the elements, with those of h.c.p. crystalline one.

2. Experimental results

The amorphous ferromagnet $Co_{75}B_{10}Si_{15}$ has been prepared by rapid quenching from the liquid state. The h.c.p. crystalline single-phase has been prepared by annealing at 350°C for about 28 days starting from amorphous material. We have confirmed the structure of amorphous and h.c.p. $Co_{75}B_{10}Si_{15}$ by means of X-ray diffraction measurements. The diffraction pattern of the amorphous material consists of a few halo with a small shoulder on the right-hand side of the second peak; such a diffraction pattern provides evidence for the presence of an amorphous structure [1]. The diffraction pattern of the h.c.p. single phase gives for the lattice constants $a = 2.52$ Å and $c = 4.08$ Å, while Co, B, Si atoms are distributed on h.c.p. sites at random.

The temperature dependence of the magnetization of amorphous and h.c.p. $Co_{75}B_{10}Si_{15}$ have been measured over the range from 4.2 to 1050 K in an applied field of 5.7 kOe. These results are shown in fig. 1. Amorphous and h.c.p. Co–B–Si have values of magnetization at 0 K and the Curie temperature not much different from each other, those of the latter being slightly lower (see table I).

The field dependence of magnetization of amorphous and h.c.p. $Co_{75}B_{10}Si_{15}$ has been measured in the range 2–8 kOe at 4.2 and 77 K. These results (fig. 2) show a larger magnetic anisotropy in h.c.p. $Co_{75}B_{10}Si_{15}$ than in amorphous material.

At low temperatures the demagnetization arises from the excitation of long wavelength spin waves. On the basis of spin wave theory such a demagnetization is described dominantly by a $T^{3/2}$ temperature dependence as follows:

$$-\Delta M(T) = M(0) - M(T) = M(0)BT^{3/2}, \quad (1)$$

with

$$B = \frac{g\mu_B}{M(0)} \left(\frac{k_B}{4\pi D}\right)^{3/2} F_{3/2}(T)$$

and

$$F_{3/2}(T) = \sum_{n=1}^{\infty} n^{-3/2} \cdot \exp-(nT_g/T) \quad (2)$$

where B shows the $T^{3/2}$ dependence, $F_{3/2}(T)$ is the Bose–Einstein integral function, g the g-factor, μ_B the Bohr magneton, k_B the Boltzmann constant, D the spin-wave dispersion coefficient and $T_g = (g\mu_B/k_B)(H + H_A)$ contains the anisotropy field H_A and applied magnetic field H. The values of D calculated from the slope of fig. 3

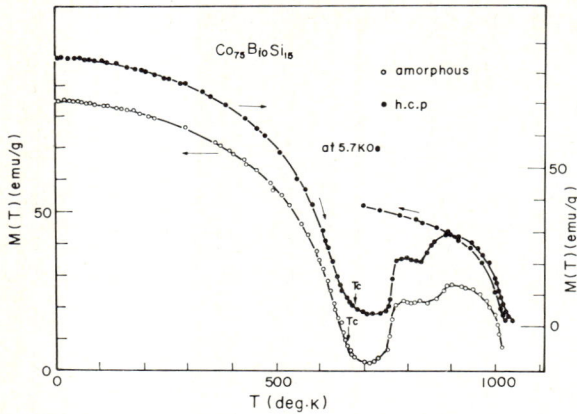

Fig. 1. Variation of magnetization with temperature. ○ amorphous $Co_{75}B_{10}Si_{15}$, and ● h.c.p. crystalline $Co_{75}B_{10}Si_{15}$.

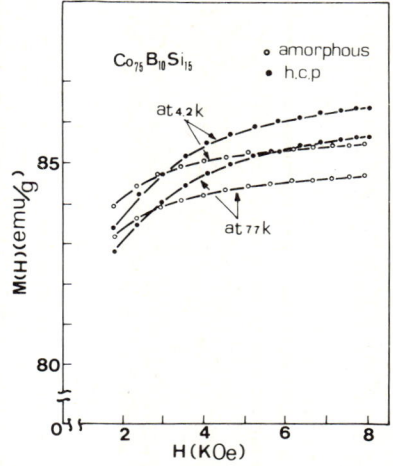

Fig. 2. Magnetization curves of amorphous and h.c.p. crystalline $Co_{75}B_{10}Si_{15}$ at 4.2 and 77 K.

using the same density of 7.34 g/cm³ for h.c.p. and for amorphous material are listed in table I.

3. Discussion

In studies of magnetic properties of amorphous ferromagnets it is difficult to distinguish the effects due to structural disorder from that of the glass former atoms such as Si, B and C.

The differences in magnetic properties of amorphous $Co_{75}B_{10}Si_{15}$, crystalline $Co_{75}B_{10}Si_{15}$ and pure Co are shown in table I. The difference between the 1st and 2nd rows gives the effect of structural disorder and the difference between 2nd and 3rd rows the effect of introducing the glass former atoms Si and B. By comparison, it can be deduced that structural disorder has little

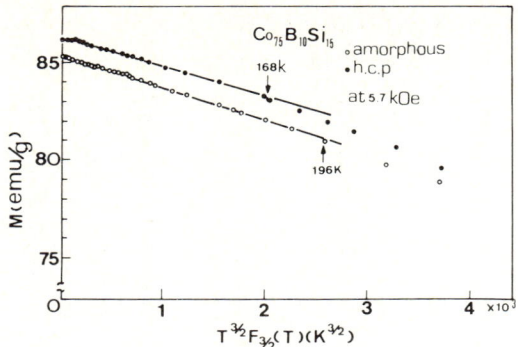

Fig. 3. Magnetization plotted against $T^{3/2}F_{3/2}(T)$ at applied magnetic field 5.7 kOe. ○ amorphous $Co_{75}B_{10}Si_{15}$ and ● h.c.p. crystalline $Co_{75}B_{10}Si_{15}$.

Table I
Values of $M_s(0)$, T_c, B and D for amorphous $Co_{75}B_{10}Si_{15}$, h.c.p. crystalline $Co_{75}B_{10}Si_{15}$ and pure Co

	$M_s(0)(\mu_B/Co)$	$T_c(K)$	$10^5 B(K^{-3/2})$	$D(meV Å^2)$
$Co_{75}B_{10}Si_{15}$ -amorphous	1.011 ± 0.005	660 ± 6	1.69 ± 0.03	198 ± 4
$Co_{75}B_{10}Si_{15}$ -h.c.p.	1.030 ± 0.020	675 ± 10	1.27 ± 0.02	235 ± 4
Co h.c.p.	1.70	1070	0.172	510 (T.A.)

T.A. = Triple axis spectrometry [3].

effect on the magnetization $M(0)$ and Curie temperature T_c but has an appreciable effect on B and D. Also, the presence of glass former atoms Si and B results in a significant decrease of $M(0)$ and T_c.

As far as $M(0)$ and T_c are concerned the present results agree with those reported by Schneider et al. [2] in the case of Fe–P–As. As shown in table I, the effect of structural disorder on the value of D is only about 14% of that of introducing the glass former atoms Si and B. It seems unlikely that this is an effect of density alone (we assumed the densities of crystalline and amorphous phases to be the same).

References

[1] Y. Waseda, Solid State Phys. 10 (1975) 17 (in Japanese).
[2] J. Schneider and H. Wiesner, Phys. Status Solidi (a) 29 (1975) 151.
[3] G. Shirane, V.J. Minkiewicz and R. Nathans, J. Appl. Phys. 39 (1968) 383.

BARKHAUSEN NOISE IN AMORPHOUS FERROMAGNETIC MATERIALS

F. FIORILLO, P. MAZZETTI, F. VINAI and G.P. SOARDO

Istituto Elettrotecnico Nazionale Galileo Ferraris, Gruppo Nazionale Struttura della Materia del CNR, UR 24, 10125 Torino, Italy

Preliminary results on the Barkhausen noise spectra in amorphous magnetic ribbons as a function of sample shape and of temperature provide information on the clustering of Bloch wall jumps in these materials. A relation is found between noise intensity and core losses, which also suggests some hints for a new approach to the problem of anomalous excess losses in magnetic materials.

1. Introduction

A well-known feature of magnetization processes in crystalline magnetic materials is the clustering of elementary jumps of domain walls into large Barkhausen groups [1]. A similar behaviour is present on amorphous magnetic ribbons, usually characterized by few large magnetization reversals, which give rise on quasi-static hysteresis loops to step-like structures. At increasing frequencies these steps are smoothed-out, because of the superposition of large magnetization jumps, and correspondingly the loop area grows rapidly [2].

In the present work the magnetization dynamics is investigated on amorphous samples by studying Barkhausen noise power spectra, taken as a function of sample shape, magnetizing frequency and temperature T. The results relative to $Fe_{40}Ni_{40}P_{14}B_6$ (Allied Chemical, Metglas 2826) show that the magnetization process is dominated by a strong coupling between elementary jumps of domain walls, which leads to large clusters, and that demagnetizing fields drastically affect both this coupling and the hysteresis loop behaviour. Measurements of the temperature dependence of B. noise on $Fe_{35}Ni_{32}Cr_{15}P_{12}B_6$ (Metglas 2826A), made up to the Curie point ($\sim -40°C$), show that the spectral density drops with increasing T more rapidly than the square of saturation magnetization $I^2(T)$. Core losses are also found to behave in a similar way, thus pointing to a relation with noise, which is briefly discussed.

2. Experimental

Barkhausen noise measurements were performed with a standard apparatus [3]. In fig. 1, curve 1 shows the power spectrum $\phi_{11}(\omega)$ obtained on a toroid of as-quenched Metglas 2826 (24 layers, 2 cm diameter). The triangular waveform exciting field had a magnetizing frequency

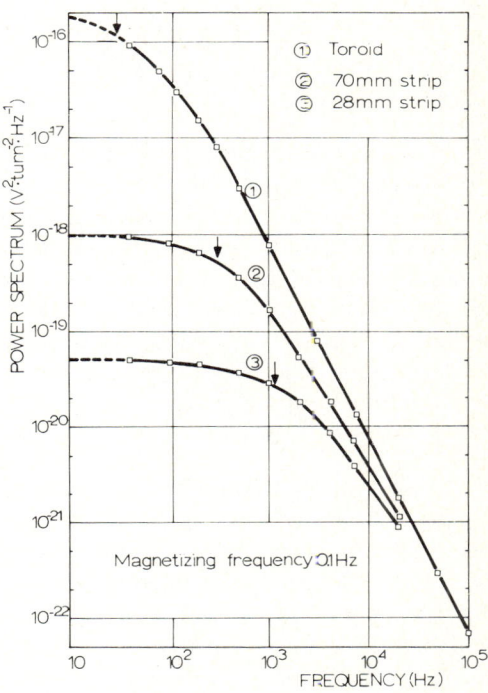

Fig. 1. Power spectra of Barkhausen noise (Metglas 2826). Arrows show cut-off frequencies.

$f_m = 0.1$ Hz, and a maximum amplitude sufficient to obtain apparent saturation. The analysis frequency f_a ranged from a lower limit at which continuous and line spectra mix-up, up to 0.1 MHz. It is found that on the whole f_a range the spectrum intensity is linearly dependent on f_m, up to $f_m = 1$ Hz. This proportionality behaviour does not usually hold in crystalline samples [3], and shows that in amorphous materials the large B. groups are better separated. This fact can actually be accounted for by the expression for the spectral density [3],

$$\phi_{11}(\omega) = \varphi(\omega)\left[1 + 2\frac{\rho(1-\nu\tau_0)^2}{\omega^2\tau_0^2\rho^2(1-\nu\tau_0)^2 + 1}\right], \quad (1)$$

which may also be used to obtain the average group duration from the experimental spectra. In eq. (1) $\varphi(\omega)$ is the power spectrum of the noise in the absence of correlation between elementary pulses; ν is the mean number of pulses per unit time; ρ is the mean number of pulses within a group; τ_0 is their average time separation. At low f_m, $\nu\tau_0 \ll 1$ [3], and eq. [1], from the measured cut-off frequency f_c, gives the mean duration of groups: in toroids $\rho\tau_0 = 1/(2\pi f_c) \sim 5 \times 10^{-3}$ s, in agreement with the value directly obtained from scope observations at extremely low f_m.

When the sample forms an open magnetic circuit, coupling between elementary pulses is strongly weakened by the building-up of a demagnetizing field: clusters become smaller, which leads, according to eq. [1], to a higher f_c and to a decreased spectral density in the low f_a range. This is shown by spectra 2 and 3 in fig. 1, taken at $f_m = 0.1$ Hz on strips respectively 70 and 28 mm long. The cut-off frequencies give a mean cluster duration $\rho\tau_0 \cong 5 \times 10^{-4}$ s for the 70 mm strip, which drops to $\sim 1.5 \times 10^{-4}$ s in the 28 mm one.

The results on the temperature dependence of the noise intensity in Metglas 2826A are presented in fig. 2, which also shows the behaviour with T of $I^2(T)$ and of the coercive field H_c. If the amplitude of each elementary pulse should simply change linearly with $I(T)$, one would expect the noise intensity to drop proportionally to $I^2(T)$. Experimentally one finds that the noise intensity falls with increasing T more rapidly than this law, which, from eq. (1), can be explained by an increase of τ_0. This increase of pulse duration can actually be expected, since H_c, which is proportional to the pressure acting on a Bloch wall during a B. pulse, decreases with increasing T. It should further be noted that the temperature dependence of the noise intensity is closely similar to the one of the static losses, which are proportional to $[I(T) \times H_c(T)]$ (curve 3, fig. 2).

3. Conclusions

The study of B. noise in amorphous magnetic ribbons is of particular interest since it permits some insights into the dynamical behaviour of Bloch walls during magnetization. The results show that these materials are characterized both by a high B. noise and by large anomalous

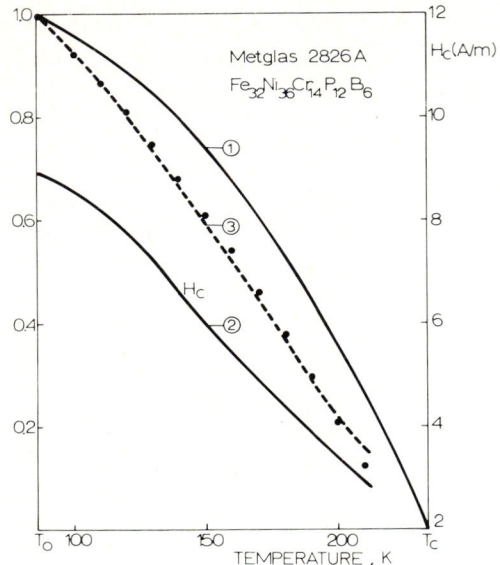

Fig. 2. Temperature dependence of B. noise intensity (black dots) compared to: 1, square saturation magnetization; 2, coercive field H_c; 3, static losses. Reduced units are used but for H_c.

dynamical losses, that is by large excess losses with respect to the ones predicted by a classical model, in which a uniform induction is assumed throughout the sample cross-section [4]. It seems plausible then to explain the loss anomaly in terms of correlation among Bloch wall movements, leading to a superposition of local eddy currents responsible for dissipative processes. Roughly speaking, one might suggest that the anomalous dynamical losses in these materials are generated by a change of the conventionally called static losses, due to a shortening of cluster duration when f_m is increased. This point seems to be supported by the fact that a reduction of the B. noise obtained by shortening the sample length (curve 3, fig. 1), is associated with a reduction of dynamical losses.

A general theory of anomalous losses in magnetic materials based on these concepts has actually been developed and will be published in another paper.

References

[1] H. Bittel, IEEE Trans. Magn. MAG-5 (1969) 359.
[2] T. Egami, P.J. Flanders and C.D. Graham Jr., Appl. Phys. Lett. 26 (1975) 128.
[3] P. Mazzetti and G. Montalenti, Proc. ICM-64 (Institute of Physics, London, 1965) p. 701.
[4] J.W. Shilling and G.L. Houze Jr., IEEE Trans. Magn. MAG-10 (1974) 195.

THE HIGH-FIELD SUSCEPTIBILITY AND THE MICROSTRUCTURE OF AMORPHOUS Ni–Fe–P–B ALLOYS

H. KRONMÜLLER and H. GRIMM

Max-Planck-Institut für Metallforschung, 7000 Stuttgart 80, W. Germany

The law of approach to saturation of quenched amorphous $Fe_{40}Ni_{40}P_{14}B_6$ alloys has been investigated as a function of the temperature and the magnetic field. These measurements were used for a study of the spin wave excitation spectrum and the microstructural properties of amorphous material.

In ferromagnetic materials deviations from ideal saturation are due to magnetocrystalline, magnetostrictive and dipolar interactions as well as thermally excited magnons. Accordingly the deviation ΔM_s from the average spontaneous magnetization for a given temperature T, and magnetic field H is composed of two terms

$$\Delta M_s(T, H) = \Delta M_1(T, H) + \Delta M_2(T, H), \quad (1)$$

where ΔM_1 is due to microstructural inhomogeneities [1] and ΔM_2 represents the effects of spin waves [2]. Because of the complex nature of the ΔM_i effects it is difficult to determine directly the theoretically interesting spin wave contribution $\Delta M_2(T, 0)$ for zero magnetic field. Therefore we have investigated the susceptibility, $\chi(T, H)$, in the vicinity of saturation. The measurements were performed with a commercial vibration magnetometer in the temperature and field ranges $4.2\,K \leq T < 300\,K$, and $0.05\,T < \mu_0 H < 4.5\,T$. It turned out that χ can be represented by the following series:

$$\chi(T, H) = \frac{A(T)}{H^2} + \frac{B(T)}{H} + \frac{C(T)}{\sqrt{H}}. \quad (2)$$

Here the first two terms are due to structural inhomogeneities and the third term results from thermally excited magnons. The A/H^2 term also has been found by Kazama and Kameda [3]. This contribution gives rise to an A/H term in $M_s(T, H)$ and was determined by plotting $M_s(T, H)$ versus $1/H$ as shown in fig. 1. For large values of $1/H$ a linear $M_s(T, H)$ vs. $1/H$ relation is obtained, and by extrapolation to $H \to 0$ $M_s(T, 0)$ can be determined. The behaviour of $\chi(T, H)$ at large magnetic fields was investigated by representing $\chi \cdot H = B(T) + C(T)\sqrt{H} + A(T)/H$ as a function of $\sqrt{H}$. At large magnetic fields, according to fig. 2, this leads to a linear relation showing that in this field range the term A/H can be neglected. The

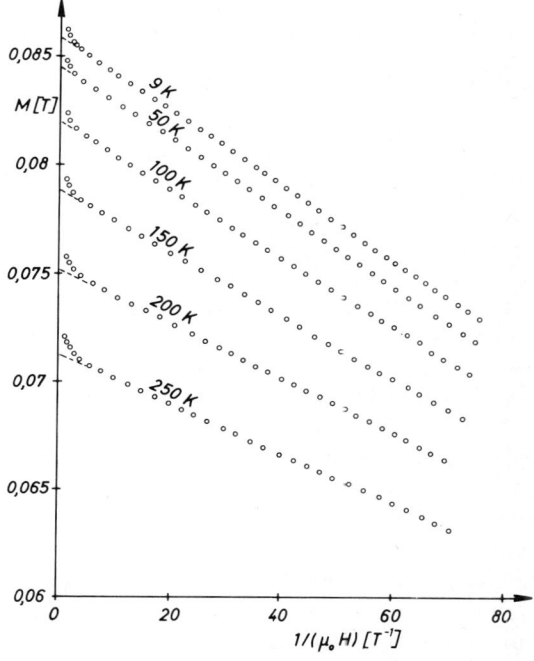

Fig. 1. Spontaneous magnetization, $M_s(T, H)$ vs. $1/H$ for different temperatures.

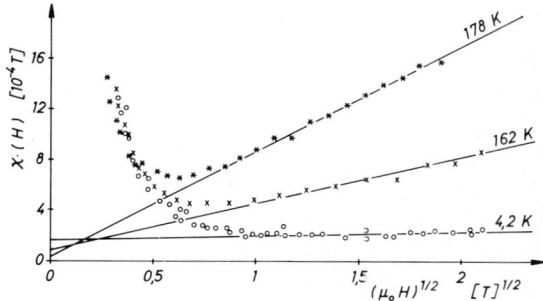

Fig. 2. $\chi(T, H) \cdot H$ vs. $\sqrt{H}$ for three different temperatures.

parameter $C(T)$ is given by the slopes of the straight lines in fig. 2, and $B(T)$ is determined from the intercept of $\chi \cdot H$ for $H = 0$ if the straight line is extrapolated to $H = 0$. The temperature dependences of $A(T)$, $B(T)$ and $C(T)$

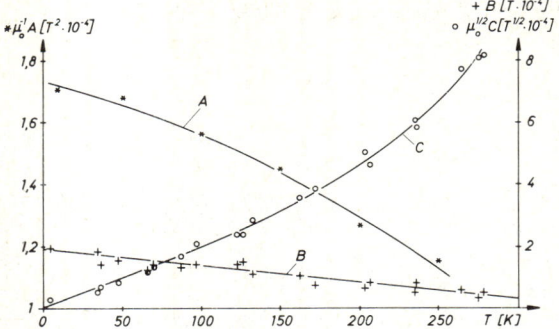

Fig. 3. The susceptibility parameters A, B, C in dependence of temperature.

are shown in fig. 3. Here we find C to be a linear function of T in the temperature range up to 150 K. The parameter $A(T)$ turns out to be proportional to $M^2(T,0)$, whereas $B(T)$ decreases linearly with increasing T. In the whole temperature range investigated $M(T,0)$ obeyed Bloch's law $M(T,0) = M(0,0)[1-(T/T_0)^{3/2}]$ with $T_0 = 805$ K.

It has been shown recently [4] that in amorphous ferromagnets extended spin waves of the type $\exp i(\mathbf{k}\cdot\mathbf{r} - \omega t)$ ($\mathbf{k}$ = wave vector, ω = eigenfrequency) as well as localized spin waves $\exp(-\kappa r)\exp i(\mathbf{k}\cdot\mathbf{r} - \omega t)$ (κ = decay length) may develop, where the range limitation term results from local fluctuations of the exchange interactions. The energy dispersion relation in the case of extended spin waves (neglecting dipolar interactions) is given by

$$\epsilon_k = Dk^2 + g\mu_B \langle S \rangle \cdot H \qquad (3)$$

(g = g-factor, μ_B = Bohr's magneton, $\langle S \rangle$ = mean spin quantum number). For the localized spin waves we have approximately [4]

$$\epsilon_k = D'k^2 + g\mu_B \langle S \rangle \cdot H + \tilde{\epsilon}. \qquad (4)$$

Here $\tilde{\epsilon}$ results from fluctuating exchange interactions. The dispersion relation given by eq. (3) leads to the $H^{-1/2}$-dependence of the parasusceptibility [2], whereas eq. (4) would give a $(H + \tilde{H}_0)^{-1/2}$-dependence with $\tilde{H}_0 = \tilde{\epsilon}/g\mu_B\langle S \rangle$. Since this latter dependence could not be detected we conclude that the localized spin waves play a minor role in the approach to saturation.

A quantitative comparison of our results for $M_s(T,H)$ with spin wave theory is possible by equating the relation

$$1 = 1.17 \left(\frac{k_B}{\mu_B}\right)^{1/2} \left(\frac{C(T)}{T}\frac{T_0^{3/2}}{M(0,0)}\right), \qquad (5)$$

which may be derived from spin wave theory for the low temperature limit [2] (k_B = Boltzmann's constant). Inserting into eq. (5) our experimental values for T_0, $M(0,0)$ and $C(T)/T$ we find a numerical value of 1.25 which agrees fairly well with the theoretical predictions.

In our measurements of χ no $1/H^3$ term could be detected showing that crystal energy and long-range internal stresses do not contribute to χ. The A/H^2 term may be due either to short range internal stresses resulting from elastic dipoles [1] or to internal stray fields due to non-magnetic precipitations or fluctuations, ΔM_s, of the average spontaneous magnetization [5]. Since A is found to be proportional to $[M(T,0)]^2$ this latter interpretation is favoured. From Néel's result [5] $A = 0.022 \cdot 4\pi M_s \Delta M_s$ we obtain at RT $\Delta M_s/M_s = 0.05$, i.e. the fluctuations of M_s correspond to $\pm 2.5\%$.

References

[1] H. Kronmüller, in: Moderne Probleme der Metallphysik, Vol. II, A. Seeger (Ed.) (Springer-Verlag, Berlin–Heidelberg–New York, 1966) p. 24.
[2] T. Holstein and H. Primakoff, Phys. Rev. 58 (1940) 1098.
[3] N. Kazama and M. Kameda, Joint MMM-Intermagn. Conf. (1976).
[4] R.G. Henderson and A.M. de Graaf, in: Amorphous Magnetism, H.O. Hooper and A.M. de Graaf (Eds.) (Plenum Press, New York–London, 1973) p. 331.
[5] L. Néel, J. Phys. Rad. 9 (1948) 184.

24 GHz TRANSMISSION STUDIES ON THE AMORPHOUS ALLOY METGLAS 2826 FROM 26 TO 270°C

J.F. COCHRAN, B. HEINRICH and R. BAARTMAN
Simon Fraser University Department of Physics, Burnaby, British Columbia, Canada V5A 1S6

Ferromagnetic anti-resonance transmission studies at 24 GHz have been used to study the temperature dependence of the magnetic properties of METGLAS 2826 from room temperatures to 270°C. The g-factor was found to be independent of temperature and to have the value 2.053 ± 0.005. Anisotropy effects at room temperature were less than 10 Oe, and this sets an upper limit of 0.2×10^4 erg/cm^3 on the anisotropy constant.

We have used the phenomenon of ferromagnetic anti-resonance transmission (FMAR) at 24 GHz to study the magnetic properties of the amorphous metal alloy METGLAS 2826 over the temperature range 26–270°C. Circular specimens 2 mm in diameter were spark cut from a ribbon of METGLAS 2826 supplied by Allied Chemical Corporation, Morristown, N.J. The results of the measurements at each temperature were fitted to the usual theory [1] which is based upon the Landau–Lifshitz equation of motion for the magnetization. The values for g, $4\pi M_s$, and the damping parameter λ, obtained from fitting the theory to the data at each temperature up to 208°C are shown in fig. 1. The g-factor of METGLAS 2826 was found to be temperature independent, within the uncertainties introduced by the fitting procedure, and had the value 2.053 ± 0.005 in good agreement with the room temperature FMR measurements reported by Hasegawa [2]. The magnetic damping parameter exhibited a different behaviour for the perpendicular and the parallel configurations. For the parallel configuration λ appeared to have the temperature independent value $(0.9 \pm 0.2) \times 10^8$ s^{-1}. For the perpendicular configuration, λ varied from $(3 \pm 1) \times 10^8$ s^{-1} at 300 K to $(0.6 \pm 0.3) \times 10^8$ s^{-1} at 480 K. This behaviour is quite different from the behaviour of the damping parameter observed in the alloy Supermalloy [3] and in pure nickel. In these materials λ was found to be temperature independent up to the Curie temperature, and λ was the same for both the parallel and perpendicular configurations. The excellent agreement between theory and experiment for the parallel configuration up to temperatures of 176°C is illustrated in fig. 2. The theoretical curves were all calculated using the same value $\lambda = 0.9 \times 10^8$ s^{-1}. Note the excellent agreement between the experimental and theoretical amplitudes. The data began to exhibit appreciable deviation from the calculated line shapes for temperatures higher than 176°C. Pure tin (m.p. = 232°C) was used as a solder to mount the specimens: the change in amplitude of the signal with time at 235°C was due to the tin creeping over the METGLAS surfaces and so partially obscuring the transmission aperture.

The external field was rotated in the specimen plane at room temperatures and no evidence was found for an anisotropy shift larger than 10 Oe. This observation places an upper limit of

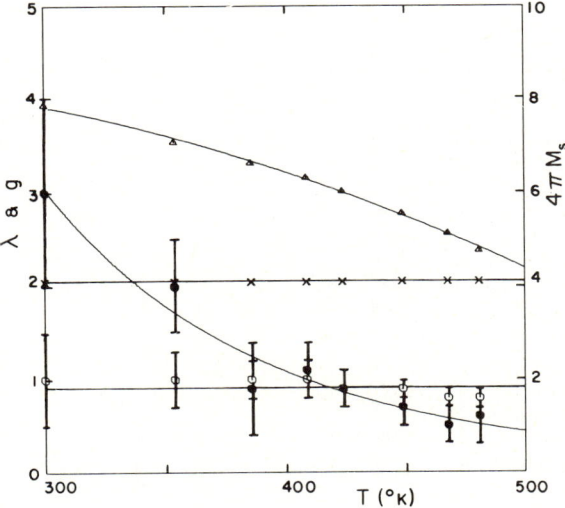

Fig. 1. The temperature variation of g, $4\pi M_s$, and λ for METGLAS 2826 as determined from transmission experiments. Crosses and triangles are used to designate the data for g and $4\pi M_s$: the estimated uncertainties are indicated by the sizes of the symbols. The solid circles represent values of the damping parameter deduced for the perpendicular configuration, the open circles represent values of the damping parameter deduced for the parallel configuration. In each case, estimated uncertainties are indicated by the vertical bars.

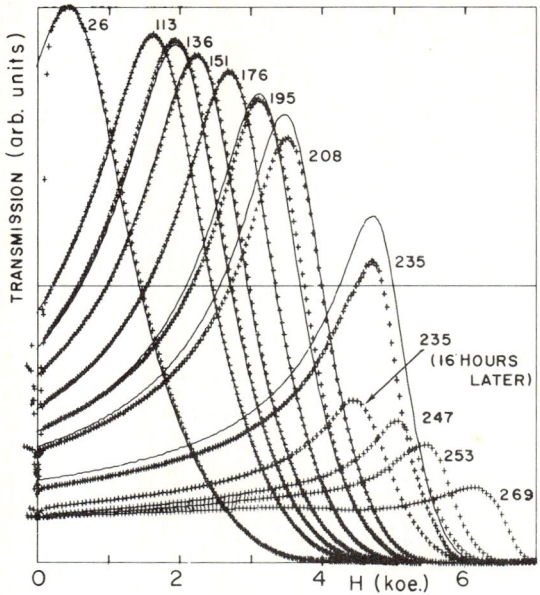

Fig. 2. Experimental transmission amplitudes observed at the indicated temperatures in °C for an 18 μm thick disc of METGLAS 2826. The solid lines are theoretical curves calculated using $g = 2.053$, $\lambda = 0.9 \times 10^8$ s^{-1}. The theory was scaled to fit the peak experimental amplitude at 26°C.

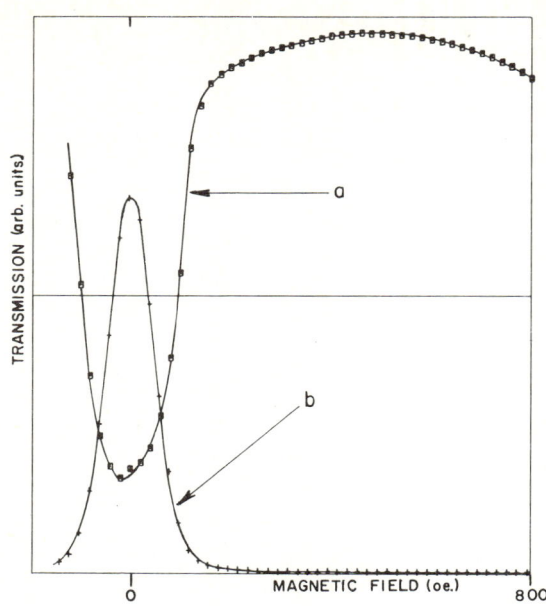

Fig. 3. Transmission amplitude vs. magnetic field at 26°C near zero field for a specimen of METGLAS 2826, 18 μm thick. The d.c. magnetic field was parallel to the specimen plane in each case, but for (a) the d.c. field was orthogonal to the RF magnetic field, whereas for (b) the d.c. field was parallel to the RF magnetic field.

0.2×10^4 erg/cm^3 on the anisotropy constant. This value is much smaller than the value $K = 1.4 \times 10^4$ erg/cm^3 reported by Hasegawa [3]. In the course of the anisotropy measurements we discovered a striking increase in transmission for external fields centered around zero field and parallel to the RF driving field (parallel–parallel configuration). This effect is shown in fig. 3, in which the transmission signal for the conventional configuration is compared with that for the parallel–parallel configuration, and indeed it can be seen from fig. 3 that for d.c. fields greater than 200 Oe the transmission amplitude becomes very small. The enhanced transmission near zero field is clearly related to the presence of domain walls in the specimen. This zero-field effect can be explained using a simple argument based upon the intrinsic stiffness of a domain wall. The RF magnetic field varies across the thickness of the specimen due to the skin-effect. This inhomogeneous RF field exerts an inhomogeneous driving force on the domain wall and as a result the wall must bow. One can write [4] for the wall displacement, y,

$$\alpha(\partial^2 y/\partial^2 z) = -h, \qquad \alpha = \xi\sqrt{KA}/2M_s,$$

where z is the direction along the specimen normal, M_s is the saturation magnetization, $\sqrt{KA}$ is the surface energy of the Bloch wall and ξ is a dimensionless constant of order unity. This equation leads to a wave-number dependent susceptibility, which when combined with Maxwell's equations gives

$$k^4 = \frac{i}{\delta_0^2} \frac{4\pi M_s}{(\alpha d)},$$

where d is an average spacing between domain walls.

In the limit of a large surface energy, the penetration depth can become very large. We hope to be able to use this zero-field effect to investigate the temperature dependence of the product $\bar{K}A$.

The authors wish to thank the National Research Council of Canada for grants which made this work possible.

References

[1] B. Heinrich and V.F. Meshcharyakov, Zh. Eksp. Teor. Fiz. 59 (1970) 424 [Soviet Physics JETP 32 (1971) 232].

[2] E. Hasegawa, in: Magnetism and Magnetic Materials – 1975, J.J. Becker, G.H. Lander and J.J. Rhyne (Eds.) (American Institute of Physics, New York, 1976) p. 216.

[3] B. Heinrich, J.F. Cochran and G. Dewar, Proc. Int. Conf. of Magnetism, ICM-73 (Nauka, Moscow, 1974).

[4] B. Heinrich and A.S. Arrott, Can. J. Phys. 50 (1972) 710.

TEMPERATURE AND COMPOSITIONAL DEPENDENCE OF MAGNETIC PROPERTIES OF AMORPHOUS FeGe FILMS

G. SURAN*, H. DAVER† and J. SZTERN*

Laboratoire de Magnétisme, CNRS, 92190 Bellevue, France
†*Groupe de Transition de Phases, 38042 Grenoble, France*

The magnetic properties of amorphous Fe_xGe_{1-x} films were studied systematically as a function of iron concentration between 4.2 and 300 K using FMR. We observed a continuous evolution of the temperature dependence of the resonance linewidth and saturation magnetization when the Fe content is changed in the film. The temperature dependence of $D(T)$ is discussed briefly.

1. Introduction

Recently there is a considerable interest in amorphous ferromagnetism. A great amount of experimental and theoretical work is being done in order to determine the influence of topological disorder on the various magnetic properties. The aim of the present paper is to show the evolution of some magnetic properties with Fe concentration for Fe_xGe_{1-x} amorphous thin films. The measurements were made using ferromagnetic resonance (FMR), so basic magnetic properties as the saturation magnetization, spin-wave spectra and their temperature dependence, as well as ferromagnetic relaxation could be studied in a systematic way. Some details of the results reported here were published partly elsewhere [1, 2].

2. Experimental

The Fe_xGe_{1-x} films were obtained by evaporation from a single crucible. The films were deposited on glass substrates cooled down to 77 K and during deposition the vacuum was 10^{-7} torr. In order to obtain a homogeneous melt a mixture of high purity Fe and Ge is melted in a BeO crucible by HF heating. The rate of evaporation of Ge being higher than that of Fe, the Fe content in the film is typically 10–15% lower than in the melt. In order to eliminate a variation of composition in the film the deposited thickness is ranged between 600 and 800 Å. Then the compositional gradient along the thickness of the film becomes negligible (less than 2%) as shown by a careful analysis [1]. The magnetic measurements were performed by FMR at 17.5 GHz in the range 4.2–300 K. The saturation magnetization and g factor are deduced using the Kittel formula from resonance fields corresponding to perpendicular and parallel orientation ($H_\perp$ and $H_\parallel$). The composition range studied was $0.4 < x < 0.7$.

3. Results and discussion

One of the most interesting results obtained is the evolution of the resonance linewidth of the main mode as a function of film composition. Fig. 1 shows the temperature dependence of $\Delta H_\perp$ for films of various compositions, the magnitude and variation of $\Delta H_\parallel(T)$ being the same. For samples with $x > 0.6$ we find $\Delta H_\perp = 60 \pm 10$ Oe (curve corresponding to $x = 0.63$). $\Delta H_\perp$ is minimum at room temperature and then increases slowly and continuously up to 80–90 Oe at 4.2 K. For $x < 0.6$ the resonance linewidth presents a well defined minimum close to $T = 100$ K. At $T = 100$ K the higher the Fe content the lower is $\Delta H_\perp$. This minimum in $\Delta H_\perp$ at 100 K increases (fig. 1) from 50 Oe for $x = 0.53$ to 500 Oe for $x = 0.433$. The large increase of ΔH between 100 and 4.2 K is related to some local anisotropy the origin of which is actually not well determined: it could be related to the fact that a part of the Fe atoms are antiferromagnetically coupled. One can remark that for $x < 0.45$ no FMR can be observed for temperatures lower than 20 K, this behaviour being similar to that observed in spin glasses with mictomagnetic clusters. The existence of this local anisotropy and its temperature dependence were also confirmed experimentally by magnetoresistance measurements [3]. For $x < 0.47$ at 4 K the resistance cannot be saturated with applied field even up to 70 kG while at 70 K saturation is attained with a few kG. The resonance linewidth can be analyzed in terms of Landau–Lifshitz (L–L) relaxation mechanism. The L–L damping

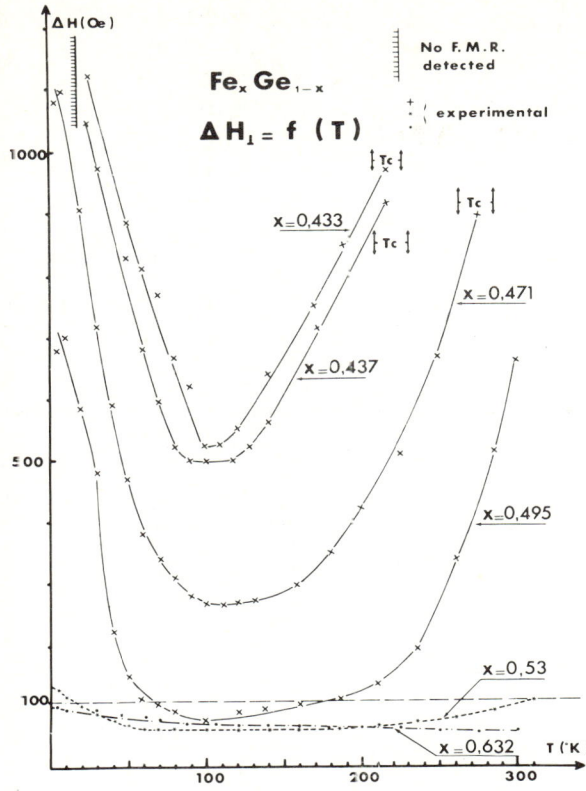

Fig. 1. Temperature dependence of the resonance linewidth $\Delta H_\perp$ for various FeGe amorphous alloys. The discontinuous line indicates the upper limit of $\Delta H_\perp$ in order that SSW spectra be resolved.

parameter λ is defined as $\Delta H = 2\omega\lambda/M\gamma^2$ where M is the saturation magnetization, $\gamma = eg/2mc$ and ΔH is the peak to peak resonance linewidth. For samples with $x > 0.6$ at room temperature λ is comprised between 8 and $8.5 \times 10^{+7}$ s^{-1}. This value is close to that obtained in various crystalline ferromagnetic films such as Py or Fe. On the other hand λ increases only slightly as a function of temperature so at 300 K λ can be considered to be close to the intrinsic relaxation term. For $0.5 < x < 0.6$ and at 100 K λ decreases with decreasing Fe content and presents a minimum for $x \simeq 0.5$ where λ (100 K) $= 3 \times 10^7$ s^{-1}. For this composition the conductivity is no longer of metallic type, so this minimum could be related to the fact that the exchange conductivity contribution to ΔH is minimal. For compositions $x < 0.5$ and at 100 K one observes again an apparent increase of λ. However, for this composition range the contribution of the anisotropy is no longer negligible even near 100 K. The increase of ΔH between 100 and 300 K can be explained by the fact that the resonance linewidth – as it is generally observed in crystalline ferromagnetic metals – presents a maximum near the Curie temperature T_c. The increase of ΔH is more gradual than in high quality crystalline materials and correspondingly a slowly increasing damping parameter λ is found. These results reflect the amorphous structure of the sample and relate to local exchange fluctuations: at certain sites the Fe atom is in a paramagnetic state while for other still in a ferromagnetic one [4].

When the resonance linewidth of the main mode is lower than 100 Oe well resolved secondary peaks – standing spin wave (SSW) modes – were detected. The temperature dependence of the SSW mode linewidth is the same as that of the main line. Variations of the same order of magnitude are found. This result shows that the main contribution to the SSW linewidth is due to the anisotropy observed.

3.1. $M_s(T)$, T_c: variation as a function of Fe concentration

For the concentration range $x > 0.6$ the saturation magnetization follows perfectly the Bloch law $M_s = M_0 (1 - BT^{3/2})$ [2] up to 300 K. For $0.5 < x < 0.6$ M_s follows the Bloch law only in the low temperature region. The observed deviations increase with decreasing Fe content. For even lower concentrations ($x < 0.5$) $M_s(T)$ tends towards a linear law [1]. This result can be explained by the fact that for high Fe concentrations the decrease of magnetization is simply due to thermal excitation of spin waves while for low Fe content the spin waves are influenced by the fluctuations in the local short range order (increase in the fluctuation of exchange constant) as was shown by a calculation of Huber et al. [5]. The evolution of the magnetic moment per Fe atom as a function of composition $\mu(x)$ is also characteristic: $\mu(x)$ is close to 2.1 μ_B/Fe for $0.6 < x < 0.66$ and then decreases linearly between $0.4 < x < 0.6$, the critical concentration being close to $x \sim 0.4$. For $x < 0.48$ T_c is lower than 300 K and so could be studied directly. It appears that at this concentration range T_c is not well determined since a smooth transition between the ferromagnetic region and the paramagnetic one is found. When the magnetic properties of amorphous samples are compared to those of various crystalline FeGe phases these properties are close to those of the cubic one [both T_c and $\mu(x)$] [6] showing that the short range order in the amorphous phase is close to that in the cubic phase. This result is similar to that observed on other amorphous materials.

3.2. $D(T)$

For $\Delta H_\perp < 100$ Oe one observes well defined SSW spectra. For $x > 0.6$ the dispersion relation is fairly well quadratic and one can determine the temperature dependence of $D(T)$. The surface anisotropy K_s of our best samples appears to be very low ($K_s < 10^2$ erg/cm^2) and careful measurements on high quality samples yield $D(T) = D(0)(1 - CT^2)$. This result is not surprising as in this composition range the conduction is metallic like and a T^2 law is predicted by an itinerant electron model. The overall variation of $D(T)$ is also close to this model. The values of D as deduced from SSW and magnetization slope agree well near 4.2 K [2].

In conclusion Fe_xGe_{1-x} amorphous films have a magnetic behaviour similar to that of other TM–MA amorphous alloys for $x > 0.6$. For lower concentrations one observes a progressive deviation from "classical" amorphous ferromagnets. Part of this deviation can be explained by the evolution of local structure the study of which is presently in progress.

References

[1] G. Suran, H. Daver and J.C. Bruyere, AIP Conf. Proc. 29 (1975) 162.
[2] G. Suran, H. Daver and J. Sztern, Joint MMM-Intermagn. Conf., Pittsburgh (June, 1976) 5D-2.
[3] H. Daver, Y. Cros, J.C. Bruyere, to be published.
[4] J. Morkowski, H. Phys. Chem. Solids 25 (1964) 1183.
[5] D.L. Huber and R.P. Siemann, Solid State Commun. 17 (1975) 769.
[6] K. Kanematsu and T. Ohoyama, J. Phys. Soc. Jap. 20 (1965) 236.

SPIN GLASSES

K.H. FISCHER†

Institut für Festkörperforschung der Kernforschungsanlage Jülich, D-517 Jülich, Germany

(Invited paper)

A critical review is presented of the various concepts and theories of spin glasses and mictomagnets, including a recent theory of Edwards and Anderson. Experimental results for the magnetization, static susceptibility, specific heat, resistivity, and neutron scattering are discussed. These results are completed by computer simulations, which show hysteresis and aftereffects in spin glasses and thus yield further insight into this fascinating subject.

1. Introduction

Spin glasses have been investigated for a long time. In order to explain a linear term in the specific heat $c_H(T)$ of Cu Mn [1] and Ag Mn [2] at low temperatures, Marshall [3] and Klein and Brout [4] developed molecular-field theories for random dilute magnetic alloys. More recently, Cannella et al. [5] discovered a sharp cusp in the static susceptibility of a series of alloys, typically noble metals with between 0.2 and 10% 3d-transition-metal impurities. The temperature of the cusp T_f (the "freezing temperature") was found to be between 0.2 and 10 K (for illustrative examples compare [6]). In Au Fe a hyperfine field splitting of the Mössbauer line into six lines was also observed roughly at T_f [7]. From the line intensities it has been concluded that the spins have random directions below T_f. Neutron scattering experiments [8] indicate the absence of any long-range order. However, magnetic short-range order has been observed in Cu Mn, and magnetic and chemical short-range order in Au Fe [7, 9]. Remanence effects in the magnetization have been found long ago in Cu Mn and Au Fe at $T < T_f$ [10, 11]. One also has aftereffects: The remanence decreases roughly logarithmically as a function of time [11]. A strongly asymmetric hysteresis loop has been observed after field cooling in Cu Mn at $T < T_f$ [12].

In contrast to the *magnetic* properties no well-defined characteristic temperature has been observed in the *thermal* properties. Instead, the specific heat shows a broad maximum at a temperature $T_m > T_f$ [13]. The maximum temperature of the derivative of the resistivity $d\rho/dT$ also does not coincide with T_f [6].

As a criterion of a good spin glass two con-

ditions have been proposed [6]:

(1) The system should possess "good" magnetic moments near T_f. This means that the characteristic temperature T_K for the vanishing (or "compensation") of the static moment *in the extreme dilute limit* (the Kondo or local spin-fluctuation temperature) should be small compared to T_f. The moment should also be fairly localized.

(2) The solubility should be good. This second criterion has been questioned since typical spin-glass properties such as remanence and a large sharp cusp in the zero field susceptibility have also been observed in *annealed* $Cu_{75}Mn_{25}$ [14]. One is therefore inclined to conclude that a random distribution of the magnetic component is not a necessary condition for a spin glass. Of course, for the theorist random systems are easier to handle, and in the following I will only consider theories based on this assumption. Systems with chemical clusters sometimes are called "mictomagnets" [12, 14]. On the other hand, sometimes [11] only systems with T_f proportional to the magnetic impurity concentration c have been called "spin glasses", for reasons to be discussed below.

2. Microscopic picture

The spin glass properties are apparently a consequence of the interactions between the magnetic moments. There are several conceivable interactions, the most important one being the Rudermann–Kittel–Kasuya–Yosida (RKKY) interaction: The spin system is described by a Heisenberg or Ising Hamiltonian ($b \equiv g\mu_B B_0$, where B_0 is an external field)

$$\mathcal{H} = -\sum_{i<j} J_{ij} \mathbf{S}_i \mathbf{S}_j + b \sum_i S_i^z \qquad (1)$$

† SFB Aachen-Köln-Jülich.

with the exchange interaction $J_{ij} = J(|\mathbf{R}_i - \mathbf{R}_j|)$

$$J(R) = J' \cos 2k_F R / R^3. \quad (2)$$

Here, k_F is the Fermi momentum of the host conduction electrons. For the Ising model one has the numbers $S_i = \pm 1$ instead of $\mathbf{S}_i$. We neglected a term proportional to R^{-4} in (2). The interaction (2) has an alternating sign and is of long range, yielding ferro- and antiferromagnetically-coupled spins. Whether or not the second property is essential for spin glasses is not yet clear. Amorphous dilute magnetic alloys, in which the RKKY-interaction is exponentially damped, also show spin glass properties [15].

The R^{-3}-dependence of the RKKY-interaction leads to some kind of corresponding states for $c \to 0$ [16]. Thus the specific heat $c_H(T, B_0, c)$ "scales" as $c_H = cf_1(T/c, B_0/c)$, the magnetization as $M(T, B_0, c) = cf_2(T/c, B_0/c)$ etc. This leads to $T_f \propto c$ (in agreement with experimental results), and one can also write

$$c_H = cf_1\left(\frac{T}{T_f}, \frac{b}{kT_f}\right), \quad M = cf_2\left(\frac{T}{T_f}, \frac{b}{kT_f}\right). \quad (3)$$

The more general relations (3) with $T_f(c)$ have been obtained [17] in the framework of a theory of Edwards and Anderson [18] (hereafter referred to as EA).

For more concentrated spin glasses one has (at least for Au Fe) $T_f \propto c^m$ with $0.55 < m < 0.75$ [6]. One expects *direct exchange interactions* to become important. For Au Fe with ferromagnetically coupled n.n. pairs [7, 20] this leads to an abrupt transition of $T_f(c)$ [19] at roughly the percolation limit (which is about 15% for f.c.c. lattices). Besides small ferromagnetic clusters, above T_f one has clusters with more or less random spin directions. This is indicated by the specific heat, which shows the freezing-in of a large number of spin degrees of freedom above T_f.

We now consider various thermal averages. For fixed impurity configuration a spin S_i will feel an internal field from its neighbours which, however, averages to zero if the average is taken over sufficiently long time ($\langle \ldots \rangle$ means thermal average) for $T > T_f$

$$\langle S_i \rangle = 0, \quad T > T_f. \quad (4)$$

The basic idea for spin glasses is that there exists a well-defined temperature T_f below which all (or at least a macroscopic part of all) spins assume a fixed local axis. Thus for fixed configuration

$$\langle S_i \rangle \neq 0, \quad T < T_f, \quad (5)$$

since the internal field at the lattice site i no longer averages to zero. This also should be true if clusters of finite size have been formed above T_f. However, if the local axes are randomly distributed (as is assumed for spin glasses, since the total magnetization is zero), the configuration average $\overline{\langle S_i \rangle} = 0$ for all temperatures.

We find the relevant parameter for spin glasses by considering the static susceptibility. For N spins it follows from the free energy that

$$\chi = \chi^0 \sum_{ij} [\overline{\langle S_i S_j \rangle} - \overline{\langle S_i \rangle \langle S_j \rangle}] / NS(S+1), \quad (6)$$

where the susceptibility of the independent spins $\chi^0 = NS(S+1)(g\mu_B)^2/3kT$. For reasons of simplicity we now consider a model [18] in which the random distribution of impurity sites is replaced by a symmetric distribution of exchange couplings J_{ij}. With $P(J_{ij}) = P(-J_{ij})$ ($P(J_{ij})$ is the probability of finding a coupling constant J_{ij}) one has

$$\overline{\langle S_i S_j \rangle} = S(S+1)\delta_{ij}, \quad \overline{\langle S_i \rangle \langle S_j \rangle} = \overline{\langle S_i \rangle}^2 \delta_{ij} \quad (7)$$

since each spin has a neighbouring spin j with up or down direction with the same probability. For a classical Heisenberg model or an Ising model with $S^2 = 1$

$$\chi = \chi^0(1-q), \quad q \equiv \overline{\langle S_i \rangle^2}, \quad (8)$$

where q is the relevant spin-glass parameter. From (4) it follows that $q(T) = 0$ for $T > T_f$. In this approach short-range correlations are neglected.

3. Theories for static properties

3.1. High-temperature expansion

One must distinguish between three temperature ranges. For $T > T_J$, where kT_J is the largest coupling energy in (2), the usual high-temperature expansion can be used. For $T_J > T > T_f$ a virial expansion of the free energy with the interaction (2) up to the second virial coefficient has been carried out [21]. One obtains the law of corresponding states eq. (3) (with $T_f \propto c$), but no reliable information about the system near T_f. An extrapolation over the whole temperature range yields a maximum in the specific heat, which is perhaps not too surprising.

3.2. Edwards-Anderson theory

A theory which explains the sharp cusp in the static susceptibility $\chi(T)$ by a sudden freezing-in of the impurity spins in random directions has been proposed by Edwards and Anderson [18] and extended by various authors [17, 22-24]. In this theory short-range correlations are neglected. From the configuration-averaged free energy a self-consistent equation for the parameter $q(T)$ eq. (8) is derived under the assumption of a symmetric distribution $P(J_{ij})$ of independent coupling constants J_{ij}. The classical Heisenberg model [18] and the Ising model [17, 24] yield fairly similar results. Both susceptibility and specific heat show a cusp at T_f, and both are proportional to T for $T \to 0$. However, since the results are based on a generalized molecular field approximation, the agreement of the low temperature behaviour with the experimental results [1, 2, 5] may be fortuitous. The same, of course, holds for the Marshall-Klein-Brout theory.

The susceptibility and specific heat have also been calculated for a finite external magnetic field. As expected, the cusp in $\chi(T)$ and $c_H(T)$ is rounded off if the field energy b becomes comparable with the thermal energy kT_f. In the case of the susceptibility the experimental data [6] indicate a considerably stronger field dependence.

A large discrepancy between theory [18, 22] and experiments [13] seems to exist for the specific heat, which shows no indication of a cusp. As compared to the susceptibility, the specific heat is more sensitive to the formation of random clusters. Furthermore, a MFA usually yields worse results for $c_H(T)$ than for $\chi(T)$ near a phase transition. Indeed, Monte-Carlo calculations [25] for the same model but avoiding the MFA yield a cusp in $\chi(T)$ and a broad maximum in $c_H(T)$. Thus, the discrepancy with the exponential results seems to be due to the MFA, and not to the model.

Recently, some doubts have arisen concerning the reliability of a trick used in the EA theory, where one calculates the configuration-averaged free energy from $-\beta F = \overline{\ln Z} = \lim_{n \to 0}(1/n)(\overline{Z^n} - 1)$. Kosterlitz et al. [26] find identical results with and without applying this trick to a spherical model of a spin glass with infinite-ranged interactions, whereas Thouless et al. [27] find different behaviour at low temperatures for the EA model in MFA and also in a low temperature expansion.

3.3. Cluster theories

A theory of spin glasses based on the formation of clusters has been proposed by D.A. Smith [28]. A magnetic cluster is defined as a connected group of spins which is coupled together by exchange interactions eq. (2) with magnitude greater than the thermal energy. With decreasing temperature these clusters grow until the percolation limit is reached at T_f. For $T > T_f$ eq. (4) holds, since the clusters are still able to rotate and the internal field at each lattice site averages to zero. For $T < T_f$ we have $\langle S_i \rangle \neq 0$ for a macroscopic contribution of all spins frozen into an infinite cluster.

Detailed calculations have been performed for a Bethe lattice. The number of "loose" spins below T_f decreases exponentially, leading to an exponential temperature dependence of $\chi(T)$ for $T \to 0$. At T_f a sharp cusp is obtained, and for $T > T_f$ and small impurity concentration one obtains (as in the EA model) the susceptibility of free spins.

In my opinion, this theory does not contradict the EA theory as long as random clusters are considered. In the EA theory local quantities are taken into account only, without specifying the way in which the spin glass freezes in.

4. Time-dependent properties

4.1. Experiments

Spin glasses show remanence effects as mentioned in the Introduction. Depending on the experimental situation, one observes an "isothermal remanent magnetization" (I.R.M.) in which a field B_0 is applied at $T < T_f$ for a sufficiently long time and then switched off, or a "thermoremanent magnetization" (T.R.M.) in which the sample is cooled down from $T > T_f$ in an external field, which then is switched off [11]. Both remanence effects saturate at sufficiently high fields and vanish for $T > T_f$.

The time dependent magnetization of $Au_{0.96}Fe_{0.04}$ in a field $B_0 = 10$ gauss has been investigated in detail by Guy [29]. The solid line in fig. 1 shows the magnetization if the field is applied for less than one minute. The magnetization with B_0 applied for about three hours (solid circles) is markedly different, indicating that after three minutes the system has not yet reached its (field-dependent) equilibrium. The cusp obtained after three minutes is similar to that obtained by Mydosh et al. by means of an ac method with a field of 17–155 Hz and about 5 gauss. Mukhopadhyay et al. [12] estimate that the true "zero field" susceptibility is only reached at a threshold field B_s below 4 gauss. Still another magnetization is obtained after field cooling (open circles in fig. 1). In some systems the "reversible" part (the short-time contribution) and the "irreversible" part of $\chi(t)$ (the additional magnetization obtained after waiting sufficiently long) have been found to add up [11] to $\chi(T)$ = constant below T_f.

The corresponding contributions to the magnetization as a function of time are seen in fig. 2. After switching on a magnetic field, the magnetization jumps in a rather short time to a value M_0 (the "reversible" part) and then increases slowly like a typical aftereffect: The system has to overcome energy barriers by thermal excitations or by tunneling processes in order to reach states with lower free energy. The decrease of remanent magnetization in Au Fe seems to follow a $\ln \cdot t$ law. For fields smaller than B_s the irreversible part of $\chi(T)$ should vanish.

4.2. Theories

Dynamical theories of spin glasses are still in a very preliminary state. The time dependence of the spin glass parameter q, starting from a non-equilibrium value, has been obtained by solving either a Fokker–Planck equation for a classical 2-dimensional Heisenberg model [30] or a master equation for an Ising system [31]. In the latter case Monte-Carlo calculations have also been performed [25]. Edwards and Anderson obtain for the time-dependent correlation function

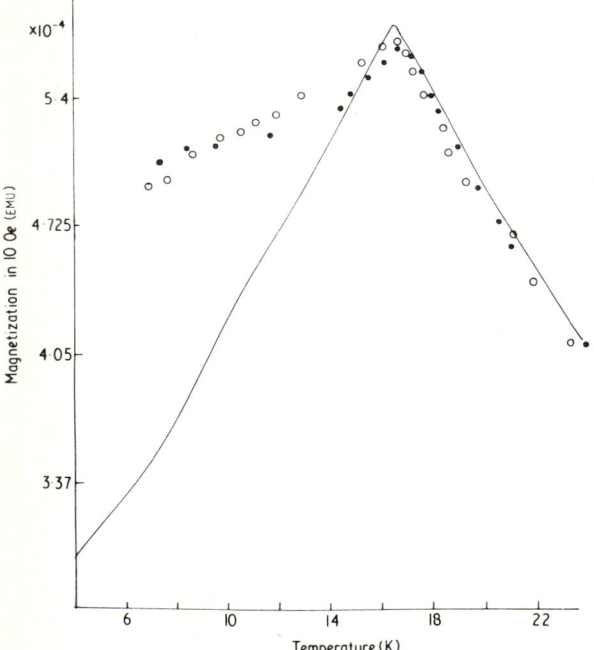

Fig. 1. Time dependent magnetization in a field $B_0 = 10$ gauss. Solid line: Magnetization with B_0 applied for less than one minute. Solid circles: B_0 applied for about three hours. Open circles: Magnetization after field cooling (after Guy [29]).

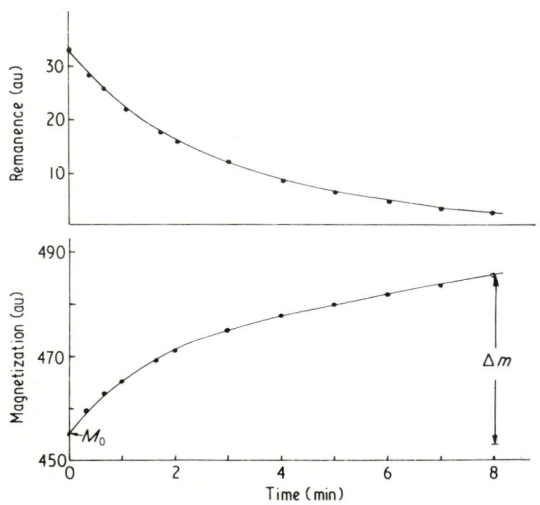

Fig. 2. Magnetization as a function of time. Upper curve: Decay of the isothermal remanence obtained after switching off a field of $B_0 = 20$ gauss. Lower curve: Development of magnetization with time in an applied field $B_0 = 20$ gauss (after Guy [29]).

$$\overline{\langle S_i(t)S_i(0)\rangle} = \exp\left\{-\left[1-\left(\frac{T_f}{T}\right)^2\right]\frac{kT}{\nu}t\right\}$$
$$\cong \exp\left[-2\left(1-\frac{T_f}{T}\right)\frac{kT}{\nu}t\right], \quad T > T_f,$$
(9)

where $\exp[-kTt/\nu]$ is the relaxation of an independent spin in the heat bath. The corresponding expression for $T < T_f$ reads

$$\overline{\langle S_i(t)S_i(0)\rangle} = q + (1-q)\exp\left[-q\frac{kT_f}{\nu}t\right], \quad T < T_f,$$
(10)

with $q \equiv \langle S_i(\infty)S_i(0)\rangle$ and $\langle S_i^2\rangle = 1$.

In the second approach one starts from the master equation for the probability $P(S_1 \ldots S_N, t)$ for a given configuration of Ising spins $(S_i = \pm 1)$

$$\frac{d}{dt}P(S_1 \ldots S_N, t) = -\sum_i W(S_i \to -S_i)$$
$$\times P(S_1 \ldots S_i \ldots S_N, t)$$
$$+ \sum_i W(-S_i \to S_i)$$
$$\times P(S_1 \ldots -S_i \ldots S_N, t).$$
(11)

Following the classical paper of Suzuki and Kubo [32], one makes the ansatz for the spin-flip transition probability $W(S_i \to -S_i)$ with $E_i \equiv \Sigma_j J_{ij} S_j$

$$W(S_i \to -S_i) = \frac{1}{2\tau}(1 - S_i \tgh \beta E_i),$$
(12)

where τ is the relaxation time for a single spin in a heat bath. The expectation value of a spin S_i is defined as

$$\langle S_i\rangle(t) = \sum_{\{S\}} S_i P(S_1 \ldots S_N, t).$$
(13)

One obtains with (11) and (12)

$$\frac{d}{dt}\langle S_i\rangle = -\frac{1}{\tau}(\langle S_i\rangle - \langle \tgh \beta E_i\rangle)$$
(14)

We apply these equations to a very crude model for spin glasses. In equilibrium the spin S_i points in the direction of the local field E_i. Thus $|\langle S_i\rangle| = \langle \tgh \beta |E_i|\rangle \approx \tgh \beta \langle |E_i|\rangle$ where the second equation holds in MFA only. For the configuration average we make the crude approximation

$$\overline{|\langle S_i\rangle|} \approx \overline{(\langle S_i\rangle^2)}^{1/2} \equiv q^{1/2},$$
$$\overline{\tgh \beta \langle |E_i|\rangle} \approx \tgh \overline{(\langle E_i\rangle^2)}^{1/2}.$$
(15)

For $P(J_{ij}) = P(-J_{ij})$ and $B_0 = 0$ one has with

$$\overline{J_{ij}\langle S_j\rangle} = 0 \text{ and } \Delta^2 \equiv \sum_j \overline{J_{ij}^2} \text{ (with a cut off)}$$

$$\overline{\langle E_i\rangle^2} = \sum_j \overline{J_{ij}^2 \langle S_j\rangle^2} \approx \Delta^2 q$$
(16)

if one neglects correlations between $J_{ij}\langle S_j\rangle$ and $J_{il}\langle S_l\rangle$ for $j \neq l$. This leads to a simple self-consistent equation for the parameter q

$$q^{1/2} = \tgh \beta \Delta q^{1/2}.$$
(17)

For $T \approx T_f$ with $kT_f = \Delta$ eq. (17) yields results which are not too different from those of the EA theory [17]. For $T \ll T_f$ the results are different.

In the same approximation one derives a dynamical equation for $q(t) = \overline{\langle S_i\rangle^2}(t)$

$$\left(1 + \tau\frac{d}{dt}\right)q^{1/2} = \tgh \beta \Delta q^{1/2}$$
(18)

with the solution for $T \approx T_f$ (compare [32])

$$q(t) = q(0)\exp\left[-2\left(1-\frac{T_f}{T}\right)\frac{t}{\tau}\right] \quad T > T_f \quad (19a)$$

$$= \left[\frac{1}{q(0)} + \frac{2t}{3\tau}\right]^{-1} \quad T = T_f \quad (19b)$$

$$= \left\{\sqrt{q} + (\sqrt{q(0)} - \sqrt{q})\exp\left[2\left(1-\frac{T_f}{T}\right)\frac{t}{\tau}\right]\right\}^2$$
$$T < T_f, \quad (19c)$$

where $q \equiv q(t = \infty)$. The results (9) and (19a) are rather similar though the quantities involved are slightly different. In both cases one obtains critical slowing down at T_f. Similar results are obtained for the self-correlation function $\overline{\langle S_i(t)S_i(0)\rangle}$ from the master equation (11) for $T > T_f$ [31].

From the fluctuation–dissipation theorem [32] and with (10) the dynamical susceptibility is

$$\chi_{ii}(\omega) = \chi^0 \frac{1-q}{1+i\omega\tau_1}, \quad T < T_f \quad (20)$$

with $\chi^0 = N(g\mu_B)^2/kT$ where the inverse relaxation time $\tau_1^{-1} = qkT_f/\nu$ goes to zero for $T = T_f$. For $\omega = 0$ eq. (20) reduces to (8). For $T > T_f$ one has $\tau_1^{-1} = 2k(T - T_f)/\nu$.

In the dynamical theories described so far no remanence effects are obtained, since the system always relaxes into its equilibrium state. Furthermore, in order to obtain remanence one has to study the system with a finite field included.

4.3. Monte-Carlo calculations

The most extensive Monte-Carlo calculations of both static and dynamical properties of spin glasses have been performed by Binder and Schröder [25] for two and three-dimensional Ising models with a symmetrical Gaussian distribution of n.n. interactions. The authors calculated the spin relaxation $\overline{\langle S_i \rangle}(t)$, energy relaxation $\langle E \rangle(t)$, and relaxations of $q(t)$ and $\overline{\langle S_i(t) S_i(0) \rangle}$ from the master equation (11). Starting with a ferromagnetic spin configuration, the system relaxes in a fairly short time ($\sim \tau$) to a metastable state with remanent magnetization $\overline{\langle S_i \rangle} \neq 0$. In agreement with the experiments [11], $\overline{\langle S_i \rangle}$ is largest for $T = 0$ and vanishes for $T = T_f$. On a "large" time scale ($t \gg \tau$) one observes aftereffects: The remanent magnetization vanishes as $\overline{\langle S_i \rangle}(t) \propto t^{-a}$, $a \approx kT/2(\Delta J)$, where ΔJ is the width of the distribution of n.n. interactions. It is remarkable that both remanence and aftereffects are obtained without involving additional uniaxial anisotropy or domain structure [11]. Of course, the mechanism of the interaction with the heat bath remains unknown in the theories discussed [25, 30, 31].

With decreasing temperatures the relaxation of $\overline{\langle S_i(t) S_i \rangle}$ slows down. For $T \approx T_f$ the relaxation becomes rather slow after an initial decrease from $\langle S_i^2 \rangle = 1$. Unfortunately, in the time available it could not be decided whether or not $\overline{\langle S_i(t) S_i(0) \rangle}$ goes to a finite value for $T \leq T_f$.

5. Recent developments

Recently the self-consistent equation for the spin-glass parameter $q(T)$, eq. (8), as obtained in the EA theory has been rederived without using the $n \to 0$ trick [33–36]. The susceptibility, eq. (8), remains the same, whereas the free energy and therewith the thermal properties such as entropy and specific heat are different [33, 34]. If one modifies the distribution $P(J)$ in a suitable way [33] the entropy for $T \to 0$ no longer becomes negative [24]. The same holds true for the free energy obtained in a spherical model for an infinite range Ising spin glass [26, 37].

Neutron scattering [38] indicates that the cusp observed in the static susceptibility with momentum transfer $q = 0$ persists also for small finite q-values ($q \leq 2 \times 10^{-2}$ Å^{-1}). However, the temperature of the cusp decreases with increasing q. This can be explained by the assumption of clusters which persist below T_f and freeze in at lower temperature only. The scattering intensity at $T_f(q)$ decreases strongly indicating a sharp decrease of the number of spins involved.

6. Conclusions

Despite considerable effort some of the most important questions about spin glasses are still unsettled.

(1) Is there a phase transition at T_f? If so, what type is it? Whereas the cusp of $\chi(T)$ can be explained by the sudden freezing-in a macroscopic number of spins, it remains unclear why no anomalous behaviour is seen in the specific heat at T_f.

(2) If one believes the specific heat experiments, there is considerable cluster formation above T_f. However, it is not yet clear whether or not percolation is important at T_f.

(3) Remanence and aftereffects seem to be essential for spin glasses. But the nature of the energy barriers involved is obscure. Is an additional anisotropic interaction essential?

(4) Well below T_f one expects collective excitations. Antiferromagnetic excitations (with a spin wave stiffness $D \propto |k|$) [30] have been proposed. The temperature dependence of the specific heat and resistivity calculated with these excitations are in disagreement with the experimental data. Therefore, a domain structure has been invoked [30], where a domain structure separates two of the many degenerate

ground states. For a Heisenberg model with $P(J_{ij}) = P(-J_{ij})$ this degeneracy is at least of the order 2^N (N is the number of spins), since a transformation $J_{ij} \to -J_{ij}$ and $S_j \to -S_j$ (for fixed S_i) leaves the energy for $B_0 = 0$ unchanged. Thus there still is no explanation of the linear temperature dependence of the specific heat for $T \to 0$ despite the fact that older [3] and more recent [18] MFA's predict such a behaviour.

I would like to express my sincere thanks to K. Binder, B. Coles, T.J. Hicks and J. Mydosh for discussions which considerably improved my understanding of spin glasses. Corresponding with P. Wells and L. Wenger is gratefully acknowledged. I would also like to thank P.A. Beck, S. Edwards, K. Binder, D.L. Huber, R.W. Tustison, and A.P. Young for sending preprints.

References

[1] J.E. Zimmermann and F.E. Hoarse, J. Phys. Chem. Solids 17 (1960) 52.
[2] J. De Nobel and F.J. Du Chatenier, Physica 25 (1959) 969.
J. Du Chatenier et al., Physica 32 (1966) 403, 561, 1097.
[3] W. Marshall, Phys. Rev. 118 (1960) 1519.
[4] M.W. Klein and R. Brout, Phys. Rev. 132 (1963) 2412.
[5] V. Cannella, J.A. Mydosh and J.I. Budnick, J. Appl. Phys. 42 (1971) 1689.
[6] J. Mydosh, A.I.P. Conf. Proc. 24 (1974) 131.
[7] C.E. Violet and R.J. Borg, Phys. Rev. 149 (1966) 540.
B. Window, Phys. Rev. B6 (1972) 2013.
[8] A. Arrott, J. App. Phys. 36 (1965) 1093.
[9] H. Sato, S.A. Werner, and R. Kikuchi, J. Phys. (Paris) 35, C4 (1974) 25.
N. Ahmed and T.J. Hicks, J. Phys. F: Metal Phys. 5 (1975) 27, 2168, and to be published.
[10] J.S. Kouvel, J. Phys. Chem. Solids 21 (1961) 57.
[11] J.L. Tholence and R. Tournier, J. Phys. (Paris) 35, C4 (1974) 229.
J.L. Tholence, thesis, University of Grenoble (1973).
[12] A. Mukhopadhyay and P.A. Beck, Solid State Commun. 16 (1975) 1067.
A. Mukhopadhyay et al., J. Less Comm. Metals 43 (1975) 69.
[13] L.E. Wenger and P.H. Keesom, Phys. Rev. B11 (1975) 3497, and to be published.
[14] R.W. Tustison, Solid State Comm. (1976).
R.W. Tustison and P.A. Beck, Solid State Comm. (1976).
[15] D. Korn, Z. Physik 187 (1965) 463.
[16] A. Blandin, thesis, University of Paris (1961).
J. Souletie and R. Tournier, R. Low. Temp. Physics 1 (1969) 95.
[17] K.H. Fischer, Solid State Comm. 18 (1976) 1515.
[18] S.F. Edwards and P.W. Anderson, J. Phys. F: Metal Phys. 5 (1975) 965.
[19] A.P. Murani, S. Roth, P. Radhakrishna, B.D. Rainford, B.R. Coles, K. Ibel, G. Götze, and F. Mezei, J. Phys. F: Metal Phys. 6 (1976) 425.
[20] H. Scheuer, M. Loewenhaupt and W. Just, J. Magnetism and Magnetic Materials 2 (1976).
[21] A.I. Larkin and D.E. Khmelnitzkii, Sov. Phys. JETP 31 (1970) 958.
K. Matho, to be published.
[22] K.H. Fischer, Phys. Rev. Letters 34 (1975) 1348.
[23] D. Sherrington and B.W. Southern, J. Phys. F: Metal Phys. 5 (1975) L49.
[24] D. Sherrington and S. Kirkpatrick, Phys. Rev. Letters 35 (1975) 1792.
[25] K. Binder and K. Schröder, Solid State Comm. (1976) and Phys. Rev. B (1976).
[26] J.M. Kosterlitz, D.J. Thouless and R.C. Jones, Phys. Rev. Letters 36 (1976) 1217.
[27] D.J. Thouless, P.W. Anderson, E. Lieb and R.G. Palm, to be published.
[28] D.A. Smith, J. Phys. F: Metal Phys. 4 (1974) L266; 5 (1975) 2148.
[29] C.N. Guy, J. Phys. F: Metal Phys. 5 (1975) L242.
[30] S.F. Edwards and P.W. Anderson, J. Phys. F: Metal Phys. 6 (1976) 1927.
[31] W. Kinzel and K.H. Fischer, to be published.
[32] M. Suzuki and R. Kubo, J. Phys. Soc. Japan 24 (1968) 51.
[33] B.W. Southern, preprint.
[34] M.W. Klein, preprint.
[35] T. Plefka, preprint.
[36] T. Kaneyoshi, J. Phys. C: Solid State Phys. 9 (1976) L289.
[37] B.W. Southern and A.P. Young, preprint.
[38] A.P. Murani, Phys. Rev. Letters 37 (1976) 450.

REMANENCE REVERSAL IN PARTIALLY ORDERED Ni₃Mn*

T. SATOH†, R.B. GOLDFARB and C.E. PATTON
Department of Physics, Colorado State University, Fort Collins, Colorado 80523, USA

Partially ordered Ni₃Mn alloys, field-cooled to 4 K, exhibit a remanent magnetization which decreases upon warming, changes sign, and becomes opposite to the direction of the initial cooling field. This phenomenon is explained in terms of Mn atoms with different local environments.

1. Introduction

Magnetic properties of Ni₃Mn alloys have been qualitatively explained in terms of competing magnetic interactions between antiferromagnetic Mn–Mn atomic pairs, and ferromagnetic Mn–Ni and Ni–Ni pairs [1]. Recently, a three phase model was proposed for the disordered alloy in which Mn atoms were classified according to their nearest-neighbor environment (NNE) [2]. One type of local aggregation, $Mn^{(A)}$, has a Mn-rich NNE, antiferromagnetic character, and a relatively high Néel temperature. The second type, $Mn^{(B)}$, has a NNE which is ~25 at.% Mn. It is exposed to a weak molecular field. The third type, $Mn^{(C)}$, has a Mn-deficient NNE and is ferromagnetic. Other theoretical and experimental work support the coexistence of ferro- and antiferromagnetic local aggregates [3, 4].

Compared to the alloy cooled in zero field, disordered Ni₃Mn, field-cooled to 4 K, is known to possess an exchange anisotropy characterized by a displaced hysteresis loop and a remanent magnetization M_r [5]. This is illustrated in fig. 1. The origin of this exchange anisotropy has been qualitatively explained by the three phase model [2].

In the present work, the remanent magnetization M_r has been measured as a function of alloy composition, heat treatment, cooling field, and temperature. For alloys near 25 at.% Mn, annealed from 1 to 30 min, M_r decreases rapidly on warming from 4 K and changes sign near 100 K. That is, the direction of the remanent magnetization becomes opposite to the direction of the initial cooling field. The data can

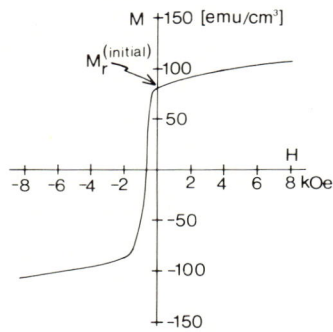

Fig. 1. Magnetization-versus-field curve at 4 K for a disordered 24.6 at.% Mn sample, field-cooled in 8 kOe. The initial remanent magnetization $M_r^{(initial)}$ is indicated.

be explained by the existence of $Mn^{(A)}$ which couples antiferromagnetically to its NNE.

2. Experiment and results

The disordered samples were prepared as described in ref. 2. Compositions analyzed were 22.1, 24.6 and 26.2 at.%Mn. Partially ordered states were achieved by annealing 1, 3, 11 and 30 min at 500°C. A heat treatment of 50 h at 500°C was used to obtain the equilibrium ordered state [6].

Remanent magnetization was measured with a commercial vibrating sample magnetometer. Considerable care was taken in measuring the relatively weak remanence. (1) Prior to measurement, the cooling-field magnet was removed. (2) Current to the temperature controller was switched off during measurement. (3) Two measurements were made, one with the sample rotated 180° to cancel any effect of asymmetry in the pick-up coils or sample position. The reversal in remanence was distinct.

In fig. 2, the remanent magnetization M_r is plotted against temperature T for the 24.6 at.% Mn alloy, annealed at 500°C for 11 min. Reversal behavior is shown for various cooling fields.

---

* Supported by the National Science Foundation, grant DMR73-02665.
† On leave from the Electrotechnical Laboratory, Tokyo, Japan.

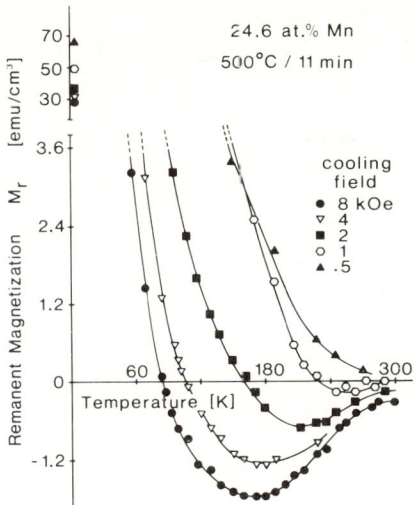

Fig. 2. Remanent magnetization M_r as a function of temperature for a 24.6 at.% Mn sample, partially ordered by annealing at 500°C for 11 min. The alloy was cooled to 4 K in various fields, as indicated. The initial remanence at 4 K for each cooling field is shown.

The initial remanence at 4 K, $M_r^{(initial)}$, and the temperature for zero remanence, T_0, decrease with increasing cooling field. The maximum reversed remanence, $M_r^{(rev)}$, increases with increasing cooling field. In fig. 3, the $M_r^{(initial)}$ and the maximum $M_r^{(rev)}$ are shown as a function of annealing time at 500°C for 24.6 at.% Mn cooled in 8 kOe. The largest reversed remanence occurs for the sample annealed for 11 min.

Several experimental facts are significant: (a) The only composition showing remanence reversal is 24.6 at.% Mn. (b) Only the partially ordered states exhibit remanence reversal. The disordered and equilibrium states show no effect. (c) Cooling fields greater than 2 kOe are required for significant remanence reversal. (d) T_0 is a sensitive function of cooling field and relatively insensitive to annealing treatment.

3. Discussion

The existence of Mn atoms with antiferromagnetic character with respect to their NNE, such as $Mn^{(A)}$, is attractive in explaining the remanence after field cooling [2] and the remanence reversal reported here. Upon field cooling, $M_r^{(initial)}$ is due to the field-aligned ferromagnetic NNE. The decrease in the $M_r^{(initial)}$ at 4 K with increasing cooling field (fig. 2) is due to an enhanced antiferromagnetic alignment of $Mn^{(A)}$ as the NNE is aligned with the field.

Now consider the role of $Mn^{(A)}$ in remanence reversal. Assume that the ferromagnetic NNE of the $Mn^{(A)}$ has a relatively low Curie temperature. Suppose that, upon warming, a small antiferromagnetic exchange survives to preserve the orientation of the $Mn^{(A)}$ opposite to the NNE and cooling field orientation. As the temperature increases, the net NNE moment will decrease. At some point it will fall below the $Mn^{(A)}$ moment and remanence reversal will occur.

The $Mn^{(A)}$ aggregates responsible for remanence reversal should be only those $Mn^{(A)}$ aggregates which are isolated, i.e. in which none of the NNE Mn atoms are also $Mn^{(A)}$. Multiple aggregates would be expected to have a rather complicated spin configuration and zero net moment. Hence, they would not contribute to the remanence reversal.

This qualitative explanation is supported by the observed maximum in the reversal effect for the partially ordered alloy. From a simple counting procedure based on a short range order (SRO) parameter related to annealing time, the number of isolated $Mn^{(A)}$ aggregates has been found to initially increase and then decrease with SRO development. Moreover, the fact that T_0 is more dependent on the cooling field than on annealing time, lends credibility to the proposition that isolated $Mn^{(A)}$ aggregates are responsible for the reversal in remanent magnetization. Based on the assumption of 1 μ_B per Mn atom [4] and a zero average moment for the matrix above 120 K, only a 0.008 existence

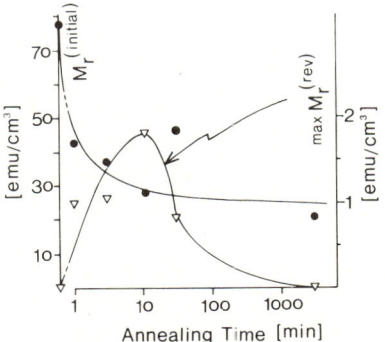

Fig. 3. Initial remanent magnetization $M_r^{(initial)}$ and maximum reversed remanence $M_r^{(rev)}$ as a function of annealing time at 500°C after an initially disordered state. The samples were 24.6 at.% Mn, field-cooled in 8 kOe.

probability for isolated Mn$^{(A)}$ is sufficient for a reversed remanence of 1.5 emu/cm^3.

Acknowledgement

The authors are grateful to Prof. H. Sato for illuminating discussions on atomic order development in Ni–Mn alloys.

References

[1] W.J. Carr, Jr., Phys. Rev. 85 (1952) 590.
[2] T. Satoh, C.E. Patton and R.B. Goldfarb, Joint MMM-Intermag. Conference, Pittsburgh, June 1976: AIP Conf. Proc. 34 (1976) 361.
[3] T. Jo, J. Phys. Soc. Jap. 40 (1976) 715; and this conference.
[4] Y. Kitaoka and K. Asayama, J. Phys. Soc. Jap. 40 (1976) 1521.
[5] J.S. Kouvel and C.D. Graham, Jr., J. Phys. Chem. Solids 11 (1959) 220.
[6] C.E. Patton, M. Vardeman and G.L. Baker, IEEE Trans. Mag. 11 (1975) 1350.

MAGNETIC ORDER IN FRESHWATER FERROMANGANESE NODULES

J.M.D. COEY

Groupe des Transitions de Phases, CNRS BP 166, 38042 Grenoble Cedex, France

and

D.W. SCHINDLER

Freshwater Institute, University of Manitoba, Winnipeg R3T 2N6, Canada

A set of eight ferromanganese concretations whose crystallinity deteriorates with increasing manganese content all order magnetically well above 4.2 K. The series presents a continuous evolution from antiferromagnetic order in the iron-rich members, composed essentially of cristalline goethite, to speromagnetic order in the manganese-rich ones, which are amorphous.

The study of magnetic order in amorphous insulating compounds has been hampered to date by the lack of suitable materials. In such compounds the interaction between a pair of magnetic atoms will be short-ranged antiferromagnetic superexchange, in contrast to the well-known metallic glasses where the exchange tends to be long-ranged and ferromagnetic or oscillatory in character.

Here, we report on the magnetic properties of an interesting class of materials [1] which have some relevance in the context of amorphous magnetism. Out of thirty ferromanganese concretations taken from five lakes in the Experimental Lakes Area of Western Ontario, eight were selected for magnetic measurements. Their Fe and Mn content are listed in table I. The total of the other metals Na, K, Mg and Ca does not exceed 1 wt.% and C is approximately 3 wt.%. The weight loss on ignition was in the range 15–20%, hence 80–90% of the inorganic fraction of the samples is hydrated iron or manganese oxide. Those with the most manganese are X-ray amorphous whereas those with relatively more iron contain progressively better-crystallized goethite.

Mössbauer spectra at 4.2 K show a broad magnetic hyperfine spectrum with H_{hf} in the range 480–510 kOe. They resemble the spectrum of an amorphous ferric gel at 4.2 K [2]. The relative intensity of the lines for L17 were invariant in applied fields of up to 50 kOe, showing there is scarcely any preferred direction for the atomic moments relative to the applied field. Spectra at 296 K show only a ferric quadrupole doublet in all samples whose splitting decreases from 0.65 mm/s for L24 to 0.55 mm/s for L28.

Magnetization curves at 4.2 K are shown in fig. 1. For the more crystalline, iron-rich samples, they are practically linear with M_s, the moment extrapolated to $M = 0$, hardly different from zero, as expected for a normal antiferromagnet. For the amorphous manganese-rich

Table I
Composition and magnetic properties of ferromanganese nodules

Nodule	Fe (wt.%)	Mn (wt.%)	M_s (emu/g*)	χ_∞ (emu/g*)	T_c (K)
L24	19.7	17.8	4.3	2.9×10^{-4}	67
L17	32.5	12.3	3.9	2.3	96
L12	39.4	9.6	3.1	2.0	117
L1	36.9	7.8	3.0	2.1	115
L18	46.2	3.0	1.3	1.3	197
L7	47.2	1.0	0.4	1.1	244
L10	48.9	0.4	1.0	1.2	213
L28	36.8	0.2	0.6	1.1	235

* Measured at 4.2 K, expressed per gram of transition metal.

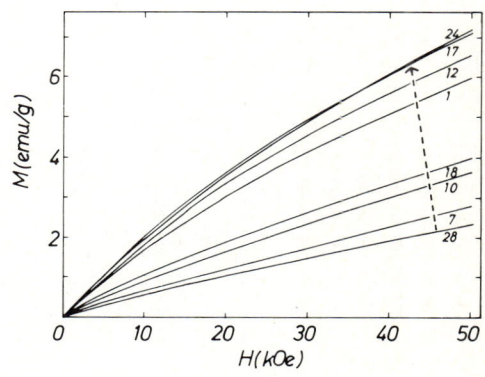

Fig. 1. Magnetization curves for eight nodules at 4.2 K. The arrow is in the sense of increasing amorphousness.

samples, they are convex and become linear only in fields >30 kOe. Similar curves have been reported for amorphous Fe(OH)$_3$ 0.9H$_2$O [2], amorphous YFe$_2$ [3] and amorphous iron-rich nodules in a polymer. Values of M_s and the high-field susceptibility χ_∞ are given in table I.

Provided the anisotropy energy in L17 is less than 3×10^5 erg/cm^3 the field invariance of its Mössbauer spectrum implies that it is not antiferromagnetic (i.e. spins within a domain all parallel or antiparallel to a given direction) or else there would have been a spin flop. Instead, it is probably *speromagnetic*, a name given [2] to a type of magnetic order found in amorphous solids, where the spins within a domain are distributed in *all* directions, with no overall correlations beyond r_0, at most a few nearest neighbour distances. Formally antiferromagnetic order is described by the conditions $\bar{S}_{iz} \neq 0$; $\bar{S}_i \cdot \bar{S}_j = \pm \bar{S}_{iz}^2$; $\langle \bar{S}_i \cdot \bar{S}_j \rangle = f\bar{S}_{iz}^2$ whereas speromagnetic order is described by the conditions $\bar{S}_{iz} \neq 0$; $\bar{S}_i \cdot \bar{S}_j = g\bar{S}_{iz}^2$; $\langle \bar{S}_i \cdot \bar{S}_j \rangle = 0$, $r_{ij} > r_0$, where $-$ is a time average and $\langle \rangle$ an average over all pairs ij the same distance r_{ij} apart. f is a rational fraction and g a number, both between 1 and -1. Speromagnetism is therefore a type of magnetic order (one of several new types found in amorphous solids [4, 5]) essentially similar to theoreticians' [6] but not experimentalists' understanding of the term "spin glass".

The evolution of magnetic order with crystallinity is illustrated schematically in fig. 2, where the sense of the dashed arrow is the same as in fig. 1. A rough estimate of T_c, the magnetic ordering temperature for the nodules is obtained from the expression for an antiferromagnetic powder $T_c = Ng^2\mu_B^2 S(S+1)/9k\chi_\infty$, which will also be valid, as a first approximation, for a speromagnet. The values of T_c listed in table I show a clear tendency to decrease with increasing manganese content and amorphousness.

Two possible reasons for speromagnetic order may be identified. The one which probably applies to the Mn-rich nodules discussed here (and perhaps to YFe$_2$ as well) is a broadly distributed range of negative or positive and negative exchange fields [2, 6]. The other is a strong, randomly directed local anisotropy together with similarly distributed exchange. However, the influence of local anisotropy on Fe^{3+} or Mn^{2+} should be small, and it is translated into a bulk anisotropy through the effect of texture. An ideal amorphous solid is isotropic, i.e. free of texture, and would have no bulk anisotropy. Since $K \simeq 10^5$ erg/cm^3 for crystalline goethite, it should be even smaller in the more amorphous nodules.

A plausible explanation for the convex shape of the $M(H)$ curves of amorphous speromagnetic samples is that it is caused by a few percent of atoms which experience a very weak exchange field and are almost paramagnetic. A small central peak in the 4.2 K Mössbauer spectrum may be identified with such atoms. However, superparamagnetism of amorphous fine particles $\simeq 100$ Å may also contribute as in the ferric gel [2] and the polymer containing iron-rich nodules [7].

In conclusion, the freshwater ferromanganese nodules are a good system for demonstrating qualitatively the influence of crystallinity on the magnetism of compounds with antiferromagnetic superexchange interactions. As natural materials, they are difficult to characterize completely and the effect of the manganese is not clearly separated from that of the amorphousness.

References

[1] W.S. Broecker, Chemical Oceanography (Harcourt, Brace & Jovanovich, New York, 1974) ch. IV.
[2] J.M.D. Coey and P.W. Readman, Earth Planet Sci. Lett. 21 (1973) 45; Nature 246 (1973) 476. See also Nature 246 (1973) 445.
[3] J.J. Rhyne, J.M. Schelleng and N.C. Koon, Phys. Rev. B10 (1974) 4672.
[4] J.M.D. Coey, J. Chappert, J.P. Rebouillat and T.S. Wang, Phys. Rev. Lett. 36 (1976) 1061.
[5] J.P. Rebouillat, A. Lienard, J.M.D. Coey, R. Arrese-Boggiano and J. Chappert, these proceedings.
[6] See, for example, D. Sherrington, AIP Conf. Proc. 29 (1975) 224.
[7] C. Meyer, J. Phys. 37 (1976) C6 in press.

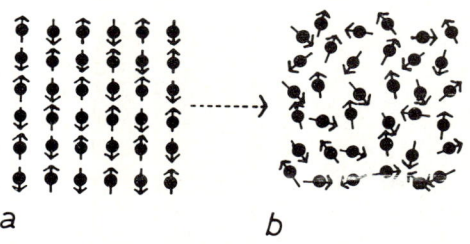

Fig. 2. Evolution of the atomic and magnetic structure for the series of nodules from (a) crystalline antiferromagnet to (b) amorphous speromagnet.

THE LOW TEMPERATURE HEAT CAPACITY OF Ni–Cu ALLOYS IN A MAGNETIC FIELD

P.C. LANCHESTER, N.F. WHITEHEAD* and P. WELLS†

Physics Department, Southampton University, Southampton SO9 5NH, England

The heat capacity of nearly ferromagnetic Ni–Cu alloys has been measured from 0.3 to 4.0 K in fields up to 6.5 T. Having subtracted the electronic term in a novel way, the field dependence of the excess heat capacity C_M for $T \geq 1.8$ K is approximately that expected for superparamagnetic clusters. However as $T \to 0$, $C_M \to 0$ linearly rather than exponentially.

1. Introduction

Near the critical concentration for ferromagnetism of alloys such as Ni–Cu and Fe–V the low temperature heat capacity is found to be of the form

$$C = \gamma T + AT^3 + C_M, \qquad (1)$$

where C_M is associated with the presence of superparamagnetic clusters. It is usually assumed that C_M arises from oscillation of these clusters in a local anisotropy field H_a for which

$$C_M = BE(T_E/T), \qquad (2)$$

where B is a constant proportional to the cluster concentration and $E(T_E/T)$ is the Einstein function with $kT_E = 2\mu_B H_a$ [1]. In this model B is not expected to vary in an applied field H, but C_M should decrease in increasing fields as H begins to dominate H_a causing the effective field H' and the Einstein temperature T'_E to rise. For $T \ll T'_E$, C_M should vary exponentially with temperature.

Although C_M is known to decrease with increasing field, the high field results for Fe–V when fitted to eqs. (1) and (2) suggest that B decreases and that γ too is field dependent [2]. In order to study these unexpected effects we have measured the high field heat capacity of Ni–Cu alloys, which are probably the best characterized alloys of this kind.

2. Experimental

Three $Ni_{1-x}Cu_x$ samples with $x = 0.558$(A), $x = 0.589$(B) and $x = 0.601$(C), have been stu-

* Post Office Development Department, 100–110 High Holborn, London, England.
† Department of Applied Physics, Caulfield Institute of Technology, Victoria 3145, Australia.

died between 0.3 and 4.0 K in fields up to 6.5 T. The samples were prepared by Metals Research Ltd., of Cambridge from materials of 5 N purity, containing <10 ppm Fe. They were homogenized by annealing in vacuo at 1100 C for 72 h and then water quenched. Measurements were made in a ^{3}He cryostat using a carbon resistance thermometer calibrated in situ against a magnetically shielded standard [3, 4].

3. Results

As the lattice term was too small to determine reliably this was calculated assuming $\Theta_D = 390$ K [5]. In high fields a small hyperfine term due to the copper was also allowed for. The resultant data C' was then fitted to $C' = \gamma T + BE(T_E/T)$ yielding results remarkably similar to those for Fe–V [2]. The cluster concentration appeared to vanish above 2 T and γ increased by as much as 30% up to 2 T, falling slightly at higher fields [3].

However, it is unlikely that the true electronic γ is so field dependent and a reduction in the cluster concentration contradicts the high field magnetization data [6] which gives cluster concentrations in reasonable accord with the zero field heat capacity. We conclude that eq. (2) is a poor representation of C_M. Further evidence of this both for the Ni–Cu and Fe–V alloys is apparent in the behaviour of C_M at temperatures well below that at which C_M peaks, which is always linear in $T(C_M = \gamma'T)$ rather than exponential.

In the absence of a satisfactory theory an empirical approach has been adopted to evaluate C_M. As the applied field $H \to \infty$, γ' for each sample is observed to tend to a limiting value $\gamma(\infty)$, which has been taken to be the true electronic γ for all fields (fig. 1). That is, we assume the electronic γ is independent of field and that

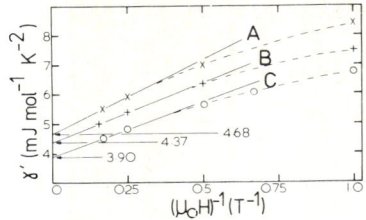

Fig. 1. γ' versus $(\mu_0 H)^{-1}$ for Ni–Cu alloys.

there is a cluster component C_M linear in T at low temperatures which is suppressed in very high fields. Values of $C_M = C' - \gamma(\infty)T$ for all samples in zero field are shown in fig. 2(a), whilst fig. 2(b) shows the field dependence of C_M for sample A which is typical of all the samples.

It is likely that the fall in C_M towards 4 K observed in the zero field results for samples A and C [fig. 2(a)] is a general feature resulting from the finite spin of the clusters [5]. The fact that this does not occur for sample B could be due to a small error in $\gamma(\infty)$ for this sample. Below 1.6 K these results fit the Einstein function reasonably well with the values of B and H_a given in table I.

An applied field increases the effective cluster concentration so that in moderate fields C_M towards 4 K increases before falling again at higher fields [fig. 2(b)]. The behaviour above about 1.5 K can be fitted very approximately to the superparamagnetic model by assuming that for $\mu_0 H \geq 1$ T the cluster concentration is constant, corresponding to a field independent value of $B' > B$ and that the effective field is $H' = H + H'_a$ where $H'_a \simeq H_a$ is also field independent. This is illustrated by the continuous curves in fig. 2(b) and the related parameters B' and H'_a are given in table I.

As might be expected the value of B' are in better agreement with the corresponding values deduced from the high field magnetization [5] than are the values of B.

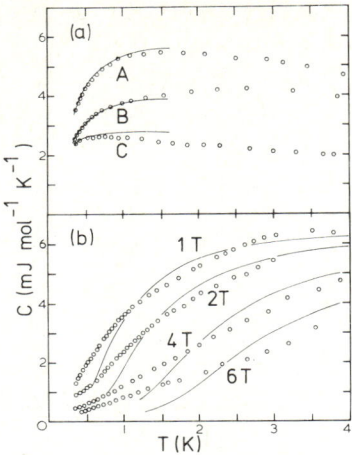

Fig. 2. (a) $C_M = C' - \gamma(\infty)T$ versus T for $H = 0$. The solid lines are fits to the Einstein function for $T < 1.6$ K. (b) C_M versus T for sample A in fields $\mu_0 H = 1, 2, 4$ and 6 T. The solid lines correspond to the Einstein function with $B' = 6.5$ mJ mol^{-1} K^{-1}, $\mu_0 H' = \mu_0 H + 1.1$ T.

Table I
Fitted parameters for Ni–Cu alloys

Sample	$\gamma(\infty)$ (mJ mol^{-1} K^{-2})	B (mJ mol^{-1} K^{-1})	$\mu_0 H_a$ (T)	B' (mJ mol^{-1} K^{-1})	$\mu_0 H'$(T) $= \mu_0(H + H'_a)$
A	4.68 ± 0.07	5.8	0.75	6.5	$\mu_0 H + 1.1$
B	4.37 ± 0.07	4.0	0.66	5	$\mu_0 H + 0.75$
C	3.90 ± 0.07	2.8	0.38	4	$\mu_0 H + 0.38$

4. Conclusion

The analysis we have used brings the interpretation of the heat capacity into line with the magnetization results, so that both broadly support the superparamagnetic model. However, the detailed behaviour and particularly the linear dependence of C_M as $T \to 0$ in high fields is not understood although it is qualitatively like that predicted for very dilute impurities by Klein [7].

References

[1] K. Schroder, J. Appl. Phys. 32 (1961) 880.
[2] W. Procter and R.G. Scurlock, 11th Int. Conf. Low Temp. Phys. St. Andrews (1968) 1320.
[3] N.F. Whitehead, Ph.D. Thesis, University of Southampton, England (1974).
[4] N.F. Whitehead, P.C. Lanchester and R.G. Scurlock, J. Sci. Inst. 7 (1974) 117.
[5] R.L. Falge, Jr. and N.M. Wolcott, J. Low Temp. Phys. 5 (1971) 617.
[6] F. Acker and R. Huguenin, Phys. Lett. 38A (1972) 343.
[7] M.W. Klein, Phys. Rev. 188 (1969) 933.

THE ROLE OF FLUCTUATIONS IN THE EVALUATION OF THE SPECIFIC HEAT IN DILUTE MAGNETIC ALLOYS

I. RIESS

Physics Department, Technion – IIT, Haifa, Israel

We have included fluctuations in the calculations for the Ising dilute magnetic alloy. It is found that the specific heat C_M exhibits a broad maximum above the spin glass transition temperature T_c. At T_c, C_M exhibits a small λ-type discontinuity. For $T \to 0$, $C_M \propto T$. The susceptibility exhibits a cusp at T_c.

The purpose of this paper is to discuss the inclusion of fluctuations in the theory of dilute magnetic alloys [1–3], and to present recent results for the magnetic specific heat C_M and susceptibility χ in these alloys. We in particular concentrate on the specific heat above the spin glass transition temperature T_c.

The self-consistent mean field theory (MRFA) by Klein [4] was most successful in deriving the low temperature ($T \ll T_c$) properties of dilute magnetic alloys [5]. This approach predicted an infinite T_c thus the existence of internal magnetic fields at all temperatures. However, it was shown by Cannella and Mydosh [6] that the susceptibility exhibits a cusp at a finite temperature T_{cusp}. The cusp was associated with a magnetic phase transition [6, 7], so that $T_c = T_{\text{cusp}}$. The MRFA was recently reconsidered by limiting the RKKY interaction *below* nearest neighbours [8]. This gives a finite T_c and a cusp in χ. However, two problems still remained with the MRFA. The main one is typical to any mean field approximation (MFA) that it gives a zero specific heat above T_c. The other problem was associated with the shape of the cusp which had a negative slope $\partial \chi / \partial T$ just below T_c rather than a positive one as seen experimentally [6].

Edwards and Anderson [9] suggested a different approach which seemed to include correlations between spins beyond the trivial ones included in the MFA. They also introduced a spin glass order parameter, the spatial average of the square of the local spin $q = \overline{\langle S_l^z \rangle^2}$. The latter however was not new since Klein [4] has considered the order parameter $m = \overline{|\langle S_l^z \rangle|}$ which had a similar meaning. The interesting result of their work was that they obtained a finite temperature above which the order parameter vanished. At that temperature (i.e. T_c) χ exhibits a cusp. This approach was then extended by others [10, 11]. We see however three main physical difficulties associated with this approach. (a) The use of a gaussian distribution $P(J)$ of the exchange interaction which is far from representing the RKKY interaction in a dilute alloy. (b) The specific heat result for $T > T_c$ is $C_M \propto T_c^{-2}$ while experimentally C_M goes through a maximum at $T_m > T_c$ with (almost?) no cusp at T_c [12]. (c) The entropy is negative at low temperatures [11].

An alternative approach was suggested by Riess [1, 2] and Riess and Mavroyannis [3]. In this approach fluctuations due to the spin–spin interaction are considered as well as mean fields. We have used a Green function technique and have decoupled the equation of motion at the second order to be able to include the fluctuations. This leads to the following modification of the quasi single spin states. In the MFA the many spin system is decoupled into individual spins having their eigenstates $\uparrow$ and $\downarrow$ shifted by an energy which depends on the interaction. The latter is proportional to the local magnetic field. We have calculated the broadening of the density of states $\rho_{l,\sigma}(\omega)$ ($\sigma = \uparrow$ or $\downarrow$, l indicates the spin site) as well as a modification in the shift in the energy of the state. Strong fluctuations are reflected by a large broadening of $\rho_{l,\sigma}(\omega)$. For $T > T_c$, $\rho_{l,\uparrow}(\omega) = \rho_{l,\downarrow}(\omega) \neq \delta(\omega)$, i.e. the states $l \uparrow$ and $l \downarrow$ have degenerate broadened density of states. For $T < T_c$ or in an external magnetic field $\rho_{l,\uparrow}(\omega)$ is shifted relative to $\rho_{l,\downarrow}(\omega)$ so that $\langle S_l^z \rangle \neq 0$.

Since $\rho_{l,\sigma}(\omega) \neq \delta(\omega)$ at $T > T_c$ the average energy of the system changes with T and the specific heat does not vanish at $T > T_c$. This is quite different from the MFA where $C_M \equiv 0$ at $T > T_c$.

We have then shown [2, 3] that to treat the specific heat and susceptibility at all temperatures one has to consider the possibility of coexistence of local magnetic fields (H) and fluctuations (V). The dilute alloy is characterized by a joint distribution function $P(H, V)$. The latter depends on the order parameter m. On the other hand the order parameter m is the average of $|\langle S_i^z \rangle|$ over the whole system, i.e. over $P(H, V)$. Thus m and $P(H, V)$ should be calculated self-consistently. We have solved for $P(H, V)$ and m. The highest temperature at which m does not vanish is T_c. $P(H, V)$ was then used to calculate C_M and χ from expressions derived before [2]. We find that the specific heat $C_M \propto T$ for $T \to 0$. It has a small λ-type discontinuity at T_c and a large rounded maximum at $T_m > T_c$. $C_M \propto T^{-1}$ for $T \gg T_m$. This is shown in fig. 1. The susceptibility was also calculated. $\chi \simeq$ constant for $T \to 0$. χ exhibits a cusp at T_c and decreases as T^{-1} for $T > T_c$. The details of the calculations and the approximations used for $P(H, V)$ will be given elsewhere [13]. Our results differ from previous ones [4, 9, 10, 11] mainly for C_M at $T > T_c$ where we get that C_M has a broad maximum as observed experimentally.

We wish to thank Professor M.W. Klein for many helpful discussions.

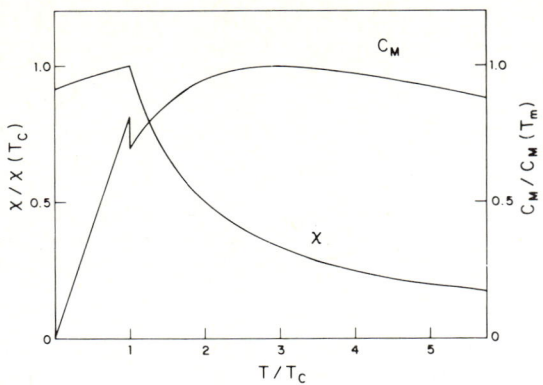

Fig. 1. $C_M/C_M(T_m)$ and $\chi/\chi(T_c)$ versus T for a dilute magnetic alloy.

References

[1] I. Riess and A. Ron, Phys. Rev. B8 (1973) 3467.
[2] I. Riess, Physica 74 (1974) 496.
[3] I. Riess and C. Mavroyannis, Physica 75 (1974) 483.
[4] M.W. Klein, Phys. Rev. 173 (1968) 552, Phys. Rev. 188 (1969) 933.
[5] J.F. Franz and D.J. Sellymer, Phys. Rev. B8 (1973) 2083.
[6] V. Cannella and J.A. Mydosh, Phys. Rev. B6 (1972) 4220.
[7] K. Adkins and N. Rivier, J. Phys. Paris 35, C4 (1974) 237.
[8] I. Riess and M.W. Klein, Phys. Rev. B (1976).
[9] S.F. Edwards and P.W. Anderson, J. Phys. F5 (1975) 965.
[10] K.H. Fischer, Phys. Rev. Letters 34 (1975) 1438.
D. Sherrington and B.W. Southern, J. Phys. F5 (1975) L49.
[11] D. Sherrington and S. Kirkpatrick, Phys. Rev. Lett. 35 (1975) 1792.
[12] L.E. Wegner and P.H. Keesom, Phys. Rev. B13 (1976) 4053.
[13] I. Reiss, to be published.

ELECTRICAL RESISTIVITY OF THE PtCr SYSTEM THROUGH THE PERCOLATION LIMIT*

R.M. ROSHKO and G. WILLIAMS

Department of Physics, University of Manitoba, Winnipeg R3T 2N2, Canada

The electrical resistivity of PtCr alloys, containing between 13 and 18 at.% Cr, have been measured between 1.4 and 300 K in an attempt to establish the influence of comparatively high single impurity characteristic temperatures (T_K or T_S) on the onset of magnetic order in such systems.

1. Introduction

The electric and magnetic properties of canonical spin glass systems (systems formed from moderate concentrations of first transition series impurities in Cu, Ag or Au) have recently been the subject of extensive experimental [1] and theoretical [2] activity. One prerequisite in a canonical spin glass (csg) is that the isolated impurity characteristic temperature (T_K or T_S) is well below the spin freezing temperature T_0, so that at this latter temperature the impurities have a well defined magnetic moment. In this way conduction electron scattering from "single" impurities has little influence on the resistivity around T_0, manifesting itself possibly only via a weak temperature dependent contribution to the resistivity above the maximum [1]. We have recently [3] begun to investigate the onset of magnetic ordering in alloy systems with substantially higher values for T_K or T_S, and in this paper we report preliminary data on PtCr alloys containing between 13 and 18 at.% Cr.

2. Experimental details

The alloys were prepared individually by arc melting from 5N's pure Pt wire and 5N's pure Cr beads; melting losses were negligibly small. Resistivity samples in the form of thin strips were cut from cold rolled buttons of the various alloys and were annealed for 24 h at 610°C. No evidence of superlattice lines were found in any of the samples.

Resistance measurements were performed using the standard four probe potentiometric technique; the sample temperatures were found from ^{4}He vapour pressure tables below 4.2 K, and by nonlinear gas thermometer techniques above 4.2 K.

* Work supported in part by the National Research Council of Canada.

3. Results and discussion

The characteristic temperature associated with isolated Cr impurities in Pt has been variously estimated [4, 5] between 50 and 400 K.

We summarize our experimental results in fig. 1, in which the resistivity $\rho(T)$ and the incremental resistivity $\Delta\rho(T)$ [$= \rho_{\text{Alloy}}(T) - \rho_{\text{Pt}}(T)$] for the alloys of lowest (13 at.%) and highest

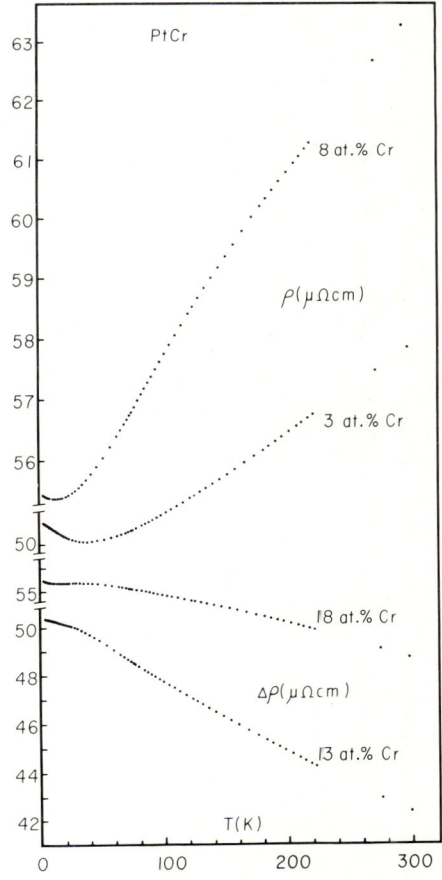

Fig. 1. The resistivity $\rho(T)$ and incremental resistivity $\Delta\rho(T)$ (both in $\mu\Omega$cm) plotted against temperature T (in K), for the 13 and 18 at.% Cr samples.

(18 at.%) Cr content are reproduced as a function of temperature up to 300 K. Minima in these resistivity versus temperature curves occur for all the alloys examined, with the position of these minima decreasing from 38 K at 13 at.% Cr to around 10 K at 18 at.% Cr. Above these minima the resistivity increases smoothly with increasing temperature, but whereas in the 13, 14, 15 and 16 at.% Cr alloys the resistivity temperature curves are concave with respect to the temperature axis ($d\rho/dT$ increasing with increasing temperature), in the 17 and 18 at.% alloys the resistivity–temperature curves exhibit an inflexion point. As can be seen from the resistivity data on the most concentrated alloy in fig. 1, the position of this inflexion point is not easily estimated, although it does occur in the vicinity of 60 K in both the 17 and 18 at.% Cr samples.

The incremental resistivity of these alloys is comparatively featureless, decreasing almost monotonically with increasing temperature. In particular there is no well-defined resistivity maximum in any of the alloys studied, a feature which one normally associates with the occurrence of magnetic order. A careful examination of the $\Delta\rho$ vs. T curve for the 18 at.% Cr sample, however, does indicate some semblance of structure at low temperature which is reminiscent of the behaviour reported [6] for the VFe system with around 27.5 at.% Fe. In addition, the difference $\Delta\rho(T = 0) - \Delta\rho(T = 300 \text{ K})$, which falls smoothly from a value of 8.1 $\mu\Omega$cm at 13 at.% Cr to 6.3 $\mu\Omega$cm at 17 at.% Cr decreases abruptly to 3.7 $\mu\Omega$cm at 18 at.%. Such behaviour has led us to suspect that 18 at.% Cr in Pt was close to the critical concentration for the onset of magnetic order.

With this point in mind we have performed a.c. susceptibility measurements (at 500 Hz in an a.c. driving field of 0.3 Oe r.m.s.) on the resistivity samples, and a preliminary account of these is given below. The specimens containing 16 at.% Cr or less exhibited no anomalous behaviour, whereas the 17 and 18 at.% Cr alloys exhibited a peak in the a.c. susceptibility at low temperature. These peaks are not sharp as in AuFe, but are rounded as in hydrogenated giant moment systems [7]; that in the 17 at.% Cr sample is very weak and peaks around 26 K, whereas that in the 18 at.% Cr specimen is much stronger and is centred at 55 K.

In any event it appears that any long range coupling of the RKKY type in PtCr is very ineffective, with interactions occurring primarily through near-neighbour d–d overlap.

References

[1] P.J. Ford and J.A. Mydosh, Phys. Rev. B (1976) in press.
[2] N. Rivier and K. Adkins, J. Phys. F; Metal Phys. 5 (1975) 1745.
[3] R.M. Roshko and Gwyn Williams, in: Proc. LT14, Vol. 3, M. Krusius and M. Vuorio (Eds.) (North-Holland, Amsterdam) p. 274.
[4] W.M. Star, E. de Vroede and C. van Baarle, Physica 59 (1972) 128.
[5] R.M. Roshko and Gwyn Williams, Phys. Rev. B9 (1974) 4945.
[6] H. Claus and J.A. Mydosh, Solid State Commun. 14 (1974) 209.
[7] J.P. Burger, D.S. MacLachlan, R. Mailfert and B. Souffaché, p. 278 of ref. 3.

ELECTRICAL RESISTIVITY OF Fe DISSOLVED IN PdAg MATRIX ALLOYS

K.V. RAO*
Clarkson College of Technology, Potsdam, New York 13676, USA

Ö. RAPP, CH. JOHANNESSON, H.U. ÅSTRÖM†
Royal Institute of Technology, Stockholm, Sweden

J.I. BUDNICK, T.J. BURCH and V. NICULESCU‡
University of Connecticut, Storrs, Connecticut 06268, USA

The temperature dependence of the resistivities of the system $Pd_{1-x}Ag_x$ up to $x = 0.5$, doped with 0.5, 1.0 and 2 at.% Fe has been measured. Clear reduction in the resistivities near the ferromagnetic and spin glass ordering temperatures are observed. The resistance minima observed in some of these alloys are discussed in terms of local s–d effects.

The $(Pd_{1-x}Ag_x)_{1-y}Fe_y$ alloy system is interesting because as the Ag concentration of the host is increased, reducing the matrix susceptibility, the alloys show a variety of magnetic behaviour [1]. For instance, studies of the temperature dependence of the low field magnetic susceptibility, $\chi(T)$ of alloys, containing 1 at.% Fe have shown that those with $0 \leq x \leq 0.25$ are ferromagnetic while those with $0.33 \leq x \leq 0.50$ have sharp susceptibility peaks thought to be characteristic of spin glass type ordering. This paper reports the salient features of a study of the effects of magnetic ordering on the electrical resistivity of these alloys including the spin glass region.

The resistivity samples for alloys with $0.025 \leq x \leq 0.25$ were wires drawn from the susceptibility samples of ref. 1 and then annealed. The samples of alloys with $x = 0.33$ and 0.5 were prepared by induction melting in alumina crucibles under argon atmospheres. All ingots were then cold worked to about half their original diameter and annealed. Bars with a cross section of about 0.009 cm² were then prepared by cold rolling and reannealed. All of the annealings were done in evacuated quartz tubes at 1000°C for at least 48 h followed by a quenching in water. X-ray diffraction showed that the final materials were single phase and had lattice parameters consistent with the nominal compositions of the alloys.

For $Pd_{1-x}Ag_x$ alloys containing 1 at.% Fe with $x = 0$, 0.025, 0.10, 0.25 and 0.50, the resistivity is significantly larger than the matrix resistivity [2–5]. The resistivity of the $x = 0.33$ alloy, however, is approximately equal to that of the matrix. The temperature dependences of the resistivity, $\rho(T)$, for the three ferromagnetic samples, $x = 0.025$, 0.10 and 0.25, show clear breaks in the $\rho(T)$ curves in the neighbourhood of the Curie temperatures obtained from the $\chi(T)$ measurements [1]. These changes in $\rho(T)$ are due to the reduction in spin-disorder scattering with ferromagnetic ordering. For the samples containing 1 at.% Fe with Ag concentrations $x = 0.5$ and 0.33, a broad shallow minimum is observed above the cusp temperature determined from the low field χ measurements. At lower temperatures in these two samples there is observed a clear reduction in $\rho(T)$. A similar observation for $x = 0.5$ at.% Fe alloy has been reported [6]. However, this reduction may be due in part to the strong variation with Fe concentration of the temperature dependence of the "matrix resistivity", as well as to the onset of the spin glass transition.

A more detailed picture of the effect of Fe impurities on the resistivity of a particular matrix, namely the system $(Pd_{0.5}Ag_{0.5})_{1-y}Fe_y$, is shown in fig. 1, where $\rho(T)$ is plotted for alloys with $y = 0$, 0.005, 0.01 and 0.02. It is clearly seen that the addition of Fe to this matrix initially increases the overall resistivity. For the 2 at.% Fe sample, the resistivity is below that of the original matrix value at low temperatures. In fig. 2, the temperature dependence of the incremental resistivity, $\Delta\rho(T) = \rho(T)_{alloy} - \rho(T)_{matrix}$ is shown. $\Delta\rho(T)$ of the $y = 0.005$ alloy is nearly linear and has a negative

* Supported by the National Science Foundation, USA.
† Supported by Statens Naturvetenskapliga Forskningsråd.
‡ Supported by the University of Connecticut Research Foundation, USA.

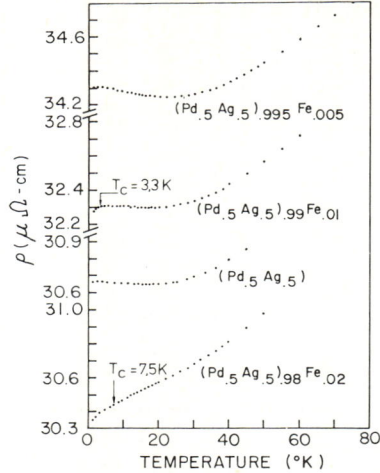

Fig. 1. The temperature dependence of the absolute resistivity of the $Pd_{0.5}Ag_{0.5}$ matrix and alloys formed by substituting 0.5, 1.0 and 2.0 at.% Fe into this matrix. The temperature, T_c, corresponds to the peak in $\chi(T)$ [1].

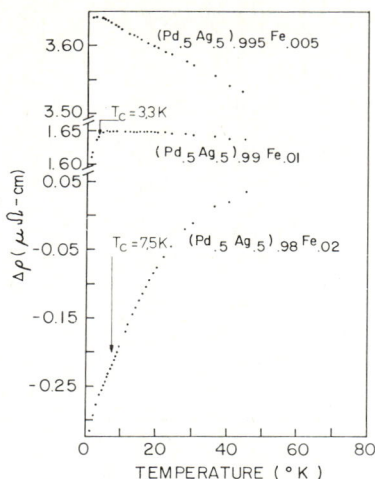

Fig. 2. The temperature dependence of the incremental resistivity $\Delta\rho(T) = \rho(T)_{alloy} - \rho(T)_{matrix}$ for the $Pd_{0.5}Ag_{0.5}$ matrix with substitutions of 0.5, 1.0 and 2.0 at.% Fe.

slope of $-2.5 \times 10^{-3}\ \mu\Omega cm/K$, at temperatures above any possible spin glass ordering temperature. Low field χ measurements down to about 1.5 K have shown no cusp. For the $y = 0.01$ alloy the slope of $\Delta\rho(T)$ is almost linear with a very small value of $6 \times 10^{-4}\ \mu\Omega cm/K$ down to about 5 K, below which $\Delta\rho$ decreases significantly. Low field measurements indicate a cusp-like behaviour at about 3.3 K. For the 2 at.% sample $\Delta\rho(T)$ decreases sharply with decreasing temperatures and falls off more sharply at the lowest temperatures. At the cusp temperature 7.5 K deduced from the low field measurements, there may be a very shallow depression in the resistivity. In the $Pd_{0.5}Ag_{0.5}$ matrix a common feature of the data is the existence of a strong decrease in $\rho(T)$ below the spin glass ordering temperatures. For the $Pd_{0.67}Ag_{0.33}$ matrix with 1 at.% Fe, not shown, a similar strong fall off in ρ with temperature is observed below the cusp temperature 4.4 K.

Even though these experiments provide some clear initial data on the resistivity in the region of the spin glass ordering temperature it is not yet possible to fully characterize the effects on $\rho(T)$ due to the onset of the spin glass transition and in the temperature region below the cusp temperature. The changes in the temperature dependence of $\rho(T)$ observed about the region of the susceptibility cusp occur over a broad temperature range. The broad minima observed in the $Pd_{0.50}Ag_{0.50}$ and $Pd_{0.67}Ag_{0.33}$ matrices and also in samples with 0.5 and 1 at.% Fe seem to be consistent with the local s-d effects on the scattering in these alloys discussed by Murani [3]. For small amounts of Fe our results can be described by a negative T^2 term in the resistivity below about 40 K whose coefficient is strongly dependent on Fe concentration. For the 2 at.% Fe alloy, however, a strong positive T^2 contribution is found to be necessary.

A more complete analysis along with additional data on these systems will be published elsewhere.

References

[1] J.I. Budnick, V. Cannella and T.J. Burch, AIP Conf. Proc. 18 (1974) 307.
[2] B.R. Coles and J.C. Taylor, Proc. Roy. Soc. Ser A267 (1962) 139.
[3] A. Murani, Phys. Rev. Lett. 33 (1974) 91.
[4] L.R. Edwards, C.W. Chen and S. Legvold, Solid State Commun. 8 (1970) 1403.
[5] W.R.G. Kemp, P.G. Klemens, A.K. Sreedhar and G.K. White, Proc. Roy. Soc. A233 (1956) 480.
[6] R.A. Levy and J.A. Rayne, Phys. Lett. 53A (1975) 329.

SPIN GLASS AND MICTOMAGNETIC REGIMES IN MoFe ALLOYS

R. CAUDRON*, P. COSTA*, B. LOEGEL**, J.L. THOLENCE† and R. TOURNIER†

Susceptibility measurements on MoFe spin-glasses show the validity of the scaling laws down to 0.05 K. The specific heat and the electrical resistivity are furthermore quantitatively strongly modified by a magnetic field. However, the general behaviour is not altered (i.e. the C/T maxima remain unchanged). No effect of "field cooling" treatments on the specific heat has been detected.

In a preceding paper [1], we have shown that the behaviour of the molybdenum–iron alloys is similar to that of the noble-metal based spin-glass systems (RKKY interactions only) in the 0.25–3 at.% Fe concentration range: the scaling laws [2] are obeyed by the magnetization and the specific heat. The values of the interaction parameter have been deduced from the low temperature behaviour of the specific heat, and from the high temperature results, on the basis of Larkin's virial development [3]. These values (2500 and 1730, respectively) are much lower than for the other spin-glass systems, and we expect to compare them with the values we can deduce from the high field behaviour of the magnetization, interpreted in a pair-interaction model due to Matho [4].

However, the Mo–Fe system differs from the other spin-glasses by the very high ratio of the temperature of the specific heat maximum to the temperature of the susceptibility maximum. In former work we were only able to give a lower estimate of about 8 for this ratio, the susceptibility maxima occurring for temperatures below the lower limit of the temperature range of the experiment (1.5 K). For Au–Fe, the value of this ratio is only of 1.3 [2].

In the mictomagnetic range, the scaling law is still followed for the specific heat, which reflects essentially the contribution of the impurities submitted to long range interactions only; when the concentration is increased, short range interactions become important and giant moments appear, leading to strong deviations of the magnetic properties from the scaling laws. The resistivity follows roughly a behaviour similar to that observed for various nickel-based alloys [5]

and Cr–Fe alloys [6] ($\rho - \rho_0 = AT^n$ $1 < n < 2$), in a fairly wide range of temperature. However, below the ordering temperature, the resistivity exponent n is equal to 1, and not to 1.5, as formerly stated by Canella and Mydosh [7]. This linear resistivity term is due to clusters, dominating the magnetic properties.

In this paper, we describe new experiments performed in the spin-glass regime, down to 0.05 K, well below the susceptibility maxima. On the other hand, we have performed specific heat experiments under a magnetic field of 55 kOe, for a spin-glass (2.45%) and a mictomagnetic alloy (7.2%). Magnetoresistance were also performed on mictomagnetic alloys.

The susceptibilities of the 0.51 and 1.42 alloys are shown in fig. 1, plotted against the reduced variable T/c. They both can be extrapolated for $T = 0$ to a value of 2.4×10^{-5} emu/g, to be compared to 1.9×10^{-5} emu/g, deduced from the low temperature specific heat coefficient and the high temperature spin value. The scaling law for susceptibility seems to be obeyed. The maxima occur at about 0.6 K/%. In the mictomagnetic regime, these temperatures increase much faster than the concentration. This behaviour is at variance with that of the classical spin-glass systems, where the ordering temperatures in-

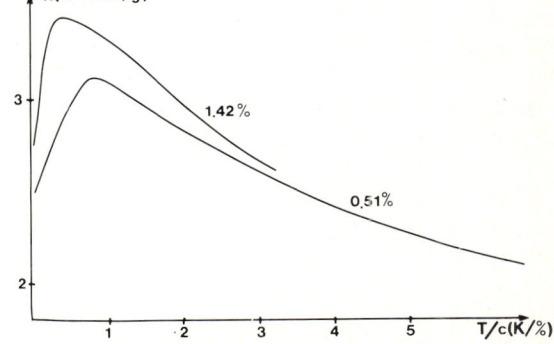

Fig. 1. Susceptibility of a 0.51 and 1.42 at.% Fe–Mo–Fe alloys, plotted versus the reduced variable T/c.

* Direction des Matériaux, ONERA, 29, avenue de la Division Leclerc, 92320 Chatillon, France.
** Laboratoire de Structure Electronique des Solides, 4, rue Blaise Pascal, 67000 Strasbourg, France.
† CRTBT, Avenue des Martyrs, BP 166, 38042 Grenoble Cedex, France.

crease more slowly in the mictomagnetic regime than in the spin-glass regime, due to the reduction of the mean free path [2].

As formerly predicted, a remanent magnetization is present in the spin-glass regime. However, in our case, at variance with the Cu–Mn case [2], the remanent magnetization follows only approximately the scaling law, and presents a plateau at temperatures lower than 0.1 K/%.

The results of the specific heat experiments under magnetic field, are plotted in fig. 2. We

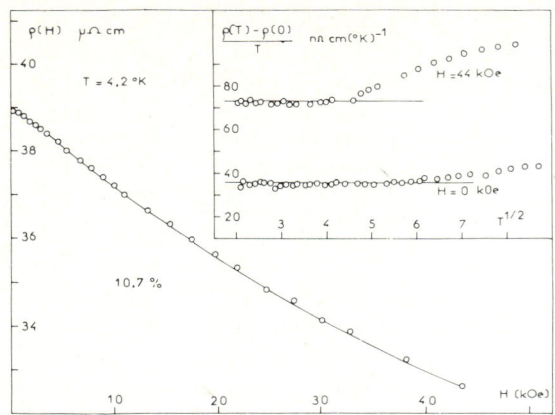

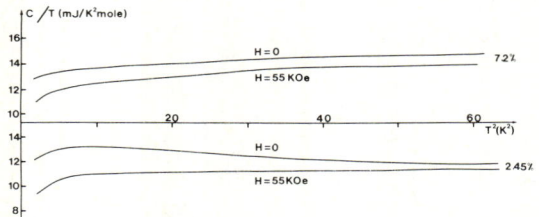

Fig. 2. Specific heat, in zero field and under a field of 55 kOe, of a 2.45 and a 7.2 at.% Fe alloys.

Fig. 3. The field dependence of the electrical resistivity at $T = 4.2$ K for an alloy containing 10.7 at.% Fe. Insert: temperature dependent part of the resistivity in zero field and for $H = 44$ kOe; the straight lines correspond to the law $\rho(T) - \rho(0) = At$.

first notice that the magnetic field strongly modifies the specific heat, a confirmation of the spin-glass origin of the specific heat. Furthermore, it is obvious that the effect of the field is less important for the more concentrated (mictomagnetic) alloy; this is a direct consequence of the scaling laws. We notice also that the maximum of C/T is not altered neither in amplitude nor in position, by the magnetic field; as formerly stated [1], the maximum is not to be explained by a local minimum of the field distribution at $H = 0$ [8], but rather by a reduction of the effective fields, due to the thermal motion of the magnetic moments [9]. Similarly, the magnetoresistance measurements in the mictomagnetic regime show that a magnetic field reduces significantly the resistivity at low temperatures (fig. 3), but the general behaviour versus temperature is not altered.

The 2.45 and 7.2% samples have been cooled in a field of 55 kOe and then their specific heat has been measured in zero field. We have detected no effect of the "field cooling" treatment, in contra-distinction at least for the mictomagnetic sample, with the magnetic properties. This is a further confirmation of the leading role of the long range interactions (RKKY) on the specific heat, the short range interactions (clustering) dominating the magnetic properties.

References

[1] A. Amamou, R. Caudron, P. Costa, J.M. Friedt, F. Gautier and B. Loegel, J. Phys. F., to be published Dec. 1976, A. Amamou, Thesis, Strasbourg (1975), R. Caudron, Thesis, Strasbourg (1976).
[2] J. Souletie and R. Tournier, J. Low Temp. Phys. 1 (1969) 95.
[3] A.I. Larkin and D.E. Khmel'Nitskii, Sov. Phys. JETP 31 (1970) 958.
[4] K. Matho, to be published.
[5] F. Gautier and B. Loegel, Solid State Commun. 11 (1972) 1205.
[6] B. Loegel, J. Phys. F. 5 (1975) 497.
[7] V. Canella and J.A. Mydosh, Phys. Rev. 136 (1972) 4220.
[8] M.W. Klein, Phys. Rev. 136A (1964) A 1156.
[9] M.W. Klein, Phys. Rev. 173 (1968) 552.

INFLUENCE OF REDUCED MEAN FREE PATH ON THE ELECTRICAL RESISTIVITY OF AuFe SPIN GLASS ALLOYS*

R. BUCHMANN, H.P. FALKE, H.P. JABLONSKI and E.F. WASSERMANN

2. Phys. Institut der RWTH Aachen, 5100 Aachen, W. Germany

The resistivity of quench-condensed films of the AuFe spin glass system has been investigated with different concentrations of structural defects above 0.3 K. The temperature of the resistance maximum is considerably lowered by a reduced mean free path. The temperature dependence of the resistivity below T_{max} is also discussed.

Although it is well established in the meantime that the properties of spin glasses are due to interactions between the magnetic impurities, which are thought to be mainly of the Rudermann–Kittel–Kasuya–Yosida (RKKY) type, there is little knowledge about how the spin glass properties (e.g. the spin freezing temperature T_0) depend on these basic interactions.

One possibility to change the interactions in a given alloy is to introduce nonmagnetic impurites or structural defects. These defects reduce the mean free path λ of the conduction electrons. As shown by de Gennes, this results (mainly) in a reduction of the strength of the RKKY-interaction by a factor of $e^{-r/\lambda}$ (damping) [1].

In this paper we present some results of our investigations on the electrical resistivity of quench condensed AuFe alloy-films with concentrations from 0.24 to 6.0 at.% Fe.

The films (800–1000 Å) are prepared in a ^{3}He-refrigerator by successive flash evaporation in uhv of about 150 pellets of the alloy onto a helium cooled quartz substrate. The electrical resistivity is measured with a relative accuracy of 1×10^{-5} down to 0.3 K after condensation and after heating the films stepwise to different annealing temperatures (T_a = 50, 150, 220 and 300 K).

In fig. 1 we have plotted the resistivity difference $\Delta\rho$ versus temperature for an Au + 0.24 at.% Fe alloy film annealed at different temperatures T_a. As can be seen the shape of the resistance curve is changed once the mean free path is changed. The most pronounced effect is a considerable shift in the temperature of the resistivity maximum T_{max}. In table I we have given the values for T_{max}, the residual resistivities $\rho_{4.2}$, and the corresponding mean

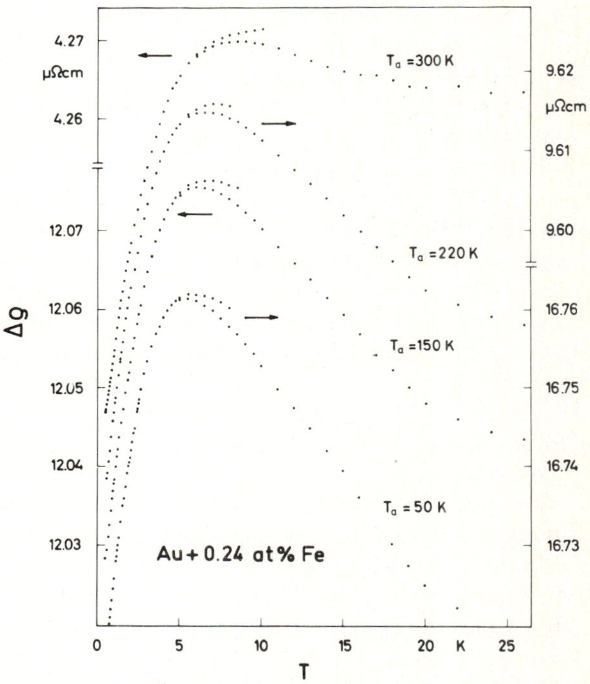

Fig. 1. Resistivity difference $\Delta\rho(T) = \rho(T) - \rho_{Ph}(T)$ versus temperature for an Au + 0.24 at.% Fe film annealed at different temperatures T_a. [$\rho_{Ph}(T)$ is the T-dependent resistivity of a pure Au film with an equal residual resistivity as the alloy film.] The upper points in each curve show $\rho(T)$ for comparison.

Table I
Temperature of the resistance maximum T_{max}, residual resistivities at 4.2 K $\rho_{4.2}$, mean free paths λ, and interaction strengths Δ_c as calculated from T_{max} for an Au + 0.24 at.% Fe alloy film (thickness 980 Å) after annealing at different temperatures T_a. Values for the corresponding bulk alloy are also given

T_a(K)	50	150	220	300	Bulk
T_{max}(K)	5.3	6.0	6.5	8.5	?
$\rho_{4.2}(\mu\Omega cm)$	16.8	12.0	9.6	4.3	1.8
λ(Å)	50	70	88	195	460
Δ_c(K)	2.6	2.8	3.0	3.7	

* Work was supported by DFG, SFB 125.

free path values λ as deduced from the relation $\rho \cdot \lambda = 8.4 \cdot 10^{-12}\,\Omega\text{cm}^2$. For the film annealed at $T_a = 50\,\text{K}$ we find $\lambda = 50\,\text{Å}$ and $T_{\max} = 5.3\,\text{K}$. After annealing the film at 300 K the mean free path has increased by a factor of 4, whereas $T_{\max}$ has increased by 60% to 8.5 K.

The only theory giving a relation between $T_{\max}$ and an interaction strength Δ_c, taking into account the Kondo-effect in higher order, has recently been given by Larsen [2]. This theory has been confirmed experimentally by Shilling et al. who investigated the pressure dependence of $T_{\max}$ for different spin glass systems [3]. These experiments show that an increase of the Kondo-temperature T_K in AuFe by a factor of two reduces $T_{\max}$ only by a few percent. From this result we conclude that the shift of $T_{\max}$ to lower values in our disordered films is due to the reduced electronic mean free path. A shift in T_K for AuFe films, similar to that found for quench condensed CuFe films [4], would be of minor influence on a shift in $T_{\max}$. The values of Δ_c as calculated from $T_{\max}$ using the theory of Larsen ($T_K = 0.24$ K) are also given in table I. We find that Δ_c is reduced by 30% in the film with the smallest λ compared to the film annealed at 300 K.

In order to look for differences in the temperature dependence of the resistivity after the different annealing steps below $T_{\max}$, we have differentiated the $\Delta\rho(T)$-curves for all films. In fig. 2 the result for Au + 0.24 at.% Fe has been plotted as an example. The main feature is that the position of the maximum in $d\Delta\rho/dT$, T_m, is independent of T_a [$T_m = (1.05 \pm 0.1)$K for Au + 0.24 at.% Fe]. Around T_m we find a temperature range where $\Delta\rho$ is proportional to T. This range is not affected by T_a. For the higher concentrated films (0.6 and 6 at.% Fe) we find a $T^{3/2}$-dependence below a temperature which is independent of T_a, too. The only change we find in $d\Delta\rho/dT$ for the differently annealed films is a second, less pronounced maximum (or shoulder) above T_m, which is reduced with increasing T_a, but whose position seems to be unchanged. For all alloys this shoulder is found at a temperature close to T_0. For the film with 0.24 at.% Fe, the temperature for this shoulder is about 2.6 K (see

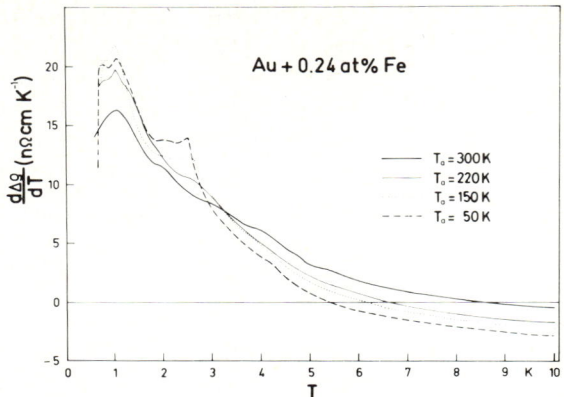

Fig. 2. Temperature dependence of the temperature derivative of the impurity resistivity $d\Delta\rho/dT$ for an Au + 0.24 at.% Fe film annealed at different temperatures T_a.

fig. 2), whereas susceptibility measurements on the same bulk alloy have given $T_0 = 2.7$ K [5].

In a more recent paper [6], Larsen calculates the concentration dependence of the spin freezing temperature T_0, taking into account a finite mean free path. In this paper T_0 is supposed to be equal to Δ_c, except for a spin-dependent constant. Following Larsen, we conclude that T_0 for the alloy with 0.24 at.% Fe should also be reduced by 30% (see table I) to 1.9 K, if λ is reduced to 50 Å. Probably, it will be hard to find this reduction of T_0 in the resistivity, because it is not yet clear if T_0 can be determined from resistivity measurements at all.

In summary, we have shown that there is a remarkable influence of the electronic mean free path on the interaction strength. From our measurements, however, we can not decide, if there is an influence of a reduced mean free path on the spin freezing temperature T_0.

References

[1] P.G. de Gennes, J. Phys. Radium 23 (1962) 630.
[2] U. Larsen, Phys. Rev., to be published.
[3] J.S. Schilling, P.J. Ford, U. Larsen and J.A. Mydosh, Phys. Rev., to be published.
[4] H.P. Falke, H.P. Jablonski and E.F. Wassermann, Solid State Commun. 19 (1976) 273.
[5] P.J. Ford, private communication.
[6] U. Larsen, preprint, to be published.

SPIN FREEZING IN THE SPIN GLASS PHASE OF PdMn

H.A. ZWEERS, W. PELT, G.J. NIEUWENHUYS and J.A. MYDOSH
Kamerlingh Onnes Laboratorium der Rijksuniversiteit, Leiden, The Netherlands

Low-field susceptibility measurements on PdMn alloys (Mn conc. > 4 at.%) exhibit sharply defined freezing temperatures T_0. A reanalysis of the specific heat data shows that the entropy rate of change has a maximum at T_0, which then gives a simple and consistent magnetic phase diagram.

Dilute PdMn alloys (Mn conc. < 2.5 at.%) are known to form a giant moment system [1, 2]. The total magnetic moment per dissolved Mn atom (7.5 μ_B) is larger than the moment of the bare atom (5 μ_B); this is the result of an induced polarization of the Pd 4d-band. For Mn concentrations of order 1 at.%, the induced polarization causes long range ferromagnetic ordering at liquid-He temperatures. Direct Mn–Mn interactions, which are antiferromagnetic, influence the ordering when the Mn concentration is further increased [2–5]. For alloys with Mn contents larger than 2.5 at.%, this antiferromagnetic direct interaction starts to dominate the long-range ferromagnetism. The present work deals with this "spin-glass" or mixed magnetic state of PdMn alloys for Mn concentrations between 4 and 9.5 at.%.

The low-field susceptibility, χ, measurements of Burger and McLachlan [6] showed a ferromagnetic transition (knee in χ at T_c), for 3 at.%, and a transition characteristic of a spin-glass at 10 at.% Mn in Pd. We recently obtained results for the intermediate concentrations: 4, 6 and 8 at.%. These results are presented in fig. 1, which shows that the freezing temperatures, T_0, are well defined from the susceptibility, and that the total response (or signal amplitude) to the a.c. magnetic field greatly decreases with increasing concentration. The χ-data at temperatures just below T_0 show a shoulder, which is not yet understood but has been found earlier for some spin-glass systems [7].

Our specific heat work [5] revealed broad maxima in the magnetic specific heat, ΔC, as a function of T at temperatures far above T_0. We re-analyzed these data by plotting $\Delta C/T$ ($= dS_m/dT$, S_m being the magnetic entropy) as a function of temperature (see fig. 2) and then found that the temperature at which dS_m/dT attains its maximum very closely coincides with T_0 deduced from the a.c. susceptibility measurements. This is illustrated in fig. 3, where

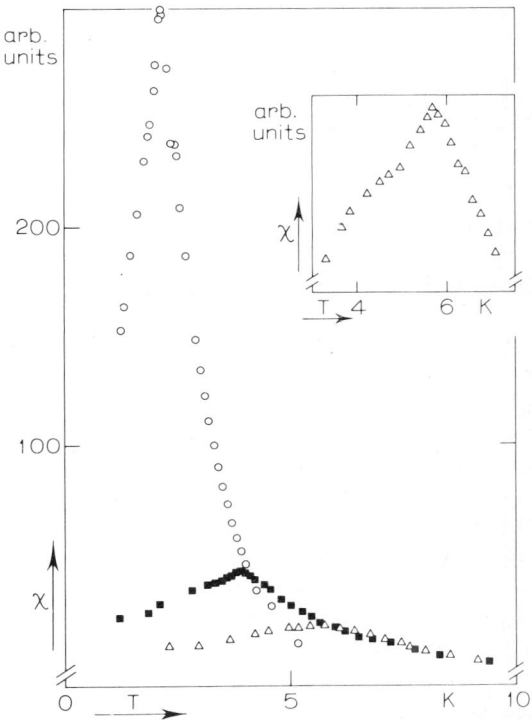

Fig. 1. Temperature dependence of the low-field a.c. susceptibility for PdMn: (○) 4 at.%, (■) 6 at.% and (△) 8 at.%. The sharp peaks in χ denote the freezing temperature.

we plot the various ordering temperatures as a function of concentration.

We wish to emphasize that the spin-glass PdMn alloys show a sharp magnetic transition indicated by the peak in the susceptibility at temperatures which coincide with the temperature where a maximum in the temperature derivative of the magnetic entropy is found. This implies that in the PdMn system the freezing is associated with a relatively large change in the magnetic entropy (revealed by $\Delta C/T$) rather than with a change in the magnetic internal energy (revealed by ΔC). By analyzing various data from the literature [8] on spin glasses in this respect, we found that the

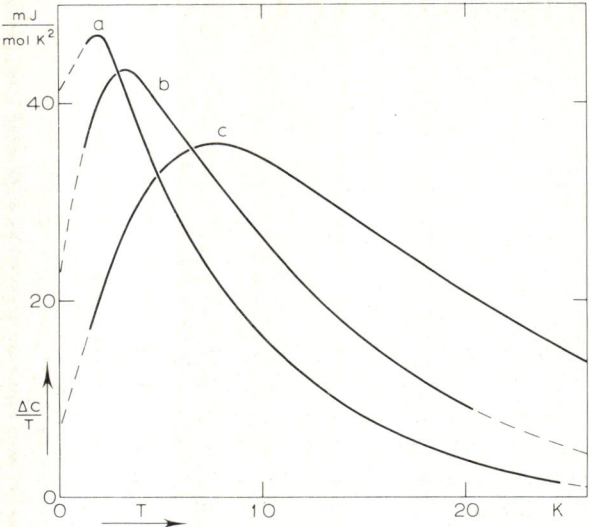

Fig. 2. $\Delta C/T$ versus T for PdMn. Curve (a) 4 at.%, curve (b) 5.5 at.% and curve (c) 9.5 at.%.

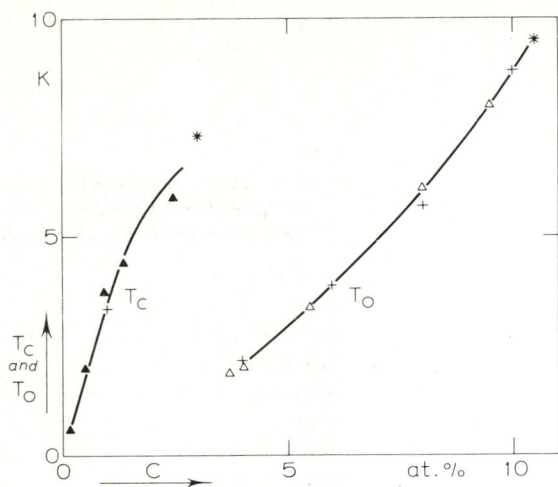

Fig. 3. Curie-temperature, T_c, and freezing temperature, T_0, for the PdMn system. ▲ sharp transition in ΔC versus T; △ maximum in $\Delta C/T$, see fig. 2; ∗ susceptibility measurement of Burger et al., see ref. 6; and + susceptibility, present work, see fig. 1.

agreement of the temperatures at which the susceptibility, and $\Delta C/T$ attain their maxima, is not as good as for PdMn. In a number of cases, even no maximum in $\Delta C/T$ could be detected. Nevertheless, we feel that a more detailed and systematic study of appropriate spin glasses is worthwhile in order to gain further insight into the connection between the thermal and magnetic properties.

The spin glass PdMn system is unique in this respect for the ratio between the different magnetic interactions can be changed as a function of concentration and the magnetic ordering temperature remains in the liquid-helium range. Another unusual feature of PdMn is the large decrease of the susceptibility magnitude as the concentration is increased. This is exactly the opposite for the other spin glass alloys. Such behaviour illustrates the role of clusters (or giant moments) in contributing to the strength of χ at T_0. In PdMn increasing the concentration statistically enhances the destruction of the giant moments and χ decreases. Similar behaviour has been produced via different heat treatments and plastic deformation at a constant concentration in CuMn [9].

The authors wish to acknowledge the interest and help with the experiments of the members of the Leiden metal physics group. This work is part of the research program of the Foundation F.O.M., supported financially by the Dutch Organisation for Basic Research (Z.W.O.), and the Dutch Organisation for Applied Physics Research (T.N.O. Metals' Institute).

References

[1] G.J. Nieuwenhuys, Advan. Phys. 24 (1975) 515.
[2] W.M. Star, S. Foner and E.J. McNiff, Jr., Phys. Rev. B12 (1975) 2690.
[3] J. Rault and J.P. Burger, C.R. Acad. Sci. (Paris) B269 (1969) 1085.
[4] B.R. Coles, H. Jameison, R.H. Taylor and A. Tari, J. Phys. F5 (1975) 565.
[5] H.A. Zweers and G.J. van den Berg, J. Phys. F5 (1975) 555, Comm. Kamerlingh Onnes Lab., Leiden, No. 412b.
[6] J.P. Burger and D.S. McLachlan, Solid State Commun. 13 (1973) 1563.
[7] V. Cannella, private communication.
[8] J.A. Mydosh, AIP Conf. Proc. 24 (1975) 131 and references therein.
[9] R.W. Tustison, Solid State Commun. 19 (1976) 1075.

^{63}Cu RESONANCE AND RELAXATION IN SPIN-GLASS CuMn

D.E. MACLAUGHLIN* and H. ALLOUL

Laboratoire de Physique des Solides,† Université Paris-Sud, Orsay, France

Host nuclear resonance measurements have been made in CuMn spin-glass alloys with Mn concentrations in the range 0.1–0.4 at.%. No large increase in linewidth was observed below the ordering temperature T_0, in apparent conflict with other spin-probe experiments. Relaxation measurements indicate impurity spin correlation times in the range 10^{-10}–10^{-8} s.

In large measure the current controversy over the nature of magnetic order in dilute magnetic alloys, or "spin glasses" [1], concerns the question of the transition from the high-temperature "paramagnetic" state to the spin-glass state of frozen-in orientational disorder. At present three kinds of measurements have been carried out which indicate an abrupt change of state at a well-defined temperature T_0: Mössbauer-effect measurements of impurity nuclear hyperfine fields [2], magnetic susceptibility studies at low applied fields [1, 3], and investigations of internal field distributions by muon spin depolarization [4]. Other properties of spin-glass systems appear not to change sharply at or near T_0 [1]. This striking dependence on the type of measurement poses a severe challenge to theories of the spin-glass state, and has provoked a search for additional experimental evidence which might help to elucidate the nature of the transition.

We have undertaken ^{63}Cu nuclear resonance and relaxation measurements in the CuMn spin-glass system, in the expectation that results would be similar to those of other studies using spin-dependent probes [2,4]; i.e. that a wide inhomogeneous field distribution would set in below T_0. Fig. 1 summarizes our linewidth measurements [5]. It can be seen that no abrupt change of linewidth is observed in the range of applied field H and temperature below T_0. The susceptibility in fields higher than a few hundred Oe behaves similarly.

As discussed below, the appearance below T_0 of a static, spatially random orientational configuration of the impurity spin system should lead to an increase of the host NMR linewidth

---

* Work supported in part by the US National Science Foundation and the UC Riverside Academic Senate. Present address: Department of Physics, University of California, Riverside, Ca. 92502, U.S.A.

† Laboratoire associé au CNRS.

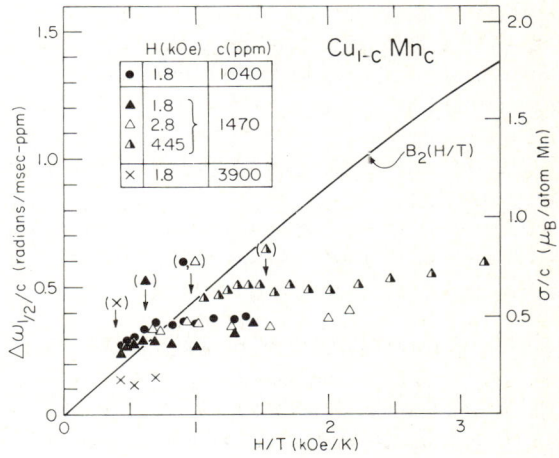

Fig. 1. Dependence of FWHM ^{63}Cu NMR linewidth $\Delta\omega_{1/2}$ on applied field H, temperature T, and impurity concentration c in CuMn spin-glass systems. The values of H/T corresponding to the zero-field spin-glass ordering temperatures T_0 are indicated by the symbols in parentheses. The Brillouin function $B_2(H/T)$ gives the linewidth dependence found in dilute alloys well above T_0 [7].

by a factor of between four and five for the concentrations employed. The absence of such an increase appears to be in conflict with mean-field theories of spin glasses [6] and also with the results of Mössbauer-effect and muon depolarization experiments.

Sample preparation and characterization, and the pulsed NMR spectroscopic techniques used in this study have been given elsewhere [5, 7]. For the 1040-ppm and 1470-ppm samples, integrated intensity measurements on the ^{63}Cu resonance spectra indicated the onset of a loss of intensity at $T \lesssim 1.3\, T_0$. Thus the linewidths given in fig. 1 are not representative of all the nuclear spins in the samples. If the linewidth of the "lost" nuclei were dominated by static inhomogeneous broadening, the spectrometer signal-to-noise ratio puts a lower limit of about 4×10^6 rad/s on their linewidth. Alternatively, a transverse spin-relaxation rate of greater than 10^5 s^{-1} would cause the signal of the lost nuclei to decay within the "dead" time of the receiver. In either case the nuclei in question would be expected to be in or near clusters of impurity ions, within which strong impurity–impurity spin correlations would be expected to form well above the freezing temperature. The discrepancy between our results and those of Mössbauer and muon-depolarization experiments may now be stated more precisely: the host NMR measurements indicate that an appreciable fraction of the spin probes experience a low value of the local field due to the impurity spin system, whereas the other methods show no evidence for such a "paramagnetic line".

In fig. 1 the Brillouin function $B_2(H/T)$ has been drawn, in units as indicated on the right-hand scale. The conversion factor

$$\sigma/c\,(\mu_B/\text{atom Mn}) = (0.77 \pm 0.04)$$
$$\times \Delta\omega_{1/2}/c \text{ (rad/ms ppm)}$$

between the impurity magnetization σ (well described by the Brillouin function), and host NMR linewidth $\Delta\omega_{1/2}$ (FWHM), was obtained experimentally from measurements by Alloul et al. [7] on a 75-ppm CuMn sample at temperatures well above T_0. The above relation was obeyed for values of H/T well into the regime of saturation of the Brillouin function. It can be seen that for the lowest values of H/T employed, and for T somewhat greater than T_0, the present values of the linewidth are larger than in the dilute sample. This discrepancy may be due to the onset at low applied field of second-order quadrupolar broadening in the relatively high-concentration samples used in the present study. The limited data available suggest that the discrepancy is reduced for higher applied fields, as would be expected from this mechanism.

In the presence of a random orientational configuration of static impurity spins the expected host NMR linewidth $\Delta\omega_r$ should be proportional to the angular average of $|S_z|$, which is $S/2$ [5]. This value corresponds approximately to the upper border of fig. 1. It should be noted, however, that this value of the linewidth will be observed only if the characteristic correlation time τ for the impurity spin system becomes longer than the inverse of the static-limit linewidth $\Delta\omega_r$. This restriction places an upper limit of about 10^{-6} s on τ, based on the present linewidth measurements for $c \sim 1000$ ppm.

Longitudinal (spin–lattice) and transverse nuclear spin relaxation rates also provide a measure of impurity-spin dynamics. If impurity spin correlation times are short compared to the inverse linewidth, as above, the simplest picture [8] for the resulting rates T_1^{-1}, T_2^{-1} gives

$$T_1^{-1},\, T_2^{-1} \simeq (\Delta\omega_r)^2 \tau.$$

This expression, evaluated using measured values of T_1 and T_2 and $\Delta\omega_r = 10^6$ rad/s, yields a range of τ between 10^{-8} and 10^{-10} s for $c \sim 1000$ ppm. While consistent with the upper limit placed by the measured linewidth, as discussed above, these figures are somewhat smaller than the lower limit of about 10^{-7}–10^{-8} s imposed by the observed hyperfine splitting in Mössbauer measurements and by the observed linewidth in muon depolarization [9]. It should be emphasized that the discrepancy concerns only a fraction of the host nuclear spins, and that the resonance properties of the remaining nuclei, if they could be observed, might well be in agreement with the other spin-probe methods.

The host NMR results were obtained for the most part on samples with magnetic impurity concentrations of the order of 0.1 at.%, whereas in all but a few Mössbauer experiments, and in the muon studies reported to date, concentrations of about 1 at.% were used. NMR experiments are presently under way to determine if the anomalous behavior reported in this paper persists to higher concentrations.

Preliminary experimental results in samples of 0.6, 0.8 and 1.0 at.% Mn concentrations, at 1.3 and 9 kOe applied fields, indicate rapid loss of NMR intensity with decreasing temperature at or near T_0. This result is consistent with muon experiments at comparable concentrations; if the muon-depolarization time is controlled by relaxation effects the NMR signal would decay too rapidly to be observed. At 9 kOe the lost intensity is recovered in a broad line (>1000 Oe) at lower temperatures ($\leqslant 0.4\, T_0$).

References

[1] See J.A. Mydosh, in: Magnetism and Magnetic Materials – 1974, C.D. Graham. Jr., J.J. Rhyne, and G.H. Lander (Eds.) (American Institute of Physics, New York, 1975) p. 131, and references contained therein.
[2] C.E. Violet and R.J. Borg, Phys. Rev. 149 (1966) 540.
[3] See, for example, V. Cannella, in: Amorphous Magnetism, H.O. Hooper and A.M. de Graaf (Eds.) (Plenum, New York, 1973) p. 195.
[4] D.E. Murnick, A.T. Fiory, and W.J. Kossler, Phys. Rev. Lett. 36 (1976) 100.
[5] A preliminary account of this work has been given in D.E. MacLaughlin and H. Alloul, Phys. Rev. Lett. 36 (1976) 1158.
[6] K. Adkins and N. Rivier, J. Phys. (Paris) Colloq. 35 (1974) C4-237.
S.F. Edwards and P.W. Anderson, J. Phys. F5 (1975) 965.
[7] H. Alloul, J. Darville, and P. Berrier, J. Phys. F4 (1974) 2050.
[8] C.P. Slichter, Principles of Magnetic Resonance (Harper & Row, New York, 1963) sec. 5.7.
[9] D. Bloyet, thèse, Université Paris-Sud, Orsay, France (1976) sec. II.7.2.

THE DYNAMICS OF MAGNETIC MOMENTS IN THE SPIN GLASS SYSTEM AuFe*

H. SCHEUER
Sonderforschungsbereich 125, Fehlordnung in Metallen, Aachen-Jülich-Köln, Universität zu Köln, D-5000 Köln 41, W. Germany

M. LOEWENHAUPT and W. SCHMATZ
Institut für Festkörperforschung der Kernforschungsanlage Jülich, D-5170 Jülich, W. Germany

The spin glass system AuFe was investigated in the concentration range of 5 to 15 at.% Fe for temperatures between 3 and 300 K by diffuse magnetic scattering of unpolarized neutrons in zero external field. The energy spectra were analyzed in terms of two quasielastic lines with different linewidths due to two species of magnetic moments with different relaxation rates.

The binary alloy AuFe with small concentrations of Fe atoms is a classical example for spin glass systems, which show as typical magnetic behaviour cusps in the low field susceptibility at distinct temperatures T_f [1]. Above the ordering temperature T_f one expects large fluctuations of the magnetic moments in the system with time. Such time dependent spin fluctuations cannot be studied by a static method such as susceptibility measurements, but neutron scattering experiments with energy analysis are suited for the determination of spin dynamics because the energy broadening of the scattered neutrons is a direct measure of the spin relaxation rates [2].

The double differential cross section per host atom for the scattering from magnetic impurities in a nonmagnetic metallic matrix is the Fourier transform of the correlations in space and time. With the assumption that the time dependent correlation function of the localized magnetic moment decays exponentially, it is given by

$$\frac{d^2\sigma}{d\Omega\,d\omega} \sim \frac{1}{\pi}\frac{1/\tau}{\omega^2 + (1/\tau)^2}|F(x)|^2, \quad (1)$$

where $\hbar\omega = E_0 - E_1$ is the neutron energy transfer and $x = k_0 - k_1$ the momentum transfer. $1/\tau$ is the spin relaxation rate and thus $2\hbar/\tau$ is the energy linewidth of the quasielastic scattering. $F(x)$ is the magnetic form factor, whose determination allows to deduce whether one has uncoupled single spins (3d form factor) or correlated regions of magnetic clusters in the system.

Our measurements were performed on the instruments D7 at the HFR Grenoble and DNS at the FRJ2-Dido in Jülich with neutrons of 3.3 meV of incident energy. We prepared $Au_{1-x}Fe_x$ samples $(20 \times 8 \times 0.5\,mm)$ with concentrations of $x = 0.05$, 0.10, and 0.15 which were at first investigated at room temperature. The samples with $x = 0.05$ and 0.15 were furthermore studied in the temperature range from 3 to 300 K. The magnetic scattering was obtained from a difference measurement of doped and undoped samples. The detailed experimental and data analysing procedure is described elsewhere [3].

For all concentrations and temperatures we could decompose the measured time of flight spectra into three individual contributions: an elastic line with the same halfwidth as the calibration scatterer vanadium and two quasielastic lines with different halfwidths $\Gamma/2$. Fig. 1 shows as an example the decomposition for the Au 15 at.% Fe sample at room temperature and a scattering angle of $\vartheta = 15°$. The dashed line represents in energy space a broad quasielastic line with a value of about 10.5 meV for $\Gamma/2$, the dash-dotted line a narrow one with $\Gamma/2 \approx 2$ meV. The narrow line vanishes rapidly with increasing scattering angle and is due to a slow relaxation process with a correlation size of about 10 Å. The fast relaxation rate of 10.5 meV is nearly independent of x and the intensity varies with a 3d form factor. This process is therefore a single site behaviour. For the lower concentrations and $T = 300$ K we observed within the experimental error the same quasielastic line-

---

* Work performed within the program of SFB 125 Aachen-Jülich-Köln.

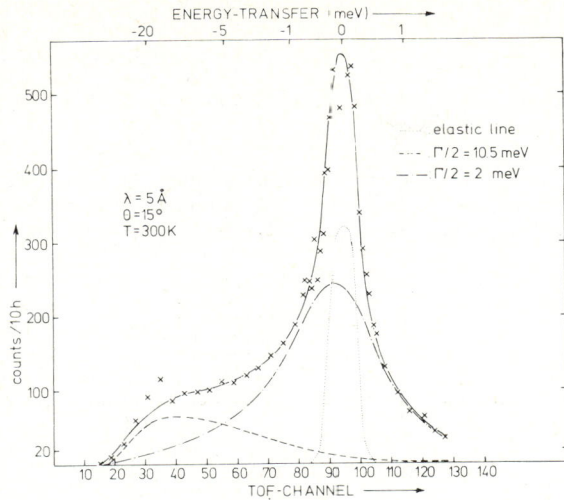

Fig. 1. Difference of TOF-spectra of Au 15 at.% Fe and pure Au for one counter (scattering angle 15°) at $T = 300$ K.

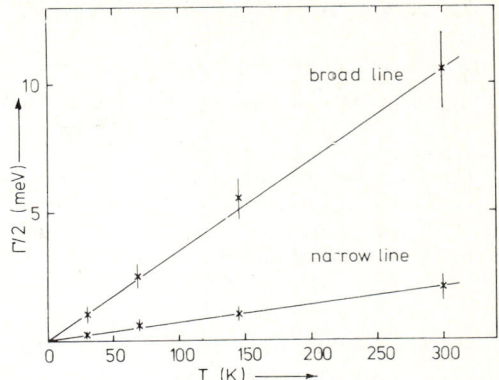

Fig. 2. Temperature dependence of the quasielastic half linewidths for the Au 5 at.% Fe sample.

widths, that means the two relaxation processes are at room temperature independent of concentration.

Fig. 2 shows the temperature dependence of the linewidths for the Au 5 at.% Fe sample. The upper line corresponds to the fast relaxation process of single spins, the lower to the slower one of spin clusters. In both cases one observes a nearly linear decrease of the linewidths with decreasing temperature at least for temperatures between 30 and 300 K. Also the Au 15 at.% Fe sample shows within the errors of the measurements the same linear temperature dependence of the linewidths, that means the dynamical behaviour of the two species of magnetic moments can be described for both concentrations by simple Korringa laws. This indicates that the spin relaxation is mainly determined by the interaction of the local magnetic moment with the conduction electrons. Further detailed investigations of the temperature dependence of the relaxation rates in the vicinity of T_f are planned.

We acknowledge the valuable help of Dr. W. Just during the neutron scattering experiments in Grenoble.

References

[1] V. Canella and J.A. Mydosh, Phys. Rev. B6 (1972) 4220.
[2] M. Loewenhaupt, H. Scheuer and W. Schmatz, Proc. Conf. on Neutron Scattering, C26, Gatlinburg (1976), to be published.
[3] H. Scheuer, M. Loewenhaupt and W. Just, J. Magn. Magn. Mater., to be published.

La(Gd)Al$_2$ – A SIMPLE SPIN GLASS?

M.H. BENNETT and B.R. COLES
Physics Department, Imperial College, London SW7, England

We compare this system with $\underline{\text{AuFe}}$ where nearest neighbour interactions of d–d overlap character might give complications. Alloys of up to 9% Gd substitution show simple spin-glass behaviour, those with 10–16% show the "cluster glass" character of Au 15% Fe, and more concentrated alloys show the magnetization character of pure GdAl$_2$ which is less simple than that of Au 30% Fe.

It has long been recognized that the spin glass behaviour of alloy systems like Au–Fe is complicated, at concentrations greater than a few atomic per cent, by nearest neighbour interactions and the formation of magnetic clusters even in statistically random alloys. Ideal spin-glasses have local moments coupled by long-range oscillatory interactions via conduction electrons; Fe nearest neighbours must couple by d–d overlap effects. It is therefore desirable to study systems in which the latter interactions are absent, i.e. we have 4f local moments, and the continuous range of solid solutions in the cubic Laves phase structure LaAl$_2$–GdAl$_2$ seems ideal for such studies. (LaGd has metallurgical complexities and YGd yields a helical structure at >3% Gd.) This system has been studied previously [1, 2] and reported to be ferromagnetic over almost the whole concentration range with a linear dependence of paramagnetic Curie temperature θ_p on concentration. Our measurements (of low field magnetization by both Faraday methods and a vibrating sample magnetometer, and of electrical resistivity) show that the higher fields used in previous measurements over-emphasize the ferromagnetic character, and that the low field behaviour for much of the composition range resembles that of Au–Fe alloys.

The magnetization in an applied field of 5 Oe is shown in fig. 1 as a function of temperature for a number of the alloys. It is clear that the 8.26% Gd alloy is a good spin glass, although the temperature of the maximum is rather lower than would be expected from the specific heat maxima of more dilute alloys [3]. The resistivity (fig. 2) of this and more dilute alloys shows the typical spin glass behaviour, with $d(\Delta\rho)/dT$ decreasing steadily as the temperature is raised. (As in some Mn containing spin glasses any maximum in this quantity lies well below T_g,

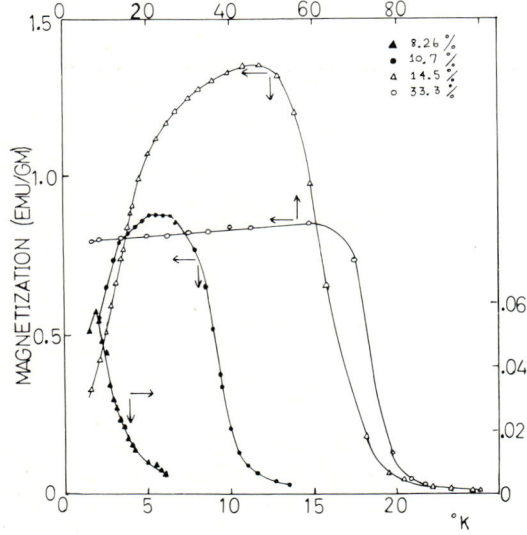

Fig. 1. Magnetization (emu/g) against temperature, in an applied field of 5 Oe, for 8.26, 10.7, 14.5 and 33.3% Gd.

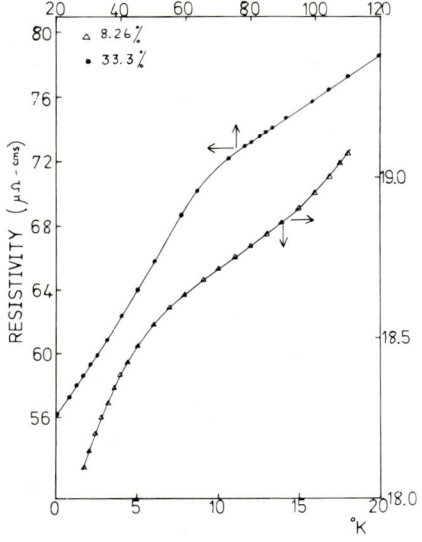

Fig. 2. Resistivity ($\mu\Omega$-cms) against temperature for 8.26 and 33.3% Gd.

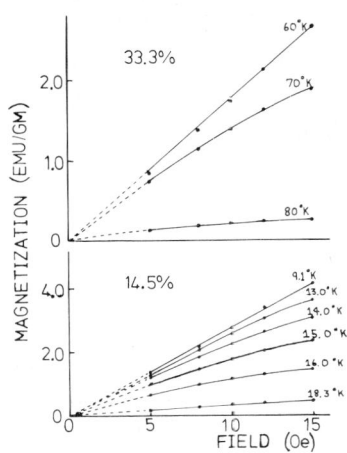

Fig. 3. Magnetisation against field for 14.5 and 33.3% Gd.

and in our case below the temperature range of our measurements.) It should be emphasized that θ_p for the 8.26% alloy is $+5.5$ K.

At somewhat higher concentrations and fields of more than 200 Oe [2] the magnetization increases steadily with decreasing temperature, but our data for the 14.5% alloy at <15 Oe show clearly the low temperature fall in M associated with the freezing in a spin-glass manner of the giant cluster moments. It is the rapid increase in alignment of these moments by small fields below θ_p (23 K) and above T_g that gives a quasi-ferromagnetic character. This is strikingly similar to the behaviour described by Murani [4] for a Au 15% Fe alloy, and like him we find remanence and time-dependence only at temperatures below that of the magnetization maximum. (The magnetization in 5 Oe reaches about $\frac{1}{3}\mu_B$ per Gd atom.) It is interesting to note that EPR data [5] for alloys in this range show significant g-shifts only below this temperature, although line broadening has begun at much higher temperatures; this again closely parallels Au-Fe behaviour (R.H. Taylor, unpublished).

At even higher concentrations of Gd ($>16.7\%$) the resistivity (fig. 2) and magnetization show temperature dependences much more like that of pure $GdAl_2$, although we are unable to specify a critical composition for the onset of ferromagnetism. This difficulty arises from the linearity at low fields of plots (fig. 3) of M vs. H, so that Arrott plots (of M^2 vs. H/M) never satisfy the criterion for ferromagnetism. Peculiarities in $M(H, T)$ and the specific heat of $GdAl_2$ are well known [6]. It thus seems that by eliminating special nearest-neighbour interactions we have simplified the character of the spin glass at the cost of complicating the character of the eventual ferromagnetism.

References

[1] K.H.J. Buschow, J.F. Fast, A.M. van Diepen and H.W. de Wijn, Phys. Status Solidi 24 (1967) 715.
[2] M.B. Maple, Thesis, Univ. California, San Diego (1969).
[3] R.J. Trainor and D.C. McCollum, Phys. Rev. B11 (1975) 3581.
[4] A.P. Murani, J. Phys. F4 (1974) 757.
[5] R.H. Taylor, J. Phys. F3 (1973) L110.
[6] H. Hacker, R. Gupta and M.L. Sheppard, Phys. Status Solidi A9 (1972) 601.

RESISTANCE MAXIMUM IN SPIN GLASSES

U. LARSEN
H.C. Ørsted Institute, University of Copenhagen, DK-2100 Copenhagen, Denmark

J.S. SCHILLING, P.J. FORD
Experimentalphysik IV, Universität Bochum, 463 Bochum, W. Germany

and

J.A. MYDOSH
Kamerlingh Onnes Laboratorium der Rijksuniversiteit, Leiden, The Netherlands

Theoretical expressions based on the s–d exchange interaction are obtained for the freezing temperature and the temperature of the resistance maximum in spin glasses. A number of recent experimental results, such as the concentration and pressure dependence of these quantities, can be interpreted within this framework.

Experimentally, there are several well defined and directly *observable characteristic temperatures* in spin glasses. Two of these are the freezing temperature, T_0, identified by a cusp or sharp maximum in the susceptibility, and the temperature, T_M, of the resistance maximum [1].

Many systems, like AuFe, are simultaneously spin glasses and Kondo alloys. There are therefore two *basic energy scales*: the single-spin Kondo temperature, T_K, and the interaction strength, Δ_c, between different spins. According to the relative magnitudes of these, the systems can be characterized as: (1) *Kondo alloys* ($T_K > \Delta_c$), and (2) *spin glasses* ($\Delta_c > T_K$).

The s–d exchange interaction, $-J\boldsymbol{\sigma}_i \cdot \boldsymbol{S}_i$, between a single spin, $\boldsymbol{S}_i$, and the conduction electron spin density, $\boldsymbol{\sigma}_i$, at site i gives rise to both the Kondo effect and to the RKKY interaction, $\mathcal{J}(R_{ij})\boldsymbol{S}_i \cdot \boldsymbol{S}_j$, between two spins a distance R_{ij} apart. The *fundamental model parameters* are therefore J and the concentration of spins, c.

Theoretical relations between these quantities are obtained below in four steps. Initially the dependence of the scales T_K and Δ_c on J and c is calculated

$$T_K = T_K(J); \qquad \Delta_c = \Delta_c(J). \tag{1}$$

Then the observables T_0 and T_M are calculated in terms of the scales

$$T_0 = T_0(\Delta_c, T_K); \qquad T_M = T_M(\Delta_c, T_K) \tag{2}$$

in the extreme spin glass regime $\Delta_c \gg T_K$. We use these relations to explain how the dependence of the observables T_0 and T_M on *experimentally accessible variables* like pressure, P, and concentration can be understood in terms of a very simple relationship with the fundamental parameters.

(1) The Kondo temperature is well known and defined by $T_K(J) = E_F \exp[-1/n(E_F)|J|]$, where $n(E_F)$ is the electron density of states at the Fermi energy E_F.

(2) The root-mean-square RKKY interaction energy of an average spin (site i) due to all the other spins (sites j) is $[\Sigma_{j\neq i} \mathcal{J}(R_{ij})^2]^{1/2}_{av}$. From this one obtains [2]

$$\Delta_c(J) = a(S)f(c,\lambda)J^2/E_F, \tag{3}$$

where f is a smoothly increasing function of c. $a(S)$ is a prefactor of order unity, depending on the spin S. Major effects determining $f(c,\lambda)$ are the statistical distribution of nearest-neighbor distances and (self-)damping of the RKKY interaction beyond the electron mean-free-path λ. Details are given in ref. 2.

(3) In the Edwards–Anderson theory [3], disregarding the Kondo effect, T_0 is proportional to the r.m.s.-(RKKY) interaction strength, and it therefore follows that

$$T_0 = b(S)\Delta_c(J) = [a(S)b(S)]f(c,\lambda)J^2/E_F \tag{4}$$

for $\Delta_c \gg T_K$. $b(S)$ is a prefactor depending on the spin S. For quantum spins [4] $[a(S)b(S)] = [(2S+1)^4 - 1]^{1/2}/12$. Also, λ can be inferred from resistance measurements [5]. Then comparing eq. (4) with the observed values in AuFe [2, 6], one finds $|J| = 0.26$ eV, in agreement with $J = -0.25$ eV obtained from the low-c Kondo resistance [7]. As can be seen in fig. 1 of ref. 6, the non-trivial c-dependence is well described by eq. (4) for $0.1\% \leq c \leq 10\%$, while the smaller observed T_0 for $c \leq 0.1\%$ can be ascribed to the breakdown of eq. (4) and to $T_0 \to 0$ when $\Delta_c \to T_K$.

(4) At temperatures above freezing the interactions produce spin-flip transitions at a rate $\Delta_c/\hbar$, manifesting themselves as "noise" in the spin system [8]. This interferes with the buildup

with decreasing temperature of the Kondo resistance and causes the maximum. In this model $T_M = T_M(\Delta_c, T_K)$ has been obtained [9]. Asymptotically, for $\Delta_c \gg T_K$ or $|J| \ll E_F$, one gets

$$T_M \sim \Delta_c \ln(\Delta_c/T_K) \sim |J| f(c, \lambda), \qquad (5)$$

where eq. (3) was used. Because $T_0 \simeq \Delta_c \propto J^2/E_F$ there are essential differences in magnitudes in this limit

$$T_M \gg T_0 \simeq \Delta_c \gg T_K, \qquad (6)$$

which is in agreement with observation [1, 5] in the extreme spin glass regime. However, as shown in figs. 1(a)–(b) for two concentrations $c_2 > c_1$, T_M goes through a maximum with increasing $|J|$ and tends to vanish at the transition (where $\Delta_c = T_K$) into the Kondo regime [9, 10].

The relation $T_M = T_M(\Delta_c, T_K)$ may be written in the form $\Delta_c = T_M Q(\ln T_M/T_K)$. Then from the measured T_M [6], using $T_K = 0.19$ K ($J = -0.25$ eV), calculating "experimental" values of Δ_c gives points well on the theoretical curve determined above to fit T_0, as is evident in fig. 1 of ref. 6. This supports eq. (4) with $b(S) = 1.0$ and is an independent check on the value of J.

Conversely, using the theoretical Δ_c from eq. (3) for $J = -0.25$ eV and $b(S) = 1$ gives a theoretical T_M-curve which describes well the non-trivial c-dependence of T_M, as is also evident in fig. 1 of ref. 6.

Considering the pressure dependence, the universally increasing T_K can be interpreted as an increase in $|J| = |J_0|(1 + \epsilon P)$ in all systems so far investigated [11]. On the other hand, as shown in fig. 2 of ref. 6, the effects of pressure on T_M are very diverse. But in terms of different values of J_0 in different systems this is explained by the behavior of T_M shown in fig. 1(b). The trend has now been observed in several spin glasses, that $dT_M/dP \sim dT_M/d|J|$ starts out positive for "small-T_K" spin glasses, like AuMn [6, 10], goes through zero (CuMn [6, 10]) and turns negative in "high-T_K" spin glasses, like AuFe [6, 10], LaCe [6, 12] and MoFe [6, 13]. This is consistent with the fact that the ratio of energy scales, T_K/Δ_c, is increasing with $|J|$, and the behavior of T_M is essentially predictable on the basis of this theory and estimates of T_K. A quantitative comparison between theory and experiment is forthcoming [10].

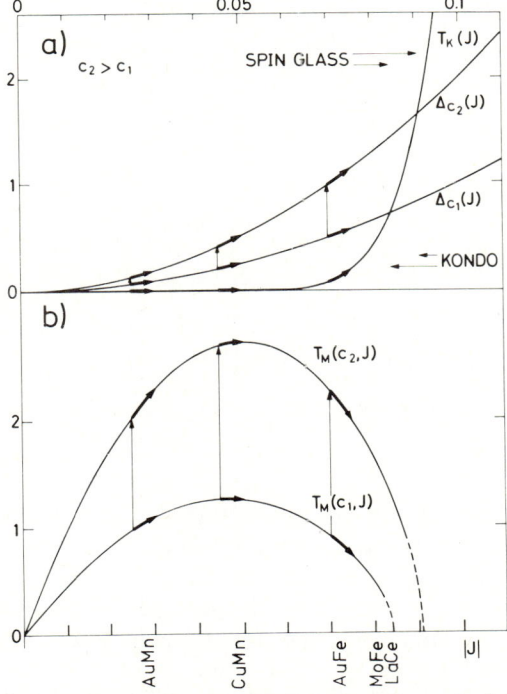

Fig. 1. Characteristic temperatures and energy scales in spin glasses, qualitatively as functions of $|J|$ for two concentrations $c_2 > c_1$. The value of $|J|$ that separates "spin glass" ($\Delta_c > T_K$) and "Kondo" ($T_K > \Delta_c$) regimes increases with concentration. Heavy arrows indicate changes with increasing pressure, light vertical arrows changes with increasing concentration. Broken curves indicate conjectured behavior in the transitional region.

References

[1] J.A. Mydosh, AIP Conf. Proc. 24 (1975) 131.
[2] U. Larsen, Solid State Commun. (1977), to appear.
[3] S.F. Edwards and P.W. Anderson, J. Phys. F5 (1975) 965.
[4] K.H. Fischer, Phys. Rev. Lett. 34 (1975) 1438.
D. Sherrington and B.W. Southern, J. Phys. F5 (1975) L49.
[5] J.A. Mydosh, P.J. Ford, M.P. Kawatra and T.E. Whall, Phys. Rev. B10 (1974) 2845.
P.J. Ford and J.A. Mydosh, Phys. Rev. B14 (1976) 2057.
[6] P.J. Ford, J.S. Schilling, U. Larsen and J.A. Mydosh, following paper (5C5) of this conference, see fig. 1 and 2.
[7] J.W. Loram, T.E. Whall and P.J. Ford, Phys. Rev. B2 (1970) 857.
[8] I. Riess and A. Ron, Phys. Rev. B8 (1973) 3467.
[9] U. Larsen, Phys. Rev. B (1976), to appear, and Phys. Lett. 40A (1972) 39.
[10] J.S. Schilling, P.J. Ford, U. Larsen and J.A. Mydosh, Phys. Rev. B (1976), to appear.
[11] See, for example, J.S. Schilling, W.B. Holzapfel, Phys. Rev. B8 (1973) 1216, and J. Crone, J.S. Schilling, Solid State Commun. 17 (1975) 791.
[12] F. Zimmer and J.S. Schilling, to be published.
[13] P.J. Ford and J.S. Schilling, J. Phys. F6 (1976) L285.

PRESSURE AND CONCENTRATION DEPENDENCES OF THE RESISTIVITY OF SPIN GLASSES*

P.J. FORD, J.S. SCHILLING

Experimentalphysik IV, Ruhr-Universität Bochum, 463 Bochum, W. Germany

U. LARSEN

H.C. Ørsted Institute, University of Copenhagen, DK-2100 Copenhagen, Denmark

and

J.A. MYDOSH

Kamerlingh Onnes Laboratorium der Rijks-Universiteit, Leiden, The Netherlands

Data are presented of the pressure and concentration dependences of the temperature of the resistance maximum T_M for some typical spin glasses. They are interpreted in terms of a recent theory due to Larsen which takes into account the relative magnitudes of the Kondo temperature T_K and the r.m.s. – RKKY interaction strength Δ_c.

Over the last few years there has been considerable interest in studying spin glasses [1]. This appears to be a new and important state of magnetism which is characterised by the spins on the impurity atoms freezing or locking into random directions below a temperature T_0. Ford and Mydosh [2] have suggested five particularly simple spin glass systems among the combinations of 3d transition metal impurities in copper, silver and gold. These are AuFe, AuCr, AuMn, CuMn and AgMn. However, spin glass behaviour seems to be becoming an increasingly widespread phenomenon and has also been observed in 4d host-3d impurity alloys as well as in rare earth systems. Measurements of the electrical resistivity are an extremely sensitive method for studying the Kondo effect as well as the interplay between spin glass and Kondo behaviour at temperatures well above T_K. One consequence of this interplay is the resistance maximum and in this paper we will be mainly discussing the way in which this varies with the concentration of the impurities and upon applying pressure. In some recent work, Larsen [3] has shown that the temperature of the resistance maximum, T_M, is a function of two fundamental parameters. These are the Kondo temperature T_K and the root-mean-square (r.m.s.) Ruderman–Kittel–Kasuya–Yosida (RKKY) interaction strength Δ_c, i.e. $T_M = T_M(\Delta_c, T_K)$.

In fig. 1 we show a plot of experimental values of T_M and T_0, taken from a compre-

* This work is supported in part by the Deutsche Forschungsgemeinschaft.

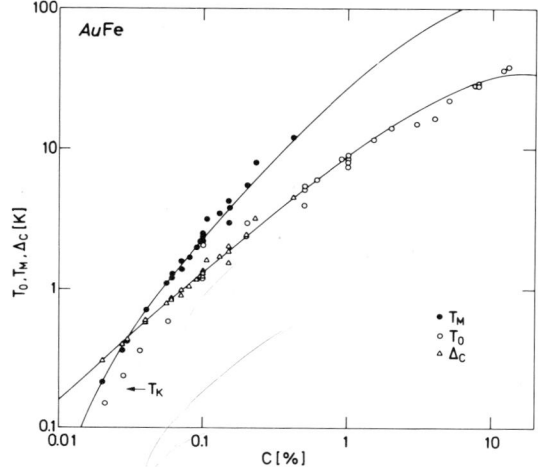

Fig. 1. Experimental data for T_M and T_0 and derived values of Δ_c as a function of concentration, c, for AuFe. T_K for AuFe at zero pressure has been taken to be 0.19 K, ($J = -0.25$ eV). The solid lines through the points represent theoretical curves for T_M and T_0 due to Larsen [3, 4].

hensive survey of the literature, and derived values of Δ_c, calculated from eq. 79 of Larsen [3], for AuFe, which can be regarded as an archetypal example of a spin glass. The solid lines through the experimental points represent theoretical curves for T_M and T_0 due to Larsen [3, 4]. Two important features can be seen from fig. 1. Firstly, contrary to widely held beliefs, T_M is never a linear function of the concentration, c, although it may approximate to this over a narrow concentration range. There are two main reasons for this. In the first place, Δ_c is affected both by the spatial distribution of the nearest neighbour impurities and also by the self damping of the RKKY interaction between the spins and it therefore has a complicated concentration dependence. Secondly, the expression $T_M = T_M(\Delta_c, T_K)$ is itself a complex function and contains for example terms which are logarithmic in c, which, although slowly varying, nevertheless diverge as c tends to zero. The second important feature of fig. 1 is that $\Delta_c = T_0$ over a wide range of concentration except at low c where T_0 is being depressed by the Kondo effect. In general one would only expect Δ_c to be proportional to T_0 and the equality may be fortuitous for the AuFe system. The concentration dependences of T_M, T_0 and Δ_c for the four remaining simple spin glass systems are currently being investigated and will be reported at a later date.

Recently a new dimension has been added to the study of the resistance maximum by pressure measurements [5-9]. Increasing the concentration of impurities only affects Δ_c, whereas increasing the pressure mainly affects T_K. It has been found that applying pressure produces a rich variety both in the magnitude and the direction of the shift in T_M. For example, for impurity concentrations of 0.1 at.%, pressure causes T_M to increase for AuMn [6, 7], to be unaltered for CuMn [6, 7] and to decrease for AuFe [5-7]. The variation of T_M with the relative volume V/V_0 for several alloys is shown in fig. 2. Larsen's theory [3] accounts for these variations by taking into consideration the relative magnitudes of the Kondo effect T_K and the r.m.s.-RKKY interaction strength Δ_c. For systems with very low values of T_K, like AuMn, $\Delta_c/T_K \gg 1$ and it can be shown [7] that $dT_M/dP > 0$. For systems like CuMn and AuFe with smaller values of Δ_c/T_K, dT_M/dP can become zero or negative in sign. Thus the magnitude of T_K is reflected in a rather subtle manner by the direction and magnitude of the pressure shift of T_M. Experiments on very dilute alloys [10, 11] have shown that T_K always increases with pressure. Two particularly interesting cases appear to be MoFe [8] and LaCe [9] where rather dramatic decreases of T_M with pressure are observed. We believe that in both systems one is probably in a region where T_K and Δ_c are comparable in magnitude. In this situation Larsen's present theory would predict that both T_M and T_0 would tend rapidly to zero when T_K tends to Δ_c, although to

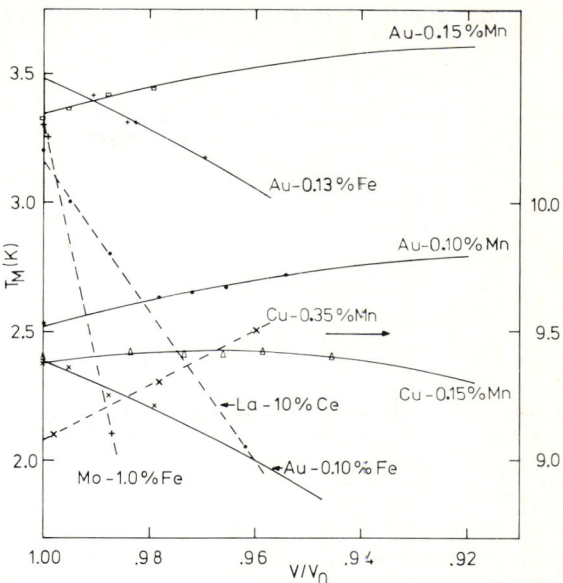

Fig. 2. T_M versus relative volume V/V_0 for several spin glasses. Cu−0.35% Mn uses temperature scale to the right. All concentrations are in at.%. The solid lines are fits using a theory due to Larsen [3]. The dashed lines are preliminary straight-line fits.

remain rigorously correct in the region where $T_K \approx \Delta_c$ the theory needs a more elaborate calculation. This region is of particular interest since it is where the Kondo effect has the greatest influence on the spin glass behaviour. More detailed accounts of this work will be forthcoming.

References

[1] J.A. Mydosh, AIP Conf. Proc. 24 (1975) 131.
[2] P.J. Ford and J.A. Mydosh, J. de Physique C4 (1974) 241, and Phys. Rev. B14 (1976) 2057.
[3] U. Larsen, Phys. Rev. B14 (1976) 4356; see also previous paper, 5C4, of this conference.
[4] U. Larsen, Solid State Commun. to appear.
[5] J.S. Schilling, J. Crone, P.J. Ford, S. Methfessel and J.A. Mydosh, J. Phys. F4 (1974) L116.
[6] J.S. Schilling, P.J. Ford, J. Crone and J.A. Mydosh, EPS Conf. Abstr. 1A (1975) 23.
[7] J.S. Schilling, P.J. Ford, U. Larsen and J.A. Mydosh, Phys. Rev. B14 (1976) 4368.
[8] P.J. Ford and J.S. Schilling, J. Phys. F6 (1976) L285.
[9] F. Zimmer and J.S. Schilling, to be published.
[10] J.S. Schilling and W.B. Holzapfel, Phys. Rev. B8 (1973) 1216.
[11] J. Crone and J.S. Schilling, Solid State Commun. 17 (1975) 791.

IMPURITY INTERACTIONS IN KONDO SYSTEMS

G. GRÜNER

Central Research Institute for Physics, Budapest, Hungary

and

E. BABIC

Institute of Physics of the University, Zagreb, Yugoslavia

The sharp distinction between effects caused by interaction between well defined impurity spins (spin glass) and that of Kondo-impurities is discussed and demonstrated on specific examples.

Interactions between magnetic impurities in a metallic matrix are usually discussed in terms of the long range coupling between well defined spins S. This coupling gives rise to a spin glass, or mictomagnet with properties extensively discussed in the literature [1]. The assumption of a spin with infinite lifetime is essential in having a spin glass state, thus one requires $T_K \ll T_c$ where T_K the Kondo temperature of the single impurity, T_c the spin glass ordering temperature. We discuss the opposite limit $T_K > T_c$ and show that the temperature and concentration dependence of the various physical properties are drastically different from the spin glass state.

In the Kondo state, below T_K, conduction electrons at energy range kT_K around the Fermi level ϵ_F are performing the screening [2, 3], thus $\alpha(T_F/T_K)$ conduction electrons screen one impurity with spin $S = \frac{1}{2}$ where α accounts for the difference of the shape of the Kondo-resonance and of the conduction band. With increasing impurity concentration not all the impurities will be completely screened, this leads to a lowering of T_K. Using a statistical picture for the distribution of the impurities we get [4] for arbitrary spin S

$$\ln\left(\frac{T_K^\circ}{T_K(c)}\right) = \left(\alpha \frac{N}{2S} \frac{T_K}{T_F}\right)^{-1} c, \quad (1)$$

for impurity concentration c where N is the number of electrons/atom in the host. The logarithmic concentration dependence of T_K is shown in fig. 1 for various dilute alloys spanning a wide range of Kondo temperatures ($T_K \sim 1200$ for AlCr, 500 for AlMn and 30 for CuFe). Fig. 2 shows the concentration dependence of T_K evaluated for several alloy systems, using plots

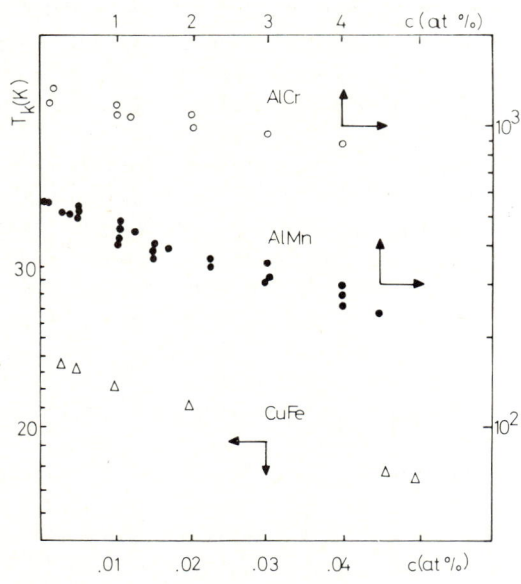

Fig. 1. Concentration dependence of T_K derived from the resistivity, using the relation for AlCr, AlMn (ref. 4) and CuFe (ref. 3).

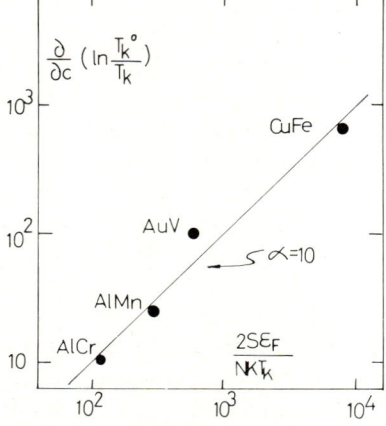

Fig. 2. Concentration dependence of T_K versus the parameter given in eq. (1). The straight line gives $\alpha = 10$.

like in Fig. 1, together with the $S\epsilon_F 2(NkT_K)^{-1}$ values. It is evident, that eq. (1) holds well for a broad range of parameters, and we obtain $\alpha \sim 10$ for the normalization factor. It is not possible to evaluate the temperature dependences of the various physical quantities in the region where interactions of type discussed above are important. We argue, however, that this region is intermediate between the single impurity Kondo-effect and the highly correlated electron gas [5] (one scatterer and many electrons in the former many scatterers and few extra electrons in the latter case). Thus we expect properties characteristic to an interacting Fermi gas, i.e. T^2 dependence of the resistivity, specific heat proportional to the temperature and susceptibility $x \sim 1 - T^2$ well below T_K. Indeed, these properties have been observed in cases where T_K is high and the impurity concentration is low.

Finally, we note that the interaction effect we described and the long range interaction which leads to a spin glass state act against each other. The Kondo effect leads a finite impurity lifetime, thus to a reduced Ruderman–Kittel–Kasuya–Yoshida interaction. On the other hand, the onset of a spin glass state reduces the probability of spin-flip due to the developing internal fields, this reduces T_K. We propose therefore a phase diagram shown in fig. 3 for the behaviour expected in general for dilute alloys. At temperatures below T_K° increasing impurity concentration leads to interactions described above, and thus to a region what we call collective Kondo state. By increasing c further T_K is reduced below T and at still higher concentrations spin glass behaviour sets in. At high concentrations direct, nearest neighbour interactions are dominating. Naturally, all different behaviours are expected to be observed only in some favourable cases, depending mainly on T_K°. With small T_K° the collective

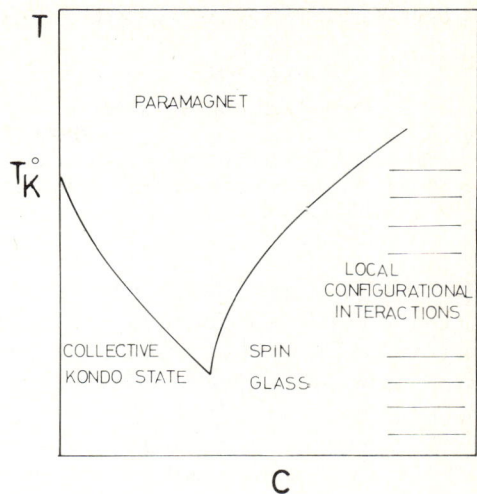

Fig. 3. Proposed phase diagram in Kondo alloys.

Kondo region is small, this situation is appropriate for AuFe. When T_K° is high, the collective Kondo state extends to high concentrations. Spin glass behaviour is not observed in these cases, increasing the concentration further nearest neighbour interactions are dominating. This the case for AlMn. CuFe and systems with similar Kondo temperatures may be good candidates for observing all the various types of impurity interactions.

Useful discussions with J. Cooper, N. Rivier and V. Zlatic are acknowledged.

References

[1] Y. Mydosh, in: Magnetism and Magnetic Materials, C.W. Graham, J.J. Phyne, G.H. Lander (Eds.), (Amer. Inst. Phys., New York, 1975) p. 131.
[2] G. Grüner and A. Zawadowski, Rep. Progr. Phys. 12 (1974) 1497.
[3] W.M. Star, PhD thesis, University of Leiden (1971).
[4] E. Babic and G. Grüner, Physica 84B (1976) 37.
[5] N.F. Mott, J. Phys. F. 4 (1974) L46.

SPIN GLASSES EXHIBIT ROCK MAGNETISM

E.P. WOHLFARTH

Department of Mathematics, Imperial College, London SW7, England

The physical properties of spin glasses are briefly discussed in terms of magnetic clusters, their anisotropy and their blocking temperatures. It is shown that the susceptibility of spin glasses has a pronounced and asymmetrical peak at the blocking temperature while in general no pronounced specific heat anomaly is expected there.

The concept "spin glasses" is fairly new and has given rise to a considerable amount of experimental and theoretical work on dilute alloys. It has become clear that there is an essential conflict between two opposing view points on this matter and that it seems difficult to reconcile these. The first view point is based on the presence of the localized spins of isolated transition metal atoms (Fe, Mn,...) in a non-transition metal matrix (Au, Cu,...). These spins interact with each other via an RKKY mechanism residing in the itinerant electrons of the matrix. Ingenious mathematical [1] and computational techniques are then used to obtain significant physical results. Comparison is made most frequently with the observations [2] that the susceptibilities of spin glasses show pronounced maxima at temperatures unrelated to any obvious phase transitions. The mathematics suffices to reproduce this result. However, it has not been found possible to use this model to describe a whole host of other observations on spin glasses, including the following:

(1) The susceptibility, temperature curves of spin glasses, including their maxima, can depend sensitively on the variables of the experiments including the metallurgy of the specimens and the frequency, time and field applied at various stages.

(2) The specific heat, temperature curves of spin glasses [3] show little evidence of a phase transition of the type described by the mathematical models [4].

(3) Neutron data (see, for example, [5], [6]) can not be interpreted solely in terms of the presence of isolated spins.

Hence the alternative view point is very rapidly gaining credence that the physical situation regarding spin glasses demands the presence of spin clusters rather than isolated spins. This view point then recalls at once the pioneering work of Néel [7] on the superparamagnetic properties of small particles and of their blocking temperatures where this type of rapid spin fluctuation with calculable relaxation time is frozen into a more stable ferromagnetic behaviour. Néel applied this concept to rock magnetism and thus we claim here that which heads this brief paper.

At and below the blocking temperature T_B the clusters thus have a susceptibility $\chi(T)$ which in the simplest case is given by [8] $\chi(T) = M_s^2(T)/3K(T)$, where M_s is the saturation magnetization and K the anisotropy constant. At and above T_B the susceptibility is given by a Langevin formula $\chi(T) \simeq vM_s^2(T)/3kT$, where v is the cluster volume. Hence, using Néel's expression for the blocking temperature, $\chi(T) \simeq 25\{M_S^2(T)/3K(T_B)\}(T_B/T)$. Hence at T_B, $\chi(T)$ is "thermally enhanced" [9] by a factor ~25. The total $\chi(T)$ curve has a very sharp and asymmetrical maximum indeed at T_B. By analogy with the behaviour of ordinary ferromagnets at their Curie points we may continue to call this a Hopkinson peak [10] (see the nomenclature of [9] and [11]). Even if there is a *spectrum* of blocking temperatures [9] this sharpness should to some extent survive. Fig. 1 [11] shows the

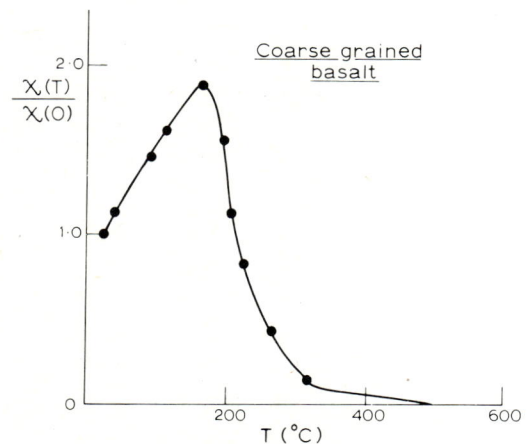

Fig. 1. Temperature dependence of the susceptibility $\chi(T)$ of a coarse grained basalt [11].

$\chi(T)$ curve of a coarse grained basalt containing magnetic oxide mineral inclusions; the sharp maximum recalls clearly that of spin glasses. Similar maxima have also been observed in cobalt and manganese aluminosilicate glasses (these are *real* glasses!) [12] and are there ascribed similarly to clustering, here of cobalt and manganese atoms. The $\chi(T)$ maxima were calculated more elaborately than above. Long before spin glasses became a fashionable concept maxima in the susceptibility were observed in dilute alloys [13] and discussed in terms of clustering effects.

What about the specific heat and the origin of K? The former has been calculated for superparamagnetic particles [14]. The contribution to the total specific heat coming from superparamagnetism was shown to be important, relative to the normal contributions, at low temperatures only. Hence, unless T_B lies in this low temperature range, no pronounced specific heat anomaly is expected at the blocking temperature. For the dilute Cu–0.7% Co alloys of [14], for example, this low temperature range is below about 1 K. Hence for alloys showing spin glass behaviour in the susceptibility the absence of a pronounced specific heat anomaly [3] is easily explained on the present basis. The anisotropy constant deduced [14] for Cu–Co from the specific heat data turns out to be anomalously large compared to the crystal anisotropy constant of Co. It was thus tentatively suggested that the anisotropy is determined partly by surface effects ("surface anisotropy") which may be non-cubic. Other types of anisotropy may arise as a result of cluster shape, stresses on clusters or interactions (dipole–dipole,...) between them ("interaction anisotropy"); a brief description of types of anisotropy is given in [15]. It is to be hoped that once the idea of clusters causing spin glass behaviour is accepted the search for the magnitude and origin of the anisotropy as being the key quantity of the whole phenomenon can begin in earnest.

I have benefited greatly from reading a recent review article on this subject [16], and helpful discussions with C.N. Guy.

Note added in proof: Time effects were introduced into the Edwards–Anderson model by these authors [17]. No numerical estimates of the expected time scales were made. However, the physics of this model is such that the observed effects, which frequently involve long times, are unlikely to be explained on this basis.

References

[1] S.F. Edwards and P.W. Anderson, J. Phys. F5 (1975) 965.
[2] V. Canella and J.A. Mydosh, Phys. Rev. B6 (1972) 4220.
[3] L.E. Wenger and P.H. Keesom, Phys. Rev. B11 (1975) 3497.
[4] H. Müske and G. Heber, J. Phys. F6 (1976) 1353.
[5] N. Ahmed and T.J. Hicks, J. Phys. F5 (1975) 2168.
[6] A.P. Murani, G. Goeltz and K. Ibel, Solid State Commun. 19 (1976) 733.
[7] L. Néel, Ann. Géophys. 5 (1949) 99; for a Fokker–Planck treatment of thermal fluctuations and magnetism, see W.F. Brown, Phys. Rev. 130 (1963) 1677.
[8] E.C. Stoner and E.P. Wohlfarth, Phil. Trans. Roy. Soc., A240 (1948) 599.
[9] D.J. Dunlop, J. Geophys. 40 (1974) 439.
[10] J. Hopkinson, Phil. Trans. Roy. Soc., A180 (1889) 443.
[11] C. Radhakrishnamurty and S.D. Likhite, Earth and Planet. Sci. Lett. 7 (1970) 389.
[12] R.A. Verhelst, R.W. Kline, A.M. de Graaf and H.O. Hooper, Phys. Rev. B11 (1975) 4427.
[13] J.S. Kouvel, J. Phys. Chem. Solids 21 (1961) 57.
P.A. Beck, Magnetism in Alloys (1972) 211.
J.L. Tholence and R. Tournier, J. de Physique 35, C4 (1974) 229.
H. Claus, Phys. Rev. Lett. 34 (1975) 26; and others.
[14] E.J. Hayes, A. Hahn and E.P. Wohlfarth, J. Phys. F2 (1972) 351.
[15] E.P. Wohlfarth, Magnetism (Rado-Suhl) 3 (1963) 351.
[16] G. Heber, updated version of J. Magn. Magn. Mater. 2 (1976) 47.
[17] S.F. Edwards and P.W. Anderson, J. Phys. F6 (1976) 1927.

THERMODYNAMICS OF DILUTE MAGNETIC ALLOYS: VIRIAL EXPANSION VERSUS T^{-1} EXPANSION

K. MATHO

Centre de Recherches sur les Très Basses Températures, CNRS, BP 166 Centre de Tri, 38042 Grenoble Cedex, France

The results of Larkin and Khmelnitzkii [1] (LK) on spin-glass susceptibility χ, entropy S_m and specific heat C_m are generalized within the second virial approximation to include the T-dependence arising from high energy cut-offs in the exchange distribution. For $T \to \infty$, results of the T^{-1} expansion are recovered.

The work reported here concerns high temperature thermodynamic properties of random dilute magnetic alloys. The basic model is the RKKY–hamiltonian in zero external field

$$\mathcal{H} = -\tfrac{1}{2} \sum W(|R - R'|) S(R) \cdot S(R') \qquad (1)$$

summed over impurity positions R, R'. The distribution of exchange energies $D(W) = \Sigma \delta[W - W(R)]$, summed over all lattice positions $R \neq 0$, is idealized by a continuous function with high energy cut-offs (fig. 1).

Low order moments of the exact RKKY-distribution $D^{(\nu)}$, $\nu = 1, 2, \ldots$, have been discussed by Kok and Anderson [2]. For a unique choice of the cut-offs (fig. 1), namely

$$T_{1f} + T_{1a} = D^{(2)}/W_1 = 2T_1 \qquad (2)$$

and

$$(T_{1f} - T_{1a})/(T_{1f} + T_{1a}) = \tanh(D^{(1)}/2W_1) = \delta, \qquad (3)$$

the continuous distribution reproduces the moments $D^{(1)}$ and $D^{(2)}$.

We now discuss second virial corrections to the leading free spin terms of various thermodynamic functions. Two well known limits are resumed in table I: for $T_1 \ll T$ (row 1), the pair corrections reduce to the T^{-1}-expansion with the moments $D^{(1)}$ and $D^{(2)}$ as coefficients [2]. For $T \ll T_1$ (row 2), the corrections [1] depend only on the central amplitude W_1 through the virial expansion parameter $\xi = CW_1/kT$. C is the atomic concentration.

In order to discuss the range $T \sim T_1$, the initial susceptibility

$$3kT \cdot \chi = (g\mu_B)^2 \Big(S(S+1) + C \int dW D(W) \langle S_1 \cdot S_2 \rangle \Big) \qquad (4)$$

is chosen for sake of argument. Fig. 1 shows the energy dependence of the exact thermal correlation function $\langle S_1 \cdot S_2 \rangle$ for a single pair, at different temperatures. It is clear that (a) samples only the first moment $D^{(1)}$. On the contrary, (d) contains strongly correlated pairs which cannot be described by any finite order of a T^{-1}-expansion. Furthermore, the symmetric central peak of $D(W)$ is sampled by the correlated quantum spins because of asymmetric saturation values. The spin dependence of the antiferromagnetic "Weiss-constant" $kT_s^* = CW_1\theta_s$ for this limit has been calculated with the exact correlation function [3] (table II).

To integrate eq. (4) between arbitrary limits $-kT_{1a}$ and kT_{1f}, the correlation function has been approximated [with $Y = (S+1)W/6kT$] by

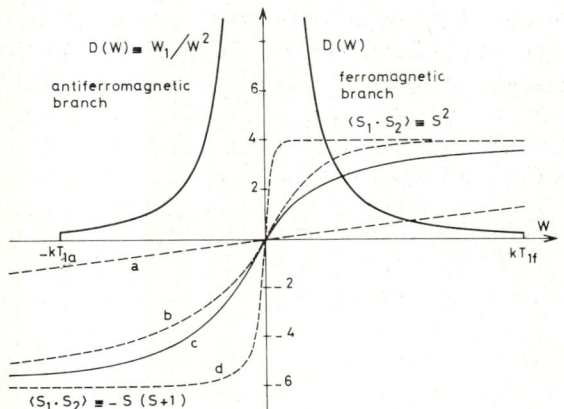

Fig. 1. Definition of a model for the exchange energies in the RKKY–hamiltonian, characterized by scaling parameter W_1 and asymmetric cut-offs $T_{1f} = (1+\delta)T_1$, $T_{1a} = (1-\delta)T_1$. Also shown, the correlation function to be summed over in eq. (4): (a) $T \gg T_1$; (b) $T \sim T_1$; (d) $T \ll T_1$; exact $\langle S_1 \cdot S_2 \rangle$ for $S = 2$. Curve (c) approximates (b) according to eq. (5).

Table I

Approximation	Susceptibility χ per spin	Entropy S_m per spin	Specific heat C_m per spin
T^{-1}-Expansion $T \gg T_1$	$\chi_{\text{Curie}}\left(1 + \dfrac{CS(S+1)D^{(1)}}{3kT}\right)$	$k\ln(2S+1) - \dfrac{CS^2(S+1)^2 D^{(2)}}{6kT^2}$	$\dfrac{CS^2(S+1)^2 D^{(2)}}{3kT^2}$
Scaling limit $CW_1 \ll kT \ll kT_1$	$\chi_{\text{Curie}}(1 - \xi\theta_s)$	$k(\ln(2S+1) - S(2S+1)\xi)$	$kS(2S+1)\xi$
Entire range $CW_1 \ll kT$; $\langle S_1 \cdot S_2 \rangle \simeq$ eq. (5)	$\xi \to \xi F_\chi$; $F_\chi = 1 + \ln^{-1}\left(\dfrac{S+1}{S}\right)$ $\times \left[\dfrac{\tau(\tau_f - \tau_a)}{2(\tau+\tau_f)(\tau+\tau_a)} + \ln\dfrac{1+\tau/\tau_f}{1+\tau/\tau_a}\right]$	$\xi \to \xi F_s$; $F_s = \dfrac{S\tau_f}{(2S+1)(\tau+\tau_f)}$ $+ \dfrac{(S+1)\tau_a}{(2S+1)(\tau+\tau_a)}$	$\xi \to \xi F_c$ $F_c = \dfrac{S\tau_f(2\tau+\tau_f)}{(2S+1)(\tau+\tau_f)^2}$ $+ \dfrac{(S+1)\tau_a(2\tau+\tau_a)}{(2S+1)(\tau+\tau_a)^2}$
Definitions	$\xi = CW_1/kT$; θ_s: table II; $\tau = T/T_1$; $\tau_f = \tfrac{1}{6}(S+1)^2(1+\delta)$; $\tau_a = \tfrac{1}{6}S(S+1)(1-\delta)$		

Table II

Quantum spin S	$\tfrac{1}{2}$	1	$\tfrac{3}{2}$	2	$\tfrac{5}{2}$
θ_s	0.50	0.64	0.74	0.81	0.86
θ'_s	0.27	0.46	0.64	0.81	0.98

$$\langle S_1 \cdot S_2 \rangle \simeq \begin{cases} S^2(S+1)Y[(1+(S+1)Y)^{-1} \\ \quad + (1+(S+1)Y)^{-2}]; & Y > 0 \\ S^2(S+1)Y[(1-SY)^{-1} \\ \quad + (1-SY)^{-2}]; & Y < 0. \end{cases} \quad (5)$$

The approximate "Weiss-constant" in the scaling limit $T \ll T_1$ is (table II) $kT_s^* \simeq CW_1 \theta'_s = \tfrac{1}{3}CW_1 S(S+1)\ln(S+1)/S$. The rest of the information on the cut-off dependence of the integral in eq. (4) and analogous integrals for the entropy and specific heat, is contained in three normalized functions F_χ, F_s and F_c (table I, row 3; fig. 2). By the substitution $\xi \to \xi F$ in the corresponding scaling formulae of row 2, one obtains the full T-dependence.

Although the effects discussed here are quite small, they reveal interesting quantum-mechanical and statistical information on the RKKY-model. They might be worth a closer experimental analysis.

References

[1] A.I. Larkin and D.E. Khmelnitzkii, Sov. Phys. JETP 31, 958 (1970). Our virial expansion parameter is $4\eta V_0/3kT$ in the (LK) notation.
[2] W.C. Kok and P.W. Anderson, Phil. Mag. 24 (1971) 1141.
[3] K. Matho, to be published.

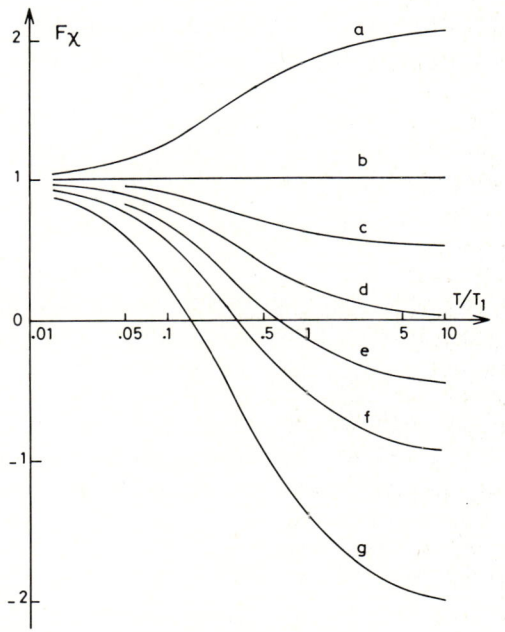

Fig. 2. Temperature dependence of the "Weiss-constant" above the scaling limit for $S = 2$. Mean cut-off T_1, eq. (2). Asymmetry-parameter $\delta = \mp(0.4(a,g); 0.2(b,f); 0.1(c,e); 0(d))$, eq. (3). Sign change occurs for $\delta > 0$. General condition for no T-dependence is $\delta = -(2S+1)^{-1}$.

STATISTICAL THERMODYNAMICS OF SPIN GLASSES AND QUENCHED MAGNETIC SYSTEMS

N. RIVIER

The Blackett Laboratory, Physics Dept., Imperial College, London SW7, England

From the information theory point of view, there is a disorder contribution to the entropy, associated with the probability distribution given a priori in quenched systems. It is observable if, as in spin glasses, this distribution is a function of the order parameter, hence of temperature. The theory is self-consistent.

Spin glasses are a particular case of quenched systems [1] in which the positions of the magnetic atoms are not in thermal equilibrium with the host metal, but given a priori by some probability distribution. The statistical thermodynamics is then unusual in that it contains a double ensemble average (the standard Gibbs average over the magnetic configurations for every given spatial configuration plus the probabilistic average over the latter) with ill-defined – and possibly unequal – a priori probabilities. It is essential therefore to construct a statistical thermodynamics based solely on our knowledge of the mechanical system, i.e. its hamiltonian and the probability distribution P associated with the randomness in the positions, and free of additional assumptions such as the nature of the ensemble(s). Such a constructive, subjective statistical mechanics has been proposed and justified by Jaynes [2] on the basis of information theory [3].

To the probability distribution P associated with the randomness of the alloy corresponds an entropy of disorder [3], $S_{dis} = -k \, \text{Tr} \, P \ln P$. The distribution P itself takes various forms in different theories of spin glasses – it can be a distribution of exchange interactions [4] or of local internal magnetic field [4–7]. If, as in refs. 5 and 7, P depends on the order parameter $q(T, H_e)$ (labelled m in AR) of the spin glass and is therefore temperature and field dependent, the disorder entropy does contribute to the specific heat. In view of considerable disagreement between theory and (computer or direct) experiment, a stable footing to evaluate the specific heat is clearly necessary. Moreover, in the EA type of theories in their present form, the spin glass state is unstable [8] relatively to the unfrozen, paramagnetic state – it is not even metastable. The aim of this paper is to construct a statistical mechanics free of unphysical assumptions which yield, for a large class of probability distributions P, a stable spin glass state.

Consider a spin glass in, for simplicity, a local mean field picture, that is described by a dynamical variable (the local spin S) and a random variable (the local dimensionless mean field $h = H/a$) according to some occupation frequency $\Pi(h, S)$. The actual distribution is that which maximizes the entropy $-k \, \text{Tr} \, \Pi \ln \Pi$ – the trace running over all continuous and discrete variables (h, S) – subject to the constraints corresponding to our knowledge of the system. These constraints are (i) the expectation value of the energy $E = \text{Tr} \, \Pi(h, S)\epsilon_s(h)$; (ii) a normalization condition $\text{Tr} \, \Pi(h, S) = 1$; and (iii) the probability distribution of the local mean field $P(h)$ which can be calculated from first principles. Quenched systems have clearly a structure of conditional probability [1], that is

$$\Pi(h, S) = P(h)\Pi(S|h), \qquad (1)$$

where $\Pi(S|h)$ gives the occupation frequency of the state S for a given mean field h. It is normalized, thus $P(h) = \text{Tr}_S \, \Pi(h, S)$. Our physical knowledge of the system consists of the proper enumeration of the states (h, S) of their energies (i.e. the hamiltonian of the system) $\epsilon_S(h)$, and of the probability distribution $P(h)$. The rest is statistical inference. Using Lagrange multipliers, we obtain

$$\Pi(h, S) = P(h) \exp[-\beta \epsilon_S(h)]/Z(h), \qquad (2)$$

where $Z(h) = \text{Tr}_S \exp[-\beta \epsilon_S(h)]$ and β is identified with $(kT)^{-1}$ following standard arguments. As a consequence, the entropy reads

$$S = -k \, \text{Tr} \, \Pi \ln \Pi = -k \int dh \, P(h) \ln P(h) + \int dh \, P(h) S(h), \qquad (3)$$

where $S(h)$ is the entropy of a spin in a fixed magnetic field. There is the expected contribution to the entropy due to the disorder. Moreover, the energy and the free energy both have the form often assumed as definition of quenched systems, namely an average of the energy instead of the partition function. Note the additional term in the free energy, associated with quenched disorder [9].

Is the theory consistent? The specific heat C_H can be calculated either as $(\partial E/\partial T)_{H_e}$ or as $T(\partial S/\partial T)_{H_e}$. They give identical results only if F is a stationary function of the order parameter $q(T, H_e)$,

$$\delta F/\delta q = 0 = kT \left\{ \int dh \left(\frac{\partial P}{\partial q} \right) [\ln P - \ln Z(h)] - \int dh\, P \left(\frac{\partial}{\partial q} \right) \ln Z(h) \right\}. \quad (4)$$

This is indeed a self-consistency equation for the order parameter q and the link with the Landau theory of phase transition is established. Consequently

$$C_H = k\beta^2 \int dh\, P(h) \frac{\partial^2}{\partial \beta^2} \ln Z(h) - \int dh \left(\frac{\partial}{\partial q} \right) \left[P \left(\frac{\partial}{\partial \beta} \right) \ln Z \right] \left(\frac{dq}{dT} \right). \quad (5)$$

The first term is standard and yields the linear T dependence of C_H at low temperatures. It is the only contribution to C_H if $q = 0$. The second term is new. Since $dq/dT \sim T$ in AR, it does not affect the linear T dependence of C_H. A recent calculation [10] of C_H using only the second term of the entropy, eq. (3), is accordingly incorrect.

The bulk magnetization $M = -(\partial F/\partial H_e)_T = -(\partial E/\partial H_e)_S$ yields no problem of self consistency. However, stationarity of F [with respect to q, eq. (4) and a similar relation with respect to M] reduces the equation for M to the simple form, $M = \int dh\, P(h) M(h)$, with $M(h) = -kT(\partial/\partial H_e) \ln Z(h)$, in agreement with standard theories [4–8]. M vanishes in the absence of an external field for a symmetrical distribution P, as it should in spin glasses.

The theory has been completely general so far. Spin glasses will be defined by their energy levels $\epsilon_S(h)$ and the local internal field distribution $P(h)$. The energy levels are given by

$$\epsilon_S(h) = -Sa(h + h_e) + \tfrac{1}{2} H_0 q^2, \quad (6)$$

with a Zeeman term and the self-energy of a cluster of magnetization q in its own magnetic field $H_0 q$ (cf. AR). This second term, an energy shift, does not affect the dynamics of the spin at fixed temperature and external magnetic field. It does, however, keep q finite in the spin glass phase. For the local field distribution, we take as a simple illustration the double lorentzian [5] at $H_e = 0$ [$P(h) = a^{-1} P(H)$],

$$P(H) = \tfrac{1}{2}[P_{\text{loc}}(q, H) + P_{\text{loc}}(-q, H)]$$
$$= \tfrac{1}{2}(\Delta/k)\{[(H - qH_0)^2 + \Delta^2]^{-1} + [-q]\} \quad (7)$$

constructed by taking $P_{\text{loc}}(q, H)$ as a conditional probability of a field H associated with a region of short range order (cluster) of definite orientation $+q$. (This, instead of eq. (4) of AR which had been somewhat arbitrarily constructed.) In (7), $\partial P/\partial q = 0$ as $q = 0$ so that C_H remains finite at the freezing temperature. All other results are qualitatively the same as those of ref. 5. The equation for q (4) reads then,

$$q - (T/H_0)(\partial/\partial q) S_{\text{dis}} = \tfrac{1}{2} \int dH\, P_{\text{loc}}(q, H) \tanh \tfrac{1}{2} \beta H. \quad (8)$$

At $T = 0$, the term involving S_{dis} vanishes, and the self-consistency equation is identical to that of AR. For $S = \tfrac{1}{2}$, the S_{dis} term should be negligible at all temperatures since $S_{\text{dis}} \sim \ln$ (width of distribution), admittedly for a gaussian distribution, and this width is a constant independent of the nature of the order for $S = \tfrac{1}{2}$ [5]. However, for eq. (7), $(\partial/\partial q) S_{\text{dis}} = kq/[(q^2 + \delta^2)(1 + \sqrt{(1 + q^2/\delta^2)})] \geq 0$ with $\delta = \Delta/H_0$ since the width of the distribution appears to increase as the wings separate. The same paradoxical inequality, at odds with the label "order parameter" for q, is found for all other distribution investigated, namely those of refs. 5 and 7. It should change sign for $S > \tfrac{1}{2}$ where anisotropy effects associated with freezing decrease the width of the distribution. Note also that a three-dimensional field distribution should be used, since the simple connection [11] between Ising and Heisenberg spin glasses does not hold for non linear functions of P like S_{dis}.

Nevertheless, the unfrozen state $q = 0$ is always a solution of (8) or (4) and a non-zero solution appears below the freezing temperature T_0. $F(q)$ has a minimum for the spin glass state solution $q \neq 0$, and the spin glass is stable below

T_0. If $(\partial/\partial q)S_{\text{dis}} > 0$, however, a parasitic $q \neq 0$ solution appears for $T > T_1 (> T_0)$ which is certainly unphysical. Below T_1, the system behaves normally. Efforts should be made in the construction of the probability distribution $P(h)$ to obtain an entropy decreasing, if at all, with the order parameter.

I am grateful to Chris Guy and John Stephenson for discussions.

References

[1] R. Brout, Phys. Rev. 115 (1959) 824.
[2] E.T. Jaynes, Phys. Rev. 106 (1957) 620.
[3] C.E. Shannon, Bell System Tech. J. 27 (1948) 379, 623.
[4] S.F. Edwards and P.W. Anderson, J. Phys. F. 5 (1975) 965.
[5] K.J. Adkins and N. Rivier (AR), J. de Physique (Paris) 35 (1974) 4-237.
[6] M.W. Klein and R. Brout, Phys. Rev. 132 (1963) 2412 and references therein.
[7] J.M. Kosterlitz, D.J. Thouless and R.C. Jones, Physica, this issue (1977).
[8] D. Sherrington and S. Kirkpatrick, Phys. Rev. Lett. 35 (1975) 1792.
[9] R.M. Mazo, J. Chem. Phys. 39 (1963) 1224.
[10] H. Muske and G. Heber, J. Phys. F 6 (1976) 1353.
[11] N. Rivier, Phys. Rev. Lett. 37 (1976) 232.

SPHERICAL MODEL OF A SPIN GLASS

J.M. KOSTERLITZ, D.J. THOULESS and RAYMUND C. JONES
Department of Mathematical Physics, University of Birmingham, Birmingham B15 2TT, England

A spherical model of a spin glass is solved in the limit of infinite-ranged interactions with a gaussian probability distribution. It is shown that for suitable values of the mean and variance of the distribution, the model can display either spin glass or ferromagnetic ordering; identical results obtain using the $n \to 0$ method. The distribution of internal fields is shown to be gaussian.

1. The model

Most of the recent theoretical work has examined the behaviour of an Ising system with random exchange interactions of infinite range [1–3]; the results obtained from such a model may be expected to have the same status as do the predictions of a Curie–Weiss theory of an Ising ferromagnet. In the model below the thermodynamic properties and the disorder are more easily disentangled.

Consider a lattice of classical spins coupled in pairs by the hamiltonian $H_1 = -\Sigma_{(ij)} J_{ij} S_i S_j$. The interactions $\{J_{ij}\}$ are chosen to be infinitely long ranged and described by a probability distribution $P(J_{ij}) = (2\pi\sigma^2)^{-1/2} \exp[-(J_{ij} - J_0)^2/2\sigma^2]$. The spins are constrained by the spherical condition of Berlin and Kac [4] which requires $\Sigma_i S_i^2 = N$. For convenience we shall use the intensive variables $\tilde{J} \equiv N^{1/2}\sigma$ and $\tilde{J}_0 = NJ_0$.

We quote first some known properties [5, 2] of a symmetric random matrix such as J_{ij}: suppose that an eigenvector $|\lambda\rangle$ has a corresponding eigenvalue J_λ, then if $\tilde{J}_0 = 0$ the averaged eigenvalue spectrum is bounded by $|2\tilde{J}|$ and has a semicircular density $\rho_0(J_\lambda)$ [5]. For $J_0 \neq 0$ the corresponding result is $\rho(J_\lambda) = \rho_0(J_\lambda)$ for $\tilde{J}_0 < \tilde{J}$ and $\rho_0(J_\lambda) + N^{-1}\delta(J_\lambda - J_m)$ for $\tilde{J}_0 > \tilde{J}$, where $J_m = J_0 + \tilde{J}^2/\tilde{J}_0$.

The matrix elements $\langle i|\lambda\rangle$ for $\tilde{J}_0 = 0$ have the property [3] that $\langle\langle i|\lambda\rangle\rangle_{av} = 0$; $\langle|\langle i|\lambda\rangle|^2\rangle_{av} = 1/N$. The partition function for the model is straightforwardly shown to be $Z = (2\pi i)^{-1} \int_{c-i\infty}^{c+i\infty} dz \times \exp\{N[z - (2N)^{-1}\Sigma_\lambda \ln(z - J_\lambda/2T)]\}$, where the contour c is taken to the right of the largest eigenvalue.

2. Thermodynamic results

Details of the derivation are given in the paper by Kosterlitz et al. [2]. Briefly, we wish to calculate the averaged free energy per site and this is done by replacing $N^{-1}\Sigma_\lambda$ by an integral over the appropriate averaged density function and then performing the standard spherical model manipulations based on a saddle point integration in Z.

Consider first the case $\tilde{J}_0 = 0$; the saddle point of the integrand "sticks" at the point where it first encounters the cut which begins at the largest eigenvalue, $2\tilde{J}$, of J_λ. This indicates a phase transition at a critical temperature $T_c = \tilde{J}$. The specific heat is then readily shown to have the value $\tilde{J}^2/2T^2$ for $T > T_c$ and $\frac{1}{2}$ for $T < T_c$. This cusp corresponds to a critical exponent $\alpha = -1$. The natural order parameter for this system is the spin corresponding to the largest eigenvalue and for this one obtains $\langle S_{2\tilde{J}}\rangle = N^{1/2}(1 - T/\tilde{J})^{1/2}$ for $T < T_c$ (with all other $\langle S_\lambda\rangle = 0$). This gives a critical exponent $\beta = \frac{1}{2}$. Again the natural response function to calculate is the staggered response to an external field $h_{2\tilde{J}}$. This susceptibility is easily calculated from $\langle S_{2\tilde{J}}^2\rangle$ which diverges at T_c with an exponent $\gamma = 2$. The scaling relation $\alpha + 2\beta + \gamma = 2$ is thus obeyed exactly for this choice of order parameter. The uniform susceptibility here has the constant value $1/\tilde{J}$ for $T < T_c$ and shows a Curie behaviour for $T > T_c$. The mean magnetization per site is zero, but the mean square value of this quantity has the value $1 - T/\tilde{J}$. This is the order parameter q of other authors [1].

When $\tilde{J} \neq 0$, but $\tilde{J}_0 < \tilde{J}$, the results are independent of $\tilde{J}_0$ and a spin glass phase exists as outlined above. However when $\tilde{J}_0 > \tilde{J}$, the system is dominated by the largest isolated eigenvalue J_m of $\rho(J_\lambda)$. The calculations are similar; one finds a critical temperature $T_c = \tilde{J}_0$ with a specific heat $\tilde{J}^2/2T^2$ for $T > T_c$ and $\frac{1}{2}$ for $T < T_c$ and with a mean field like discontinuity of $\frac{1}{2}(1 - \tilde{J}^2/\tilde{J}_0^2)$ across the critical isotherm, corresponding to $\alpha = 0$. The natural order parameter is again $\langle S_{J_m} \rangle \equiv (1 - T/\tilde{J}_0)^{1/2}$ (with all other $\langle S_\lambda \rangle = 0$) and using the result [2]

$$\left\langle \frac{1}{\sqrt{N}} \sum_i \langle i|J_m\rangle \right\rangle_{av} = \left(1 - \frac{\tilde{J}^2}{\tilde{J}_0^2}\right)^{1/2}, \quad (1)$$

one finds that there is a mean magnetization per site whose value is $(1 - \tilde{J}^2/\tilde{J}_0^2)^{1/2}(1 - T/\tilde{J}_0)^{1/2}$. The system is thus ferromagnetically ordered and the uniform susceptibility diverges with an exponent $\gamma = 1$, characteristic of mean field behaviour.

3. The internal field distribution

We now calculate the distribution of internal "molecular" fields for our model. Consider a set of uncoupled spins, each acted upon by a magnetic field h_i and subject to the spherical constraint. The hamiltonian is $H_2 = \Sigma_i h_i S_i$, and the partition function, readily shown to have a saddle point at

$$z^* = \left[1 + \left(1 + \frac{4}{N}\frac{\Sigma}{T^2} h_i^2\right)^{1/2}\right]/4.$$

From standard thermodynamic arguments one finds a magnetization per site $m_i = h_i/2Tz^*$. We shall proceed by finding the distribution of $\{h_i\}$ which gives the same value of $\langle m_i^2 \rangle_{av}$ (or q) as in the spin glass of section 2. Now we have seen that $m_i = \langle i|2\tilde{J}\rangle\langle S_{2\tilde{J}}\rangle$ in the spin glass, and that $\langle \Sigma_i |\langle i|2\tilde{J}\rangle|^2 \rangle_{av} = 1$. Denote the element $\langle i|2\tilde{J}\rangle$ by x_i, then if for convenience we work in a microcanonical ensemble, the distribution of any one element, say x, is given by

$$P(x_1) = \text{constant} \int \delta\left(x_1^2 + \sum_2^N x_i^2 - 1\right) \prod_2^N dx_i, \quad (2)$$

which, for large N, gives $P(x) = (N/2\pi)^{1/2} e^{-Nx^2/2}$. Thus we see that $m_i = h_i/2Tz^* = x_i N^{1/2} q^{1/2}$. We may now eliminate the h_i by means of the saddle point condition and for consistency we find that $z^* = 1/2(1-q)$: and hence that $x_i = h_i(1-q)/2T\sqrt{N}q^{1/2}$. The distribution of internal fields is thus the gaussian

$$p(h) = (2\pi s^2)^{-1/2} \exp{-h^2/2s^2}. \quad (3)$$

with $s^2 \equiv qT^2/(1-q)^2$.

In the ferromagnetic phase $\tilde{J}_0 > \tilde{J}$ the calculation is similar: one simply calculates an integral similar to eq. (2) but with $x_i \equiv \langle i|J_m\rangle$ and with the additional constraint (1). The self consistency arguments are similar and lead to a gaussian distribution of fields centred not about zero as in eq. (3) but about a finite value $\bar{h} = Tq^{1/2}\langle \Sigma_i \langle i|J_m\rangle\rangle_{av}/(1-q)$, and with a variance $\Sigma^2 \equiv (\tilde{J}^2/\tilde{J}_0^2)[T^2 q/(1-q)^2]$, where $q = 1 - T/\tilde{J}_0$, now.

It should be noted that these field distributions do not in fact give the averaged free energy correctly because of neglect of fluctuations.

References

[1] S.F. Edwards and P.W. Anderson, J. Phys. F. 5 (1975) 965.
D. Sherrington and S. Kirkpatrick, Phys. Rev. Lett. 35 (1975) 1792.
K.H. Fischer, Phys. Rev. Lett. 34 (1975) 1438.
[2] J.M. Kosterlitz, D.J. Thouless and R.C. Jones, Phys. Rev. Lett. 36 (1976) 1217.
[3] D.J. Thouless, P.W. Anderson and R.B. Palmer, to be published.
[4] T.H. Berlin and M. Kac, Phys. Rev. 86 (1952) 821.
[5] M.L. Mehta, Random Matrices and the Statistical Theory of Energy Levels (Academic, New York, 1967) p. 240.
S.F. Edwards and R.C. Jones, J. Phys. A 9 (1976) 1595.
[6] W. Marshall, Phys. Rev. 118 (1960) 1520.

COMPUTER SIMULATION OF SPIN GLASSES AND MICTOMAGNETS

F.A. DE ROZARIO, D.A. SMITH

Physics Department, Monash University, Clayton, Victoria 3168, Australia

and

C.H.J. JOHNSON

Division of Chemical Physics, CSIRO, P.O. Box 160, Clayton, Victoria 3168, Australia

Simulation on finite samples of 200, and 1000 spins suggests that the ground state of a classical spin glass is basically a random 3D spin structure but with slight ferromagnetic alignment. The internal field distribution gives a T^3 specific heat term. Spin correlation functions for mictomagnets are also obtained.

We have generated approximate ground states for classical spin glasses of n spins for $n = 200$ (9 samples) and 1000 (3 samples). The energy function is the RKKY form

$$E = -\sum_{\langle ij \rangle} J(R_{ij}) \boldsymbol{S}_i \cdot \boldsymbol{S}_j, \quad (1)$$
$$J(R) = A \cos 2k_F R / (2k_F R)^3.$$

For each sample, the spins are randomly sited on a s.c. lattice of 10^{36} sites within a unit cube. From a random initial spin configuration, the energy is minimized by holding one spin fixed and successively aligning all others in their internal fields. Satisfactory convergence in the distributions was normally achieved after 200–300 repetitions. The distribution of $\cos \theta$ between pairs is almost isotropic (fig. 1) but peaks slightly at $\theta = 0$. The total magnetic moment is 15 ± 6 units ($n = 200$) and 48 ± 14 ($n = 1000$), i.e. approximately equal to $\sqrt{n}$.

The internal field distributions are very sensitive to sample size. For $n = 1000$ the distribution of the magnitude of $\boldsymbol{H}$ can be fitted by the form

$$P(H) = (4H_0/\pi) H^2 / (H^2 + H_0^2)^2 \quad (2)$$

appropriate for an isotropic 3D distribution [1]. In terms of the concentration p, where $k_F^3 = 3\pi^2 n/p$ for a monovalent host, the reduced width $H_0/pAS \approx 0.03$ compared to $1/36\pi = 0.00884$ for the independent-random-spin model.

The low-temperature specific heat in this model is

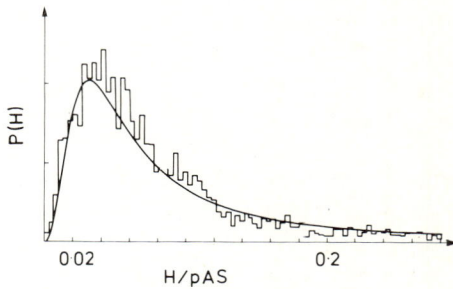

Fig. 2. Distribution of the internal field in units of pAS for $n = 1000$ (3 samples), fitted by least-squares to the function given by eq. (2).

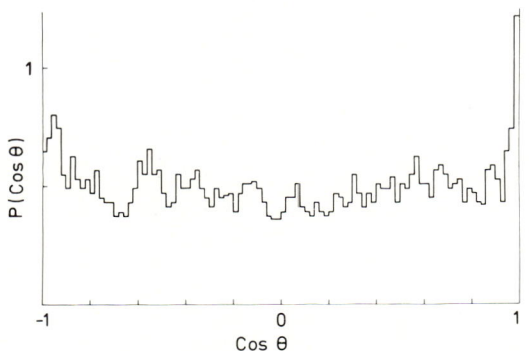

Fig. 1. The distribution of $\cos \theta$, θ = angle between two spins, over all pairs for $n = 200$ (9 samples).

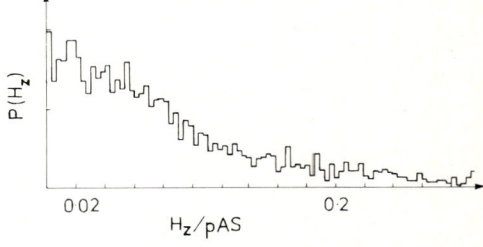

Fig. 3. Distribution of the z component of the internal field, where z is an arbitrary axis.

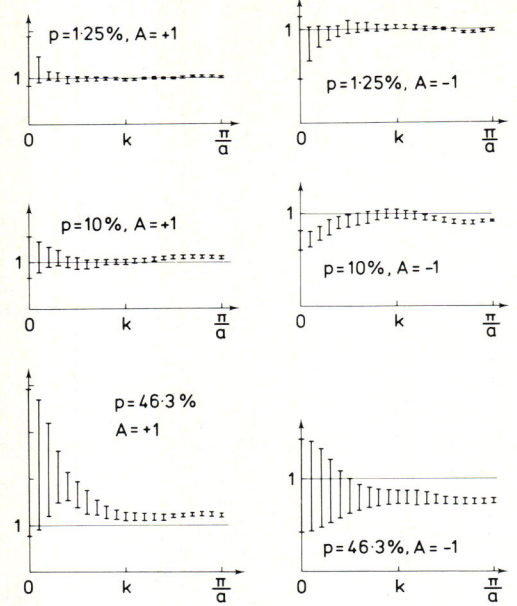

Fig. 4. $S(k)$ for $n = 100$ (15 samples), various concentrations and $A = \pm 1$ (ferro or antiferromagnetic n.n. exchange).

$$\Delta C = (nk_B^2 T/2S) \int_0^\infty x^2 B_S'(x) P(k_B T x/S) \, dx, \quad (3)$$

where $B_S(x)$ is the Brillouin function. Collective excitations are not included. For $k_B T \ll H_0$, eq. (3) gives a T^3 law

$$\Delta C = (8\pi^3/15)[1 - (2S + 1)^{-3}](k_B T/H_0)^3 n k_B \quad (4)$$

in contrast to the observed linear variation [2]. Measurements at lower temperatures are required.

The same algorithm was applied to systems of $n = 100$ in the mictomagnetic region, for $p = 1.25, 4.55, 10, 19.5$ and 46.3%. We used 15 samples; 1500 repetitions were required. The

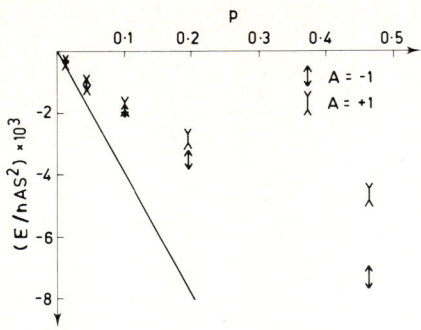

Fig. 5. Energy per spin is reduced units for $n = 100$. The slope near the origin is from runs at $p = 24 \times 10^{-6}$.

spin configuration is sensitive to the nearest-neighbour exchange integrals $J_1, J_2 \ldots$. We chose $A = +1 (J_1 + ve, J_2 - ve)$ and $A = -1 (J_1 - ve, J_2 + ve)$. The mictomagnetic state appears to be less stable than the spin-glass (fig. 5). Fig. 4 shows the influence of nearest neighbours on the directionally-averaged correlation function

$$S(k) = \frac{1}{n} \sum_{ij} \frac{\sin kR_{ij}}{kR_{ij}} \langle S_i \cdot S_j \rangle. \quad (5)$$

In the dilute or spin-glass limit $S(k) \to 1$. Neutron diffraction results on 5 and 10% polycrystalline CuMn [3] are qualitatively described by our results.

Additional results are available from the authors.

References

[1] C. Held and M.W. Klein, Phys. Rev. Lett. 35 (1975) 1783.
[2] J. Souletie and R. Tournier, J. Low Temp. Phys. 1 (1969) 95.
F.W. Smith, Phys. Rev. B9 (1974) 942.
[3] N. Ahmed and T.J. Hicks, J. Phys. F5 (1975) 2168.

SPIN DEPENDENT CURVATURE OF THE LOW TEMPERATURE SUSCEPTIBILITY OF RKKY SPIN-GLASSES

B. FISCHER and M.W. KLEIN
Department of Physics, Bar-Ilan University, Ramat-Gan, Israel

The low temperature magnetic susceptibility of spin-glasses is calculated in the Ising model MRF-Approximation. We find that the curvature of $\chi(T)$ as a function of temperature is S-dependent, where S is the impurity spin. The curvature is negative for $S < \frac{3}{2}$, is approximately zero for $S = \frac{3}{2}$, and becomes positive for $S > \frac{3}{2}$.

In a recent paper the authors [1] examined the low temperature thermodynamic properties of RKKY [2] spin glass system, using an Ising-like and Heisenberg-like probability distribution of internal fields. For the Ising-like distribution they used the lorentzian derived from the mean random field approximation [3], whereas for the Heisenberg-like distribution they used the form derived by Adkins and Rivier [4] for $T = 0$, and Held and Klein [5] for general temperatures. In each case they found that whereas the Ising-like distribution gave reasonably good agreement with the very low temperature measurements of the magnetization, the Heisenberg-like distribution did not.

Thus it seems that the Ising-like distribution describes the very low temperature properties of some spin glasses quite well (the reason for which is not yet well understood). Hence in this paper we use an Ising-like distribution to obtain the spin dependent magnetization for very small and large externally applied fields at very low temperatures. In particular we show that the model predicts that the second derivative of the magnetization with respect to the temperature is strongly spin dependent, and may be negative, positive or close to zero depending upon the impurity spin.

The probability distribution $P(H)$ of the Ising internal field H for small values of H ($H \ll \Delta$) in the MRF-approximation is [3]

$$P(H) = \pi^{-1}\Delta(B)/[\Delta^2(B) + H^2], \quad (1)$$

where $\Delta = \gamma m(B)c$, $\gamma = (2\pi^2/3)n_0|a|$, n_0 is the number of sites per unit cell, $|a|$ is the strength of the RKKY interaction at a distance of lattice constant and c is the impurity concentration. $m(B)$ is the spin-glass order parameter in external magnetic field B defined by

$$m(B) = \int_{-\infty}^{\infty} P(H)S|B_S[\beta(H+B)]|\,dH, \quad (2)$$

where B_S is the Brillouin function for spin S, and the vertical brackets in eq. (2) indicates absolute values.

The magnetization M of the system is

$$M = N_0 c \int_{-\infty}^{\infty} P(H)SB_S[\beta(H+B)]\,dH, \quad (3)$$

where N_0 is the number of sites in the solid ($g = \mu_B = k_B = 1$). We rewrite m and M as follows:

$$m = S + S\int_{-\infty}^{\infty}[B_S(\beta H) - 1](P^-(H) + P^+(H))\,dH, \quad (3a)$$

$$M = N_c S \int_0^{\infty} [1 + (B_S(\beta H) - 1] \times [P^-(H) - P^+(H)]\,dH, \quad (3b)$$

where $P^{\pm}(H) = \pi^{-1}\Delta/[\Delta^2 + (H \pm B)^2]$. This enables the calculation of m and M for low temperatures:

$$m = S\left\{1 - \frac{2\ln(2S+1)(S\gamma c)}{\pi[(S\gamma c)^2 + B^2]}T \right.$$
$$\left. - \left(\frac{2\ln(2S+1)(S\gamma c)^2}{\pi[(S\gamma c)^2 + B^2]}\right)^2 T^2 + O(T^3)\right\} \quad (4)$$

$$M = N_0 c S\left\{\frac{2}{\pi}\tan^{-1}\left(\frac{B}{\Delta}\right) - \frac{4\pi}{3}\frac{B\Delta T^2}{(2S+1)(\Delta^2 + B^2)^2}\right.$$
$$\left. - \frac{\pi^3 B^3 \Delta}{(\Delta^2 + B^2)^4}[1 - (2S+1)^{-3}]T^4\right\}. \quad (5)$$

For high external field, $B \gg \Delta$, we obtain

$$M = N_0 cS \left\{ 1 - \frac{2\gamma Sc}{\pi B} \left[1 + \frac{2\pi^2}{3(2S+1)} \left(\frac{T}{B}\right)^2 \right] \right.$$

$$+ \ln(2S+1) \left(\frac{2}{\pi}\right)^2 \left(\frac{\gamma cS}{B}\right)^2$$

$$\left. \times \left[\left(\frac{T}{B}\right) + \left(\frac{\pi}{2}\right)^2 \frac{8}{3(2S+1)} \left(\frac{T}{B}\right)^3 \right] \right.$$

$$\left. + 0(c^3 T) + 0(cT^4) \right\}. \qquad (6)$$

The lowest order in c of this expression is similar to the result of Hou and Coles [6] for spin $\frac{1}{2}$, and thus agrees with experiment. For a general spin S our results differ somewhat from that of Hou and Coles [6].

For very low fields, $B \ll \gamma cS$ and low temperatures we keep in eq. (5) only terms linear in B and obtain

$$\chi = \frac{M}{B} = \frac{N_0}{\gamma} \left\{ \frac{2}{\pi} + \frac{4\ln(2S+1)}{\pi^2} \left(\frac{T}{\gamma cS}\right) \right.$$

$$+ \left(\frac{16[\ln(2S+1)]^2}{\pi^3} - \frac{4\pi}{3(2S+1)} \right) \left(\frac{T}{\gamma cS}\right)^2$$

$$\left. + 0\left(\frac{T}{\Delta}\right)^3 + 0(B^2) \right\}. \qquad (7)$$

Thus we can conclude for the susceptibility:

(a) In the limit as $T \to 0$, χ is independent of S, independent of c and inversely proportional to the strength of the RKKY interaction a. The concentration independence of $\chi(T=0)$ was found to be in agreement with experiments of Franz and Sellmyer [7].

(b) The term linear in T is inversely proportional to the impurity concentration.

(c) The curvature of χ with respect to $y = (T/\gamma cS)$ is given by

$$K = \frac{\gamma}{2N_0} \frac{\partial \chi^2}{\partial y^2} = \frac{16}{\pi^3} [\ln(2S+1)]^2 - \frac{4\pi}{3(2S+1)}. \qquad (8)$$

We find that for $S = \frac{3}{2}$, K is close to zero. For $S < \frac{3}{2}$, K is negative ($K = -1.85$ for $S = \frac{1}{2}$), and K is positive for $S > \frac{3}{2}$ ($K = +0.5$ for $S = 2$, and $K = 0.96$ for $S = \frac{5}{2}$). Examining the curvature as measured by Mydosh [8] we find positive curvature for Ag-Mn. On the other hand the curvature of the susceptibility in Au-Fe measured by Guy [9] is almost zero. This would be in accordance with our predictions if we assume that the iron spin in Au-Fe is $\frac{3}{2}$. It would be interesting to measure the curvature of samples containing different impurities with a view to compare our predictions with experiment.

The results obtained so far apply to the case in which the distance of closest approach between the impurities is allowed to go to zero. It is only for this case that the probability distribution of fields is a lorentzian. However, in a real physical situation the distance of closest approach between the impurities is restricted to a near neighbor distance. For this case there are corrections to the lorentzian. For very low concentrations, for example, we find that if the distance of closest approach is limited to r_{nn}, where r_{nn} is the near neighbor distance, that $P(H=0) = (\pi \Delta)^{-1} \exp(4\sqrt{2}\pi c/3)$, instead of $(\pi \Delta)^{-1}$ when the cutoff distance is zero. Thus for low concentrations there is a correction arising from the cutoff in the interaction, and the correction is sensitive to the cutoff.

Finally we mention that whereas using a lorentzian gives an infinite transition temperature for the system, however when one introduces a near-neighbor cutoff the transition temperature is finite. For this case one obtains [10] a cusp in the magnetic susceptibility at the spin-glass transition temperature T_c, where $T_c \propto c^\alpha$ and $\alpha \simeq 0.66$.

References

[1] B. Fischer and M.W. Klein, to be published.
[2] M.A. Ruderman and C. Kittel, Phys. Rev. 96 (1954) 99.
[3] M.W. Klein, Phys. Rev. 173 (1968) 552; Phys. Rev. 188 (1969) 933.
[4] K. Adkins and N. Rivier, J. de Physique 35 (1974) C4-237.
[5] C. Held and M.W. Klein, Phys. Rev. Lett. 35 (1975) 1783.
[6] P.W. Hou and B.R. Coles, Phys. Rev. Lett. 35 (1975) 1655.
[7] J.M. Franz and D.J. Sellmyer, Phys. Rev. B8 (1973) 2083.
[8] J.A. Mydosh, AIP Conf. Proc. on Magnetism 24 (1974) 131.
[9] C.N. Guy, J. Phys. F5 (1975) L242.
[10] I. Riess and M.W. Klein, Phys. Rev., Dec. 1976.

MAGNETIC PROPERTIES OF Pr–Nd AND Pr–Tb ALLOYS

B.V.B. SARKISSIAN

Physics Department, Imperial College, London SW7, England

Resistivity and magnetization data for Pr–Nd and Pr–Tb alloys are reported. For the Pr–Nd system they suggest conventional antiferromagnetism at very low fields, but at moderate fields ferromagnetic couplings seem to appear. In contrast the Pr–Tb alloys show the behaviour usually associated with spin glass alloys.

Because praseodymium has a singlet ground state it should be interesting to study the magnetic effects produced by alloying it with a light rare-earth (Kramer's) element like Nd, and with a heavy rare-earth (non-Kramer's) element like Tb. We report investigations of the magnetic susceptibilities and electrical resistivities of some Pr–Nd and Pr–Tb alloys in the double hexagonal crystal structure. Conventional potentiometric methods were used for the resistivity measurements, Faraday-type force measurements in a solenoid with gradient coils for the low field magnetization measurements, and similar measurements in the fringe field of an electro-magnet for fields greater than 30 Oe. Measurements of magnetization in low fields (5 Oe) show a sharp susceptibility peak for alloys of 6–30% Nd (see lowest curve in fig. 1).

This behaviour has some of the characteristics of a conventional antiferromagnet, but it should be noted that time dependence and isothermal remanence is found even in these low fields. The resistivity of these alloys shows fairly well defined (but not sharp) maxima and minima in $d\rho/dT$ vs. T plots, the temperatures of which are shown in fig. 2, which also includes the temperatures of the sharp cusp found in low field (5 Oe) magnetization. For an alloy of 2% Nd, however, no such anomalous behaviour in the resistivity is observed even down to 0.25 K indicating a composition limit for the onset of such magnetic ordering. The field dependence of the magnetization–temperature curves is very anomalous. The sharp cusp seems to disappear in moderate fields, and to give way to ferromagnetic behaviour in high fields (> 500 Oe) but the saturation magnetization is much less than one would expect for a simple ferromagnet. Similar measurements have been made on Pr–Tb alloys (0.5–20% Tb). The magnetization–temperature curves of these alloys in fields of 500–1000 Oe show maxima (fig.

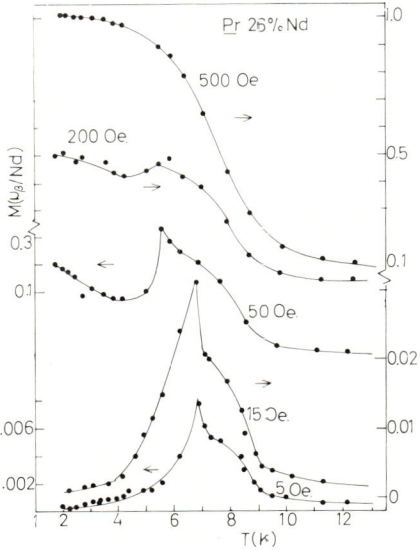

Fig. 1. Magnetization M of Pr 26% Nd alloy as a function of the temperature T in different constant applied fields. Some curves are displaced vertically; note the changes in the vertical scale for some curves. The number by each curve indicates the applied field.

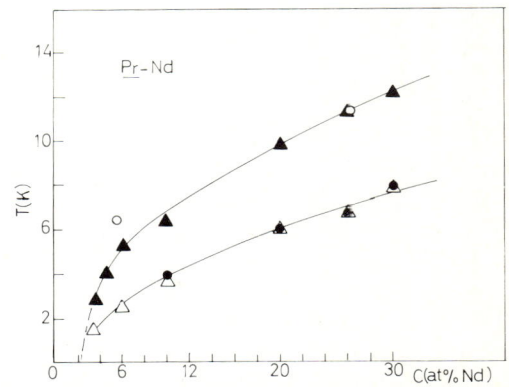

Fig. 2. Deduced temperatures from the resistivity anomalies and the sharp cusp in the low field magnetization–temperature curve for Pr–Nd alloys. △ maximum in $d\rho/dT$, ● cusp in magnetization, ▲ minimum in $d\rho/dT$, ○ neutron scattering Néel points [3].

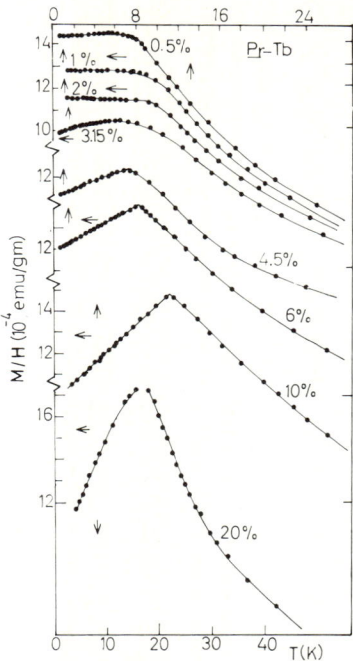

Fig. 3. Susceptibility (M/H) for Pr–Tb alloys as a function of temperature in the concentration range 0.5–20% Tb; the number by each curve indicates solute concentrations. The magnetization is a linear function of the applied field up to 8 kOe. Note that the temperature scale for the first alloy is that shown at the bottom of the figure.

3) the temperatures of which are given in fig. 4. The temperature dependence of the magnetization for alloys with less than 3% Tb is characteristic of a singlet ground state, as observed in pure Pr. The resistivity of these alloys does not show any well-defined anomalies but is however complicated by the presence of crystal field and phonon scattering terms. The field dependence of the magnetization–temperature curves is not anomalous as in Pr–Nd alloys, for temperatures less than that of the maximum in the magnetization the magnetization-field curves are linear. In addition the magnetization was observed to exhibit time dependent effects, isothermal remanent magnetization, and other properties characteristic of spin glass ordering. The magnetization maxima in these alloys are much sharper than those of normal spin glasses in corresponding fields. This behaviour is similar to what is observed in Y–RE and Sc–RE alloys

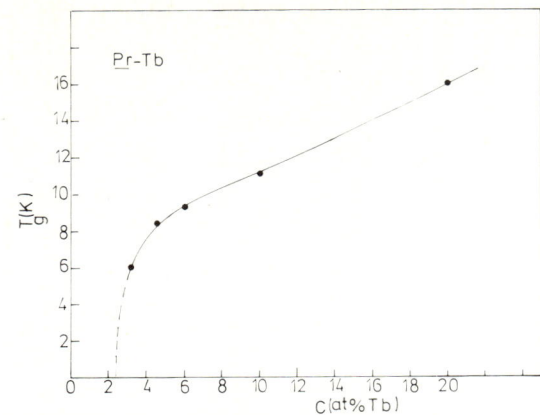

Fig. 4. Spin freezing temperature T_g for Pr–Tb alloys plotted against Tb concentration.

[1] (excluding Gd alloys) in their spin glass regions. The observed diffuse peak in the neutron scattering [2] could also be associated with spin glass ordering in Pr–Tb alloys.

On the basis of their study of neutron scattering by Pr–Nd alloys Lebech et al. [3] concluded that antiferromagnetic periodic structures are formed in the fairly dilute impurity concentration region, but the magnetization measurements were not straightforward in their character. Our measurements appear to confirm the presence of antiferromagnetic ordering but the complex field dependence of the magnetization-temperature curves suggests a rather complex magnetic structure. Field-dependent neutron scattering measurements will be useful in clarifying the observed magnetic behaviour.

The author thanks Professor B.R. Coles and Dr. B.D. Rainford for helpful discussions and comments. Financial support from the Science Research Council is also acknowledged.

References

[1] B.V.B. Sarkissian and B.R. Coles, Commun. Phys. 1 (1976) 17.
[2] D. Chatterjee, K.N.R. Taylor and M.W. Stringfellow, Proc. Conf. Rare Earth and Actinides, Durham (1971) pp. 22–23.
[3] B. Lebech, K.A. McEwan and P.A. Lindgard, J. Phys. C.: Solid State Phys. 8 (1975) 1684.

SPIN GLASS PROPERTIES AND ANISOTROPY IN $(Ti_{1-x}V_x)_2O_3$ SINGLE CRYSTALS

J. DUMAS, C. SCHLENKER

Groupe des Transitions de Phases, CNRS, BP 166, 38042 Grenoble Cedex, France

J.L. THOLENCE and R. TOURNIER

Centre de Recherches sur les Très Basses Températures, CNRS, BP 166, 38042 Grenoble Cedex, France

The freezing temperature, the susceptibilities χ_{rev}, χ_{irr} and the TRM have been measured as a function of x in $(Ti_{1-x}V_x)_2O_3$. The observed anisotropy is well accounted for by a Néel model of magnetic clusters with a non uniform distribution of the anisotropy axis. The dipolar interaction might be responsible for this anisotropy.

While pure Ti_2O_3 is, at low temperatures, a semiconductor which shows a small Van Vleck type paramagnetism, it has been established that the incorporation of V in this oxide induces both a Curie–Weiss behaviour for the susceptibility and, for vanadium concentrations $x > 1\%$, a metallic conductivity [1]. We have reported previously that the effective magnetic moment per V atom corresponds roughly to a spin $\frac{1}{2}$ for $x > 4\%$ [1] and that NMR measurements performed on V nuclei for $x = 10\%$ show a temperature dependent Knightshift indicating that the magnetic moments are on the V ions [1, 2]. These results are not in agreement with those of ref. [3]. $(Ti_{1-x}V_x)_2O_3$ has also been shown to exhibit spin glass properties for $x > 1\%$ [4]. We have proposed band schemes which account for both the electrical and magnetic data; the local moment is due to a V virtual bound state decoupled through the Coulomb repulsion and the spin glass behaviour to RKKY interactions [1]. The uniaxial corundum structure, with pairs of cations located along the c axis, leads to a strong anisotropy of the spin glass properties [4]. We now report a study of these properties and of their anisotropy as a function of x.

Fig. 1 shows the dependence on x of the freezing temperature T_B obtained from the maximum of the curve of the reversible susceptibility χ_{rev} versus temperature (see fig. 1, insert). T_B is found to be roughly proportional to x and independent of the orientation of the magnetic field. The saturated thermoremanent magnetization (TRM) σ_{sr}, obtained from the curve of the TRM versus the cooling field h (shown in the insert of fig. 2) and extrapolated to 0 K, is found to increase with x as x^2 for $x > 2\%$ (fig. 1). While the TRM is anisotropic for low h, σ_{sr} is the same for both field orientations (fig. 2, insert). Typical

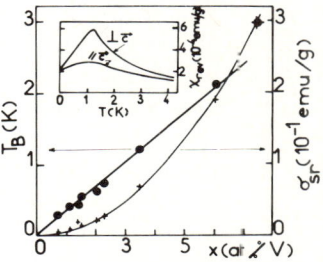

Fig. 1. Freezing temperature (obtained from the maximum of $\chi_{rev}(T)$ and saturated TRM extrapolated to 0 K, vs. x. Insert: reversible susceptibility vs. T for $H \parallel C$ and $H \perp C$, $x = 3.5\%$.

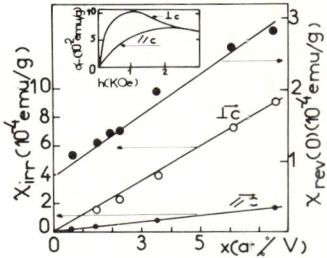

Fig. 2. Reversible susceptibility $\chi_{rev}(0)$ and irreversible susceptibility $\chi_{irr}(0)$ deduced from the initial slope of σ_r vs. h, extrapolated to 0 K. $\chi_{irr}^{\parallel}$ and $\chi_{irr}^{\perp}$ correspond to $h \parallel$ and $\perp C$. Insert: TRM vs. the cooling field h for $h \parallel$ and $\perp C$, $x = 3.5\%$, $T = 0.06$ K.

values for the inverse magnetic field which destroys σ_{sr} are 800 Oe for $H \parallel C$ and 560 Oe for $H \perp C$ at 0.05 K for $x = 2\%$. The irreversible susceptibility $\chi_{irr}(0)$ calculated from the initial slope of the curve of the TRM vs. h and extrapolated to 0 K, is shown as a function of x in fig. 2. $\chi_{irr}^{\perp}$ obtained with $h \perp c$ is found $\sim 5\chi_{irr}^{\parallel}$; $\chi_{irr}^{\perp}(0)$, $\chi_{irr}^{\parallel}(0)$ and $\chi_{rev}(0)$ are found roughly linear with x (fig. 2); no anisotropy is detected for $\chi_{rev}(0)$. For $T \gg T_B$, in the paramagnetic regime, the anisotropy of χ is found to be only $\sim 10\%$ for $x = 2\%$ and smaller at larger x.

The linearity of the freezing temperature with the concentration of the magnetic impurities is usually ascribed to the dependence of the RKKY interactions with the third power of the distance r between the moments. In the case of $(Ti_{1-x}V_x)_2O_3$, at small x, both the density of states at the Fermi level and the magnetic moment depend on x and one would not expect in this model, the law $T_B \sim x$ to be valid. We therefore propose that this law is due to a different mechanism. In the Néel model for fine particles [5], the system goes from the superparamagnetic regime to the frozen one when the time constants τ of the grains become of the order of magnitude of the experimental times. τ is determined by the anisotropy energy W of the grain $\tau \sim \exp(W/kT)$ and T_B depends essentially on W. The anisotropy observed for χ suggests in our case a model of magnetic regions with a uniaxial anisotropy; the distribution function of the orientation of their axis is uniform inside the plane $\perp C$ and much lower along C; then, in the case $H//C$, for most of the clusters, H is // to the hard axis and $\chi_{irr}^{//}$ is small as experimentally observed. The variation of σ_{sr} with x^2 together with the increase of $\chi(0)$ with x indicates that the presence of ferromagnetic pairs, possibly of nearest neighbour V with a direct exchange coupling, becomes predominant at large x.

The origin of the anisotropy is one of the most delicate problems of the spin glass. In our case, the *local single-ion anisotropy* is expected to be very small as $S = \frac{1}{2}$; the g-factor anisotropy can account for the high-temperature behaviour but probably not for the spin glass anisotropy. The *RKKY coupling* could lead to some anisotropy if one takes into account the *spin–orbit interaction*; but this contribution should be negligible for a spin $\frac{1}{2}$ [6]. One cannot exclude the contribution of the *pseudodipolar anisotropy* of the direct exchange between nearest neighbour V, but this effect should become important only for large x. In the case of the corundum structure, if one assumes a predominant ferromagnetic exchange (as indicated by the positive Curie temperatures [1]), one can show that the *dipolar coupling* leads to an anisotropy with a hard axis $//C$ as experimentally observed. In the Néel model, the r^{-3} dependence of this coupling then accounts for the linear dependence of T_B with x. The anisotropy energy W of a region can be evaluated from the width of the dipolar field distribution $\langle\Delta Hd^2\rangle^{1/2} \sim \langle\mu^2/r^6\rangle^{1/2}$ and from the average number N of magnetic ions in the region. Assuming that σ_{sr} is due to the uncompensated moment of the clusters and is $\sim \sqrt{N}$ [5], one obtains from the ratio σ_{sr}/σ_{sat} for $x = 2\%$, $N \simeq 1600$, $\langle\Delta Hd^2\rangle^{1/2} \simeq 130$ Oe and $W/k \simeq 14$ K in good agreement with the expected value $W \simeq 20\,kT_B$. We therefore conclude that the most likely mechanism for the anisotropy in $(Ti_{1-x}V_x)_2O_3$ is the dipolar interaction which may also determine the values of the freezing temperatures.

References

[1] J. Dumas and C. Schlenker, Colloque C.N.R.S. Transitions Métal-Nonmétal, Autrans, 28 June–1st July 1976, J. de Physique (to be published).
[2] M. Minier (private communication).
[3] Y. Miyako and T. Ito, J. Phys. Soc. Jap. 39 (1975) 1212 and this conference.
[4] J. Dumas, C. Schlenker, J.L. Tholence and R. Tournier, Solid State Commun. 17 (1975) 1215 and AIP Conf. Proc. 29 (1976) 431.
[5] J.L. Tholence and R. Tournier, J. de Physique 35 (1974) C4-229.
[6] D.A. Smith and G.P. Haberkern, J. Phys. F 3 (1973) 856.

MAGNETISM OF V-DOPED Ti_2O_3

Y. MIYAKO, S. SIMIZU and Y. KIMISHIMA
Department of Physics, Faculty of Science, Hokkaido University, Sapporo, Japan

Adding V-atoms introduces localized spins in Ti_2O_3. The NMR frequency shift of ^{51}V does not follow the susceptibility, which might suggest that V atom is nonmagnetic. The NMR absorption intensity decreased with increasing V concentration. At low temperatures, V-doped Ti_2O_3 shows spin glass behavior.

With regard to the origin of the magnetic moment, fig. 1 shows the susceptibility χd of a $(Ti_{0.99}V_{0.01})_2O_3$ single crystal [Weiss constant $H = (-0.8 \pm 0.5)$ K was determined from measurements in liquid helium temperatures as described below]. The susceptibility was measured by torsion balance, and the measured susceptibility was analyzed by a formula $\chi_m = \chi_d + \chi_c$, where χ_d indicates the susceptibility which obeys the Curie–Weiss law and χc is the paramagnetic susceptibility independent of temperature. The anisotropy of χd was not observed for the applied field H parallel and perpendicular to the c-axis of the crystal, but χc was anisotropic, that is $\chi_\parallel (H \parallel c\text{-axis}) = 3.8 \times 10^{-4}$ emu/mol and $\chi_\perp (H \perp c\text{-axis}) = 3.2 \times 10^{-4}$ emu/mol, respectively. From the Curie constant a localized spin of $S \sim 1$ is estimated assuming that the number of the localized spins is the same as that of doped vanadium and the Zeeman splitting factor $g = 2$. According to the susceptibility measurement of $(Ti_{0.995}V_{0.005})_2O_3$, $(Ti_{0.99}V_{0.01})_2O_3$ and $(Ti_{0.9}V_{0.1})_2O_3$, the effective magnetic moment seemed to vary from $(3.3 \pm 0.3)\mu_B$ to $(2.5 \pm 0.3)\mu_B$ with increasing V concentration.

To study the magnetic state of V atom in Ti_2O_3, NMR of ^{51}V was measured for single crystals and for powdered samples. In the measurement of the Knight shift, as given in fig. 1, the correction was made for the additive shift due to quadrupole interaction for powdered samples. The negative Knight shift obtained for ^{51}V in Ti_2O_3 is independent of temperature, while the susceptibility obeys the Curie–Weiss law. The Knight shift was also independent of V concentration within our experimental accuracy and no anisotropy of the Knight shift was obtained for a $(Ti_{0.99}V_{0.01})_2O_3$ single crystal at room and liquid helium temperatures. However, NMR absorption intensity of ^{51}V decreased with increasing V concentration. The intensity decreased rapidly above about 2 at.% and the intensity of $(Ti_{0.9}V_{0.1})_2O_3$ was about one-seventh compared with that of $(Ti_{0.99}V_{0.01})_2O_3$.

As the NMR frequency shift of ^{51}V does not follow the susceptibility, ^{51}V observed in the NMR is considered to be nonmagnetic. Especially, it has been confirmed by comparing the intensity with V metal that all of ^{51}V has no localized moment on the atom in Ti_2O_3 for the concentration less than 1 at.%.

Van Zandt [6] has proposed a band model to explain the specific heat data of V-doped Ti_2O_3 measured by Sjöstrand and Keesom [7]. As noticed by Van Zandt, it is known to exist in a very narrow band near the Fermi level. However, he has not analyzed the magnetic properties. According to his model, it seems difficult to interpret the magnetic properties.

There is no apparent evidence in our present experiment concerning the electronic ground state of the V atom in Ti_2O_3. We assume that the electronic ground state of the V atom lies with

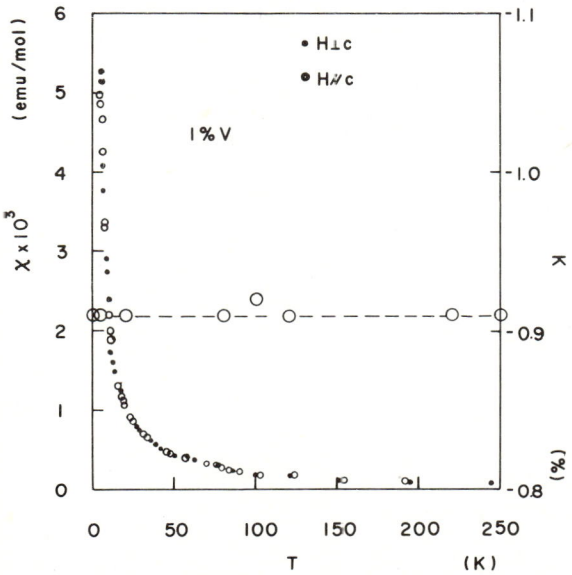

Fig. 1. The susceptibility of a $(Ti_{0.99}V_{0.01})_2O_3$ single crystal at the applied magnetic field parallel and perpendicular to the c-axis and the Knight shift by larger open circles.

up and down spin below the Fermi level. In this case, a paired configuration of the Ti atoms along the c-axis will be substituted by Ti–V configuration by adding V atoms. Then, the near neighbour Ti atoms of the V atom will have an unpaired electron as the paired Ti–Ti configuration will be broken. By adding V_2O_3 into the host Ti_2O_3, a very narrow energy band of near neighbour Ti atoms to the doped V atom will be produced near the Fermi level. By the existence of the spin polarized band of near neighbour Ti atoms, spin polarization of the bonding a_{1g} will be induced and, as a result, total spin polarization becomes larger than $\frac{1}{2}$. With regard to spin glass, the magnetic interaction of V-doped Ti_2O_3 was suggested to be of the RKKY type from the temperature dependence of the NMR line width. V-doped Ti_2O_3 shows spin glass behavior as first reported by Dumas et al. [8].

Fig. 2 shows the inverse of susceptibility of $(Ti_{0.995}V_{0.005})_2O_3$ and $(Ti_{0.99}V_{0.01})_2O_3$ as a function of temperature. The arrows in the figure indicate the transition temperature T_0 from paramagnetic state to spin glass. Magnetic susceptibility was measured by the a.c. bridge method and the a.c. field was less than 1 G. Samples were cooled by the adiabatic magnetic cooling method. The Weiss constant of $(Ti_{0.995}V_{0.005})_2O_3$, $(Ti_{0.99}V_{0.01})_2O_3$ and $(Ti_{0.9}O_{0.1})_2O_3$ is -1.2 ± 0.5 K, -0.8 ± 0.5 K, and $+10 \pm 5$ K, respectively. It seems that the susceptibilities have different Curie constants in two temperature regions. The effective magnetic moment for $(Ti_{0.995}V_{0.005})_2O_3$ and $(Ti_{0.99}V_{0.01})_2O_3$ is 3.3 μ_B in the temperature region of $T > 2$ K and becomes 20% smaller in the lower temperatures of $T_0 \leqslant T < 1$ K. When the temperature is decreased to the T_0, the Weiss constant tends to decrease to zero more rapidly than the decreasing magnetic moment. Sjöstrand and Keesom have observed an anomaly due to the magnetic moment in the specific heat measurement of $(Ti_{0.981}V_{0.019})_2O_3$ [7]. The anomaly may be due to the spin freezing in V-doped Ti_2O_3. To make it clear, we are now undertaking the specific heat measurement of

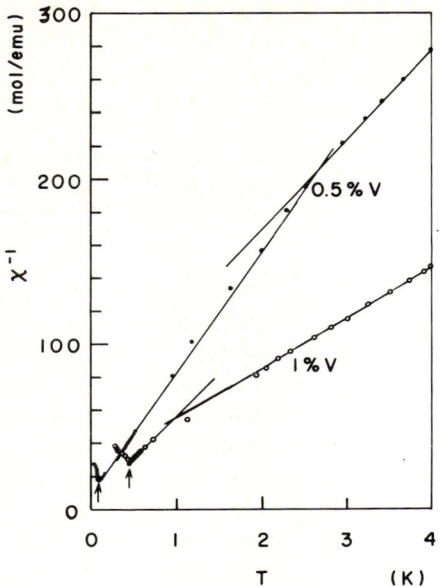

Fig. 2. The inverse of susceptibilities of $(Ti_{0.99}V_{0.01})_2O_3$ and $(Ti_{0.995}V_{0.005})_2O_3$.

$(Ti_{0.99}V_{0.01})_2O_3$ in the temperature region down to 0.2 K.

One of the authors (Y.M.) expresses his deep thanks to Professor J.M. Honig of Purdue University for the supply of good single crystals and to Professor J.B. Goodenough of Oxford University for valuable discussions.

References

[1] R.E. Loehman, C.N.R. Rao and J.M. Honig, J. Phys. Chem. 73 (1969) 1781.
[2] J.M. Honig, L.L. Van Zandt, T.B. Reed and J. Sohn, Phys. Rev. 182 (1969) 863.
[3] G.V. Chandrashekhar, Q. Won Choi, J. Moyo and J.M. Honig, Mater. Res. Bull. 5 (1970) 999.
[4] J. Dumas, C. Schlenker and R.C. Natoli, Solid State Commun. 16 (1975) 493.
[5] Y. Miyako and T. Itō, J. Phys. Soc. Jap. 39 (1975) 1212.
[6] L.L. Van Zandt, Phys. Rev. Lett. 31 (1973) 598.
[7] M.E. Sjöstrand and P.H. Keesom, Phys. Rev. B7 (1973) 3558.
[8] J. Dumas, C. Schlenker, J.L. Tholence and R. Tournier, Solid State Commun. 17 (1975) 1215.

COMPUTER SIMULATION OF SPIN GLASSES

K. BINDER

Fachrichtung 11.1, Universität des Saarlandes, 66 Saarbrücken, W. Germany

For two- and three-dimensional Ising model spin glasses we compute by Monte Carlo both a local order parameter and a nonlocal one, whose "susceptibility" diverges at the freezing temperature. The distribution of effective fields seems to have a shallow minimum at $H_{\text{eff}} = 0$. Symmetric hysteresis loops are found.

In order to investigate the Edwards–Anderson (EA) model [1], where it is even controversial if a sharp freeze-in transition exists [2,3], Monte-Carlo studies have been performed [3,4]. Ising spins with random exchange J_{ij} (distributed according to $P(J_{ij}) \propto \exp\{-J_{ij}^2/2(\Delta J)^2\}$ between nearest neighbors on square [3] and simple cubic [4] lattices were treated. Here we extend this work, concentrating upon hysteresis loops, and the appropriate definition and properties of spin glass order parameters.

Fig. 1 shows an example of the hysteresis loops obtained in two dimensions. The loops are (nearly) symmetric on the scale of fields which produces rounding of the cusp in the susceptibility χ [3], in contrast to what is found experimentally [5].

Fig. 2 shows the temperature variation of the *local* order parameter $q = \langle\langle\mu_i\rangle^2\rangle$ [the inner bracket being a thermal average over the spin states $\mu_i = \pm 1$, the outer an average over $P(J)$] and the "global" order parameter $\langle\psi\rangle$, which is defined [3] as $\psi = \Sigma\phi_i^{(l)}\mu_i/N^d$. N^d is the total number of spins in the lattice, and $\phi_i^{(l)}$ the *phase function* of the *l*th ground state. In a simple (anti-) ferromagnet, one has only two such functions $[\phi_i^{(1,2)} = \pm \exp(i\mathbf{r}_i\mathbf{Q})$, where $\mathbf{Q}$ is the wavevector associated with the ordering, while

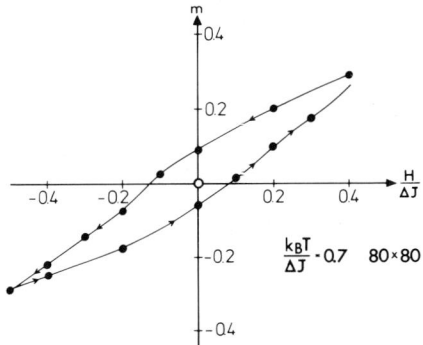

Fig. 1. Magnetization plotted vs. magnetic field. Each point is based on averages over 400 Monte-Carlo steps/spin of an 80×80 lattice.

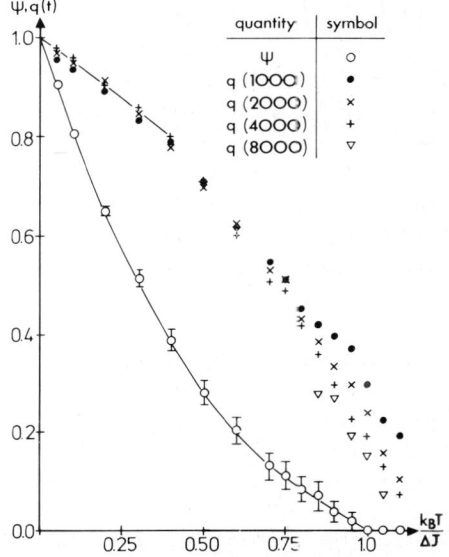

Fig. 2. Order parameters plotted vs. temperature, based on runs for 34×34 and 68×68 lattices. The cusp of the susceptibility occurs at $k_B T_f/\Delta J \approx 1.0$ while in three dimensions it occurs at $k_B T_f/\Delta J \approx 1.5$.

in the spin glass there may be many of them, and no simple formula for their $\mathbf{r}_i$-dependence can be given. We determined ψ by finding *one* $\phi_i^{(l)}$ by doing first a $T = 0$ calculation, using the resulting spin configuration both as $\phi_i^{(l)}$ and as initial state for the $T \neq 0$ run. The advantages of this definition of the order parameter are: (i) q depends very strongly on the time t up to which the run is extended, while ψ does not. (ii) q cannot have critical (i.e. divergent) fluctuations, while ψ does. In fig. 3 the generalized susceptibility $\chi_\psi \equiv (\langle\psi^2\rangle - \langle\psi\rangle^2)N^d\Delta J/k_B T$ is plotted vs. T. $\chi_\psi \gg \chi$ in the vicinity of T_f, implying strong spatial correlations of $\phi_i^{(l)}\mu_i$ at various sites although $\langle\mu_i\mu_j\rangle = 0$ for $i \neq j$. This result contradicts [1] where it was argued that no spatial correlation whatsoever exists, but only a correlation in time.

Rewriting the hamiltonian as $\mathcal{H} = -\Sigma\mu_i H_i^{\text{eff}}$, the distribution of effective fields $P(H^{\text{eff}})$ may

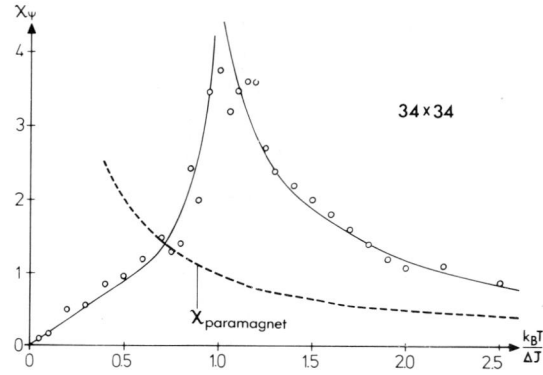

Fig. 3. χ_ψ plotted vs. temperature. Note that $\chi = \chi_{\text{ideal paramagnet}}$ for temperatures above the freezing temperature.

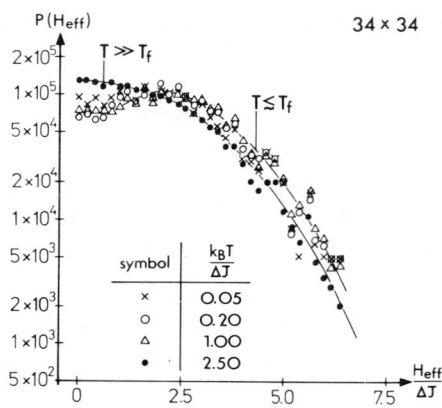

Fig. 4. Distribution of effective fields at various temperatures (in arbitrary units).

be found (fig. 4). $P(H_{\text{eff}})$ has no strong variation with T, in contrast to [6].

References

[1] S.F. Edwards and P.W. Anderson, J. Phys. F5 (1975) 965.
[2] J.K. Matho, preprint; A.D. Young, preprint.
[3] K. Binder and K. Schröder, Solid State Commun. 18 (1976) 1361; Phys. Rev. B14 (1976) 2141.
[4] K. Binder and D. Stauffer, Phys. Lett. 57A (1976) 177.
[5] J.A. Mydosh, AIP Conf. Proc. 24 (1974) 131.
[6] M.W. Klein, preprint.

REMANENT MAGNETIZATION OF SPIN GLASSES

J.L. THOLENCE and R. TOURNIER
Centre de Recherches sur les Très Basses Températures, CNRS, BP 166X, 38042 Grenoble, Cédex, France

A spin glass is an assembly of clouds containing n spins with uncompensated moments ($\sqrt{M^2g} = \mu\sqrt{n}$) each one frozen only below its own blocking temperature T_b. The potential barriers may be due to the dipolar coupling. The largest clouds begin to freeze at T_B (temperature of the maximum of the reversible susceptibility). The progressive freezing down to 0 K is characterized by a saturated remanent magnetization σ_{rs} varying as $e^{-\alpha T/c}$.

We recall that the saturated remanent magnetization σ_{rs}/c of a spin glass depends only on T/c at low concentrations [1]. The existence of this scaling law eliminates an explanation due to clustering effects in a solid solution out of equilibrium. A large number of moments are retained in the direction of the field previously applied, by anisotropy phenomena related to a $1/r^3$ interaction. The irreversible properties look like the properties of an assembly of fine antiferromagnetic grains [2] in the sense that the perfectly disordered solid solution behaves like an assembly of independent regions. Each region has an anisotropy energy E_a depending on its number of spins n:

$$E_a = \tfrac{1}{2} M_g H_a = (Q + \ln \tau) kT, \qquad (1)$$

where M_g is the uncompensated moment of the cloud [$(n^+ - n^-)\mu$ in an Ising model], H_a its anisotropy field, τ its relaxation time which is equal to the time of measurement at the blocking temperature T_b. Formula (1) can be written $\tau = \tau_0 e^{E_a/kT}$. Q is equal to log $1/\tau_0$.

In agreement with the Néel model:
(1) Q has been experimentally determined [3, 4] equal to 15 in some spin glasses ($\tau_0 \simeq 3 \times 10^{-7}$ s).
(2) The remanent magnetization varies [3] as log t.
(3) The thermoremanent magnetization (TRM) is higher than the isothermal remanent magnetization (IRM) obtained in the same field h. In low fields, the TRM varies as h; the IRM begins to vary as h^2 and as h in higher fields. The saturation of the two quantities is the same and is a small fraction of the total saturation of the alloy [4].

We want, in this paper, to focus attention on the distribution of the uncompensated moment of clouds which is assumed to follow a gaussian law:

$$P(|M_g|) = \frac{2}{M_{g0}\sqrt{2\pi}} e^{-M_g^2/2M_{g0}^2}, \qquad M_{g0} = \mu\sqrt{n_0}, \qquad (2)$$

where n_0 is the number of spins in the clouds. We may determine n_0 from the saturated remanent magnetization σ_{rs} at 0 K and from the total saturation of the alloy: $\sigma_s = N_0 n_0 \mu_0$:

$$\sigma_{rs}(T=0) = \frac{N_0}{2}\left(\frac{2n_0}{\pi}\right)^{1/2} \mu_0, \qquad n_0 = \frac{1}{2\pi}\left(\frac{\sigma_s}{\sigma_{rs}}\right)^2. \qquad (3)$$

N_0 is the number of clouds ($N_0 = Nc/n_0$); n_0 is independent of c when σ_{rs} and $\sigma_s \sim c$. If the distribution follows a gaussian law we must find again [1] the Curie constant C of all impurities knowing only σ_{rs} since

$$C = \frac{M_{g0}^2}{3k}\left(\frac{\mu_{eff}}{\mu_0}\right)^2 = Nc\frac{\mu_{eff}^2}{3k} = \frac{2\pi}{N_0}\frac{\sigma_{rs}^2}{3k}\left(\frac{\mu_{eff}}{\mu_0}\right)^2.$$

For <u>Cu</u> Mn, n_0 is nearly independent of c and equal to 260. For <u>Au</u>–Fe alloys $350 < n < 500$ when $0.01 > c > 0.001$. The calculated Curie constants[4] are in very good agreement with the experiment. Then the gaussian law [2] is a good representation of the M_g distribution. In order to respect the scaling laws, E_a is assumed to be proportional to the number n of spins in each cloud: $E_a = 20\, kT_b = n\mu_0\Delta_a$. Δ_a is the width of the distribution of the local anisotropy fields proportional to c. The clouds which contain n spins have a blocking temperature T_b; therefore, the mean moment of the clouds which have their T_b between T and $T + dT$ increases with T_b. Assuming $\overline{M_g(T)} = \mu_0(\sqrt{2n}/\sqrt{\pi})$ we can write

$$\frac{\overline{M_g(T)^2}}{2M_{g0}^2} = \frac{n}{\pi n_0} = \frac{\alpha T}{c},$$

where α is independent of c. Using (2) we calculate

$$\sigma_{rs}(T) = \tfrac{1}{2} N_0 \int_{M_g(T)}^{\infty} M_g P(M_g)\, dM_g = \sigma_{rs}(0)\, e^{-\alpha T/c}. \tag{4}$$

In fig. 1 we see that this law is obeyed. The width of the anisotropy field per impurity is

$$\Delta_a = 20kc/\pi n_0 \mu \alpha. \tag{5}$$

For AuFe, $n_0 = 330$, $\Delta_a = 135$ Oe for $c = 0.01$. In order to determine the origin of Δ_a we calculate the width Δ_d of the distribution of dipolar fields using the width Δ_{KB} of the distribution of RKKY molecular fields calculated by Klein and Brout [5]. The RRKY field is written $A\mu \cos(2k_F r)/r^3$. We do not calculate the mean square of the anisotropic part of the dipolar interaction. We assume that the dipolar field is μ/r^3, so

$$\Delta_{KB} = 19.6 \frac{A\mu c}{d^3}, \qquad \Delta_d = 27.7 \frac{\mu c}{d^3},$$

where d is the lattice constant. We obtain $\Delta_d \simeq 82$ Oe for AuFe with $c = 0.01$. Δ_a and Δ_d are not largely different inside the rough approximation we have used. Δ_d is perhaps a little too low. Up to now, from this example the anisotropic dipolar coupling seems to be sufficient to explain the presence of potential barriers. We may have an idea of the phenomenon which determines the number n_0. An n-cloud has an uncompensated moment which can freeze if its potential barrier is equal or larger than the coupling of the n spins with the other spins outside the cloud. We put a spin at the origin. We consider a sphere around it containing n spins, with a radius r_c proportional to $c^{-1/3}$. The distribution of the molecular field created at the origin by the spins located outside the sphere has a width $\delta_{r > r_c}$.

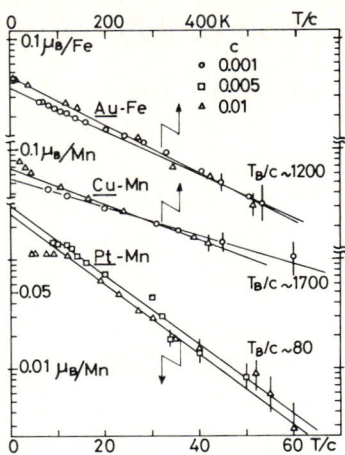

Fig. 1. The saturated remanent magnetization σ_{rs} of several systems is shown in the reduced diagram $\log(\sigma_{rs}/c) = f(T/c)$. Data for Cu Mn and Pt Mn are from refs. 1, 6 and 7.

$$E_a = n_0 \mu \Delta_d > 2\mu \sqrt{n_0}\, \delta_{r > r_c}.$$

We obtain n_0 nearly equal to A and independent of c. The mean number of spins inside the clouds is of the order of the ratio of the amplitudes of the RKKY and dipolar couplings, in agreement with the experimental results on Cu–Mn and Au–Fe alloys.

Then the irreversible properties will be significantly enhanced in systems with a low RKKY interaction. It would be the case for gadolinium diluted in normal metals.

References

[1] J. Souletie and R. Tournier, J. Low Temp. Phys. 1 (1969) 95.
[2] L. Néel, Cours de Physique Théorique, Les Houches (Presses Universitaires de France, Paris, 1961).
[3] R. Tournier, Thesis, University of Grenoble (1965). Y. Ishikawa, R. Tournier and J. Fillippi, J. Phys. Chem. Solids 26 (1965) 1727.
[4] J.L. Tholence and R. Tournier, J. de Physique 35 (1974) C4-229, J.L. Tholence, Thesis, University of Grenoble (1973).
[5] M.W. Klein and R. Brout, Phys. Rev. 132 (1963) 2412.
[6] J.A. Careaga, Thesis, University of Grenoble (1967).
[7] J.L. Tholence and E.F. Wassermann, this Conference, and to be published.

LOW TEMPERATURE SUSCEPTIBILITY OF DILUTE PtMn ALLOYS*

J.L. THOLENCE

Centre de Recherches sur les Très Basses Températures, Centre National de la Recherche Scientifique, BP 166 Centre de Tri, 38042 Grenoble-Cedex, France

and

E.F. WASSERMANN

2. Physikalisches Institut der RWTH Aachen, 5100 Aachen, W. Germany

The temperature dependence of the reversible susceptibility and thermoremanent magnetization TRM of dilute PtMn alloys indicates that there is a tendency from typical spin glass to Kondo-behaviour at concentrations around 0.1% Mn.

In an earlier paper [1] we reported about low temperature magnetization measurements on PtMn alloys with Mn-concentrations between 0.5 and 2.5 at.%. In the meantime alloys with higher concentrations (5–15%) have also been studied. The susceptibility of all samples shows the typical spin glass behaviour: a pronounced peak at the spin glass freezing temperature T_f. The effective magnetic moment as taken from the Curie–Weiss law above T_f decreases with increasing concentration. In the dilute range we find for the 0.5% alloy $p_{eff} = 6.0$, corresponding to a spin value $S = 2.5$. For the concentrated alloys $p_{eff} = 5.3$ has been measured. The absolute values of the susceptibility decrease with increasing concentration, indicating antiferromagnetic ordering in PtMn. Similar results have been reported by Pou-Wei-Hou [2] for PtMn in the concentration range 0.1–12 at.%. In comparison to other spin glass systems, the absolute values of the susceptibility for $T \to 0$, $\chi(0)$ are an order of magnitude higher in PtMn [$\chi(0) = 9$–10×10^{-5} emu/g] than, for example, in CuMn or AuMn. This indicates that there is some enhancement present due to the Pt-matrix.

In this paper we report magnetic measurements on more dilute PtMn samples (0.05, 0.1, and 0.25 at.% Mn). The magnetization has been determined by an extraction method in a low d.c. field (~50 Oe) in a temperature range between 0.05 and 4.2 K. Fig. 1 shows the incremental, reversible susceptibility as a function of the reduced temperature T/c. We observe a maximum (no sharp peak because of a small

* Work was supported by Sonderforschungsbereich 125 der Deutschen Forschungsgemeinschaft.

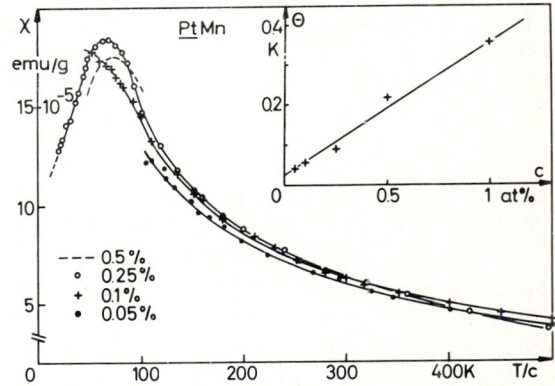

Fig. 1. Impurity contribution of the susceptibility as measured in low field χ versus normalized temperature T/c. The insert shows the Weiss-temperature as a function of the impurity concentration.

measuring field) for the 0.25% alloy at $T_f = 0.17$ K and an overall susceptibility behaviour similar to that observed in the other samples [1]. Yet, compared to the 0.5% alloy, the 0.25% sample still has a higher absolute susceptibility at low temperatures (for the 0.5% alloy the position of maximum is indicated by the dotted line in fig. 1), and the effective moment rises to $p_{eff} = 6.2$.

Cooling of the samples in a weak field h from above T_f results in the occurence of a remanence, the thermoremanent magnetization TRM after reducing h to zero. Fig. 2 (right-hand part) shows the temperature dependence of the TRM for the 0.25% sample for three different fields h. The TRM vanishes at T_f, and from a plot of the TRM versus h for $T = $ const. one obtains the irreversible susceptibility χ_{irr}, given by the initial slope $\Delta(\text{TRM})/\Delta h|_{h \to 0}$. The TRM saturates in small fields, the saturation value is

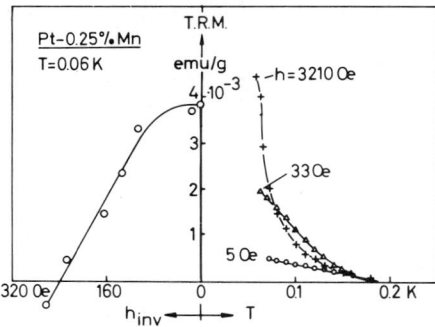

Fig. 2. Right-hand part: temperature dependence of the TRM for a Pt-0.05% Mn alloy. The sample is cooled in fields h from $T > T_f$ to $T \ll T_f$ and then $h \to 0$. Left-hand part: TRM as a function of the inverse field h_{inv}, opposite to h. h_{inv} is stepwise increased and reduced to zero and the remaining TRM measured. The area under the curve is a measure for the anisotropy energy of the frozen in spin glass domains.

$\sigma_{rs} = 4.2 \times 10^{-3}$ emu/g at $T = 0.06$ K for the 0.25% alloy. The saturation field is proportional to c (~ 3 kOe/at.%). The remanent magnetization in spin glasses can be understood in terms of the domain model by Néel [3] in which the randomly frozen-in moments form magnetic domains for T approaching zero. The resulting moment of such a domain lies in one direction of the anisotropy axis of the domain. Its mean value is $|\overline{M_{gl}}| = \sqrt{(2n/\pi)}\,\mu$ where n is the number of frozen single moments μ of such a domain. $n = (1/2\pi)(\sigma_s/\sigma_{rs})^2$ can be calculated [3] from

$$\sigma_{rs}(T=0) = \frac{1}{2}\frac{N_L c}{Mn}\sqrt{\frac{2n}{\pi}}\,\mu.$$

As for most of the spin glass systems the ratio $\sigma_s/\sigma_{rs} \approx 50\text{–}70$, resulting in $n = 400\text{–}800$.

One can get an idea of the size of the anisotropy energy by applying a field h_{inv} in opposite direction to h. At $T = $ const, h_{inv} is increased stepwise and then reduced to zero. The resultant remanence as a function of h_{inv} is shown in fig. 2 (left-hand part). The area under this curve is a measure for the anisotropy energy. One can see that at $h_{inv} = 200$ Oe ($T \approx 0.06$ K) the remanence for the 0.25% alloy has vanished.

The more dilute 0.1 and 0.05% alloys show deviations from the universal behaviour in reduced diagrams discussed so far. The effective moment rises to $p_{eff} = 6.44$ for both samples. For the 0.1% alloy the TRM is very small, and at $T = 0.07$ K a remanence is no longer measurable, indicating a T_f value of about 0.07 K. Since in fig. 1, however, we no longer find a maximum in $\chi(T)$, although the minimum reachable temperature is not low enough, it is obvious that with decreasing impurity concentration T_f/c is shifted to lower and lower temperatures. This means that the number of magnetic impurities decreases more rapidly than c. We think that this is an indication that the system tends more and more from the spin glass to Kondo behaviour in the dilute concentration range. On the other hand, the low temperature resistivity [4] behaviour of the same samples is not very different from what is observed in the high concentration spin glass regime. Minima in $\Delta\rho(T)$ are always present.

One can get an estimate of the Kondo-temperature of the PtMn system from an extrapolation to $c \to 0$ of the Weiss-temperatures θ versus impurity concentration c. This is shown in the inset of fig. 1. If θ for $c \to 0$ corresponds to T_K there will be a T_k of the order 25 mk for PtMn. A comparison of $T_K = 25$ mk and $T_f/c \approx 70\text{–}80$ results in a value $c_K \approx 300$ ppm, separating the spin glass from the low concentration Kondo regime. Below this concentration c_K ordering temperatures proportional to c^2 should be observed. The discussion demonstrates that more dilute alloys at extremely low temperatures have to be studied to get more information about the Kondo behaviour of PtMn.

References

[1] E.F. Wassermann and J.L. Tholence, Proc. 21st Ann. Conf. on Magnetism and Magnetic Materials, Philadelphia (1975); AIP Conf. Proc. 29 (1976).
[2] Pou-Wei-Hou, Thesis, Imperial College, London (1975).
[3] J.L. Tholence and R. Tournier, J. de Physique C 4 (1974) 229.
[4] J. Kästner, K. Matho, J.L. Tholence and E.F. Wassermann, Solid State Commun., to be published.

THE BEHAVIOUR OF SPIN GLASSES IN LOW d.c. FIELDS

C.N. GUY

Physics Department, Imperial College, London SW7, England

The susceptibility, thermoremanence, isothermal remanence and magnetic viscosity of Au–Fe alloys are investigated using low d.c. fields. The susceptibility of a Cu–Mn alloy is presented for comparison. An attempt is made to interpret this data in terms of the Néel blocking model, applied to fine magnetic particles.

1. Introduction

Nearly all the recent theoretical investigations (see ref. 1) of spin glasses have been concerned primarily with the sharp peak in the a.c. susceptibility. The static properties of the frozen state ($T < T_G$) on the other hand remain unexplained within these new theoretical models. Several authors have pointed out the strong analogy which exists between these latter properties of spin glasses and those of single domain particles [2, 3]; this is strengthened by the existence of very sharp peaks in the initial susceptibility $\chi(T)$ of nearly all magnetic rocks [4]. In this paper we follow this approach by presenting low static field measurements of Au–Fe and Cu–Mn spin glasses and attempting to interpret the results using the phenomenology of rock magnetism.

All the data was obtained using a vibrating sample magnetometer. All the alloys were prepared by arc melting, homogenizing at 900°C and quenching to room temperature.

2. Results and discussion

The low field d.c. susceptibilities of all the alloys studied are shown in fig. 1 plotted against T/T_G. Fig. 2 shows the results of field cooling experiments [$H_{ext} = 20$ Oe] for Au–Fe alloys. Fig. 3(a) shows the variation of incremental isothermal remanence (Δ(IRM)) obtained from zero field-cooled samples after the application of 20 Oe for 2 min.

From fig. 1 it can be seen that $\chi(T)$ has a smooth maximum at T_G and that the Au–Fe alloys show a pronounced shoulder close to $0.6T_G$ where Δ(IRM) has a maximum [5] [fig. 3(a)]. In recent work [7] it has been shown that $\chi(T) = (\partial M/\partial H)_{H=H_0}$ is independent of applied cooling fields up to $H_0 = 60$ Oe and that the thermoremanence TRM, fig. 2, is directly proportional to the cooling field in the range studied (5–40 Oe). The field cooled state of these spin glasses is time independent as long as the

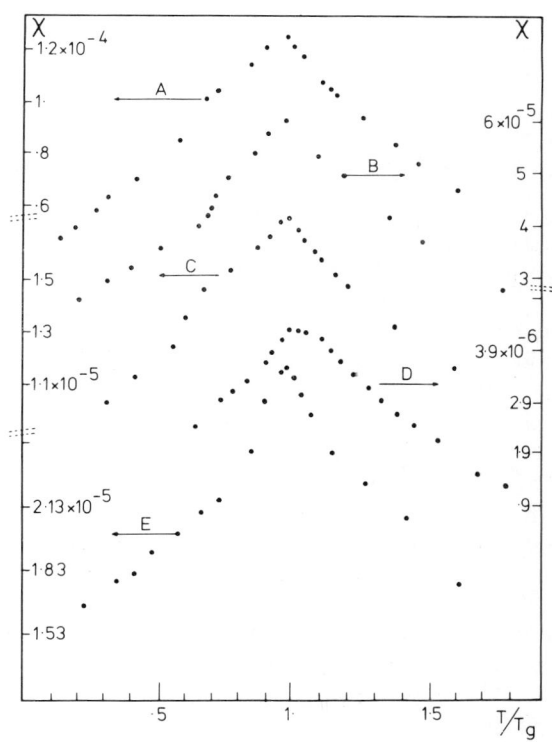

Fig. 1. The susceptibility (emu/gm) for (A) Au 7 at.% Fe, (B) Au 4 at.% Fe, (C) Au 2 at.% Fe, (D) Au 0.25 at.% Fe, (E) Cu 2 at.% Mn plotted against reduced temperature T/T_g.

cooling field is maintained. The thermoremanence is, however, time dependent approximately following.

$$M(t, T) = \text{const} - S(T) \ln t, \qquad t < 30 \text{ s.} \qquad (1)$$

In fig. 2 the initial ($t = 10$ s) value of TRM is plotted. The rapid initial time decay is responsible for the relatively large error bars in fig. 3(a). Preliminary measurements of $S(T)$ show a similar peaked behaviour to that of Δ(IRM), fig. 3(b).

To account for these observations within the fine particle model we assume that spin glasses contain, for whatever reason, large magnetic clusters each with a net spontaneous moment $m(T)$ and each having an associated anisotropy energy E. The magnetic behaviour of the en-

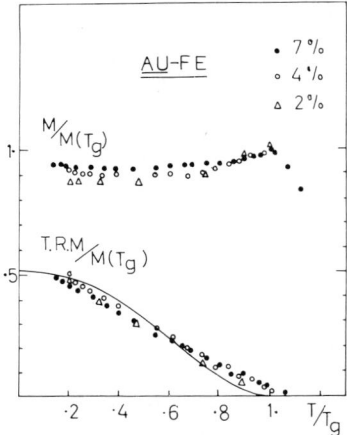

Fig. 2. Total magnetization and TRM of Au–Fe alloys scaled to $M(T_g)$ and plotted against T/T_g. Solid line is the calculated variation of TRM (see text).

semble of clusters will depend on the distribution of anisotropy energies $f(E)$ and $m(T)$. The Néel model relates the blocking temperature T_B to be particle anisotropy by

$$KT_B = [2E \pm mH_{ext}/V]^2/4E \ln C\tau_m, \quad (2)$$

with V the particle volume, τ_m the measuring time and $C \simeq 10^9$ s^{-1}. At T_B the particle moment ceases to respond to external field within a time τ_m. We argue that fig. 3(a) essentially represents $f(E) \equiv f(T_B)$ in the Au–Fe spin glasses since at a particular temperature $T < T_G$, particles with $T_B > T$ do not respond to the 2 min pulse and particles with $T_B < T$ respond but decay too rapidly to contribute any remanence. Using the phenomenological description [7] of remanent decay it can be shown that the total TRM of the ensemble is

$$M(T, \tau_m) \propto m(T) \int_{KT' \ln C\tau_m}^{KT'_{max} \ln C\tau_m} f(E) dE. \quad (3)$$

Using the data of fig. 3(a) and eq. (3), assuming $m(T)$ to be constant and $T'_{max} = T_G$, we calculate the expected variation of TRM with T shown in fig. 2 for 2, 7 at.% Au–Fe. The same procedure yields the time dependence of the remanence for $t > \tau_m$, at a fixed $T < T_G$. Preliminary measurements [7] show qualitative agreement with this calculation.

This simple model naturally leads us to an association of T_G with the maximum blocking temperature in the distribution $f(T_B)$. If this were correct we would not expect to have any sharp specific heat anomaly at T_G; the abrupt change in magnetic susceptibility would result simply from the strong dependence of relaxation time on temperature ($1/\tau \approx C e^{-E/KT}$). The data of fig. 2 suggests, on the other hand, that T_G has a deeper significance than merely the highest blocking temperature; both the TRM and the total magnetization appear to be determined by the net magnetic alignment obtaining as the sample cools through T_G.

We conclude that much of the spin glass phenomenon can be interpreted qualitatively using a simple phenomenological blocking model by analogy with rock magnetism [4]. It remains to be shown however that magnetic clusters are necessary for spin glass behaviour.

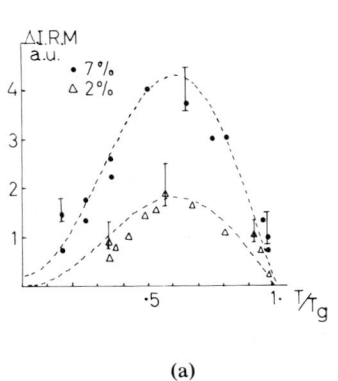

(a)

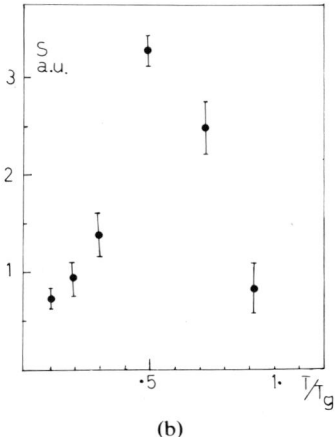

(b)

Fig. 3. (a) The variation of ΔIRM for 2 and 7 at.% Au–Fe plotted against T/T_g. (b) The variation of $S(T)$, the magnetic viscosity coefficient, with T/T_g for Au–4 at.% Fe.

References

[1] K.F. Fischer, ICM '76, Paper 7B12.
[2] J.L. Tholence, and R. Tournier, J. de Physique 35 (1974) C4-229.
[3] P. Beck, Less Common Metals 28 (1972) 193.
[4] E.P. Wohlfarth, ICM '76, Paper 5C7.
[5] T. Nagata, Rock Magnetism (Maruzen, Tokyo, 1961).
[6] J.S. Kouvel, J. Phys. Chem. Solids 21 (1961) 57.
[7] C.N. Guy, in preparation.
[8] L. Néel, Advan. Phys. 4 (1955) 191.

LONG-TIME RELAXATION EFFECTS IN SPIN-GLASSES: THE HEAT CAPACITY OF Au Fe

G.J. NIEUWENHUYS and J.A. MYDOSH

Kamerlingh Onnes Laboratorium der Rijksuniversiteit, Leiden, The Netherlands

Heat-capacity measurements on a Au Fe (4 at.%) spin glass reveal long-time relaxation processes due to the magnetic history of the sample. We present measurements of the heat flow from or to the sample induced by changes of an external magnetic field. A $1/t$ dependence is observed, in agreement with Monte-Carlo simulations.

Spin-glass systems have been receiving much attention from both the theoretical as well as the experimental point of view. For quite recent reviews one is referred to the work of Anderson [1], Fisher [2] and Mydosh [3]. Most of the studies have concerned the static properties of spin glasses. However, present interest is shifting to the dynamical properties and the relaxation effects coupled with the application of external magnetic fields. Motivated by the intriguing magnetization measurements of Guy [4], we have studied the dynamical behaviour of the heat capacity for a Au Fe (4 at.%) alloy.

The measurements were performed using the adiabatic method as described by Boerstoel et al. [5] in which a constant rate of heat is applied to the sample during a time Δt, causing a temperature increase of ΔT, from which the heat capacity is deduced according to $C = \dot{Q} \Delta t/\Delta T$. In fig. 1 three recorder traces (temperature versus time) obtained from this method are shown. Focusing attention on the middle graph; the vertical lines are the temperature versus time curves without heating the sample; the horizontal ones result from the application of heat at a constant rate to the sample. This measurement was carried out in zero field for all temperatures (including the cooling of the sample). No relaxation effects can be seen except for the intrinsic experimental ones. The excess specific heat of the alloy, deduced in this way, was directly proportional to temperature, with a constant of 1 mJ/K²mole. This result for temperatures much less than the freezing temperature, T_f, is in good agreement with those found by Wenger and Keesom [6].

The left and right recorder traces were obtained when cooled in zero field (left curve) and then measured in an external field of 20 kOe, or cooled in 20 kOe (right curve) and then measured in zero field. Long timescale (> 100 s)

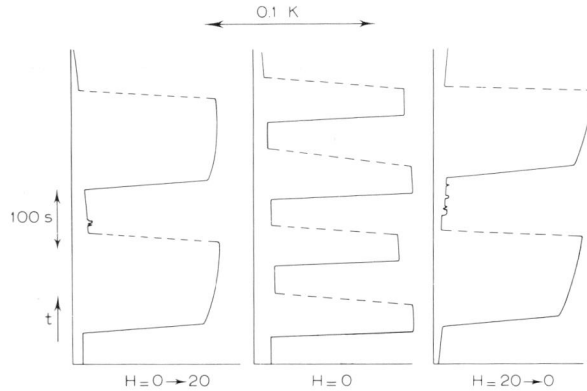

Fig. 1. Recorder traces at about 2 K of the temperature versus time obtained from the adiabatic heat-capacity measuring method applied to AuFe 4 at.%. The middle curve is obtained by cooling and measuring the sample in zero external field; the left curve by cooling zero field and measuring in 20 kOe, and the right one by cooling in 20 kOe and measuring in zero field.

relaxation effects are evident from the graphs. Clearly no exact specific heat values could be deduced from the two curves. Nevertheless, if one neglects the long-time relaxation, the value obtained agrees with that of the middle graph.

Analogous, but much smaller, relaxation effects were obtained when the sample was rapidly cooled from high temperature to $T = 1.2$ K. Since such long-time relaxation effects should be explained by the presence of weakly coupled spin–lattice systems, we repeated the heat-capacity measurements, now using a constant heating method in such a way that the rate of the temperature increase of the sample was nearly constant. Unfortunately, the differences in the values for the heat capacity for different rates of temperature increase were not much larger than the experimental error.

We therefore tried to investigate the relaxation effects by connecting the sample via a heat link to the liquid He-bath, and then monitoring

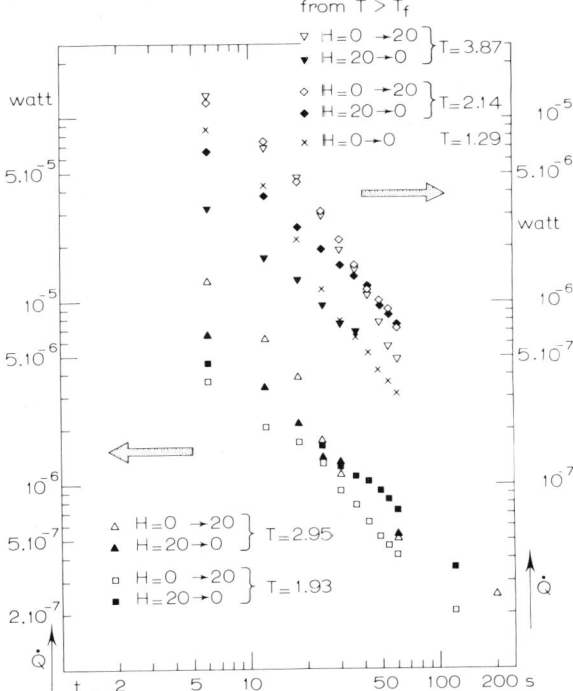

Fig. 2. Heat flow versus time for AuFe 4 at.% at the indicated temperatures and changes of the external field.

the rate of the heat flow to or from the sample (i.e. by detecting the temperature difference across the heat link), due to a change of the external field. Care was taken with the dimensions of the sample and the heat link so as to maintain an intrinsic experimental relaxation time less than 1 s.

In fig. 2 a series of heat flow, $\dot{Q}$, versus time curves are shown. The upper curve was obtained after heating the sample before each change of the field to a temperature well above T_f. The lower curves were obtained without this "temperature cycling". As can be seen the relaxation cannot be described by a single relaxation time, since $\dot{Q}$ varies roughly as t^{-1}. This type of relaxation is indeed very much slower than the normal exponentional one. If one wishes to describe an inverse-time relaxation process by an effective exponential constant, it should contain "log t" rather than "t".

It should be noted that, decreasing the field ($H = 20$ kOe $\rightarrow 0$) as well as increasing the field ($H = 0 \rightarrow 20$ kOe) always causes the heat to flow from the sample to the liquid helium bath.

Theoretical investigations of the dynamical behaviour of spin glasses have been begun by Edwards and Anderson [7], Rivier [8] and Fisher [9]. However, we feel that the recent Monte-Carlo simulations by Binder et al. [10] are more workable to compare with our experiments. These authors found that the dynamical behaviour of a number of magnetic properties (including the energy) of a spin glass (treated within the Edwards and Anderson model) can be described by a t^{-a} function. We repeated this kind of calculation, especially applied to Au Fe(4 at.%), and found similar results. This is in agreement with the t^{-1} dependence of the heat flow

In conclusion we remark that:

(1) The derivative of the energy after a large change of the external magnetic field is proportional to the inverse of the time; in agreement with Monte-Carlo simulations.

(2) The Monte-Carlo simulations are based on the Ising model; no anisotropic exchange is included artificially. We therefore think that the long-time relaxation processes are intrinsic with spin glasses (and other diluted magnetic systems), and may be qualitatively explained by considering the large number of spin flips needed to go from one to another state in the glass and taking into account the decreasing number of ways in which this is possible when equilibrium is almost reached.

(3) "log t"-scaling instead of "t"-scaling has already been predicted for a number of other magnetic systems (see e.g. Stauffer [11]). A more extended investigation (not restricted to spin glasses) may be worthwhile.

The authors acknowledge valuable discussions with Drs. N. Rivier and K.H. Fisher, and thank the members of the Leiden Metal Physics group for their interest and help with the experiments. This work is part of the research programme of the Foundation F.O.M., supported financially by Z.W.O. and T.N.O. (Metals' Institute).

References

[1] P.W. Anderson, in: Amorphous Magnetism-II (Plenum Press, New York, 1977).
[2] K.H. Fisher, in: Proc. Int. Con. on Magnetism; Amsterdam (1976).

[3] J.A. Mydosh, AIP Con. Proc. 24 (1975) 131.
[4] C.N. Guy, J. Phys. F 5 (1975) L242.
[5] B.M. Boerstoel, W.J.J. van Dissel and M.B.M. Jacobs, Physica 38 (1968) 287; Commun. Kamerlingh Onnes Lab., Leiden No. 363a.
[6] L.E. Wenger and P.H. Keesom, Phys. Rev. B11 (1975) 3497.
[7] S.F. Edwards and P.W. Anderson, J. Phys. F 6 (1976) 1927.
[8] N. Rivier, Amorphous Magnetism-II (Plenum Press, New York, 1977).
[9] K.H. Fisher, private communication.
[10] K. Binder and K. Schröder, Solid State Commun. 18 (1976) 1361. K. Binder and K. Schröder, Phys. Rev. B 14 (1976) 2142. B.K. Binder and D. Stauffer, Phys. Lett. 57A (1976) 177.
[11] D. Stauffer, Phys. Rev. Lett. 35 (1975) 394.

CHAPTER 6

INSULATORS AND SEMICONDUCTORS
EXCHANGE

Transition metal compounds; Magnetic structure (*Session 5F*)	885
Magnetic structure I; Insulators (*Session 3W*)	899
Magnetic structure II; Oxides (*Session 8X*)	919
Magnetic oxides (*Session 9E*)	934
Magnetite (*Session 2C*)	948
Magnons and phonons (*Session 3C*)	968
Magnetic semiconductors; Transport (*Session 8U*)	982
Metal–nonmetal transitions; Magnetic semiconductors (*Session 3D*)	1001
Exchange interactions in insulators (*Invited paper 9AI1*)	1012
Prospects for *ab initio* calculations and models of superexchange interactions (*Invited paper 9AI2*)	1018
Exchange I (*Session 8D*)	1025
Exchange II (*Session 9D*)	1037

MAGNETIC STRUCTURES OF $Mn_{2-x}Fe_xSb$

C. BLAAUW, G.R. MACKAY and W. LEIPER
Department of Physics, Dalhousie University, Halifax, Nova Scotia, Canada*

$Mn_{2-x}Fe_xSb$ ($x \leq 0.2$) has been studied by ^{57}Fe Mössbauer spectroscopy. The spin flop transition of Mn_2Sb at T_s is observed through the T dependence of the quadrupole interaction. The corresponding change in the hyperfine magnetic field from 49.6 kOe at $T < T_s$ to 79.4 kOe at $T > T_s$ is attributed to crystalline anisotropy. T_s increases with increasing values of x.

1. Introduction

The intermetallic compound Mn_2Sb orders in the tetragonal Cu_2Sb (C38) structure and is ferrimagnetic with $T_c = 550$ K. Two distinct Mn sites exist in the lattice, Mn_I and Mn_{II}, carrying different magnetic moments, with μ_{Mn_I} aligned antiparallel to $\mu_{Mn_{II}}$ in the magnetically ordered state [1]. An interesting feature is the presence of a spin flop transition at $T_s = 240$ K. The spins are aligned parallel to the crystallographic c-axis at $T > T_s$ and perpendicular to it at $T < T_s$.

The magnetic properties of Mn_2Sb are sensitive to impurity contamination; e.g. doping the material with small amounts of Cr leads to an antiferromagnetic structure, first noted by Swoboda et al. [2]. Our purpose in this investigation was to determine the influence of doping with Fe on the magnetic properties of Mn_2Sb.

2. Experimental procedure

Samples were prepared by melting the constituent elements in an r–f furnace in an argon atmosphere. All samples were shown by X-ray analysis to be single phase. Samples of composition $Mn_{2-x}Fe_xSb$ were studied, with $x = 0.01$, 0.02 and 0.2, using iron enriched in ^{57}Fe where necessary. In addition a sample with $x = 0.02$ and one of pure Mn_2Sb were doped with ^{57}Co and studied as Mössbauer sources.

Spectra were taken of all samples at room temperature (RT) and liquid nitrogen temperature (LNT). The results obtained from these measurements are presented in table I.

3. Discussion

No significant concentration dependence could be detected in spectra taken at low impurity concentrations ($x \leq 0.02$). Also, spectra taken from the ^{57}Co doped samples used as Mössbauer sources were not different from the ^{57}Fe absorption spectra of the other samples. Thus all spectra at $x \leq 0.02$ can be represented with one set of data points. In fig. 1 some characteristic spectra are shown.

Two conclusions may be directly inferred from these results. One is that Co occupies the same position as Fe in the Mn_2Sb lattice. The crystallographic sites in Mn_2Sb are sufficiently distinct to expect observable differences between emission and absorption spectra if Co and Fe were to occupy different lattice positions. A second conclusion is that, because the results remain unchanged in the limit of zero impurity concentration (the ^{57}Co concentration in the pure Mn_2Sb sample ≈ 1 ppm), the difference between the RT and LNT spectra is directly related to the spin flop transition of pure Mn_2Sb. Our discussion will focus on how the Möss-

Table I
^{57}Fe Mössbauer data for different samples of $Mn_{2-x}Fe_xSb$

$Mn_{2-x}Fe_xSb$	$T = 80$ K			$T = 300$ K		
	$|H|$(kOe)	QS(mm/s)	IS(mm/s)	$|H|$(kOe)	QS(mm/s)	IS(mm/s)
$x \leq 0.02$	49.6 ± 0.5	-0.10 ± 0.02	0.6 ± 0.1	79.4 ± 0.5	0.12 ± 0.02	0.5 ± 0.1
$x = 0.2$	57.3 ± 0.5	-0.10 ± 0.02	0.6 ± 0.1	56.3 ± 0.5	-0.07 ± 0.02	0.5 ± 0.1

* This research has been supported by the National Research Council of Canada and by the Faculty of Graduate Studies at Dalhousie University.

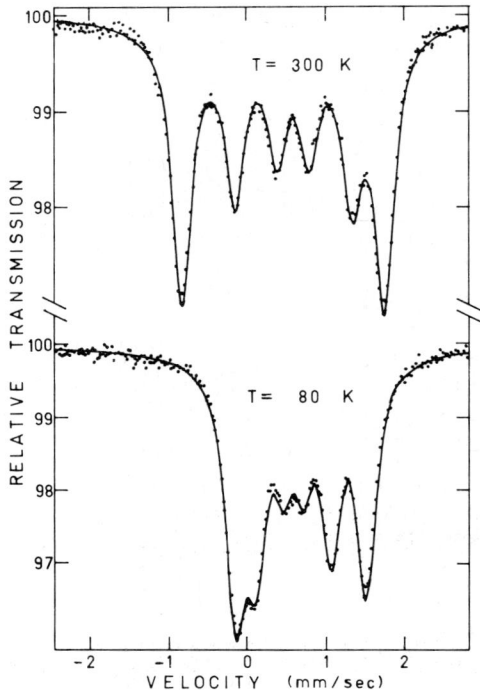

Fig. 1. ^{57}Fe Mössbauer spectra of $Mn_{1.98}Fe_{0.02}Sb$. The isomer shift is relative to metallic iron.

bauer properties in the RT and LNT spectra relate to this transition.

It would be interesting to see whether a change in the isomer shift (IS) occurs at T_s. Unfortunately this information cannot be obtained from our results. In our spectra we are in the limit of validity of the approximation that the quadrupole interaction is small compared with the magnetic interaction. When this approximation is not valid, the relative positions and the relative intensities of the lines are not given by simple relationships but depend on the ratio of these two interactions. This makes an accurate determination of the IS extremely difficult. The difference in IS between RT and LNT can be fully attributed to a second order Doppler shift.

The quadrupole splitting (QS) is in agreement with the picture of a spin flop transition. Because the Mn positions have axial symmetry, the axis of symmetry being the c-axis, one expects to observe a QS. The transition of the spins from being parallel to the c-axis at $T > T_s$ to being parallel to the basal plane at $T < T_s$ should correspond to a change in the apparent QS by a factor of -0.5. Though this is not observed directly, such a change is within the experimental error, if we assume a general decrease of the QS with increasing T of the same order as is seen in the $x = 0.2$ sample, which does not exhibit a spin transition between LNT and RT.

Probably the most interesting result is the change in the value of the magnetic hyperfine field (H) from $|H| = 49.6$ kOe at $T < T_s$ to $|H| = 79.4$ kOe at $T > T_s$. To make sure that this difference is not the result of measuring at two different reduced temperatures in different magnetic phases, we also took spectra of a $Mn_{1.98}Fe_{0.02}Sb$ sample at 5 K and at 400 K. The results confirm that the LNT and RT results are approximately saturation values of H. If we assume that no major changes in the electronic structure occur at T_s, then the change indicates a crystalline anisotropy, to be attributed to anisotropy in dipolar and orbital contributions to H. Crystalline anisotropy has been observed before, e.g. in Fe_2O_3 [3], but not to our knowledge of the magnitude observed here. The effect is presently being investigated in more detail.

The spectra taken of the $Mn_{1.8}Fe_{0.2}Sb$ sample show Mössbauer parameters of both spectra are not significantly different from those at LNT in the other samples, indicating that increasing the Fe content raises T_s. In this sample, Fe occupies 20% of the sites in one of the Mn sublattices. A different value of H and a distribution in values caused by different neighbouring atomic configurations, which were observed in the spectra, are therefore to be expected. The data in table I for this sample represent an average obtained by fitting the spectra with one sextet.

References

[1] M.K. Wilkinson, N.S. Gingrich and C.G. Shull, J. Phys. Chem. Solids 2 (1957) 289.
[2] T.J. Swoboda, W.H. Cloud, T.A. Bither, M.S. Sadler and H.S. Jarrett, Phys. Rev. Lett. 4 (1960) 509.
[3] F. Van der Woude, Phys. Status Solidi 17 (1966) 417.

TEST OF THE "3c" AND "4c" Fe$_7$Se$_8$ OKAZAKI'S SUPERSTRUCTURES BY MÖSSBAUER EFFECT

G.A. FATSEAS, J.L. DORMANN, R. DRUILHE, L. BROSSARD and P. GIBART

Laboratoire de Magnétisme, CNRS, 1, Place Aristide Briand, 92190 Meudon, France

High statistics ^{57}Fe Mössbauer spectra have been obtained at 300 K and below 77 K, on both "3c" and "4c" absorbers of NiAs-type ferrimagnetic iron selenide Fe$_7$Se$_8$. It is shown that the results obtained at the low temperature region are not consistent with the Okazaki's "3c" and "4c" superstructures model.

1. Introduction

The ferrimagnetic iron selenide ☐ Fe$_7$Se$_8$ exists in two types of crystallographic superstructures [1], "3c" and "4c", attributed to an ordered arrangement of vacancies ☐. Both superstructures have identical iron sites; three sites, A, B, C, with different environments in number and positions of vacancies (see, for example ref. 2).

The relative populations of these iron atoms are for both-superstructures at 300 K, A:B:C = 3:2:2, and they have been reported to be constant even at [1] (or down to at least [3]) liquid nitrogen temperature for "3c" and down to at least 20 K for "4c" [3].

In this paper we report Mössbauer spectroscopy results in the low-temperature region (4.2–77 K), in order to test the relative intensities cited above. Such data are lacking in all earlier Mössbauer data [2, 4, 5].

2. Experimental and preliminary results

High statistics Mössbauer spectra were obtained on powder "3c" and "4c" Fe$_7$Se$_8$, with a high counting rate spectrometer, a 50 mC ^{57}Fe (Rh) source, and a symmetrical saw-tooth signal vibrator. All our samples contain 10 mg natural iron per cm^2 corresponding to a thin Mössbauer absorber. The least squares fitting routine uses lorentzian line-shapes with equal line width for all peaks and all sites and equal Mössbauer absorption coefficient for all sites.

The initial material was a single crystal prepared by the Bridgman method at 1200°C and checked by X-ray Laue photography and by micrographic examination. Platelets cut in this crystal were annealed by heating to 1000°C at 4°/h, kept at this temperature for two weeks and then cooled down at room temperature at 4°/h.

Preliminary spectra obtained on these samples at 300 K ($T_c \simeq 450$ K) showed the presence of a central paramagnetic doublet of 8% intensity overlapping the hyperfine Zeeman spectrum. In an attempt to reduce this paramagnetic component, a platelet of the initial single crystal was again annealed at 800°C with the same, as above, heating and slow cooling rate. With these new samples the spectra were fitted with $\simeq 3.5\%$ paramagnetic component intensity. Then, for further possible reduction of this component, a third annealing was attempted as follows. Platelets of the initial single crystal (and also powder separately), were heated to 800°C at 15°/h and kept at this temperature for three days. Then they were cooled down to 400°C at 4°/h and quenched to room temperature for the "3c" superstructure, or slowly cooled successively from 400 to 250°C at 2.5°/h and from 280 to 20°C at 2°/h for "4c" superstructure.

All our final spectra, whose only some at them [6] are given in fig. 1, were obtained on these last samples and they were fitted, at 300 K and at all temperatures below 77 K, with 2% ($\pm 0.5\%$) paramagnetic intensity superposed to several six-line hyperfine Zeeman paterns.

3. Results and discussion

3.1. Results at 300 and 77 K

Three iron sites A, B, C and four sites A$_1$, A$_2$, B, C were detected in "3c" and "4c", respectively, in good agreement with the results of Boumford et al. [5] for both "3c" and "4c" absorbers.

The agreement is also good with the results of H. Nam Ok et al. [2] for "3c" absorber and concerns not only the number of iron sites but also their relative intensities and the hyperfine parameters H, δ and ϵ. According to the correspondence between the Mössbauer sites and the Okazaki's iron-sites proposed by Boumford

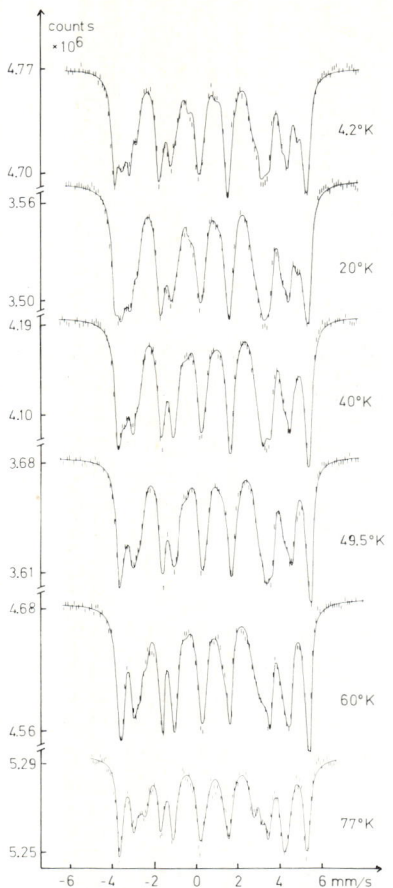

Fig. 1. Fitted spectra obtained on "3c" absorber between 77 and 4.2 K. The fitting corresponds to the solution no. 1.

et al. [5], the Mössbauer intensity-ratios are very near to the Okazaki's ratios $A:B:C = 3:2:2$ leading to the conclusion that the Mössbauer results are consistent with the Okazaki's model at 300 and 77 K.

3.2. Low-temperature results (4.2–60 K)

At the temperature region between 60 and 4.2 K two good different fitting of each experimental spectrum have been obtained, giving two different solutions (nos. 1 and 2).

The most important result of these solutions is that both of them revealed important perturbations of the intensities and the hyperfine fields of several sites in this low-temperature region. These perturbations begin at approximately 60 K and give rise to the following features:

(1) An apparition of a low-field site ($\simeq 110$ kOe) (D-site) below 60 K growing at the expense of both B and C sites. It is argued [6] that this site cannot be attributed to the presence of a parasitic phase or to experimental anomalies and it can not be eliminated by other fitting-solutions.

(2) Very large temperature intensity-variations for almost all sites, particularly in "4c" superstructure.

(3) Large deviations of the Mössbauer intensity-ratios $I_A:I_B:I_C$ from the Okazaki's ratios $A:B:C = 3:2:2$, in all temperatures below 60 K. At 4.2 K and for "3c" for example, we have $I_A:I_B:I_C:I_D = 3:1.4:1.90:0.7$ for solution 1 and $3.5:1.6:1.20:0.7$ for solution 2.

(4) Subdivision of the A-site in A_1, A_2 for both superstructures and both solutions. This subdivision is not justified, for the "3c" superstructure, in the light of the explanation given for these sites by Boumford et al. [5].

All these variations lead to the conclusion that our low-temperature Mössbauer results, unlike those at 300 and 77 K, are not consistent with the Okazaki's model for "3c" and "4c" superstructures. This means, in other words, that with the results obtained here and at the present step of understanding there is not any reasonable correspondence between the Mössbauer sites and the crystallographic sites of the Okazaki's model.

3.3. Isomer shifts (δ) and quadrupole splittings (ϵ)

These parameters are not affected at all by the intensity and hyperfine fields-perturbations cited above. δ varies, relative to metallic iron, from 0.86 (77 K) to 0.89 (4.2 K) for all sites. (All values for both δ and ϵ with 0.03 mm/s error.) The quadrupole splittings $\epsilon = \frac{1}{4}(D_{21} - D_{65})$ [5] is negative for all sites and both superstructure and solutions and equal, at all temperatures, to -0.05 mm/s for A-sites, -0.14 for B and -0.12 for C-sites. The D-site has a positive ϵ-value varying from 0.36 (77 K) to 0.25 (4.2 K).

References

[1] A. Okazaki, J. Phys. Soc. Jap. 16 (1961) 1162.
[2] H. Nam Ok and S. Won Lee, Phys. Rev. B 8 (1973) 4267.
[3] M. Kawaminami and A. Okazaki, J. Phys. Soc. Jap. 29 (1967) 924.
[4] G.A. Fatseas, C.R. Acad. Sci. Paris, 265 (1967) 1073.
[5] C. Boumford and A.H. Morrish, Phys. Stat. Sol. (a) 22 (1974) 435.
[6] More extended results will be published elsewhere.

$FeCr_{2-x}In_xS_4$: STRUCTURAL AND MAGNETIC PROPERTIES

L. GOLDSTEIN, L. BROSSARD, M. GUITTARD and J.-L. DORMANN

Laboratoire de Magnétisme, CNRS, 1, Place A. Briand, 92190 Bellevue, France

The solid solutions $FeCr_{2-x}In_xS_4$ remain spinel type for all x. Fe is moved towards B sites when x is increased. It is shown that the transition from the ferrimagnetic state ($FeCr_2S_4$) to the AF state ($FeIn_2S_4$ at low temperature) occurs in a relatively broad range of composition.

1. Introduction

$FeCr_2S_4$ and $FeIn_2S_4$ [1] are normal and inverse spinels, respectively, the first one being a ferrimagnet ($T_c = 180$ K) and $FeIn_2S_4$ is AF. The solid solution $FeCr_{2-x}In_xS_4$ remains spinel for all compositions. The compounds were prepared for different values of x by heat treatment under H_2S atmosphere of mixed oxides issued from co-decomposition of the corresponding nitrates.

The a parameter deduced from X-ray measurements for different x shows that the solid solution follows the Vegard law ($a = 9.995$ Å for $x = 0$, $a = 10.610$ Å for $x = 2$). Paramagnetic Mössbauer spectra at room temperature were done for each composition. The quadrupole splitting of Fe^{2+} in the octahedral site is large due to the trigonal deformation ($u > 0.375$), very different from Fe^{2+} at A sites. So it is possible to evaluate the repartition of Fe^{2+} among A and B sites with a rather good approximation. The Cr^{3+} ions being only at B, it is possible to deduce the In concentration at each of the sites. Fig. 1 shows that for small x In substitutes in B and for higher concentrations In starts to occupy the tetrahedral sites. So it results that the In affinity for the A site increases with x. The intensities of the X-ray peaks (400, 422, 444, 333, 531, 800) are in good agreement with Mössbauer data [2].

2. Magnetic measurements

The magnetic moments were determined between $T = 4.2$ K and the transition temperature as a function of the applied magnetic field for different compositions $x = 0.4, 0.6, 1.0, 1.2$ and 1.6. Curves $M(T)$ are plotted in fig. 2 for different x. For the compositions $x \geq 1.3$ the compounds are antiferromagnetic. As shown later the transition from the ordered ferrimagnetic structure to the antiferromagnetic is relatively broad and make possible to distinguish three regions with different magnetic order as a

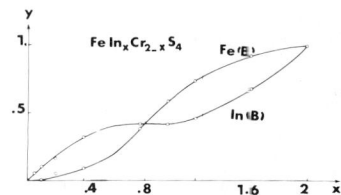

Fig. 1. Concentration of In and Fe in B site as a function of x.

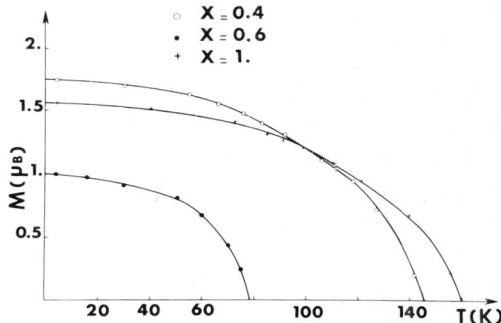

Fig. 2. $M(T)$ curves for different x.

function of x.

(a) $0 < x \leq 0.8$ where $|Fe|_B < |Cr|_B$
     and $|In|_A < |Fe|_A$

The compounds are ferrimagnetic. M_s can be calculated assuming a Néel collinear structure where Fe^{2+} in A sites are antiparallel to Fe^{2+} and Cr^{3+} in B sites ($M_s = 1.40 \mu_B$ for $x = 0.4$, $M_s = 1.65 \mu_B$ for $x = 0.6$). These results are in good agreement with the experimental determination (fig. 2). The Curie temperature decreases slowly in this range of compositions ($T_c = 160$ K for $x = 0.4$, $T_c = 145$ K for $x = 0.6$). This is explained by the localization of iron atoms mainly in A sites. The strong negative interaction between Fe^{2+} in A and Cr^{3+} in B (~ 10 K) is predominant in these compounds and gives a ferrimagnetic order with a smooth decrease of T_c.

(b) $1.3 \leq x < 2$ where $|Cr|_B < |Fe|_B$
and $|Fe|_A < |In|_A$

The compounds are antiferromagnetic, the Néel temperature is relatively low ($T_N = 20$ K for $x = 1.6$) Fe is now in B so that A–B exchange interaction is negligeable. As encountered in other thiospinels of the same kind, such as Fe_2GeO_4 [3], Fe_B^{2+}–Fe_B^{2+} interaction is weak and negative. The influence of the remaining Cr^{3+} ions is rather complex due to the competition of n.n. positive interaction and n.n.n. negative interaction. As in other $MInCrS_4$ thiospinels it is assumed that Cr favours the AF order. Finally, the dilute Fe^{2+} in A sites must be coupled antiferromagnetically to the B moment ($J_{AB} < 0$) and so must give an AF contribution when integrated over a wide volume. Mössbauer spectra for $FeIn_2S_4$ at 4.2 K clearly show the superposition of different magnetic site contributions related with the probability of finding different configurations of Fe neighbours for a specified Fe in B site. In all cases the quadrupole effect is positive. So, the $^5T_{2g}$ state is split by the trigonal crystal field in a doublet ground state and an excited singlet. It is assumed that this effect is due to the predominant trigonal crystal field in relation with the sulfur position, instead of the next cation effect as observed in oxides spinels [3,4]. The angle between $V_{zz}\|\langle 111\rangle$ and the spin direction is $\theta = 55°$. It indicates, that the $\langle 100\rangle$ is an easy axis of magnetization.

(c) $0.8 \leq x \leq 1.3$

In this relatively broad range of composition the compounds change from the ferrimagnetic to the AF state. Special studies have been made of the composition $FeInCrS_4$ [5] ($T_c = 75$ K). The resulting moment (0.8 μ_B) is in reasonable agreement with the neutron diffraction data* ($\bar{S}_A = 0.8$ μ_B, $\bar{S}_B = 0.4$ μ_B). The $M(H)$ curves show the presence of magnetocrystalline anisotropy. Fe^{2+} in A and B sites give an easy direc-

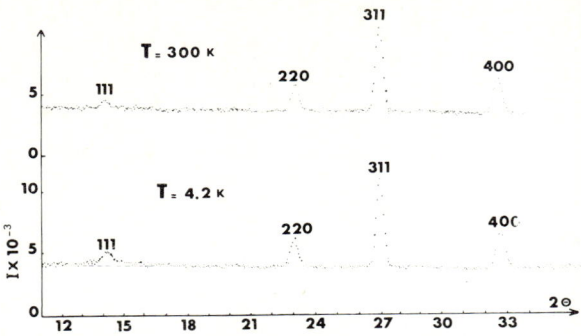

Fig. 3. Neutron diffraction spectra for $FeInCrS_4$ at temperatures of 300 and 4.2 K.

tion along $\langle 100\rangle$. Neutron diffraction data (fig. 3) at 4.2 K indicate typical short-range order. In particular there is an increase of the background near the $\langle 111\rangle$ peak; the mean values of the moments at A and B sites are low, although T_c is well above 4.2 K.

3. Discussion

The high anisotropy of Fe^{2+} in A and B sites ($K_1 > 0$) stabilizes the moment along the $\langle 100\rangle$ direction and maintains a collinear structure. It must be pointed out that even at low temperature for $T \ll T_c$ there is no long-range order, probably due to the competition between exchange interactions of the same magnitude. On the other hand for a definite B atom it is rather difficult to have his n.n. in antiparallel position, the order diminishes going to the n.n.n. and so on. This situation is to compare with a short-range order as can be observed near T_c [3].

References

[1] W. Schlein and A. Wold, J. Solid State Chem. 4 (1972) 286.
[2] L. Brossard, L. Goldstein and M. Guittard, J. Phys., to be published.
[3] M. Eibschutz, Ganiel and S. Shtrikman, Phys. Rev. 151 (1966) 245.
[4] J.C. Slonczewski, JAP 32 (1961) 253S.
[5] Y. Mimura, M. Shimada and M. Koizumi, Solid State Commun. 15 (1974) 1035.

---

* Neutron diffraction was done at the Leon Brillouin Institute, Saclay.

MAGNETIC PROPERTIES OF THE $Cr_2Se_{3-x}Te_x$ ($x = 0 \sim 3$) AND $Cr_3Se_{4-x}Te_x$ ($x = 0 \sim 4$) SYSTEMS

M. YUZURI and K. SEGI

Department of Physics, Faculty of Engineering, Yokohama National University, Yokohama 240, Japan

The magnetic properties of $Cr_2Se_{3-x}Te_x$ ($x = 0 \sim 3$) and that the transition from antiferromagnetism to ferromagnetism took place at about $x = 2.5$ in the $Cr_2Se_{3-x}Te_x$ system and at $x = 0.8$ in the $Cr_3Se_{4-x}Te_x$ system and that the lattice parameters discontinuously changed at about the magnetic transition points.

1. Introduction

It has been known that Cr_2Se_3 has a NiAs structure with an ordered arrangement of vacant chromium sites and is antiferromagnetic with a Néel temperature of about 43 K [1] and that Cr_2Te_3 has also the same structure as Cr_2Se_3 and is ferromagnetic with a Curie temperature of about 180 K [2]. If the $Cr_2Se_{3-x}Te_x$ ($x = 0 \sim 3$) system forms a solid solution, we can expect the transition from antiferromagnetism to ferromagnetism at some value of x. At the same time we can expect a discontinuity of lattice constants at the same value of x mentioned above by Kittel's [3] exchange inversion theory. Similar results are also expected in the $Cr_3Se_{4-x}Te_x$ system with monoclinic structure. It is known that Cr_3Se_4 [1] is antiferromagnetic with a Néel temperature of about 82 K and Cr_3Te_4 is ferromagnetic with a Curie temperature of about 320 K [2]. Hence, in the present studies magnetic and crystallographic analyses of these compounds were undertaken. In the following pages the results of the measurement and a discussion of the nature of the magnetic transition are given.

2. Experimental results and discussion

The specimens used were prepared by the usual ceramic method. The magnetic properties of the specimens were measured by means of an automatically recording magnetic balance. The results are shown in figs. 1 and 2. As seen from the figures, it was found that the transition from antiferromagnetism to ferromagnetism took place at about $x = 2.5$ in a $Cr_2Se_{3-x}Te_x$ system and at about $x = 0.8$ in a $Cr_3Se_{4-x}Te_x$ system. The temperature dependence of the inverse susceptibility of the $Cr_2Se_{3-x}Te_x$ system is almost parallel with each other at all tellurium concentrations. From the Curie constant ob-

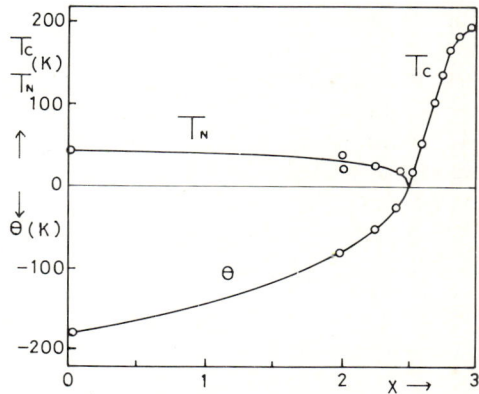

Fig. 1. Curie, Néel and paramagnetic Curie temperature, T_c, T_N and θ versus content x in $Cr_2Se_{3-x}Te_x$.

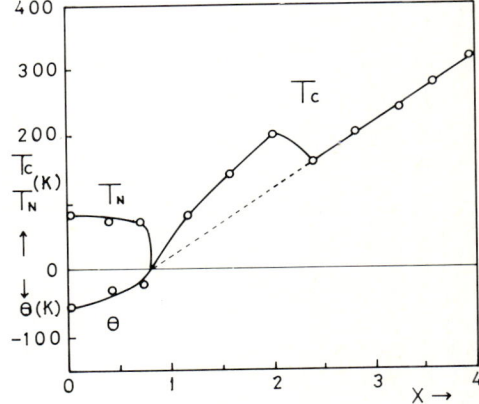

Fig. 2. Curie, Néel and paramagnetic Curie temperature versus content x in $Cr_3Se_{4-x}Te_x$.

tained from the slope of $1/\chi - T$ curve, the effective magneton number per chromium ion was calculated to be about 3.8. Assuming a Landé factor $g = 2$, the mean spin quantum number S became about 3/2 for each value of x. Therefore, spins are considered to be canted and the canting angle becomes smaller with increasing content x and spins reach a collinear arrangement at Cr_2Te_3 in the ferromagnetic range.

The X-ray data, using a diffractometer at room temperature, are shown in fig. 3. The parameters of the crystal lattice of $Cr_2Se_{3-x}Te_x$ system, a and c, increase monotonously with increasing content x, but they show a discontinuity at about $x = 2.5$ corresponding to the magnetic transition point. It is considered that the crystallographic discontinuity is related to Kittel's exchange inversion theory. However, the ratio c/a decreases monotonously from $x = 0$ to $x = 3$.

As shown in fig. 2, the $Cr_3Se_{4-x}Te_x$ system shows a similar behavior as the $Cr_2Se_{3-x}Te_x$ system. In the ferromagnetic range, the canting of spin is also considered to be the same as in the $Cr_2Se_{3-x}Te_x$ system. However, the change of Curie temperature with concentration showed a discontinuity at $x = 2.0$.

The X-ray study in this system showed that the concentration dependence of the crystal lattice parameters changed almost monotonously; only a slight change was shown at $x = 0.8$ and 2.0. Cr_3Te_4 has originally an anomaly at 80 K in its thermomagnetic curve. This anomaly has been considered to be the intrinsic antiferromagnetic Néel temperature. This phenomenon maybe affects the magnetic and crystallographic properties of the $Cr_3Se_{4-x}Te_x$ system. Therefore, the whole $Cr_3Se_{4-x}Te_x$ system is considered to become somewhat more complex compared with the $Cr_2Se_{3-x}Te_x$ system and to be more interesting.

The magnetic properties of the $Cr_2Se_{3-x}Te_x$ and $Cr_3Se_{4-x}Te_x$ systems are fundamentally explained by Hirone and Adachi's [4] theory related to the magnetic properties of NiAs structure, and the discontinuity of the crystal lattice parameters would be explained by Kittel's exchange inversion theory. More precise calculations will be done by application of exchange interactions, taking into account the vacancies of chromium ions as in the studies of Hashimoto et al. [2] related to Cr_2Te_3 and Dwight et al. [5] related to Cr_5S_6.

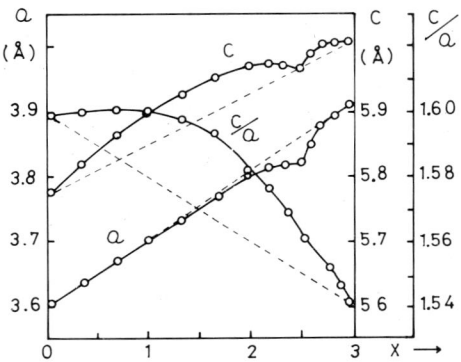

Fig. 3. Lattice parameters a, c and c/a versus content x in $Cr_2Se_{3-x}Te_x$.

References

[1] M. Yuzuri, J. Phys. Soc. Jap. 35 (1973) 1252.
[2] T. Hashimoto, K. Hoya, M. Yamaguchi and I. Ichitsubo, J. Phys. Soc. Jap. 31 (1971) 679 and J. Phys. Soc. Jap. 32 (1972) 635.
[3] C. Kittel, Phys. Rev. 120 (1960) 335.
[4] T. Hirone and K. Adachi, J. Phys. Soc. Jap. 12 (1957) 156.
[5] K. Dwight, N. Menyuk and J. A. Kafalas, Phys. Rev. B2 (1970) 3630.

MAGNETIC STRUCTURES OF THE SPINELS COMPOUNDS $MnCr_{2-x}V_xS_4$

L. GOLDSTEIN, P. GIBART, M. MEJAI

Laboratoire de Magnétisme, CNRS, 92190 Bellevue, France

and

M. PERRIN

Institut Léon Brillouin, Centre d'Etudes Nucléaires Saclay, BP 2, 91120 Gif-sur-Yvette, France

The compounds $MnCr_{2-x}V_xS_4$ exist in the spinel phase up to $x = 0.6$. Vanadium substitutes Cr in B sites. V^{3+} couples antiferromagnetically to Cn^{3+}. The transition temperature from the Yafet–Kittel to the Néel ferrimagnet decreases with x.

1. Introduction

The ferrimagnetic thiospinel $MnCr_2S_4$ ($T_c = 74$ K) exhibits a Yafet–Kittel (Y.K.)–Néel transition at 5.5 K, the canting occurring on A-site (Mn atoms) [1, 2]. Nauciel-Bloch et al. [3, 4] showed that reliable agreement for the temperature dependence of the sublattice magnetization can be obtained when exchange striction terms are included. The compounds $MnCr_{2-x}V_xS_4$ were studied in order to understand the effect of V substitution on the magnetic properties and on the different magnetic transitions.

2. Experimental

$MnCr_2S_4$ exhibits a transition from spinel to defect NiAs at $T_K = 1800$ K. The substitution of V atoms results in lowering the transition temperature ($T_K = 1200$ K for $x = 0.4$). The spinel phase exists up to $x = 0.6$. The slight increase of the lattice parameter a with x does not show a transition from localized to delocalized V states as observed in $CuCr_{2-x}V_xS_4$ [5].

3. Ionic configuration

The compound $MnCr_{1.8}V_2S_4$ was studied in detail using neutron diffraction. The room temperature spectrum ($T \gg T_c$) was best fitted with a reliability factor $R = 3.6\%$ assuming a small degree of inversion: 0.04 V are in A site and 0.06 Mn in B site. Mn in B site occurs as Mn^{3+}. This explains the magnetocrystalline anisotropy ($K_1 > 0$) deduced from the field dependence of the magnetic peaks. The existence of two kinds of Mn has been confirmed by NMR.*

At 4.2 K, where the compound is a Néel ferrimagnet, it is found that $S_A = 4.25\ \mu_B$, $S_B = 2.45\ \mu_B$, values which are in good agreement with the magnetization data, $M_s = 0.6\ \mu_B$. A reliable formula consistent with all these features must account for a small amount of A-site vacancies. The formula could be

$$Mn^{II}_{0.93}\ V^{III}_{0.04}\ \square_{0.03}\ Cr^{III}_{1.79}\ V^{III}_{0.16}\ Mn^{III}_{0.06}\ S_4.$$

4. High field measurements and magnetic structures

The magnetization of $MnCr_{2-x}V_xS_4$ was measured between 1.5 and T_c and up to 150 kG using SNCI† high fields facilities. Fig. 1 shows

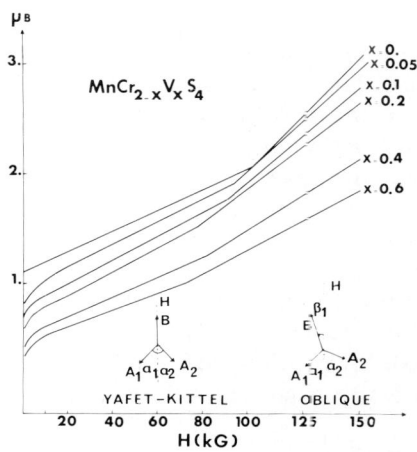

Fig. 1. Magnetization vs. H at 4.2 K for $MnCr_{2-x}V_xS_4$.

* The NMR data were provided by Dr. Le Dang Khoi, Institut d'Electronique, Orsay.
† SNCI: Service National des Champs Intenses, Grenoble.

the magnetization at 4.2 K vs. H for all compositions. The main experimental features are the following:

(1) $M_s(0)$ decreases with vanadium substitution.

(2) The magnetization varies linearly with H, and there is a change in the slope at a critical field $H_c = 75$ kG for $x = 0.6$ and 110 kG for $x = 0$.

(3) The zero field extrapolation of the high field part of the curves is close to the origin; this is true for all compositions.

Plumier et al. [6] showed that $MnCr_2S_4$ exhibits at least three magnetic structures: Néel ($T > 5.5$ K, $H = 0$), Yafet–Kittel ($T < 5.5$ K, $H = 0$), oblique ($T < 20$ K, $H > 110$ kG). He wrote the energy of the system as follows:

$$E = -4J_{AA}S_{A1}S_{A2}\cos(\alpha_1 + \alpha_2)$$
$$- 12J_{AB}S_B|S_{A1}\cos(\beta + \alpha_1) + S_{A2}\cos(\alpha_2 - \beta)|$$
$$+ 4p(S_{A1}S_{A2})^2\cos^2(\alpha_1 + \alpha_2) - 12J_{BB}S_B^2$$
$$- H(4S_B\cos\beta + S_{A1}\cos\alpha_1 + S_{A2}\cos\alpha_2), \quad (1)$$

where p is the parameter of the biquadratic exchange, the other terms are defined in fig. 1.

From eq. (1) Plumier deduced that in the oblique region $H + 3J_{Cr-Mn}M = 0$. Assuming that V atoms are in B sites, the slope b of the magnetization vs. H is calculated in a molecular field approximation, when x is not too large

$$b = -\frac{1}{3J_{Cr-Mn}}\left(\frac{(2-x)S_{Cr} + xS_v}{(2-x)S_{Cr} + xS_v[(J_{v-Mn})/(J_{Cr-Mn})]}\right). \quad (2)$$

Eq. (2) is only valid at low temperature, J_{V-Mn} is the interaction between A site Mn and B site V, and is assumed to be weak. The slope of the curve is directly related to the orientation of V moments with respect to Cr moments. The observed decrease of the slope as a function of V concentration indicated that V moment couples antiferromagnetically to Cr moment.

The rapid decrease of $M_s(0)$ is consistent with antiparallel V–Cr coupling for $x \leq 0.1$. For $x > 0.1$, small inversion of Mn^{III} in B site results in slightly decreasing the total moment.

5. The Yafet–Kittel–Néel transition (T_{Y-K})

In $MnCr_2S_4$, in neutron diffraction, the (200) peak appear for $T < 5.5$ K [7] and $M(H)$ is constant up to 5.5 K. The substitution of V atoms results in dropping T_{Y-K}. In $MnCr_{1.8}V_{0.2}S_4$, at 1.5 K the (200) peak does not appear. On the other hand, $M(H)$ does not vary with T up to 2.2 K. As pointed out by Plumier [6] the Y–K structure can occur at temperatures higher than T_{Y-K} under applied magnetic field.

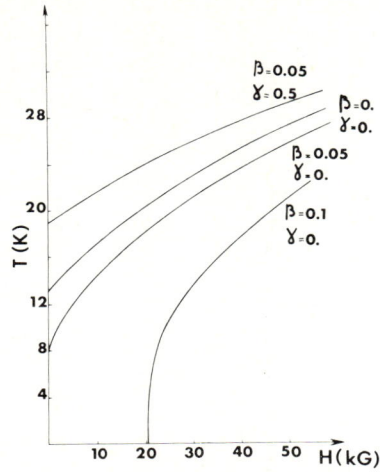

Fig. 2. Theoretical Yafet–Kittel–Néel transition vs. field in $MnCr_2S_4$ for different values of the parameter β and γ. The Néel phase lies above the curves.

In order to derive an approximate magnetic phase diagram, we calculated T_{Y-K} as a function of H and T, starting from the equations derived by Nauciel-Bloch [3] and Castets [4]:

$$E = +12J_{AB}\left(1 + \frac{\beta\Delta}{J_{AB}}\right)m_A m_B \cos\alpha$$
$$- 4J_{AA}m_A^2\cos 2\alpha - 3J_{BB}\left(1 + \frac{\gamma\Delta}{J_{BB}}\right)m_B^2 + \tfrac{1}{2}C\Delta^2, \quad (3)$$

where $\Delta = 3(\Delta a/a)$ is the isotropic volume change, and β, γ adjustables coefficients, $m_A = -\langle S_A\rangle$, $m_B = 2\langle S_B\rangle$.

We introduced the magnetic field H, this adds a term $-H(4m_B - 2m_A \cos\alpha)$ to (3). The transition T_{Y-K} is assumed to be second order, T_{Y-K} was computed from the minimum of the free energy. T_{Y-K} was calculated as a function of H and T. It appears from these calculations that T_{Y-K} increases with H as experimentally observed. In fig. 2 the $T_{Y-K}(H)$ curves calculated for different values of β and γ are plotted. It can be seen from this figure that a MFA can account for the magnetic phase diagram when temperature dependence of the exchange integral is taken into account. The calculated transition temperature is very sensitive to the value of β. The reduction of T_{Y-K} with vanadium substitution could be due to a weak increase of β.

In a further extension, the whole magnetic phase diagram will be established.

References

[1] F.L. Lotgering, J. Phys. Chem. Solids 29 (1968) 2193.
[2] J. Denis, Y. Allain and R. Plumier, J. Appl. Phys. 41 (1970) 1091.
[3] M. Nauciel-Bloch, A. Castets and R. Plumier, Phys. Lett. 39A (1970) 311.
[4] A. Castets, Thesis 3è cycle, Paris (1973).
[5] M. Robbins, A. Menth, M.A. Miksovsky and R.C. Sherwood, J. Phys. Chem. Solids 31 (1970) 423.
[6] R. Plumier, R. Conte, J. Denis, M. Nauciel-Bloch, Journal de Physique 31 (1971) C1-55.
[7] R. Plumier and M. Sougi, C.R. Acad. Sci. 268 (1969) 1549.

MAGNETIC STRUCTURES OF CrUS$_3$ RELATED COMPOUNDS SHOWING DELOCALIZED MOMENTS

P. WOLFERS

Institut Laue-Langevin and Laboratoire des Rayons X, CNRS, 38042 Grenoble Cedex, France

and

G. FILLION

Laboratoire de Magnétisms, CNRS, 38042 Grenoble Cedex, France

The magnetic structures of VUS$_3$ and CrUSe$_3$ are determined via powder neutron diffraction techniques. Like a previous discovery in CrUS$_3$, a significant spin density located more than 2 Å away from any atom is found. It can be attributed to an extended $6d_{z^2}$ orbital of uranium.

1. Crystal structure

The compounds VUS$_3$ and CrUSe$_3$ are isomorphous to the prototype material CrUS$_3$ which crystallizes [1] in the orthorhombic space group Pnam-D_{2h}^{16} with Cr in octahedral site 4a, U and S$_I$ in 4c (symmetry m) and S$_{II}$ in 8d general position. Uranium is located in a somewhat distorted right trigonal prism of six sulfurs with two of the rectangular faces capped with additional more distant sulfurs neighbors (fig. 1).

2. Magnetic structures and delocalized moments

For both compounds, refinements with only localized moments do not lead to a reasonable fit of the data. Therefore we used the same simple delocalized model as for CrUS$_3$ [2] with four additional magnetic sites of the same form factor as the uranium atom. Then, the same refinement techniques give the components of both localized and additional moments with good veracity factors. The best choice for the additional positions is always found to be along the US$_6$ prism axis at intervals of about 1 Å. The results are summarized in table I and the VUS$_3$ structure is shown in fig. 2. The relation with the uranium neighboring appears in fig. 1 for the CrUS$_3$ case.

3. Magnetic properties and localized moments

All the three compounds have nearly the same magnetic behavior, that is: (1) a weak ferromagnetic component below a Curie point, (2) a nearly temperature-independant superimposed constant susceptibility at high fields, and (3) the presence of narrow domain walls associated with high propagation threshold fields and magnetization after effect [2]. This last feature indicates a large magnetocrystalline anisotropy energy at least of same order as the exchange interactions [3]. A straightforward calculation of the crystal field effect on the levels of 5f^2 and 5f^3 configurations suggests that this anisotropy is due to crystal field effects on preponderant 5f-character orbitals of uranium and large spin–orbit coupling. In fact, though the extension of the radial part of the 5f-orbitals enhances the fourth and sixth order terms, the low symmetry (m) of the uranium site leads to preponderant second order terms and removes all non-Kramers degeneracies. This yields the uranium moments to be sharply bound to some easy direction. However, the calculated angles do not fit the experimental ones. This is pri-

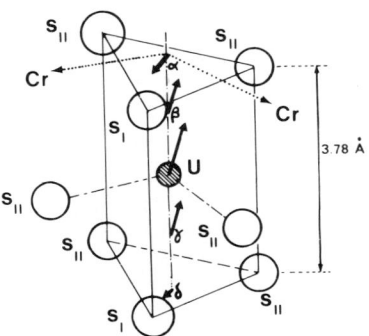

Fig. 1. Neighboring of uranium showing the US$_6$ prism, the two capping sulfurs and the four additional magnetic sites α, β, γ, δ of the delocalized model of CrUS$_3$.

Table I
Magnetic structures of VUS_3, $CrUSe_3$ and $CrUS_3$ for comparison. (magnetic modes are indicated in bracket)

			VUS_3	$CrUSe_3$	$CrUS_3$
	Curie temp. (K)		230	105	110
	Operating temp. (K)		100	7	4.2
	Schubnikov group		Pn'am'	Pna'2'$_1$	Pn'am'
Components of magnetic sites	V or Cr	x	−2.1 (C)	−2.1 (F)	−2.4 (C)
		y		−1.7 (C)	
		z		0.5 (A)	
	α	x			−0.1 (C)
		y	0.3 (F)	0.4 (C)	−0.5 (F)
		z		−0.5 (A)	
	β	x	0.3 (C)		0.5 (C)
		y		0.1 (C)	0.4 (F)
		z			
	U	x	2.1 (C)	2.3 (F)	0.9 (C)
		y	0.5 (F)	0.8 (C)	0.7 (F)
		z		−1.4 (A)	
	γ	x	0.1 (C)	−0.1 (F)	0.5 (C)
		y			0.4 (F)
		z			
	δ	x	−0.1 (C)	0.2 (F)	0.2 (C)
		y		0.2 (C)	
		z		−0.3 (A)	

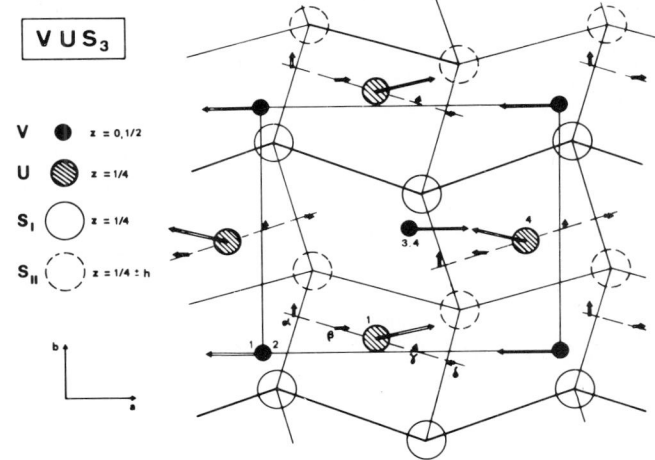

Fig. 2. Projection of the magnetic structure of VUS_3 on the plane $z = \frac{1}{4}$.

marily due to the wrong evaluation of crystal field parameters by the point charge model used but the arguments for a large expected anisotropy will remain.

4. Bonding and delocalized moments

On the other hand, we made a semi-empirical study of chemical bonding in $CrUS_3$ [4] from the combined approach of Pauling's metallic radii and the bidirectional orbital approximation (BOA) [5]. The first method permits an estimate of bond orders from the observed interatomic distances and the second one gives a distribution of orbitals employed in bonding. The most significant results of this work are the introduction of strong f-character (20–30%) into the BOA approach for the bonding of uranium in

$CrUS_3$ and the determination of the available orbitals for the ≈ 3 unpaired electrons of uranium. The following situation is suggested in reasonable agreement with neutron diffraction results: one electron occupies the $5fy(3x^2 - y^2)$ orbital well localized at the U site, the second occupies a combination of the $5fx(5z^2 - r^2)$ and $5fy(5z^2 - r^2)$ orbitals somewhat spacially separated from the first at β and γ sites and the third occupies an extended $6d_{z^2}$ orbital at greater distance (~ 2 Å) distributed between the α and δ sites. Some similar distribution would be expected for the other compounds of the same structure.

We are grateful to Prof. F.L. Carter for very helpful discussions and to the staff of the Laue-Langevin Institute for technical support of the experiments.

References

[1] H. Noel, J. Padiou and J. Prigent, C.R. Acad. Sci. Paris 280 (1975) C-123.
[2] P. Wolfers, G. Fillion, M. Bacmann and H. Noel, J. de Physique 37 (1976) 233.
[3] B. Barbara, C. Becle, R. Lemaire and D. Paccard, J. de Physique Colloq. 32 (1971) C1-299.
[4] F.L. Carter, P. Wolfers and G. Fillion, Int. R.E. Research Conf., Vail, Cororado (1976) (unpublished), (to be published in J. Phys. C: Solid State Phys.).
[5] F.L. Carter, in: Density of States, L. Bennet (Ed.) (NBS Special Publication, 323, 1970) p. 358.

MÖSSBAUER INVESTIGATION OF THE MAGNETIC BEHAVIOUR OF HEXAHALIDES OF Ir(IV) AT VERY LOW TEMPERATURES*

W. POTZEL, F.E. WAGNER, W. GIERISCH, E. GEBAUER and G.M. KALVIUS

Physik Department, Technische Universität München, D-8046 Garching, W. Germany

The low temperature magnetic behaviour of K_2IrF_6, $Li_2IrCl_6 \cdot xH_2O$, Rb_2IrCl_6 and Cs_2IrCl_6 has been studied in a $^3He/^4He$ dilution refrigerator with the 73 keV resonance of ^{193}Ir. Longitudinally magnetic fields up to 56 kOe could be applied. Around 100 mK all compounds show hyperfine spectra indicating the existence of a magnetically ordered state. The hyperfine fields at the Ir nuclei below 100 mK are similar to those found in other hexahalides of Ir(IV) ($5d^5$). They are interpreted as the sum of orbital and contact contributions of opposite sign. Cs_2IrCl_6 and K_2IrF_6 show an increase of the magnetic hyperfine splitting with external field. In K_2IrF_6 this increase amounts to more than 50% of the zero field hyperfine splitting.

1. Introduction

The magnetic behaviour of octahedral complexes of Ir(IV) has been studied by various authors [1-9]. Tetravalent iridium has a $5d^5$ electron configuration. In the presence of a crystalline electric field the odd number of electrons leads to Kramers-degenerate electronic levels. At higher temperatures Ir(IV) hexahalides are paramagnetic [10, 11], at temperatures below 4.2 K antiferromagnetic transitions have been observed for K_2IrCl_6, $(NH_4)_2IrCl_6$ and several others [1-9]. The relative magnitudes of the contact and orbital contributions to the hyperfine fields at the Ir site have been estimated for several Ir-complexes from the hyperfine anomaly of ^{193}Ir [1, 2, 12].

2. Experimental

The Mössbauer experiments were performed with the 73 keV γ-rays of ^{193}Ir. The ^{193}Os ($T_{1/2}$ = 30 h) sources were obtained by neutron activation of either ^{192}Os metal or an $^{192}Os_{0.6}V_{99.4}$ alloy. The Os metal sources exhibit a small electric quadrupole splitting which has to be taken into account in the evaluation of the data. The OsV sources give a narrower linewidth [$W = (0.72 \pm 0.02)$ mm/s against Ir metal extrapolated to zero absorber thickness]. The experiments were performed in a $^3He/^4He$ dilution refrigerator [13] with the polycrystalline absorbers (~100 mg/cm^2 of natural Ir) mounted inside the mixing chamber. A longitudinal field up to 56 kOe could be applied. It acts on source and absorber simultaneously.

* Supported by the Bundesministerium für Forschung und Technologie.

3. Results and discussion

For all compounds hyperfine patterns typical for a magnetically ordered state were observed at temperatures below 0.5 K. Fig. 1 shows spectra for Rb_2IrCl_6 and K_2IrF_6, taken at temperatures around 100 mK without external

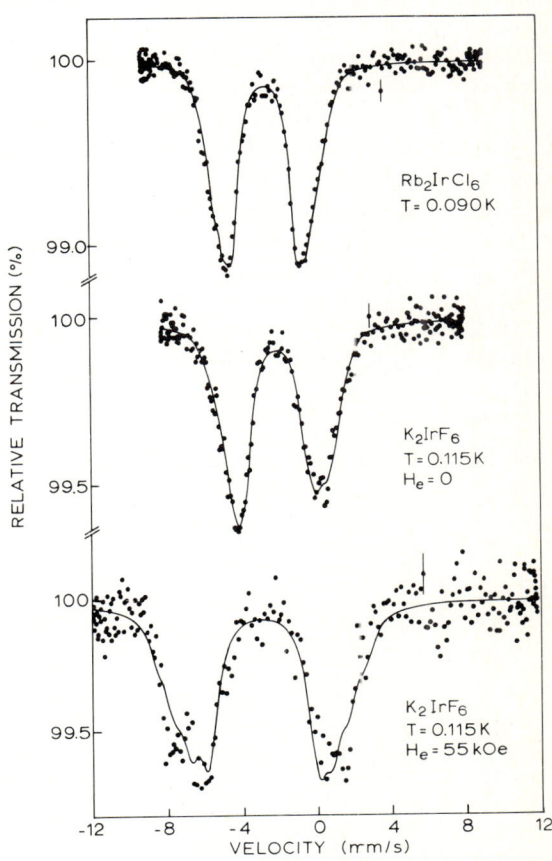

Fig. 1. Mössbauer spectra of Rb_2IrCl_6 and K_2IrF_6 in zero external field and of K_2IrF_6 in an applied field of 55 kOe. The zero field spectrum of K_2IrF_6 was taken with an Os metal source, the other two with a source of Os in V.

magnetic field. The hyperfine fields were obtained by fitting a superposition of eight lorentzian lines to the data as has been described in ref. 1. The hyperfine fields obtained at the lowest temperatures are given in table I. These values are close to those found earlier [1] in other hexahalides of Ir(IV). They can be considered as the sum of an orbital contribution of approximately +600 kOe and a contact contribution of approximately −200 kOe [2].

Table I also summarizes the information obtained on the magnetic transition temperatures T_M of the studied compounds. The value for K_2IrF_6 is the lowest found so far for any iridium hexahalide.

Fig. 2 shows the magnetic hyperfine field at the iridium nuclei as a function of reduced temperature for K_2IrF_6 and Rb_2IrCl_6. In both cases we find deviations of the hyperfine fields from the spin $\frac{1}{2}$ Brillouin function predicted by molecular field theory [14].

The hyperfine field in Rb_2IrCl_6 remains nearly constant as the temperature is raised to $0.9 \times T_M$ and then rapidly drops to zero. In the region $0.9 < T/T_M < 1$ an additional single line appears in the Mössbauer spectra, whose relative intensity increases fast when the ordering temperature is approached. A similar behaviour has previously been observed in other Ir-hexahalides [1] and indicates a first-order phase transition.

In K_2IrF_6 one finds an electric quadrupole splitting of $\Delta E_Q = 0.55(4)$ mm/s at 4.2 K. This causes additional complexities in the hyperfine spectra below T_M since the angle between the electric field gradient and the magnetic hyperfine field is not known. It is possible that this angle changes upon cooling the compound and the deviations from the Brillouin function around $T/T_M \approx 0.7$ may have this origin.

Polycrystalline absorbers of Cs_2IrCl_6 and K_2IrF_6 have also been studied in external fields up to 56 kOe (see fig. 1). In both cases an increase of the magnetic hyperfine splitting with the external field was observed, indicating that the internal field is positive [2]. In K_2IrF_6 an increase of more than 50% was seen, which cannot be explained by a simple superposition of the hyperfine field of 446 kOe and the applied field of 55 kOe. It indicates that the spin structure of K_2IrF_6 changes drastically when an external field is applied, but the data on this compound are still too scarce for an attempt of a detailed interpretation.

We thank Prof. W. Preetz and Dr. K. Rössler for making K_2IrF_6 available for us, and Dr. U. Wagner for preparing the samples of Rb_2IrCl_6, Cs_2IrCl_6 and $Li_2IrCl_6 \cdot xH_2O$.

Table I

Compound	H(T) (kOe)	T (K)	T_M (K)
K_2IrF_6	446 ± 8	0.068	0.460 ± 0.010
$Li_2IrCl_6 \cdot xH_2O$	337 ± 8	0.081	≈ 1.2
Rb_2IrCl_6	418 ± 8	0.090	1.85 ± 0.10
Cs_2IrCl_6	452 ± 8	0.088	$0.088 < T_M < 1.85$

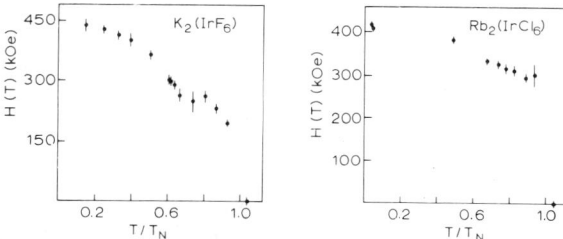

Fig. 2. Internal magnetic fields at the iridium nuclei as a function of reduced temperature for K_2IrF_6 and Rb_2IrCl_6.

References

[1] F.E. Wagner and Ursel Zahn, Z. Phys. 233 (1970) 1.
[2] F.E. Wagner and W. Potzel, in: Hyperfine Interactions in Excited Nuclei, Vol. 2, G. Goldring and R. Kalish (Eds.) (Gordon and Breach, New York, 1971) p. 681.
[3] A.H. Cooke, R. Lazenby, F.R. McKim, J. Owen and W.P. Wolf, Proc. Roy. Soc. A250 (1959) 97.
[4] C.A. Bailey and P.L. Smith, Phys. Rev. 114 (1959) 1010.
[5] A.J. Lindop, J. Phys. C3 (1970) 1984.
[6] G.L. Baker and R.L. Armstrong, Can. J. Phys. 48 (1970) 1649.
[7] R.G. Wheeler, F.M. Reames and E.J. Wachtel, J. Appl. Phys. 39 (1968) 915.
[8] V.J. Minkiewicz, G. Shirane, B.C. Frazer, R.G. Wheeler and P.B. Dorain, J. Phys. Chem. Solids 29 (1968) 881.
[9] M.T. Hutchings and C.G. Windsor, Proc. Phys. Soc. London 91 (1967) 928.
[10] V. Norman and J.C. Morrow, J. Chem. Phys. 31 (1959) 455.
[11] B.N. Figgis, J. Lewis and F.E. Mabbs, J. Chem. Soc. (1959) 3138.
[12] G.J. Perlow, W. Henning, D. Olson and G.L. Goodman, Phys. Rev. Lett. 23 (1969) 680.
[13] G.J. Enholin, T.E. Katila, O.V. Lounasmaa, and P. Reivari, Cryogenics 8 (1968) 136.
G.M. Kalvius, T.E. Katila and O.V. Lounasmaa, in: Mössbauer Effect Methodology, Vol. 5, I.J. Gruverman (ed.) (1970) p. 231.
[14] P.W. Anderson, Phys. Rev. 79 (1950) 705.

MAGNETIC STRUCTURE OF NiS$_2$ (PYRITE TYPE)

T. MIYADAI, K. KIKUCHI

Department of Physics, Faculty of Science, Hokkaido University, Sapporo, Japan

and

Y. ITO

The Institute for Solid State Physics, University of Tokyo, Tokyo, Japan

The susceptibility (χ_w) in the expression $M = M_0 + \chi_w H$ and the magnetic field effect on the neutron diffraction intensity were measured on single crystals of NiS$_2$ below $T_c(=30\,\text{K})$. χ_w was found to be almost isotropic and temperature independent ($\chi_w = 5 \times 10^{-6}$ emu/g). Any field effect on the intensity of antiferromagnetic reflections could not be detected. An improved model of magnetic structure is proposed.

1. Introduction

NiS$_2$ shows two types (M1 and M2) of antiferromagnetic (AF) reflections of neutron diffraction (ND) below 30 K; M1 corresponds to ordering of the first kind (f.c.c. lattice) and M2 to the ordering of the second kind [1, 2]. Furthermore, below 30 K ($= T_c$) a weak-ferromagnetic (WF) moment appears. The magnetic structure at low temperatures, however, has not yet been well established. Previously, we proposed a model (model-1) [2] for the magnetic structure which explains the observed intensities of ND. After that, Nishihara et al. [3], from the Mössbauer study on ^{57}Fe in Ni$_{0.995}$Fe$_{0.005}$S$_2$, suggested that the arrangement of the spin component contributing to M1 (abbreviated as M1-component) is non-collinear, whereas our model-1 is collinear with respect to each of M1- and M2- components. Czjzek et al. [4], from the Mössbauer study on ^{61}Ni in NiS$_2$, suggested that there are two kinds of Ni sites with respect to the direction of the spin. This is compatible with model-1.

2. Experiments

Magnetization curves along [100] and [110] directions were measured from 4.2 K up to T_c on a single-crystal which had been cooled in a field of 15 kOe parallel to [100] or [110] direction, respectively. The magnetization curve is approximately expressed by a linear relation $M = M_0 + \chi_w H$. The values at 4.2 K are $M_0 = 0.61$ emu/g for [100] direction and 0.47 emu/g for [110]. $\chi_w \cong 5 \times 10^{-6}$ emu/g, almost the same for both directions. The anisotropy of M_0 was, in our previous work [5], explained as being due to the change in WF domain distribution; after magnetic annealing along [100], the sample is in a single ([100]) domain state in which all WF moments are parallel to the [100] axis, while after magnetic annealing along [110], the sample is in "two" ([100] and [010]) domain states even in a field of 20 kOe, leading to a reduction in the net moment by a factor of $1/\sqrt{2}$. Taking account of the domain distribution, the fact that the same value was obtained for χ_w along both directions, indicates that χ_w is isotropic within the experimental errors. Further, χ_w was found to be isotropic in the whole temperature region and also to be temperature independent except near T_c.

Next, we examined the magnetic field effect on the intensity of the AF reflections of ND. But we could not find any difference between intensities without and with a magnetic field of about 10 kOe. The examined reflections are (200)M1 with a field along [001] or [01$\bar{1}$] direction, and (111)M2 with a field along [01$\bar{1}$] direction. (The index is referred to twice the chemical unit cell.) From model-1, including the mechanism of weak ferromagnetism, it is expected for the above field directions that (111)M2 should remain unchanged, but that (200)M1 should vanish. So, the present observation is contradictory with Model-1.

3. Magnetic structure (Model-3)

Here, we propose a new model (model-3) which explains the ND intensities, Mössbauer study, the field effect on ND, the behavior of χ_w

and the displacement of WF 90°-wall [5]. Model-3 is essentially non-collinear with respect to each of M1- and M2-components. For the M1-associated structure, three degenerate modes [6] are combined linearly. This structure is a modified one of $MnTe_2$ type [7]. For the M2-associated structure, four degenerate modes [6] are linearly combined. This structure consists of four interpenetrating (independent) AF simple cubic lattices. The whole magnetic structure is constructed from the above M1- and M2-associated structures, and it consists of eight sublattices. Strictly speaking, the whole structure has an orthorhombic symmetry (the magnetic unit cell is twice the chemical cell), so that there may exist six kinds of AF domains. In the calculation of the intensity of ND, a random distribution of these domains were assumed.

The direction of each spin was determined so as to obtain the best fit to the observed intensities under two constraints; (1) M1-component makes a unique angle ($\theta = 21°$) with the electric field gradient axis ($= \langle 111 \rangle$) [3] and (2) it is perpendicular to M2-component. An example of z_a-domain is shown in Fig. 1. Model-3 is compatible with both the data of Nishihara et al. [3] and of Czjzek et al. [4].

The WF moment can appear if we assume a slight canting of each spin towards, say, the [100] direction. The direction of canting may be either of the $\langle 100 \rangle$ directions, depending on circumstances such as the applied field-direction. This mechanism seems to be compatible with the observation described above, although its theoretical basis is not clear. By this model, 90°-wall displacement [5] can take place through a switching of the direction of canting, leaving the AF domains unchanged. So, no field effect on the intensity of ND is expected. The non-

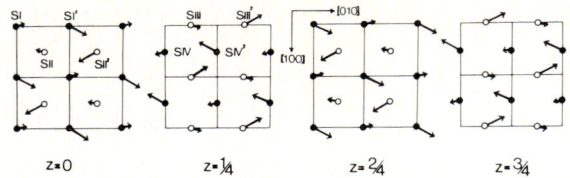

Fig. 1. Projection of model-3 structure (Z_a-domain) onto (001) plane. The solid or open circle denotes the positive or negative sign of Z-component of the spin, respectively. Si and Si' have the same M1-component but different M2-components. The direction cosines of each spin are $S_I = (\bar{a}bc)$, $S_{I'} = (def)$, $S_{II} = (\bar{a}\bar{b}\bar{c})$, $S_{II'} = (d\bar{e}\bar{f})$, $S_{III} = (ab\bar{c})$, $S_{III'} = (\bar{d}ef)$, $S_{IV} = (a\bar{b}c)$, $S_{IV'} = (\bar{d}\bar{e}f)$ with $a = 0.04$, $b = 0.35$, $c = 0.94$, $d = 0.47$, $e = 0.87$ and $f = 0.17$. The canting for WF moment is not shown.

collinear AF structure will give rise to the isotropic and temperature independent susceptibility (χ_w).

The authors thank Dr. J. Akimitsu for ND measurement and helpful discussions, and K. Takizawa, T. Fukui and K. Uchino for providing them with single crystals. Two of the authors (T.M. and K.K.). also thank Prof. S. Miyahara for valuable discussions.

References

[1] J.M. Hastings and L.M. Corliss, IBM J. Res. Develop. 14 (1970) 227.
[2] T. Miyadai, K. Takizawa, H. Nagata, H. Ito, S. Miyahara and K. Hirakawa, J. Phys. Soc. Jap. 38 (1975)115.
[3] Y. Nishihara, S. Ogawa and S. Waki, J. Phys. Soc. Jap. 39 (1975) 63.
[4] G. Czjzek, J. Fink, H. Schmidt, G. Krill, F. Gautier, M. F. Lapierre, and C. Robert, J. de Physique 35 (1974) C6–621; J. Phys. C9 (1976) 761.
[5] H. Nagata and T. Miyadai, J. J. Appl. Phys. 15 (1976) 1507.
[6] Y. Yamamoto and T. Nagamiya, J. Phys. Soc. Jap. 32 (1972) 1248.
[7] J.M. Hastings, L.M. Corliss, M. Blume and M. Pasternak, Phys. Rev. B1 (1970) 3209.

MAGNETIC PROPERTIES OF $NiBr_2 \cdot 6H_2O$

S.N. BHATIA and R.L. CARLIN
Department of Chemistry, University of Illinois at Chicago Circle, Chicago, Illinois 60680, USA

$NiBr_2 \cdot 6H_2O$ is quite similar magnetically to $NiCl_2 \cdot 6H_2O$. With an antiferromagnetic T_c of 8.30 ± 0.02 K, the susceptibilities at higher temperatures display small anisotropy, and may be fit by the usual spin hamiltonian. The energy gap in the spin-wave spectrum at $\kappa = 0$ corresponds to $T_{AE} \approx 7.0$ K.

The monoclinic compound $NiBr_2 \cdot 6H_2O$ is isomorphic to $NiCl_2 \cdot 6H_2O$. Continuing our studies on magnetic ordering in hydrated halides [1, 2], the zero-field susceptibilities and specific heat of $NiBr_2 \cdot 6H_2O$ have been measured between 1 and 30 K. The single-ion anisotropy is small, and the principal axis susceptibilities are readily fit by the usual [1, 3] spin-Hamiltonian for Ni^{2+} in octahedral coordination (fig. 1). An antiferromagnetic transition is observed at 8.30 ± 0.02 K, from the inflection in the susceptibilities in the ac plane. The preferred axis was determined as lying $11 \pm 1.5°$ away from the a axis, in the ac plane (fig. 2).

The magnetic parameters are

$g_{a_{\parallel}} = 2.28 \pm 0.05$, $D/k = -1.5 \pm 0.5$ K,
$g_b = 2.25 \pm 0.05$, $E/k = 0.0 \pm 0.05$ K,
$g_{c\perp} = 2.28 \pm 0.05$, $zJ/k = -11.4 \pm 0.2$ K,

where the exchange parameter has been determined by a molecular field approximation [1].

The specific heat behavior is consistent with these results (fig. 3). A λ-shaped peak is observed at 8.30 ± 0.02 K, and the high temperature portion is readily fit by the sum of a 3-d Debye function ($\theta_D = 125.7 \pm 0.5$ K) and a B/T^2 magnetic term, with $B = 91 \pm 3$ cal-K/mole. The anticipated entropy change of $R \ln 3$ is accounted for and there is relatively little short-range order; $S_c = 67\%$. Thus, like $NiCl_2 \cdot 6H_2O$ [3], $NiBr_2 \cdot 6H_2O$ appears to be a normal, two-sublattice, 3-d antiferromagnet. The two compounds are quite similar, magnetically and crystallographically, with the only difference being that the exchange interactions are stronger in the bromide.

Following standard procedures [3, 4], the anisotropy field (8 kOe) and exchange field

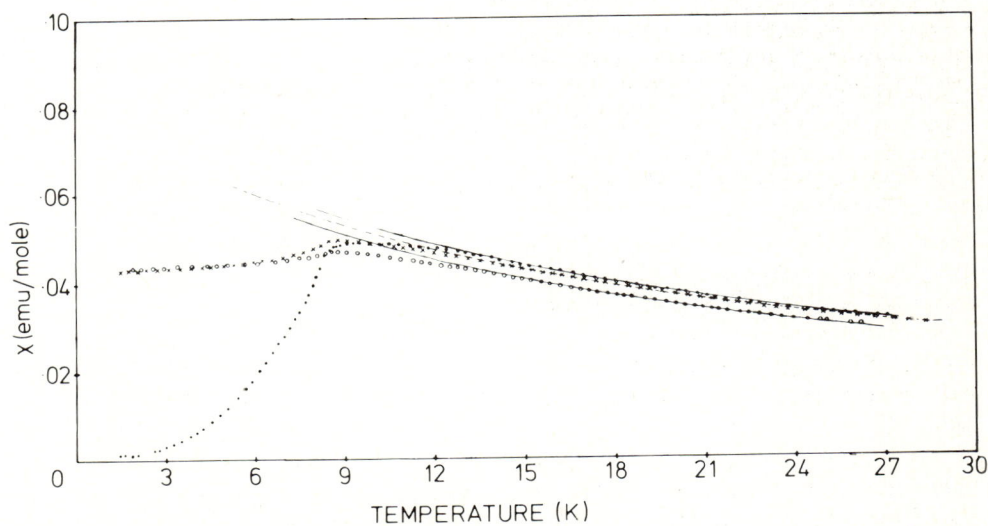

Fig. 1. Crystal axes susceptibilities of $NiBr_2 \cdot 6H_2O$. The closed circles (●) are $\chi_{a\parallel}$, open circles (○) for χ_b and the crosses (X) for $\chi_{c\perp}$. The solid and the dashed curves represent the fitted susceptibilities for the values of the parameters given in the text.

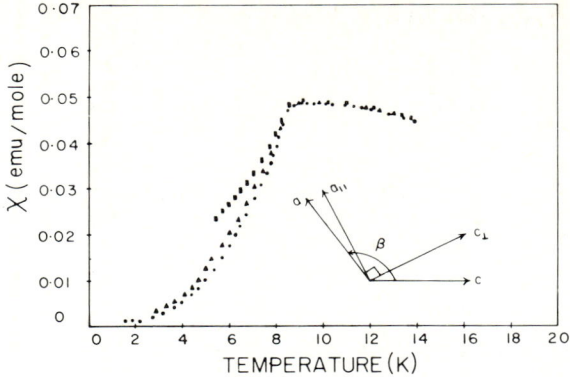

Fig. 2. Magnetic susceptibilities measured to locate the orientation of the preferred axis of spin alignment. The closed rectangles (■) represent the measurements along the a-axis. The triangles (△) and closed circles (●) respectively are measured at 14° and 11° from the a-axis in the ac plane. The inset shows the orientations of the various axes.

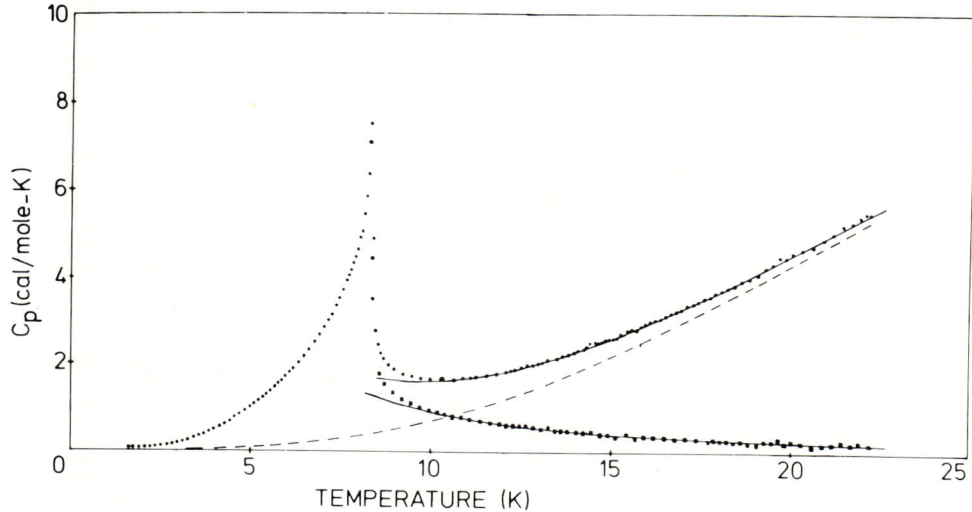

Fig. 3. Experimental heat capacity showing a sharp λ transition at 8.30 K (T_c). The solid curve through the experimental data [closed circles (●)] represents the fit to a Debye function and a T^{-2} magnetic term. The magnetic heat capacity (obtained by subtracting the lattice term from the measured data) is shown separately [closed rectangles (■)] along with the fitted curve.

(147 kOe) have been evaluated, and the parallel susceptibility and specific heat have been fit between 3 and 7 K ($\sim 0.8\ T_c$) to the spin-wave relationships. We find $T_{AE} \approx 7.0 \pm 0.7$ K, where kT_{AE} is the gap in the spin-wave energy spectrum corresponding to $\kappa = 0$. The calculated $\chi_\perp(T=0) = 0.0423$ emu/mole compares well with the extrapolated experimental value of 0.043 emu/mole.

Acknowledgement is made to the donors of the Petroleum Research Fund administered by the American Chemical Society, for the support of this research.

References

[1] J.N. McElearney, D.B. Losee, S. Merchant and R.L. Carlin, Phys. Rev B7 (1973) 3314.
[2] R.L. Carlin, Accts. Chem. Res. 9 (1976) 67.
[3] A.I. Hamburger and S.A. Friedberg, Physica 69 (1973) 67.
[4] F. Keffer, in: Handbuch der Physik, Vol. 18, Pt. 2, S. Flügge (ed.) (Springer, Berlin, 1966) p. 1.

MAGNETIC TRANSITIONS IN YbIG AT LOW TEMPERATURES

J. LOOS

Institute of Solid State Physics, Czechoslovakian Academy of Science, Prague, Czechoslovakia

The stability of possible magnetic phases, the existence of soft modes and the limits of validity of mean field approach are discussed for a model with one saturated Fe-sublattice and six rare earth sublattices in the garnet structure.

The model with one saturated Fe-sublattice and six sublattices of mutually independent Yb-ions was used in [1] as the basis of a numerical procedure determining the possible magnetic phases of YbIG. The same model is used in the present contribution to discuss the stability of these phases and to deduce the free energy expansion describing their transformations.

Let us investigate small oscillations of the Fe-sublattice magnetization M about its equilibrium direction. Considering the low-frequency modes only, the ground state doublet splitting Δ_j of the Yb-ion in the jth sublattice is determined by the external field H and the instantaneous direction of M. The following total effective field acting on M is found:

$$H_{\text{eff}} = H + \sum_{j=1}^{6} \tfrac{1}{2} n_j (\partial \Delta_j / \partial M). \quad (1)$$

Here, the instantaneous difference in populations n_j of the doublet levels takes the canonical quasi-equilibrium value at the given temperature T, if the soft mode frequency $\omega \to 0$. Under these assumptions, the linearized equations of motion for M were solved and the formulae for ω and the susceptibility tensor were obtained for H and M both lying in the same easy plane [2]. The equation $\omega = 0$ defines in the T–H plane the instability curves of corresponding phases and these curves deliminate the possible metastability regions of neighbouring phases. This is shown on fig. 1(a) and (b) for $H\|\langle 100\rangle$ and $H\|\langle 111\rangle$, respectively.

The series expansion of the free energy may be used in the neighbourhood of the continuous transition points. The order parameter η for the transition between the collinear phase and the canted one is defined to be proportional to the component of M perpendicular to H in the easy plane, i.e. $\eta = M_y/M$. Because the stability with respect to the deviations from the easy plane

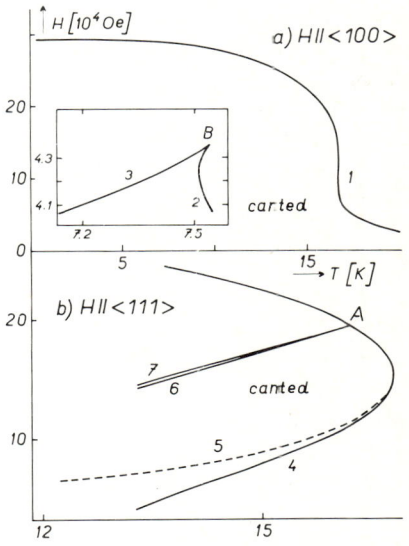

Fig. 1. The $\omega = 0$ lines. In particular, curve 4 adheres to the canted phase and curve 5 to the collinear phase. These two lines coincide at point A which is the end-point of curves 6 and 7, too. Curves 2, 3 and 4, 5, respectively, deliminate the metastability regions, whereas both the neighbouring phases are instable in the region between the curves 6, 7. Probably, an intermediate phase exists in this latter region.

should be examined, the free energy is developed not only in powers of η, but also in powers of the component of M perpendicular to the easy plane, namely, $\xi = M_x/M$. The corresponding expansions of the magnetic Gibbs' free energy up to fourth order in ξ, η for $H\|\langle 100\rangle$ and for $H\|\langle 111\rangle$, respectively, are as follows:

$$G_{\langle 100\rangle} = a(\xi^2 + \eta^2) + c(\xi^4 + \eta^4) + d\xi^2\eta^2, \quad (2)$$

$$G_{\langle 111\rangle} = a'(\xi^2 + \eta^2) + b(\eta^3 - 3\eta\xi^2) + c'(\xi^2 + \eta^2)^2. \quad (3)$$

The coefficients of these polynomials are given as functions of T, H depending only on the known parameters of the Yb-ion spin hamiltonian. The series expansion (2) has four minima of G for $a < 0$ corresponding to four equivalent directions of M in easy planes going through $\langle 100 \rangle$ directions. The second order transition line $a = 0$ between the canted phase and the collinear one is identical with the $\omega = 0$ line 1 in fig. 1. The expansion (3) describes the first order transition between the phases with $\eta = 0$ and $\eta \neq 0$. These two phases are identical at the critical point A which fulfils the equations $a' = 0$, $b = 0$ simultaneously. The coexistence curve of these two phases is defined by the equation $b^2 - 4a'c' = 0$. The equilibrium conditions deduced from the expansion (3) for $4a'c' < b^2$ lead to six directions of M non-collinear to H but always only three of them are stable. These two triads of solutions correspond to two types of canting in the three easy planes going through the $\langle 111 \rangle$ direction. The rotation of M but one type of canting to the other occurs at the line $b = 0$ where both these canted phases become instable with respect to deviations of M perpendicular to the easy plane. Of course, the non-approximated treatment gives two instability curves, namely the curves 6, 7 on fig. 1(b).

The continuous transition at the point B was already discussed phenomenologically in [3]. Fig. 1(a) shows the existence of a metastability region bounded by curves 2 and 3. Two minima of G at given T, H in this region may be again obtained from the series expansion of G with respect to the direction of M adhering to the maximum of G at the same values of T, H. However, the latter direction depends now on T, H and it should be determined from the stationarity condition of G. The deviations from the direction of M determined in this way will be characterized by ξ, η as previously. The expansion of G in powers of ξ, η has the following form:

$$G = a_1\xi^2 + a_2\eta^2 + b_1\xi^2\eta + b_2\eta^3 + c_1\xi^4 + c_2\eta^4 + f\xi^2\eta^2. \quad (4)$$

The coexistence curve of the two considered canted phases with the order parameter $\eta_{1,2} = \pm(-a_2/2c_2)^{1/2}$ is given by the condition $b_2 = 0$.

The known dependence of the expansion coefficients on T, H allows us to estimate the limits of the mean field approach used if the average quadratic fluctuation inside the correlation volume or the relative importance of the specific heat contribution due to fluctuations [4] are examined. These estimates restrict the critical region along the second order transition line 1 to the temperature interval $\Delta T \approx 10^{-5}$–10^{-4} K and on the coexistence line at the point B with temperature T_B to $|T - T_B| \leq 10^{-6}$ K. The fluctuations at the point A with temperature T_A may be important for $|T - T_A| \leq 10^{-1}$ K, measured on the coexistence curve of the collinear phase and the canted one.

References

[1] R. Alben, Phys. Rev. B2 (1970) 2767.
[2] J. Loos, Czech. J. Phys. B26 (1976) in print.
[3] R. Alben, Phys. Rev. Lett. 24 (1970) 68.
[4] V.L. Ginzburg, Fiz. Tverd. Tela 2 (1960) 2031; A.P. Levanyuk, Fiz. Tverd. Tela 5 (1963) 1776.

EIGHT SUBLATTICES FeI$_2$ MODEL

D. BERTRAND, A.R. FERT, J. GÉLARD, J. LÉOTIN and J.C. OUSSET

Laboratoire de Physique des Solides, Université Paul Sabatier – 31077, Toulouse Cédex, France

An eight sublattices model describes reasonably well the ferrous iodide magnetic behaviour: in zero magnetic field three AFMR uniform-modes are observed; in a magnetic field parallel to the spin direction FeI$_2$ presents a four-step metamagnetic behaviour. We determine a set of exchange coupling parameters between different sublattices.

1. Introduction

At low temperature FeI$_2$ orders antiferromagnetically ($T_N = 9.3$ K). By neutron diffraction experiments, we have observed [1] that the spins are oriented along the c axis, but contrary to FeCl$_2$ and FeBr$_2$, there is an antiferromagnetic order in each layer with two rows of up spins followed by two rows of down spins and so on (fig. 1).

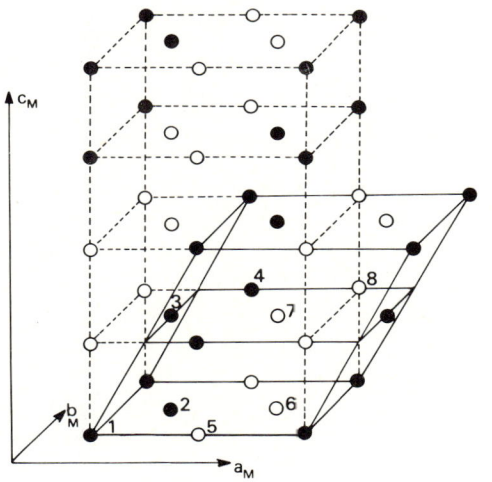

Fig. 1. Magnetic primitive cell of FeI$_2$. Black and white circles represent magnetic ions with moments in opposite direction. Each of the eight sublattices are labelled by a number 1, 2, ..., 8. The magnetic orthorhombic cell is represented by dashed lines.

In a magnetic field parallel to the spin direction we have observed [2] four successive transitions at $H = 46, 60, 92, 122$ kOe corresponding roughly to a four-step metamagnetic behaviour. Mössbauer experiments have been performed by de Graaf and Trooster who propose a twelve sublattices model [3].

To investigate the properties of FeI$_2$, we have undertaken far infrared AFMR experiments, the results of which will be presented later. At the same time, Petitgrand and Meyer obtained, by means of a Fourier transform spectrometer, three absorption peaks corresponding to different uniform mode excitations in zero magnetic field [4].

Using a theoretical model in which the spins are divided into eight different sublattices, we try to explain the observed metamagnetic behaviour and the spin-wave spectrum at the center of the Brillouin zone.

Fig. 1 shows the primitive cell of FeI$_2$, Fe^{2+} ions of each sublattice being labelled by a number p ($p = 1, 2, 3, 4$ up spins, $p = 5, 6, 7, 8$ down spins).

2. Theoretical model

The magnetic hamiltonian appropriate to the Fe^{2+} ions in FeI$_2$ is a simple $S = 1$ hamiltonian with anisotropy and exchange terms:

$$\mathcal{H} = \sum_i D(S_i^{z2} - \tfrac{2}{3}) - \sum_i g_\| \mu_B H_0 S_i^z$$
$$+ \sum_{i,j} [-g_\|'^2 \mathcal{J}_{ij} S_i^z S_j^z - g_\perp'^2 \mathcal{J}_{ij}(S_i^x S_j^x + S_i^y S_j^y)],$$

where D is the cristalline anisotropy parameter, $\mathcal{J}_{ij}$ the isotropic exchange constant between real spins, $\mathbf{g}(g_\|, g_\perp)$ the anisotropic g factor connecting magnetic moment operator and effective spin operator and $\mathbf{g}'(g_\perp', g_\|')$ the anisotropic g' factor connecting real spin operator and effective spin operator.

2.1. Spin wave spectrum

In order to study the spin waves generated from the ordered array of spins, we introduce the Fourier transforms of the exchange parameters between ions on p and q sublattices:

$$J_{pq}(\mathbf{k}) = \sum_{j(q)} 2J_{i(p)j(q)} \exp[i\mathbf{k}(\mathbf{r}_{i(p)} - \mathbf{r}_{j(q)})],$$

$$J_{ij} = g_\perp'^2 \mathcal{J}_{ij}.$$

Taking into account the symmetry of FeI_2 the set of exchange parameters $J_{pq}(\mathbf{k})$ may be expressed in terms of only five parameters $J_{1p}(\mathbf{k})$, $p = 1, 2, \ldots, 5$.

By a spin wave treatment, we find eight magnon branches with:

$$\hbar\omega_1(\mathbf{k}) = \{h(\mathbf{k})[h(\mathbf{k}) - 2J_{15}(\mathbf{k}) - 2J_{12}(\mathbf{k}) - 2J_{13}(\mathbf{k}) - 2J_{14}(\mathbf{k})]\}^{1/2} \pm g_\parallel \mu_B H_0,$$

the other three energies being obtained by introducing two positive signs among J_{12}, J_{13} and J_{14}.

where

$$h(\mathbf{k}) = h_A - [J(\mathbf{k}) - J(0)] + [J'(\mathbf{k}) - J'(0)],$$

$$h_A = g_\parallel \mu_B H_A = -D + \frac{1-\eta}{\eta}[J(0) - J'(0)],$$

$$\eta = \frac{g_\perp'^2}{g_\parallel'^2}; \qquad J(\mathbf{k}) = \sum_{p=1}^{4} J_{1p}(\mathbf{k});$$

$$J'(\mathbf{k}) = \sum_{p=5}^{8} J_{1p}(\mathbf{k}).$$

In FeI_2, the only unknown parameters are h_A and $J_{1p}(\mathbf{k})$ with $p = 1, 2, 3, 4, 5$.

2.2. Step metamagnetism in a magnetic field parallel to the spin direction

In a magnetic field parallel to the spin direction, we may observe intermediate phases corresponding to the reversal of some of the down spins sublattices. If we suppose that spin-flop phases does not occur this lead to a step by step metamagnetic behaviour.

The energies of different phases are:

antiferromagnetic phase (4 spins up) (FeI_2 structure in zero field)

$$E_{AF} = -\frac{1}{2\eta}(J_{11} - J_{15}),$$

ferrimagnetic state I (5 spins up)

$$E_{FI} = -\frac{1}{2\eta}(J_{11} - \tfrac{1}{2}J_{15}) - \tfrac{1}{4}g_\parallel \mu_B H_0,$$

ferrimagnetic state II (6 spins up)

$$E_{FII} = -\frac{1}{2\eta}(J_{11} + K) - \tfrac{1}{2}g_\parallel \mu_B H_0,$$

with $K = J_{12}$ (5, 6 down) or J_{13} (5, 7 down) or J_{14} (5, 8 down).

ferrimagnetic state III (7 spins up)

$$E_{FIII} = -\frac{1}{2\eta}[J'(0) + J_{11} - \tfrac{1}{2}J_{15}] - \tfrac{3}{4}g_\parallel \mu_B H_0,$$

paramagnetic phase (8 spins up)

$$E_p = -\frac{1}{2\eta}[J(0) + J'(0)] - g_\parallel \mu_B H_0,$$

and the transition field between two successive states are

$$H_1 = -\frac{1}{\eta g_\parallel \mu_B} J_{15}; \qquad H_2 = -\frac{1}{\eta g_\parallel \mu_B}[J_{15} + 2K],$$

$$H_3 = -\frac{1}{\eta g_\parallel \mu_B}[2J'(0) - J_{15} - 2K];$$

$$H_4 = -\frac{1}{\eta g_\parallel \mu_B}[2J'(0) - J_{15}].$$

3. Discussion of the case of FeI_2

From the experimental value of H_1 (this first magnetization discontinuity, obtained by magnetization measurements [1], is about a quarter of magnetization at saturation), we obtain $J_{15} = -9.7$ K.

According to Petitgrand and Meyer, only three infrared absorption peaks are observed at 21.6, 29.2 and 32.2 cm^{-1} with relative strengths respectively equal to 2, 1 and 1 suggesting that the peak observed at 21.6 cm^{-1} corresponds to two uniform modes of lower energy.

In our model, the exchange coupling constants J_{12}, J_{13}, J_{14} and the anisotropy field h_A may be calculated from the absorption peaks. We obtain:

$h_A = 29.6$ K, $\qquad J_{12} = -1.6$ K,
$J_{14} = -8.4$ K, $\qquad J_{13} = -1.6$ K.

From these values of parameters we compare theoretical predictions of our model with other experimental results 46, 60, 92 and 122 kOe. The transition fields are 46, 62, 141 and 156 kOe, and experimental results 46, 60, 92 and 122 kOe. The perpendicular susceptibility is 0.064 cgs mole^{-1} (exp.: 0.071). If we introduce exchange parameters between spins, we can show that the predominant interactions are between nearest and next nearest neighbours along the c axis.

In conclusion, in spite of some disagreement with experimental Mössbauer results (De Graaf and Trooster give $\Delta M/M_S = \frac{1}{3}$ for the first magnetic discontinuity) our eight sublattices model gives a good interpretation of the step metamagnetic behaviour, the infrared absorption spectra, and the principal other magnetic properties of FeI$_2$.

References

[1] J. Gelard, A.R. Fert, P. Meriel and Y. Allain, Solid State Commun. 14 (1974) 187.
[2] A.R. Fert, J. Gelard and P. Carrara, Solid State Commun. 13 (1973) 1219.
[3] H. de Graaf and J.M. Trooster, Solid State Commun. 12 (1975) 1387.
[4] D. Petitgrand and P. Meyer, J. Phys. 37 (1976) 5.

MÖSSBAUER EFFECT STUDIES OF βFe_2O_3

E.R. BAUMINGER, L. BEN-DOR, I. FELNER, E. FISCHBEIN, I. NOWIK and S. OFER

The Hebrew University, Jerusalem, Israel

Mössbauer studies have been performed on thin films of βFe_2O_3 deposited on fused silica or polycrystalline alumina. All the Mössbauer spectra are composed of two subspectra with relative intensities of 3:1. The parameters of the hyperfine interactions in ^{57}Fe below and above the ordering temperature (106.5 ± 1.5 K) were determined.

Several investigations of the magnetic behaviour of solid solutions of Fe_2O_3 and Mn_2O_3 with the cubic bixbyite structure have been performed [1–6]. Studies on pure cubic Fe_2O_3 have not been published. Recently, pure cubic Fe_2O_3, which was identified as βFe_2O_3, was prepared as thin films, by chemical vapor deposition from trifluoroacetylacetonate iron at 300°C in the presence of oxygen, on various backings [7]. X-ray analysis of the films showed that they were well crystallized in the bixbyite structure, which is identical to the C structure of the rare earth oxides. The lattice constant was found to be $a_0 = 9.393$ Å. 24 Fe^{3+} ions in the cubic unit cell have C_2 symmetry ("d" site) and 8 ions have C_{3i} symmetry ("b" site).

Recoilless absorption spectra of the 14.4 keV gamma ray emitted from a ^{57}Co in Cu source in $\beta^{57}Fe_2O_3$ thin films were measured between 4.1 and 400 K. The films were deposited on fused silica or polycrystalline alumina. All spectra above 108 K showed pure quadrupole splittings. Least squares computer fits to the experimental spectra, showed that all spectra were composed of two subspectra with relative intensities of 1:3. A typical spectrum of this kind is shown in fig. 1. The relative intensities obtained, show unambiguously that the subspectrum with relative intensity 3, stems from the Fe nuclei occupying the d sites, whereas the spectrum with relative intensity 1, stems from the Fe nuclei occupying the b sites. The parameters obtained from the least squares fits are: (a) in the d sites, the quadrupole interaction parameter $eqQ = 1.44$ mm/s [2], the isomer shift (at 295 K relative to iron metal) 0.324(8) mm/s, and the full line width at half height $\Gamma_d = 0.280(5)$ mm/s; (b) in the b site: $eqQ = 1.944(20)$ mm/s, $(I.S.)_b = 0.331(8)$ mm/s and $\Gamma_b = 0.38(2)$ mm/s. The quadrupole splittings and line widths were temperature independent above 108 K. Below 105 K

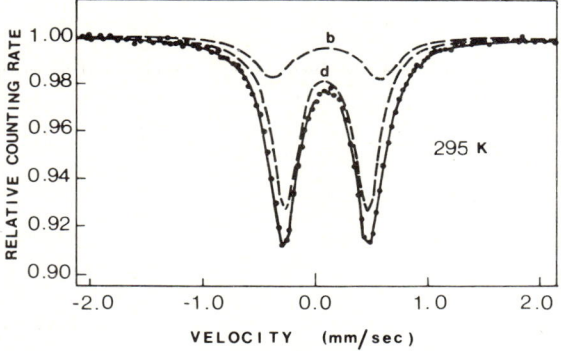

Fig. 1. Recoilless absorption spectrum of the 14.4 keV gamma ray of ^{57}Fe in βFe_2O_3 at 295 K. The solid line is the theoretical curve obtained by a least squares computer fit to the experimental points assuming two sites with pure quadrupole interactions. The dashed lines are the theoretical curves for the b and d sites.

magnetically split spectra were obtained. The ordering temperature of this compound is therefore 106.5 ± 1.5 K. The purest and best resolved magnetic spectra were obtained with the samples prepared on alumina backings and the hyperfine interaction parameters were therefore deduced from measurements carried out on these samples.

Typical spectra obtained at 4.1 and 90 K are shown in fig. 2. The two split subspectra can clearly be seen in the figure. The individual lines in the spectrum obtained at 4.1 K are narrow and well defined, whereas at higher temperatures the individual lines are broadened. This broadening may be due to relaxation effects. The effective magnetic fields obtained at the various temperatures are summarized in table I. The values determined for the effective quadrupole interaction parameters (eqQ) are 0.37(2) mm/s for the Fe ions in the d sites and 1.80(1) mm/s for the Fe ions in the b sites.

The sharpness of the absorption lines observed at 4.1 K indicates that in both sites the spins of the Fe^{3+} ions point in a definite direc-

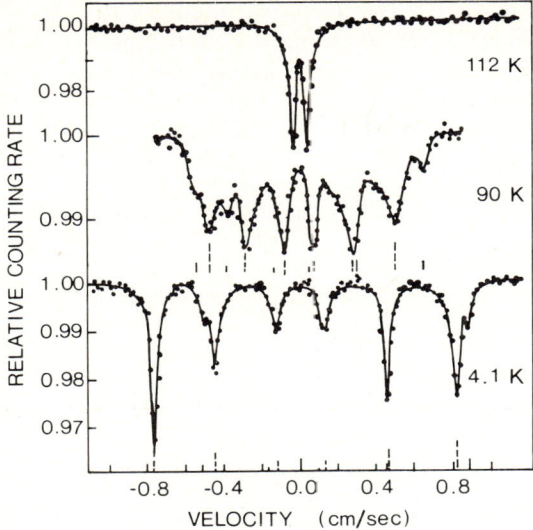

Fig. 2. Recoilless absorption spectra of the 14.4 keV gamma ray of ^{57}Fe in βFe$_2$O$_3$ at various temperatures. The dashed lines below the spectra show the positions and relative intensities of the lines corresponding to the d site, the smaller full lines below the spectra show the positions and relative intensities of the lines corresponding to the b sites.

Table I
Values of H_{eff} in the d site and b site of βFe$_2$O$_3$

T (K)	H_{eff}(kOe) d site	H_{eff}(kOe) b site
4.1	496(2)	519(3)
30	474(3)	502(5)
60	429(4)	479(6)
80	355(4)	410(6)
90	305(4)	373(6)
95	271(5)	312(6)
103	186(7)	
107	0	0

tion relative to the principal axes of the electric gradients (EFG). The fact that the quadrupole interaction parameters observed in the b site are the same above and below the magnetic transition temperature, proves that at this site the spins of the Fe^{3+} ions point in the direction of the local symmetry axis, that is, in the local [111] direction.

In the d site, one of the principal axes of the EFG lies along the [100] direction, which is the local symmetry axis (C_2). The other two principal axes are in the plane perpendicular to this direction. From the fact that the quadrupole interactions measured above and below the ordering temperatures differ by a factor of about 4, it is clear that in this site the spins do not point in the direction of the largest component of the EFG.

Point charge calculations of the EFGs in the d and b sites of βFe$_2$O$_3$ were performed for various values of the parameters u, x, y, z, which determine the positions of the iron and oxygen ions in the crystal unit cell [8]. These calculations show that the EFGs are extremely sensitive to the values of u, x, y and z. Because of the low accuracy of the determination of the parameters, it is impossible to make a quantitative comparison between the experimental and the calculated value of the quadrupole interaction parameters. The only conclusion that can be drawn is, that if it is assumed that the Fe spins in the d sites point along the [100] direction, as is the case for Er$_2$O$_3$ [9], then the experimental results are consistent with the theoretical calculations.

The temperature dependence of the hyperfine fields in the two Fe sites, is quite different (fig. 3). Assuming the molecular field approximation, we have tried to fit the temperature dependence of the hyperfine fields in the two sites. If we denote by σ_d and σ_b the reduced hyperfine fields in the d and b site, respectively, we may write

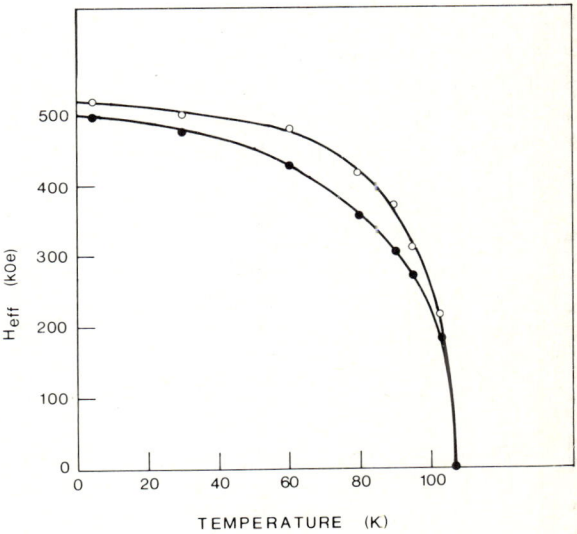

Fig. 3. Hyperfine fields as functions of temperature in the d site (full points) and in the b site (open circles). The curves are the theoretical fits to the experimental data, assuming the molecular field approximation.

the following equations:

$$\sigma_d(\tau) = B_{5/2}[\{W_{dd}\sigma_d(\tau) + W_{db}\sigma_b(\tau)\}/\tau], \quad (1)$$

and

$$\sigma_b(\tau) = B_{5/2}[\{W_{bd}\sigma_d(\tau) + W_{bb}\sigma_b(\tau)\}/\tau], \quad (2)$$

where τ is the reduced temperature and W_{ij} are the inter- and intra-sublattice exchange energies. The contribution of a j ion to W_{ij} acting on an i ion, is determined by the strength of the exchange interaction between them, and the relative orientations of their spins. (W_{bd} may therefore be different in sign and magnitude from W_{db}).

A least squares fit procedure to the experimental points in fig. 3 leads to the solid curves shown in the figure and the following parameters:

$W_{dd} = 2.67(5) \ T_c$, $W_{db} = -0.30(5) \ T_c$,

$W_{bd} = 1.86(5) \ T_c$, and $W_{bb} = 0.84(8) \ T_c$.

The relative magnitudes obtained for the W_{ij} energies seem reasonable, taking into account the number of closest neighbours of each ion and the distances between them.

References

[1] E. Banks and E. Kostiner, J. Appl. Phys. 37 (1966) 1423.
[2] E. Banks, E. Kostiner and G.K. Wertheim, J. Chem. Phys. 45 (1966) 118.
[3] S. Geller, R.W. Grant, J.A. Cape and G.P. Espinosa, J. Appl. Phys. 38 (1967) 1457.
[4] R.R. Chevalier, G. Roult and E.F. Bertant, Solid State Commun. 5 (1967) 7.
[5] R.W. Grant, S. Geller, J.A. Cape and G.P. Espinosa, Phys. Rev. 175 (1968) 686.
[6] S. Geller, Acta Cryst. B27 (1971) 821.
[7] L. Ben-Dor, E. Fischbein, I. Felner and Z. Kalman, to be published in J. Electro-Chem. Soc.
[8] R.W.G. Wykoff, Crystal Structures, 2nd ed., Vol. 2 (Interscience, 1964) p. 161.
[9] G.M. Kalvius, G.K. Shenoy and B.D. Dunlap, in: Proc. Conf. Les Elements des Terres Rares, Paris–Grenoble (1969) p. 477.

NEUTRON DIFFRACTION STUDY OF THE MAGNETIC STRUCTURE OF THE BASIC IRON SULPHATE FeOHSO$_4$

D. SCHEERLINCK and E. LEGRAND

Materials Science Department, SCK/CEN, B-2400 Mol, Belgium

Low temperature neutron-diffraction measurements of FeOHSO$_4$ show that the modulation of the moments can be described by means of two wave vectors k_1 $[2\pi, 2\pi, 0]$ and k_2 $[2\pi, 2\pi, 2\pi]$. Magnetic intensities described by the $[2\pi, 2\pi, 0]$ mode decrease slower with increasing temperature.

The crystal structure of FeOHSO$_4$ has been determined from X-ray diffraction measurements on single crystals by Johanson [1]. This compound is orthorhombic with four molecules in the unit cell. The space group is P$_{nma}$ and the lattice parameters are $a = 7.33$ Å, $b = 6.42$ Å and $c = 7.14$ Å. The magnetic properties have been extensively studied by Rumbold and Wilson [2]. From Mössbauer and susceptibility measurements these authors deduced the existence of linear chain antiferromagnetism along the crystallographic a direction. As they notice, this is indeed supported by consideration of the crystal structure since the superexchange paths between the iron ions are much shorter along the a-axis than in other directions. The abrupt decrease of the susceptibility below 56 K indicates that below this temperature a three-dimensional ordering of the magnetic moments appears. Since the complete magnetic structure can hardly be deduced from such measurements we performed neutron diffraction measurements to obtain more information.

Neutron diffraction patterns, with a neutron wavelength of 1.263 Å, were taken at 4.2, 48.6, 55.6, 77.3, 106 and 293 K. The position of the hydrogen atoms was determined by a least square profile analysis of the room temperature measurement, the other parameters being taken from the structure determination by Johanson. Especially the (101) reflection and the (011) + (110) reflections depend strongly on the choice of the H-coordinates. Unfortunately the calculated intensity of the (111) reflection remains always higher than the measured one. Possibly it is necessary to include the other position parameters in the refinement; it is also possible that some of the diffraction lines are somewhat enlarged due to mechanical distortions of the grains introduced by the sifting of the powder which may disturb a line profile analysis of the data. An analysis with integrated measured intensities was not possible due to the strong overlap of the diffraction peaks.

At 4.2 K the cell parameters have decreased to $a = 7.28$ Å, $b = 6.33$ Å and $c = 7.10$ Å. Moreover, superstructure lines appear and the intensities of other diffraction lines increases (fig. 1). The presence of these superstructure lines indicates that the magnetic structure is not simple antiferromagnetic since the chemical unit cell already contains 4 Fe atoms. The first superstructure line can be indexed by doubling the b-axis of the chemical unit cell. To index the second superstructure line the b-axis as well as the c or a-axis (or both) must be doubled.

Since the a and c axes have nearly the same length no distinction can be made between them with the available experimental resolution of the diffraction patterns. However, from structural considerations it seems reasonable to suppose that the magnetic moments in the a-direction are antiferromagnetically coupled. It is therefore plausible to describe the magnetic structure in a: $a, 2b, 2c$ unit cell.

Looking for possible spin configurations following the Fourier method of Bertaut [3] one finds that a spiral spin structure has the lowest energy, however, the stability conditions are not satisfied. Other possible wave vectors to describe the magnetic modulation are

$$k_1 = [2\pi, 2\pi, 0] \quad \text{and} \quad k_2 = [2\pi, 2\pi, 2\pi].$$

The first superstructure can be explained with the wave vector k_1, but not the second one. On the other hand the latter can be explained with a vector k_2 but not the first one. Since no unique wave vector k could be found which describes both superstructure lines at the same time, we tried to describe the structure with a combination of the two wave vectors k_1 and k_2.

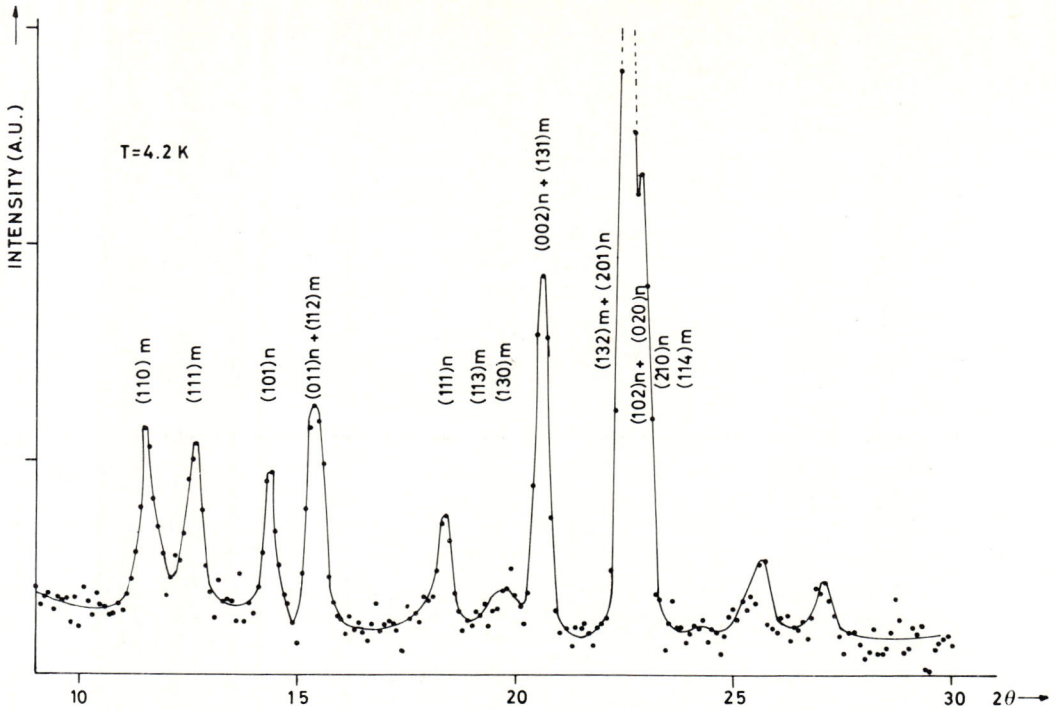

Fig. 1. 4.2 K neutron (powder) diffraction diagram of FeOHSO$_4$: (m) magnetic superstructure reflections indexed in the $(a, 2b, 2c)$ unit cell; (n) nuclear intensities.

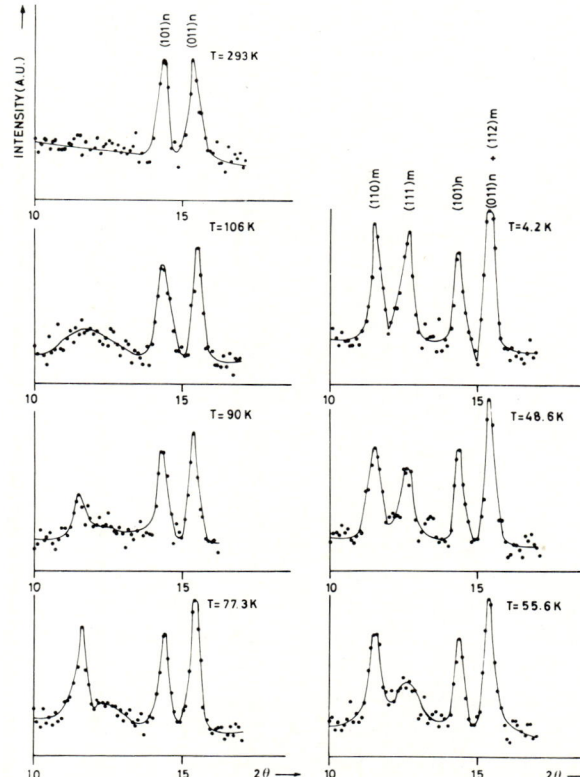

Fig. 2. Temperature evolution in the region of the magnetic superstructure lines intensities at $T = 293, 106, 90, 77.3, 55.6, 48.6$ and 4.2 K.

Therefore the unit vector $\sigma(i)$ of the ith iron atom was decomposed into $\sigma(i) = \sigma_1(i) + \sigma_2(i)$, with $\sigma_1(i)$ perpendicular to $\sigma_2(i)$. The modulation of the components $\sigma_1(i)$ and $\sigma_2(i)$ is then described by the modes k_1 and k_2, respectively. The best fit between calculated and measured values is found for:

$$\sigma_1(1) = [1, 1, 0], \qquad \sigma_2(1) = \left[\frac{1}{\sqrt{2}}, \frac{-1}{\sqrt{2}}, 1\right],$$

$\mu = gS = 4.55$ Bohr magneton,

which gives a reliability factor of 3%.

In fig. 2 the scattered intensity in the region of the magnetic superstructure lines is shown at different temperatures. It is notable that the superstructure line and other magnetic contributions in the diffraction patterns, which are described by the wave vector k_2, decrease much faster with increasing temperature than the line described by k_1.

The value found for the magnitude of the magnetic moment is somewhat smaller than the value which is to be expected for an Fe^{3+} ion. The temperature dependence of the magnetic diffraction lines indicates that the magnitude of the exchange interactions in various crystallographic directions are very different. However, the broad hump which is present at 106 K indicates that at this temperature the interchain interaction is still not completely negligible.

References

[1] G. Johanson, Acta Chem. Scand. 16 (1962) 1234.
[2] B.D. Rumbold and G.V.H. Wilson, J. Phys. Chem. Solids 35 (1974) 241.
[3] E.F. Bertaut, in: Magnetism III, Rado and Shul (Eds.) (Academic Press, 1963) p. 150.

MAGNETIC STRUCTURE OF THE PEROVSKITE-LIKE COMPOUND TbMnO₃

S. QUEZEL, F. TCHEOU, J. ROSSAT-MIGNOD, G. QUEZEL* and E. ROUDAUT

DRF/DN, Centre d'Etudes Nucléaires, 85 X, 38041 Grenoble Cedex, France
* Laboratoire de Magnétisme, CNRS, 166 X, 38042 Grenoble Cedex, France

Magnetic and neutron diffraction experiments on TbMnO₃ powder and single crystals are reported. Below 40 K a sine-wave ordering of the Mn^{3+} moments is found with the wave-vector along the b-axis $k_{Mn} = (0\;0.28\;0)$. Below 7 K a short-range ordering of the Tb^{3+} moments takes place with a different wave-vector $k_{Tb} = (0\;0.415\;0)$.

1. Introduction

The rare-earth orthoferrites and orthochromites have been extensively studied. Much less work has been done on manganites and especially on heavy rare-earth orthorhombic compounds. This paper deals with the magnetic and neutron diffraction studies of terbium orthomanganite. It crystallizes in an orthorhombically distorted perovskite structure (space group Pbnm) with 4 formula units/elementary cell.

2. Experimental

Investigations were made on both powder sample and single crystals (size $2 \times 2 \times 2.5$ mm). Magnetic susceptibilities in the temperature range 1.5–300 K and magnetization in field up to 120 kOe have been measured. Neutron experiments were carried out using a double axis spectrometer operating at the reactor Siloe at temperatures from 1.5 to 300 K.

3. Magnetic measurements

Magnetic susceptibility, measured on powder sample follows a Curie–Weiss law down to about 28 K ; however, a weak decrease of the Curie constant takes place at about 40 K. This temperature may correspond to the ordering of the Mn^{3+} moments. At 28 K a small anomaly has been detected and the reciprocal susceptibility shows a broad minimum at about 7 K which may be attributed to an antiferromagnetic ordering of the Tb^{3+} moments. The susceptibility measured along the three crystallographic axes of a single crystal shows a large anisotropy : χ_a and χ_b have about the same value but χ_c is ten times lower. This anisotropy is due to the Tb^{3+} moments and indicates that they are located in the a–b plane.

This fact is confirmed by high field magnetization measurements at 1.5 K [fig. 1 (a)]. The weak magnetization along the c-axis increases linearly with field; it may be mainly due to the Mn^{3+} contribution. In the a–b plane, an important anisotropy is observed (for $H = 120$ kOe, $\sigma_a = 6.8\ \mu_B$ and $\sigma_b = 4.7\ \mu_B$/formula unit). From the 120 kOe magnetization vs. field directions, we deduced that the terbium moments lie symmetrically with respect to the a-axis with an angle of about 33° [fig. 1 (b)] comparable to those found in other terbium perovskites [1]. The transitions observed at low field result from the metamagnetic behaviour of terbium moments.

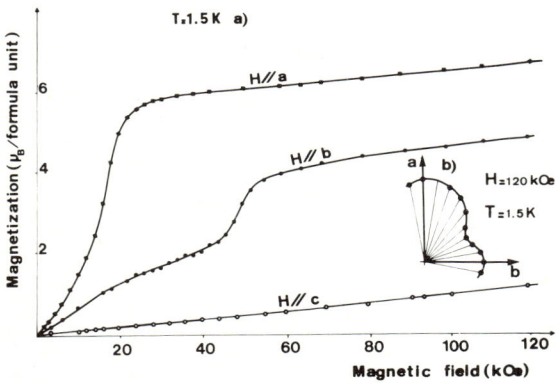

Fig. 1. Magnetization of TbMnO₃ single crystal, in μ_B/formula unit (a) for applied fields along the a, b, c axes; (b) vs. field directions within the a–b plane, in polar coordinates.

4. Neutron diffraction results

Neutron diffraction experiments have been made to precise the magnetic order of terbium and manganese moments. The diagrams obtained from a powder sample at low temperatures show magnetic superlattice peaks indicating an incommensurate structure which propagates along the **b**-axis. The value and the thermal variation of the propagation vector has been determined by scanning the reciprocal space along lattice lines parallel to the **b**-axis. Fig. 2 represents scans along $(0\,k\,1)$ and $(1\,k\,1)$ directions at 1.8 and 10 K, respectively; the same superlattice peaks observed at both temperatures can be attributed to the manganese moments ordering with a propagation vector $\boldsymbol{k}_{Mn} = (0\,0.28\,0)$. At $T = 1.8$ K additional weaker broad peaks are found and can be associated to the terbium ordering with a wave vector $\boldsymbol{k}_{Tb} = (0\,0.415\,0)$. No satellites associated to $\boldsymbol{k}_{Mn}$ were observed in Brillouin zones of type F $(h + k = 2p,\ l = 2p)$ and C $(h + k = 2p + 1,\ l = 2p)$ whereas strong and weak peaks, respectively, exist in Brillouin zones A $(h + k = 2p,\ l = 2p + 1)$ and G $(h + k = 2p + 1,\ l = 2p + 1)$. It must be noted that the G-type satellites disappear above 20 K. The $\boldsymbol{k}_{Mn}$ wave vector keeps the same value up to 30 K and increases to $k = 0.295$ until T_N [fig. 3 (b)]. The thermal variation of the $(001)^+_{Mn}$ intensity [fig. 3 (a)] gives a Néel temperature of about 40 K and shows an increase below 25 K up to 10 K. On the powder sample the same behaviour occurs but below 10 K a decrease of the intensity has been observed; moreover the magnetic peaks due to Tb^{3+} order are far more intense than in single crystals and give an ordering temperature of about 7 K (fig. 3).

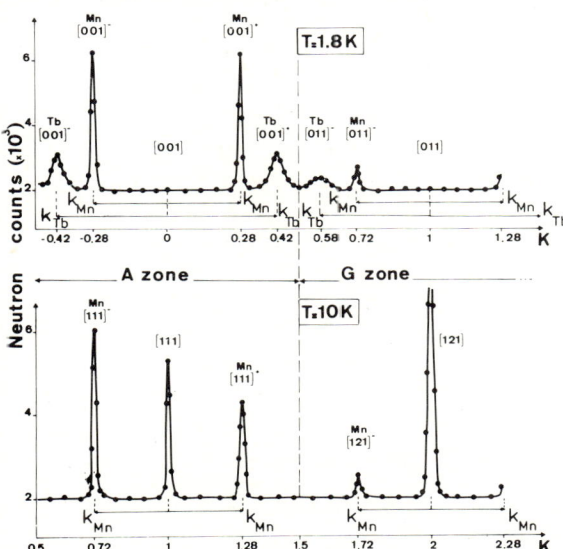

Fig. 2. $(0\,k\,1)$ scan at 1.8 K – nuclear peaks are absent. Manganese and terbium magnetic satellites are present. $(1\,k\,1)$ scan at 10 K – nuclear peaks and manganese magnetic satellites are observed.

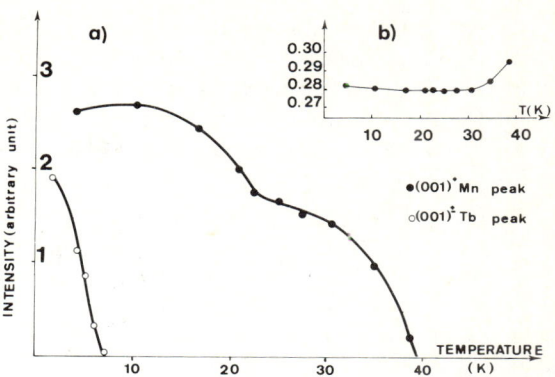

Fig. 3. Thermal variation of the $(001)^-$ intensities – the Mn^{3+} propagation vector value vs. temperature is given in inset.

5. Magnetic structure

Since the magnetic ordering in this compound appears to be quite complex the structure above 25 K is reported in a first step. The magnetic intensities measured at 30 K were difficult to interpret without taking extinction effects into account. In fact, nuclear intensities show considerable extinction: for the most intense peak, the measured value is only 32% of the theoretical one. The parameters g and r of the extinction theory [2] have been determined using a program of nuclear structure refinement ($g = 0.8'$, $r = 7\ \mu m$). The reliable factor for intensities is 5.5% with 92 observations.

In TbMnO$_3$ (space group Pbnm) the Mn^{3+} ions define four Bravais lattices labelled as : 1, $(\frac{1}{2}\,0\,0)$; 2, $(\frac{1}{2}\,0\,\frac{1}{2})$; 3, $(0\,\frac{1}{2}\,\frac{1}{2})$; 4, $(0\,\frac{1}{2}\,0)$. The absence of any satellite in F and C Brillouin zone implies the following relations between the Fourier components : $\boldsymbol{m}_k^{(2)} = -\boldsymbol{m}_k^{(1)}$ and $\boldsymbol{m}_k^{(3)} = -\boldsymbol{m}_k^{(4)}$. Moreover, the vanishing of the G-type satellites means that the phase between $\boldsymbol{m}_k^{(4)}$ and $\boldsymbol{m}_k^{(1)}$ is close to $\pi k_{Mn} = 50.4°$. The presence of only one Fourier component leads to a spiral or a sine-wave structure. Taking these conditions into account, all possible structures have been investigated and refined, applying the previously determined extinction corrections. The only reliable model ($R \approx 9\%$) consists of a sine-wave structure with the moment direction parallel to the propagation vector (**b**-axis). The sine-wave amplitude is 2.6 $\mu_B \pm 0.1$. This arrangement can only be understood if a large anisotropy confined the Mn^{3+} moment along the **b**-axis; it must be noted that the same direction has been observed in light rare-earth manganites [3–5]. At low temperatures, the structure still remains a sine-wave, the increase of the intensities may be interpreted either as a terbium polarization or as a decoupling and a rotation of the moments of both (1) and (4) Bravais lattices.

Both magnetic and neutron experiments reveal that terbium moment ordering occurs below 7 K with a sine-wave structure: the moments lie along two symmetrical directions with respect to the b-axis (57°) within the $a-b$ plane. This order is quite complex (existence of satellites in A–G–C zones) and of short-range as shown by the broad and weak superlattice peaks.

References

[1] J. Mareschal, J. Sivardiere and G.F. de Vries, J. Appl. Phys. 29 (2) (1968) 1364.
[2] P.J. Becker and Ph. Coppens, Acta Cryst. A30 (1974) 129; Acta Cryst. A31 (1975) 417.
[3] E.O. Wollan and W.C. Koehler, Phys. Rev. 100 (2) (1955) 545.
[4] G. Matsumoto, J. Phys. Soc. Jap. 29 (3) (1970) 606.
[5] S. Quezel-Ambrunaz, Bull. Soc. Fr. Min. Crist. 91 (1968) 339.

ON THE MAGNETIC PROPERTIES OF THE BASIC COMPOUNDS IN THE Fe_2O_3–Bi_2O_3 SYSTEM

J. DE SITTER, C. DAUWE, E. DE GRAVE, A. GOVAERT
and G. ROBBRECHT

Laboratory of Magnetism, University of Ghent, Ghent, Belgium

The Mössbauer spectrum of $BiFeO_3$ shows the presence of two non-equivalent octahedral sites for the ferric ion, having different quadrupole splittings below the Néel temperature. The hyperfine parameters in $Bi_2Fe_4O_9$ point to a greater degree of covalency for the ferric ions in tetrahedral coordination than in octahedral coordination.

1. Introduction

The Mössbauer effect has been used to study the compounds $BiFeO_3$ and $Bi_2Fe_4O_9$. In a previous paper [1], we mentioned the influence of the preparation on the room temperature Mössbauer spectrum of $BiFeO_3$. In all samples, prepared from the stoichiometric mixture of Bi_2O_3 and Fe_2O_3, the Mössbauer spectrum showed the presence of additional central absorption lines, due to small segregations of $Bi_2Fe_4O_9$. Single phase $BiFeO_3$ could only be obtained using the preparation method, described by Achenbach [2].

A $Bi_2Fe_4O_9$ sample was obtained by the usual ceramical techniques and showed some segregation of α-Fe_2O_3. This segregation was taken into account in analysing the Mössbauer spectra.

2. $BiFeO_3$

The Mössbauer spectra of $BiFeO_3$ were recorded between 77 and 800 K. Above its Néel temperature (645 K), $BiFeO_3$ shows a spectrum, consisting of two lines with equal intensity, and having a line width of about 0.27 mm/s. This indicates that at this temperature all Fe^{3+}-sites are equivalent.

The quadrupole coupling constant is about 0.42 mm/s at 700 K, which is in good agreement with literature data[3]. At increasing temperature the quadrupole splitting decreases slowly, having a value of 0.38 mm/s at 800 K. Below the Néel point the spectrum consists of a strongly broadened Zeeman pattern. Below 475 K some splitting of the absorption lines becomes visible and remains constant down to 77 K. The spectra can only be satisfactorily analysed using a superposition of two six-lines spectra. The linewidth of each component is less than 0.33 mm/s, while for a single six-lines pattern it is more than 0.50 mm/s, indicating that indeed not all the Fe-positions are equivalent. In fig. 1 the spectrum at 100 K is shown, together with the best fitting sum of twelve lorentzian lines. Because the $BiFeO_3$ sample was obtained by quenching from 750°C, the effect of an annealing process at 600°C was checked. Within experimental error, there is no difference between the hyperfine parameters obtained for the quenched and for the annealed sample. The relative intensity of both subspectra is also the same in the two samples. In our opinion this means that the line broadening and splitting are not due to defects, introduced by the quenching, but must be seen as a structural property of $BiFeO_3$.

An analysis with two subspectra seems

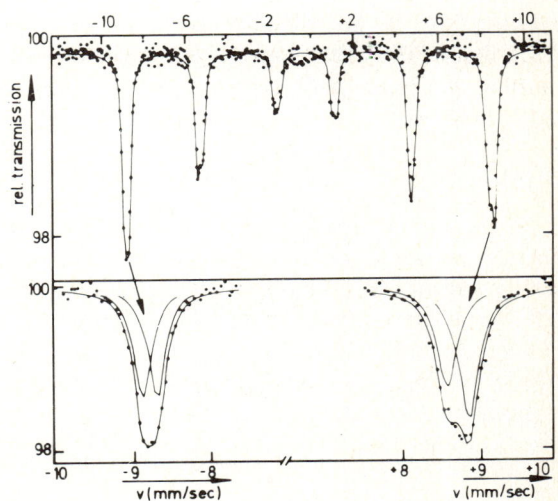

Fig. 1. The Mössbauer spectrum at 100 K of $BiFeO_3$. The fully drawn curve is the sum of twelve lorentzian lines fitted to the experimental data.

acceptable to us because of the fine structure observed in all spectra recorded below T_N. All lines were fitted without constraints, and the fitting led to a normal and unambiguous behaviour of both sets of hyperfine parameters.

The temperature dependence of the hyperfine parameters for both subspectra is studied. The isomer shift and the hyperfine field are equal on both sublattices over the whole temperature range. The only distinction is found between the quadrupole parameters 2ϵ, which are -0.03 mm/s and 0.44 mm/s, respectively. From the relative intensity of both subspectra we can conclude that the two Fe^{3+}-lattice sites are equally populated. From our measurements we can propose the following structure for $BiFeO_3$.

Each ferric ion is antiferromagnetically coupled to its six nearest neighbours, which all belong to the sublattice generating the other hyperfine pattern. Because no indication of any electrostatic transition at the Néel temperature has been found, the difference between the 2ϵ-values below the Néel temperature should be due to different angles between the main EFG-axis and the direction of the hyperfine field. These angles should be about $58 \pm 5°$ and $0 \pm 10°$ respectively. Because Fe^{3+} has no orbital angular momentum, a non-collinear spin-structure is very unlikely in this compound, so that we may conclude that the EFG-axis makes different angles with the trigonal axis for both sublattices. This implies that the point symmetry of a Fe^{3+}-site should be lower than trigonal, so that the EFG-tensor is not strictly axially symmetric. This means also that the oxygen displacements cannot be explained as rotations of the octahedra around the trigonal axis [4].

3. $Bi_2Fe_4O_9$

In the orthorhombic compound $Bi_2Fe_4O_9$, the ferric ions are equally distributed over tetrahedrally and octahedrally coordinated lattice sites. The hyperfine parameters of this compound are reported by Kostiner [5] and Bokov [6]. Both authors claim a very unusual temperature dependence for the isomer shift at tetrahedral sites, being smaller at 77 K than at 300 K. We

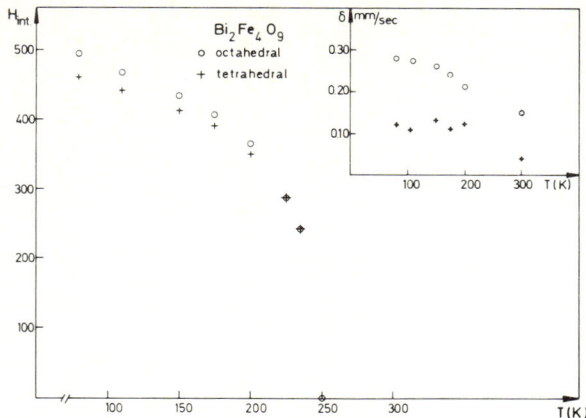

Fig. 2. Magnetic hyperfine field and isomer shift (vs. $^{57}Co/Pd$) for both ferric sites in $Bi_2Fe_4O_9$.

recorded Mössbauer spectra of $Bi_2Fe_4O_9$ between these temperatures and found a normal behaviour for all hyperfine parameters. Below 200 K the lines, belonging to both lattice sites can well be separated. In fig. 2 the temperature dependence of the hyperfine fields and the isomer shifts is illustrated. The ratio between the Mössbauer fractions at the tetrahedral and the octahedral sites is about 1.1. All these data point to a greater degree of covalency at the tetrahedral sites. We also found a greater value for the quadrupole coupling constant at tetrahedral sites, which cannot be explained by a simple point-charge model.

The authors thank IWONL and FKFO for financial support.

References

[1] J. De Sitter, C. Dauwe, E. De Grave and A. Govaert, Solid State Commun. 12 (1973) 645.
[2] G.D. Achenbach, W.J. James and R. Gerson, J. Amer. Ceram. Soc. 50 (1967) 437.
[3] C. Blaauw and F. van der Woude, J. Phys. C6 (1973) 1422.
[4] J.M. Moreau, C. Michel, R. Gerson and W.J. James, Acta Cryst. B26 (1970) 1425.
[5] E. Kostiner and G.L. Shoemaker, J. Solid State Chem. 3 (1971) 186.
[6] V.A. Bokov, G.V. Novikov, V.A. Trukhtanov and S.I. Yushchuk, Sov. Phys. Solid State 11 (1970) 2324.

NON-COLLINEAR MAGNETIC STRUCTURE IN MANGANESE-ZINC FERRITES

A.H. MORRISH and P.J. SCHURER

Department of Physics, University of Manitoba, Winnipeg, Canada R3T 2N2

The non-collinear magnetic structure of $Mn_{1-x}Zn_xFe_2O_4$, where $x = 0.7$ and 0.8, has been investigated as a function of temperature by Mössbauer spectroscopy with a 50 kOe magnetic field. For $x = 0.8$ the canting angle decreases as the temperature is raised above 4.2 K, and becomes essentially zero when $T = 30$ K. For $x = 0.7$, the canting is zero at $T = 20$ K.

1. Introduction

Over the past three or four years the magnetic structure of the disordered ferrites, $(M_{1-x}Zn_x)Fe_2O_4$ where $M = Mn$, Fe, Co, and Ni, has been studied in several laboratories by Mössbauer spectroscopy that utilizes large magnetic fields. Almost all these experiments have been conducted at 4.2 K, and sometimes the external magnetic field, usually produced by a superconducting solenoid, is varied up to as high as 90 kOe. In the zinc-rich region a non-collinear structure is found which is dependent on the strength of the external field. A program has commenced in our laboratory to investigate these systems at temperatures above that of liquid helium.

The $(Mn_{1-x}Zn_x)Fe_2O_4$ system with $x > 0.5$ is of special interest because all, or almost all, the ferric ions are reported to occupy octahedral or B sites in the spinel lattice [1, 2]. When iron cations also occupy the tetrahedral or A sites, as for the other mixed zinc ferrites, the overlapping Mössbauer absorptions obscure the analysis. The present communication reports on the Mössbauer spectra of $Mn_{0.3}Zn_{0.7}Fe_2O_4$ and $Mn_{0.2}Zn_{0.8}Fe_2O_4$ in an external longitudinal magnetic field of 50 kOe at liquid helium and higher temperatures.

2. Results and discussion

The Mössbauer spectrometer consisted of a ^{57}Co in Cr source, a constant-acceleration transducer, a proportional counter, and a multichannel analyzer operated in the time mode. The samples were placed inside a variable temperature offset cryostat inserted into a room temperature 5.1 cm. diameter bore of a superconducting solenoid. The two manganese-zinc ferrite samples were the same ones used earlier [2].

2.1. $(Mn_{0.2}Zn_{0.8})Fe_2O_4$

The Mössbauer spectra for $(Mn_{0.2}Zn_{0.8})Fe_2O_4$ in a magnetic field of 50 kOe applied parallel to the γ-ray direction at $T = 4.2$, 15, and 30 K are displayed in fig. 1. The spectrum at $T = 4.2$ K can be adequately fitted with three overlapping six-line patterns [2]. As the temperature is raised two features are obvious. One, the areas of the 2 and 5 lines, $A_{2,5}$, decrease compared to that of the 1 and 6 lines, $A_{1,6}$. Two, the absorption lines broaden, and hence possess more structure. In the analysis it is reasonable, at least as a first step, to base the number of patterns used on the possible configurations for the nearest-neighbor

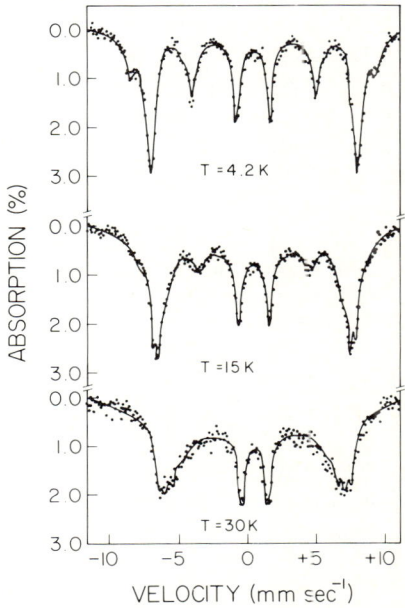

Fig. 1. Mössbauer spectra of $Mn_{0.2}Zn_{0.8}Fe_2O_4$ in an external logitudinal magnetic field of 50 kOe at $T = 4.2$, 15, and 30 K. The points indicate the experimental data and the curves the sum of four ^{57}Fe patterns obtained by a least-squares computer fit.

cations. For a ferric ion in a B site, the probability that the six nearest A-site neighbors will all be Zn ions is 0.26. Similarly, the probabilities for the other A-site arrangements are 0.39 for 5Zn1Mn, 0.24 for 4Zn2Mn, 0.08 for 3Zn3Mn, and 0.02 for 2Zn4Mn. If the last two are grouped together (for a total probability of only 0.10), the implication is that four patterns should be used.

The sum of the four patterns, fitted by computer with a minimal number of constraints, and yielding excellent values for χ^2, are drawn in fig. 1. The hyperfine fields for each pattern, 1 to 4, in decreasing magnitude, are listed in table I. Pattern 1 is identified with ferric ions with 6 Zn nearest A-site neighbors. These ferric ions have reversed spins, that is, a magnetic moment with a component antiparallel to the applied field direction. The relative areas of this pattern, compared to the total, are 28, 20 and 22% at $T = 4.2$, 15, and 30 K, respectively, and are reasonably close to the 26% expected. The average canting angles obtained from the ratio $A_{2,5}/A_{1,6}$, are 25(7), 18(5) and 0(8) degrees at $T = 4.2$, 15, and 30 degrees, respectively (the errors are given in parentheses).

The magnetic moments for the rest of the ferric ions have a component parallel to the applied field direction. Therefore, pattern 4 is presumably associated with the 5Zn1Mn configuration. The relative areas of the patterns 2 to 4 are not in as good accord with the configuration probabilities; by imposing more constraints, the relative areas can be brought into respectable agreement, but at the expense of an increase in χ^2. The average canting angles for the three patterns, combined, (2, 3, and 4), are 36(6), 21(4), and 0(8) degrees at 4.2, 15, and 30 K, respectively.

At $T = 30$ K, it is less certain that pattern 1 represents reversed spins. Further, the width of pattern 4 becomes very large, and corresponds to a wide range in hyperfine fields.

2.2. $Mn_{0.3}Zn_{0.7}Fe_2O_4$

For this composition, the average canting angle is smaller; it is $\sim 25°$ at $T = 4.2$ K, and $\sim 0°$, at $T = 20$ K. As the temperature is raised, the linewidths increase, consistent with a significantly larger range in the hyperfine fields. The persistence of small outer 1 and 6 lines at higher temperatures seems to imply that about 3% of the ferric ions occupy A sites, contrary to earlier suppositions [1, 2].

3. Conclusions

Although more structure appears in the Mössbauer spectra at temperatures above $T = 4.2$ K, the resolution obtained was insufficient to reveal the details in the magnetic structure. More data on the line shapes may permit the development of a local canting model. The increase in the range of hyperfine fields above 4.2 K for $H = 0$ appears to imply that neighbors more distant than the nearest ones are important. It has been established that the average canting angle, determined in an external magnetic field of 50 kOe, decreases with increasing temperature.

A grant from the National Research Council of Canada made this research possible.

References

[1] U. Konig and G. Chol, J. Appl. Crystallogr. 1 (1968) 124.
[2] A.H. Morrish and P.E. Clark, Phys. Rev. B11 (1975) 278.

Table I
Hyperfine fields (in kOe) for the four patterns fitted to the Mössbauer spectrum of $Mn_{0.2}Zn_{0.8}Fe_2O_4$ in a 50 kOe magnetic field

Temperature (K)	H_1	H_2	H_3	H_4
4.2	558	490	468	446
15	507	462	438	407
30	441	408	375	331

SPIN REORIENTATION IN TmCrO$_3$

T. TAMAKI, K. TSUSHIMA

Broadcasting Science Research Laboratories of Nippon Hōsō, Kyōkai, Kinuta, Setagaya-ku, Tokyo 157, Japan

and

Y. YAMAGUCHI

Electro-Technical Laboratory, Tanashi, Tokyo 188, Japan

A new type of the temperature- and field-induced spin reorientation was studied for TmCrO$_3$ by the magnetization measurements with the method of sample motion technique and with the magnetic torque meter from 1.63 to 300 K. At the low temperature, field induced spin reorientation and Tm^{3+} reordering temperature, T_{N2}, were observed.

TmCrO$_3$, in conformity with other rare earth orthochromites (RCrO$_3$), is an orthorhombically distorted perovskite structure [space group $P_{bnm}(D_{2h}^{16})$] with four formula units per unit cell [1]. RCrO$_3$ has been known as the weak ferromagnetic crystal which shows the spin reorientation (SR) transition, in general, between the temperature, T_2 and T_1, below the first Néel temperature, T_{N1}. They also have another transition temperature, T_{N2}, below which rare earth spins begin to reorder by the magnetic interaction between rare earth ions. By this time, three kinds of the SR have been observed in RCrO$_3$: from $\Gamma_4(G_x, A_y, F_z; F_z^R)$ type spin structure above T_2 to $\Gamma_2(F_x, C_y, G_z; F_x^R, C_y^R)$ below T_1, Γ_4 to $\Gamma_1(A_x, G_y, C_z; C_z^R)$ and Γ_2 to Γ_1, respectively. The magnetic study of TmCrO$_3$ has been done by the neutron diffraction [2], specific heat [3] and magnetization measurement [4], but its magnetic property is not clear yet. Therefore, we measured carefully the magnetizations particularly at low field and low temperature ($T < 10$ K) by the magnetometer. The magnetization, principal susceptibilities and magnetic torque for flux-grown single crystals of TmCrO$_3$ were measured by the magnetometer of sample motion type and by the magnetic torquemeter down to 1.63 K. The lattice constants of our crystals were determined by X-ray diffraction as $a = 5.206$ Å, $b = 5.504$ Å, $c = 7.497$ Å. The single crystal was shaped into sphere and was oriented to its proper crystal axis by X-ray Laue method, then transferred and cemented as it was to the measuring-sample holder. In this crystal, the weak ferromagnetism has been observed by Hornreich et al. [4] along the a-axis until 2.2 K, but they did not find T_{N2} nor temperature induced SR. Our measurements down to 1.63 K show that the weak ferromagnetic moment (WFM) rises very steeply below the compensation temperature 29 K until $T_{N2} \simeq 5$ K where it drops again as temperature decreases. Fig. 1 shows the temperature dependence of spontaneous magnetic moment along the a and c-axes ($\sigma_a, \sigma_{a(0)}$ and $\sigma_{c(0)}$) and susceptibility (χ_c). σ_a indicates values of WFM from the magnetic torque, $\sigma_{a(0)}$ and $\sigma_{c(0)}$ are those from the magnetization data, respectively, at zero field. Each magnetization datum was corrected for demagnetizing field. Susceptibility, χ_c rises very steeply near $T_{N2} \simeq 5$ K. In fig. 2 are

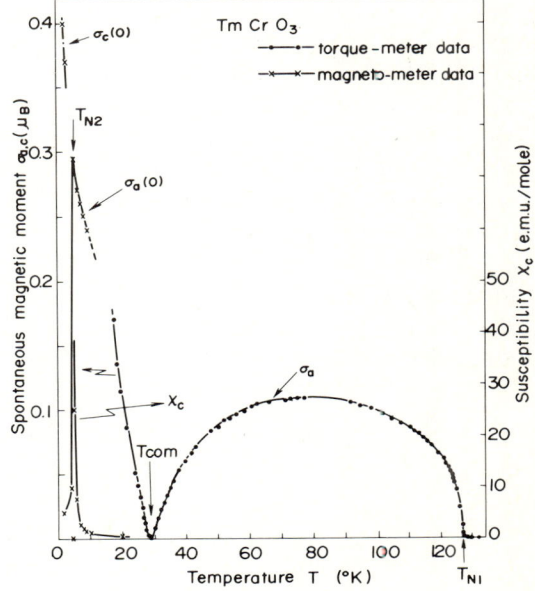

Fig. 1. Spontaneous magnetic moment ($\sigma_a, \sigma_{a(0)}$ and $\sigma_{c(0)}$), and magnetic susceptibility (χ_c) of TmCrO$_3$ as a function of temperature (●●● is from the magnetic torque data (σ_a), ×××× from magnetometer data. $\sigma_{a(0)}$ and $\sigma_{c(0)}$ indicate remanent magnetizations.)

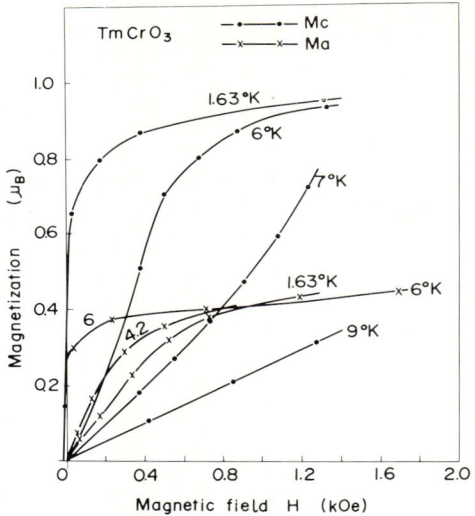

Fig. 2. Field dependence of the magnetization of TmCrO$_3$ along the a and c-axes.

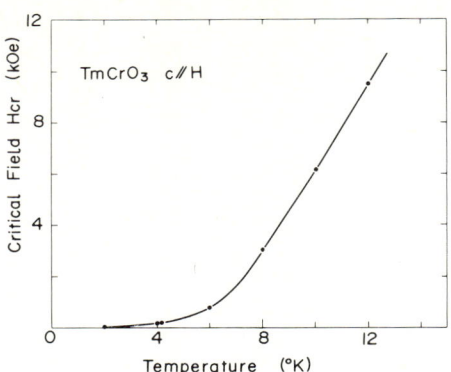

Fig. 3. Critical field (H_{cr}) of field induced spin reorientation as a function of temperature in TmCrO$_3$ along the c-axes.

shown the magnetization versus external field obtained with field parallel to the a and c-axes. At 1.63 K, the remanent magnetization of $\sigma_{c(0)} = 0.4\,\mu_B$ and $\sigma_{a(0)} \doteq 0.02\,\mu_B$ was clearly observed. For both cases the coercive force was approximately 20 Oe. The values, $\sigma_{a(0)}$, $\sigma_{c(0)}$ and χ_c in fig. 1 were plotted using those of fig. 2. From our results, the spin configuration of TmCrO$_3$ for $T_2 < T < T_{N2}$ is identified to be $\Gamma_{2i(i=6\,or\,7)}$ (F_x, C_y, G_z; F_x^R, C_y^R, A_z^R or F_x^R, C_y^R, G_z^R) which holds from T_{N2} to T_2 and it will presumably be $\Gamma_{2i} + \Gamma_{4i}$ (G_x, A_y, F_z; F_z^R, A_z^R or F_z^R, G_z^R) or Γ_{24i} [5] below T_2 ($2 < T_2 < 4.2$ K). This type of SR is new which is also consistent with the recent optical absorption measurement of Cr^{3+} excitons by Aoyagi et al. [6]. Note that the applied field required to rotate the WFM from the a to the c-direction at 4.2 K is as small as 200 Oe. The critical field, H_{cr}, of the field induced SR versus temperature is shown in fig. 3.

From this result, TmCrO$_3$ is very soft magnetic material near 4 K, and its derivative dH_{cr}/dT is obtained as 1.6 (kOe/deg.) between 7 and 12 K. It would be necessary to consider the magnetic interaction between the Cr^{3+} ions and Tm^{3+} ions which have the easy axis along the c-direction.

The authors thank Dr. T. Teranishi for his help and guidance in the X-ray measurement, and H. Nishimura and T. Matsui for considerable assistance.

References

[1] S. Geller and E.A. Wood, Acta Cryst. 9 (1956) 563.
[2] E.F. Bertaut, J. Mareschal, G. de Vries, R. Aleonard, R. Pauthenet, J.P. Rebouillat and V. Zarubicka, IEEE Trans. Mag. 2 (1966) 453.
[3] J.B. Ayasse, A. Berton and J. Sivardiere, Compt. Rend. 271B (1970) 1220.
[4] R.M. Hornreich, B.M. Wanklyn and I. Yaeger, Int. J. Magn. 4 (1973) 313.
[5] T. Yamaguchi and K. Tsushima, Phys. Rev. 8B (1973) 5187.
[6] K. Aoyagi, M. Kajiura, T. Morishita and K. Tsushima, Abstr. 31st Ann. Conf. Phys. Soc. Jap. 2 (1976) 60.

MÖSSBAUER STUDY OF ATiNdO$_5$ (A = Cr, Mn, Fe)

P.J. SCHURER and A.H. MORRISH
Department of Physics, University of Manitoba, Winnipeg, Canada R3T 2N2

The covalency contributions to the isomer shift, δ, and the magnetic hyperfine field, H_{hf}, for Fe^{3+} ions in octahedral and square-pyramid sites have been calculated and compared with experimentally determined values for FeTiNdO$_5$. In addition the distribution of A^{3+} and Ti^{4+} ions in ATiNdO$_5$ (A = Cr, Mn, Fe) has been investigated.

1. Introduction

The compounds ATiNdO$_5$ (A = Cr, Mn, Fe) belong to the orthorhombic space group Pbam and are antiferromagnets [1]. The A^{3+} and Ti^{4+} ions occupy the 4f and 4h sites in this space group which have octahedral and square-pyramid oxygen environments, respectively. A neutron-diffraction study showed that the Cr^{3+} ions have a strong preference for the octahedral site, whereas the Fe^{3+} ions are almost equally distributed over the 4f and 4h sites. The Nd^{3+} ions occupy the 4g positions. No experimental information on the distribution of Mn^{3+} and Ti^{4+} ions in the Mn compound is available.

We have studied the compounds Cr$_{0.95}$Fe$_{0.05}$TiNdO$_5$, Mn$_{0.95}$Fe$_{0.05}$TiNdO$_5$, and FeTiNdO$_5$ using the Mössbauer effect (ME) technique. These measurements can, among others, also give information about the distribution of A^{3+} ions over the two available lattice positions.

2. Experiments and analyses

The compounds, FeTiNdO$_5$, Cr$_{0.95}$Fe$_{0.05}$TiNdO$_5$, and Mn$_{0.95}$Fe$_{0.05}$TiNdO$_5$, the last two with iron enriched to 70% in the isotope ^{57}Fe, have been prepared in the same way as described by Buisson [1]. ME spectra at 4.2 K show that the Fe^{3+} spins are magnetically ordered in all three compounds. For the Fe compound the Néel temperature, T_N, of 33 K is substantially higher than for the two other compounds. The magnetic transition temperature for the Fe doped Cr compound is $T_N = 17.5$ K; it is slightly higher than the value of $T_N = 13$ K found for the undoped Cr compound [1]. No magnetic ordering was observed for MnTiNdO$_5$ at 1.5 K using neutron diffraction, whereas for the Fe doped compound T_N is found to be 16.2 K.

The ME spectrum of FeTiNdO$_5$ at 4.2 K, shown in fig. 1, consists of six absorption peaks with very broad line widths (FWHM ≈

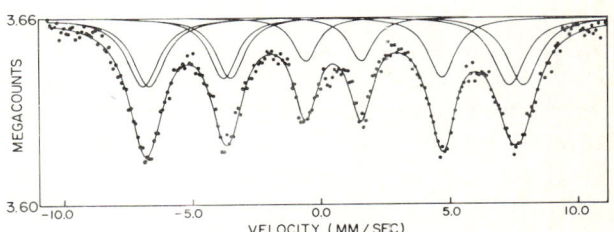

Fig. 1. Mössbauer spectrum of FeTiNdO$_5$ at 4.2 K. The points represent the experimental data and the curves the two individual patterns and their sum as obtained by a computer fit.

1.4 mm s^{-1}). From the neutron-diffraction results, the ME spectrum is expected to consist of two six-line absorption patterns of almost equal intensity which correspond to Fe^{3+} ions situated at 4f and 4h sites, respectively. At this point it is useful to estimate the values of the hyperfine fields expected for these two sites. It is well known that the differences in H_{hf} and δ measured for Fe^{3+} ions in various oxygen surroundings (octahedral, tetrahedral, etc.) can be understood by taking into account covalency effects [2, 3], that is

$$H_{hf} = H_{free} + H_{cov} + H_{red} + H_{sthf}$$

and

$$\delta = \delta_{free} + \delta_{cov} + \delta_{red}.$$

The terms, H_{cov} and δ_{cov}, have their origin in the overlap of doubly occupied Fe^{3+} ns orbitals with oxygen p-electron orbitals and in the transfer of p-electrons into empty 4s orbitals. Likewise, H_{red} and δ_{red} are contributions from the overlap and transfer effects involving the Fe^{3+} 3d orbitals. The supertransferred hyperfine field, H_{sthf}, has its source in the transfer of p-electrons into empty $d_{3z^2-r^2}$ orbitals of the neighbouring Fe^{3+} ions in the superexchange bonds. In the calculations of these covalency contributions for $FeTiNdO_5$, only σ-bonding has been considered. The average Fe–O distances, determined from X-ray diffraction measurements, are 2.11 Å for the 4f and 1.92 Å for the 4h sites in $FeTiNdO_5$. For these distances the values of the single orbital overlap integrals $S'_{ns} = \langle p | \phi_{ns} \rangle$ and $S_\sigma = \langle p^\uparrow | d^\uparrow_{3z^2-r^2} \rangle$ were determined using the values given by Sawatzky et al. [2]. For octahedral sites, transfer integral values of $a'_{4s}(p_z \rightarrow \phi_{4s}) = 0.125$ and $\beta_\sigma(p^\downarrow_z \rightarrow d^\downarrow_{3z^2-r^2}) = 0.320$ have been determined experimentally for a Fe^{3+}–O^{2-} distance of 2.011 Å. If $B_\sigma \propto S_\sigma$ and $a'_{4s} \propto S'_{4s'}$ then $B_\sigma(4f) = 0.286$ and $a'_{4s}(4f) = 0.121$ are reasonable to use in the calculation. Furthermore, for tetrahedral sites with a Fe^{3+}–O^{2-} distance of 1.89 Å values of $a'_{4s} = 0.181$ and $B_\sigma = 0.466$ have been reported [2]; these values have also been used in the calculations for the square-pyramid site. The average superexchange bond angle is $\theta \approx 120°$. Finally, H_{free} has been taken equal to -630 kOe.

The results of the calculation are shown in table I. The average supertransferred hyperfine field, $\bar{H}_{sthf}$, has been calculated assuming a random distribution of Fe^{3+} and Ti^{4+} ions over 4f and 4h sites. Such a distribution is expected to cause a line broadening in the two absorption patterns since there will then be a dispersion in H_{sthf} and also in the electric field gradient. As a result the two absorption patterns corresponding to 4f and 4h sites are not well resolved. From a least-squares fit of the $FeTiNdO_5$ spectrum, the average values of the isomer shift and the hyperfine field for the two sites were obtained (table I). In addition an area ratio for the two sites was determined to be close to unity, which confirms that the Fe^{3+} ions, and therefore also the Ti^{4+} ions, are almost equally distributed over the 4f and 4h sites. Furthermore, there is good agreement between the measured and calculated values of H_{hf} for the 4h site. On the other hand, $|H_{hf}(4f)|$ and also the difference $\delta(4f) - \delta(4h)$ is smaller than expected from the calculation. These discrepancies may occur because the separation of the 4f sites along the c axis is only about 2.9 Å. As a consequence, the overlap between orbitals of neighbouring 4f ions probably produces a decrease in $|H_{hf}(4f)|$ and in $\delta(4f)$.

In $Cr_{0.95}Fe_{0.05}TiNdO_5$ the Cr^{3+} ions prefer the 4f site [1]. Since the Fe^{3+} ions have no preference for either of the two available sites, the majority of the Fe^{3+} ions are expected to occupy the 4h sites. The spectrum at 4.2 K shows two components; the largest (95%) can be associated with 4h sites and the smallest either with 4f sites or perhaps with another phase. The hyperfine fields extrapolated to 0 K, are $H_{hf}(4f) = -491 \pm 5$ kOe and $H_{hf}(4h) = -401 \pm 5$ kOe. The difference between these two values is very close to that calculated for the $FeTiNdO_5$ compound.

For $MnTiNdO_5$, Buisson [1] has proposed that the Mn^{3+} ions prefer the 4h sites. Then the Fe^{3+} ions in $Mn_{0.95}Fe_{0.05}TiNdO_5$ are expected to occupy 4f sites preferentially. From a computer fit we determine $H_{hf}(4f) = -455 \pm 5$ kOe and $H_{hf}(4h) = -399 \pm 5$ kOe and the ratio of the areas of the two components is $A(4f)/A(4h) = 1.9$. This indicates that the Mn^{3+} ions indeed prefer the 4h site.

Table I
Parameters deduced for $FeTiNdO_5$

Site	δ_{cov}	δ_{red}	δ (meas)	H_{cov}	H_{red}	$\bar{H}_{sthf}$	$H_{hf}(0)$ (calc)	$H_{hf}(0)$ (meas)
4f	−0.606	0.096	0.44(5)	73	43	−22	−536	−468(3)
4h	−0.875	0.196	0.43(5)	111	79	−8	−448	−444(3)

The isomer shifts, δ, with respect to iron, are in mm s^{-1}. The magnetic fields, H, are in kOe. The figures in parentheses are the probable errors.

References

[1] G. Buisson, J. Phys. Chem. Solids 31 (1970) 1171.
[2] G.A. Sawatzky, C. Boekema and F. van der Woude, Proc. Conf. on Mössbauer spectrometry (Phys. Soc. DDR, Dresden, 1971) p. 238.
[3] C. Boekema, F. van der Woude and G.A. Sawatzky, Int. J. Magn. 3 (1972) 341.

MAGNETIC STRUCTURES OF SOME [AC$_3$](B$_4$)O$_{12}$ COMPOUNDS WITH A PEROVSKITE-LIKE STRUCTURE

A. COLLOMB, D. SAMARAS, B. BOCHU*, J. CHENAVAS, M.N. DESCHIZEAUX*,
G. FILLION†, J.C. JOUBERT* and M. MAREZIO

Laboratoire des Rayons X, CNRS, 166 X, 38042 Grenoble Cedex, France

The magnetic structures as determined by neutron diffraction powder data of three compounds belonging to the series [ACu$_3$](B$_4$)O$_{12}$ are described. ThCu$_3$Mn$_4$O$_{12}$ is ferrimagnetic (T_C = 430 K), Ca(Cu$_{0.78}$Mn$_{0.22}$)$_3$Mn$_4$O$_{12}$ is ferromagnetic (T_C = 350 K), while CaCu$_3$Ti$_4$O$_{12}$ is antiferromagnetic (T_N = 27 K).

The [AC$_3$](B$_4$)O$_{12}$ compounds have a perovskite-like structure, they are cubic with a cell parameter of about 7.4 Å, and the space group is Im3 [1]. The stability of this structure is due to the presence of a Jahn-Teller ion, such as Cu^{2+} or Mn^{3+}, on the C site. The particularity of the [Ca(Cu$^{2+}_{0.78}$Mn$^{3+}_{0.22}$)$_3$]Mn$_4$O$_{12}$ compound is to contain, on the C site, both Cu^{2+} and Mn^{3+} ions.

We shall describe the magnetic structures of the [CaCu$^{2+}_3$](Ti$_4$)O$_{12}$, [ThCu$^{2+}_3$](Mn$^{4+}_{0.5}$Mn$^{3+}_{0.5}$)$_4$O$_{12}$, [Ca(Cu$^{2+}_{0.78}$Mn$^{3+}_{0.22}$)$_3$](Mn$^{4+}_{0.835}$Mn$^{3+}_{0.165}$)$_4$O$_{12}$ compounds, studied by neutron diffraction.

1. [CaCu$^{2+}_3$](Ti$_4$)O$_{12}$

The susceptibility measurements, at low temperature, show an antiferromagnetic behaviour with a Néel temperature T_N = 27 K. The presence, in neutron diffraction powder pattern, of the magnetic intensities with $h+k+l = 2n+1$ shows that the crystallographic translation ($\frac{1}{2}\frac{1}{2}\frac{1}{2}$) is now a magnetic anti-translation. The Schubnikov group is R$\bar{3}$'. The magnetic moments of the Cu^{2+} ions, related to each other by the three fold rotation, are on a cone around the [111] axis and are situated in the (110) plane. The angle between the spin direction and the [001] axis is 33°. The experimental moment of the Cu^{2+} ion is 1 μ_β.

2. [ThCu$^{2+}_3$](Mn$^{4+}_{0.5}$Mn$^{3+}_{0.5}$)$_4$O$_{12}$ [2]

This compound is ferrimagnetic with a Curie temperature of 430 K. The magnetic structure refinement based on neutron diffraction powder data (table I) leads to a collinear configuration of the Cu^{2+} and of ⟨Mn$^{4+}_{0.5}$Mn$^{3+}_{0.5}$⟩ moments along

*Laboratoire de Génie Physique, INPG, 38040 Grenoble Cedex, France.

†Laboratoire de Magnétisme, CNRS, 166 X, 38042 Grenoble Cedex, France.

the [111] axis (fig. 1). These orientations lead to a spontaneous magnetization σ_S = 9.4 ± 0.5 μ_β/mol which is in good agreement with the value obtained from the magnetic measurements (σ = 9.0 ± 0.5 μ_β/mol) (fig. 2).

Table I
Observed and calculated magnetic intensities of the [ThCu$_3$](Mn$_4$)O$_{12}$ compound

h	k	l	$I_{obs.}$	$I_{cal.}$
2	0	0	190	194
2	2	0	111	92
2	2	2	(a)	137
4	0	0	(b)	20
4	2	0	270(b)	224
4	2	2	54(b)	38

(a) Not measurable because of the large value of the nuclear intensity.

(b) Badly measured because of the indetermination of the background.

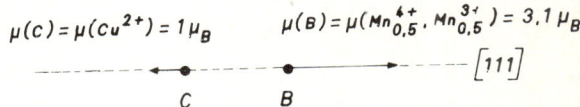

Fig. 1. Magnetic structure of [ThCu$_3$](Mn$_4$)O$_{12}$.

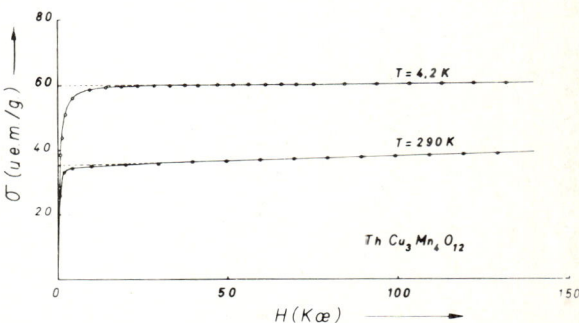

Fig. 2. Magnetization of [ThCu$_3$](Mn$_4$)O$_{12}$ versus magnetic field.

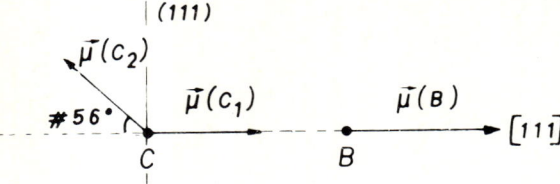

Fig. 3. Magnetic structure of $[Ca(Cu_{0.78}^{2+}Mn_{0.22}^{3+})_3](Mn_4)O_{12}$.

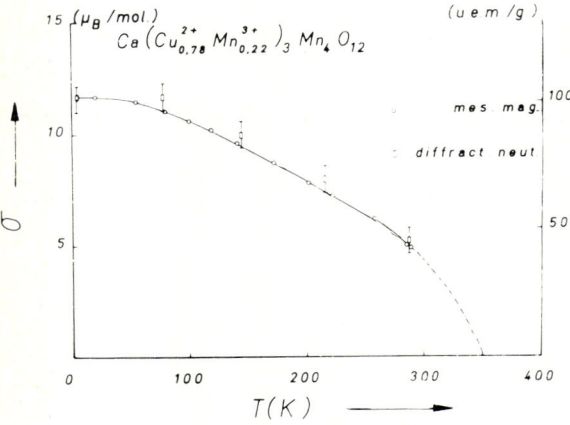

Fig. 4. Magnetization of $[Ca(Cu_{0.78}^{2+}Mn_{0.22}^{3+})](Mn_4)O_{12}$ versus temperature.

3. $[Ca(Cu_{0.78}^{2+}Mn_{0.22}^{3+})_3](Mn_{0.835}^{4+}Mn_{0.165}^{3+})_4O_{12}$ [3]

The ordering temperature of this compound is $T = 350$ K. The magnetic structure is shown in fig. 3. The magnetic moment situated on the B site corresponds to an average moment $\mu(Mn_{0.835}^{4+}Mn_{0.165}^{3+}) = 2.75\ \mu_B$, while that situated on the C site is due to two contributions namely, $\mu(C_1) = \mu(Cu_{0.78}^{2+}) = 0.78\ \mu_B$ and $\mu(C_2) = \mu(Mn_{0.22}^{3+}) = 0.77\ \mu_B$. The spin arrangement found in this compound is consistent with the trigonal symmetry. For instance, the angles between the projections on the (111) planes of the $\mu(C_1)$ or of the $\mu(C_2)$, are 120°. The magnetizations obtained by neutron diffraction and magnetic measurements, in the temperature range 4.2–300 K, are shown on the fig. 4.

References

[1] B. Bochu, J. Chenavas, A. Collomb, M.N. Deschizeaux, G. Fillion, J.C. Joubert and M. Marezio, Proc. Int. Conf. on Magnetism, Amsterdam (1976).
[2] M.N. Deschizeaux, J.C. Joubert, A. Vegas, A. Collomb, J. Chenavas and M. Marezio, J. Solid State Chem. 19 (1976) 45.
[3] A. Collomb, Thèse d'Etat, Université de Grenoble, France (1976).

SYNTHESIS, STRUCTURE CHARACTERISATION AND MAGNETIC PROPERTIES OF SOME NEW MAGNETIC PEROVSKITE-LIKE OXIDES

B. BOCHU*, J. CHENAVAS†, A. COLLOMB†, M.N. DESCHIZEAUX*, G. FILLION‡, J.C. JOUBERT† and M. MAREZIO†

*Laboratoire de Génie Physique, INPG, 38040 Grenoble Cedex, France
†Laboratoire des Rayons X, CNRS, 166 X, 38042 Grenoble Cedex, France
‡Laboratoire de Magnétisme, CNRS, 166 X, 38042 Grenoble Cedex, France

The synthesis conditions of several $AC_3B_4O_{12}$ perovskite-like compounds belonging to the $ACu_3^{2+}B_4O_{12}$ subseries are given and discussed. These compounds have remarkable magnetic properties. Chemical formulae as determined by neutron diffraction powder data are discussed.

1. Introduction

The $[AC_3](B)_4O_{12}$ compounds have a perovskite-like arrangement when A is a large monovalent, divalent, trivalent or tetravalent cation such as Na^+, Ca^{2+}, Cd^{2+}, Y^{3+}, RE^{3+}, Th^{4+}, B a small cation suitable for octahedral coordination such as Mn^{4+}, Mn^{3+}, Cr^{4+}, Cr^{3+}, Ge^{4+}, Ti^{4+}, Ta^{5+}... and C a Jahn–Teller cation [1]. Two subseries have been prepared so far: $ACu_3^{2+}B_4O_{12}$ and $AMn_3^{3+}B_4O_{12}$. Their prototypes are $CaCu_3Mn_4O_{12}$ and $NaMn_3Mn_4O_{12}$, respectively. When the B cations are Mn^{4+} or Mn^{3+} or both, the compounds of the copper subseries are strongly ferro- or ferrimagnetic whereas the compounds of the manganese subseries are antiferromagnetic.

2. Synthesis

Some of the compounds which contain Mn^{3+} on both the B and C sites, e.g. $La^{3+}Mn_3^{3+}Mn_4^{3+}O_{12}$, must be synthesised under very high pressure conditions because they are indeed high pressure phases. On the contrary, compounds which contain copper on the C-sites and, for example, Ti^{4+} on the B ones, can easily be prepared at normal pressure. In this case the temperature of the synthesis must not exceed $\sim 1000°C$ because at this temperature the Cu^{2+} cations begin to reduce to Cu^{1+}. A much lower temperature should be used for the synthesis of the compounds containing Mn^{4+} or Cr^{4+} on the B sites because these cations are not stable in air above 400°C. Unfortunately this temperature is too low for a solid-state reaction. It is therefore necessary, for preparing these compounds, to use either very high pressure and temperature conditions, or conventional hydrothermal synthesis under a high oxidizing atmosphere. For example, $Th^{4+}Cu_3^{2+}(Mn_2^{4+}Mn_2^{3+})O_{12}$ can be prepared from the stoichiometric mixture $Th(OH)_4 + 3CuO + 2MnO_2 + Mn_2O_3$ either in a belt-type apparatus at 80 kbar and 1200°C or by hydrothermal synthesis at 2 kbar and 600°C. Other oxidizing agents such as $KClO_3$ or $HClO_3$ can be used. Chlorides or sulfates are other possible starting materials. Table I shows the synthesis conditions of some of these com-

Table I
Synthesis conditions and cell parameters of some $AC_3B_4O_{12}$ compounds

Compounds	Starting materials	Synthesis conditions	Cell parameters
$Ca^{2+}Cu_3^{2+}Mn_4^{4+}O_{12}$	$CaCl_2 + CuCl_2 + KMnO_4 + H_2O$ $Ca(OH)_2$, CuO, MnO_2	1.5 kb 600°C 24 h 40 kb 1000°C 1 h	$a = 7.241$ Å
$Y^{3+}Cu_3^{2+}Mn_3^{4+}Mn^{3+}O_{12}$	$YCl_3 + CuSO_4, 5H_2O + KMnO_4 + H_2O$	3 kb 500°C 24 h 50 kb 800°C 1 h	$a = 7.253(2)$ Å
$Th^{4+}Cu_3^{2+}Mn_2^{4+}Mn_2^{3+}O_{12}$	$Th(OH)_4 + {}_3CuO + {}_2MnO_2 + Mn_2O_3$	2 kb 600°C 80 kb 1200°C 1 h	$a = 7.359(1)$ Å

pounds together with their lattice parameters. These values were obtained by least-squaring the X-ray powder data taken with a 360 mm circumference Guinier focusing camera, FeKα radiation (λ = 1.9373 Å) and KCl as an internal standard.

3. Crystal structure of $CaCu_3Mn4O_{12}$

The crystal symmetry of $CaCu_3Mn_4O_{12}$ is cubic, space group Im3 with a cell parameter of 7.241 Å which is twice the parameter of the simple ABO_3 perovskite. The structural arrangement consists of a tilted network of oxygen octahedra sharing corners, the Mn^{4+}–O–Mn^{4+} angle is 140° instead of 180°. This tilt gives rise to two types of polyhedra (the A and C sites) around the calcium and copper cation, respectively. The calcium cations are surrounded by quite regular icosahedra with 12 Ca–O distances equal to 2.563(3) Å. The copper cations are surrounded by very distorted 12-oxygen polyhedra which consist of three mutually perpendicular rectangles of different size, the smallest and the largest of which are almost squares. The three sets of Cu–O distances are 1.942(3), 2.707(3) and 3.181(3) Å. Fig. 1 shows the A, C and B polyhedra.

4. Discussion

$YCu_3Mn_4O_{12}$ and $ThCu_3Mn_4O_{12}$ are ferrimagnetic; the magnetizations and the Curie temperatures are 9.4 μ_B/mole, 9 μ_B/mole and 400 K, 430 K, for the two compounds, respectively. On the contrary, $CaCu_3Mn_4O_{12}$ is ferromagnetic. The magnetization and the Curie temperature are 11.5 μ_B/mole and 350 K. The magnetic properties of $ThCu_3Mn_4O_{12}$ and $CaCu_3Mn_4O_{12}$ along with their magnetic structures as determined by neutron diffraction data are reported in the following article.

Structural refinements based on nuclear neutron diffraction data show that the chemical

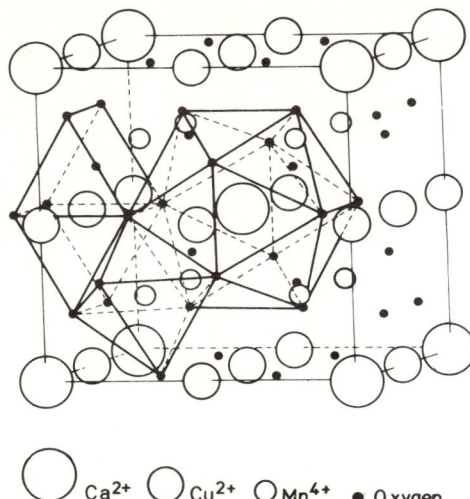

Fig. 1. A, C and B polyhedra in $CaCu_3Mn_4O_{12}$.

formula of the calcium–copper manganese compound is $Ca^{2+}[Cu^{2+}_{2.3}Mn^{3+}_{0.7}](Mn^{4+}_{3.3}Mn^{3+}_{0.7})O_{12}$ and that of the thorium and yttrium compounds are instead $Th^{4+}[Cu^{2+}_3](Mn^{3+}_2Mn^{4+}_2)O_{12}$ and $Y^{3+}[Cu^{2+}_3](Mn^{3+}Mn^{4+}_3)O_{12}$, respectively. A large number of hydrothermal runs were done at different pressures and temperatures in order to synthesize the stoichiometric $CaCu_3Mn_4O_{12}$. It has been found by X-ray diffraction powder data that the cell parameter increases when the pressure decreases and/or the temperature increases. Such experimental conditions favor the reduction of the manganese cations, that is, some of the Mn^{3+} replace the Mn^{4+} in the B sites and, as a consequence, the same amount of Cu^{2+} are replaced by Mn^{3+} on the C-sites. This corroborates qualitatively the chemical formula deduced from the neutron diffraction data.

References

[1] M.N. Deschizeaux, J.C. Joubert, A. Vegas, A. Collomb, J. Chenavas and M. Marezio, J. Solid State Chem. 19 (1976) 45, and references therein.

COLLECTIVE ELECTRON STATES IN $Zn_xFe_{3-x}O_4$ AND $Cd_xFe_{3-x}O_4$ FOR $0 \leq x \leq 0.3$

B.J. EVANS and HANG NAM OK

Department of Chemistry, The University of Michigan, Ann Arbor, Michigan 48109, USA

^{57}Fe Mössbauer measurements of $Zn_xFe_{3-x}O_4$ and $Cd_xFe_{3-x}O_4$ have been made at 300 K in zero applied field and at 180 K in zero and 55 kG applied fields. The compositional dependence of the isomer shifts, B to A area ratios and hyperfine fields, which are sensitive to and diagnostic of the conduction mechanism, have been determined under all three experimental settings. The decreasing isomer shift difference between the A and B site, and increasing area ratio are in good agreement with the predictions of band conduction. The hyperfine field at the B site having no Zn or Cd A site neighbors is also observed to increase for small x values, again, in good accord with band conduction.

1. Introduction

In recent years, it has become apparent that significant insights into the mechanism of electron delocalization in binary transitions metal oxides can be obtained from physical properties measurements on materials containing a second, substitutional metal ion [1–4]. In a previous ^{57}Fe Mössbauer study of magnetite with Zn and Cd substitutions, i.e., $Cd_xFe_{3-x}O_4$ and $Zn_xFe_{3-x}O_4$, it was concluded that a band model for the conduction mechanism in Fe_3O_4 is in best agreement with the critical Mössbauer parameters [5]. The Mössbauer spectra were obtained, however, in the absence of an applied magnetic field; and due to line broadening and decreased resolution of the A and B site patterns for $x > 0.1$, the conclusions reached depended heavily on least squares computer analyses of the spectra. Therefore, in the present study we have performed ^{57}Fe Mössbauer measurements on $Zn_xFe_{3-x}O_4$ and $Cd_xFe_{3-x}O_4$ in an applied magnetic field of 55 kG at 180 K with improved spectral resolution. In addition to confirming the salient conclusions of the earlier study, some new spectral features were discovered which are interesting in terms of recent results of others regarding line broadening mechanisms in Fe_3O_4 at 300 K [6].

2. Experimental

The $M_xFe_{3-x}O_4$ (M = Zn, Cd; x = 0.1, 0.2, and 0.3) are identical to those used previously and were prepared by standard ceramic [5] techniques. Precise stoichiometry was assured by firing intimate mixtures of spectroscopic grade ZnO, or CdO, Fe_2O_3 and Fe metal in evacuated quartz tubes with small free volumes at 1273 and 1223 K for Zn and Cd doped Fe_3O_4, respectively. X-ray powder diffraction patterns were obtained with a focused Guinier camera and indicated the presence of only a single phase spinel in every case. The lattice constants have been reported previously and show no unexpected variations [5]. The Mössbauer spectra were obtained with an electromechanical transducer operating in a constant-acceleration mode in conjunction with 1024 channel analyzer. A ^{57}Co source in Pd metal was used for the 180 K applied field measurements and a ^{57}Co/Rh source was used for the room temperature measurements. The absorber density was approximately 10 mg/cm^2 of natural Fe.

3. Results

^{57}Fe Mössbauer spectra were obtained at 300 K in zero applied magnetic field and at 180 K in a 0 and 55 kG external field. The $Cd_xFe_{3-x}O_4$ spectra at 180 K and in a 55 kG field are shown in fig. 1. The $Zn_xFe_{3-x}O_4$ spectra are quite similar except for somewhat narrower B site lines.

The solid line is the result of fitting to the spectra a local molecular field model in which the magnetic hyperfine field at a B site is determined by the number of A site Zn or Cd neighbors, with the hyperfine field experiencing a linear decrement, ΔH, for each Zn or Cd neighbor. The intensity of each pattern is computed assuming a random intra-site distribution of Cd and Zn; neither Zn nor Cd occupies the B sites. The parameters in such a model are H_0, the field at a B site having no diamagnetic A site neighbor, and ΔH, the decrement per A site diamagnetic neighbor. The A site hyperfine field, $H(A)$, and the isomer shifts of all patterns are completely free parameters. Completely satis-

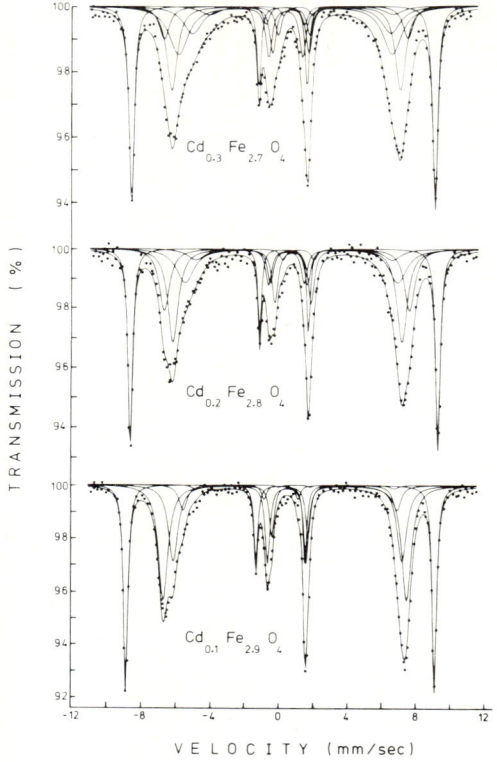

Fig. 1. ^{57}Fe Mössbauer spectra of $Cd_xFe_{3-x}O_4$ at 180 K in 55 kG field. The ^{57}Co source temperature is 200 K. The more widely split pattern is from A site Fe^{3+} ions and the pattern with the smaller splitting and broader lines is due to B site Fe ions. The solid lines through the data points are the results of a least squares fit of several sets of 4-line (lorentzian) patterns to the experimental data.

factory fits are obtained with the above model as indicated by the solid line in fig. 1. The relevant Mössbauer parameters resulting from the fits are shown in table I.

4. Discussion

The area ratios of the A and composite B site subspectra, as fitted in this study, are in good agreement with the expectation that *all* B site Fe ions contribute *only* to the B site subspectrum, i.e. there is little or no tendency for the formation of localized B site Fe^{3+} ions which would contribute to the A site pattern. As shown in table I, the isomer shift difference between Fe(B) with no Zn or Cd neighbors and Fe^{3+}(A) is proportional to the conduction electron concentration. This result is in agreement with that expected if the conduction electrons interact equally, more or less, with all B site ions as in the case of a band description of the electron itinerancy. This result is, however, inconsistent with a pair-wise, localized hopping model since the average charge state within Fe^{2+}–Fe^{3+} pairs would be independent of x and would result in a nearly constant isomer shift. It is to be noted that the isomer shifts variations in zero applied field at 300 K and in a 55 kG field at 180 K are in excellent accord.

Since the conduction electrons are polarized opposite to the Fe^{3+}(B) 3d electrons, the slight increase in $H_0(B)$ is in accord with the de-

Table I
Results obtained in least squares fits of 4 to 6 patterns to the Mössbauer spectra of $Zn_xFe_{3-x}O_4$ and $Cd_xFe_{3-x}O_4$ at 300 K in zero applied field and 180 K in a 55 kG field

x	$H^a(A)$ (kG)	$H_0(B)^a$ (kG)	ΔH^a (kG)	ΔH^b (kG)	$S_0(B) - S(A)^a$ (mm/sec)	$S_0(B) - S(A)^b$ (mm/sec)	$(A_\beta/A_\alpha)^a$	$(A_\beta/A_\alpha)^b$
				$Zn_xFe_{3-x}O_4$				
0.1	557$_1$[c]	433$_2$	11$_2$	7.2$_1$	0.34$_1$	0.32$_1$	2.48$_5$	1.91$_1$
0.2	552$_1$	446$_2$	22$_2$	16.3$_6$	0.21$_1$	0.28$_1$	2.49$_5$	2.1$_1$
0.3	551$_1$	445$_2$	19$_2$	18.4$_5$	0.16$_1$	0.18$_1$	2.69$_5$	2.7$_1$
				$Cd_xFe_{3-x}O_4$				
0.1	554$_2$	437$_2$	26$_2$	18.6$_1$	0.32$_1$	0.35$_1$	2.39$_5$	1.91$_1$
0.2	552$_2$	441$_2$	29$_2$	22$_1$	0.19$_1$	0.26$_1$	2.78$_5$	2.5$_1$
0.3	545$_2$	434$_2$	27$_2$	27$_1$	0.15$_1$	0.20$_4$	2.84$_5$	3.1$_2$

[a] At 180 K in a 55 kG external field.
[b] At 300 K in no external field.
[c] Subscript below each number indicates estimated error in last digit.

creasing conduction electron concentration as x goes from 0.1 to 0.2 and provides further support for a band description of the conduction electrons.

It has also been noticed in the applied field spectra that the hyperfine fields and external fields are not rigorously additive and this discrepancy is believed to be related to a small anisotropy in the B site hyperfine field as discussed recently [6].

Thus, the good accord between the ^{57}Fe Mössbauer spectra of $Cd_xFe_{3-x}O_4$ and $Zn_xFe_{3-x}O_4$ at 300 K in zero applied field and at 180 K in a 55 kG field and the good agreement between the parameter values observed and those expected for band conduction indicate that despite a narrow bandwidth the collective electron state in Fe_3O_4 is a rather stable one.

Support of this study by the National Science Foundation is gratefully acknowledged. It is also a pleasure to acknowledge the assistance of Dr. L. J. Swartzendruber in obtaining the applied field spectra.

References

[1] D.C. Dobson, J.W. Linnett and M.M. Rahman, J. Phys. Chem. Solids 32 (1970) 2727.
[2] D.B. McWhan and J.P. Remeika, Phys. Rev. B2 (1970) 3734.
[3] D.B. McWhan, M. Marezio, J.P. Remeika, and P.D. Dernier, Phys. Rev. B10 (1974) 490.
[4] B.J. Evans, AIP Conf. Proc. 24 (1975) 73.
[5] Hang Nam Ok and B.J. Evans, Phys. Rev. B14 (1976) 2956.
[6] A.M. van Diepen, Phys. Letters (1976), In press.

MAGNETIC PROPERTIES OF THE SPINEL SERIES $Co_{2-y}Zn_yTiO_4$ $(0 \leq y \leq 1)$

K. DE STROOPER and G. ROBBRECHT

Laboratory of Magnetism, University of Ghent, B-9000 Ghent, Belgium

Magnetization measurements were carried out on the spinels $Co_{2-y}Zn_yTiO_4$ from 4.2 to 300 K in magnetic fields up to 28 kOe. The exchange integrals J_{AA}, J_{AB} and J_{BB} were deduced and qualitatively explained. A model for the magnetic structure is proposed. An empirical formula is fitted to the coercive force energy.

According to Sakamoto et al. [1], the cobalt-zinctitanates show unusual magnetic properties. We have explained these phenomena for Co_2TiO_4 in a previous paper [2]. We have performed more accurate and more extensive measurements on the spinel series $Co_{2-y}Zn_yTiO_4$ ($y = 0.0$; 0.2; 0.4; 0.6; 0.8 and 1.0) from 4.2 to 300 K in magnetic fields up to 28 kOe. The samples were prepared by the oxide sintering technique; only slowly cooled samples were studied. Nearly all Zn^{2+} ions are situated on the tetrahedral sites (A sites) [3]; all Ti^{4+} ions occupy the octahedral sites (B sites) [3]. Only the Co^{2+} ions possess a magnetic moment.

The inverse of the paramagnetic susceptibility as a function of temperature shows a hyperbolic behaviour for all samples. Because these curves nearly follow a Curie–Weiss law from 77 to 300 K, the Curie constant C and the asymptotic Curie temperature θ_{as} may be accurately determined (table I). The magnetic moments of the Co^{2+} ions on A and B sites were deduced from the Curie constants. The Co^{2+} ion on the A sites has a second order orbital contribution to the magnetic moment [4]; the magnetic moment equals $3.60 \pm 0.03 \mu_B$. The orbital momentum of the Co^{2+} ion on the B sites is not completely quenched by the crystal field [4]; the magnetic moment equals $3.87 \pm 0.07 \mu_B$.

From magnetic measurements below T_c, the behaviour of the spontaneous magnetization σ_0, the differential susceptibility at high fields χ_d, and the coercive force H_c was deduced as a function of temperature. The shape of the experimental (σ_0, T) curves indicates a collinear spin structure [5]. A difference of two Brillouin functions was fitted to these curves, assuming that the AB interaction is antiferromagnetic and dominating. The fittings were performed by the SIMPLEX method [6]. The values of the magnetic moments, as determined from the paramagnetic susceptibility measurements, were used in the argument of the Brillouin functions. The best values obtained for the exchange integrals J_{AA}, J_{AB}, J_{BB} [7] and the sublattice magnetizations $M_A(O)$ and $M_B(O)$ are listed in table I (see also fig. 1). From the values of $M_A(O)$, we can conclude that this fitting procedure is wrong for $y = 0.8$ and $y = 1.0$, because the BB interactions become dominating and determine in first approximation the values of T_c and θ_{as}. From T_c and θ_{as} (table I) we could calculate the exchange integrals: $J_{B_1B_2} = -6.0$ K and $J_{B_1B_1} = 2.3$ K for $CoZnTiO_4$, following the molecular field theory.

J_{AB} and J_{AA} are due to an antiferromagnetic superexchange interaction, and remain constant because all materials show the same lattice parameter (8.445 ± 0.003 Å). J_{BB} (and $J_{B_1B_1}$ for

Table I
Results determined from the para- and ferrimagnetic region

y	$C \left(\dfrac{cm^3 K}{mole}\right)$	θ_{as}(K)	T_c (K)	$M_A(O)$ $\left(\dfrac{emu}{mole}\right)$	$M_B(O)$ $\left(\dfrac{emu}{mole}\right)$	J_{AA} (K)	J_{AB} (K)	J_{BB} (K)
0.0	5.36 ± 0.10	-130 ± 5	53 ± 1	20 460	19 760	-4.6 ± 0.3	-6.3 ± 0.3	-5.5 ± 0.3
0.2	4.82 ± 0.10	-104 ± 5	48 ± 1	17 400	17 830	-4.4	-6.0	-4.9
0.4	4.28 ± 0.08	-72 ± 5	40 ± 1	13 250	14 140	-4.4	-6.0	-3.8
0.6	3.73 ± 0.08	-49 ± 5	31 ± 1	8480	9820	-4.6	-6.4	-2.0
0.8	3.19 ± 0.06	-20 ± 5	30 ± 1	8400	9150			
1.0	2.67 ± 0.05	-5 ± 5	28 ± 1	6750	7070			

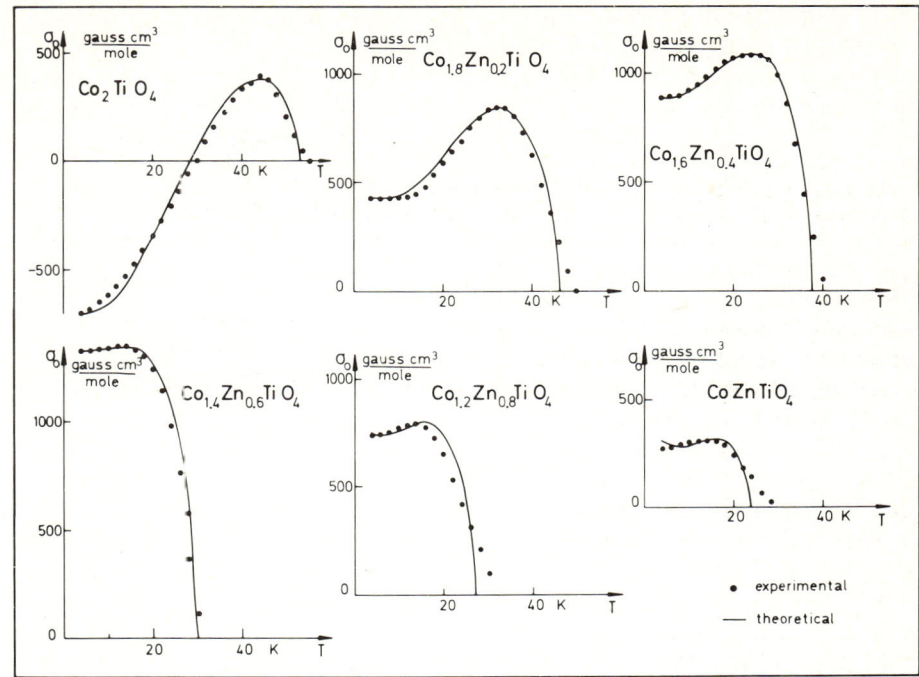

Fig. 1. Spontaneous magnetization vs. temperature for $Co_{2-y}Zn_yTiO_4$.

$CoZnTiO_4$) is due to a competition between a ferromagnetic superexchange interaction and an antiferromagnetic direct exchange interaction. The direct exchange is dominating for Co_2TiO_4, but the ferromagnetic part becomes stronger with increasing Zn^{2+} content. The antiferromagnetic long-range superexchange interaction ($J_{B_1B_2}$), arising from the covalent properties of the Zn^{2+} ions on the tetrahedral sites, is responsible for this, giving two neighbouring Co^{2+} ions on the B sites the same spin directions.

The differential susceptibility χ_d (fig. 2) increases with increasing Zn^{2+} content. The maximum at the Curie temperature completely vanishes for $y = 1.0$, indicating that $CoZnTiO_4$ is an antiferromagnet with a spin structure perpendicular to the applied field.

The Curie temperature T_c (see table I) as a function of the Zn^{2+} content shows a discontinuity at $y = 0.6$, because the BB interaction becomes dominating above $y = 0.6$. The following model for the magnetic structure of this series is proposed. Two nearest Co^{2+} neighbours on the A sites give parrallel spin configuration for their nearest Co^{2+} neighbours on the B sites (cf. Co_2TiO_4). Two nearest Zn^{2+} neighbours on

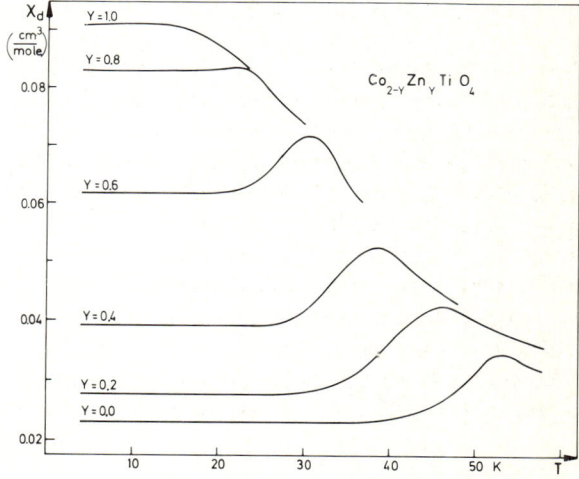

Fig. 2. Differential susceptibility vs. temperature for $Co_{2-y}Zn_yTiO_4$.

the A sites give an antiparallel spin configuration for their nearest Co^{2+} neighbours at the B sites, perpendicular to the applied field (cf. $CoZnTiO_4$). In the other cases, there is a probability of two-thirds that the B spins are parallel. The values of σ_0 and χ_d at 4.2 K, calculated from this model, are in good agreement with the experimental values.

The hysteresis curves are not rectangular and show a small relaxation effect. An empirical formula can be fitted to the behaviour of the energy arising from the coercive force:

$$\sigma_0 H_c = \frac{N_A k T_c^2}{6 \times 460} \exp\left(-\frac{460\,T}{T_c^2}\right),$$

in which N_A is the Avogadro number and k is the Boltzmann constant; the factor six arises from the number of magnetic B neighbours of an A ion; 460 K is of the same order of magnitude as the spin–orbit coupling energy.

The authors thank the FKFO for financial support.

References

[1] N. Sakamoto, J. Phys. Soc. Jap. 17 (1962) 99.
[2] K. De Strooper, C. Henriet-Iserentant, G. Robbrecht and V. Brabers, C.R. Acad. Sci. Paris 277B (1973) 75.
[3] P. Poix, Ann. Chim. 10 (1965) 42.
[4] F. Varret and F. Hartmann-Boutron, Ann. Phys. 3 (1968) 157.
[5] L. Néel, Ann. Phys. 12 (1948) 137.
[6] C. Dauwe, M. Dorikens and L. Dorikens-Vanpraet, Appl Phys. 5 (1974) 45.
[7] J.S. Smart, in: Magnetism III, Rado and Suhl, ed. (Academic Press, New York, 1963) ch. II.

MAGNETIC PROPERTIES OF A NEW SERIES OF MIXED ALCALINE EARTH AND RARE EARTH FERRITES WITH FORMULA SrLn$_2$Fe$_2$O$_7$

D. SAMARAS, A. COLLOMB, J.C. JOUBERT

Laboratoire des Rayons X, CNRS, 166 X, 38042 Grenoble Cedex, France

and

R. CHEVALIER

Groupe d'Interactions Hyperfines, DRF-CEN-G, 85 X, 38041 Grenoble Cedex, France

Magnetic structures and spin reorientations in the SrLn$_2$Fe$_2$O$_7$ compounds have been studied by neutron diffraction and Mössbauer spectroscopy. The results as a whole show that the rare earth's order is induced by the exchange interactions between the two sublattices; their anisotropy allows the observed spin reorientations to be explained.

1. Introduction

The mixed strontium or barium and rare earth ferrites, with the general formula SrLn$_2$Fe$_2$O$_7$, crystallize in a tetragonal cell with parameters $a \simeq 5.5$ Å (except BaLa$_2$Fe$_2$O$_7$ where $a \simeq 3.95$ Å) and $c \simeq 20$ Å. The crystal structure is built up of perovskite blocks, infinite in two dimensions and two cells in width in the third dimension which coincides with the c axis [1]. In the real structure the Fe^{3+} cation has five oxygen neighbors placed at the corners of a trigonal bipyramid [2] instead of six neighbors forming an octahedra of the idealized Sr$_3$Ti$_2$O$_7$ structure type [3].

These compounds are all antiferromagnetic, with Néel temperatures of about 550 K [4]. In some cases (Ln = Nd, Tb) a ferromagnetic component appears at low temperatures [4].

2. Study by neutron diffraction

At room temperature only the Fe^{3+} ions are ordered. The investigation by neutron diffraction does not provide unambiguously the magnetic structure. The possible magnetic modes are $G_x^- \pm A_y^-$ in SrTb$_2$Fe$_2$O$_7$ [5] and $A_x^- \pm G_y^-$ in the compounds of Pr and Nd. The magnetic moments are aligned either along the [110] direction or along the [1$\bar{1}$0] one. The two modes give the same intensities of magnetic reflections.

Above 235 K, in BaLa$_2$Fe$_2$O$_7$, one observes a continuous rotation of the magnetic moments in the basal plane from the base diagonal direction to the [100] one.

At 4.2 K, in the Nd and Tb compounds the magnetic moments are along the c axis, the magnetic structure being described by the A_z^- mode. The rare earth's spin configuration is described by the mode $-A_z^-$ for Nd and by the A_z^-, F_x^+, C_y^+ ones for Tb. In the Pr compound neither spin reorientation nor ordering of the Pr has been observed by neutron diffraction down to liquid helium.

3. Study by Mössbauer spectroscopy

The spin reorientation was observed by the use of a SrNd$_2$Fe$_2$O$_7$ single crystal with the c axis parallel to the γ rays. At room temperature, the ratios of the lines intensity are 3:4:1, which correspond to a spin orientation perpendicular to the direction of the γ photons. On the other hand, at 4.2 K the $\Delta m = 0$ transitions disappear, the magnetic moments are aligned along the c axis.

The quadrupole coupling measured from a powder spectrum in the magnetic-order region is related to the orientation (θ, Φ) of the internal magnetic field with respect to the principal axes of the EFG by the formula

$$\Delta \epsilon_m = (-1)^m \frac{e^2 qQ}{2} (3\cos^2\theta - 1 + \eta \sin^2\theta \cos 2\Phi). \quad (1)$$

The study of the single crystals paramagnetic spectrum gives a 14° angle between the major EFG principal axis and the c one, and an asymmetry parameter of $\eta = 0.40$. For $\Phi = \frac{1}{2}\Pi$ the formula (1) yields a calculated value for ϵ_m which is in agreement with the observed one (the spins are along the [110] diagonal).

For the Sm, Eu, Gd and Tb compounds, the room temperature values of ϵ_m is almost equal to the ϵ_m value found for $SrNd_2Fe_2O_7$. This indicates that the magnetic structures are similar, that is the spins are aligned along the [110] diagonal.

At 4.2 K, for the same compounds the ϵ_m values are highly modified. The change of sign observed for the Nd and Tb compounds can be explained by spin reorientation from the [110] direction to the c axis. At low temperature, the value of ϵ_m for the Gd compound varies but not in sign, and this is explained by a rotation of the spins in the basal plane from the [110] direction to the [1$\bar{1}$0] one. For the Pr, Sm and Eu compounds, no modification of the ϵ_m values has been observed.

4. Discussion

In all compounds investigated, when only the Fe^{3+} ions are ordered, the magnetic moments are aligned in the basal plane. Since one can calculate that the spins configurations along the c axis are favored by the dipolar energy, one must admit then that, as in $BaFe_{12}O_{19}$, the crystal field anisotropy of the Fe^{3+} ions align the spins along the bipyramid's axis.

Crystal field calculations at the rare earth's site show that the rare earth's spins direction is not induced by the crystal field's anisotropy. In fact, for Nd as for Tb as well, the crystal field would favor directions close to the [110] one, whereas experimentally one observes configurations along the c axis. One must admit an anisotropy due to the exchange between the two sublattices. Indeed, an order of the rare earth and a reorientation of the Fe^{3+} spins along the c axis at low temperature are observed when the rare earth's ground state is a doublet (a Kramers doublet for Nd, a pseudo-doubled consisting of two singlets $3\,cm^{-1}$ spaced for Tb) which allows strong interactions with a high anisotropy.

The Fe^{3+} spins rotation within the basal plane, observed in the La and Gd compounds is not obviously influenced by the rare earth. It can be explained by assuming a thermal variation of the first order anisotropy constant for the Fe^{3+} ion.

References

[1] J.C. Joubert, D. Samaras, A. Collomb, G. le Flem and A. Daoudi, Mat. Res. Bull. 6 (1971) 341.
[2] D. Samaras, A. Collomb and J.C. Joubert, J. Solid State Chem. 7 (1973) 337.
[3] S.N. Ruddlesden and P. Popper, Acta Crystallogr. 11 (1958) 54.
[4] D. Samaras, A. Collomb and J.C. Joubert, Mat. Res. Bull. 9 (1974) 693.
[5] D. Samaras, A. Collomb, J.C. Joubert and E.F. Bertaut, J. Solid State Chem. 12 (1975) 127.

INCORPORATION OF IRIDIUM IN GARNETS ON TETRAHEDRAL SITES

B. ANDLAUER
Institut für Angewandte Festkörperphysik der Fraunhofer–Gesellschaft, 7800 Freiburg, W. Germany

and

W. TOLKSDORF
Philips Forschungslaboratorium, 2000 Hamburg, W. Germany

Optical absorption measurements in $Y_3Fe_5O_{12}$:Ir indicate that a certain fraction of iridium is incorporated on tetrahedral sites. From the investigation it is concluded that the observed iridium centre contributes to the magnetic anisotropy of $Y_3Fe_5O_{12}$:Ir.

1. Introduction

The magnetic anisotropy and the magnetostriction of $Y_3Fe_5O_{12}$ can be drastically changed by doping with iridium, via the strong spin–orbit coupling exhibited by this ion [1]. For a microscopic interpretation of these effects, it has been assumed that iridium is incorporated mainly on octahedral sites, in the valence state Ir^{4+} [1]. In contrast, we have found that a certain fraction of iridium occupies also the tetrahedral garnet sites.

2. The absorption spectrum

Single crystals of $Y_3Fe_5O_{12}$:Ir exhibit a group of intense absorption lines near $4900\,cm^{-1}$, see fig. 1. The strongly anisotropic exchange gives rise to energy shifts of as much as $200\,cm^{-1}$, by rotating the magnetization within the {100} or {110} plane, see fig. 2. The absorption can be ascribed to a single electronic transition of an iridium centre being incorporated on 24 magnetically inequivalent sites. From the pronounced polarization dependence it may be inferred that the transition is electric dipole-allowed, being σ-polarized to almost 100% relative to a $\langle 100 \rangle$ axis. The E1 character together with the high oscillator strength, $f > 10^{-4}$, reveal that the iridium ions reside on sites without inversion symmetry.

The angular dependence of the ground state $|g\rangle$ and the excited state $|e\rangle$ may be described on the basis of effective spin hamiltonians of the type [2]

$$\mathcal{H} = \mu_B \cdot \boldsymbol{H}_{ex} \cdot \boldsymbol{g} \cdot \boldsymbol{S} \qquad (1)$$

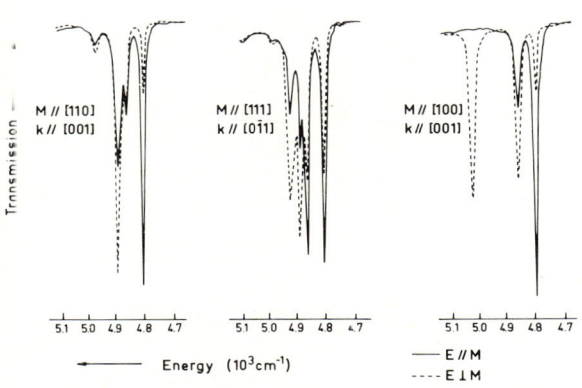

Fig. 1. Energy and polarization dependence of the absorption in $Y_3Fe_5O_{12}$:Ir near $4900\,cm^{-1}$, for three different geometries of the magnetization, at $T = 12\,K$. The thickness of the sample is 0.2 mm, the iridium concentration 0.0056 per formula unit. k and E are the wave vector and electric field vector of incident light, respectively.

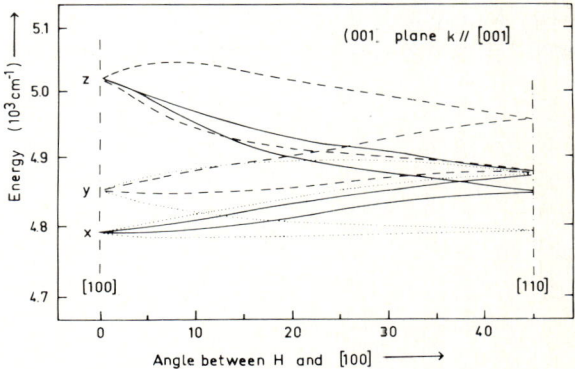

Fig. 2. Angular dependence of the absorption in $Y_3Fe_5O_{12}$:Ir, as obtained by rotating the external saturation field H within the (001) plane. There are 12 magnetically inequivalent positions for this geometry. By rotating H through 180°, a particular iridium centre occupies four different positions that have been characterized by an identical designation of the spectral lines.

Here, μ_B is the Bohr magneton, H_{ex} the exchange field, **g** the effective g tensor and **S** the effective spin. In the case of anisotropic exchange, H_{ex} and the magnetization **M** are related by an exchange tensor, $H_{ex} = \Lambda \cdot M$ [3]. An analysis of the angular dependence of the spectra has been performed on the assumption that the magnetic principal axes (i, j, k) of **g** and Λ coincide and that they are the same for $|g\rangle$ and $|e\rangle$. Fig. 3 shows the position of the magnetic axes (i, j, k) relative to the $\langle 100 \rangle$ axes (x, y, z). The k axis is tilted from z within the (z, y) plane by $\theta = \pm 9.5 \pm 0.5°$; the i axis is rotated relative to x within the plane $\perp k$ by $\phi = \pm 21.5 \pm 0.5°$. The energies of the transition, for **M** lying parallel to the three principal axes, are (in cm^{-1}): $E_i = 4779 \pm 4$, $E_j = 4776 \pm 10$, $E_k = 5042 \pm 2$. The coincidence of E_i and E_j, within the error limits, indicates an axial symmetry of the centre, the unique character of the k axis being underlined by the considerably higher transition energy and the polarization dependence, see above.

3. Conclusion

The results suggest strongly that the iridium centre is localized on a tetrahedral site, the effective tetragonal axis being tilted by $\pm 9.5°$ from the original S_4 axis, the crystallographic rotation angle β being increased from $\pm 15.6°$ [4] to $\phi = \pm 21.5°$. Further investigations of the exchange splitting up to $T_c = 559$ K revealed a temperature-dependent spin-canting [5]. The tilt and the spin-canting both prove the effectiveness of spin-orbit coupling within the electronic states $|g\rangle$, $|e\rangle$. Hence, it is concluded that the iridium centre should also contribute to the magnetic anisotropy of $Y_3Fe_5O_{12}$:Ir. As is suggested by Zeeman measurements on the same transition in $Y_3Ga_5O_{12}$:Ir [5], the charge state of the centre is probably Ir^{4+}.

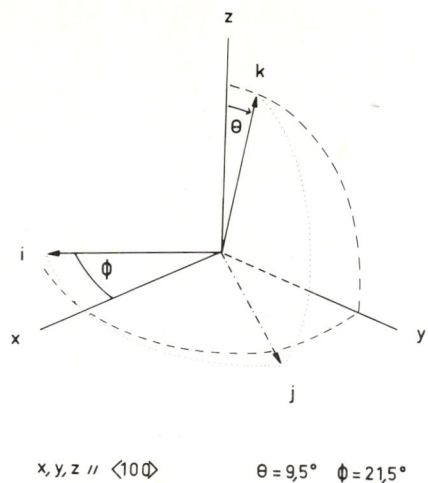

Fig. 3. The position of the magnetic symmetry axes (i, j, k) relative to the $\langle 100 \rangle$ axes (x, y, z), see text.

References

[1] P. Hansen, J. Schuldt and W. Tolksdorf, Phys. Rev. B8 (1973) 4274.
[2] See, e.g., A. Abragam and B. Bleaney, Electron Paramagnetic Resonance of Transition Ions (Clarendon Press, Oxford, 1970) p. 131.
[3] K.A. Wickersheim, Phys. Rev. 122 (1961) 1376.
[4] S. Geller, Acta Cryst. 12 (1959) 944.
[5] B. Andlauer and W. Tolksdorf, to be published.

MAGNETIC PROPERTIES OF ALUMINIUM SUBSTITUTED Zn_2–W HEXAGONAL FERRITES

G. ALBANESE, M. CARBUCICCHIO
Istituto di Fisica, Università di Parma, Parma, Italy

F. BOLZONI, S. RINALDI
Lab. MASPEC del CNR, Parma, Italy

and

G. SLOCCARI and E. LUCCHINI
Istituto di Chimica Applicata, Università di Trieste, Trieste, Italy

The temperature dependence of the saturation magnetization and of the anisotropy field in $BaZn_2Fe_{16-x}Al_xO_{27}$ compounds for $x = 1, 2, 3, 4$ has been measured. Mössbauer absorption spectra of ^{57}Fe γ-rays have been also measured for all the samples in the temperature range from 78 to 700°C. These data indicate that up to $x = 4$ aluminium preferentially substitutes the iron in the octahedral sites of the spinel block. No remarkable effects on the equilibrium of superexchange interactions have been detected.

The crystalline structure of W compounds is built up by the superposition of the so called R-block and two S blocks [1]. The cations are distributed among different sublattices denoted by the same symbols as in Zn_2–W ferrite [2].

The $BaZn_2Fe_{16}O_{27}(Zn_2$–W$)$ ferrite presents the largest saturation magnetization among the known hexagonal ferrimagnetic oxides. A detailed study of the intrinsic properties of this compound has been reported recently [2]. In this paper we present and discuss Mössbauer results and magnetic measurements on the Al substituted compound $BaZn_2Fe_{16-x}Al_xO_{27}$ for $x = 1, 2, 3, 4$. The polycrystalline samples have been prepared by standard ceramic techniques and controlled by means of micrographic and X-rays tests. The Mössbauer, magnetization and anisotropy measurements have been done by means of previously described techniques [2, 3].

The Mössbauer absorption spectra of the aluminium substituted Zn_2–W strongly resemble the spectrum of the pure compound previously studied [2]. In fig. 1 the spectra for the compounds with $x = 1$ and 3 at room temperature are reported. These spectra can be interpreted as the superposition of three Zeeman sextets. We attribute sextet I to iron ions in the K sublattice and to part of the iron ions in the a sublattice. The sextet II is due to the f_{IV} and to the remaining part of a sublattice while to the sextet III contributes the f_{VI} sublattice. The

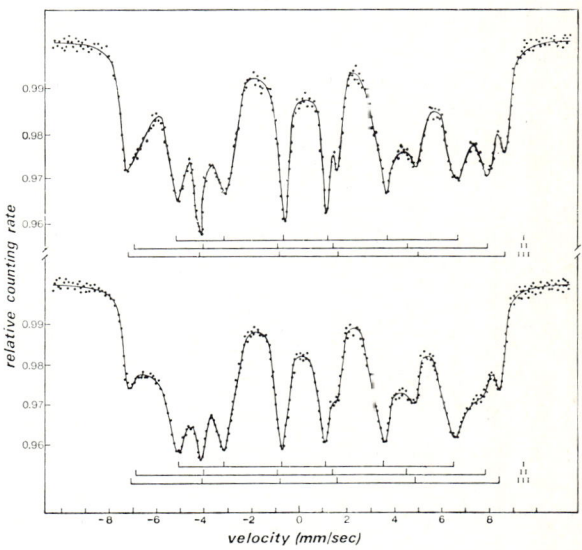

Fig. 1. Mössbauer spectra at room temperature for $BaZn_2Fe_{16-x}Al_xO_{27}$ compounds for $x = 1$ (a) and $x = 3$ (b); I, sublattice K and part of sublattice a; II, sublattice f_{IV} and part of sublattice a; III, sublattice f_{VI}.

sextet originated by b sublattice has not been detected because of its low intensity.

The obtained spectra put in evidence a systematic decrease of the relative intensity of sextet II with increasing x indicating the preferential entrance of Al in the cationic sublattices inside the spinel blocks.

From the Mössbauer spectra the temperature

dependence of the magnetic hyperfine fields (H_{hf}) relative to the various sextets has been determined. We may notice that the dependence of $H_{hf}(T)/H_{hf}(0)$ vs. the reduced temperature T/T_c looks very similar to the compounds with different aluminium content.

By measuring the temperature dependence of the line width, we obtained the Curie temperatures that turned out to be $T_c = 648$, 613, 588, 558 and 533 ± 5 K for $x = 0, 1, 2, 3$ and 4, respectively; these data agree with those obtained from magnetization measurements.

In fig. 2 the temperature dependence of the saturation magnetization for the various compounds is reported. The values at 0 K turns out to be $\sigma(0) = 123, 108, 94, 80$ and 66 emu/g for $x = 0, 1, 2, 3$ and 4, respectively. We can notice that both the saturation magnetizations at 0 K and the Curie temperatures vary linearly with the Al content.

The anisotropy constant K_I is reported in fig. 3 for the various x as a function of the reduced temperature T/T_c. We can notice that the behaviour for $x = 0, 1, 2$ and 3 is the same while the data for $x = 4$ deviate slightly from the others.

As regards the order of substitution of Fe^{3+} by Al^{3+} in the various sublattices, the Mössbauer data, as we noticed before clearly indicate that Al^{3+} enter only the lattice sites of the spinel block. We can now distinguish between the octahedral a and tetrahedral f_{IV} sublattices taking into account the magnetic order of Zn_2–W ferrite and the dependence of the saturation magnetization at 0 K on the order of substitution. As already known [2] the magnetic order of Zn_2–W

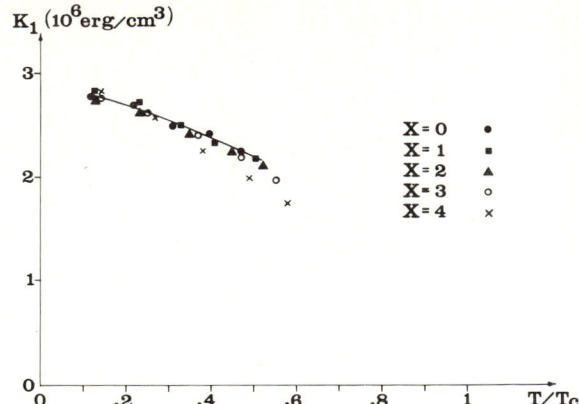

Fig. 3. Anisotropy constant K_I vs. reduced temperature for the compounds $BaZn_2Fe_{16-x}Al_xO_{27}$.

compound deviates from the Gorter scheme [1]: in fact, the value of $35\mu_B$ measured at 0 K for Zn_2–W can be explained only by admitting that 10% of the Fe^{3+} ions in the a sublattice reverse their spin due to the weakening of the tetrahedral–octahedral superexchange interactions caused by the presence of Zn-ions in tetrahedral sublattice. The dependence of the saturation magnetization at 0 K for the various x agrees very well with the values calculated assuming that all the Al^{3+} ions enter statistically the a sites. The temperature dependence of the anisotropy constant also agrees with the assumed order of substitution. In fact, it is known [4] that the main contribution to the anisotropy of these compounds comes from the b and the K sublattices which are not interested to the substitution.

All the obtained data indicate that up to $x = 4$ the substitution of Fe^{3+} by Al^{3+} in Zn_2–W compound does not affect the equilibrium of the superexchange interactions. In fact for various x no differences in the curve of the sublattice magnetization vs. reduced temperature have been observed. Moreover, also the $\sigma(T)/\sigma(0)$ vs. T/T_c curves are the same for all x.

References

[1] J. Smith and H.P.J. Wijn: Ferrites (Philips Technical Library, Eindhoven, 1959).
[2] G. Albanese, M. Carbucicchio and G. Asti, Appl. Phys. 11 (1976) 81.
[3] G. Asti and S. Rinaldi, J. Appl. Phys. 45 (1974) 3400.
[4] G. Asti and S. Rinaldi presented to Joint MMM-Intermag. Conf., Pittsburg (1976).

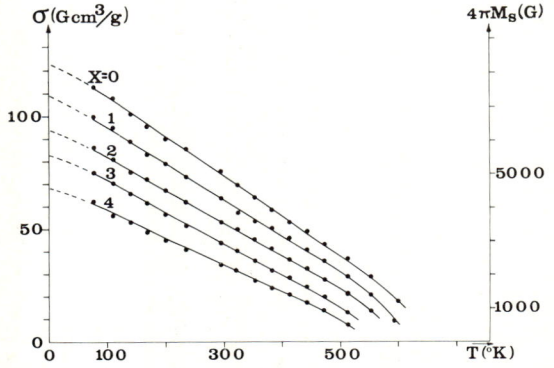

Fig. 2. Temperature dependence of the saturation magnetization for the compounds $BaZn_2Fe_{16-x}Al_xO_{27}$.

UNIVERSAL HYSTERESIS LOOP FOR SOFT FERRIMAGNETIC POLYCRYSTALS

A. GLOBUS

Equipe de Recherche Matériaux Magnétiques, CNRS, 92190 Meudon-Bellevue, France

The author presents the achievement of a work upon the magnetization mechanisms based on the concept of the total anisotropy K and on a model for hysteresis loop previously proposed. Initial magnetization curves of ferrites and YIG are reduced to a universal curve as a function of K and of the mean grain size of the sample. The same procedure allows the author to obtain a universal hysteresis loop.

The present work shows that the initial magnetization curves plotted for different compositions – spinels and a garnet – and for samples with various grain sizes can be reduced to a single intercomposition curve by taking into account the total anisotropy field and the grain size. The compositions involved in this paper are the following: YIG; $NiFe_2O_4$; $Ni_{0.9}Zn_{0.1}Fe_2O_4$; $Ni_{0.9}Cd_{0.1}Fe_2O_4$; $Ni_{0.8}Cd_{0.2}Fe_2O_4$ and $Ni_{0.7}Cd_{0.3}Fe_2O_4$.

Some properties of ferrimagnetic ceramics have permitted – better than in the case of metals – the parameters responsible for the magnetization mechanisms to be separated and fundamental laws to be established. For example, the existence of a total anisotropy K specific to polycrystals and non-dependent from technology has allowed us to elaborate a model [1, 2, 3] in which the analysis of the wall diameter variations during its displacement inside a spherical grain leads to a description of the magnetization mechanisms. The principle of this model is summarized in fig. 1, while figs. 2 and 3 show the experimental results and the reduced curves which are obtained.

Fig. 3 presents hysteresis loops obtained from materials of different compositions, plotted for samples with different grain size and for different

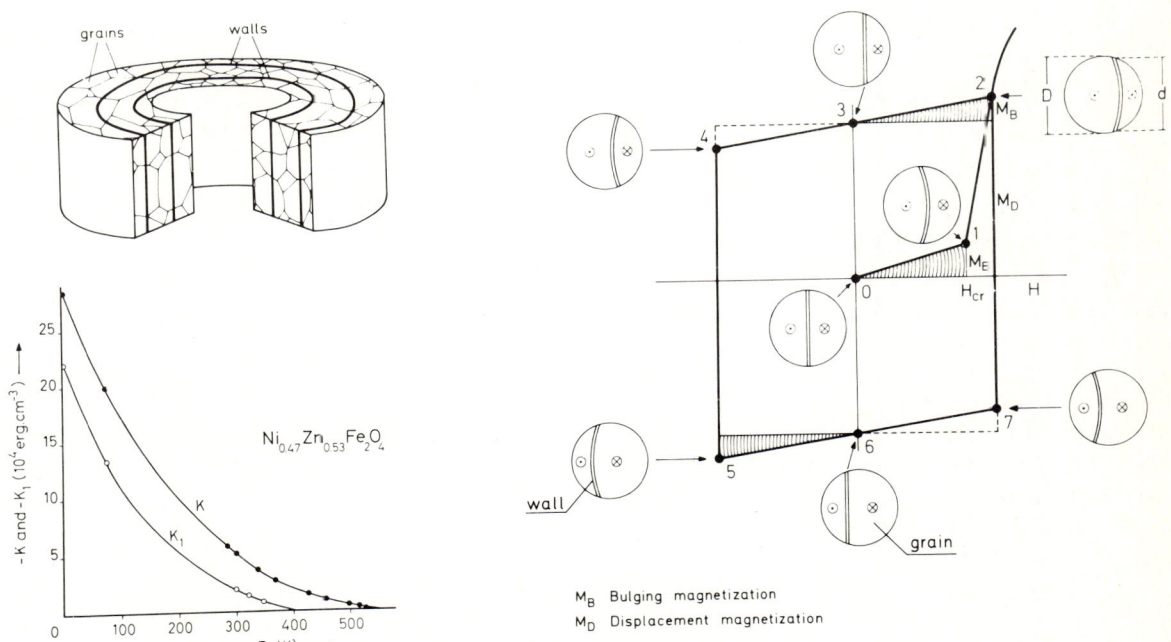

Fig. 1. Magnetization mechanisms in a polycrystalline toroid. Top left, one of the valid domain wall configurations; bottom left, the predominance of the total anisotropy K over the magnetocrystalline anisotropy K_1 in the temperature range below the Curie temperature T_c; right, the model for the domain wall motion during the initial magnetization and hysteresis processes.

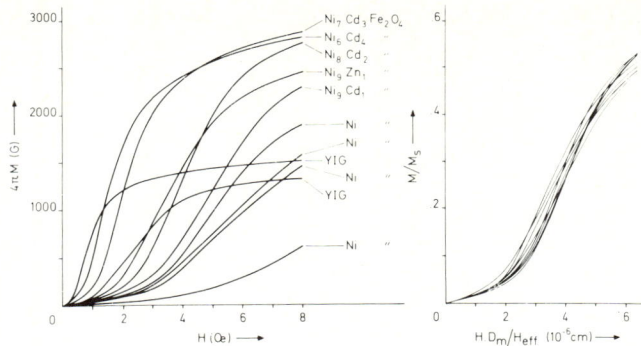

Fig. 2. Initial magnetization curves for some compositions (left) reduced to a universal curve (right) as a function of the mean grain size D_m and the total anisotropy field H_{eff}.

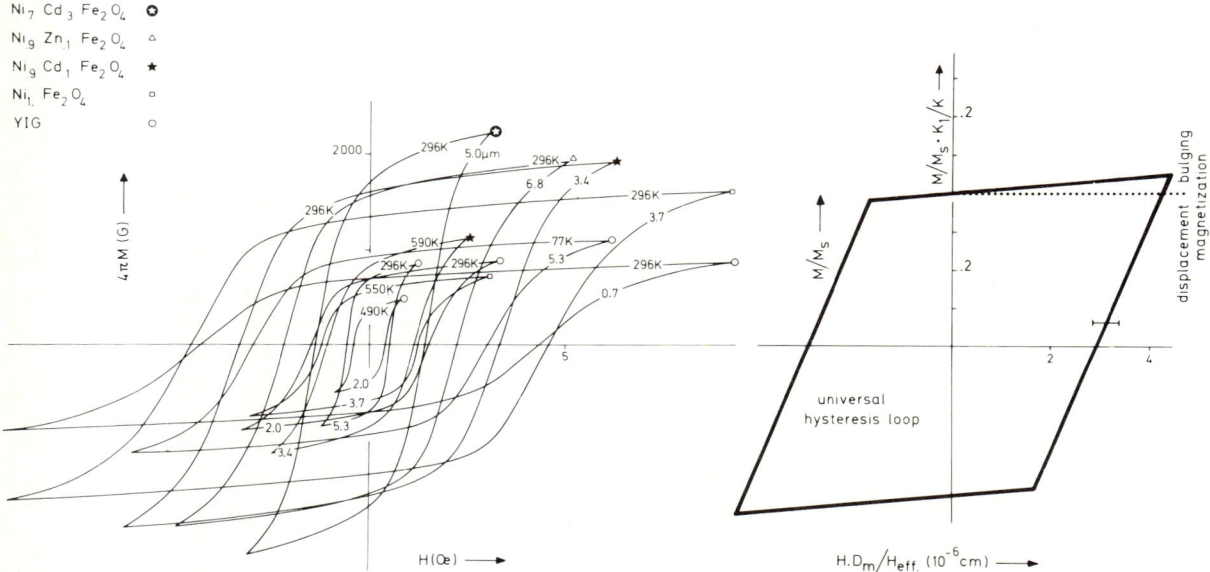

Fig. 3. Universal hysteresis loop (right) obtained from the hysteresis loops of samples of different compositions, with various grain sizes, and plotted at different measurement temperatures. The possible error for coercive force is indicated.

measurement temperatures. This set of loops can be reduced to a single one if taking into account the previously established parameters: total anisotropy K and grain size D_m.

Of course it is necessary to take into account the other related to the model parameters: the ratio M/M_s and, for the reversible part only the ratio K_1/K. Fig. 3 shows such a universal hysteresis loop.

References

[1] A. Globus, C.R. Acad. Sci. 255 (1962) 1709.
[2] A. Globus, Proc. S.M.M. 2, Cardiff (1975), to be published.
[3] A. Globus, supplement au J. de Physique 1977, Int. Conf. of Ferrites JCF 2 (to be published).

CHEMICAL SITE PREFERENCE IN GARNETS

S. METHFESSEL

Ruhr-University, D-4630 Bochum, BRD

J.C. SUITS

IBM Research, San Jose, California 95193, USA

> The dodecahedral coordination of oxygen ions around the c-sites is suggested to be important for the preferential distribution of RE ions in garnet surfaces growing from the melt. Comparison with the situation in cubic, monoclinic and hexagonal RE_2O_3 indicates that the ionic size ratio should control the site preference.

Garnets crystals or films containing a mixture of rare earth ions show a noncubic magnetic anisotropy which is useful for bubble devices. This anisotropy depends in a very complicated manner on the type and concentration of the ions incorporated in the garnets and on the conditions during preparation. The situation is simplest when the anisotropy is produced by two different RE ions in the dodecahedral c-sites of iron garnets. The local symmetry of the 24 c-sites is orthorhombic, but the local orthorhombic axes are arranged in such a way that overall cubic properties result when the RE ions are distributed at random. However, as soon as different RE ions prefer sites with different orientations of the local axes the cubic degeneration is lifted and orthorhombic effects appear in a strength which depends on the difference of both RE ions with respect to their properties and distribution [1]. This is, first of all, a pure symmetry effect and the resulting noncubic anisotropy can be described by group theoretical considerations in a two parameter formula when the site preferential distribution of the RE ions and the mechanism producing the local anisotropy are known [2]. The difficulties in the comparison with experimental results come mainly from uncertainties in this two aspects.

This paper is concerned with the parameters which let the rare earth ions prefer special c-sites during the growth of the crystal from the melt. In principle, it is the reduction of the 24-fold degeneracy of the c-sites in special subgroups with different orientations of their local symmetry axes relative to the growing surface. One has to find interactions of the RE ions with other lattice neighbors at the surface which represent the local symmetry and are strong enough to persist against the thermal agitation at a melting temperature of about 1500 K. Callen [1] has suggested that the superexchange interaction of the RE spins with the spins of the tetrahedral Fe ions may strain the surface in a characteristic way, which depends on the relative position of the Fe ions to the growing surface and can be felt as variation of the "sticking coefficients" by the RE ions coming from the melt. We believe that the exchange fields acting on the RE ions have not sufficient strength to persist as significant parameters up to the melting temperature. Moreover, site preference on the c-sites seems to occur also with nonmagnetic ions.

Since the chemical bonding of the RE ions to the neighboring oxygen ions is by far the strongest interaction in the garnet, it seems to be more reasonable to make the O^{2-}-coordination around the c-sites responsible for the preferential growth. Since the 2:3 oxides of the trivalent rare earth are very stable compounds with melting temperatures around 2500 K one might even expect that RE-O coordinations typical for 2:3 oxides should be present in the garnet melt at 1500 K.

In a garnet such as $8 \cdot [Y^c(Fe^dO_4)]_3Fe_2^a$, the Y^c ions alternate with Fe^dO_4-tetrahedra along 3 sets of chains X, Y, Z parallel to the crystal axes x, y, z, respectively.

The oxygen ions form around the c-sites an orthorhombic "pseudocube" with three perpendicular two-fold axes. The axis $\epsilon_1 = x$ is parallel to the chain direction, the axes $\epsilon_2 = \frac{1}{2}\sqrt{2}(y-z)$, $\epsilon_3 = \frac{1}{2}\sqrt{2}(y+z)$ lie under 45° between the y and z directions. If one moves from the c-site $(\frac{1}{8}0\frac{1}{4})$ in the first octant along x to the c-site $(\frac{5}{8}0\frac{1}{4})$ in the second octant, the local symmetry is reflected at the xy-plane and the oxygen coordination changes from X_A to X_B. Table I gives the positions and local axes for all 8 atoms at X and X^+ chains which are related to one another by inversion*. The data for the corresponding c-sites at the chains Y, Z in y and z direction is obtained by rotation around the suitable diagonal axis in each octant.

The surface of a growing crystal cuts different oxygen dodecahedra open in different ways depending on the orientation of the local axes ϵ_1, ϵ_2, ϵ_3 relative to the growth direction. In fig. 1, as an example, the growth direction is assumed to be parallel to the (111) direction of the crystal. The part of the oxygen dodekahedron which is already formed below the substrate when the RE-ion arrives from the melt has different form for c-sites in X_A, Y_A, Z_A and X_B, Y_B, Z_B chains. The degeneration of the 24 c-sites is lifted into 2 subgroups with 12 sites each. The coordination of the X_B, Y_B and Z_B sites is very similar to one half of the irregular octrahedron found in R_2O_3 with cubic Ia3 structure. The flat square coordination below the X_A, Y_A, Z_A sites resembles much better the situation in RE_2O_3 with monoclinic $C2/m$ or hexagonal $C\bar{3}m$ structure [3]. The hexagonal structure is more stable in RE_2O_3 with ratios of ionic radii RE^{3+} to O^{2-} larger than 0.87, for smaller RE ions the monoclinic and finally the cubic form is preferred [4]. From this coordinational preference in RE_2O_3 one can expect a preference of large RE ions for the A-type sites on a growing (111) plane of garnet leaving the narrower B-type sites to the smaller RE ions. Therefore, the important parameter for site preference should be the differences in the ionic radii.

We used the situation of growth in the (111) direction as an example to demonstrate the

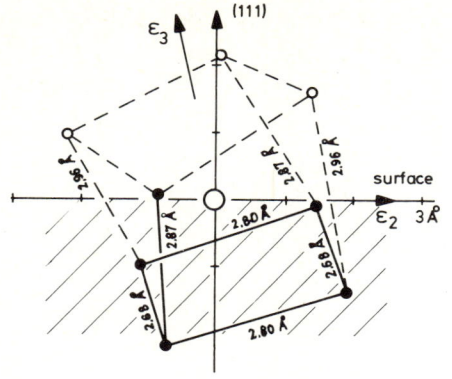

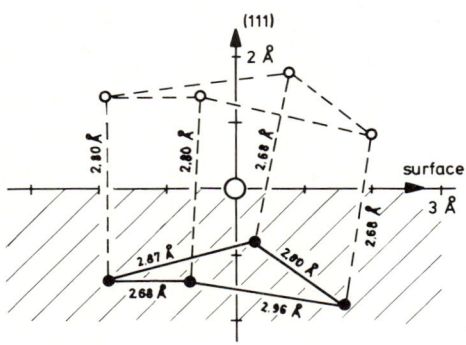

Fig. 1. Coordination of oxygen ions below a (111) surface at the moment the RE ion arrives at the center. (a) for c-sites of type X_B, Y_B, and Z_B; (b) for c-sites of type X_A, Y_A, and Z_A.

possibility of the chemical mechanism of site preference. It can be extended to other growth directions by the application of Callen's procedure to the oxygen instead to the Fe coordination of the c-sites [5]. The magnetic anisotropy produced by exchange interactions should not depend on the mechanism of site preference. Since, however, the chemical site preference is sensitive to ionic radii, it will distribute ions of different size in a regular manner through the lattice. The regular occupation of A-sites by large and B-sites by smaller RE-ions will, as an example, distort the environment of the a-site ions as well as of the d-sites, and can produce there a single ion anisotropy which fluctuates a little from site to site but produces an average noncubic symmetry which is related to the growth direction. The chemical site preference works, of course, for nonmagnetic RE ions also.

Table I
Coordination of the oxygen ions at c-sites in X-chains

	X_A	X_A^+	X_B	X_B^+
Type:	$\frac{1}{8}0\frac{1}{4}$	$\frac{3}{8}\frac{1}{2}\frac{1}{4}$	$\frac{5}{8}0\frac{1}{4}$	$\frac{3}{8}0\frac{3}{4}$
c-site:	$\frac{5}{8}\frac{1}{2}\frac{3}{4}$	$\frac{7}{8}0\frac{3}{4}$	$\frac{1}{8}\frac{1}{2}\frac{3}{4}$	$\frac{7}{8}\frac{1}{2}\frac{1}{4}$
ϵ_1	100	$\bar{1}00$	100	$\bar{1}00$
ϵ_2	$0\bar{1}\bar{1}$	011	$01\bar{1}$	$0\bar{1}1$
ϵ_3	$0\bar{1}1$	$0\bar{1}1$	$0\bar{1}\bar{1}$	011

* In Callen's paper is $X_1 = X_B$, $X_2 = X_B^+$, $X_3 = X_A$, $X_4 = X_A^+$.

This paper has been stimulated by discussion with several colleagues during a visit in K. Lees Group at the IBM research center in San Jose.

References

[1] H. Callen, Appl. Phys. Lett. 18 (1971) 311.
A. Rosencwaig, W.J. Tabor and R.D. Pierce, Phys. Rev. Lett. 26 (1971) 775.
A. Rosencwaig and W.J. Tabor, AIP Conf. Proc. 5 (1971) 51.
[2] E.M. Gyorgy, A. Rosencwaig, E.I. Blount, W.J. Tabor and M.E. Lines, Appl. Phys. Lett. 26 (1971) 779.
[3] R.W.G. Wyckhoff, Crystal Structures, Vol. 2 (Interscience, New York 1966).
[4] M.M. Schieber, Experimental Magnetochemistry (North Holland, Amsterdam, 1967) p. 304.
[5] H. Callen, Nat. Res. Bull. 7 (1971) 931.

ANOMALIES IN THE TEMPERATURE DEPENDENCE OF THE HYPERFINE FIELDS IN Sn DOPED AND PURE MAGNETITE

C. BOEKEMA, J. DE JONG and F. VAN DER WOUDE

Solid State Physics Laboratory, Materials Science Center, University of Groningen, Groningen, The Netherlands

and

G.A. SAWATZKY

Laboratory for Physical Chemistry, Materials Science Center, University of Groningen, Groningen, The Netherlands

A Mössbauer study of the behavior of the hyperfine interactions at the ^{57}Fe and ^{119}Sn nuclei in Fe$_3$O$_4$ and Sn doped Fe$_3$O$_4$ has revealed anomalies in the temperature dependence of these hyperfine fields in the Verwey transition temperature region.

Recent experimental studies [1–3] of the physical properties of magnetite above and in the region of the Verwey temperature (T_V) of about 120 K give evidence of a complex nature of the Verwey phase transition, showing a series of transitions with electronic and ionic characteristics. In the neighborhood of T_V two closely spaced transitions were observed by means of heat capacity measurements [1]. The lower temperature anomaly at 113 K is attributed to ionic ordering, while the high temperature anomaly is assigned to electronic ordering. Spontaneous magnetization reversals in magnetite were detected [2] at up to six discrete temperatures between 110 and 190 K. Because also line intensity rearrangements of Mössbauer spectra of Fe$_3$O$_4$ in this Verwey temperature region were observed and because the rotation characteristics are due to spin–orbit coupling, it was suggested that spin and charge density changes at the B sites in magnetite are involved. We have revealed in a Mössbauer study [4] of slightly doped magnetite the occurrence of spin and charge density oscillations in the B sublattice of magnetite above T_V, reflecting a nonlocal part in the character of the conduction electron response in magnetite. The behavior of the magnetocrystalline anisotropy [3] above T_V shows a temperature dependence, together with other physical properties such as the elastic modulus C_{44}, indicating some local lattice distortions or charge localization. Considering the complex nature of the Verwey transition it is evident that the peculiar phenomena show electronic and ionic features. In order to elucidate some aspects of this complexity, we have studied the behavior of the hyperfine interactions at the ^{57}Fe and ^{119}Sn nuclei in Fe$_3$O$_4$ and (Fe)[Sn$_x$Fe$_{2-x}$]O$_4$ with $x = 0.01$ and 0.05 as a function of temperature. Magnetite can be characterized as (Fe^{3+})$_A$[Fe$_2^{3+}$e^{-1}]$_B$O$_4$. Above T_V there are crystallographically two equivalent A sites and four equivalent B sites per unit cell in the spinel structure. Magnetically there are two kinds of B sites at a ratio 3:1. The resulting anisotropy plays an important part in the behavior of the hyperfine fields in magnetite.

The samples of pure magnetite and Sn doped magnetite were prepared using a ceramic technique [4] until a single spinel phase was obtained. The Mössbauer spectrometer [4] was calibrated with absorbers of Fe and powdered α-Fe$_2$O$_3$. Commercial sources of ^{57}Co in a Cr or Rh matrix and a Ba ^{119}SnO$_3$ source were used.

The ^{119}Sn Mössbauer spectra of (Fe)[Sn$_x$Fe$_{2-x}$]O$_4$ ($x = 0.01$ and 0.05) show below the ferrimagnetic Néel temperature an asymmetrically broadened six line pattern, caused by an effective magnetic field at the ^{119}Sn nuclei. No doublet of possible Sn^{2+} or SnO was observed. The isomer shift (0.195 mm/s relative to the Ba ^{119}SnO$_3$ source) at room temperature is indicative for a 4+ valency of the Sn impurities in magnetite. No quadrupole effect was measured for all temperatures.

Below the ferrimagnetic Néel temperature the ^{57}Fe Mössbauer spectra show two six-line patterns. The width of the absorption lines due to Fe^{3+} ions at the A sites is narrow, while the absorption lines due to the Fe ions at the B sites are broadened. This broadening is due to the fast electron hopping and the anisotropy at the

B sites. Due to this anistropy which causes a quadrupole effect and a magnetic anisotropy effect [1, 5], the first B line shows some asymmetric broadening, while in the second B line both effects cancel. Above T_V the ^{57}Fe spectra of magnetite can be succesfully interpreted in terms of one A pattern and two B patterns with an intensity ratio 3:1. In fig. 1(a) we have plotted as a function of T/T_{FN} the absolute difference in magnetic hyperfine field at the ^{57}Fe nuclei at the two inequivalent B sites, due to the anisotropy (ΔH_{an}). The width of the B lines is shown in fig. 1(b). Its behavior as a function of temperature can be interpreted as a relaxation effect due to the electron exchange at the B sites. We have found the temperature region above about 240 K the A linewidth is 0.28 ± 0.02 mm/s; between T_V and 240 K the A linewidth is 0.32 ± 0.02 mm/s. Only near the Verwey transition the A lines broaden somewhat.

The effective magnetic hyperfine field at the ^{119}Sn nuclei in Sn doped magnetite is due to supertransfer (A-B) and direct transfer (B-B) effects, which oppose each other. Its direction is parallel to the magnetic moment of the Fe^{3+} ion at an A site, if we assume, in analogy with Sn-doped $NiFe_2O_4$ and $MnFe_2O_4$, that the supertransfer hyperfine contribution is greater than the direct contribution. The asymmetric broadening of the ^{119}Sn absorption lines in Sn doped Fe_3O_4 can have two causes: (1) the magnetic anisotropy effect, or (2) the effect of a different number of diamagnetic Sn ions around the absorbing Sn ion. For the second effect we would expect a broadening towards higher fields, which is not observed. The measured asymmetric broadening towards lower fields can be explained by means of magnetic anisotropy effects, coming from the lattice. Analysing the ^{119}Sn spectra of Sn doped Fe_3O_4 in this way yields the following results. In fig. 1(c) the absolute difference between the hyperfine field at the two inequivalent B sites due to this anisotropy effect (ΔH_{an}) is given as a function of T/T_{FN}. Also in fig 1(d) the linewidth behavior is given. Below 200 K, towards T_V there is a drastic increase of this linewidth. For $Sn_{0.01}Fe_{2.99}O_4$ the room temperature value of $\Delta H_{an}(=7.3$ kOe) is in excellent agreement with the NMR result [5] and Mössbauer results [fig. 1(a)] on pure magnetite. A calculation of ΔH_{an} for $T = 0$ K

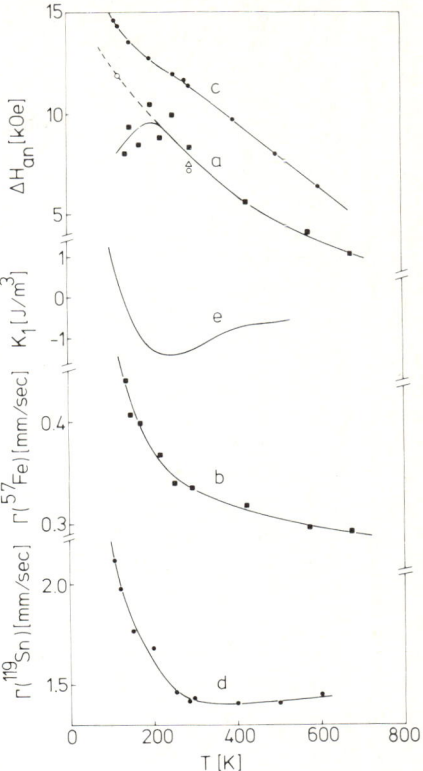

Fig. 1. The temperature dependence of the absolute difference in magnetic anisotropic hyperfine fields at the ^{57}Fe nuclei in Fe_3O_4 (a) and ^{119}Sn nuclei in $Sn_{0.05}Fe_{2.95}O_4$ (c) and $Sn_{0.01}Fe_{2.99}O_4$ (○) at the B sites, together with the NMR result (△, ref. 5) and of the linewidth of the B lines in Fe_3O_4 (b) and the ^{119}Sn absorption lines in Sn doped magnetite (d), and of the magnetocrystalline anisotropy constant K_1 of magnetite (e).

taking dipole–dipole interactions into account yields 17 kOe. The values of ΔH_{an} for $Sn_{0.05}Fe_{2.95}O_4$ are somewhat higher because of the above mentioned second effect, which also broadens the ^{119}Sn absorption line.

In contrast with the results of pure magnetite [fig. 1(a)] there is no maximum in $\Delta H_{an}(Sn)$ [fig. 1(c)] because only a lattice contribution to the magnetic anisotropy is possible for the ^{119}Sn nuclei. For the ^{57}Fe nuclei there exists also a small ionic contribution to the anisotropic hyperfine field, which is the reason for the maximum in $\Delta H_{an}(Fe)$ [fig. 1(a)]. Applying one ion crystal field theory and assuming for the Fe^{2+} ions a trigonal field splitting of 1000 cm^{-1}, a typical value, with the singlet state as ground state we would expect differences in the magnetic anisotropic fields in the order of 100 kOe. With the obtained results the trigonal field splitting is of the order of the electron hopping energy or even smaller. Because of the same characteristics [fig. 1(e)] one may relate this behavior to the behavior of the magnetocrystalline anisotropy constant K_1 of Fe_3O_4 [3]. Chikazumi [3] explains the change in sign of K_1 by means of formation of anisotropic ions such as Fe^{2+} or Fe^{1+} ions, due to local lattice distortion or local charge ordering. The broadening of the ^{119}Sn absorption lines in Sn doped magnetite near the Verwey transition may show an onset of localization of the conduction electrons in magnetite. At room temperature the conduction electrons are moving relatively free between the B sites. While going down in temperature towards T_V, the extra 3d electrons slow down in their movement, getting more and more a localized character in their behavior and making an onset of localization possible.

This work is part of the research program of the "Stichting voor Fundamenteel Onderzoek der Materie" (Foundation for Fundamental Research on Matter – F.O.M.) and has been made possible by financial support from the "Nederlandse Organisatie voor Zuiver Wetenschappelijk Onderzoek" (Netherlands Organization for the Advancement of Pure Research – Z.W.O.).

References

[1] B.J. Evans, AIP Proc. 24 (1975) 73.
[2] R.A. Buckwald et al., Phys. Rev. Lett. 35 (1975) 878.
[3] S. Chikazumi, Tech. Rep. ISSP Ser. A. (1975) 737 (to be published in AIP Proc. 1976).
[4] C. Boekema et al., J. Phys. C: Solid State Phys. 9 (1976) 2439.
[5] M. Rubinstein, G.H. Stauss and F.J. Bruni, AIP Proc. 10 (1973) 1384.

MAGNETOELECTRIC EFFECTS ARISING FROM THE ELECTRIC-FIELD-DEPENDENT MACROSCOPIC MAGNETIC ANISOTROPY ENERGY IN MAGNETITE

G.T. RADO and J.M. FERRARI

Naval Research Laboratory, Washington, D.C. 20375, USA

The electric-field-dependent macroscopic magnetic anisotropy energy, previously introduced by the authors, is shown to produce linear as well as two kinds of bilinear magnetoelectric susceptibilities in a magnetically biased ferromagnet. For the linear and one bilinear susceptibility, the calculated and measured values agree in magnetite (Fe_3O_4) at 4.2 K.

Having introduced the concept of an electric-field-dependence of the macroscopic magnetic anisotropy energy [1], we now show that in a ferromagnetic crystal, biased by means of an external static magnetic field H_0, such a dependence gives rise to linear as well as to two kinds of bilinear magnetoelectric (ME) effects described by the terms $-\alpha_{ij}e_ih_j$, $-\frac{1}{2}\beta_{ijk}e_ih_jh_k$ and $-\frac{1}{2}\gamma_{ijk}e_ie_jh_k$ in the appropriate free energy density $\tilde{F}$. After augmenting our expression [1] for $\tilde{F}$ with a Zeeman term containing the components of h, we expand $\tilde{F}$ to third order in the perturbing electric and magnetic fields, $e = E$ and $h = H - H_0$, and require the total magnetization M to be in equilibrium in each order of the perturbation. In this way we calculate the dependences on H_0 of the linear ME susceptibility α_{ij} and of the bilinear ME susceptibilities β_{ijk} and γ_{ijk}. To measure these dependences, we use 1 kHz fields for e and h and observe the 2 kHz as well as the 1 kHz parts of both M $(=-\partial\tilde{F}/\partial H)$ and of the total polarization P $(=-\partial\tilde{F}/\partial E)$ in a magnetically annealed, disk-shaped crystal of synthetic magnetite (Fe_3O_4) at 4.2 K. The faces of the disk [1] are, in orthorhombic notation, perpendicular to the crystallographic a (magnetically hard) axis and thus parallel to the plane defined by the b (magnetically intermediate) and the c (magnetically easy) axes. For α_{ac}, α_{ab} and β_{abb}, but not for γ_{aab}, the calculated and measured magnitudes as well as H_0-dependences are in good agreement whenever the magnetization process is expected to be dominated by domain rotations. We also note that the (magnetic-anneal-dependent) value of our $|\beta_{abb}|$ at $H_0 = 0$ is at least 10^{-6} Oe^{-1} and thus exceeds by two orders of magnitude the largest $|\beta_{ijk}|$ ever observed in any material. Gaussian units are used throughout this paper.

In expanding on the above statements, we find that space limitations prohibit the displaying as well as the derivation of our formulae for the ME susceptibilities. Furthermore, we are obliged to confine the discussion to just one of our two experimental situations, namely a one-domain configuration in which H_0 is applied at some angle ψ with respect to the c axis in the bc plane.

To determine α_{ac} experimentally, we measured both the magnetization m_c induced by an applied e_a and the polarization p_a induced by an applied h_c. Experimental curves of $|\alpha_{ac}|/|\partial K'_b/\partial E_a|_0$ as a function of H_0 and ψ are shown in fig. 1. The reason for plotting $|\alpha_{ac}|/|\partial K'_b/\partial E_a|_0$ rather than $|\alpha_{ac}|$ itself is that $(\partial K'_b/\partial E_a)_0$ varies [1] from one magnetic anneal to the next. Here K'_b is one of the previously introduced [1] electric-field-dependent anisotropy constants K'_b and K'_{bb}, and the subscript zero denotes $e = h = 0$. The parameters $(\partial K'_b/\partial E_a)_0$ and $(\partial K'_{bb}/\partial E_a)_0$ had been measured previously [1] (by means of a static method) and

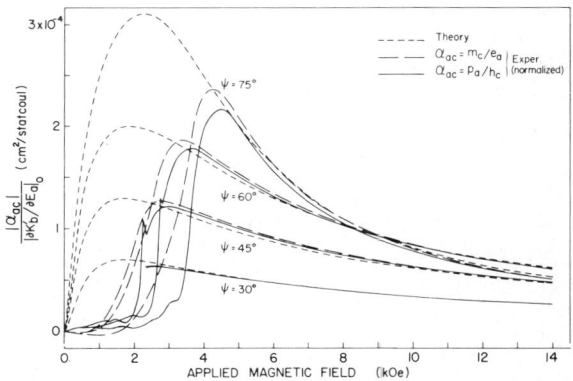

Fig. 1. Theoretical and experimental curves of the linear ME susceptibility $|\alpha_{ac}|$ as a function of H_0 and ψ. The theoretical values of $|\alpha_{ac}|$ are plotted as ratios of $|\alpha_{ac}|$ to $|\partial K'_b/\partial E_a|_0$, where the latter quantity is an H_0-independent parameter. The experimental values of $|\alpha_{ac}|$ are normalized to the theoretical values at $H_0 = 14$ kOe.

$(\partial K'_b/\partial E_a)_0$ was again measured after each magnetic anneal.

Also shown in fig. 1 are theoretical curves of $|\alpha_{ac}|/|\partial K'_b/\partial E_a|_0$ as a function of H_0 and ψ. The ordinates are expressed in units of cm^2/statcoul and do not involve any adjustable parameters. Our formula for α_{ac} predicts that $|\alpha_{ac}|/|\partial K'_b/\partial E_a|_0$ depends solely on H_0 and ψ, which are directly measurable, on the anisotropy fields $2K_b/M$ and $4K_{bb}/M$, which we take from the experiments of Palmer [2] as having the values 0.888 and 1.868 kOe, respectively, and on the ratio of $(\partial K'_{bb}/\partial E_a)_0$ to $(\partial K'_b/\partial E_a)_0$ which we take from our previous work [1] to have the value 0.08. Fig. 1 shows, among other things, that at low values of H_0 the theoretical and experimental curves disagree. This is undoubtedly due to the fact that in this range of H_0 values the domain wall displacements predominate over the domain rotations described by our theory. More interesting, perhaps, is the fact that at low values of H_0 the α_{ac} measured via m_c/e_a disagree with those measured via p_a/h_c. A possible explanation of this fact is that in the region of domain wall displacements the effect of h_c is so large that $\tilde{F}$ cannot be expanded in powers of e_a and h_c and hence m_c/e_a cannot be expected to equal p_a/h_c. At sufficiently high values of H_0, on the other hand, the domain rotations are expected to predominate so that the power series expansion of $\tilde{F}$ and our thermodynamic theory are both applicable.

To investigate the numerical value of $|\alpha_{ac}|$, we consider, both experimentally and theoretically, an annealing condition characterized by $|\partial K'_b/\partial E_a|_0 = 3.03$ statcoul/cm^2. We find, for example, that at $H_0 = 14$ kOe and $\psi = 75°$ the measured $|p_a/h_c|$ is about 1.4×10^{-4} which agrees reasonably well with the calculated value $|\alpha_{ac}| = 1.60 \times 10^{-4}$. Thus we conclude, as stated above, that not only the H_0-dependence but even the numerical value of $|\alpha_{ac}|$ is explained satisfactorily by the electric-field-dependent anisotropy mechanism of the present theory.

Included in a full account of this work, to be published in Physical Review B, 1977, are the detailed theory for the one-domain configuration as well as for a two-domain configuration, the experimental data on α_{ab}, β_{abb} and γ_{aab}, and experimental evidence for the proportionality of $|\alpha_{ab}|$ and $|\beta_{abb}|$, but not of γ_{aab}, to the quantity $|\partial K'_b/\partial E_a|_0$ measured after each of numerous magnetic anneals.

References

[1] G.T. Rado and J.M. Ferrari, Phys. Rev. B12 (1975) 5166. Erratum: Phys. Rev. B14 (1976) 4239.
[2] W. Palmer, Phys. Rev. 131 (1963) 1057.

SELECTIVE EXCITATION DOUBLE MÖSSBAUER STUDIES (SEDM) OF ELECTRON HOPPING IN MAGNETITE (Fe_3O_4)*

B. BALKO† and G.R. HOY

Physics Department, Boston University, Boston, Massachusetts, USA

Using the SEDM technique, we have established that the electron hopping time at the octahedral (B) sites in magnetite is zero or less than 10^{-11} s at room temperature. Electrical conductivity measurements have predicted such a value. In addition, unexpected relaxation effects were observed at the tetrahedral (A) sites.

Magnetite, Fe_3O_4, above the Verwey transition ($T_v = 119$ K) is a cubic inverse spinel with the tetrahedral (A) sites occupied by Fe^{3+} ions and the octahedral (B) sites by Fe^{2+} and Fe^{3+} ions. The usually high electrical conductivity ($250\,\Omega^{-1}\,cm^{-1}$ at 300 K) results from a rapid electron exchange between the ferrous and ferric ions on the octahedral (B) sites. Previous Mössbauer measurements [1] have found that the electron exchange relaxation time is 10^{-9} s at room temperature. This result was obtained by examing the line broadening in the hyperfine Mössbauer pattern associated with the octahedral (B) sites.

However, electrical conductivity results suggest that the relaxation time should be about two orders of magnitude shorter. This discrepancy stimulated our interest in applying the SEDM technique to magnetite. SEDM procedures have previously been applied to detect the presence of relaxation processes [2]. Fig. 1 gives a schematic diagram of the experimental set-up. In this technique a single line source is driven at a predetermined constant velocity. The source radiation impinges on the scatterer which is made of the material under investigation, i.e. magnetite in this case. Some of the nuclei in the scatterer became resonantly excited. The radiation subsequently emitted, when the nuclei decay, is subjected to analysis by using a resonant single line absorber moving at constant acceleration. In this way the energy spectral distribution of the radiation emitted by the scatterer can be measured. These SEDM procedures have certain advantages over conventional Mössbauer spectroscopy. First, the resulting SEDM spectrum is usually quite simple containing one or perhaps two peaks. Se-

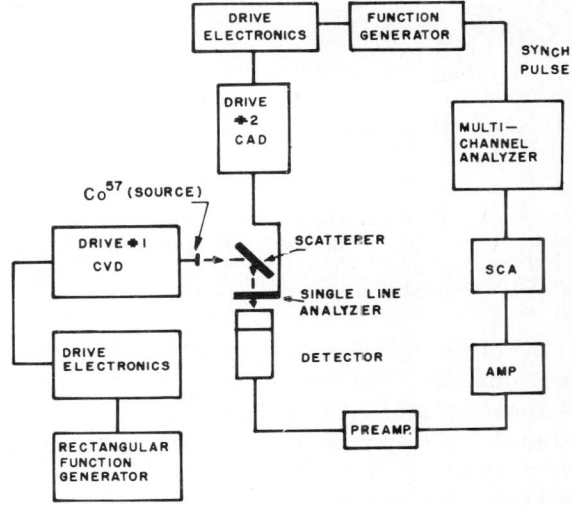

Fig. 1. A schematic representation of the experimental configuration used in SEDM. CVD stands for constant velocity drive, and CAD for constant acceleration drive.

condly, relaxation effects can be observed more clearly. This is so, because in conventional Mössbauer spectroscopy thickness and inhomogeneous field effects produce line broadening which can be mistaken for relaxation.

In fig. 2 we show a typical Mössbauer spectrum of magnetite at room temperature. The dips labeled B correspond to the hyperfine pattern due to iron nuclei on the octahedral (B) sites, while those labeled A correspond, similarly, to the tetrahedral (A) sites. We have concentrated our studies on the region in the neighborhood of the 1A and 1B dips. In fig. 3 we show our SEDM results using a magnetite scatterer at room temperature. The locations of the 1A and 1B dips are shown by the arrows at the top of the figure and the two dashed vertical lines. The arrows at the bottom of each spectrum locate the setting of the constant velocity drive in each case.

---

* Supported by the National Science Foundation, Grant no. DMR 73-07665A03.
† Present address: National Institutes of Health, Bethesda, Maryland.

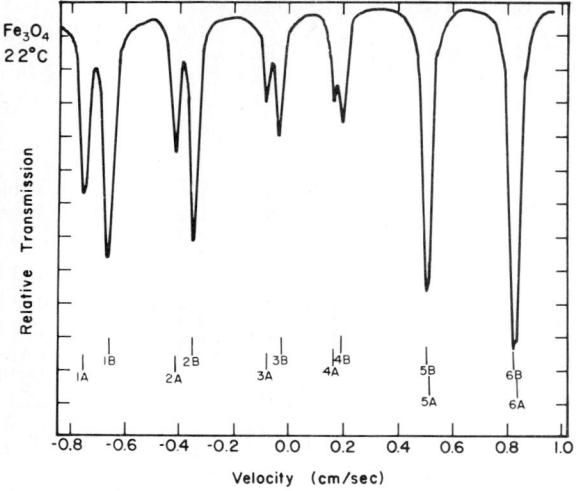

Fig. 2. A typical Mössbauer absorption spectrum using magnetite at room temperature. The dips labeled B arise from the contribution due to the octahedral sites, and those labeled A are due to the tetrahedral sites.

In order to analyze these data we used methods previously developed [2]. To briefly summarize the general features of such an analysis, we can state the following results. When there is no relaxation occurring in the scatterer, the experimental peak appears at the excitation energy with perhaps some asymmetry. However, when relaxation is present, the experimental peak is shifted toward the resonance position and the asymmetry can be quite pronounced. The solid curves shown in fig. 3 are calculated assuming *no* relaxation. The only free parameter in these calculations is a single normalization. In this case, the calculations were normalized to the second spectrum from the top. Thus the other calculated curves have no free parameters. These calculations were made using a ratio of B to A sites equal to two. The sample thickness parameter $\beta = 336$.

Perhaps the most important conclusion we can draw from these results is the following. Since the top three spectra are fitted rather well with an SEDM non-relaxating calculation, the electron hopping at the B sites must be sufficiently rapid that for these parameters one can not see relaxation efforts. Preliminary cal-

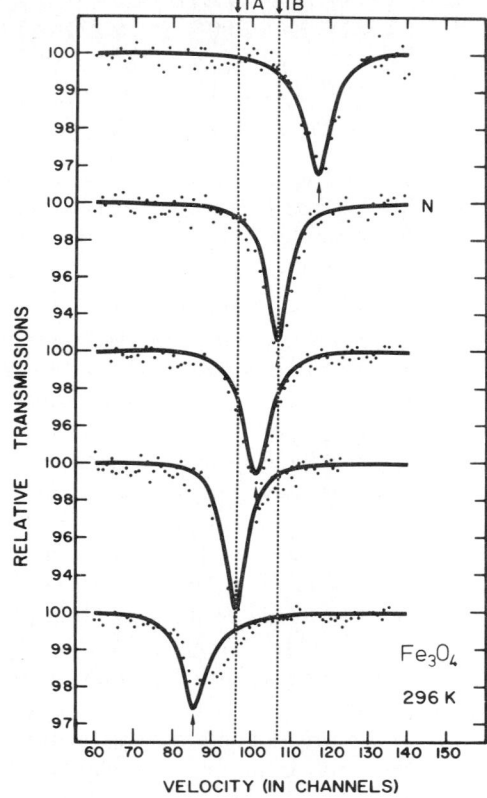

Fig. 3. Our SEDM results using a magnetite scatterer at room temperature are shown in this figure. The arrows labeled 1A and 1B and the dashed vertical lines locate the positions of those resonances (see fig. 2). The small arrows locate the excitation energy (i.e. CVD setting) for each of the five cases. The dots are the data and the solid curves give the theoretical SEDM results assuming *no* relaxation.

culations indicate that the electron exchange time must be zero or less that 10^{-11} s.

The most surprising result is seen in the bottom spectrum of fig. 3. It appears that there is relaxation occurring at the A sites at room temperature. We know of no other experimental result or theoretical model that predicts this effect. We observed this result over several months on the same sample.

References

[1] W. Kundig and R.S. Hargrove, Solid State Commun. 7 (1969) 223.
[2] B. Balko and G.R. Hoy, Phys. Rev. B10 (1974) 36.

MÖSSBAUER LINEWIDTH OF OCTAHEDRAL IRON IN MAGNETITE

A.M. VAN DIEPEN

Philips Research Laboratories, Eindhoven, The Netherlands

In the study of Fe^{2+}–Fe^{3+} valence exchange on octahedral sites in Fe_3O_4 the line broadening of the Mössbauer spectrum of the B sites with respect to the A sites has frequently been considered experimental proof for an electron hopping model. The present investigation shows, however, that this broadening is the result of electric quadrupolar and magnetic dipolar effects.

The electrical conduction of magnetite above the Verwey transition (119 K) is ascribed to the presence of nominally equal amounts of Fe^{2+} and Fe^{3+} on the octahedral (B) sites in the spinel structure. The linewidth of the Mössbauer spectrum of B-site Fe, as compared to that of A-site Fe^{3+}, and its temperature dependence have frequently been considered experimental proof for electron hopping as the conduction mechanism [1, 2], opposing the band model which is, however, supported by several other experiments [3]. In the analysis of doped magnetite also important conclusions are derived from the Mössbauer line shape [4]. The purpose of the present study is to investigate the effect of other mechanisms on the line broadening for B-site Fe in Fe_3O_4.

As a result of the trigonal symmetry at the B sites non-zero nuclear quadrupole and magnetic dipole splittings occur that may manifest themselves as broadening of the lines in the Mössbauer spectrum. This was first noted by Evans [5], who decomposed the B-site pattern into two subspectra with 3:1 intensity and determined a value of 0.100 mm/s for $|e^2qQ|$ at the B site, in fact including also the dipolar field. The resulting contribution to the width of the line B1 (0.44 mm/s), the outermost negative-velocity peak of the B-site subspectrum, was considered negligible by the authors of ref. 4.

In the present study a new estimate is made based on Mössbauer spectra of a synthetic single crystal of Fe_3O_4 while the direction of the hyperfine field is changed by means of an external magnetic field. The magnetic dipolar field and the nuclear quadrupole interaction to first order produce a splitting of the B1 line proportional to $(3\cos^2\theta - 1)$, where θ is the angle between the hyperfine field and the local symmetry axis ([111]). By varying θ the sum of these interactions thus can be obtained.

The experiment was done at room temperature on a $\langle 1\bar{1}0 \rangle$ single-crystal platelet, 85 μm thick, of Bridgeman-grown Fe_3O_4. An external magnetic field H of 8 kG was applied in the plane of the platelet. Mössbauer spectra were taken with H parallel to the [001], [110], and [111] directions. A field of this magnitude is amply sufficient to saturate the magnetization, i.e. to turn the hyperfine field completely in the desired crystallographic direction.

Since there are four body diagonals in the cube, for an arbitrary direction of the hyperfine field there are four angles θ and thus four subspectra for the B site. If H_{hf} is along an [001] direction they coincide so that only one B-site spectrum is observed. For $H_{hf}//[110]$ there are two subspectra of equal intensity, while with $H_{hf}//[111]$ there are two subspectra with relative intensities 3:1. Fig. 1 gives the low-velocity

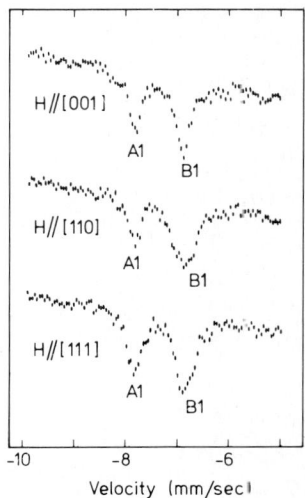

Fig. 1. Low velocity parts of room temperature Mössbauer spectra of Fe_3O_4 with a field of 8 kG applied as indicated in the three main crystallographic directions of a $\langle 1\bar{1}0 \rangle$ single-crystal platelet.

parts of the three spectra. It is immediately seen that the line shape of the B1 line reflects the $(3\cos^2\theta - 1)$ dependence: for $H//[001]$ only one line is observed of the same width as the A1 line, for $H//[110]$ it is about twice as wide but still symmetric, and for $H//[111]$ it is asymmetrically broadened. The A site is perfectly cubic so that its Mössbauer spectrum is not affected by turning the magnetization. Effective linewidths at half height are: $\Gamma_{A1} = 0.28$ mm/s for all three directions, $\Gamma_{B1}^{[001]} = 0.30$ mm/s, $\Gamma_{B1}^{[110]} = 0.57$ mm/s, and $\Gamma_{B1}^{[111]} = 0.45$ mm/s. From $\Gamma_{B1}^{[110]}$ as compared to $\Gamma_{B1}^{[001]}$ a splitting in the former of about 0.27 mm/s is derived which implies a splitting of 0.36 mm/s in $\Gamma_{B1}^{[111]}$. This should be compared with the earlier estimates of about 0.10 mm/s [4, 5]. It is noted here that the small value of the quadrupole coupling already by itself makes it unlikely that the Fe on the B site is in a state intermediate between ionic Fe^{2+} and Fe^{3+}.

It is thus demonstrated that the line broadening observed in the B-site Mössbauer spectrum of Fe_3O_4 can strongly be suppressed by turning the hyperfine field in the [001] direction by means of an externally applied field. The B-site line broadening is shown to be the result of electric quadrupolar and magnetic dipolar effects which are zero for $(3\cos^2\theta - 1) = 0$. In order to draw conclusions from B-site lineshapes, for instance regarding the electron hopping model, these dipolar and quadrupolar effects have to be taken into account. This can unambiguously be done only by using single crystals in relatively small external fields such that the direction of the magnetization is known for all of the B sites.

The author is indebted to J.J.P. Verheijden for technical assistance with the Mössbauer measurements and to J.P.M. Damen for growing the crystal.

References

[1] W. Kündig and R.S. Hargrove, Solid State Commun. 7 (1969) 223.
[2] G.A. Sawatzky, J.M.D. Coey and A.H. Morrish, J. Appl. Phys. 40 (1969) 1402.
[3] For a review see: B.J. Evans, AIP Conf. Proc. 24 (1975) 73.
[4] C.I. Nistor, C. Boekema, F. van der Woude and G.A. Sawatzky, Proc. 5th Int. Conf. on Mössbauer Spectrometry (Bratislava, 1973) p. 99.
[5] B.J. Evans, AIP Conf. Proc. 5 (1971) 296.

DETAILED STUDIES ON THE LOW TEMPERATURE PHASE OF Fe_3O_4

S. IIDA, K. MIZUSHIMA, M. MIZOGUCHI, J. MADA, S. UMEMURA, J. YOSHIDA and K. NAKAO

Department of Physics, University of Tokyo, Bunkyo-ku, Tokyo, Japan

A part of the recent study on the low temperature phase of Fe_3O_4 is presented. Emphasis is placed on the single crystal X-ray results, Mössbauer results and NMR results. They support our previously proposed model which has an emphasis on the $d\epsilon_{xy}$ orbital of BFe^{2+}-II ions.

A systematic study on the low temperature phase of Fe_3O_4 [1–4], has been extended further in detail. By using single crystals and counter-type automatic recording X-ray spectrometer [5], the reflection intensities on the two-dimensional reciprocal lattice planes were observed. One of the typical result is shown in fig. 1, in which the reflection intensities near the {444} line at 84 K are shown. Here, we have seven different reflection peaks with three different scattering angles 2θ. In fig. 2 we show another example of the single crystal X-ray diffractions in which half integer reflections $(8\ 0\ \bar{\frac{1}{2}})$ and $(8\ 0\ \frac{1}{2})$ are clearly observed at 84 K. We have concluded that the low temperature phase of Fe_3O_4 has a monoclinic unit cell with $a = 11.888$ Å, $b = 11.847$ Å, $c = 16.773$ Å, and $\beta = 89.76_2°$ at 84 K. From the extinctions we have concluded also that the crystal symmetry of the phase is either $C_s^4 - C_c$ or $C_{2h}^6 - C2/c$, with a c-glide plane whose normal is along b axis, in agreement with the result of neutron diffraction [6, 7]. The

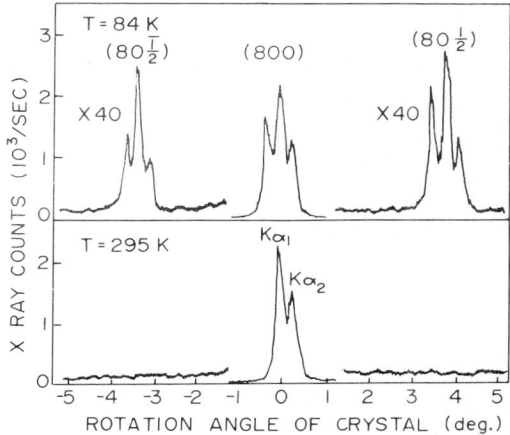

Fig. 2. Reflection intensity profile around (800) line at below and above T_v. A magnetic field was applied along [001] during cooling [Yoshida].

interrelation between this result and our proposed model will be clarified later.

From the Mössbauer observation [8, 9, 2, 4], we have proposed already that Fe^{2+}-II will have a $d\epsilon_{xy}$ orbital.

The NMR spin echo experiment [9, 10, 4], has been extended further. We show two of the results in figs. 3(a) and (b). They show the hyperfine field anisotropy of the BFe^{3+}-I and II ions. For fig. 3(a), a magnetic field of 6 kOe was applied along [113] during cooling and the magnetic field during observation is in $(12\bar{1})$ plane. For fig. 3(b), the magnetic field was along [213] during cooling and that during observation is in $(2\bar{1}\bar{1})$ plane. As a whole, a 180° symmetry is present for fig. 3(b) but it is of a 360° symmetry for fig. 3(a). The characters can be reproduced by electronic computation accurately. In conclusion, the 360° symmetry can appear only when non-saturation of the magnetization is present and this is due for fig. 3(a) but is not for fig. 3(b), because the $(12\bar{1})$ plane is close to ac-plane which includes the hard direction for

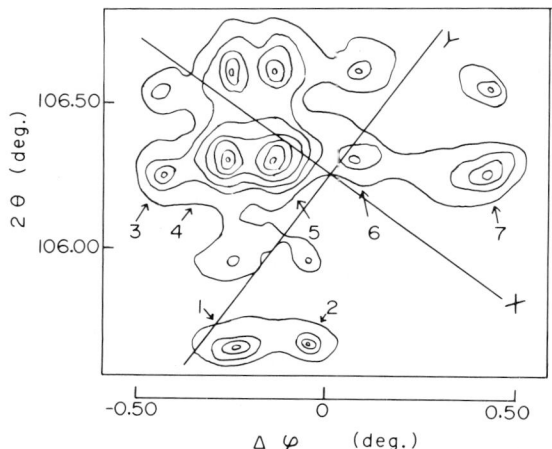

Fig. 1. Reflection intensity of {444} lines of Fe_3O_4 at 84 K in a two-dimensional reciprocal lattice space [Yoshida].

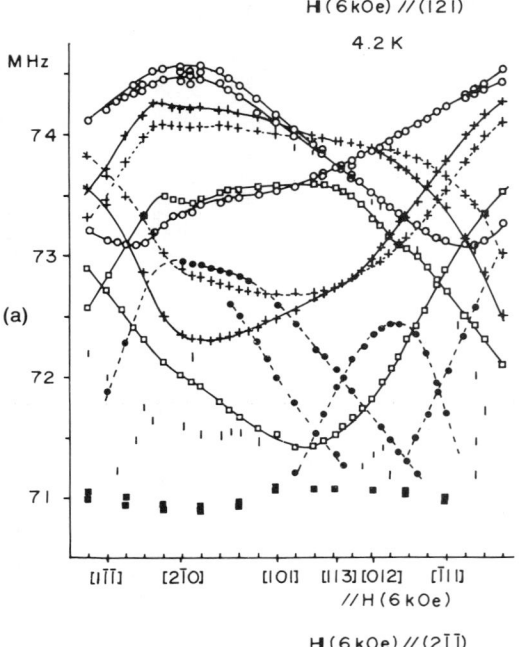

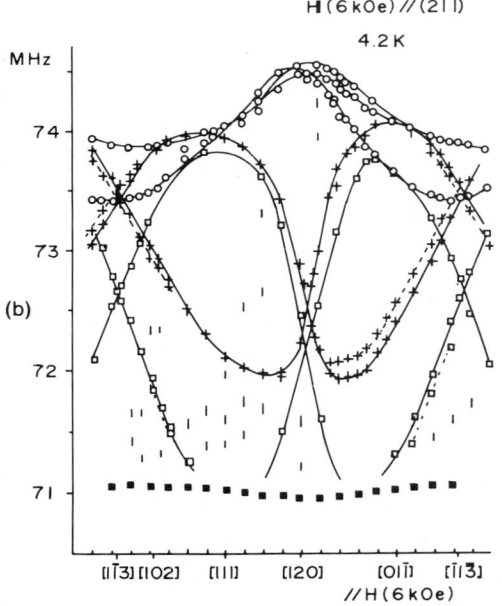

Fig. 3. Variation of the NMR peak positions of Fe_3O_4 as functions of the direction of applied magnetic field. The crystal was cooled with a magnetic field along [113] in (a) and along [213] in (b) [Mizoguchi].

the magnetization but $(2\bar{1}\bar{1})$ plane is close to the bc-plane which includes only the intermediate and easiest directions. We have a hyperfine field anisotropy of about 20 kOe for BFe^{3+} ions, which is not the case for AFe^{3+} ions, an example of which is shown by solid squares in the figures. The ferrimagnetic structure works nicely to create a uniaxial dipolar field with $\langle 111 \rangle$ symmetry on B sites.

We found a merging of the hyperfine fields of BFe^{2+}-II and BFe^{2+}-I in the same experiment with fig. 3. This is a strong evidence that the large lowering of the hyperfine field of BFe^{2+}-II is due to the spin dipolar field of the extra electron which has the $d\epsilon_{xy}$ major orbital. It confirms also that BFe^{2+}-I will have the $d\epsilon_{xz}$ and/or $d\epsilon_{yz}$ major orbital.

At the present stage of the experimental results, there is no evidence which requests to change our original model of the low temperature phase [4]. Based on our model, we have made a certain theoretical study for the origin of the presence of $d\epsilon_{xy}$ orbital for Fe^{2+}-II. Our conclusion is that, in addition to the presence of two BFe^{3+}-I in the xy plane, BFe^{2+}-II has four nearest neighbor A site Fe^{3+} ions near the xy-plane, which react cooperatively either to shift downwards or upwards resulting in the two upper or lower A site Fe^{3+} ions approaching to the xy-plane. The result definitely stabilizes the $d\epsilon_{xy}$ orbital of the BFe^{2+}-II ion. Obviously this shift makes the crystal magneto-electric [11] and it may become ferroelectric in a certain temperature range.

References

[1] S. Iida, M. Yamamoto and S. Umemura, AIP Conf. Proc. 18 (1974) 913.
[2] J. Mada and S. Iida, J. Phys. Soc. Jap. 39 (1975) 1627.
[3] S. Umemura and S. Iida, J. Phys. Soc. Jap. 40 (1976) 679.
[4] S. Iida, K. Mizushima, M. Mizoguchi, J. Mada, S. Umemura, K. Nakao and J. Yoshida, AIP Conf. Proc. 29 (1976) 388.
[5] J. Kobayashi, N. Yamaka and T. Nakamura, Phys. Rev. Lett. 11 (1963) 410.
[6] G. Shirane, S. Chikazumi, J. Akimitsu, K. Chiba, M. Matsui and U. Fuzii, J. Phys. Soc. Jap. 39 (1975) 947.
[7] M. Iizumi and G. Shirane, Solid State Commun. 17 (1975) 433.
[8] R.S. Hargrove and W. Kündig. Solid State Commun. 8 (1970) 303.
[9] M. Rubinstein and D.W. Forester, Solid State Commun. 9 (1971) 1675.
[10] N.M. Kovtun and A.A. Shamyakov, Solid State Commun. 13 (1973) 1345.
[11] G.T. Rado and J.M. Ferrari, Phys. Rev. B12 (1975) 5166.

EFFECT OF MAGNETOELASTIC COUPLING ON THE ANISOTROPY OF MAGNETITE BELOW THE TRANSITION TEMPERATURE

N. TSUYA, K.I. ARAI and K. OHMORI

Research Institute of Electrical Communication, Tohoku University, Sendai, Japan

The temperature dependence of a complete set of the magnetostriction constants in magnetite from liquid N_2 to room temperature were given. The considerable parts of the anisotropy constants measured by torque method at liquid N_2 temperature were considered to be induced by the magneto-elastic coupling with the lattice distortion.

Magnetite is well known to exhibit at 119 K a Verwey transition from cubic to a lower symmetry phase. Recently, it was found by the neutron and electron X-ray [1] diffraction that the magnetite below the transition temperature is nearly rhombohedral crystallographically with an elongation of about 0.2% along one of the [111] directions. The magnetic anisotropy constants were measured by many investigators [2–5] and reported as referring to orthorhombic symmetry. The magnetostriction and the switching of the magnetic axes were observed by Bickford [6] and by us [7] below the transition temperature. In this paper, we report, the magnetostriction constants of the magnetite single crystals by using a larger number of the specimens from liquid N_2 under the transition temperature.

Single crystals of Fe_3O_4 were grown by Iida in a floating zone type infrared ray image furnace in CO_2 atmosphere. Four spherical samples of Fe_3O_4 were made of the above single crystals. The magnetostriction of the spherical samples was measured by using a three-terminal capacitance method reported previously [8]. In this measurement, the magnetic field of 10 kOe was rotated in the space which was formed between the cubic (100) and (010) planes, in which the switching was not observed. During cooling through the transition, the magnetic field 14 kOe was applied to the [112] direction of the crystal in order to establish the orthorhombic axes. The applied magnetic field direction, observation directions of the magnetostriction and observable magnetostriction constants defined by Mason [9] in this experiment were shown in table I. Special attention was paid to remove as far as possible the effect of slight twin formations which mixed with the magnetostriction. The data were reproduced within ± several 10^{-6}.

In the observed magnetostriction constants

Table I
Directions of observation, and magnetic field rotation

Obs. H	$[1\bar{1}0]$	[110]	[001]	[100]	$[11\sqrt{2}]$	$[1\bar{1}0]$	[110]	[001]	[011]
from	[110]	[110]	[110]	[101]	$[11\sqrt{2}]$	[101]	[011]	[011]	[011]
to	[001]	[001]	[001]	[001]	[001]	[112]	[001]	[001]	[101]
Result	λ_2	λ_4	λ_6	$\lambda_{7/2}$	λ_9	complicated			

defined by Mason [9] the values of λ_2, λ_4, λ_6 and λ_9 were large and the residual constants were comparatively small. The temperature dependence of all constants became steeper toward the transition. To represent the crystallographic symmetry clearly, the expression of the magnetostriction is rearranged as follows:

$$\Lambda = \Lambda_1[2\alpha_3^2(\beta_3^2 - \tfrac{1}{3}) - \alpha_1^2(\beta_1^2 - \tfrac{1}{3}) - \alpha_2^2(\beta_2^2 - \tfrac{1}{3}) - 2(\beta_3^2 - \tfrac{1}{3})] + \Lambda_2[\alpha_1^2(\beta_1^2 - \tfrac{1}{3}) - \alpha_2^2(\beta_2^2 - \tfrac{1}{3})]$$
$$+ \Lambda_3(2\alpha_1\alpha_2\beta_1\beta_2 - \alpha_1\alpha_3\beta_1\beta_3 - \alpha_2\alpha_3\beta_2\beta_3)$$
$$+ \Lambda_4(\alpha_1\alpha_3\beta_1\beta_3 - \alpha_2\alpha_3\beta_2\beta_3)$$
$$+ 3\Lambda_5[\alpha_1^2(\beta_2^2 - \tfrac{1}{3}) - \alpha_2^2(\beta_1^2 - \tfrac{1}{3})]$$
$$+ \omega_a\alpha_1^2/3 + \omega_b\alpha_2^2/3$$
$$+ \text{cubic parts rotated } \pi/4 \text{ arround [001]},$$

where ω_a and ω_b are volume magnetostriction referred to the a and b-axes in orthorhombic plane. The temperature dependence of these constants are shown in fig. 1, where λ_{poly} and Λ_{poly} correspond to polycrystalline states above and below the transition, respectively.

To estimate the magnetoelastically induced anisotropy at a low temperature, we assume that the elastic constants remain unchanged and the change along the a, b and c-axes to be 0.0036, −0.0040, and 0.0003, respectively, using the thermal expansion coefficient to be 7.7×10^{-6} and the data of lattice parameters taken by Abraham and Calhoun [3], and Iida [1] at 77 K. The results were expressed as follows:

$$K_a^\lambda = K_{aa}^\lambda(\alpha_1^2 - \tfrac{1}{3}) + K_{bb}^\lambda(\alpha_2^2 - \tfrac{1}{3}) + K_{ab}^\lambda\alpha_1\alpha_2 + K_{ac}^\lambda\alpha_1\alpha_3 + K_{bc}^\lambda\alpha_2\alpha_3,$$

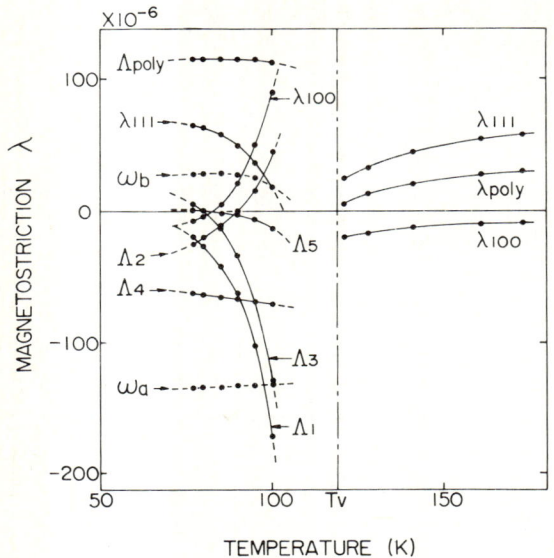

Fig. 1. Temperature dependence of the magnetostriction constants.

where these values of K_{aa}^λ, K_{bb}^λ, K_{ab}^λ and K_{bc}^λ in 10^5 erg/cm^3 were -5, 8, 0 and 0, respectively. The corresponding values measured by Bozorth [2], Calhoun [3], Palma [4] and Chikazumi [5] using the torque were ranging in 19.6–24.8, 7.7–5.5, 0 and 0, respectively. K_{ac}^λ contributes to the uniaxial term K_u, which yielded to -3 and 0.5 in 10^5 erg/cm^3, respectively. K_{bb}^λ seems to contribute remarkably.

The authors thank Professor S. Iida for supplying us with the magnetite single crystal.

References

[1] S. Iida, M. Yamamoto and S. Umemura, AIP Conf. Proc. 18 (1974) 913.
[2] H.J. Williams, R.M. Bozorth and M. Goertz, Phys. Rev. 91 (1953) 1107.
[3] B.A. Calhoun, Phys. Rev. 94 (1954) 1577.
[4] W. Palmer, Phys. Rev. 131 (1963) 1057.
[5] S. Chikazumi, private communication.
[6] L.R. Bickford, Jr., J. Pappis and J.L. Stull, Phys. Rev. 99 (1955) 1210.
L.R. Bickford, Jr., Rev. Mod. Phys. 25 (1953) 75.
[7] K.I. Arai, K. Ohmori, N. Tsuya and S. Iida, Phys. Status Solidi (a), 34 (1976) 1.
[8] N. Tsuya, K.I. Arai, K. Ohmori and Y. Shiraga, Jap. J. Appl. Phys. 13 (1974) 1808.
N. Tsuya, K. Ohmori, K.I. Arai and T. Wakiyama, Rep. Res. Inst. Commun. Tohoku Univ. 27 (1975) 1.
[9] W. P. Mason, Phys. Rev. 96 (1954) 302.

ELECTRON EXCHANGE BETWEEN Fe^{2+} AND Fe^{3+} ON OCTAHEDRAL SITES IN SPINELS

A.M. VAN DIEPEN and F.K. LOTGERING

Philips Research Laboratories, Eindhoven, The Netherlands

Thermally activated electron exchange ("hopping") between Fe^{2+} and Fe^{3+} is observed through Mössbauer spectroscopy in the system of spinels $Zn^{2+}|Zn^{2+}_{(1-x)/2}Ti^{4+}_{(1+x)/2}Fe^{3+}_{1-x}Fe^{2+}_x|O_4$. The results are discussed in terms of a hopping model based on an activation energy $q \simeq 0.12$ eV and non-equivalence of the octahedral sites. From the dependence of the asymptotic Curie temperatures on composition ferromagnetic exchange between Fe^{2+} and Fe^{3+} is derived, $J_{12}/k = +1.6$ K.

By virtue of a strong sensitivity for differences in valence state, the Mössbauer effect of ^{57}Fe is well-suited for studying Fe^{3+}–Fe^{2+} compounds. In this paper we give results for the system represented by the general formula $Zn^{2+}|Zn^{2+}_{(1-x)/2}Ti^{4+}_{(1+x)/2}Fe^{3+}_{1-x}Fe^{2+}_x|O_4$, in which the only magnetic ions are Fe^{2+} and Fe^{3+} occupying part of the B sites. This system was chosen on the basis of the following considerations. (i) The local trigonal symmetry of the octahedral sites allows a clear distinction between Fe^{2+} and Fe^{3+} based on the strong difference in quadrupole splitting and isomer shift. (ii) There is no iron on tetrahedral sites and the materials are paramagnetic down to low temperatures, resulting in simple Mössbauer spectra. (iii) Asymptotic Curie temperatures can be measured accurately, providing a reliable determination of the magnetic interaction parameters.

Mössbauer spectra taken at 78 and 300 K are given in fig. 1. The iron content is indicated and the samples are numbered 1–6. It is seen that at 78 K the Mössbauer spectra are simple superpositions of a ferric and a ferrous spectrum in accordance with the amounts of Fe^{2+} and Fe^{3+} present. At 300 K there is, in addition, continuous absorption near +1 mm/s for the mixed compounds (samples 2–5) which is absent for the end members (samples 1 and 6 with Fe^{3+} and Fe^{2+} only, respectively). For sample 3 Mössbauer spectra were also taken at 198 and 400 K which shows that this additional absorption increases in intensity with temperature. A sample having the composition of sample 3 with Fe^{3+} replaced by Ga^{3+} showed a room temperature spectrum similar to sample 6 with very little absorption at +1 mm/s. Sample 4 with $\tfrac{2}{3}Fe^{3+} + \tfrac{2}{3}Fe^{2+}$ on the B sites shows more absorption than sample 3 with $\tfrac{1}{2}Fe^{3+} + \tfrac{1}{2}Fe^{2+}$. This

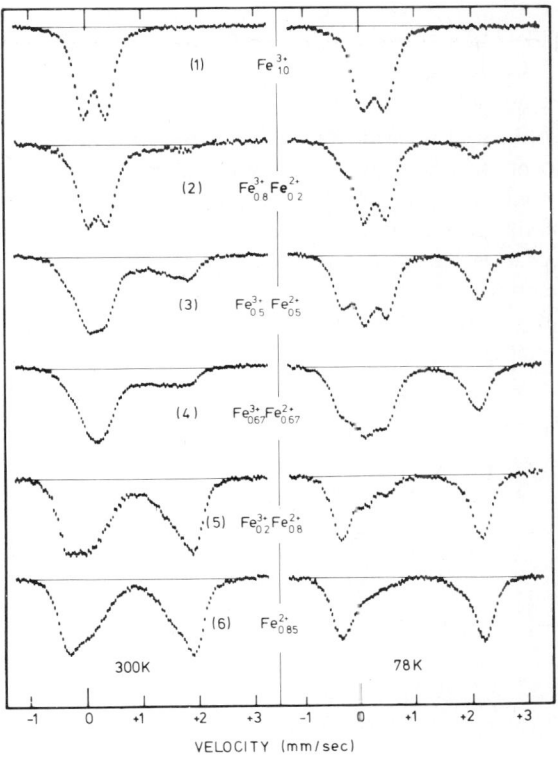

Fig. 1. Paramagnetic Mössbauer spectra taken at 300 and 78 K for the system of spinels $Zn^{2+}|Zn^{2+}_{(1-x)/2}Ti^{4+}_{(1+x)/2}Fe^{3+}_{1-x}Fe^{2+}_x|O_4$. The Fe^{3+} and Fe^{2+} contents are indicated. At 78 K superposition of a ferric and a ferrous spectrum is observed, but at 300 K there is for the mixed compounds in addition a continuous absorption near +1 mm/s which is attributed to Fe^{3+}–Fe^{2+} hopping.

continuous absorption is attributed to thermally activated electron exchange ("hopping") between Fe^{2+} and Fe^{3+} ions, which occurs for only part of the Fe^{2+}–Fe^{3+} pairs.

Due to the presence of different ions (Zn^{2+}, Ti^{4+}, Fe^{3+}, and Fe^{2+}) non-equivalence of the B sites is expected. Considering an Fe^{3+}–Fe^{2+} pair as two Fe^{3+} cores with an itinerant electron, the non-equivalence of the two sites can be represented by a potential difference U_0 for this electron that varies from pair to pair. In addition, there is the potential barrier q representing the self-trapping energy by local lattice deformation or electrical polarization. The hop frequency of the itinerant electron increases exponentially with the temperature T according to

$$\nu_{hop} = \nu_0 \exp[-(U_0+q)/kT]/[1+\exp(-U_0/kT)]. \quad (1)$$

The Mössbauer spectrum of an Fe^{3+}–Fe^{2+} pair depends on ν_{hop} with respect to a critical frequency of 10^8 s^{-1}. If $\nu_{hop} \ll 10^8 \text{ s}^{-1}$ a distinction between Fe^{3+} and Fe^{2+} is possible and the spectrum consists of superimposed Fe^{3+} and Fe^{2+} spectra. If $\nu_{hop} \gg 10^8 \text{ s}^{-1}$ a distinction cannot be made and a mixed spectrum with a thermally averaged quadrupole splitting and isomer shift occurs. Although for every Fe^{3+}–Fe^{2+} pair the mixed-spectrum lines are sharp, their positions depend on U_0. The observation of continuous absorption can thus be attributed to U_0 variation. According to eq. (1) ν_{hop} increases with T and first passes the critical frequency of 10^8 s^{-1} for pairs with $U_0 = 0$. This condition gives $q \simeq 0.12 \text{ eV}$, adopting $\nu_0 = 5 \times 10^{12} \text{ s}^{-1}$, a value that describes the electrical conduction of Fe^{2+}-containing spinels. With increasing temperature ν_{hop} for other pairs passes 10^8 s^{-1}, so that the intensity of the continuous absorption increases.

Paramagnetic Curie temperatures θ were obtained from straight χ^{-1}–T curves. The dependence of θ on x is far from linear and points to positive Fe^{2+}–Fe^{3+} interaction. The exchange interaction parameters derived are $J_{11}/k = -1.4 \text{ K}$, $J_{22}/k = -3.3 \text{ K}$, and $J_{12}/k = +1.6 \text{ K}$ for the Fe^{3+}–Fe^{3+}, Fe^{2+}–Fe^{2+}, and Fe^{3+}–Fe^{2+} interactions, respectively.

We are indebted to P.F. Bongers, B. Hoekstra, and R.P. van Stapele for suggestions and critical remarks, and to J.J.P. Verheijden and P.H.G.M. Vromans for technical assistance.

HYPERFINE FIELDS OF ^{119}Sn IN Fe_3O_4, Mn_3O_4 AND Co_3O_4

H. SEKIZAWA, T. OKADA and F. AMBE

The Institute of Physical and Chemical Research, Wako-shi, Saitama 351, Japan

^{119}Sn Mössbauer measurements were made. In Fe_3O_4, abrupt broadening was observed below the electronic transition temperature. In Mn_3O_4, a small H_{hf} with a wide distribution was observed below T_c. In Co_3O_4, H_{hf} was almost negligible even below T_N.

1. Introduction

Information about the hyperfine fields at diamagnetic ions in magnetically ordered oxides and compounds is important in understanding the mechanisms of magnetic interactions. There are numbers of such studies for spinel type ferrites [1]. In this paper, the Mössbauer hyperfine magnetic fields at ^{119}Sn in spinel (or distorted spinel) type simple oxides, Fe_3O_4, Mn_3O_4 and Co_3O_4 are reported. In these oxides, Sn ions occupy octahedral sites preferentially. The specimens were prepared by firing each oxide doped with 1–2% ^{119}Sn enriched tin in an appropriate atmosphere.

2. Experimental results

2.1. Fe_3O_4

At room temperature, the obtained spectrum was rather normal as shown in fig. 1, with an isomer shift $\delta = +0.27$ mm/s (against $BaSnO_3$) and a hyperfine magnetic field $H_{hf} = 207$ kOe; the electric quadrupole effect was almost negligible. A rather broad linewidth was observed which can be ascribed to slightly distributed H_{hf}. Down to the electronic transition temperature T_v (which was found to be 106 K for our specimen), no appreciable change occurred except the gradual increase in H_{hf}. Just below T_v, a rapid broadening of the spectra was observed, and at 5 k the broadening was so severe that almost no structure could be seen. As this drastic broadening seemed to be due to a wide distribution of H_{hf}, such a distribution function was determined by means of the Fourier expansion procedure [2]. The obtained distribution for the spectra are also shown in the figure. The width of the distribution curves, defined as the spread of H_{hf} at half maximum, is also shown in fig. 2 as a function of the temperature. The sign of H_{hf}, determined by applying an external magnetic field of 35 kOe, was found to be negative.

2.2. Mn_3O_4

Mn_3O_4 has a high Jahn–Teller distortion temperature of 1443 K and a ferrimagnetic Curie temperature $T_c = 42$ K, and the spin structure is fairly complicated [3]. At room temperature, δ was +0.27 mm/s. The profile of the spectrum

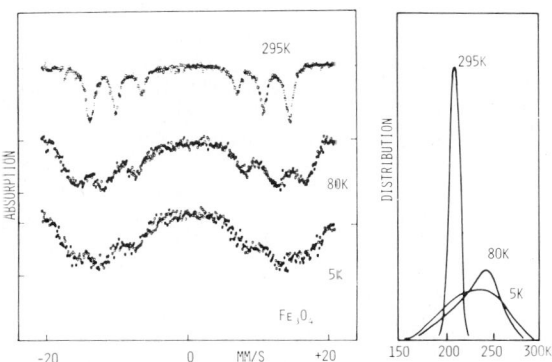

Fig. 1. Mössbauer absorption patterns of ^{119}Sn doped in Fe_3O_4, and the distributions of H_{hf} for them.

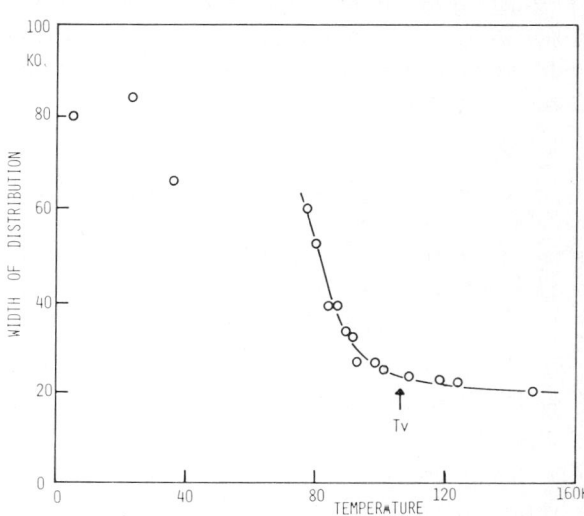

Fig. 2. Width of the distribution curves of H_{hf} at ^{119}Sn doped in Fe_3O_4.

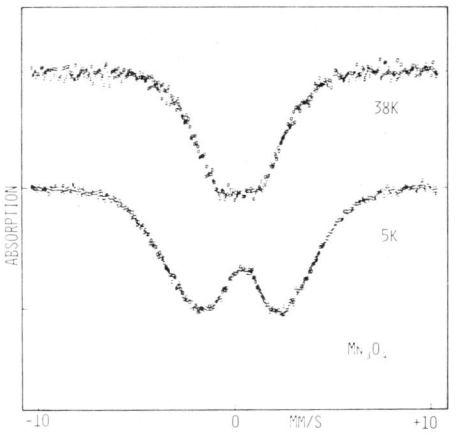

Fig. 3. Mössbauer absorption pattern of ^{119}Sn doped in Mn_3O_4.

shown in fig. 3, which is wide with flat top, suggests an electric quadrupole splitting due to the Jahn–Teller distortion of the matrix, but quantitative determination of its value was found to be difficult, probably because of the distributed values of the gradient. A remarkable widening of the spectrum was observed below T_c. As the widening was much more pronounced compared with that above T_c, the same procedure applied for Fe_3O_4 was tried assuming a fictitious line-width compatible with the room temperature spectrum. The distribution of H_{hf} was very wide but the temperature change of which confirmed the assumption. At 5 K, the peak value of the distribution was 46 kOe. The sign of H_{hf} was difficult to determine because of the complicated spin structure, but H_{hf} parallel to the net moment seems to be consistent with the spectrum with an applied magnetic field.

2.3. Co_3O_4

In this normal spinel oxide, the octahedral site is occupied by diamagnetic low spin Co^{3+}, and the tetrahedral Co^{2+} ions become antiferromagnetic below 40 K. As the room temperature spectrum with $\delta = +0.25$ mm/s showed a slight assymmetry, the possibility for the existence of not single ionic sites for tin in this oxide can not be eliminated. A broadening was observed below T_N, and a rough estimation gave H_{hf} of 8 kOe at 5 K.

3. Discussion

The observed value of H_{hf} at ^{119}Sn in Fe_3O_4 (around 250 kOe) is fairly common to various spinel ferrites. Supertransferred hyperfine (STHF) magnetic field from the A-site Fe^{3+} ion to the 5s orbitals of Sn^{4+} via oxygen ion seems to be the dominant mechanism [1]. The drastic broadening of the distribution of H_{hf} below T_v is undoubtedly connected with the transition. Sn^{4+} ions modify the complicated charge distribution in the low temperature phase, thus resulting in a number of sites with different ionic configuration environment with much different values of STHF magnetic field.

A small H_{hf} around 50 kOe at ^{119}Sn in Mn_3O_4 is comparable to the one observed in $MnFe_2O_4$ [4]. The net magnetic moment in $MnFe_2O_4$ is parallel to the B-site moment, whereas that of Mn_3O_4 is parallel to the A-site moment. This is consistent with the observed sign of the field in the latter.

The STHF magnetic field at ^{119}Sn on the B-site of the antiferromagnetic Co_3O_4 should be zero, because among the six nearest neighbor A-site Co^{2+} ions three have up and the other three have down spins [5]. The small observed H_{hf} of several kOe should be caused by some secondary effect.

References

[1] B.J. Evans and L.J. Swartzendruber, Phys. Rev. B6 (1972) 223.
[2] B. Window, J. Phys. E4 (1971) 401.
[3] G.B. Jensen and O.V. Nielsen, J. Phys. C7 (1974) 409.
[4] I.S. Lyubutin, Toshie Ohya, T.V. Dimitrieva and Kazuo Ōno, J. Phys. Soc. Jap. 36 (1974) 1006.
[5] W.L. Roth, J. Phys. Chem. Solids 25 (1964) 1.

INFLUENCE OF Fe^{2+} CONCENTRATION ON THE MÖSSBAUER SPECTRA OF IRON–COBALT AND IRON–NICKEL FERRITES

H. FRANKE and M. ROSENBERG

Ruhr-Universität Bochum, NB 03/34, Postfach 10 21 48, 4630 Bochum 1, BRD

The investigation of $Fe_{3-x}Me_xO_4$ (M = Co or Ni) with Mössbauer spectroscopy gives evidence for substitutional disorder in these compounds and for Fe^{2+}–Fe^{3+} electronic relaxation with frequencies higher as 10 MHz at room temperature.

As reported by us in a previous paper [1] the Mössbauer spectra of cobalt–iron ferrites $Fe_{3-x}Co_xO_4$ with $0.1 \leq x \leq 0.7$ could be splitted in an A-site Fe^{3+}-sextet and starting with $x = 0.1$ in an increasing number of B-site Fe^{3+}-sextets, thus allowing us to ascribe up to 5 different hyperfine fields at the B-site Fe-nuclei for $x = 0.5$ in contrast to only two sextets mentioned very recently in a paper by Murray and Linnett [2].

The experimentally observed Mössbauer B-spectra have to be related to the occurrence of five types of surroundings of the B-site Fe-ions with probabilities which obviously depend on the Co^{2+}-concentration. We have tried to interpret the appearance of the five different surroundings of the B-sites as a result of a random (statistical) distribution of the Co^{2+}-ions over the B-sublattice, i.e. assuming that the $Fe_{3-x}Co_xO_4$ system at room temperature in the composition range $0 \leq x \leq 0.7$ behaves as a substitutional alloy with compositional disorder. Both the ions Co^{2+} and Fe^{2+} have about the same radius (0.74 and 0.76 Å, respectively) a statistical distribution of the Co^{2+} on the B-sites is very probable.

For each composition a Co^{2+} distribution is established and the Fe ions have to adapt their valencies to this charge distribution as to fulfil the condition of electrical neutrality (Co^{2+} cannot change its valency by electron transfer!). Thus the Co^{2+} statistical distribution gives rise to a distribution of the Fe^{2+} ions with, on average, more Fe^{2+} ions in the Co^{2+}-poor regions and less Fe^{2+} ions in the Co^{2+}-rich regions.

Assuming a fast electron transfer with the neighbouring Fe^{3+} or Fe^{2+} ions the valency state of Fe-ion at a given B-site will fluctuate in time between the +3 and +2 states in order to preserve locally the average charge and spin $\langle n_d \rangle$ necessary to fulfil the electrical neutrality condition, i.e. a higher $\langle n_d \rangle$ in the Co-rich regions and a lower $\langle n_d \rangle$ in the Co-poor regions. The same mechanism holds for other substitutional systems as for instance $Fe_{3-x}Ni_xO_4$.

Assuming further that the hyperfine field at a given ^{57}Fe-nucleus will depend on the number of B-nearest neighbours with which the electronic exchange may take place and that up to $x = 0.7$ the hopping occurs so fast that only single valued average hyperfine fields can be observed with the Mössbauer technique for every type of surrounding of the B-ions, it was possible to assign four of the B-lorentzian fitting sextets to four different cationic configurations in groups of

(1) $3 Fe^{3+} + 3 Co^{2+}$ (B_5-sextet),
(2) $3 Fe^{3+} + 2 Co^{2+} + 1 Fe^{2+}$ (B_4-sextet),
(3) $3 Fe^{3+} + 1 Co^{2+} + 2 Fe^{2+}$ (B_3-sextet),
(4) $3 Fe^{3+} + 3 Fe^{2+}$ (B_2-sextet).

The fifth B_1-sextet had a stronger Fe^{2+} character as the other ones.

Under the same experimental conditions as reported in [1] we have prepared and investigated the Mössbauer spectra of six compositions in the system $Fe_{3-x}Ni_xO_4$ with $x = 0.1$; 0.3; 0.5; 0.7; 0.8; 0.9; 1.0 and nine compositions in the system $Fe_{3-x}Co_xO_4$ with $x = 0.5$; 0.6; 0.7; 0.75; 0.8; 0.85; 0.9; 0.95; 1.0.

The B-part of the Mössbauer spectra of the nickel–iron ferrites could be fitted over the whole compositional range by up to five lorentzian-sextets as in the case of iron–cobalt ferrites as shown in fig. 1 where the intensities of the fitting spectra for four of them (B2–B5) are plotted as function of x.

Assuming as in the case of cobalt–iron ferrites that in $Fe_{3-x}Ni_xO_4$ the Ni^{2+}-ions are statistically distributed over the B-sites and that the ob-

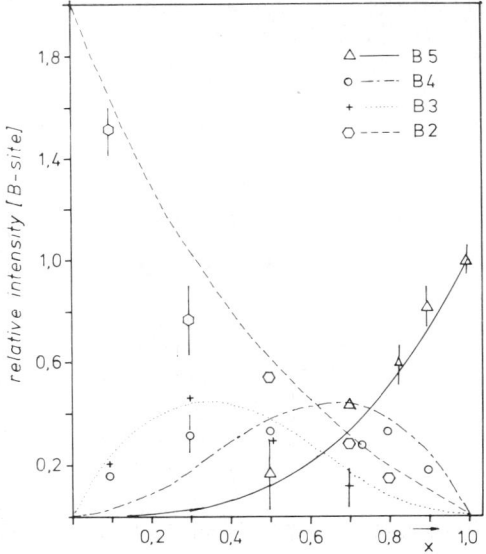

Fig. 1. Measured (B2-○, B3-+, B4-○, B5-△) and calculated (B2 ---, B3 +++, B4 -·-·-, B5 ———) intensities of the B2–B5 sextets in the $Ni_xFe_{3-x}O_4$ Mössbauer room temperature spectra.

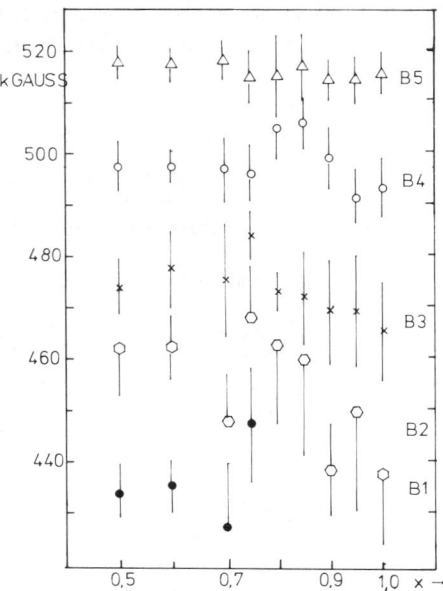

Fig. 2. B1–B5 hyperfine fields versus Co concentration in the $Co_xFe_{3-x}O_4$ system and corresponding linewidth of the Mössbauer fitting spectra (bars).

served values of the B_2–B_4 hyperfine fields originates in the above-mentioned distributions (1)–(4) with Ni^{2+} instead of Co^{2+} we obtain by using the binomial distribution the curves plotted in fig. 1 which are in qualitative agreement with the experimental data. At $x = 0.8$ it is possible to separate three of the fitting sextets B_2, B_4 and B_5 but only two (B_4 and B_5) for $x = 0.9$ in agreement with the calculated probabilities given by the binomial distribution. As shown in fig. 2 the results for $Fe_{3-x}Co_xO_4$ with $x \geqslant 0.7$ does not agree any more with the assumption of the statistical disorder of Co^{2+} over B-sites only. In the concentration range $0.7 < x \leqslant 1.0$ four different values of the B–Fe^{5+} hyperfine field could be found, lying between about 440 and 520 kOe. In the case of pure Fe_2CoO_4 the observed splitting was first reported by Sawatzky et al. [4] and attributed to the A-site Co^{2+} which occurs because, as compared to Fe_2NiO_4, cobalt ferrite is not a totally inverted spinel.

Unfortunately, there are no available data about the Co^{2+} ionic distribution on A and B sites for $x < 1.0$, and about the influence of x on the reduction of the B-site ^{57}Fe-hyperfine field owing to the presence of Co^{2+} ions on A-sites. However, we tried to apply the binomial distribution to both the six B-site and A-site

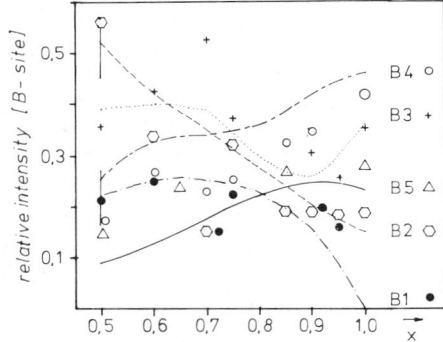

Fig. 3. Measured (B1-●, B2-○, B3-+, B4-○, B5-△) and calculated (B1 -·-·-, B2 ---, B3 +++, B4 ———, B_5 ———) intensities of the B1–B5 sextets in the $Co_xFe_{3-x}O_4$ Mössbauer room temperature spectra.

nearest-neighbour cation surroundings of a given B-site ^{57}Fe nucleus under the simplifying assumptions that 24% of the Co^{2+}-ions are on A-sites in Fe_2CoO_4 and $0.24\,x$ in $Fe_{3-x}Co_xO_4$. The calculated curves are plotted against x in fig. 3 and compared with the measured intensities of the fitting lorentzian-sextets. Only a qualitative agreement could be reached, as expected, in view of the crude approximations used in our computation model. In conclusion our investigation of Fe–Co-ferrites and Fe–Ni-ferrites with the Mössbauer spectroscopy at room temperature has given strong evidence for substitutional disorder in these compounds owing to the statistical distribution of Ni^{2+} ions on the B-sites and of the Co^{2+} ions on both the A- and B-sites.

The dependence of the hyperfine fields on the statistical configurations of the nearest neighbours cations in the concentration range $x > 0.6$ where, according to electrical properties, electronic hopping occurs, means that the hopping frequency is higher as the nuclear Larmor frequency at room temperature in agreement with the transport properties in the Ni-Fe-ferrite system [5]. An investigation of the low-temperature Mössbauer spectra of both Co-Fe and Ni-Fe-ferrites is now in progress.

References

[1] H. Franke and M. Rosenberg, Verh. der D P G 4 (1976) 303, to be published in J. Magn. Magn. Mater.
[2] P. J. Murray and J. W. Linnett, J. Phys. Chem. Solids 37 (1976) 619.
[3] J.W. Linnett and M.M. Rahman, J. Phys. Chem. Solids 33 (1972) 1465.
[4] G.A. Sawatzky, F. Van Der Woude and A.H. Morrish, Phys. Rev. 187 (1969) 747.
[5] P. Nicolau, I. Burget, M. Rosenberg and I. Belciu, IBM J. Res. Develop. 14 (1970) 249.

TWO-DIMENSIONAL MAGNON AND PHONON THERMAL CONDUCTIVITY IN HIGH MAGNETIC FIELDS

L.H.M. COENEN, H.N. DE LANG, J.H.M. STOELINGA, H. VAN KEMPEN and P. WYDER

Physics Laboratory and Research Institute for Materials, University of Nijmegen, Toernooiveld, Nijmegen, The Netherlands

The thermal conductivity of the two-dimensional Heisenberg ferromagnets $(C_nH_{2n+1}NH_3)_2CuCl_4$, $n = 1, 2$ have been measured for temperatures between 1.5 and 25 K, and in fields up to 6.5 T. The heat current consists of a small phonon and a large (between 70 and 90%) magnon contribution, which is completely quenched for fields above 4.5 T.

1. Introduction

In 1955 Sato [1] suggested the possibility of thermal conduction by magnons. Since then evidence of magnon heat conduction was indeed observed in several magnetic materials [2], with the most pronounced magnon contribution being found in YIG (70% of the total heat transport) [3].

In the present experiments the thermal conductivity of the layered copper-compounds $(C_nH_{2n+1}NH_3)_2CuCl_4$, $n = 1, 2$ have been measured as a function of temperature over the range 1.5–25 K and as a function of magnetic field from zero to 6.5 T. These crystals consist of layers of magnetic $CuCl_2$-complexes widely separated by two non-magnetic $(C_nH_{2n+1}NH_3Cl)$-groups, so that their magnetic behaviour is essentially two-dimensional. They can be regarded as excellent approximants to a two-dimensional, $S = \frac{1}{2}$, Heisenberg ferromagnet [4].

2. Results and discussion

A typical example of the thermal conductivity of the $(CH_3NH_3)_2CuCl_4$ single crystal as a function of magnetic field at a constant temperature is shown in fig. 1. After an initial increase in the low field region, the total thermal conductivity decreases, until at fields between 4 and 5 T it reaches a saturation value of about 10–20% of the original zero field magnitude.

Clearly at zero field the heat flow consists of a large magnon contribution, of at least the same order of magnitude as that present in YIG, and a much smaller phonon current. When a magnetic field is applied the magnon conductivity is reduced until at sufficiently high fields it is completely quenched and what one measures is the lattice conductivity only.

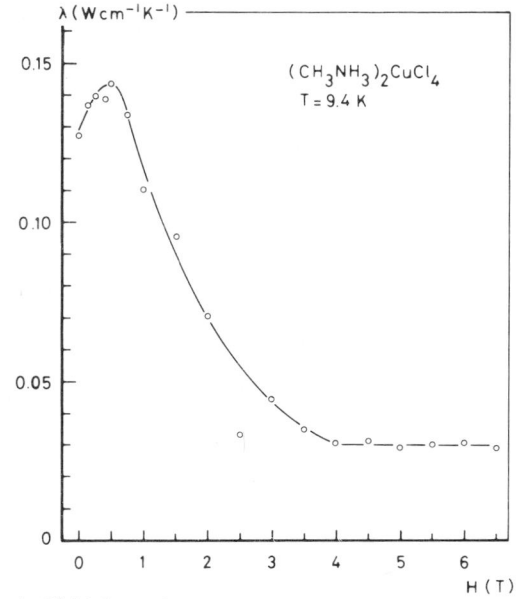

Fig. 1. Field dependence of the thermal conductivity of the methyl-compound at a constant temperature of 9.4 K.

Figs. 2 and 3 show the temperature dependence of the thermal conductivity of the methyl- and ethyl-crystal, respectively, at zero and at the highest measured field (6.5 T), the high field values representing the lattice thermal conductivity. The thermal conductivity values of the methyl-crystal, both at zero and at high field, are about an order of magnitude smaller than those of the ethyl-compound. This is due to the difference in the quality of the crystals. On examination after the experiments the ethyl-crystal was still found to be of excellent quality, but the methyl-crystal showed some cracks. The difference in crystal quality is also reflected in the temperature-dependent behaviour of the respective compounds.

Assuming non-interacting magnons and a

constant mean free path l_s, the magnon conductivity for a two-dimensional ferromagnet at low temperatures and for zero field can be written [5]

$$\lambda_s = \frac{k_B l_s T^{1.5}}{a_0 c_0 h (J/k_B)^{0.5}} \int_0^\infty \frac{x^{2.5} e^x}{(e^x - 1)^2} dx, \quad (1)$$

where k_B is Boltzmann's constant, h is Planck's constant, a_0 and c_0 are unit cell parameters, and J/k_B is the exchange parameter. As can be seen from figs. 2 and 3 the ethyl-crystal shows a far better $T^{1.5}$-behaviour than the methyl-compound. For the temperature region between 2 and 4 K a magnon mean free path $l_s = 1.3 \times 10^{-3}$ cm for the methyl- and $l_s = 1.8 \times 10^{-2}$ cm for the ethyl-crystal are estimated, which are reasonable values [5, 6].

For temperatures below 2 K, the curves drop below the $T^{1.5}$-line, due to a reduced magnon–phonon contact. Due to enhanced magnon–phonon and magnon–magnon interactions, the conductivity of the ethyl-compound decreases above 3 K. In the methyl-compound this decrease occurs at a much higher temperature, due to the lesser quality of the crystal. Above 6 K the thermal conductivity is completely dominated by phonons in the ethyl-compound. As can be seen from fig. 3 the magnitude of the high field values is greater than that of the zero field values, the magnons now interfering with the phonon transport and decreasing the conductivity rather than contributing to it.

Not only do the $(C_n H_{2n+1} NH_3)_2 CuCl_4$-crystals ($n = 1, 2, \ldots$) constitute a two-dimensional magnon system, but from far infrared measurements on these crystals evidence was obtained that above ~ 50 cm^{-1} also the optical phonon system in these compounds is two-dimensional [7–9]. Our lattice conductivity data can now provide information about the dimensionality of the acoustical phonon spectrum.

Taking into account boundary scattering and scattering at line dislocations only, the lattice conductivity of a layered system can be written [10]

$$\lambda_p = \frac{3}{c} N k_B v^2 \left(\frac{T}{\theta}\right)^2 \int_0^\infty \frac{x^3 e^x}{(e^x - 1)^2} (\tau_1^{-1} + \tau_2^{-1}) dx, \quad (2)$$

where c is the distance between two magnetic

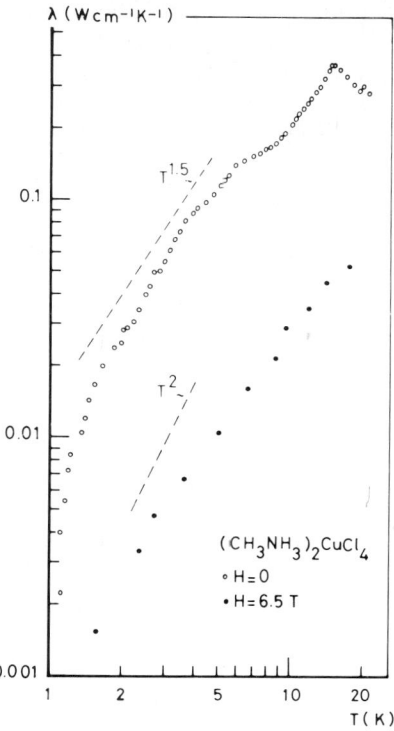

Fig. 2. Temperature dependence of the total ($H = 0$) and lattice ($H = 6.5$ T) thermal conductivity of the methyl compound ($T_c = 8.91$ K). The dashed line represents a $T^{1.5}$- and a T^2-dependence, respectively.

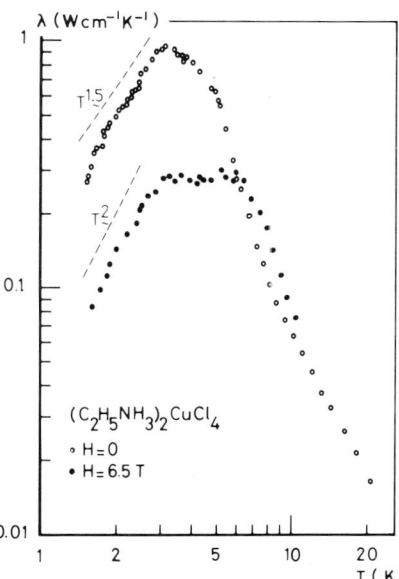

Fig. 3. Temperature dependence of the total ($H = 0$) and lattice ($H = 6.5$ T) thermal conductivity of the ethyl compound ($T_c = 10.20$ K). The dashed line represents a $T^{1.5}$- and T^2-dependence, respectively.

layers, N is the number of atoms per unit area, v is the sound velocity, θ is the Debye temperature, τ_1 is the relaxation time from boundary scattering and τ_2 the relaxation time from scattering at line dislocations. When only a relatively small amount of dislocations are present, the lattice conductivity should follow a T^2- and a T^3-dependence for a two- and a three-dimensional lattice, respectively, while in the limit of a large amount of dislocations, these dependences should be linear and quadratic. The measured temperature dependence of the lattice conductivity of both crystals lies between a T- and a T^2-dependence, thus following the behaviour of a two-dimensional lattice. Quantitatively, this was confirmed by computer fits of the measured data to eq. (2). From these fits a phonon mean free path $l_p = 1.2 \times 10^{-4}$ cm and a dislocation density $D = 1.5 \times 10^{12}$ cm^{-2} for the methyl- and $l_p = 2.5 \times 10^{-2}$ cm and $D = 1.5 \times 10^{10}$ cm^{-2} for the ethyl-crystal, respectively, was obtained.

References

[1] H. Sato, Progr. Theor. Phys. 13 (1955) 119.
[2] D. Walton, Proc. Int. Conf. on Phonon Scattering in Solids, ed. H.J. Albany (La Documentation Francaise, Paris, 1972) p. 295.
[3] R.L. Douglass, Phys. Rev. 129 (1963) 1132.
[4] L.J. De Jongh and A.R. Miedema, Advan. Phys. 23 (1974) 1.
[5] F.W. Gorter, L.J. Noordermeer, A.R. Kop and A.R. Miedema, Phys. Lett. 29A (1969) 331.
[6] A.R. Miedema, P. Bloembergen, J.H.P. Colpa, F.W. Gorter, L.J. De Jongh and L.J. Noordermeer, AIP Conf. Proc. 18 (1974) 806.
[7] J.H.M. Stoelinga and P. Wyder, J. Chem. Phys. 61 (1974) 478.
[8] J. Holvast, J.H.M. Stoelinga and P. Wyder, Ferroelectrics 13 (1976) 543.
[9] J.H.M. Stoelinga and P. Wyder, J. Chem. Phys. 64 (1976) 4612.
[10] L.H.M. Coenen, H.N. De Lang, J.H.M. Stoelinga, H. Van Kempen and P. Wyder, Solid State Commun. 20 (1976) 713.

MAGNON LIFETIME IN MnO

K. MOTIZUKI and T. NIWA
Faculty of Engineering Science, Osaka University, Toyonaka, Japan

Magnon lifetime in MnO arising from magnon–magnon interaction is studied by the spin wave approach using diagrammatic technique. The energy and lifetime of magnons with wave vectors along [100], [111], and [110] are calculated as functions of temperature.

1. Formulation

MnO is a type II antiferromagnet with the Néel temperature at 117 K and below T_N the f.c.c. crystal undergoes a small trigonal distortion. The hamiltonian is assumed to be the sum of a uniaxial anisotropy energy $DS^2_{[111]}$ ($D > 0$) and exchange interactions between n.n. and n.n.n. spins. We denote the exchange constant for n.n.n. spin pairs by J_2. As for n.n. spin pairs, we denote by $J_1 + j$ the exchange constant for n.n. parallel spins in the same (111) plane and by $J_1 - j$ that for n.n. antiparallel spins on adjacent (111) planes. j is a function of temperature and is related to spin correlation functions of the n.n. parallel and antiparallel spin pairs as shown in [1]. In order to write the hamiltonian in terms of spin deviation operators, we introduce a local coordinate system [2] and use the Dyson–Maleev transformation. Fourier-transforming the spin deviation operators and diagonalizing the quadratic part of the hamiltonian, we obtain

$$H = \sum_k \Omega_k \alpha_k^+ \alpha_k + H_4 + H_6, \tag{1}$$

where

$$H_4 = \sum_{k_1 - k_4} [I_1(k_1 k_2 k_3 k_4) \alpha_1^+ \alpha_2^+ \alpha_3^+ \alpha_4^+ \delta(k_1 + k_2 + k_3 + k_4)$$
$$+ \cdots + I_5(k_1 k_2 k_3 k_4) \alpha_1^+ \alpha_2^+ \alpha_3 \alpha_4$$
$$\times \delta(k_1 + k_2 - k_3 - k_4)]. \tag{2}$$

Expressions for I_i are given in [3]. H_6 (which we neglect) is six-magnon terms which give higher order corrections.

We define the thermal Green function

$$G_k(\tau) = \langle T_\tau [\exp(\tau H) \alpha_k^+ \exp(-\tau H) \alpha_k] \rangle. \tag{3}$$

Since H consists of the unperturbed hamiltonian H_2 and the perturbing H_4, we obtain $G_k(\tau)$ by the use of the diagrammatic method in the expanded form $G_k(\tau) =$ $G_k^0(\tau) + G_k^1(\tau) + G_k^2(\tau) + \cdots$. The Fourier transform of $G_k^0(\tau)$ is obtained as $G_k^0(\omega_n) = 1/(\Omega_k - i\omega_n)$. In order to calculate $G_k^1(\tau)$ and $G_k^2(\tau)$, we take up only those terms which are proportional to I_5, among several terms in H_4. This is because only these terms contribute to $G_k^1(\tau)$ and the main part of $G_k^2(\tau)$. $G_k^1(\tau)$ corresponds to the diagram shown in fig. 1 and $G_k^2(\tau)$ corresponds to the sum of three diagrams (a), (b), and (c) in fig. 2. First we consider diagrams (a) and (b). Since these can be reduced to the first order diagram, we can obtain the Dyson-equation for $G_k(\tau)$. Performing the Fourier transformation and solving the equation for $G_k(\omega_n)$, we obtain

$$G_k^{HF}(\omega_n) = 1/(\bar{\Omega}_k - i\omega_n), \tag{4}$$

where $\bar{\Omega}_k$ is the renormalized magnon energy given by

$$\bar{\Omega}_k = \Omega_k + \sum_{k'} n_{k'} \tilde{I}_5(kk'kk'). \tag{5}$$

Since $n_{k'} = 1/[\exp(\beta \bar{\Omega}_{k'}) - 1]$, eq. (5) is the self-consistent equation for $\bar{\Omega}_k$. Next we take account of diagram (c). This cannot be reduced to

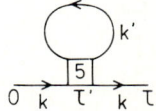

Fig. 1. Diagram of $G_k^1(\tau)$.

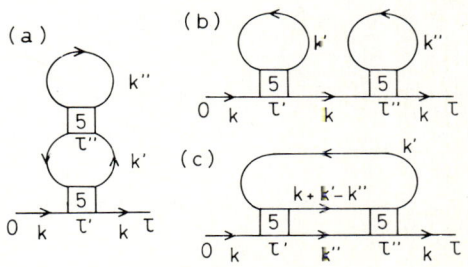

Fig. 2. Diagrams of $G_k^2(\tau)$.

the first order diagram, so we approximate the true Green function appearing in diagram (c) by $G_k^{HF}(\omega_n)$ of eq. (4). Then we obtain $G_k(\omega_n)$ in the form

$$G_k(\omega_n) = 1/(\bar{\Omega}_k - i\omega_n - \Sigma_k^2), \qquad (6)$$

where Σ_k^2 is the second order self-energy given by

$$\begin{aligned}\Sigma_k^2 = \sum_{k'k''} &\tilde{\tilde{I}}_5(k, k', k'', k + k' - k'') \\ &\times \tilde{I}_5(k'', k + k' - k'', k, k') \\ &\times \frac{(n_{k''} + n_{k+k'-k''} + 1)n_{k'} - n_{k''}n_{k+k'-k''}}{\bar{\Omega}_{k''} - \bar{\Omega}_{k'} + \bar{\Omega}_{k+k'-k''} - i\omega_n}.\end{aligned} \qquad (7)$$

We replace $i\omega_n$ by $\bar{\Omega}_k + i\delta$. The imaginary part of Σ_k^2 corresponds to the inverse of magnon lifetime, Γ_k, and the real part gives a correction to $\bar{\Omega}_k$.

2. Results

We measured energy in units of $4|J_1|S$ and we used values of $J_2/J_1 = 1.20$ and $D/|J_1| = 0.10$. If we use $J_1 = -4.44$ K [1], T_N will correspond to $t(= k_B T/4S|J_1|) = 2.64$. The magnon energies given by eq. (5) are calculated and their temperature variations are found to be similar to those obtained in [1]. In fig. 3 we show the calculated Γ_k. In the same figure we also show the calculated four-magnon joint density of states, ρ_k, which is defined by

$$\sum_{k',k''} \delta(\bar{\Omega}_{k''} - \bar{\Omega}_{k'} + \bar{\Omega}_{k+k'-k''} - \bar{\Omega}_k).$$

From comparison of the curves of Γ_k with those of ρ_k, it is apparent that magnons for large ρ_k have large Γ_k. We note that the temperature variation of Γ_k depends remarkably on k-values.

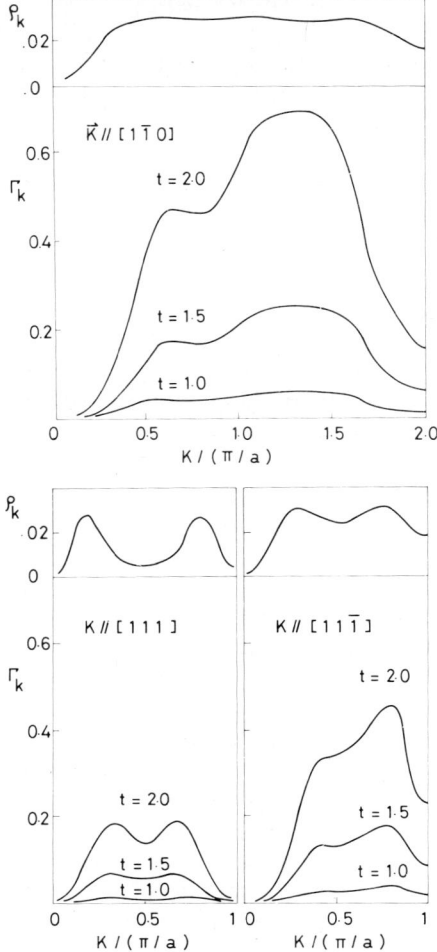

Fig. 3. The calculated magnon damping as a function of k-value along [1$\bar{1}$0], [111], and [11$\bar{1}$] at $t = 1.0$, 1.5, and 2.0 ($t = k_B T/4S|J_1|$). ρ_k is the calculated four-magnon joint density of states at $t = 1.5$.

References

[1] M. Kohgi, Y. Ishikawa, I. Harada and K. Motizuki, J. Phys. Soc. Jap. 36 (1974) 112.
[2] I. Harada and K. Motizuki, J. Phys. Soc. Jap. 32 (1972) 927.
[3] T. Niwa and K. Motizuki, J. Phys. Soc. Jap. 41 (1976).

ELEMENTARY EXCITATIONS OF ANTIFERROMAGNETIC $CoCl_2$

J.E. KARDONTCHIK, A. SORGEN and E. COHEN

Technion-Israel Institute of Technology, Haifa, Israel

A unified approach to the problem of the elementary excitations within the $^4T_{1g}$ ground term in $CoCl_2$ is presented. The hamiltonian includes single ion terms and isotropic exchange interactions between the ionic spins. The dispersion of magnons and excitons is obtained by projecting the hamiltonian on a molecular field space corresponding to $S' = 11/2$. The calculations account well for spin wave dispersion, antiferromagnetic resonance and electronic Raman scattering.

Magnetic crystals containing transition metal ions with unquenched orbital angular momentum in the ground term have been the subject of many experimental and theoretical studies. In some cases great difficulties are encountered in explaining the experimental results, particularly the elementary excitations, owing to the coupled motion of both ionic spin and orbit.

$CoCl_2$, which orders antiferromagnetically at $T_N = 24.9$ K, is an example of such a system. For this crystal there is a bulk of experimental data: inelastic neutron scattering [1], antiferromagnetic resonance [2] and electronic Raman scattering [3, 4]. The Co^{2+} ions in the ordered phase form ferromagnetic layers, with adjacent layers coupled antiferromagnetically, but the interlayer coupling constant is much smaller than the intralayer one.

In this work we have carried out a full calculation of the elementary excitations (magnons and excitons) within the $^4T_{1g}$ ground term, assuming an isotropic exchange interaction between the ionic spins [5] and taking into consideration all six Kramers doublets of the $^4T_{1g}$ term. As in the case of Co^{2+}: $CdCl_2$ [6] (which is isomorphic to $CoCl_2$) we use the following single ion hamiltonian for the $^4T_{1g}$ term

$$h_0 = -\tfrac{3}{2}\lambda \mathcal{L} \cdot S - \Delta[\mathcal{L}_z^2 - \tfrac{1}{3}\mathcal{L}(\mathcal{L}+1)]$$
$$- \frac{15}{4}\frac{\lambda^2}{E_0(^4T_{2g}) - E_0(^4T_{1g})} \quad (1)$$
$$\times [2(\mathcal{L}_\xi^2 S_\xi^2 + \mathcal{L}_\eta^2 S_\eta^2 + \mathcal{L}_\zeta^2 S_\zeta^2) - (\mathcal{L} \cdot S)^2],$$

where $\mathcal{L} = 1$ and $S = 3/2$. Since the single ion parameters depend primarily on the Cl^- octahedron, we adopted for $CoCl_2$ the values of λ and Δ obtained by fitting the energy levels of Co^{2+} in $CdCl_2$: $\lambda = -159$ cm^{-1} and $\Delta = -405$ cm^{-1}. Based on the assumption of isotropic exchange, the hamiltonian for the system of coupled Co^{2+} ions in $CoCl_2$ is given by

$$\mathcal{H} = \sum_i h_0(i) - \sum_{i,j} J_{ij} S(i) \cdot S(j). \quad (2)$$

We further assume that only two exchange parameters are significant; the intralayer (ferromagnetic) J_1 and the interlayer (antiferromagnetic) J_2. They correspond to nn and nnnn coupling, respectively. We calculate the self-consistent molecular field states for the whole $^4T_{1g}$ term, corresponding to a pseudospin space of $S' = 11/2$. Then we project the hamiltonian [eq. (2)] on the molecular field space, expressing it in terms of unit spherical tensor operators. These operators create (or annihilate) single ion excitations within the molecular field space. Fourier transforming the unit tensor operators and introducing the Bose creation and annihilation operators for all the elementary excitations branches, the hamiltonian is transformed into a bilinear form in the Bose operators. It is then diagonalized yielding the dispersion relations for all the 22 elementary excitation branches.

There are two parameters to fit the experimental results: J_1 and J_2. Since it is expected that the hamiltonian of eq. (2) will reproduce all the excitation branches of $^4T_{1g}$, we chose to fit the energy values at $q = 0$. These are given by the second magnon branch observed by antiferromagnetic resonance and the exciton lines observed by electronic Raman scattering. The exchange parameters thus obtained are $2J_1 = 5.11 \pm 0.28$ cm^{-1} and $2J_2 = -0.36 \pm 0.02$ cm^{-1}. The spin wave dispersion curves together with the data obtained by neutron scattering are shown in fig. 1. Table I summarizes the main results obtained in both the low and high energy regions. The lines at 577 cm^{-1} and 984 cm^{-1} (see

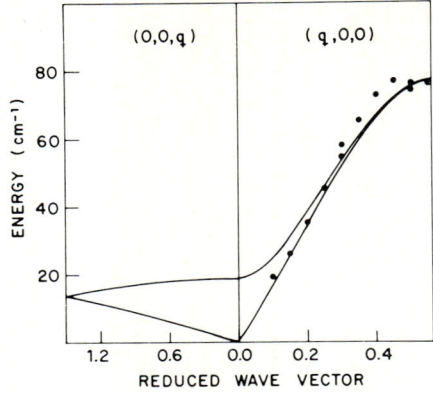

Fig. 1. Spin wave dispersion curves of $CoCl_2$. Experimental points from Ref. 1; calculation – solid curves.

Table I
Observed and calculated magnon and exciton energies (in cm^{-1}).

Excitation branch	$E(q=0)$ (exp.)	$E(q=0)$ (calc.)	$E(q_{max}, 0, 0)$ (calc.)
1	0.	0.	78
2	19.0 ± 0.2	19	78
3,4,5,6	233 ± 6	236	254
7,8	unobserved	488	504
9,10	548 ± 6	551	552
11,12	unobserved	906	906
13,14	960 ± 4	954	954
15,16	1016 ± 7	1013	1014
17,18		1021	1021
19,20,21,22	1150 ± 7	1151	1151

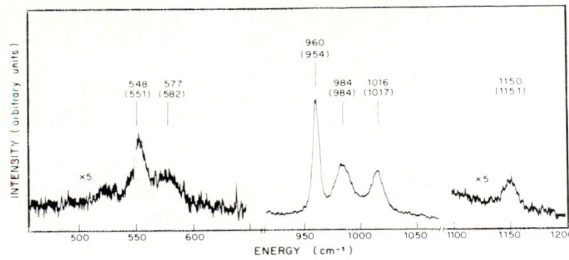

Fig. 2. Portion of the electronic Raman scattering spectrum of $CoCl_2$ at 2 K. The values in parentheses are the calculated peak positions.

fig. 2) are identified as simultaneous magnon–exciton Raman transitions involving zone-edge excitations [(78 + 504) and (78 + 906) cm^{-1}, respectively]. It should be noted that selection rules for electronic Raman scattering indicate that if a transition to a given $q = 0$ exciton is allowed then the transition to the other exciton branch originating from the same single ion Kramers doublet is forbidden. However, the latter will give rise to a simultaneous exciton–magnon scattering. The reasoning is similar to that used in the case of two-magnon scattering [7].

Finally we note that the magnetic field dependence of the antiferromagnetic resonance is well reproduced by our model.

References

[1] M.T. Hutchings, J. Phys. C6 (1973) 3143.
[2] I.S. Jacobs, S. Roberts and P.E. Lawrence, J. Appl. Phys. 39 (1968) 816.
[3] J.H. Christie and D.J. Lockwood, in 2nd Int. Conf. on Raman Scatt. in Solids, M. Balkansky, ed. (Flamarion, Paris, 1971).
[4] J.E. Kardontchik, E. Cohen and J. Makovsky, to be published in Phys. Rev.
[5] M.E. Lines, Phys. Rev. 131 (1963) 546.
[6] J.E. Kardontchik, E. Cohen and J. Makovsky, Phys. Rev. B13 (1976) 2955.
[7] P.A. Fleury and R. Loudon, Phys. Rev. 166 (1968) 514.

THE LIFETIME OF COUPLED SPIN–PHONON EXCITATIONS IN A PARAMAGNETIC CRYSTAL

C.M. CARE and J.W. TUCKER
Department of Physics, University of Sheffield, Sheffield S3 7RH, England

A coupled-mode theory of spin–phonon excitations in a paramagnetic crystal containing ions having $S > \frac{1}{2}$ is presented. The calculation is facilitated by the introduction of a Wick's theorem for operators of higher order than linear in the spin components. The theory is applied to Fe^{2+} ions in MgO.

In an insulating crystal containing paramagnetic ions the phonon scattering from different ions leads to coupled spin–phonon modes. Although these modes have been studied extensively over the last few years (e.g. [1, 2]) very little work has been done on systems with $S > \frac{1}{2}$. Ions with $S \geq 1$ differ primarily from the two-level system in that the coupling of the ion to the lattice is described by a hamiltonian quadratic rather than linear in the spin components. This paper reports results for the dispersion relation and the lifetime of the excitations in a multi-level spin system described by the hamiltonian

$$H = \sum_{qj} \hbar\omega_{qj} a^+_{qj} a_{qj} - \sum_n \hbar\omega_0 S^z_n + \sum_{qjn\gamma} \hbar \epsilon^\gamma_{qj} T^\gamma_n \phi_{qj} \exp(i\boldsymbol{q} \cdot \boldsymbol{R}_n), \tag{1}$$

with $\phi_{qj} = (\omega_{qj}/\alpha V)^{1/2}(a_{qj} + a^+_{-qj})$. a^+_{qj} and a_{qj} are the phonon creation and annihilation operators for phonons of wave vector $\boldsymbol{q}$ in the jth branch. ϵ^γ_{qj} are the spin–phonon coupling parameters with γ taking integer values in the range $-2 \leq \gamma \leq 2$. The T^γ_n operators are defined in terms of the spin components for the nth site by

$$T^0_n = (S^z_n)^2 - \tfrac{1}{3}S(S+1); \qquad T^{\pm 1}_n = S^z_n S^\pm_n + S^\pm_n S^z_n; \qquad T^{\pm 2}_n = (S^\pm_n)^2. \tag{2}$$

The dispersion relation of the coupled spin–phonon excitations is obtained by considering the poles of the phonon Green function

$$A_{j_1 j_2}(\boldsymbol{q}\lambda) = \frac{1}{\beta} \int_0^\beta \int_0^\beta d\tau_1 d\tau_2 \exp\{i\lambda(\tau_1 - \tau_2)\} \langle T_\tau \tilde{\phi}_{qj_1}(\tau_1) \tilde{\phi}_{qj_2}(\tau_2) \rangle. \tag{3}$$

It is convenient to introduce a generalized cumulant Green function $G^{\gamma_1 \gamma_2}(\boldsymbol{q}\lambda)$ similar to that of [3] but with the spin operators $\tilde{S}^\alpha$ replaced by the $\tilde{T}^\gamma$ operators of eq. (2). As in [3] **G** can be expanded diagrammatically and the poles of the phonon Green function for the coupled spin–phonon system depend on the polarization matrix **M**. In each order in perturbation theory **M** is expressed in terms of semi-invariants which we have evaluated using a generalized Wick's theorem. Our theorem, which may be regarded as an extension of that proposed by Izyumov and Kassan-Ogly [4], is useful for dealing with operators that consist of products of the spin operators S^+, S^- and S^z. Let γ denote the difference in the number of the S^+ and S^- operators in the product. Then if R^γ is any linear combination of operators with the same γ the theorem (for equidistant spin levels) is ($\gamma \neq 0$)

$$\langle T_\tau \{R^{\gamma_1}(\tau_1) R^{\gamma_2}(\tau_2) \ldots R^\gamma(\tau) \ldots R^{\gamma_n}(\tau_n)\} \rangle_0 = g_\gamma(\tau - \tau_1) \langle T_\tau \{[R^{\gamma_1}, R^\gamma]_{\tau_1} R^{\gamma_2}(\tau_2) \ldots R^{\gamma_n}(\tau_n)\} \rangle_0$$
$$+ \ldots + g_\gamma(\tau - \tau_n) \langle T_\tau \{R^{\gamma_1}(\tau_1) R^{\gamma_2}(\tau_2) \ldots [R^{\gamma_n}, R^\gamma]_{\tau_n}\} \rangle_0$$
$$g_\gamma(\tau - \tau_i) = \exp\{-\gamma\omega_0(\tau - \tau_i)\}[\pm \exp(\mp \gamma\beta\omega_0) \mp 1]^{-1}, \quad \text{if } \tau \gtrless \tau_i. \tag{4}$$

In the present application the R-operators are taken to be the T-operators defined in eq. (2).

In the phonon regime, the dispersion relation to second-order in the coupling constant is obtained by the inclusion of just the zeroth-order term for the polarization matrix. In this approximation the excitations are undamped. The lifetime comes with the addition of the second-order contribution, M_2, requiring the evaluation of fourth-order semi-invariants. These have been evaluated using the Wick's theorem of eq. (4). Detailed calculations have been made for ions in sites of octahedral symmetry in a cubic crystal with the axis of quantization along a four-fold symmetry axis. For transverse wave propagation, with the propagation and polarization vectors respectively parallel to the x and z cubic axes, the dispersion relation is

$$(\omega^2 - \omega_{qj}^2)(\omega^2 - \omega_0^2) - \frac{NG_{44}^2 \omega_0 \omega_{qj}^2 b_1}{2V\hbar\rho v_t^2} + \frac{i\pi NG_{44}^2 \omega_{qj}^2 \{\Gamma(\omega) - \Gamma(-\omega)\}}{16V\hbar\rho v_t^2(\omega^2 - \omega_0^2)} = 0, \tag{5}$$

with

$$\Gamma(\omega) = 2C_1 b_2(\omega^2 + \omega_0^2)\omega^3 + C_2 b_3(\omega - 3\omega_0)^3(\omega + \omega_0)^2\{\coth(\tfrac{3}{2}\beta\omega_0) - \coth[\tfrac{1}{2}\beta(3\omega_0 - \omega)]\}$$
$$+ \{C_2 b_4(\omega - \omega_0)^2 + C_0 b_5(\omega + \omega_0)^2\}(\omega - \omega_0)^3\{\coth(\tfrac{1}{2}\beta\omega_0) - \coth[\tfrac{1}{2}\beta(\omega_0 - \omega)]\}, \tag{6}$$

where

$$b_1 = 2(1 - 4f)\langle S_z \rangle_0 + 16\langle S_z^3 \rangle_0, \qquad b_2 = \langle (16S_z^3 - 8fS_z + 2S_z)^2 \rangle_0 - \langle 16S_z^3 - 8fS_z + 2S_z \rangle_0^2,$$
$$b_3 = 96\langle 3f^2 S_z - f(6S_z^3 + 12S_z) + 3S_z^5 + 17S_z^3 + 4S_z \rangle_0,$$
$$b_4 = 8\langle 28f^2 S_z - 12f(10S_z^3 + 4S_z) + 9(12S_z^5 + 12S_z^3 + S_z) \rangle_0,$$
$$b_5 = 2\langle -8f(4S_z^3 + S_z) + 48S_z^5 + 32S_z^3 + S_z \rangle_0, \tag{7}$$

in which $f = S(S+1)$, $C_0 = 9KG_{11}^2/8$, $C_1 = KG_{44}^2/3$ and $C_2 = (C_0 + 6C_1)/12$. G_{11} and G_{44} are the phenomenological spin–phonon coupling constants [5] for crystals having cubic symmetry and a Debye model for the elastic properties has been adopted. $K = (2/v_l^5 + 3/v_t^5)/10\pi^2\hbar\rho$, v_l and v_t being the velocities of the longitudinal and transverse branches of the phonon spectrum, respectively, and ρ is the density. This result is for a fully concentrated crystal. In a dilute crystal each thermal average, $\langle \ldots \rangle$, in eq. (7) has to be replaced by $c_s\langle \ldots \rangle$ where c_s is the fractional concentration of the paramagnetic ions. Analysis shows that in the phonon regime the term in b_2 can be interpreted as a resonant elastic phonon scattering process while the terms proportional to b_3 and b_4 arise from inelastic phonon scattering. The scattering due to b_5 is also inelastic but non-resonant. At high frequencies $\omega \gg \omega_0$ and $1/\beta$ the inverse lifetime is proportional to ω_{qj}^2 whereas in the opposite limit the elastic and inelastic scattering are proportional to ω_{qj}^4 and ω_{qj}^2, respectively. Which process dominates at a particular temperature depends crucially on the details of the spin–phonon interaction. As an example, the case of Fe^{2+} ions in MgO in the presence of a magnetic field can be considered. The low-lying Zeeman triplet can be described by the static spin hamiltonian above with $S = 1$. For this value of spin b_3 identically vanishes. Using [5], $G_{11} = 647$ cm^{-1}, $G_{44} = 407$ cm^{-1}, $v_l = 9.25 \times 10^5$ cm/s, $v_t = 6.68 \times 10^5$ cm/s and $\rho = 3.59$ gm cm^{-3} it is found that in the phonon regime quite close to resonance the inelastic and elastic scattering dominate respectively at high and low temperatures for small concentrations of paramagnetic ions, the two being equal at $\beta\omega_0 \approx 3$.

References

[1] J.W. Tucker, J. Phys. C: Solid State Phys. 5 (1972) 2064.
[2] G.A. Toombs and F.W. Sheard, J. Phys. C: Solid State Phys. 6 (1973) 1467.
[3] C.M. Care and J.W. Tucker, J. Phys. C: Solid State Phys. 9 (1976) 2681.
[4] Y.A. Izyumov and F.A. Kassan-Ogly, Fiz. Metal. Metal. 30 (1970) 225.
[5] F.B. Fidler and J.W. Tucker, J. Phys. C: Solid State Phys. 4 (1971) 2583.

MAGNON RENORMALIZATION IN $Mn_{1.88}Cr_{0.12}Sb$

J. TODOROVIĆ

Boris Kidrič Institute, Laboratory of Solid State Physics, 11001 Beograd, Yugoslavia

This intermetallic compound has the first order antiferro-ferrimagnetic phase transition at $T_s = 319$ K. The softening of the ferromagnon in the second half of the Brillouin zone can be expressed in the form $E^2(T, q) = K(q) \cdot |T - T_s| + A(q)$. In the antiferromagnetic phase the deviation from this relation is found to be considerable.

It is a well-known idea that the mechanism of structure phase transition is bonded to the instability in a normal vibrational mode of the lattice, i.e. for the "softening" of the phonon mode called the soft mode. The square of the soft mode frequency is a linear temperature function which has the value of zero or near zero, at the temperature of phase transition, depending on whether the phase transition is of second or first order. Experimentally, it is found that the first order antiferro-ferrimagnetic phase transition is followed by the appearance of the magnon soft mode [1] in the intermetallic compound $Mn_{1.88}Cr_{0.12}Sb$. Its appearance in the high temperature ferromagnetic phase (FM) on the boundary of the Brillouin zone, produces the doubling of unit cell, i.e. the formation of low temperature antiferromagnetic phase (AF).

In this paper the effect of this phase transition on the renormalization of spin waves dispersion relation in the direction of tetragonal c-axis is presented.

The intermetallic compound $Mn_{2-x}Cr_xSb$ has a tetragonal structure and each tetragonal cell contains two formula units. There are sets of three-layer sheets of strongly coupled spins in the baseplane. The effective interaction between the sets is weak and it changes the sign by discontinual change of the lattice parameters [2, 3] at the temperature T_s. This phase transition is followed by the appearance of latent heat and considerable changes in the equilibrium and transport properties [4].

The results of measurements on the sample with 12% Cr for both magnetic phases and the results of Kazama et al. [2] on the sample with 5% Cr, only for the antiferromagnetic phase, are given in fig. 1. It is found in the ferrimagnetic phase that, besides earlier measured soft mode, there exists a strong softening of magnon energy in the second half of Brillouin zone, approaching the phase transition of cooling. The

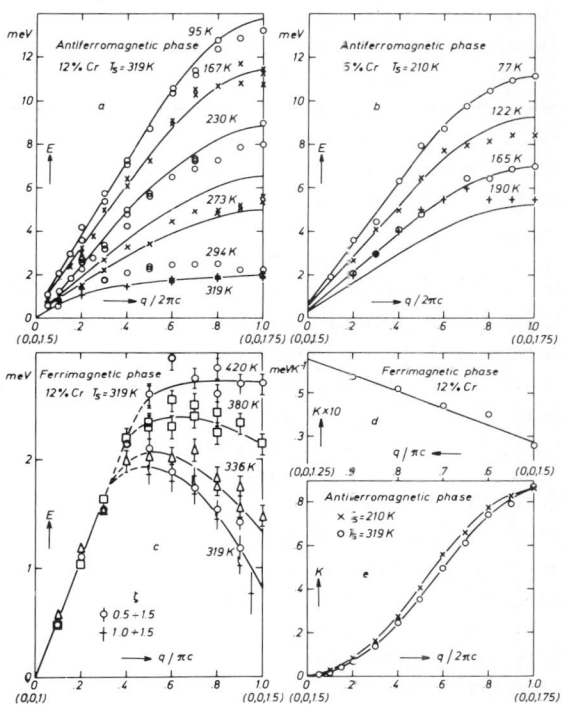

Fig. 1. Spin-wave dispersion relation in $Mn_{2-x}Cr_xSb$ for q parallel to C^*. (a) $x = 12\%$, the antiferrimagnetic phase; (b) $x = 5\%$, the antiferromagnetic phase; (c) $x = 12\%$, the ferrimagnetic phase; (d) The K coefficient for $x = 12\%$ Cr, in a function of reduced wave vectors $\zeta = q/2\pi c$, for the ferrimagnetic phase; (e) The K coefficient for $x = 12\%$ Cr, in a function of reduced wave vectors $\zeta = q/2\pi c$, for the antiferromagnetic phase. The full line is calculated according to eq. (1) and empirical values for the K coefficient.

softening is an indication for the phase transition to begin at 420 K, i.e. about $0.84\, T_c$, where T_c denotes the Curie point. In the antiferromagnetic phase, from 95 K to $T_s = 319$ K for 12% Cr and from 77 to 190 K ($T_s = 210$ K) for 5% Cr the softening of magnons includes the whole Brillouin zone of this phase. The centre of softening region of the dispersion branches is the superlattice point (001.5).

The softening of magnon energy in the ferrimagnetic phase can be expressed by

$$E^2(T, \zeta) = K(\zeta) \cdot |T - T_s| + A(\zeta), \qquad (1)$$

where E is the energy of acoustic magnons, A is the square of magnon energy at the temperature of phase transition T_s, and the $K(\zeta)$ the coefficient in the function of reduced wave $\zeta = q/2\pi c$, is given in fig. 1(d).

In the antiferromagnetic phase the square relation of the soft mode is valid in the whole range of Brillouin zone for the values of normalized temperature $|T - T_s|/T_s$ greater than about 0.5 for 12% Cr and 0.22 for 5% Cr. For the region of temperatures closer to the phase transition this relation tends to be valid in the vicinity of the superlattice point (001.5).

For the region of temperature from 273 K to T_s and for the values $\zeta > 0.5$ the disagreement from the square relation is considerable. In the region of validity of the square relation the $K(\zeta)$ coefficient has the following properties.

(a) It is practically independent on the percentage Cr, i.e. on the temperature of phase transition, for antiferromagnetic phase, see fig. 1(e).

(b) As can be seen in fig. 2, the coefficient ratio of the antiferromagnetic phase, with respect to the ferrimagnetic phase for the sample of 12%, is proportional to the square of reduced wave ζ. It allows us to accept the ratio of relative changes of magnon energy in both phases for the same $|T - T_s|$ is given by a simple expression:

$$\epsilon_{AF}^2 / \epsilon_{FM}^2 = 35 \cdot \zeta^2,$$

where ϵ is the relative change of magnon energy with respect to measured energy at the temperature of phase transition.

In conclusion, it can be said that the softening

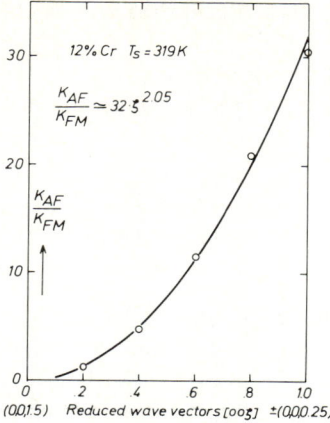

Fig. 2. The $K_{AF}(\zeta)$ coefficient ratio of the antiferromagnetic phase in comparison with the $K_{FM}(\zeta)$ coefficient for the ferrimagnetic phase. The full line is obtained by fitting two parameters.

of the spin waves branch, as a herald of the appearance of a new magnetic phase, begins further from the temperature of phase transition, but relatively close to the Curie point. It occurs in the wide range around the boundary of Brillouin zone. The instability of the new phase endures even $|T - T_s|/T = 0.7$ which is in agreement with the phenomenological model by Kittel [3].

I am greatly thankful to Dr. Sava Milioševič for many valuable discussions. I am also indebted to Dr. H.S. Jarrett for the loan of the monocrystal sample.

References

[1] J. Todorović, to be published.
[2] N. Kazama, S. Funahashi, Y. Hamaguchi and H. Watanabe, Proc. Conf. of Magnetism, Moscow, V (1973) 581.
[3] C. Kittel, Phys. Rev. 120 (1960) 335.
[4] N.P. Grazhdankina, Sov. Phys.-Uspekhi 11 (1969) 727.

EFFECTIVE ANHARMONICITY OF ELASTIC SUBSYSTEM IN ANTIFERROMAGNETS

V.I. OZHOGIN
Kurchatov Institute of Atomic Energy, Moscow, USSR

and

V.L. PREOBRAZHENSKII
Moscow Institute of Radioengineering, Electronics and Automatics, USSR

In weakly anisotropic antiferromagnets (AF), the effective anharmonicity turns out to be considerable for the elastic subsystem – due to exchange enhanced magnetostrictive coupling between elastic and magnetic subsystems (the third order effective elastic moduli can be 1.5–2 order larger than in a pure elastic system). The treatment of various nonlinear ultrasonic effects is undertaken for rombohedral AF with an easy plane anisotropy.

1. Introduction

The elastic part of the solid crystal energy is related to the strains $u_{ij} \equiv \mathbf{u}$ by the expression

$$\mathcal{F}_e = \tfrac{1}{2}\mathbf{C}^{(2)}\mathbf{u}\mathbf{u} + \tfrac{1}{6}\mathbf{C}^{(3)}\mathbf{u}\mathbf{u}\mathbf{u}. \tag{1}$$

With practically attainable $\mathbf{u}(t)$, the values $C^{(3)}u$ are so small compared with $C^{(2)}$, that the experimental observation of nonlinear effects in solids is difficult at present [1].

In magnetics, the interaction between the magnetic and elastic subsystems is substantially nonlinear in its nature. In AF the static and linear dynamic display of magnetoelastic (m.e.) coupling is enhanced by the exchange interaction [2–7]. It can be suggested that the AF exchange interaction can enhance the nonlinear display of m.e. coupling as well.

2. Qualitative analysis

The origin of the effective elastic anharmonicity and the evaluation of its magnitude can be demonstrated for the case of the simplest model of an easy plane AF with two magnetic sublattices M_1 and M_2. For this model the energy density is written as:

$$\mathcal{F} = 2M_0\{\tfrac{1}{2}Em^2 - D(m_x l_y - m_y l_x) + \tfrac{1}{2}A(\mathbf{n}\mathbf{l})^2 - \mathbf{m}\mathbf{H}\} + \tfrac{1}{2}\mathbf{C}^{(2)}\mathbf{u}\mathbf{u} + \mathbf{l}\mathbf{B}\mathbf{l}\mathbf{u} - \boldsymbol{\sigma}\mathbf{u}. \tag{2}$$

Here and below $\mathbf{m} \equiv (\mathbf{M}_1 - \mathbf{M}_2)/2M_0$, $\mathbf{l} \equiv (\mathbf{M}_1 - \mathbf{M}_2)/2M_0$, $|\mathbf{M}_1| = |\mathbf{M}_2| = M_0$; $E \equiv 2H_E$, $D \equiv H_D$, and $A \equiv H_A$ are the effective fields of exchange, Dzyaloshinskii and anisotropy interactions respectively; $\mathbf{B}$ is the m.e. moduli tensor; $\boldsymbol{\sigma}$ is the external stress tensor; $\mathbf{H} = (H_0, 0, 0)$ is the external in-plane d.c. magnetic field; $\mathbf{n} = (0, 0, 1)$ is the unit vector perpendicular to the easy plane ($A > 0$).

If $\boldsymbol{\sigma} = 0$, the magnetic subsystem equilibrium is determined by the relations $\mathbf{m} = (m_0, 0, 0)$, $\mathbf{l}_0 = (0, l_0, 0)$ with $l_0 \approx 1(T \ll T_N)$ and $m_0 \approx (H_0 + D)/E$. We shall consider only the so called "quasiferromagnetic" (QF) mode of the magnetic oscillations. In this mode, the variables (μ_y, μ_z, λ_x) take part ($\boldsymbol{\mu} \equiv \mathbf{m} - \mathbf{m}_0$, $\boldsymbol{\lambda} \equiv \mathbf{l} - \mathbf{l}_0$; $\mu(t) \ll m_0$, $\lambda(t) \ll l_0$).

For the qualitative evaluations, the m.e. part of $\mathcal{F}$ can be written schematically as

$$\mathcal{F}_{m.e} = \mathbf{l}\mathbf{B}\mathbf{l}\mathbf{u} \Rightarrow B_\alpha l_0^2(u_0 + u_\alpha^{(1)}) + B_\beta l_0 \lambda_x u_\beta^{(1)} + B_\delta \lambda_x^2(u_0 + u_\delta^{(1)}),$$

where $u^{(1)}(t) \ll u_0$ are some components of the strain tensor. From the Landau–Lifschits equations, we get

$$\begin{aligned}
\gamma^{-1}\dot{\mu}_y &= -(H_0 + D)\mu_z; & \gamma^{-1}\dot{\lambda}_x &= El_0\mu_z; \\
\gamma^{-1}\dot{\mu}_z &= H_0\mu_y + 2\tilde{B}_\delta l_0 \lambda_x(u_0 + u_\delta^{(1)}) - \tilde{B}_\beta l_0^2 u_\beta^{(1)},
\end{aligned} \tag{3}$$

Physica 86–88B (1977) 979–981 © North-Holland

where $u_0 \Rightarrow -B_\alpha l_0^2/C^{(2)}$ is a spontaneous AF striction [3–5] and $\tilde{B} \equiv B/2M_0$. We shall consider the m.e. oscillations only of a small frequencies $\Omega \ll \omega_{10}$ where ω_{10} is the frequency of QF mode of AF resonance: $(\omega_{10}/\gamma)^2 = H_0(H_0 + D) + 2H_E H_{mes}$; $H_{mes} \Rightarrow 2B_\alpha B_\delta l_0^4/2M_0 C^{(2)}$ [3–5] [H_{mes} can be modulated by $u_\delta^{(1)}(t)$]. In this case one can put $d/dt = 0$ in eq. (3) and obtain that the time dependent strains excite the oscillations of λ_x. Taking into account the modulating of ω_{10}^2, this can be expressed by the relation:

$$\lambda_x = -u_\beta^{(1)} E\tilde{B}_\beta l_0^3/[\omega_{10}(t)/\gamma]^2 \approx -A_\lambda u_\beta^{(1)} - G_\lambda u_\beta^{(1)} u_\delta^{(1)}, \tag{4}$$

with $A_\lambda \approx E\tilde{B}_\beta l_0^3/(\omega_{10}/\gamma)^2$, $G_\lambda \sim 2E^2 \tilde{B}_\beta \tilde{B}_\delta l_0^5/(\omega_{10}/\gamma)^4$. By substituting eq. (4) in the equation of motion for elastic displacement vector $\boldsymbol{R}$, one concludes that the m.e. interaction leads not only to the renormalisation of the second order elastic constants [6, 7]:

$$\Delta C_{m.e.}^{(2)} \sim EB^2 l_0^4/2M_0(\omega_{10}/\gamma)^2,$$

but to the appearance of the third order effective elastic moduli

$$C_{eff}^{(3)} \sim 2B_\delta A_\lambda^2 \sim B_\beta G_\lambda \sim 2E^2 B^3 l_0^6/(2M_0)^2 (\omega_{10}/\gamma)^4. \tag{5}$$

For hematite (α-Fe$_2$O$_3$), $E \sim 2 \times 10^7$ Oe, $B \sim 1.5 \times 10^7$ erg cm^{-3}, $M_0 \sim 10^3$ e.m.u.; $l_0 \sim 1$ at $T \sim 300$ K; $(\omega_{10}/\gamma)^2 \sim 3 \times 10^7$ Oe^2 at $H_0 \sim 1$ kOe, therefore $C_{eff}^{(3)} \sim 10^{15}$ erg cm^{-3} that is 1.5 order of magnitude larger than in a pure elastic system. The magnitudes of $C_{eff}^{(3)}$ depend on H_0 and σ via ω_{10}.

3. The effective anharmonicity in rombohedral AFEP's (α-Fe$_2$O$_3$, FeBO$_3$ etc.)

In detailed calculations we used the following expression for the energy density of a crystal belonging to D_{3d}^6 space group: $\mathscr{F} = \mathscr{F}_m + \mathscr{F}_e + \mathscr{F}_{m.e.} - 2M_0 \boldsymbol{mH} - \boldsymbol{\sigma u}$.

$$\mathscr{F}_m = 2M_0\{\tfrac{1}{2}Em^2 - D(\boldsymbol{n}[\boldsymbol{ml}]) + \tfrac{1}{2}A(\boldsymbol{nl}) + \tfrac{1}{2}\alpha(\partial \boldsymbol{l}/\partial x_i)^2\},$$

$$\mathscr{F}_e = \tfrac{1}{2}C_{11}(u_{xx}^2 + u_{yy}^2) + \tfrac{1}{2}C_{33}u_{zz}^2 + C_{12}u_{xx}u_{yy} + C_{13}(u_{xx} + u_{yy})u_{zz}$$
$$+ (C_{11} - C_{12})u_{xy}^2 + 2C_{44}(u_{xz}^2 + u_{yz}^2) + 2C_{14}[(u_{xx} - u_{yy})u_{yz} + 2u_{xy}y_{xz}].$$

$$\mathscr{F}_{m.e.} = B_{11}(l_x^2 u_{xx} + l_y^2 u_{yy}) + B_{12}(l_x^2 u_{yy} + l_y^2 u_{xx}) + 2(B_{11} - B_{12})l_x l_y u_{xy}$$
$$+ B_{33}l_z^2 u_{zz} + 2B_{44}(l_y l_z u_{yz} + l_x l_z u_{xz}) + B_{41}[l_y l_z(u_{xx} - u_{yy}) + 2l_x l_z u_{xy}]$$
$$+ 2B_{14}[(l_x^2 - l_y^2)u_{yz} + 2l_x l_y u_{xz}]. \tag{6}$$

The x-axis is directed along the twofold axis U_2 of the crystal (as in [4, 8], with some difference from [3, 6]). For simplicity we took $\sigma_{ij} = -p\delta_{xx}$ (the compression along $\boldsymbol{H}_0$). Then we:

(1) calculate the equilibrium $u_{ij}^{(0)}$, $\boldsymbol{m}_0$ and $\boldsymbol{l}_0$ for $p < p_c$ and $p > p_c$, where $p_c(H_0)$ is the critical pressure which causes the reorientation of the sublattices [7b, 9];

(2) solve the equations of motion for $\boldsymbol{m}$ and $\boldsymbol{l}$ with $d/dt = 0$, taking into account not only $u_{ij}^{(0)}$ but also $u_{ij}^{(1)}(t)$ as a small addition to u_{ij}^0 and using the condition $|\boldsymbol{M}_1| = |\boldsymbol{M}_2| = M_0$ for calculating the second order correction to $\boldsymbol{\lambda}(\boldsymbol{u})$;

(3) by substituting $\boldsymbol{\lambda}(u_{ij})$ in the equations of motion for the displacement vector $\boldsymbol{R}$, obtain the nonlinear equations with the effective third order terms. Based on these, some concrete processes are treated with the help of slowly varying amplitude method [10].

4. The sound frequency doubling

Let us consider the normal quasielastic wave with $\boldsymbol{R}_1 \| z$. In a rombohedral crystal, all three components of the wave vector $\boldsymbol{k}_1$ for this wave are nonzero. The synchronous process of generation of the wave with $\Omega_2 = 2\Omega_1$, $\boldsymbol{k}_2 \| \boldsymbol{k}_1$ and $\boldsymbol{R}_2 \| z$ seems to be possible in this case ($\boldsymbol{H}_0 \| x \| U_2$). The stationary amplitude R_{2a} of the second harmonic is described by the equation (in the approximation of the preset primary field):

$$(\partial R_{2a}/\partial \zeta) + \alpha_2 R_{2a} - \Psi R_1^2 e^{-2\alpha_1 \zeta} = 0, \tag{7}$$

where ζ is a length variable in the direction of k_1; α_1 and α_2 are the respective attenuation coefficients, and

$$\Psi \equiv \tfrac{3}{2}(E/M_0)^2 B_{14}^3 k_1 k_{1x}^2 k_{1y}/\rho \Omega_1^2 (\omega_{10}/\gamma)^4.$$

At the distance $\zeta_m = (\alpha_2 - 2\alpha_1)^{-1} \ln(\alpha_2/2\alpha_1)$ from the source (at $\zeta = 0$) of the primary wave, the amplitude R_{2a} achieves its maximum which is equal to $R_1^2 \Psi/8\alpha_1$ if we adopt for wave quality factor $Q_k = k/2\alpha = \text{const.} \, \Omega^{-1}$. For hematite at $H_0 \sim 1$ kOe, $\Omega_1 \sim 2\pi \times 10^8 \, \text{s}^{-1}$ and $Q_1 \sim 10^3$ one gets: $\zeta_m \sim 0.5$ cm, $R_{2\max}/R_1 \sim 10^5 (k_1 R_1)$.

5. The parametric interaction of collinear waves

If the crystal is excited by the plane elastic wave with the frequency Ω_p, $R_p \| y$ and $k_p \| z$, the synchronous process seems to be possible for parametric excitation of two facely running waves with $k_1 \| (-k_2) \| k_p$ and $R_1 \| R_2 \| x$. Solving the equations in the preset pumping field approximation with boundary conditions $R_1(z=0) \to 0$, $R_2(z=L) = 0$, where L is the length of the crystal along z-axis, we obtain for the threshold strains in the pumping wave:

$$\epsilon_{th} = \tfrac{1}{2} R_p k_p = 2\pi |B_{14}|(1-\kappa)/C_{44}(k_p L)\kappa^{5/2}; \quad \kappa \equiv 2E B_{14}^2/M_0 C_{44}(\omega_{10}/\gamma)^2. \tag{8}$$

For α-Fe_2O_3 one gets $\epsilon_{th} \sim 2 \times 10^{-6}$ if $\Omega_p \sim 2\pi \times 10^8 \, \text{s}^{-1}$, $L \sim 1$ cm, $H_0 \sim 1$ kOe. The account of attenuation with $Q \sim 10^3$ approximately doubles the value of ϵ_{th}.

The effects treated in sections 4 and 5 can become more pronounced if $p \to p_c(H_3)$ in which point $(\omega_{10}/\gamma)^2$ diminishes to the magnitude $2H_E H_{mes}$ [9a]. However, the attenuation of quasisound becomes larger in this case and must be taken into account more strictly.

References

[1] L.K. Zarembo and V.A. Krasil'nikov, Uspekhi Fiz. Nauk 102 (1970) 549.
[2] M.A. Savchenko, Fiz. Tverd. Tela 6 (1964) 864.
[3] S. Iida and A. Tasaki, in: Proc. Int. Conf. Magnetism (Nottingham, 1964) p. 583.
[4] A.S. Borovik-Romanov and E.G. Rudashevskii, Zh. Eksp. Teor. Fiz. 48 (1965) 74.
[5] E.A. Turov and V.G. Shavrov, Fiz. Tverd. Tela 7 (1965) 217.
[6] M.H. Seavey, Solid State Commun. 10 (1972) 219.
[7] V.I. Ozhogin and P.P. Maximenkov (a) in: Digests of Intermag Conf. (Kyoto, 1972) rep. 49–4; (b) Zh. Eksp. Teor. Fiz. 65 (1973) 657.
[8] R.Z. Levitin, A.S. Pakhomov and V.A. Shchurov, Zh. Eksp. Teor. Fiz. 56 (1969) 1242.
[9] I.E. Dikshtein, V.V. Tarasenko and V.G. Shavrov (a) Fiz. Tverd. Tela 16 (1974) 2192; (b) Zh. Eksp. Teor. Fiz. 67 (1974) 816.
[10] S.A. Ahmanov and R.V. Hohlov, The Problems of Nonlinear Optics (Academie of Sciences USSR, Moscow, 1964).

ANOMALOUS HALL EFFECT IN THE FERROMAGNETIC COMPOUND Cr_3SeTe_3

D. BABOT, M. CHEVRETON,
Laboratoire d'Etude des Matériaux (E.R.A. n° 602), 303, INSA, 69621 Villeurbanne Cédex, France

M. WINTENBERGER and B. LAMBERT-ANDRON
CENG, 85X, Centre de Tri, 38041 Grenoble Cédex, France

The spontaneous Hall coefficient R_s is determined for Cr_3SeTe_3 ($T_c = 214$ K). It is assumed that the relatively large anomalous effect ($R_s = -3.6 \times 10^{-9}$ V m/AG at 280 K) is based on an intrinsic spin–orbit coupling of the 3d delocalized electrons and on a predominant spin disorder scattering.

1. Introduction

The electrical resistivity ρ vs. T curve of the ferromagnetic compound Cr_3SeTe_3 shows a clear change in slope near the Curie temperature $T_c = 214$ K [1]. The Hall effect has been investigated with a four point probe method in the ranges 5–320 K and 0–50 kOe on sintered powders. The Hall resistivity ρ_H of ferromagnetic materials has been treated as follows [2]:

$$\rho_H = R_0 H_{int} + R_1 M, \quad (1)$$

where R_0 is the ordinary Hall coefficient, H_{int} the magnetic field inside the sample, R_1 the extraordinary Hall coefficient and M the magnetization.

2. Hall effect in Cr_3SeTe_3

Hall effect measurements are carried out on the basis of field and current reversal method in order to cancel the thermomagnetic effects. The Hall voltage V_H is then obtained and the isotherms ρ_H vs. B (B = induction outside the sample) are plotted (fig. 1). For Cr_3SeTe_3 the susceptibility χ is small and in low fields the applied field is not negligible as compared to the magnetization. Taking into account that

$$M = \chi H_{int} = \chi (H_{ext} - NM) = \frac{\chi}{1+\chi N} H_{ext}, \quad (2)$$

a new relation is obtained:

$$\rho_H = H_{ext}\left[R_0 + \frac{\chi}{1+\chi N}(R_1 - NR_0)\right], \quad (3)$$

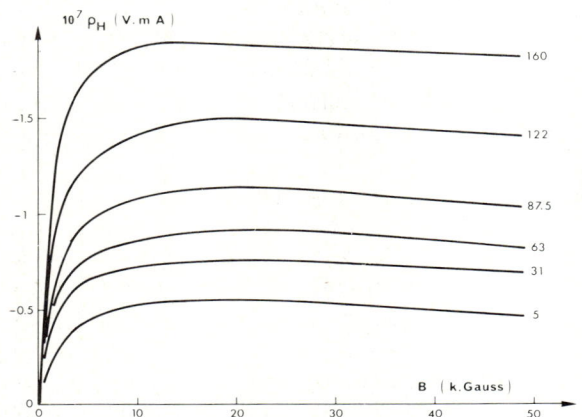

Fig. 1. Isotherms ρ_H vs. B (only drawn for $T < 160$ K).

where H_{ext} is the outside field and N the demagnetizing factor ($N \ll 1$). Taking into account that $NR_0 \ll R_1$ and $\chi N \ll 1$, the slope of the ρ_H vs. B curve is in low fields:

$$p = R_0 + \chi R_1. \quad (4)$$

This relation gives the R_1 coefficient. In order to determine χ we also need the M vs. H curves in low fields.

A second determination of R_1 is obtained from the extended part of high fields data which intercepts the ordinate axis at $\rho_s = M_s(R_1 - R_0)$ [2] and from the magnetization at saturation M_s. The $R_s = R_1 - R_0$ coefficient is called the spontaneous Hall coefficient ($R_s \approx R_1$). The same value of R_1 obtained by both methods (fig. 2) is a check that sufficiently high fields have been reached.

In the paramagnetic state ($T > 214$ K) R_1 can be obtained by two ways: in low fields as described before and also in intermediate fields

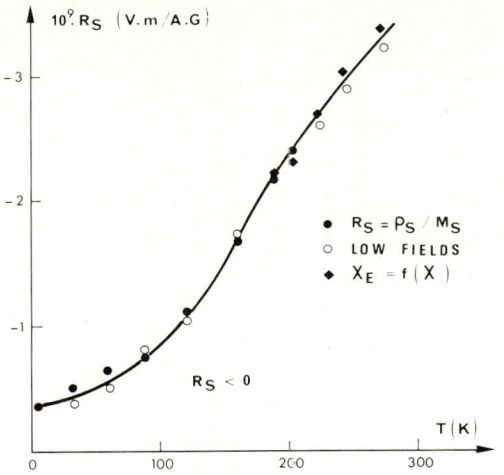

Fig. 2. The spontaneous Hall coefficient R_S as a function of temperature ($T_c = 214$ K).

where $\chi(H) = dM/dB$ is field dependent; then the following expression is obtained:

$$\chi_E = \frac{d\rho_H}{dB} = R_0 + R_S \chi(H). \quad (5)$$

A straight line is obtained when plotting χ_E against χ. The intercept on the ordinate axis is R_0 and the slope is R_S. Agreement is established between low fields and high fields data.

The spontaneous coefficient R_S increases with T in the whole investigated range. No saturation is observed above T_c. The ordinary coefficient remains temperature independent: $R_0 = +7.9 \times 10^{-13}$ Vm/AG.

3. Discussion

The relevant theories of the anomalous Hall effect are based on the existence of mixed or intrinsic spin–orbit coupling and of a scattering mechanism. The initial relation $R_S = a\rho_t^2$ proposed by Karplus and Luttinger [3] is not confirmed. The linear dependence of R_S with ρ_{ph}^2 proposed by Irkhin et al. [4] considering both spin–orbit coupling and a phonon scattering, as well as the linear dependence of R_S with ρ_m obtained by Irkhin and Abelskii [5] in case of spin disorder scattering for both spin–orbit coupling are examined (fig. 3). The ρ_t vs. T curve of Cr_3SeTe_3 allows us to determine the phonon ρ_{ph} and the magnetic ρ_m contributions. Large discrepancies are observed for these two laws. We have tried to check the relation $R_S = a\rho_m^2$ given by Kagan and Maksimov [6], obtained by using a spin wave approximation. Agreement is shown between 80 and 180 K. When T approaches T_c the molecular field approximation is more valid and this relation tends to the ρ_m form. At very low temperatures ($T < 60$ K) the $a\rho_m^2$ relation is not well obeyed.

A more realistic model has been proposed by Lazarev [7] who expected to explain the

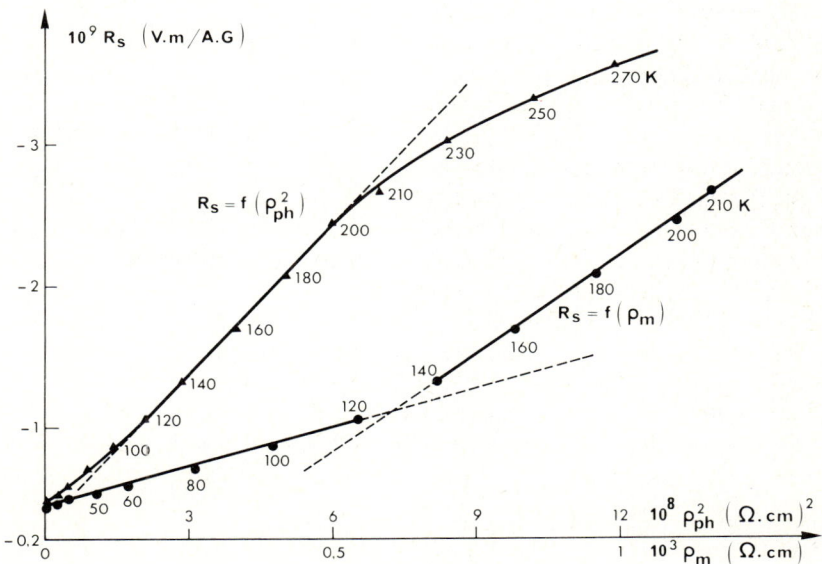

Fig. 3. The dependences of R_S on ρ_{ph}^2 and on ρ_m.

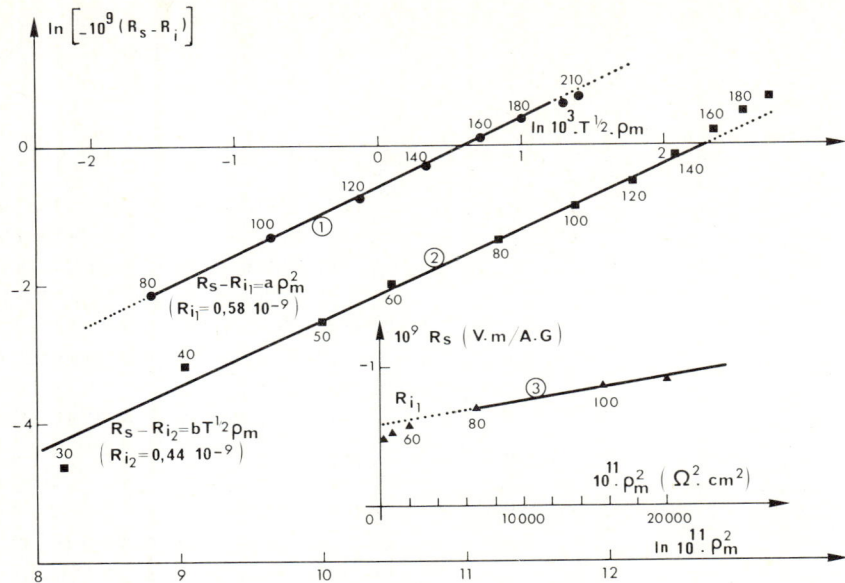

Fig. 4. The dependences of R_s on ρ_m^2 ①, ③ and on the product $T^{1/2} \cdot \rho_m$ ② as predicted for a spin polaron treatment [7].

anomalous Hall effect with the displacement study of a "spin polaron", when the mean free path is strongly reduced. This model leads to the following relation:

$$R_S(T) = bT^{1/2} \cdot \rho_m(T). \qquad (6)$$

We found that this relation is best obeyed in the low temperature range 30–150 K (fig. 4).

The Hall mobility $\mu_H = R_0/\rho_t$ has been found equal to 2.5 cm^2/Vs at 300 K. The high value of R_S is probably due to an intrinsic spin–orbit coupling of the partially delocalized 3d electrons. As described by Rhyne [8] a mixed coupling gives a high anomalous effect if μ_H is large. Spin disorder scattering predominates at low temperatures. The $Cr_3Se_{4-x}Te_x$ compounds provide a propitious way to study the anomalous Hall effect theoretically: ρ_m can be obtained for each compound, a ferro ($x \geqslant 1$) or antiferromagnetic ($x < 1$) arrangement can be studied in a selected range of temperature because T_c or T_N are tellurium concentration dependent [1].

References

[1] D. Babot, M. Wintenberger, B. Lambert-Andron and M. Chevreton, J. Solid State Chem. 8 (1973) 175.
[2] J. P. Jan, in: Solid State Physics, Vol. 5 (1957) p. 1.
[3] R. Karplus and J. M. Luttinger, Phys. Rev. 95 (5) (1954) 1154.
[4] Yu. Irkhin and V. G. Shavrov, Sov. Phys. JETP 15 (1962) 854.
[5] Yu. Irkhin and Sh. Sh. Abelskii, Sov. Phys. Solid State 6(6) (1964) 1283.
[6] Yu. Kagan and L. A. Maksimov, Sov. Phys. Solid State 7 (2) (1965) 422.
[7] G. L. Lazarev, Sov. Phys. Solid State 14 (1) (1972) 22.
[8] J. J. Rhyne, Phys. Rev. 172 (2) (1968) 523.

HALL MOBILITY OF PHOTOELECTRONS IN MnO*

Tashiro USAMI and Taizo MASUMI
Department of Pure and Applied Sciences, University of Tokyo, Komaba, Tokyo, 153 Japan

Hall mobility of photoelectrons in MnO was studied at 70–300 K *by the Redfield technique*. Experimental results indicate that the absorption edge is associated with the transition from *p–d* overlapping valence band to *s*-like conduction band and the mobility of photoelectrons *in a wide band* is ruled by the *spin-disorder scattering*.

1. Introduction

Transition metal oxides are the magnetic semiconductors characterized by the mixed band-localized states for the electrons near the Fermi energy [1]. However, the assignment of the optical transitions seems rather subtle. Besides, because of their high resistivity, there have been little reliable data available on the transport phenomena in undoped crystals. Here, we report the first successful measurement of the Hall mobility of photoelectrons in undoped MnO single crystals ($T_N = 118$ K) from 77 to 300 K by applying *the pulsed Redfield technique*. Thus, we provide a new aspect on the energy level scheme for the electrons and also on the transport mechanisms in the transition metal oxides.

2. Experimental method

Several single crystals, grown by Verneuil's method at the Nakazumi Crystal Corp., were annealed in a well-controlled H_2–CO_2 mixture of various rates and then polished in H_3PO_4. Because of their high resistivity ($\rho \geq 10^9$ Ωcm at 300 K), one inevitably has to adopt the fast pulse technique with blocking electrodes [2] at the measurement of the transient photocurrent Q to avoid serious difficulties such as the non-ohmic character of contact electrodes and the space charge effect. Configurations of three electrodes A, B and C are illustrated in fig. 1. In the Q_x-measurement, a pulsed electric field E_x is applied as shown in fig. 1(a), while in the measurement of the Hall mobility, $\mu_H = (c/H_z) \cdot (Q_y/Q_x)$, E_x is applied as shown in fig. 1(b) to compose the Redfield arrangement. Pulsed photosignals due to the induced charge in the electrode C were amplified and fed to a

---

* Based on the thesis submitted by Toshiro Usami in partial fulfilment of the requirements for the D.Sc. degree at the University of Tokyo, 1976.

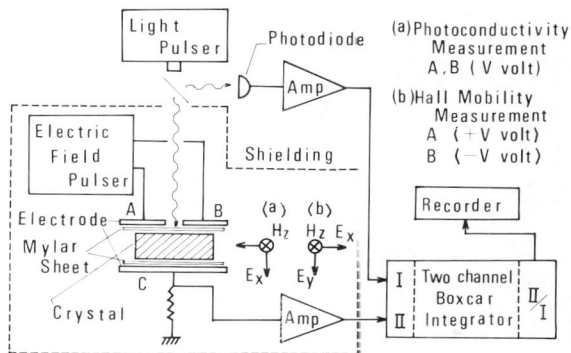

Fig. 1. Schematic diagram of the electrode arrangement and the signal detection system.

Boxcar integrator. Electric fields up to 7 kV/cm and an incident light flux of 10^{14} photons/pulse with duration of 1 μsec were used to obtain adequate signals.

3. Experimental results

Photosignals Q_x have a linear dependence on E_x and also on the light intensity. Dependence of Q_x on the wavelength of the exciting light λ, $Q_x(\lambda)$, confirmed to be insensitive to the surface state, is shown in fig. 2. At $\lambda > 390$ nm, a broad peak is observed in $Q_x(\lambda)$ only for the crystal 2-G with more excess oxygen, but this peak decreases remarkably as the temperature T decreases. At $\lambda < 390$ nm, $Q_x(\lambda)$ rises rapidly as λ decreases regardless of the degree of oxidation of crystals.

A typical signal of the Hall current Q_y is shown in fig. 3(a). The polarity of Q_y indicates that the photocarriers are electrons. The value of μ_H is about 10 cm^2/V sec at 300 K and independent of λ. Temperature dependence of $\mu_H(T)$ is illustrated in fig. 3(b). From 300 K down to about 200 K, μ_H is proportional to $T^{-3/2} \cdot \bar{\chi}(T)^{-1}$ where $\bar{\chi}(T)$ is the magnetic susceptibility [3]. At lower temperatures, $\mu_H(T)$ becomes lower for the crystals with less oxidation. No longitudinal magneto-resistance effect has been observed.

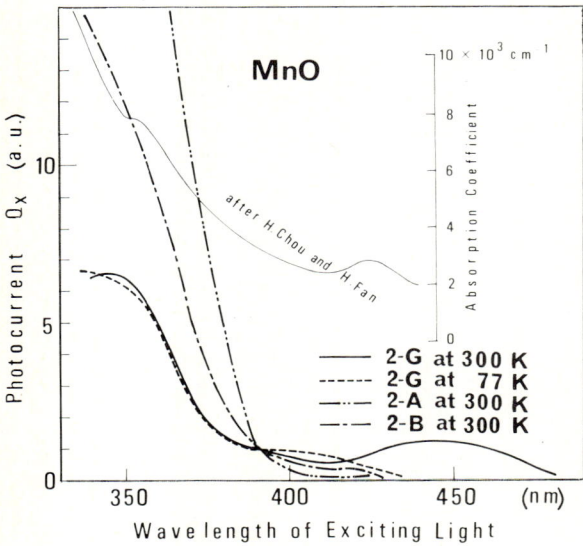

Fig. 2. Photocurrent Q_x versus wavelength of the exciting light λ. Q_x was normalized at $\lambda = 390$ nm for all samples. The crystal 2-G is as grown. Annealing condition is as follows: $P(CO_2)/P(H_2) = 0.1$ and 1 for the crystals 2-A and 2-B, respectively. For the absorption data, refer to H-h. Chou and H.Y. Fan, Phys. Rev. B10 (1974) 901.

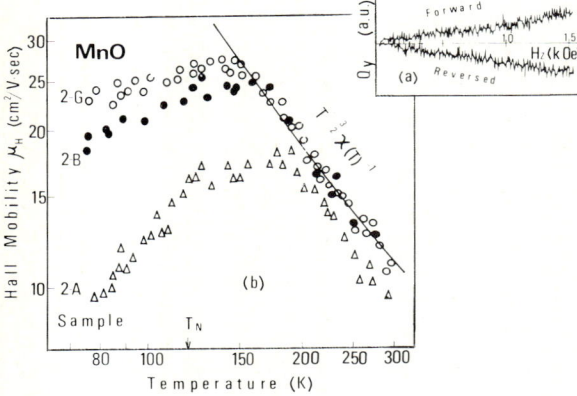

Fig. 3. (a) A recorder trace of magnetic field dependence of the Hall current $Q_y(H_z)$. Inversion of the polarity of H_z yields two branches. (b) Temperature dependence of the Hall mobility of photoelectrons in MnO crystals of various grades of oxidation.

4. Discussion and conclusion

Data of $Q_x(\lambda)$ suggest that (i) at $\lambda > 390$ nm, the transition is related to extrinsic states such as Mn-vacancy while (ii) at $\lambda < 390$ nm, the excitation involves the intrinsic interband transition. For the intrinsic region, there exist two alternative interpretations, i.e., (1) $(2p)^6(3d)^5 \to (2p)^5(3d)^6$ or (2) $(3d)^5(4s)^0 \to (3d)^4(4s)^1$. Relatively large values of μ_H, their temperature dependences and also the lack of longitudinal magnetoresistance effect indicate that the photoelectrons move in a wide band presumably due to the Mn-4s states. Then, the initial valence band may be ascribed to the overlapping $p-d$ states [4]. Temperature dependence of μ_H above 200 K is explained by the Haas model [5] where the scattering time τ is ruled by the spin-disorder. At lower temperatures, τ is determined by the charged impurity produced by partial compensation. Around T_N, the anomalies of μ_H reported for NiO and MnTe were not observed for MnO.

Thus, we conclude that the optical band edge is associated with the transition from the $p-d$ overlapping valence band to the s-like conduction band and the mobility of conduction electrons *in a wide band* is ruled by *the spin-disorder scattering*. Note that the transient photoconductivity measurement is substantial for studying the transport phenomena in magnetic semiconductors.

The authors would like to acknowledge stimulating discussions with Drs. K. Kajita and S. Komiyama. This work was supported in part by the Grants-in-Aid for Scientific Research (B).

References

[1] D. Adler and J. Feinleib, Phys. Rev. B2 (1970) 3112.
[2] K. Kajita and T. Masumi, Appl. Phys. Lett. 21 (1972) 332.
K. Kajita, T. Masumi, and T.B. Reed, Proc. ICM 73 5 (1974) p. 143.
[3] H. Bizette, C.F. Squire, and B. Tsui, Introduction to Solid State Physics, C. Kittel, 3rd ed. (John Wiley & Sons, Inc., New York, 1966), p. 483.
[4] D.E. Eastman and J.L. Freeouf, Phys. Rev. Lett. 34 (1975) 395.
[5] C. Haas, Phys. Rev. 168 (1968) 531.

LIGHT SCATTERING BY MAGNONS IN MAGNETIC SEMICONDUCTORS*

S. COUTINHO and M.D. COUTINHO-FILHO
Departmento de Física, Universidade Federal de Pernambuco, 50000 Recife, Brazil

The contributions of the s–d (or s–f) interaction on the light scattering by magnons in magnetic semiconductors are discussed. We analyze two processes involving magnon density fluctuations in the ferromagnetic case and the scattering from two-magnons in the antiferromagnetic one. The frequency dependence of the extinction coefficient is obtained and applications to $CdCr_2Se_4$ and EuTe are made.

Inelastic light scattering has proved to be a very useful technique as a probe of elementary excitations in solids. In magnetic insulators a number of scattering mechanisms involving spin waves (magnons) have been proposed [1]. Very recently, however, some calculations of resonant Raman effects involving magnetic excitations have been performed with applications for antiferromagnetic insulators and ferromagnetic semiconductors [2]. These carrier-mediated mechanisms contain resonant terms (incident light frequency approaching band gap) which make these processes rather distinct from the earlier proposed ones [1].

In this communication we shall analyze the most simple possible scattering contributions due to the s–d interaction for the magnon resonant Raman scattering in ferro- and antiferromagnetic semiconductors. Our model for the magnetic semiconductor is the following. There is a full valence band, whereas the conduction band is empty. We assume a two parabolic band model in which the electron and hole have reduced effective mass m^*, wave vector p and forbidden energy gap $\hbar\omega_g$. The magnetic aspect of the semiconductor is dictated by localized spins which we shall assume to be ferro- or anti-ferromagnetically arranged. The carriers are supposed to interact with the localized spins via the s–d interaction. Below the ordering temperature it leads in the ferromagnetic case to a spin splitting of the bands, i.e. the carrier energies are given by $E_{n,p,\sigma} = E_{n,p} - \sigma J_{sd}^{(n)} S/2$ [3]. Here $J_{sd}^{(n)}$ is the exchange integral between carrier of band n and the N localized spins of magnitude S. In the antiferromagnetic case the magnetization vanishes and in first order the band spin splitting disappears. The magnetic system is described by spin wave excitations with energies given by the usual ferromagnetic and antiferromagnetic dispersion relations.

The quantity to be evaluated is the extinction coefficient, defined as the fraction of light scattered per unity of path length per unit solid angle. The scattering processes are diagramatically represented in fig. 1. As is seen they correspond to either light scattering by magnon density fluctuations (ferromagnetic case) or two-magnon scattering (antiferromagnetic case). The two-magnon process involving magnon production at the same vertex is negligible because the main features in this scattering is caused by magnons close to the Brillouin zone boundary. The transition probability per unit time of these events is evaluated by using standard perturbation theory. In the course of these calculations the following assumptions have been made [4]. We have considered a non-interacting Bloch picture for the intermediate state, i.e., the energy of the electron-hole pair was taken to be

$$E_i = \hbar\omega_g + \frac{\hbar^2 p^2}{2m^*} - \sigma\hbar\omega_s, \qquad (2)$$

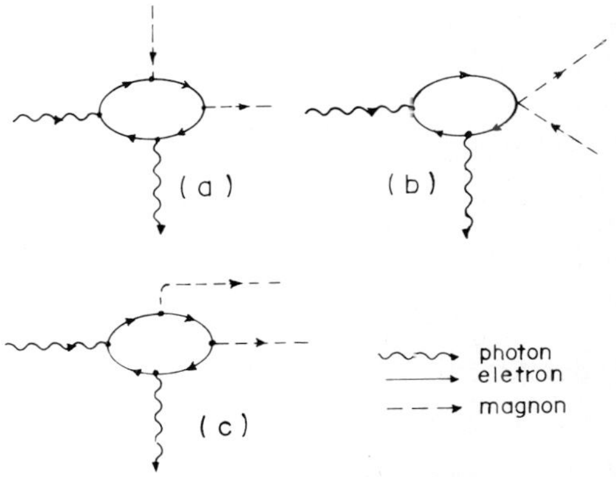

Fig. 1. (a) and (b) magnon density fluctuation processes for the ferromagnetic case; (c) two-magnon process for the antiferromagnetic case.

---

* Work supported by BNDE, CNPq and CAPES (Brazilian Government).

where $\hbar\omega_s = (J_{sd}^{(c)} - J_{sd}^{(v)})S/2$. The electron wave vector matrix elements were replaced by $\hbar/a$, where a is the lattice constant. Between the various terms which appear when changing the order of the interactions in fig. 1, we have taken the most resonant ones. The number of scattered photons in the initial state was taken to be zero. These assumptions were common to both magnetic cases. Different considerations, however, have been made when dealing with the statistical factors associated with the magnon populations, which were dictated by the possible applications in mind. Thus for the ferromagnetic case we have replaced the magnon thermal population by $k_B T/\hbar\omega_k \gg 1$, with ω_k given by the usual quadratic magnon dispersion relation. On the other hand, the antiferromagnetic magnons are assumed to have sufficiently high energy compared with thermal excitation. Finally, we have assumed the incident light directed along the z-axis (magnetization axis) and polarized along the x-axis, and that the scattered radiation is collected in a small solid angle in the back scattering geometry. There is no restriction to its polarization.

The calculations yield:

$$h_a = 4C\hbar^2(k_B T)^2 k_0^{-1}\left[\pi - 2\tan^{-1}\left(\frac{\gamma H_0}{Dk_0^2}\right)^{1/2}\right]$$
$$\times \left|\sum_\sigma \frac{(\omega_g - \omega_0 + 2\sigma\omega_s)}{(\omega_g - \omega_0 + \sigma\omega_s)^{1/2}}\right|^2, \quad (3a)$$

$$h_b = C(J_{sd}S)^2(k_B T)^2 k_0^{-1}\left[\pi - 2\tan^{-1}\left(\frac{\gamma H_0}{Dk_0^2}\right)^{1/2}\right]$$
$$\times \left|\sum_\sigma \sigma(\omega_g - \omega_0 + \sigma\omega_s)^{-1/2}\right|^2, \quad (3b)$$

$$h_c = 2\pi^2 CD^2(J_{sd}S)^4 a^{-3}|(\omega_g - \omega_0)^{-3/2}|^2, \quad (3c)$$

where $C = e^4 a^2 m^{*3}/2^9 \pi^3 \hbar^5 c^4 m^4 D^2 S^2$, D is the magnon stiffness parameter, c is the light velocity, H_0 is the external magnetic field, $\gamma = g\mu_B/\hbar$ is the gyromagnetic ratio, and m the electron mass.

It is clear from eqs. (3) that in the ferromagnetic case one has two resonant frequencies $\omega_0 = \omega_g \pm \omega_s$ due to the s–d band splitting. Using a (continuous frequency) laser in an appropriated range it will be possible to determine the J_{sd} exchange integral. Notice also that in the antiferromagnetic case the resonant factor ($\omega_0 \to \omega_g$) goes with the third power due to the absence of spin splitting in the bands. For a typical ferromagnetic chalcogenide spinel, such as $CdCr_2Se_4$ one may take $\hbar\omega_g = 1.3$ eV, $a = 10^{-7}$ cm, $S = \frac{3}{2}$, $D = 7 \times 10^{-2}$ cm^2 s^{-1}, $m^* = 0.1$ m, $J_{sd}^{(c)} - J_{sd}^{(v)} \simeq J_{sd}^{(c)} = 0.1$ eV. At $T = 100$ K, $H_0 = 1$ kG, and using the $\lambda_0 = 10\,150$ Å YAG laser line for the incident radiation one obtains $h_a \simeq h_b \simeq 10^{-8}$. The shift on the scattered light is $\Delta\omega \simeq D(2k_0)^2 \simeq 10^{10}$ s^{-1}. This shift is very small but we believe it may be resolved with the modern Brillouin spectroscopic techniques [5]. On the other hand, for an antiferromagnetic Eu-chalcogenide, such as EuTe, one may take $\hbar\omega_g = 2.1$ eV, $a = 6.6 \times 10^{-8}$ cm, $S = \frac{7}{2}$, $m^* = m$, $J_{sd}^{(c)} - J_{sd}^{(v)} \simeq J_{sd}^{(c)} = 0.1$ eV. At $T \ll T_c$, $H_0 = 0$ and using the 5.145 Å argon-ion laser line one has $h_c \simeq 10^{-7}$. The frequency shift is $\Delta\omega \simeq 2\gamma H_E \sim 10^{12}$ s^{-1}, where H_E is the exchange field.

Finally, we should point out that one may use the exciton picture or the magnetic polariton picture for the intermediated states. These changes lead to dramatic effects as was discussed by Birman [6], and may be directly incorporated in our approach.

References

[1] R.J. Elliot and R. Loudon, Phys. Lett. 3 (1963) 189.
P.A. Fleury, S.P. Porto, L.E. Cheesman and H.J. Guggenheim, Phys. Rev. Lett. 17 (1966) 84.
P.A. Fleury and R. Loudon, Phys. Rev. 166 (1968) 514.
[2] A. Stasch, Phys. Lett. 44A (1973) 291.
A. Stasch and T. Wesolek, Acta Phys. Polonica A45 (1974) 469.
S.G. Coutinho and L.C.M. Miranda, Solid State Commun. 18 (1976) 785.
[3] C. Haas, C.R.C. Critical Rev. in Solid State Sci. 1 (1970) 47.
[4] R. Loudon, Proc. Roy. Soc. A275 (1963) 218.
[5] J.R. Sandercock and W. Wettling, Solid State Commun. 13 (1973) 1729.
[6] J.L. Birman, Proc. Int. Conf. on Light Scattering in Solids, M. Balkanski, ed. (Paris, 1971) p. 15.

NEWLY OBSERVED ANOMALIES IN THE MAGNETIC AND ELECTRICAL PROPERTIES OF $Li_xMn_{1-x}Se$

M. KASAYA

Department of Physics, Faculty of Science, Tōhoku University, Sendai, Japan 980

Magnetic measurements on the $Li_xMn_{1-x}Se$ system revealed abrupt changes in the paramagnetic Curie points and the magnetization with the applied magnetic field in the range $0.031 \leq x \leq 0.053$. These anomalies along with the electrical properties are explained qualitatively by introducing molecules which contain two Mn^{3+} valency states.

In a recent paper [1], we have reported a typical anomaly in the variation of the resistivity with the Li content in the substitutional alloy $Li_xMn_{1-x}Se$. The resistivity for $x = 0.053$ is 3×10^3 times as large as that for $x = 0.031$ in the impurity conduction region. The present paper describes our detailed studies on the magnetic properties of the same system with the aim to explain the observed anomaly.

Fig. 1 shows the Curie constants per mole of Mn ion and the paramagnetic Curie points obtained from experiments above room temperature. As for the pure MnSe, the value of the Curie constant corresponds to Mn^{2+} ion with $d^5(t_{2g}^3 e_g^2)$ configuration within the experimental error. However, the Curie constants obtained from the experiment in $Li_xMn_{1-x}Se$ are lower than those expected on the consideration that $(1-2x)Mn^{2+}$ and xMn^{3+} in $Li_xMn_x^{3+}Mn_{1-2x}^{2+}Se$ have the magnetic moments of $5\mu_B$ and $0\mu_B$, respectively. This is shown in fig. 1(a), the calculated curve based on the above consideration being shown by a dotted line. It is to be noted that zero moment for Mn^{3+} is lower limit in cubic crystalline field. Heikes et al. [2] have explained the small Curie constants by assuming that each cation has an effective magnetic moment of $(1-2x)2\sqrt{\frac{5}{2}(\frac{5}{2}+1)}$ Bohr magnetons on the average. Fig. 1(b) shows the variation of the paramagnetic Curie point, θ_p, with the Li content x. Up to $x = 0.031$, the value of $\Delta\theta_p/\Delta x$ is 41×10^2 K, while after this it abruptly changes to 26×10^2 K. At $x = 0.11$, θ_p is $+50$ K.

In fig. 2 magnetization M vs. magnetic field H curves at 4.2 K are shown for various values of x. As x increases, the M vs. H curves deviate from linearity and show a field-dependence of the susceptibility. However, at $x = 0.053$, the M vs. H curve is nearly linear for the magnetic fields used. As x increases further, the M vs. H

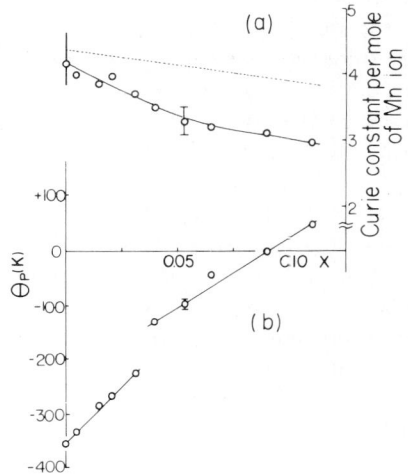

Fig. 1. (a) Curie constants vs. Li concentration x. The calculated curve based on the formula $C_x = [(1-2x)/(1-x)](C_{Mn}^{2+})$ is shown by the dotted line. (b) Paramagnetic Curie point vs. Li concentration x.

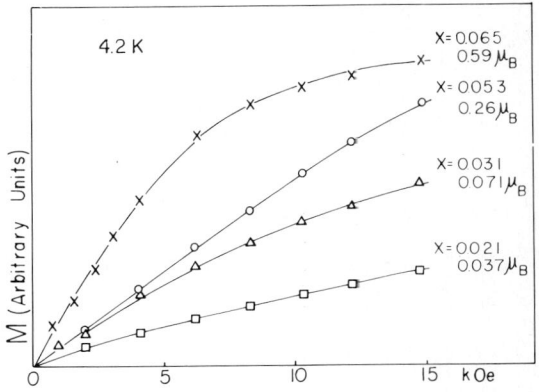

Fig. 2. Magnetization vs. magnetic field at 4.2 K. The magnetic moment listed below each curve is the value at 14.8 kOe.

curve again shows a field dependence of the susceptibility. With 6.5% Li, the magnetization per Mn ion turns out to be $0.59\mu_B$ at 14.8 kOe.

The external field dependence of the resistivity ratio, $[\rho(H)-\rho(0)]/\rho(0)$, changes at the same value of x. For 3.1% Li, this ratio increases with the square of the external field and becomes $+0.3$ at 4.2 K with 80 kOe, while the ratio for 5.3% Li is -0.02 and is nearly independent of the external field.

On the basis of the experimental facts, we will discuss the origin of the anomalous dependence of the resistivity on Li content in the impurity conduction region. For this purpose we have to estimate half the average distance between the neighbouring Li$^+$ centres. For $x = 0.005$, its value is 13 Å. On the other hand the distance between nearest-neighbour cations, which is a measure of the extension of the wave function of the hole, i.e., the Mn^{3+} valency state bound to the Li$^+$ centre, is 3.9 Å. Therefore, it is reasonable to assume that for $x = 0.005$, the Li$^+$Mn^{3+} clusters are isolated. In each of these clusters, 12 nearest neighbour Mn ions for Li$^+$ will be coupled by the charge carrier on Mn^{3+}. As x increases, these clusters interact and result in a metallic conduction and field-dependence of susceptibility at low temperatures.

At $x = 0.053$, the M vs. H curve is nearly linear for the magnetic fields used. It appears as if the interaction among the clusters disappears at this concentration of Li. At $x = 0.053$, half the average distance between the neighbouring Li$^+$ centres is 5.7 Å and this is comparable to the extension of the wave function of Mn^{3+}. From these facts, we suggest that the two clusters, which are nearest neighbours, combine together to make a molecule. At $x = 0.053$, such molecules are isolated. We think that the impurity conduction via these isolated molecules may exhibit a large resistivity, similar to that via isolated clusters.

Finally let us discuss the nature of this molecule. A schematic picture of this molecule is presented in fig. 3. In the figure, the electronic configuration of the Mn ions near Li$^+$ is the low-spin state, $t_{2g}^5 e_g^0$. This is based on the experimental fact that the Curie constants are lower than those deduced from the assumption of zero magnetic moment for the Mn^{3+} ions. On theoretical considerations, three configurations, $t_{2g}^5-t_{2g}^3-t_{2g}^5$, $t_{2g}^4-t_{2g}^4-t_{2g}^5$ and $t_{2g}^5-t_{2g}^4-t_{2g}^4$, are likely to mix within this molecule. An abrupt change of θ_p at $x = 0.053$ can be explained by considering that there is an increase in the ferromagnetic exchange interactions, i.e., e_g^0–Se–e_g^2 superexchange and $t_{2g}^3-t_{2g}^4$, $t_{2g}^4-t_{2g}^5$ double exchange [3].

The author thanks Professor J.B. Goodenough for stimulating discussions. Thanks are also due to Dr. M.M. Bajaj for his helpful suggestions during this work.

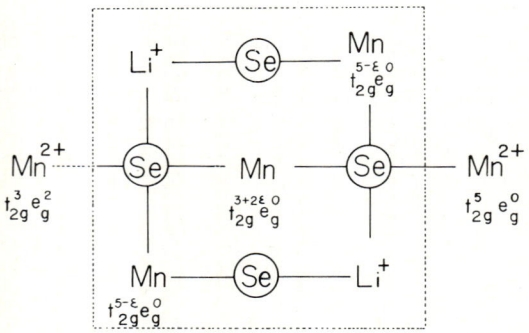

Fig. 3. Two holes trapped in t_{2g} orbitals at Li$^+$—Mn$\genfrac{}{}{0pt}{}{\text{Mn}}{\text{Mn}}$Mn—Li$^+$ molecule.

References

[1] M. Kasaya, Phys. Lett. 55A (1976) 365.
[2] R.R. Heikes, T.R. McGuire and R.J. Happel, Jr., Phys. Rev. 121 (1961) 703.
[3] J.B. Goodenough, Magnetism and the Chemical Bond (Interscience, Wiley, New York, 1963) p. 174.

INFLUENCE OF THE MAGNETIC FIELD ON THE ELECTRICAL RESISTIVITY OF U_3As_4 IN THE VICINITY OF THE CURIE POINT

Z. HENKIE and J. KLAMUT

Institute for Low Temperature and Structure Research, Polish Academy of Sciences, P.O. Box 937, 50-950 Wrocław, Poland

The influence of the magnetic field up to 150 Oe, applied along the direction [111], on the longitudinal resistivity of U_3As_4 has been investigated. Results show that the electrical resistivity depends strongly on the domain structure and at lower magnetic fields the spin critical fluctuation induces sharp maximum of the resistivity at T_c.

The electrical resistivity of magnetic materials in the vicinity of the transition temperature depends, first of all, on the magnetic critical fluctuations. Among various other factors which apparently influence the magnetic part of electrical resistivity, the topology of the Fermi surface is also considered to be of some importance. Therefore, a large variety of behaviour of the electrical resistivity in the critical region is observed. In spite of a large number of experimental and theoretical papers devoted to this problem, the influence of the spin fluctuations on the scattering of conduction electrons requires further investigation.

A peculiar dependence of the resistivity on the temperature has been found for U_3As_4. This compound reveals two resistivity maxima: the first below T_c and the second in the close vicinity of T_c. Further investigations have shown that an external magnetic field larger than 150 Oe very strongly influences the dependence $\rho(T)$ [1] in the neighbourhood of T_c, and moreover, in an anisotropic way. In the present paper we study the influence of a magnetic field smaller than 150 Oe on the longitudinal resistivity of U_3As_4 in the [111] direction.

U_3As_4 crystallizes in a b.c.c. lattice of the Th_3P_4-type structure. The dependence of the magnetization on the temperature below 198 K is characteristic for a ferromagnet [2]. This compound shows an extremely large magnetic anisotropy with the effective easy axis of magnetization along the direction [111] [3, 4]. Both the electrical and magnetic properties point to the validity of the hypothesis that this ferromagnet forms three magnetic sublattices with mutually perpendicular magnetic axis [4, 5]. This hypothesis has not been completely confirmed by neutron diffraction data.

The measurements of the resistivity were carried out with the conventional four-point d.c. method. Single crystals of the dimensions $0.3 \times 0.3 \times 4$ mm^3 with the longest dimension along the [111] axis were fastened to a copper block through an electrically insulating layer. The temperature of the block was measured with a copper-constantan thermocouple. The external field was produced by the solenoid and the earth field was compensated with Helmholtz coils.

The temperature dependence of the resistivity for various field strengths of weak magnetic fields is shown in fig. 1. It is seen that the maximum in the resistivity (below T_c) is strongly dependent on the magnetic field strength; the value of the resistivity at the maximum increases and moves towards low temperature when the field strength increases. Fig. 2 shows the dependence of the resistivity on the field strength below T_c. The resistivity increases with the increase of the field strength and tends to a saturation. It seems that the obtained dependence is related to magnetic domain processes. Two distinct maxima are

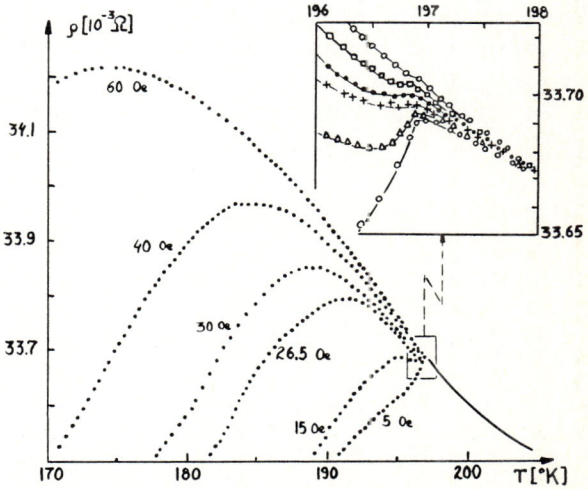

Fig. 1. Temperature dependence of the longitudinal resistivity for various field strengths for a single crystal of U_3As_4.

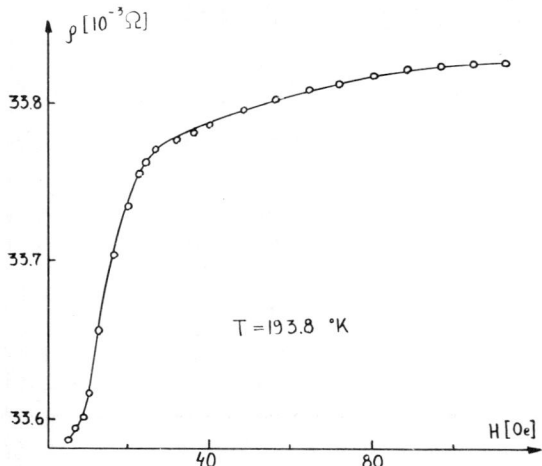

Fig. 2. Dependence of the resistivity on the field strength at the constant temperature ($T = 193.8$ K).

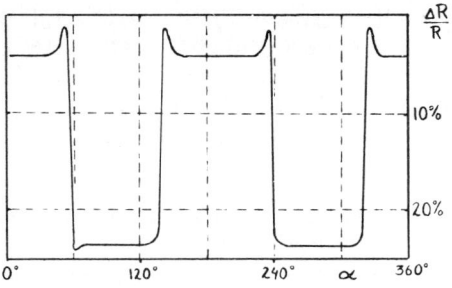

Fig. 3. Dependence of the magnetoresistance [$(\Delta R/R) = (R(0) - R(H)/R(0)) \cdot 100\%$] on the angle ($\alpha$) between the current ($i$) and the magnetic field ($\boldsymbol{H}$). $\boldsymbol{H}$ is rotating in the plane $(\bar{1}10)$; $i\|[111]$; $T = 78$ K; $H = 0.69$ T.

seen in the curves $\rho(T)$ for the fields of the strength 15, 26.5, 30 and 40 Oe. One of them occurs at the temperature corresponding to the maximum in the field-free case. The latter maximum is usually observed in ferromagnetics as a consequence of scattering of the conduction electrons by the critical magnetic fluctuations.

The important role of the domain structure in the interpretation confirms, in our opinion, the results of the measurement of the dependence of the magnetoresistance on the angle between the vectors of the current and that of magnetic field. Fig. 3 shows this dependence for the case when the vector of the current intensity is parallel to the direction [111] and that of the field is rotating in the plane $(\bar{1}10)$ at the constant temperature ($T = 78$ K) and the constant field strength (0.69 T).

The latter results indicate that the electrical resistivity of U_3As_4 is strongly anisotropic in a domain. If the magnetic moments are directed along the easy axis [111], the resistivity in this direction is about 20% greater than in the case with the magnetic moments directed in a different direction. As we mentioned, it is the non-collinear magnetic structure which is, amongst other factors, responsible for such considerable anisotropy in the domain. On the basis of results shown in figs. 2 and 3 it seems that apart from the possible causes of the anisotropy in the domain, the domain structure is of prime importance for the behaviour of electrical resistivity below T_c and at weak fields.

References

[1] Z. Henkie and J. Klamut, Phys. Status solidi (a) 20 (1973) K69.
[2] W. Trzebiatowski, A. Sępichowska and A. Zygmunt, Bull. Acad. Polon. Sci., Sér. Sci. Chim. 12 (1963) 661.
[3] C. Buhrer, J. Phys. Chem. Solids 30 (1969) 1273.
[4] K.P. Belov, Z. Henkie, A.S. Dmitrievsky, R.Z. Levitin and W. Trzebiatowski Zh. Eks. Teor. Fiz. 64 (1973) 1552.
[5] Z. Henkie and C. Bazan, Phys. Status Solidi (a) 5 (1971) 259.

MAGNETIC SUSCEPTIBILITY, ELECTRICAL RESISTIVITY AND ELASTIC CONSTANTS OF ANTIFERROMAGNETIC UN SINGLE CRYSTALS

P. DE V. DU PLESSIS and C.F. VAN DOORN*

Rand Afrikaans University, Johannesburg, South Africa

Susceptibility and electrical resistivity measurements on UN indicate $T_N \simeq 53$ K. The spin-disorder resistivity is mainly proportional to $1 - m_n^2$ (m_n is the reduced sublattice magnetization). The elastic constant C_{44} shows a renormalization proportional to m_n^2, whereas C_{11} exhibits an anomalous softening of 10% well below T_N at 47 K.

Face centered cubic uranium mononitride orders in a type I antiferromagnetic structure with the spins aligned along $\langle 100 \rangle$ axes [1] below its Néel temperature $T_N \simeq 53$ K [1, 2]. Our results on single crystal UN [3] extend previous susceptibility [2] and resistivity [4] data taken on polycrystalline and sintered samples, as well as room temperature values of single crystal elastic constants [5].

The susceptibilities (in $H = 20$ kOe) for the [100] and [110] directions are plotted in fig. 1.

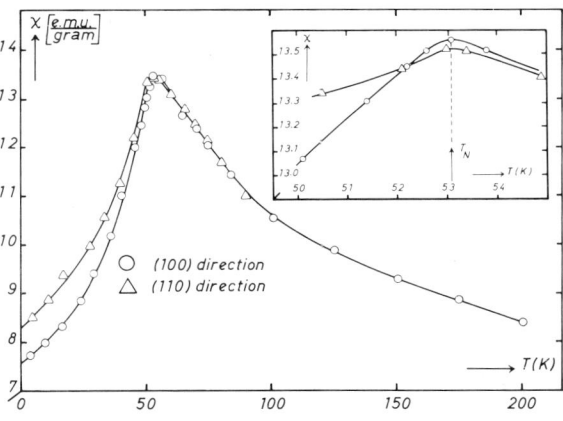

Fig. 1. Temperature dependence of the susceptibility.

The curves for both directions peak at 53.1 ± 0.2 K which is taken as T_N and agrees with previous values [1, 2]. Our paramagnetic susceptibility values between 100 and 1000 K confirm previous observations [2] that a modified Curie–Weiss law $\chi = \chi_0 + C/(T - \theta)$ is followed, but with a smaller value of $\chi_0 = 8 \times 10^6$ emu/cm^3 as compared with Troć's value of 22×10^6 emu/cm^3. Furthermore, $\theta = -247$ K and the paramagnetic moment is 2.66 μ_B per U atom.

* Now at Atomic Energy Board, Pelindaba.

Resistivity values along the [100] direction, after subtraction of the residual value $\rho_0 = 1.18 \, \mu\Omega$ cm at 4 K, are depicted in fig. 2. It is shown by a dotted line that the spin-disorder resistivity is proportional to $1 - m_n^2$ (where m_n is the reduced sublattice magnetization obtained

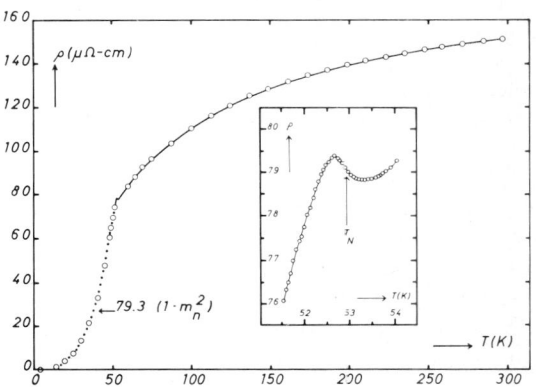

Fig. 2. Temperature dependence of the resistivity along the [100] direction.

from [1] over nearly the whole ordered region). T_N is taken as that temperature where $d\rho/dT$ attains its largest negative value [6], i.e. at 52.95 ± 0.05 K. A small but distinct peak is observed at 0.25 K below T_N and its occurrence is usually ascribed to the effect of new Brillouin zone boundaries that are introduced as a result of antiferromagnetic ordering [7]. Previous authors [4] observed either only a change in positive slope at 52 K or an indication of a peak at 59 K. The large non-linear increase in resistivity in the paramagnetic region confirms previous observations [4], and was ascribed by de Novion [8] to crystal field effects.

Values of the elastic constants C_{11} and C_{44} as determined respectively from longitudinal and transverse ultrasonic velocities along the [100] direction, are depicted in fig. 3. Renormalization of C_{44} starts upon antiferromagnetic ordering at

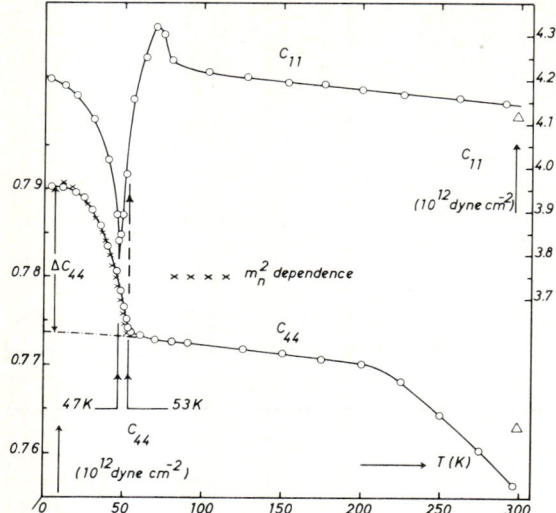

Fig. 3. Temperature dependence of the elastic constants, showing the step in C_{11} at 47 K (i.e. below T_N) and the m_n^2 dependence of ΔC_{44} (normalized to the experimental value at 35 K). Δ: Values by Guinan and Cline.

T_N, the change ΔC_{44} being proportional to m_n^2 which indicates its exchange magnetostrictive origin. In contrast the longitudinal constant C_{11} exhibits a pronounced softening of 10% at 47 K. It is known [9] that a small distortion from f.c.c. to tetragonal ($c/a = 0.99935$ at 4 K) sets in at $T_N \simeq 53$ K and measurements on our own sample yield a peak in the thermal expansion coefficient also at 53 K. Therefore neither structural changes, nor critical effects (which are also centered around T_N), should be responsible for the unexpected anomaly in C_{11} well below T_N. It is of interest that similar but smaller differences between the longitudinal velocity minimum and the transition temperature were previously reported for the antiferromagnet MnTe and the rare-earths [10].

The South African Atomic Energy Board and Council for Scientific and Industrial Research are thanked for financial support.

References

[1] N.A. Curry, Proc. Phys. Soc. 86 (1965) 1193.
[2] R. Troć, J. Solid State Chem. 13 (1975) 14 (and references cited within).
[3] Obtained from Battelle Memorial Institute, Ohio. U/N ratio 1.018 ± 0.003, oxygen 290 p.p.m. wt and carbon 20 p.p.m. wt.
[4] J.P. Moore, W. Fulkerson and D.L. McElroy, J. Amer. Ceram. Soc. 53 (1970) 76.
T. Ohmichi, T. Kikuchi and S. Nasu, J. Nucl. Sci. Tegnol. 9 (1972) 77.
[5] M. Guinan and C.F. Cline, J. Nucl. Mater. 43 (1972) 205.
[6] S. Alexander, J.S. Helman and I. Balberg, Phys. Rev. B13 (1976) 304.
[7] R.J. Elliott and F.A. Wedgwood, Proc. Phys. Soc. 81 (1963) 846.
[8] C.-H. de Novion, C.R. Acad. Sci. Paris 273 (1971) 26.
[9] J.A.C. Marples, C.F. Sampson, F.A. Wedgwood and M. Kuznietz, J. Phys. C: Solid State Phys. 8 (1975) 708.
[10] K. Walter, Solid State Commun. 5 (1967) 399.
B. Luthi, T.J. Moran and R.J. Pollina, J. Phys. Chem. Solids 31 (1970) 1741.

TEMPERATURE DEPENDENCE OF CARRIER CONCENTRATION AND MAGNETIC SUSCEPTIBILITY OF EuO DOPED WITH Eu

M. LUBECKA, J. SPAŁEK and A. WEGRZYN

Department of Solid State Physics, Academy of Mining and Metallurgy, 30–059 Cracow, Poland

The temperature dependence of the carrier concentration is calculated by taking into account the s–f exchange interaction. The results are used to obtain an expression for the static paramagnetic susceptibility in the paramagnetic regime.

EuO doped with Eu exhibits the metal-insulator transition below T_c accompanied by a strong temperature dependence of the carrier concentration in the conduction band [1]. For a deeper insight into this problem one needs to take into consideration the nature of the dope as a double donor, Eu^{2+}. We assume that the second electron is not thermally excited from the donor and therefore it is trapped and plays no role in the conduction process. But even being localized it can magnetically polarize its neighborhood as well as the other electron with the smaller binding energy. In fact, they form a triplet in the localized state [2]. The aim of this paper is to relate the carrier concentration with the static susceptibility in the paramagnetic region for the triplet state. We will summarize here our final results only.

(1) The carrier concentration in the conduction band with the inclusion of the s–f type exchange interaction in the molecular field approximation is

$$n_c = N_c \sum_{k=1}^{\infty} \frac{(-1)^{k+1}}{k^{3/2}} \exp[k\beta(\mu - \epsilon_c^0)] \cosh(kH_e), \quad (1)$$

where the chemical potential μ for the parabolic band is determined from the neutrality condition

$$n_c = \frac{N_t}{1 + 2\cosh H_\gamma \cdot \exp[-\beta(\epsilon_d^0 - \mu)]} + N_{ex}. \quad (2)$$

In these formulae:

$N_c = 2(2\pi m^* k_B T/h)^{3/2}, \quad H_e = \beta(J_c\langle S^z \rangle + \mu_B H),$

and

$H_\gamma = \beta(\gamma J_c \langle S^z \rangle + \mu_B H).$

N_t is the concentration of double donors Eu^{2+}, N_{ex} is the concentration of "extra" electrons (includes influence of acceptors and localized level at the bottom of conduction band). The conduction band edge and the donor level are located at ϵ_c^0 and ϵ_d^0, respectively; m^* is the effective mass of electrons in the conduction band; J_c is the s–f exchange constant; $\gamma = J_d/J_c$ is the relative exchange constant of interaction between localized and 4f electrons; and $\beta = (kT)^{-1}$. For the case of nondegenerated conduction we have

$$n_c = (\tfrac{1}{2}N_c N_t)^{1/2} e^{\beta/2(\epsilon_d^0 - \epsilon_c^0)} \left(\frac{\cosh H_e}{\cosh H_\gamma}\right)^{1/2}$$
$$\times \left[B + \sqrt{B^2 + \frac{N_{ex} + N_t}{N_t}} \right], \quad (3)$$

where

$$B = \frac{N_{ex}}{\sqrt{2N_c N_t}} e^{-(\beta/2)(\epsilon_d^0 - \epsilon_c^0)} \left(\frac{\cosh H_\gamma}{\cosh H_e}\right)^{1/2}$$
$$- \frac{1}{2}\sqrt{\frac{N_c}{2N_t}} \left(\frac{\cosh H_e}{\cosh H_\gamma}\right)^{1/2} e^{(\beta/2)(\epsilon_d^0 - \epsilon_c^0)}. \quad (4)$$

For $N_{ex} = 0$ and $N_t \gg N_c$ one obtains the result of Haas [3]. In fig. 1 is plotted the dependence of $n_c(T)$ for various values of N_t.

(2) To obtain information about a magnetic state of the trapped pair of electrons localized near an oxygen vacancy we have calculated the static magnetic susceptibility in the paramagnetic region for the triplet state. The magnetic state of our system is composed of the Pauli paramagnetism of electrons in the conduction 5d–6s band, the Langevin paramagnetism of trapped electrons and paramagnetism of 4f electrons which are coupled with the former ones by the interaction of s–f type, as previously. The calculations were performed in the framework of a self-consistent molecular field approximation. The susceptibility χ_f of 4f elec-

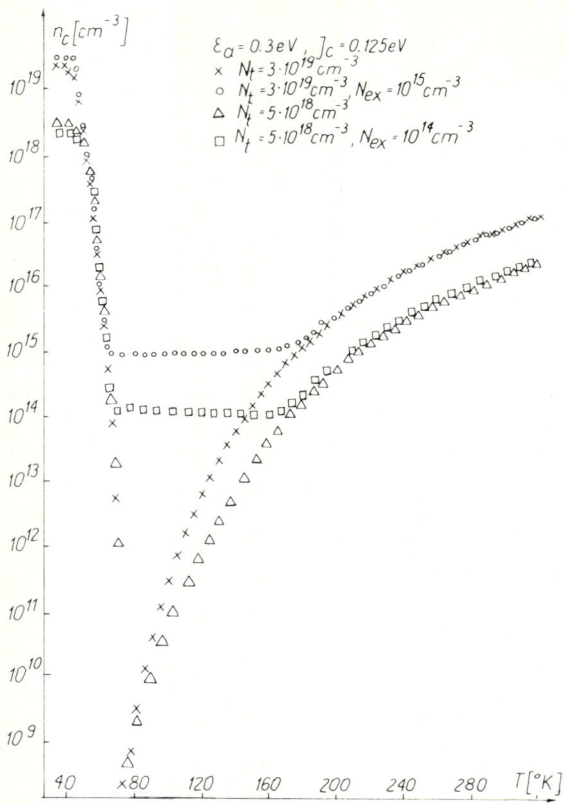

Fig. 1. The temperature dependence of carrier concentration $n_c(T)$ for single donor level.

trons is

$$\chi_f = C_M \left[1 + J_c \tilde{\chi}_P \cdot \frac{n_c}{N_0} \middle/ (g\mu_B)^2 \right.$$
$$+ \gamma \left(\frac{2N_t - n_c}{N_0} \right) \frac{J_c}{2k_B T} \right]$$
$$\times \left\{ T - \theta \left[1 + \frac{J'C_t}{J} + \gamma^2 \left(\frac{2N_t - n_c}{N_0} \right) \cdot \frac{J_c^2}{2Jzk_B T} \right. \right.$$
$$\left. \left. + \frac{J_c^2 \tilde{\chi}_P \cdot n_c}{N_0 (g\mu_B)^2 Jz} \right] \right\}^{-1}. \tag{5}$$

The total susceptibility χ_{tot} is

$$\chi_{tot} = \chi_f \left[1 + \tilde{\chi}_P \frac{J_c}{(g\mu_B)^2} \cdot \frac{n_c}{N_0} + \gamma \frac{2N_t - n_c}{N_0} \cdot \frac{J_c}{2k_B T} \right]$$
$$+ \tilde{\chi}_P n_c + \frac{(g\mu_B)^2}{k_B T} (2N_t - n_c), \tag{6}$$

where $C_M = S(S+1)(g\mu_B)^2 N_0/3k_B$, $\theta = JzS(S+1)/3k_B) \cdot \tilde{\chi}_P \cdot n_c$ is the Pauli susceptibility $\chi_P = n_c \mu_B^2/k_B T_F$, and J' is the exchange constant for the interaction of the electron in the trap level with the 4f spin of nearest neighbours. To visualize our results we have plotted in fig. 2 the

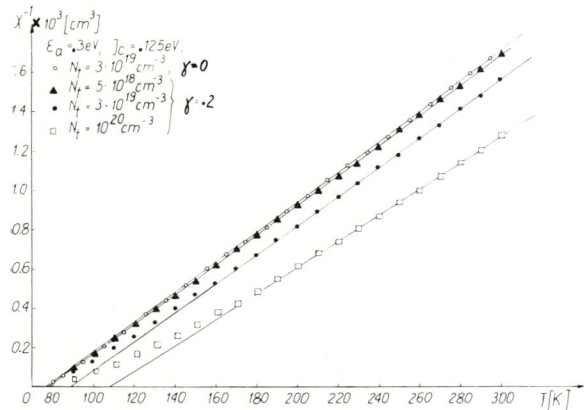

Fig. 2. The temperature dependence of inverse total static susceptibility $\chi_{tot}^{-1}(T)$.

$\chi_{tot}^{-1}(T)$ for few concentrations of traps N_t in the range of metal-insulator transition. These curves are quite similar to the experimental ones obtained in [2].

References

[1] M.R. Oliver et al., Phys. Rev. B5 (1972) 1078.
[2] J.B. Torrance, M.W. Shafer and T.R. McGuire, Phys. Rev. Lett. 29 (1972) 1168.
[3] C. Haas, in: New Developments in Semiconductors, P.R. Wallace et al., eds. (Noordhoff, Leiden, 1972).
[4] P. Leroux-Hugon, Phys. Rev. Lett. 29 (1972) 939.

SPECIFIC HEAT AND ELECTRICAL RESISTIVITY OF 3d TRANSITION METAL DISULFIDES AND DISELENIDES WITH PYRITE STRUCTURE

S. OGAWA

Electrotechnical Laboratory, Tanashi, Tokyo 188, Japan

As characteristics of a d-electron system in a very narrow band, a smaller magnetic entropy than expected for the localized moment and a large T^2 contribution to the low temperature resistivity have been found in $M_{1-x}M'_xS_2$ (M, M' = Fe, Co, and Ni) and in $NiS_{2-x}Se_x$.

The transition-metal disulfides and diselenides with the Pyrite structure appear to provide a good example of d electrons in very narrow bands [1, 2]. The metal-semiconductor transition of $NiS_{2-x}Se_x$ around $x \simeq 0.5$ is considered to be due to the effect of strong electron correlation [4, 5]. The specific heat of FeS_2–CoS_2–NiS_2 solid solutions and the low temperature resistivity of $NiS_{2-x}Se_x$ have been measured to know the thermal and electrical properties of such system.

The specific heat has a contribution of magnetic origin, as seen in fig. 1. Assuming a Debye specific heat function for the lattice contribution and a γT term for the electronic contribution, the magnetic contribution at low temperature has been deduced, and the magnetic entropy S_M has been computed. Fig. 2 shows the S_M as a function of average number of 3d e_g electrons n. From the Curie constant of the paramagnetic susceptibility of CoS_2–NiS_2, the paramagnetic moments of Co and Ni atoms are 1 and 2 μ_B, respectively, although the antiferromagnetic moment of Ni in NiS_2 is 1.2 μ_B [3]. The observed S_M values are smaller than those expected for mixtures of Co^{2+} and Ni^{2+} with localized moments of 1 and 2 μ_B.

$NiS_{2-x}Se_x$ is semiconducting for $x < 0.45$ and is metallic for $x > 0.45$ [4, 5]. The electrical resistivity at low temperature of metallic

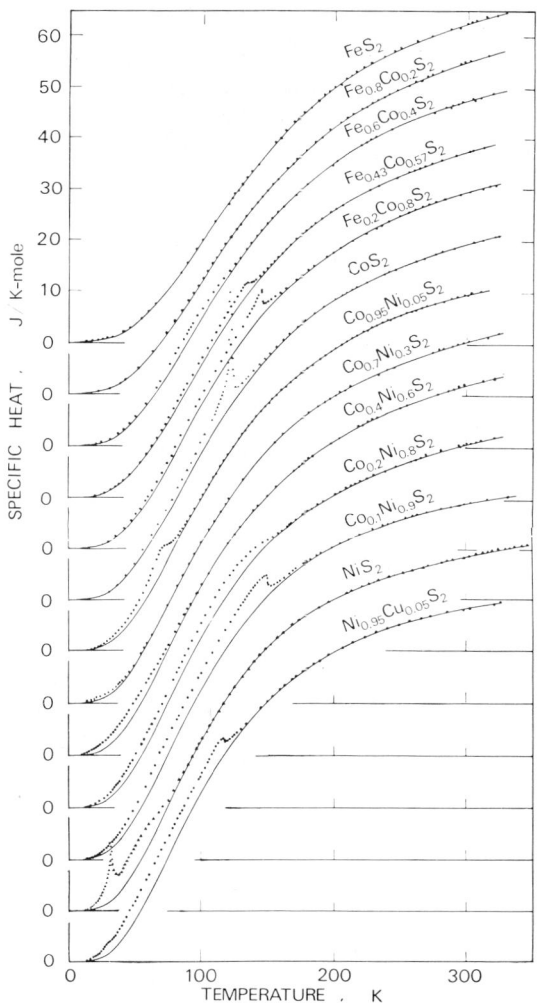

Fig. 1. Specific heats of FeS_2–CoS_2–NiS_2–$Cu_{0.05}Ni_{0.95}S_2$ solid solutions. Solid curves are those computed for lattice and electronic contributions.

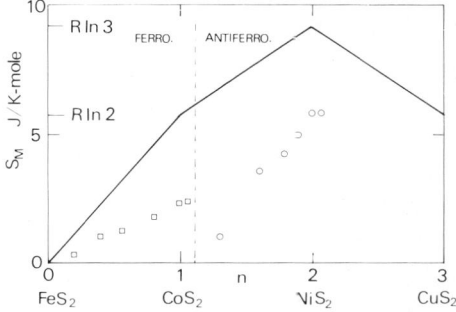

Fig. 2. Magnetic entropies, S_M, of ferromagnetic (open square) and antiferromagnetic (open circle) disulfides. The curve is the magnetic entropy calculated for the mixtures of Fe^{2+}, Co^{2+}, Ni^{2+} and Cu^{2+} ions in low spin states.

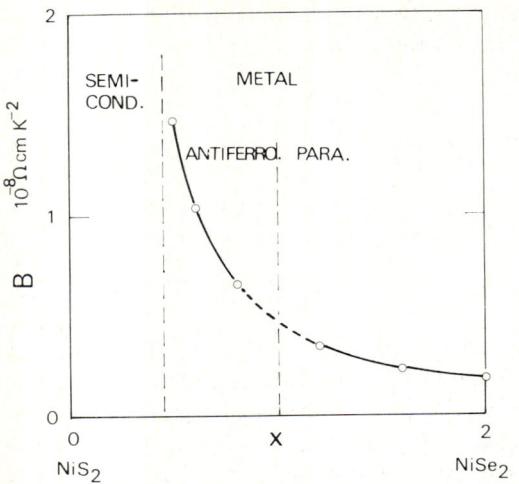

Fig. 3. The coefficient B of T^2 contribution to the electrical resistivity at low temperature, $R = R_0 + BT^2$, of metallic $NiS_{2-x}Se_x$.

$NiS_{2-x}Se_x$ was measured and found to have a large T^2 contribution except where the concentration x is close to the antiferromagnetic to paramagnetic transition ($x \approx 1$). The coefficient of the T^2 term, B, is plotted in fig. 3. The value of B is about 10^3 times as large compared with that of nickel and iron metals. B increases as the concentration x approaches the boundary of the metal-semiconductor transition.

The reduced magnetic moment and the large T^2 contribution in resistivity [6] seem to be the characteristics of d electrons in very narrow bands.

References

[1] H.S. Jarrett, W.H. Cloud, R.J. Bouchard, S.R. Butler, C.G. Frederick and J.L. Gillson, Phys. Rev. Lett. 21 (1968) 617.
[2] S. Ogawa, S. Waki and T. Teranishi, Int. J. Magn. 5 (1974) 349.
[3] J.M. Hasting and L.M. Corliss, IBM J. Res. Develop. 14 (1970) 227.
[4] H.S. Jarrett, R.J. Bouchard, J.L. Gillson, G.A. Jones, S.M. Marcus and J.F. Weiher, Mater. Res. Bull. 8 (1973) 877.
[5] G. Czjzek, J. Fink, H. Schmidt, G. Krill, M.F. Lapierre, P. Panissod, F. Gautier and C. Robert, J. Magn. Magn. Mater. 3 (1976) 58.
[6] A. Kawabata, Progr. Theor. Phys. 54 (1975) 45.

THE INFLUENCE OF COBALT IONS ON THE ANISOTROPY AND MAGNETOSTRICTION OF CdCr$_2$Se$_4$ AND HgCr$_2$Se$_4$ SINGLE CRYSTALS

T. JAGIELIŃSKI and H. SZYMCZAK,

Institute of Physics, Polish Academy of Sciences, Warsaw, Poland

By means of ferromagnetic resonance method the anisotropy and magnetostriction constants of CdCr$_2$Se$_4$ and HgCr$_2$Se$_4$ doped with cobalt ions were measured. The measurements were carried out in the temperature range from 4.2 K up to the Curie point. The obtained results are discussed within the framework of single-ion theory of anisotropy and magnetostriction.

The anisotropy and magnetostriction caused by impurity ions have been intensively investigated in recent years. Most attention has been paid to the magnetic oxides (spinels, garnets). It has been shown that the addition of the cobalt (Co^{2+}) ions to spinels or garnets increases their anisotropy and magnetostriction by several orders of magnitude. Lastly, a great interest was concerned with a new class of magnetic spinels MCr$_2$X$_4$ (where M = Cd, Hg, and X = S, Se) having interesting electrical transport properties. The purpose of the present investigations was to study the influence of Co^{2+} ions on the anisotropy and magnetostriction of CdCr$_2$Se$_4$ and HgCr$_2$Se$_4$ single crystals.

The details of crystal growth of Cd$_{1-x}$Co$_x$Cr$_2$Se$_4$ have been described elsewhere [1]. In the present work single crystals of Cd$_{1-x}$Co$_x$Cr$_2$Se$_4$ ($x = 0$, 0.01, 0.03, 0.05) and Hg$_{0.99}$Co$_{0.01}$Cr$_2$Se$_4$ were investigated. The investigations have been carried out by the ferromagnetic resonance method (FMR). The samples were polished spheres about 0.5 mm in diameter. FMR was measured at X band in the temperature range 4.2 K–T$_c$. Magnetostriction constants λ_{100} and λ_{111} were calculated using the value of the resonance field shift under the influence of stress load applied along the [110] axis. Additionally the rectangular pulse modulation of the load was applied [2]. This modulation enabled the FMR curves to be recorded simultaneously, with and without load. Similar measurements were carried out also for Hg$_{0.99}$Co$_{0.01}$Cr$_2$Se$_4$ single crystal to check whether the effects observed for cadmium spinels appear in another semiconductor spinel.

On the basis of the numerical analysis of the resonance field angular dependence it was proved, that for all investigated samples the magnetocrystalline anisotropy has to be described by three anisotropy constants. At 4.2 K the following values of magnetocrystalline anisotropy and magnetostriction constants of Co^{2+} ions in CdCr$_2$Se$_4$ were obtained: $K_1 = 0.2$ cm^{-1}/ion, $K_2 = -2.2$ cm^{-1}/ion, $K_3 = 0.8$ cm^{-1}/ion, $\lambda_{100} = -1.8 \times 10^{-24}$/ion, $\lambda_{111} = 0.8 \times 10^{-24}$/ion. Magnetocrystalline anisotropy constants as well as the magnetostriction constants were interpreted in terms of a single-ion model. This model was verified by measurements of the anisotropy and magnetostriction as a function of x. The dependences of anisotropy as well as magnetostriction on x have a linear character. The additional verification of the single-ion model for the examined crystals has been done by measurements of the temperature dependence of anisotropy and magnetostriction coefficients.

On the basis of a single-ion model the magnetocrystalline anisotropy and magnetostriction for Co^{2+} ions in tetrahedral sites of the spinel lattice were calculated. In these sites the ground state of Co^{2+} ions is $^4A_2$; thus, the generalized spin hamiltonian has the following form:

$$\mathcal{H}_s = g\beta(\mathbf{HS}) + \mu\beta\{H_xS_x^3 + H_yS_y^3 + H_zS_z^3\} - \tfrac{1}{5}(\mathbf{HS})[3S(S+1) - 1]\}, \qquad (1)$$

where $S = \tfrac{3}{2}$, $\mathbf{H}$ is the exchange field and β the Bohr magneton. The anisotropy coefficients (per spin) at 0 K can be evaluated from eq. (1) as follows:

$$K_1 = \beta H(1.59\, u - 0.88\, u^2 g^{-1}),$$
$$K_2 = 0, \qquad K_3 = 3.5\, \beta H u^2 g^{-1}. \qquad (2)$$

From the ESR data results that for Co^{2+} ions in tetrahedral sites one can assume $u = -0.01$ cm^{-1}. Presuming $H = 100$ cm^{-1} a significant contribution was obtained only for the K_1 constant ($K_1 = -1.6$ cm^{-1}). Therefore, in order to describe correctly the experimental results and especially the large values of K_2 and K_3 in comparison with K_1, it was assumed that some of the Co^{2+} ions occupy the octahedral sites. The analysis of the anisotropy constants in this case has been carried out by diagonalization of the complete hamiltonian (including crystal field, electrostatic, spin–orbit and molecular field interactions) within the complete d^7 configuration. The numerical analysis revealed that the anisotropy constants depend mainly on the Δ_t splitting of the ground state $^4T_1$:

$$\Delta_t = -0.9\,v - 1.7\,v', \tag{3}$$

where v, v' are trigonal crystal field parameters. For that reason the final fitting of the experimental data was carried out taking into account the ^4F state only. The best fitting was obtained for the following parameters: $Dq = 900$ cm^{-1}, $\Delta_t = -200$ cm^{-1}, $\xi = 400$ cm^{-1}, $H = 150$ cm^{-1}, where ξ is the spin–orbit constant. For these parameters the anisotropy constants have the following values: $K_1 = 7.6$ cm^{-1}/ion, $K_2 = -17.7$ cm^{-1}/ion, $K_3 = 3.5$ cm^{-1}/ion. The best agreement of the proposed model with the experimental results is obtained when the amount of cobalt ions occupying the octahedral sites is equal to 15–20% of total amount of cobalt.

In order to calculate the magnetostriction constants it was necessary to determine the hamiltonian of orbit–lattice interactions. For that purpose the point charge model was applied, taking into account only the nearest neighbours and treating the oscillations of XY$_4$ and XY$_6$ complexes as the oscillations of ideal tetrahedrons and octahedrons, respectively. In this approximation the hamiltonian of orbit–lattice interactions has the form

$$V_{o-1} = \sum_{\Gamma,\gamma} C_\gamma(\Gamma) V_\gamma(\Gamma) e_\gamma(\Gamma), \tag{4}$$

where $e_\gamma(\Gamma)$ are the strain-tensor components, having the transformation properties of the γ subvector of Γ cubic-group representation ($\Gamma = E, T_2$); $V_\gamma(\Gamma)$ is a function of electron coordinates, $C_\gamma(\Gamma)$ is a constant which can be evaluated by applying a point charge model. $C_\gamma(\Gamma)$ is expressed by the effective charge on the neighbouring ligands. The values of the effective charge can be calculated from the relations describing the Dq parameter.

For the tetrahedral sites $Dq = -370$ cm^{-1} (calculated on the basis of the optical spectrum of Cd$_{1-x}$Co$_x$Cr$_2$S$_4$ [3]. Finally, for the tetrahedral positions $\lambda_{100} = -1.5 \times 10^{-24}$/ion, $\lambda_{111} = -9.4 \times 10^{-24}$/ion, and for the octahedral positions $\lambda_{100} = -2.7 \times 10^{-24}$/ion, $\lambda_{111} = -3.7 \times 10^{-24}$/ion.

It was derived from the above calculations that the presented model, together with the assumption that 15–20% of Co^{2+} ions occupy octahedral positions, leads to a theoretical value of λ_{100} close to the value which results from experiments.

We did not read such an agreement for the λ_{111} constant. The reason for this discrepancy remains unclear.

References

[1] T. Jagieliński and M. Berkowski, Phys. Status Solidi (a) 27 (1975) K17.
[2] T. Jagieliński, Acta Phys. Polonica A49 (1976) 799.
[3] T.J. Coburn, D. Pearlman, E. Carnall, Jr., F. Moser, T.H. Lee, S.O. Lyu and T.W. Martin, AIP Conf. Proc. 10 (1972) 740.

PRESSURE EFFECT ON METAL–NONMETAL TRANSITION IN NiS

S. ANZAI

Faculty of Engineering, Keio University, Hiyoshi, Yokohama-shi, 223, Japan

and

K. OZAWA

Japan Atomic Energy Research Institute, Takai-mura, 319-11, Japan

Lidiard's model for itinerant antiferromagnet is extended to involve the strain energy and the volume dependence of the gap between both spin bands. The model is simulated to fit the pressure diagram and its first-order properties. The collapse of moment at T_t is explained with the thermal excitation of electrons to the other spin band.

NiAs-type NiS undergoes a first-order antiferromagnetic-to-paramagnetic transition [1] with the collapse of static-local moment [2], a sharp change in resistivity [3, 4] and cell volume [1] at a transition temperature T_t. White and Mott [5] have proposed that the transition is a semimetal-to-metal transition. The antiferromagnetic and semimetallic phase is suppressed at pressures above 20 kbar [6]. In this paper, we investigate the compressibility of NiS, and simulate an improved Lidiard model.

Changes in a- and c-parameters were measured with strain gauges pasted up on a single crystal ($T_t = 234$ K). As shown in fig. 1(a), the pressure changes in the parameters are monotonically enhanced with increasing in hydrostatic pressure at 298 K.

The Ni-sublattice in the antiferromagnetic NiS [1], in which the alternative c-plane are coupled antiparallel, is divided into two interlocking sublattices (A and B) such that they have up and down-spins as their majority spins, respectively. The A and B orbitals are mainly concentrated on the A and B sublattices, respectively. N' is the total number of d-electrons concerned with the transition. Lidiard [7] has defined a relative magnetization ζ_A (or ζ_B) such that $N'\zeta_A/2$ (or $N'\zeta_B/2$) is the excess number of A (or B) electrons having up- (or down-) spins. The definitions for θ_1, θ_{12}, η, β and F_k are the same as those of Lidiard. The abbreviated form: $k\theta'\zeta = (k\theta_1\zeta_A + k\theta_{12}\zeta_B) = (k\theta_2\zeta_B + k\theta_{12}\zeta_A)$ is used at nonexternal field. The Lidiard gap $2k\theta'\zeta$ is the difference in energy for the spin parallel and antiparallel to the majority spin in the sublattice. N is the number of electrons per cm^3 which are concerning to the transition. The electronic energy E_{el} per cm^3 and the electronic entropy S_{el} per cm^3 for these electrons are $(E_{el}/N\epsilon_0) = (kT/\epsilon_0)^3[F_2(\eta + \beta) + F_2(\eta - \beta)] - (k\theta'/\epsilon_0)(\zeta^2/2)$ and $(S_{el}/N\epsilon_0) = (kT/\epsilon_0)^3\{-(\eta + \beta)F_1(\eta + \beta) - (\eta - \beta)F_1(\eta - \beta) - (3/2)[F_2(\eta + \beta) + F_2(\eta - \beta)]\}/T$. Here, a linear form of band scheme is assumed, because the observed scheme [8] seems to be a linear one. We assume the volume-dependent relation: Ξ (in ϵ_0 unit) $\equiv 2(k\theta'/\epsilon_0)\zeta = 2(k\theta'/\epsilon_0)_0\zeta \times [1 + \xi\Delta V]$, where $\xi = [1/(k\theta'/\epsilon_0)_0][\partial(k\theta'/\epsilon_0)/\partial\Delta V]$. Here, ΔV is $(V - V_0)/V_0$ where V is the equilibrium volume of the unit cell and ϵ_0 would be the volume in absence of Ξ at $T = 0$ for $P = 0$. Interactions among inner cores and between the cores and itinerant electrons are taken into account as the strain energy $E_{st} = (1/\kappa)[(\Delta V)^2/2]$ [9]. The lattice entropy is taken to be $S_l = -(\alpha T\Delta V/\kappa) + C_v \ln(T)$ [9], where κ, α and C_v are the volume compressibility, the volume thermal-expansion coefficient and the specific

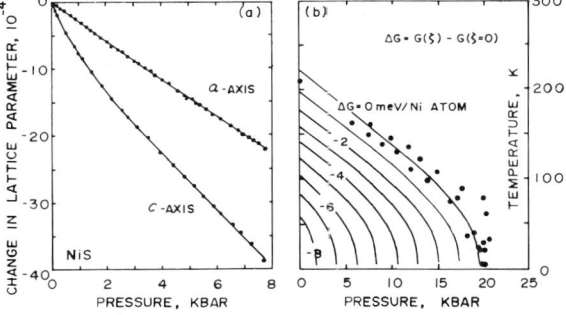

Fig. 1. (a) Isothermal curves for changes in lattice parameters ($[a(P) - a(0)]/a(0)$ and $[c(P) - c(0)]/c(0)$) at 298 K. (b) T_t–P diagram for NiS. Experimental points are taken from fig. 3 in ref. 6. The transition line is defined as the line where $\Delta G = 0$.

heat at constant volume when the virtual local moment disappears. The Gibbs free energy per cm^3 at a value of ζ is given by $G(\zeta) = E_{el}(\zeta) - T[S_{el}(\zeta) + (\alpha/\kappa)\Delta V(\zeta)] + P\Delta V(\zeta) + (1/2\kappa)\Delta V(\zeta)^2$. Here, the second term in S_l is omitted because it is assumed to be independent of ζ. ζ varies from 0 to 1 as $(k\theta'/\epsilon_0)$ changes from 0.5 to $1/\sqrt{2}$. $\partial G/\partial \Delta V$ gives the equilibrium volume change: $\Delta V(\zeta) = -P\kappa + \alpha T + (\frac{1}{2}) \times (k\theta'/\epsilon_0)_0 N\kappa\epsilon_0 \xi \zeta^2$, where $\partial\eta/\partial\Delta V = 0$. κ is deduced to be 1.1×10^{-3} kbar^{-1} from fig. 1(a). By fitting eq. (1) in ref. 10 with the values of the a and c-parameters at temperatures between 270 and 400 K [1], one obtains $\alpha = 1.5 \times 10^{-5}$ K^{-1} as the value averaged in temperatures between 0 and 250 K. Here (C_v/R) in eq. (1) is estimated from the Debye temperature 157 K [11], and the absolute moment 1.7 μ_B [13] (corresponding to $N\epsilon_0 = 3.24$ electron · kbar) is taken into account. By using the boundary conditions: (i) T_t–P diagram [6] and (ii) sudden change in volume (1.6%) at T_t for $P = 0$ [1], one obtains the most fittable parameters: $\epsilon_0/k = 380$ K, $\xi = 22$ and $(k\theta'/\epsilon_0)_0 = 0.675$. Here, the Rhodes' [12] table and expressions are used. This set of parameters gives the transition line [free energy contour in fig. 1(b)] and the change in volume (1.65%).

The calculated magnetization slowly decreases with increasing temperature and suddenly diminishes from 1.35 μ_B just below T_t to 0 μ_B above T_t. This critical value is in good agreement with that of experimental value (1.50 ± 0.10) μ_B [13]. The heat of transition at T_t is calculated to be 188 cal mol^{-1} at $T_t = 220$ K for 0 kbar, which is in fair agreement with the experimental values: 194 cal mol^{-1} for the sample with $T_t = 228$ K [14] and 160 cal mol^{-1} for $T_t = 210$ K [15]. It is plausible that the present model describes the transition in NiS. We will discuss the picture of this transition.

The electrons would move freely throughout the crystal because no gap exists between the adjacent basal planes above T_t. When the sample is cooled, calculated Ξ jumps to 50 meV per electron at T_t for $P = 0$ and has a maximum value 70 meV around 100 K. The gap formation has been observed by Barker and Remeika [16] who have found a dip of reflectivity at 140 meV for 80 K only in the antiferromagnetic phase. $\Xi/2\zeta$ is calculated to have a value larger than $1/\sqrt{2}$ in all the region below T_t on the T–P plane. Then, there is an energy interval between the Fermi level and the bottom of the other spin band at 0 K. When an electron with up-spin transfers from an atomic layer on A sublattice to that on B, it requires a finite value of excitation energy to reverse its spin direction. The electron in its ground state can move easily in the same basal plane even below T_t. NiS below T_t would have the characteristic of a semimetal as a three-dimensional lattice. This feature appears on the Lidiard model with larger value of $(k\theta'/\epsilon_0)$, but not on the Stoner's model because the latter consists of only one magnetic sublattice.

One of the authors (S.A.) thanks Takeda Science Foundation for a grant.

References

[1] J. T. Sparks and T. Komoto, J. Appl. Phys. 34 (1963) 1191.
[2] J. T. Sparks and T. Komoto, J. de Physique 25 (1964) 567.
[3] J. T. Sparks and T. Komoto, Phys. Lett. A25 (1967) 398.
[4] S. Anzai and K. Ozawa, J. Phys. Soc. Jap. 24 (1968) 271.
[5] R. M. White and N. F. Mott, Phil. Mag. 24 (1971) 845.
[6] D. B. McWhan, M. Marezio, J. P. Remeika and P. D. Dernier, Phys. Rev. B5 (1972) 2552.
[7] A. B. Lidiard, Proc. Roy. Soc. A224 (1954) 161.
[8] S. Hüffner and G. K. Wertheim, Phys. Lett. A44 (1973) 133.
[9] C. P. Bean and D. S. Rodbell, Phys. Rev. 126 (1962) 104.
[10] S. Anzai and Y. Hamaguchi, J. Phys. Soc. Jap. 38 (1975) 400.
[11] T. Ohtani, K. Adachi, K. Kosuge and S. Kachi, J. Phys. Soc. Jap. 36 (1974) 1489.
[12] P. Rhodes, Proc. Roy. Soc. A204 (1950) 396.
[13] J. T. Sparks and T. Komoto, Rev. Mod. Phys. 40 (1968) 752.
[14] R. F. Koehler, Jr. and R. L. White, J. Appl. Phys. 44 (1973) 1682.
[15] J. M. D. Coey and R. Bursetti, Phys. Rev. B11 (1975) 671.
[16] A. S. Barker, Jr. and J. P. Remeika, Phys. Rev. B10 (1974) 987.

LOCAL MOMENT FORMATION IN THE VANADIUM SULFIDES $V_{1+\delta}S_2 (\frac{1}{3} \leq \delta \leq \frac{1}{4})$

B.G. SILBERNAGEL and A.H. THOMPSON

Corporate Research Laboratories, Exxon Research and Engineering Co., Linden, N.J. 07036, USA

The process of moment localization in vanadium sulfides is examined by X-ray, magnetic susceptibility and NMR techniques. The electronic character and environment of the moments are independent of composition. Delocalization of the vanadium electrons occurs in the presence of neighboring vanadium atoms and models of the structural order are discussed.

Vanadium sulfides form a class of compounds varying almost continuously in composition from VS to V_5S_8 [1]. Alternating vanadium and sulfur layers form the basis of the structure, with every second vanadium layer being completely occupied, while the remainder are depleted. DeVries and Haas [2] successfully accounted for the magnetic properties of these materials by assuming that isolated vanadium atoms in these depleted layers bore localized moments. For V_5S_8, one-quarter of the sites in the depleted layers are occupied ($V_{1+1/4}S_2$) and no vanadium atom has a vanadium near neighbor in the layer; a maximum magnetic response is observed. For V_3S_4, one-half of the sites are occupied ($V_{1+1/2}S_2$), the atoms lie in chains, and no paramagnetic species remain. To understand the process by which the addition of vanadium atoms destroys the magnetism, we have synthesized a series of $V_{1+\delta}S_2$ compounds in the composition range from $\delta = \frac{1}{4}$ (V_5S_8) to $\delta = \frac{1}{2}$ (V_2S_3). The resulting products have been examined by X-ray, magnetic susceptibility, and NMR techniques. NMR is particularly useful since previous observations of V_5S_8 have shown that ^{51}V nuclei *residing in atoms bearing localized moments* can be observed [3]. The shifts, widths and intensities of the NMR spectra for these magnetic species probe the electronic character of the moment, its environment, and the mechanism for moment destruction.

Appropriate amounts of vanadium and sulfur were slowly heated to 850°C. The resulting products were then slowly cooled (one month) to 400°C where they were annealed. A Foner magnetometer was used for temperature-dependent susceptibility studies while NMR observations were made on a Varian WL-112 wideline spectrometer. X-ray, susceptibility, and NMR studies suggest a continuous variation of properties throughout the composition range. Evidence for antiferromagnetic order is seen in all samples for $\frac{1}{4} \leq \delta \leq 0.316$. The NMR signal associated with the localized moments shows a room temperature shift of -3.2%, which is independent of composition. The width of the resonance line, 95 ± 7 G at 13.4 K, is also independent of composition. Field dependence studies indicate that the intrinsic width is ~ 80 G.

The intensity of this resonance signal has been measured by comparing the peak heights of equimolar amounts of the sample with an ^{27}Al metallic standard signal. The resulting intensity data is shown in fig. 1(a), where the normalized intensity is plotted as a function of δ. The error flags shown are the result of standard deviations of repeated measurements and indicate accuracies of ± 5–10%. An initial linear decrease in total moment is observed for $0.25 \leq \delta \leq 0.29$. At $\delta = 0.299$ there is a distinct change in magnitude

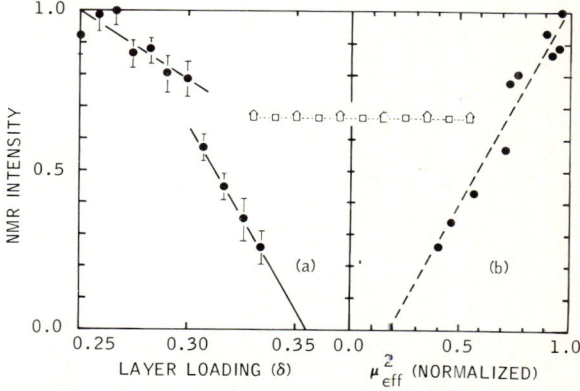

Fig. 1. Normalized intensity of the varadium NMR signal associated with localized moments plotted as a function of (a) atomic loading in depleted vanadium layers (δ), and (b) measured susceptibility. The vanadium signal intensity falls as additional atoms are added, showing a sharp break at $\delta \sim 0.30$. The inset shows the schematic arrangement of vanadium atoms on chains for V_5S_8.

and slope. The moment then delocalizes much more rapidly with the addition of more vanadium. A second gauge of the number of moments is extracted from the slope of the $1/\chi$ vs. T curve of the susceptibility observation. A plot of NMR signal intensities vs. the normalized values of μ_{eff}^2 from the χ data [fig. 1(b)] shows good correlation.

These data provide the following picture of the system. As the samples become more metal-rich, the NMR shifts and widths suggest little change in the environment or electronic character of the remaining localized moments. The loss of magnetic species is clearly reflected in the data of fig. 1, and the details of delocalization can be inferred from structural considerations. For V_3S_4 ($\delta = \frac{1}{2}$), vanadium atoms lie on chains, while for V_5S_8 ($\delta = \frac{1}{4}$), occupied sites and vacancies alternate on the chains, as indicated schematically in fig. 1. For $\delta > \frac{1}{4}$, some of the vacant sites become occupied by vanadium atoms. The slope of the intensity vs. δ curve for $0.25 \leq \delta \leq 0.30$ shows that one localized vanadium moment is lost for each excess vanadium atom added. If a vanadium nearest neighbor is sufficient to delocalize a moment, random atom placement should remove two moments for each excess atom. The present data can be understood if the atoms arrange themselves in *strings* along the chain sites, since the ratio of excess atoms to moments will be one for a string. Thus for $\delta < 0.3$, these magnetic data suggest the coexistence of strings of vanadium atoms and isolated sites. The dramatic change in moment variation with δ for $\delta > 0.30$ might well reflect the onset of an order–disorder transition among strings and residual moments on the chain sites. However, this is also near the terminus of the V_2S_3 phase region, variously reported to be $\delta \sim 0.32$ [1] and $\delta \sim 3.07$ [4], and may also reflect a structural change.

We gratefully acknowledge the assistance of C.R. Symon in sample preparation, L.A. Gebhard in collection and analysis of NMR data, and J.R. Schrieffer for helpful discussions.

References

[1] For a recent review, see A.B. de Vries and F. Jellinek, Rev. Chim. Miner. 11 (1974) 624.
[2] A.B. de Vries and C. Haas, J. Phys. Chem. Solids 34 (1973) 651.
[3] B.G. Silbernagel, R.B. Levy and F.R. Gamble, Phys. Rev. B11 (1975) 4563.
[4] R. Grønvold, H. Haraldsen, B. Pedersen and T. Tufte, Rev. Chim. Miner. 6 (1969) 215.

MAGNETIC PROPERTIES OF $(Cr_{1-x}Ti_x)_5S_6$ AND $(Cr_{1-x}V_x)_5S_6$

S. ANZAI, O. INOUE and A. KATORI

Faculty of Engineering, Keio University, Hiyoshi, Yokohama, 223, Japan

Nonmagnetic dilution effects of Ti and V atoms for Cr in $CrS_{1.18}$ are investigated by means of magnetic measurements. Magnetic phase diagrams and molar Curie constants C_{mole} are determined. A considerable difference in x-dependences of C_{mole} for both systems is briefly discussed in view of localization–delocalization concept of 3d electrons.

Cr_5S_6 has a collinear-ferrimagnetic structure between $T_c \simeq 300$ K and T_t (Haas point) $\simeq 160$ K (F-phase) [1]. Below T_t, it changes to a canted helimagnetic structure (H-phase). 3d electrons in V_5S_6, and Ti_5S_6 are mainly delocalized [2]. This paper describes the effect of nonmagnetic dilutions with Ti and V for Cr atoms in Cr_5S_6. Crystallographic properties for these systems were presented in previous papers [2–4].

As seen in the inset of fig. 1, the maximum magnetization in the F-phase decreases with increasing in x for $(Cr_{1-x}Ti_x)_5S_6$, and is extrapolated to zero at $x \simeq 0.2$. This behaviour is also observed on $(Cr_{1-x}V_x)_5S_6$. Such a feature is explained when the preferred substitution of nonmagnetic atoms for Cr in the B-sublattice is provided. Here, the formula unit is taken to be $[Cr(\uparrow)_{x_a}]_A[Cr(\downarrow)_{x_b}]_B S_{1.18}$ with $(1-x) = x_a + x_b$ where the maximum values for x_a and x_b are 0.4 and 0.6, respectively. Representative temperature-dependences of the reciprocal gram susceptibility $1/\chi_g$ are shown in fig. 1. Curves for $x(Ti) = 0$–0.5 are well described with the Néel law: $1/\chi_{mole} = (T/C_{mole}) + (1/\chi_0) - [s/(T-\nu)]$. Curves for $x(Ti) = 0.90$, $x(V) = 0.60$ and 0.80 follow the Curie–Weiss law: $\chi_{mole} = [C_{mole}/(T-\theta)] + \chi_c \times$ (formula weight). Those for $x(V) = 0.3$–0.6 (A-phase) have critical humps in χ_g at T_N. $1/\chi_g$–T curves for $(Cr_{1-x}V_x)_5S_6$ follow the Néel law below $x = 0.15$. As seen in fig. 2(D), χ_g for $x(V) = 0.90$ is mainly Pauli-paramagnetic. Computed parameters are tabulated in table I. The parameters obtained for Cr_5S_6 are in good agreement with those which are computed from Kamigaichi's data up to the higher temperature 570 K [5]. The magnetic interaction parameters n, α and β are also calculated to be 462 K, -0.50 and -0.49, respectively, from the observed values of $1/\chi_g$, s, ν and C_{mole} through eq. 2–7 in [6]. α slightly decreases and β remarkably increases with

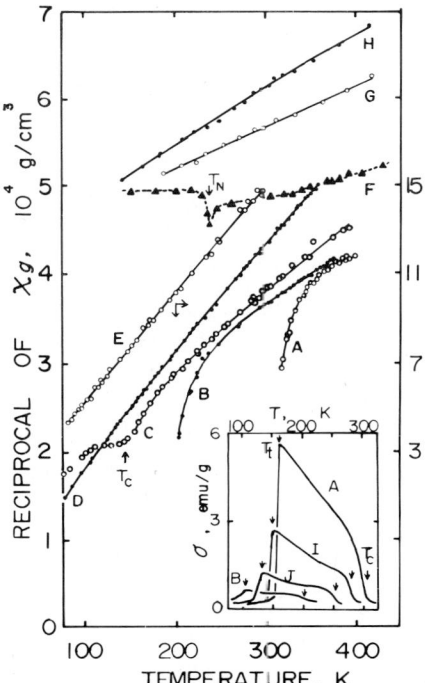

Fig. 1. $1/\chi_g$ vs. T curves for (A) $CrS_{1.18}$; (B) $x(Ti) = 0.20$; (C) $x(Ti) = 0.30$; (D) $x(Ti) = 0.50$; (E) $x(Ti) = 0.90$; (F) $x(V) = 0.30$, $x(V) = 0.60$; (H) $x(V) = 0.80$; (I) $x(V) = 0.05$; (J) $x(Ti) = 0.10$. Solid curves are drawn with the parameters on table I. The inset shows magnetization σ vs. T curves for $(Cr_{1-x}Ti_x)_5S_6$ at 7.7 kOe.

Table I
Parameters for the Néel law: $x[C_{mole}, s, 1/\chi_0, \nu]$ and for the Curie–Weiss law: $x\{C_{mole}, \theta, \chi_c$ in 10^{-5} cm^3/g$\}$

0.0 [2.72, 1625, 339, 306]	0.05(V)[2.37, 3287, 400, 288]
0.0* [2.66, 1631, 388, 306]	0.10(V)[2.55, 3505, 402, 297]
0.10(Ti)[2.48, 1515, 313, 246]	0.15(V)[2.13, 3070, 382, 309]
0.20(Ti)[1.59, 2245, 242, 186]	0.60(V){1.30, −685, 2.90}
0.30(Ti)[1.12, 2015, 172, 119]	0.80(V){0.71, −384, 4.60}
0.40(Ti)[1.00, 2815, 158, 56]	0.90(V){0.0088, +23, 3.55}
0.50(Ti)[0.77, 2758, 111, 17]	V_5S_6† {0.0091, +4, 3.43}
0.60(Ti){0.66, −29, 2, 18}	0.3(Ti)†{0.0061, +32, 3.90}
0.90(Ti){0.16, −1, 0.40}	0.5(Ti)†{0.0098, +20, 3.85}

* From Kamigaichi's data; †from the data for $(V_{1-x}Ti_x)_5S_6$ [2].

increasing in $x(\text{Ti})$ up to $x = 0.3$ but α decreases more and β increases less with $x(\text{V})$ up to $x(\text{V}) = 0.15$. Decrease in absolute value of β extends the F-phase along the x axis at lower temperatures in contrast with the V-case. n slightly increases with increasing in $x(\text{V})$, but remarkably decreases for the Ti-case. T_c were calculated from n, α and β through the relation

$$T_c = [nC_{\text{mole}}/2(x_a + x_b)] \\ \times [x_a\alpha + x_b\beta + \sqrt{(x_a\alpha - x_b\beta)^2 + 4x_ax_b}].$$

The calculated solid curves in figs. 2(A) and 2(B) are well fitted by the observed T_c below $x(\text{Ti}) = 0.40$ and $x(\text{V}) = 0.15$.

As shown in fig. 2(C), $C_{\text{mole}}[x(\text{Ti})]$ is considerably lower than expected for the simple nonmagnetic dilution model: $C_{\text{mole}}(x) = (x_a + x_b)C_{\text{mole}}(x = 0)$ [dashed curve]. For the V-case it is nearly close to those for the simple model, except for $x = 0.90$. Observed $C_{\text{mole}}[x(\text{V}) = 0.90]$ is considerably lower than that expected for the simple model. C_{mole} and χ_c for $x(\text{V}) = 0.90$ are as large as found for $(\text{V}_{1-x}\text{Ti}_x)_5\text{S}_6$ below $x = 0.5$. The change to this state seems to be steep at $x(\text{V}) \sim 0.85$, because χ_g at 298 K jumps at $0.8 < x(\text{V}) < 0.9$. The anomalous weakening effect on C_{mole} would come from the partial or nearly full collapse in Cr moments due to the invading of Ti or V electrons on Cr atoms. It is considered that the metallic electrons are in a more narrow band in V_5S_6 than in Ti_5S_6 because χ_c in V_5S_6 is considerably larger than that in Ti_5S_6 [2]. It is plausible that the intra-atomic exchange interactions between magnetic electrons on a Cr atom is more easily weakened by the conduction electrons

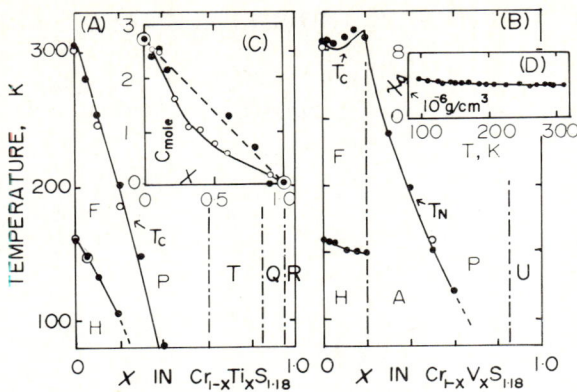

Fig. 2. Magnetic phase diagrams for (A) Ti and (B) V modified Cr_5S_6. ● and ○ signs represent magnetic and DSC measurements, respectively. T: two phases; Q: $(\text{Cr}_{0.1}\text{Ti}_{0.9})_5\text{S}_6$; R: Ti_5S_6; U: V_5S_6; P: paramagnetic region for Cr_5S_6-type of structure. (C) $C_{\text{mole}}(x)$ for Ti- (○) and V- (●) modified systems. (D) χ_g vs. T for $x(\text{V}) = 0.90$.

from Ti atoms so that the collapsing effect occurs only in the higher $x(\text{V})$.

The authors express their thanks to Matsunaga Science Foundation for a grant.

References

[1] B. van Laar, Phys. Rev. 156 (1967) 654.
[2] S. Anzai, O. Inoue, K. Kamiya and K. Sakaguchi, Jap. J. Appl. Phys. 5 (1976) 551.
[3] S. Anzai and Y. Hamaguchi, J. Phys. Soc. Jap. 38 (1975) 400.
[4] A. Katori, S. Anzai and T. Yoshida, J. Inorg. Nucl. Chem. 37 (1975) 323.
[5] T. Kamigaichi, J. Sci. Hiroshima Univ. A24 (1960) 371.
[6] E.W. Gorter, Philips Res. Rep. 9 (1954) 321.

EFFECT OF CARRIERS' STATISTICS ON THE INDIRECT (RKKY) EXCHANGE IN MAGNETIC SEMICONDUCTORS

A. MAUGER and P. LEROUX HUGON

Laboratoire de Physique des Solides, CNRS, 1, Place Aristide Briand, 92190 Meudon, France

Indirect exchange in heavily doped europium chalcogenides is calculated in the mean field approximation. We have taken into account the finite bandwidth, and carriers' statistics (splitting of spin subbands, thermal degeneracy). Our predictions, both in the paramagnetic and ferromagnetic configurations agree with experimental data on EuO.

In europium chalcogenides, the importance of the indirect coupling between localized moments through the conduction electrons introduced by doping has been recognized for some time [1]. However, the common expression of this indirect (RKKY) exchange for metals must be modified for these semiconductors, partly because the width of the d-type conduction band is small, and mostly because the exchange splitting is no longer negligible with respect to the Fermi energy, so that the strength of the indirect exchange depends on both the polarization of the electron gas and its degeneracy.

Our model for the europium chalcogenides relies on the following assumptions: (a) the exchange constant J_{df} is local; (b) alloying effects are neglected; and (c) the calculations are made in the mean field approximation. The exchange interaction between localized and free electrons is

$$-\sum_i J_{df} S_i \cdot s, \qquad (1)$$

where S_i is the spin of the localized moment at site R_i, s is the spin of a conduction electron. The carriers are in t_{2g} subbands split by the spin–orbit interaction. For mathematical convenience, we have likened the lowest subbands to a four-fold band split by the exchange in two two-fold up and down spin subbands. The values of J_{df} are $J_{df} = 0.13$ eV for EuO and 0.06 eV for EuS [2]. We have chosen the most tractable dispersion law, accounting of the finite width W of the conduction band:

$$E(k) = \tfrac{1}{2}W[1 - \cos(ka)], \qquad (2)$$

where $a = (\pi/3)^{1/2} a'/2$ in which a' is the lattice parameter. According to ref. 2, $W = 1.1$ eV (EuO) and 0.6 eV (EuS). A second quantization formalism allows us to write the second order hamiltonian deduced from eq. (1) in the usual form: $\tfrac{1}{2}\sum_{i,j} J_{eff}(R_{ij}) S_i^z S_j^z$ with

$$J_{eff}(R_{ij}) = -\frac{8}{W}\left(\frac{VJ_{df}}{4\pi^2 N R_{ij}}\right)^2 \int_0^{\pi/a} dk \left\{ f\left(E(k) + \frac{J_{df}S\sigma}{2}\right) \right.$$
$$\left. + f\left(E(k) - \frac{J_{df}S\sigma}{2}\right)\right\} k \sin(kR_{ij}) \int_0^{\pi/a} dk'$$
$$\times \frac{k' \sin(k'R_{ij})}{\cos(k'a) - \cos(ka)}. \qquad (3)$$

V is the volume of the crystal, N the number of localized spins, σ the reduced magnetization and $f(E)$ the Fermi function. In the ferromagnetic configuration, we must determine simultaneously in a self-consistent way J_{eff} and σ given by

$$\sigma = B_s\left[\frac{\sigma S^2[I(0) + J_{eff}(0)]}{k_B T}\right] \qquad (4)$$

where B_s is the Brillouin function $I(0)$ and $J_{eff}(0)$ are the Fourier transform at $q = 0$ of the direct and indirect exchange, and k_B is the Boltzmann constant.

We have plotted in fig. 1 the variation of the paramagnetic Curie temperature $\theta = S(S+1)[I(0) + J_{eff}(0)]/(3k_B)$ as a function of the composition x for $Eu_{1-x}Gd_xS$ and $Eu_{1-x}Gd_xO$, supposing that each Gd atom provides one electron. Our results are in good agreement with experimental data [3] for $Eu_{1-x}Gd_xS$. For a carrier concentration higher than 50%, the calculation becomes meaningless because alloying effects are important.

In fig. 2, we show the thermal variations of the magnetization for a typical carrier concen-

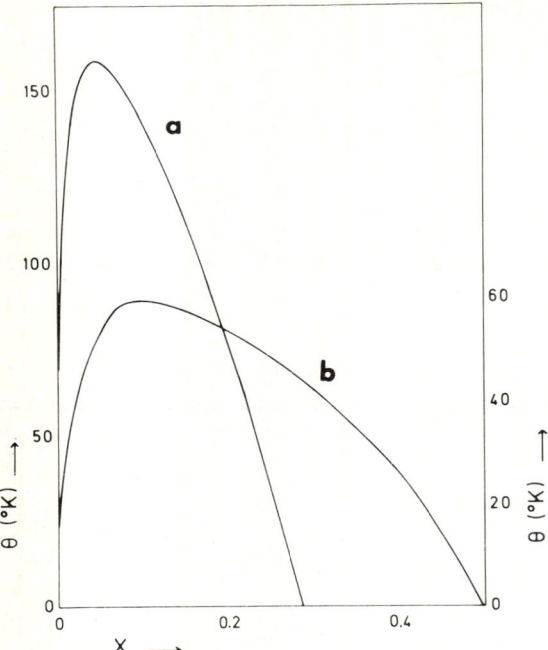

Fig. 1. Paramagnetic Curie temperature θ vs. composition x for $Eu_{1-x}Gd_xO$ (curve a, left scale) and $Eu_{1-x}Gd_xS$ (curve b, right scale).

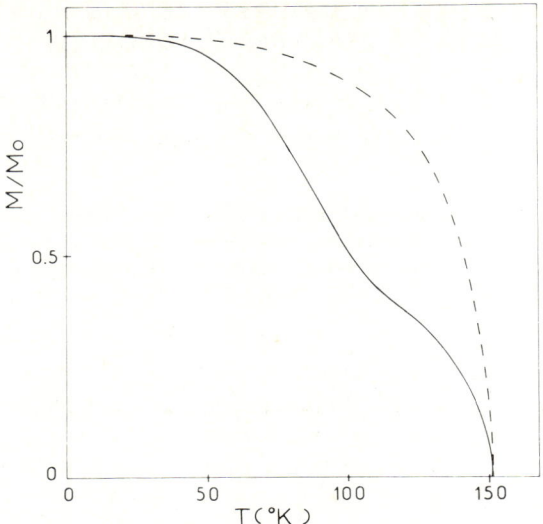

Fig. 2. Magnetization curve for EuO with a free electron concentration $n \simeq 5 \times 10^{20}$ cm^{-3} (full curve). The broken curve is the magnetization predicted by a classical Brillouin law.

tration $n = 5 \times 10^{20}$ cm^{-3} in EuO. We find a large departure from the Brillouin curve owing to two factors. (a) When free electrons are only in the spin up subbands, a thermal variation of J_{eff} is associated with that of a mean value of the wave vector q of the electrons; (b) in the vicinity of the ferro-paramagnetic transition, the spin down conduction band states become also populated. This gives rise to a characteristic anomaly of the magnetization curve, the location of which depends on the carrier concentration. Such a magnetization curve has been observed, for example, by Ferré et al. [4], Massenet et al. [5], and Schoenes et al. [6]. This would indicate that the particular shape of the magnetization curve, which is systematically below the Brillouin curve, is a characteristic property of this kind of magnetic semiconductor rather than an effect of amorphous magnetism, as suggested in ref. 6.

References

[1] F. Holtzberg, T. McGuire, S. Methfessel and J. Suits, Phys. Rev. Lett. 13 (1964) 18.
[2] P. Wachter, Crit. Rev. Solid State Sci. 3 (1972) 189.
[3] T.R. McGuire and F. Holtzberg, Amer. Inst. Phys. Conf. Proc. 10 (1973) 10.
[4] J. Ferré, J.P. Badoz, C. Paparoditis and R. Suryanarayanan, IEEE Trans. Mag. 7 (1971) 388.
[5] O. Massenet, Y. Capiomont and Nguyen van Dang, J. Appl. Phys. 45 (1974) 3593.
[6] J. Schoenes and P. Wachter, Phys. Rev. B9 (1974) 3097.

MAGNETIC AND STRUCTURAL PHASE TRANSITIONS IN SODIUM INTERCALATES Na_xVS_2 AND Na_xVSe_2

G.A. WIEGERS and C.F. VAN BRUGGEN

Laboratory of Inorganic Chemistry, Materials Science Center of the University, Groningen, The Netherlands

Non-stoichiometric compounds $Na_xVS_2(Se_2)$ were synthesized by reaction from the elements. They crystallize in two different structures: a metallic form with Pauli-paramagnetic behaviour and a semiconducting form with paramagnetic behaviour at room temperature. The semiconducting form shows a sharp paramagnetic–antiferromagnetic phase transition at ~50 K with a simultaneous first order co-operative Jahn–Teller lattice distortion.

The non-stoichiometric compounds $Na_xVS_2(Se_2)$ were synthesized by direct heating from the elements [1]. They crystallize in two different structures: a metallic (m.) form (type I, space group R3m, Z = 3; for Na_xVS_2 with $0.3 \leq x \leq 1$, and for Na_xVSe_2 with $x \simeq 0.5$–0.6) with a trigonal prismatic coordination of Na, and a semiconducting (s.c.) form (type II, space group $R\bar{3}m$, Z = 3; for $Na_xVS_2(Se_2)$ with $x \simeq 1$) with a trigonal antiprismatic coordination of Na. Both structure types (fig. 1) have V in trigonally distorted octahedra of S(Se), which are elongated along the trigonal axis in type I, but compressed in type II.

The m. type I compounds reveal a slightly temperature-dependent Pauli-paramagnetism of the conduction electrons at high temperatures (HT); at low temperatures (LT) several anomalous changes in χ occur. Fig. 2 gives the result for $NaVS_2$, but the isotypic compounds

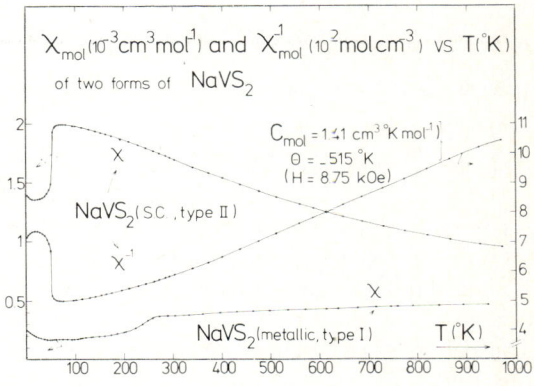

Fig. 2. Molar magnetic susceptibilities (left axis) and reciprocal susceptibility (right axis) versus temperature of two different forms of $NaVS_2$; see text.

Na_xVS_2 ($0.3 \leq x < 1$) and Na_xVSe_2 ($x \simeq 0.5$–0.6) behave almost the same. The magnetic and electrical properties of the "host" compound VSe_2 will be reported elsewhere [2].

The behaviour of the s.c. type II compounds $NaVS_2(Se_2)$ is completely different; see figs. 2 and 3. At HT χ^{-1} versus T obeys Curie–Weiss:

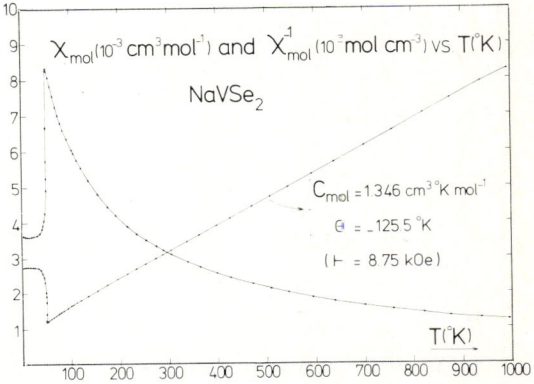

Fig. 3. Molar magnetic susceptibility and reciprocal susceptibility versus temperature of $NaVSe_2$; see text.

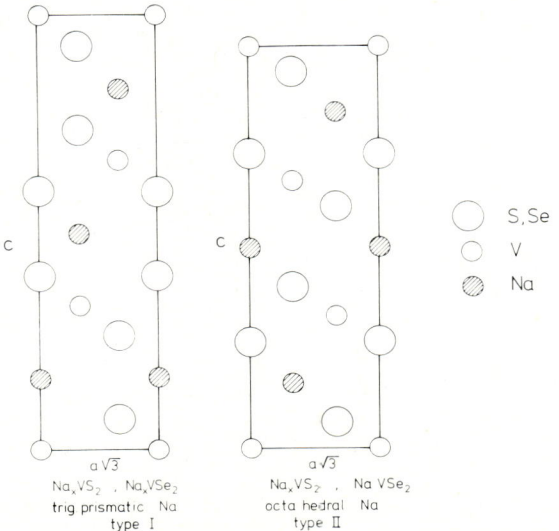

Fig. 1. Sections through the (11$\bar{2}$0) planes of the type I (left) and type II (right) structures of $Na_xVS_2(Se_2)$.

$\chi_{mol} = C_{mol}/(T - \theta)$. However, at LT these type II compounds show sharp magnetic phase transitions (ph. tr.) at ~50 K; they become antiferromagnetic (AF). No drastic change of the electrical conductivity (σ) shows up at the ph. tr. This ph. tr. is attended by a simultaneous structure change. The rhombohedral cell (space group $R\bar{3}m = D_{3d}^5$, hexagonal setting, $Z = 3$) undergoes a first order co-operative Jahn–Teller (JT) distortion in the hexagonal planes and becomes C-centered monoclinic (space group $C2/m = C_{2h}^3$, $Z = 2$); see fig. 4. For NaVSe$_2$ holds:

$T \geq 50$ K $\rightarrow a_{hex} = 3.721$ Å; $c_{hex} = 20.53$ Å;

$T \leq 50$ K $\rightarrow a_{mon} = a_{hex} \times \sqrt{3} \times 1.007 = 6.489$ Å;

$b_{mon} = b_{hex} \times 0.993 = 3.695$ Å;

$c_{mon} = [(\frac{2}{3}a_{mon})^2 + (\frac{1}{3}c_{hex})^2]^{1/2} = 8.096$ Å; $\beta = 122°19'$.

No volume change can be observed by X-ray analysis. For NaVS$_2$ the orthorhombic distortion amounts to 10‰ instead of 14‰ for NaVSe$_2$.

The HT magnetic behaviour of the s.c. type II compounds can be understood qualitatively as follows. V^{3+} ($3d^2$) has spin $S = 1$ and an unquenched orbital angular momentum, and is in the rhombohedral HT form ($T > 50$ K) situated in trigonally compressed octahedral sites of c.c.p. anions (fig. 1, right). This means that both spin–orbit coupling λ and a low-symmetry crystal field D_{3d} contribute in its Van Vleck (VV)-type paramagnetic behaviour [3, 4]. In this lattice of exchange-coupled V^{3+} ions, however, also exchange J will contribute to θ, making the experimental $\theta = \theta_{VV} + \theta_{exch.}$ in the Curie–Weiss relation $\chi_{mol} = C_{mol}/(T - \theta)$ for HT powder susceptibilities. Fig. 5 gives schematically the splitting diagram in terms of D_{3d} and λ (A = matrix element for orbital angular momentum).

The LT distortion is driven by the coupling between the localized orbital electronic state ^3E_g (fig. 5) and the crystal lattice strain ϵ_g (in D_{3d}): $^3E_g \otimes \epsilon_g$ [5]. It lowers both the group- and point-symmetry from D_{3d} to C_{2h}, and involves the simultaneous splitting of the electronic states of V^{3+} (d^2) (fig. 5). The structural ph. tr. corresponds to $k = 0$, and is of first order because it involves the decrease by a factor 3 of the number of symmetry elements [6]. These type II compounds have magnetic exchange, spin–orbit coupling and JT interaction; this results at $T_d \simeq 50$ K in a first order AF ordering of the – by spin–orbit coupling strongly reduced – magnetic moments in the direction of the – by distortion shortened – monoclinic b-axis. The magnetic

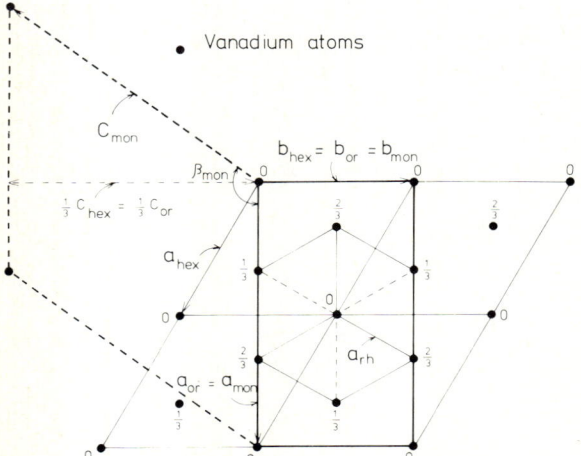

Fig. 4. Relationship for type II NaVS$_2$(Se$_2$) between the rhombohedral cell in the hexagonal setting and the orthorhombic distorted cell in the monoclinic setting (only the V atoms are shown).

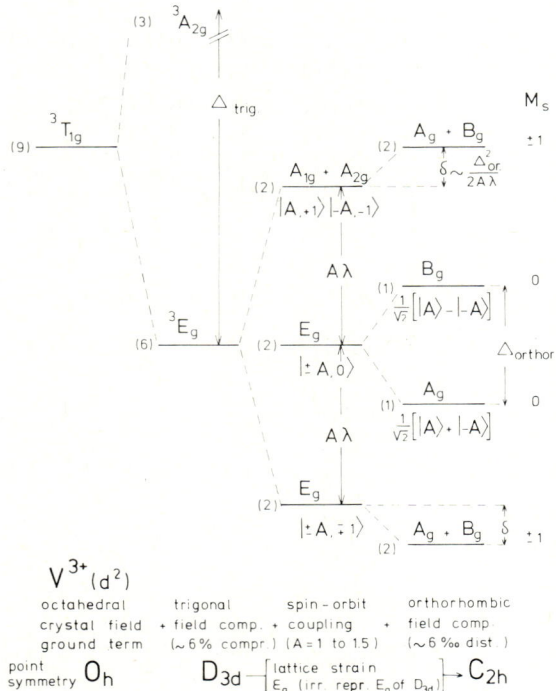

Fig. 5. Schematic splitting diagram for the octahedral crystal-field ground term $^3T_{1g}$ of V^{3+}(d^2) under the action of low-symmetry crystal fields and spin–orbit coupling in type II NaVS$_2$(Se$_2$); see text.

structure of $NaVSe_2$ was determined by neutron powder diffraction. The magnetic cell is twice the C centered monoclinic crystallographic cell, i.e. $2a_{mon} \times b_{mon} \times c_{mon}$. The basal-plane structure corresponds to the $MnBr_2$-type two-dimensional AF structure with two rows of parallel moments in the monoclinic b direction, followed by two rows of parallel moments in the opposite direction. Successive AF vanadium metal layers are ferromagnetically coupled in the direction of the monoclinic c axis. From a profile refinement of the neutron diagram follows a magnetic moment $\mu_V = 0.90 \mu_B$.

References

[1] G.A. Wiegers, R. van der Meer, H. van Heiningen, H.J. Kloosterboer and A.J.A. Alberink, Mater. Res. Bull. 9 (1974) 1261.
[2] C.F. van Bruggen and C. Haas, Solid State Commun. 20 (1976) 251.
[3] B.N. Figgis, J. Lewis, F.E. Mabbs and G.A. Webb, J. Chem. Soc. A. (1966) 1411.
[4] E. König and S. Kremer, Ber. Bunsenges. Phys. Chem. 79 (1975) 192.
[5] G.A. Gehring and K.A. Gehring, Rep. Prog. Phys. 38 (1975) 1.
[6] L.D. Landau and E.M. Lifshitz, Statistical Physics (Pergamon, London, 1962), ch. 14.

EXCHANGE INTERACTIONS IN INSULATORS

L. JANSEN and R. BLOCK

Institute of Theoretical Chemistry, University of Amsterdam, Amsterdam, The Netherlands

(Invited paper)

Exchange perturbation theory of the Rayleigh–Schrödinger type is applied to the problem of magnetic ordering in 3d-cation solids at low temperatures, on the basis of an effective-electron model with Slater- or Gaussian-type spherically symmetric orbitals in zeroth order. The effect of point-group symmetry is not explicitly considered, whereas permutation symmetry is fully taken into account. Results are presented for the following systems: (a) MnS in its three solid modifications, (b) the 180°-superexchange in the 3d-metal fluorides XMF_3 and X_2MF_4, (c) the difference in magnetic ordering for the manganese pyrites MnS_2, $MnSe_2$ and $MnTe_2$ and (d) the different Cu–Cu exchange interactions in solids of $KCuF_3$, K_2CuF_4 and $[C_nH_{2n+1}NH_3]_2 CuCl_4$ ($n = 1, 2, \ldots$).

1. Introduction

As is well known, Kramers [1] in 1934 was the first to suggest that exchange interactions between two paramagnetic 3d-cations in an insulating solid (such as solid NiO, MnS, etc.) involve a diamagnetic anion (O^{2-}, S^{2-}, etc.). At that time, there existed already some experimental evidence of this *indirect* exchange, from measurements of magnetic properties of the mineral tysonite (a fluoride of La and Ce, with Nd and Pr as impurities) by Becquerel, de Haas and van den Handel (1934) [2]. In 1946, de Klerk [3] obtained further evidence from measurements of the specific heat of copper potassium sulfate at temperatures below 1 K. The specific heat was much larger than that due to *isolated* unpaired spins of the Cu^{2+}-ions, pointing to some kind of indirect coupling mechanism between them. As magnetostatic coupling is much too weak at the distance between neighboring Cu^{2+}-ions, the indirect interaction must be of exchange type. In 1949, Shull and collaborators [4] published the first results of a series of neutron-diffraction analyses on non-conducting 3d-solids at low temperatures. The evidence of magnetic ordering of spins on different cations was conclusive proof for existence of the effect (which was then already called "*superexchange*"). Since about 1950, magnetic order and superexchange have formed a major domain of solid-state research and have found many technical applications.

Kramers' intuitive perturbational approach to superexchange based on a consideration of two cations and one anion, has, in the course of time, been replaced by more elaborate and pretentious schemes. The most detailed analysis was developed by Anderson (1959) [5] on the basis of a two-step procedure. First, he considers one magnetic electron in the field of all the anions (ligands) and all the other cations in the solid, while leaving out exchange between the unpaired electron considered and those of the other cations, in a Hartree–Fock scheme, i.e. in a one-electron approximation. If we start from a state in which the spins of all unpaired electrons are parallel, then the Hartree–Fock operator for the electron considered has the periodicity of the (cation-) lattice, so that the solutions are running Bloch waves. These can be transformed to Wannier functions, localized around the cation positions (those centered at different lattice points are thus orthogonal). In the second step, Anderson considers exchange interactions between two electrons on different Wannier orbitals, taking those orbitals to represent the unperturbed state. Because of orthogonality of Wannier functions, first-order exchange between the two electrons favors parallel alignment of their spins ("potential exchange"). In second order, with a "delocalization part" of the Hartree–Fock operator as the perturbation, a gain in energy is obtained only if the spins of the two electrons are antiparallel ("kinetic exchange").

The Anderson method does have the advantage that it can lead to either parallel or antiparallel coupling of cation spins. Rigorous calculations on this basis are, however, hardly feasible, whereas "simplified" versions [6] of Anderson's method are hopelessly confusing regarding reliability of the approximations made. We mention here also a configuration-interaction analysis based (as with Kramers) on a unit of two cations and one anion by Keffer and Oguchi (1959) [7]. Many other authors have contributed to the theory of superexchange in non-conducting solids; for excellent reviews we refer to the literature [8, 9].

It is indeed disappointing to observe that, in spite of all efforts, the accurate evaluation of superexchange interactions has turned out to be a problem of formidable complexity; for general systems neither the order of magnitude nor the sign of such interactions can be reliably extracted from such analyses. D.J. Newman discusses in these Proceedings [10] the prospects for *ab initio* calculations and models of superexchange interactions. Instead, semi-empirical rules and classification schemes have taken over where theory failed to provide a rigorous basis for the interpretation of observed regularities which have accumulated from the enormous number of accurate experiments carried out on a great many different solids. Such rules were formulated by Goodenough [11] and Kanamori [12]; for a review, see also Anderson [8]. But even on this basis, present "explanations" lag very far behind the available experimental data.

We have approached the interpretation of magnetic interactions in non-conducting (3d-) solids extrapolating from earlier work on crystal stability of rare-gas crystals and on crystal stability and elastic constants of ionic solids (closed-shell cations and anions) in terms of three-atom and three-ion exchange interactions [13]. The model for magnetic interactions, to be outlined in the next Section, has recently also been applied to the stability of rare-gas halides [14] and to the explanation of rotational barriers in simple molecules [15].

2. Effective-electron model for magnetic interactions [16]

Since an *a priori* evaluation of exchange interactions even for a simple system of, say, two paramagnetic cations and one diamagnetic anion, lies far beyond the capabilities of present first-principle theory, we adopt a model based on a number of drastic approximations. The validity of these assumptions is in part hypothetical, in part empirical and must be ultimately established by comparison with experiment. We adopt the following simplifications:

(A) *Weak* exchange interactions may be evaluated by using perturbation methods, starting from free-ion wavefunctions in zeroth order of approximation;

(B) *Permutation symmetry* of the Hamiltonian determines the essential characteristics of weak exchange interactions, whereas spatial symmetries are only of secondary importance;

(C) Properties of weak exchange interactions are not *primarily* dependent upon the number of unpaired electrons per paramagnetic cation. As a consequence we replace, on each cation, its unpaired electrons by one "*effective*" electron. The number of unpaired electrons is reflected by the "magnetic size" of the orbital of the effective electron.

Assumption (A) implies that the experimental facts to be explained must be reproduced already in the lowest orders of the perturbation series which yield a non-zero contribution to the exchange. We choose a Rayleigh–Schrödinger type of exchange perturbation theory [17]; the adoption of free-ion wavefunctions in zeroth order implies automatically that those at different centers are *not* orthogonal. Assumption (B) implies, first of all, that translational symmetry of the crystal is but of secondary importance for the phenomena of weak exchange coupling in insulating solids. Further we remark that, in spite of nonorthogonality of the free-ion wavefunctions, we can always (Schmidt-) orthogonalize the zeroth-order cation functions to those of the ligand anions in a valence-bond determinantal total wavefunction; the resulting cation functions then do have the symmetry of the ligand environment.

Finally, assumption (C) of the model may be made plausible by observing that the stable magnetic patterns of solids with 3d-cations are *all* antiferromagnetic; moreover, they all are of the so-called *second kind* (except for CrN). In addition, the orders of magnitude of their Néel temperatures are the same (a few hundred K). This indicates that the essential aspects of magnetic coupling may be described by means of one magnetic electron per cation and that the number of unpaired electrons may be incorporated as a "size"-parameter of the effective-electron orbital. In accordance with this description, we replace the electrons of a closed-shell anion by *two*, spin-paired, effective electrons. A simplest system of two cations and one anion is then described by the three-center, four-electron model proposed by Kramers [1] forty years ago.

3. Applications of the model

3.1. Angular dependence of superexchange interactions

Consider the simplest system affording superexchange interactions, i.e. two paramagnetic cations A and B, each with one effective electron (wavefunctions ϕ_A, ϕ_B), and one diamagnetic anion, C, with two spin-paired effective electrons (wavefunction ϕ_C). Starting from an assignment of electron 1 to A, 2 to B, electrons 3 and 4 to C, the zeroth-order wavefunction is

$$\psi_\sigma^{ABC} = \mathcal{A}\phi_A(1)\phi_B(2)\phi_C(3)\phi_C(4)\sigma \equiv \mathcal{A}\phi\sigma; \quad (1)$$

$\mathcal{A}$ is the antisymmetrizer and σ denotes the triplet or singlet spin-eigenfunction. In view of the fact that there are only two unpaired electrons, the first-order interaction energy [17] can always be expressed in the form

$$E_\sigma^{(1)} = M + N\langle S_1 \cdot S_2\rangle\sigma, \quad (2)$$

where M and N consist of contributions due to the pure interatomic permutations of electrons (containing as well the identity permutation). Employing the method of double-coset decomposition [18] of the antisymmetrizer, we have evaluated $E_\sigma^{(1)}$ adopting Slater functions $r^{n-1} \exp(-pr)$ for each of the effective-electron wavefunctions. Actually, the results are very similar for $n = 1, 2, 3$ and 4, so that the numerical results are given for $n = 1$ only. We consider only those geometric configurations (cation–anion–cation) in which the two cation–anion distances are equal. The opening angle θ, at the site of the anion C, was varied from 40° to 180°. The cations A and B, at fixed distance $R_{AB} = 7$ au, are assumed to be of the same kind. The choice of the orbital parameters p was $p_A = p_B = 1$ au^{-1}, $p_C = 0.75$ au^{-1}; these values roughly correspond to the unit Ni–Cl–Ni in an application [19] of the effective-electron model to the analysis of 180°-superexchange.

In fig. 1 we present the contributions to the coefficient N in eq. (2) due to the permutations P_{12} (cation–cation exchange), P_{123} and P_{132} (the

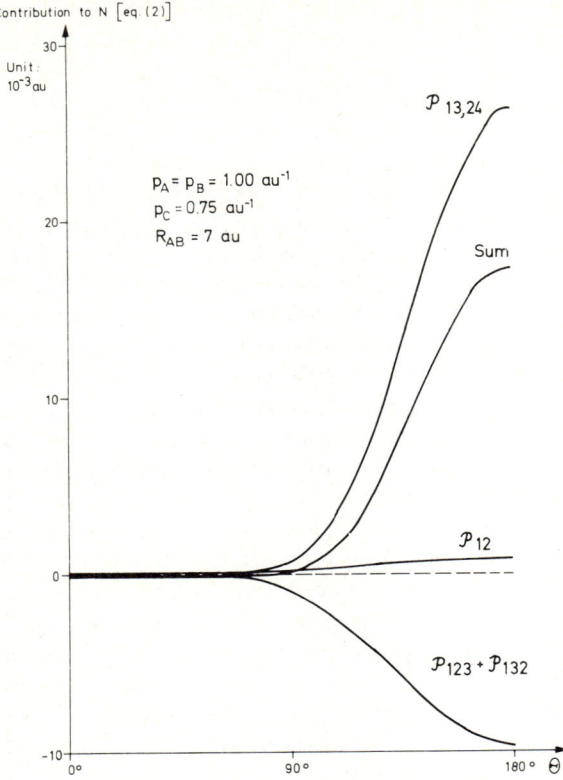

Fig. 1. Contributions to singlet–triplet splitting in the three-center, four-electron model, from permutations involving both electrons with unpaired spin (1, 2), as well as their sum, as a function of the opening angle θ at the anion. Parameters p_A, p_B (cations) and p_C (anion) of the 1s-Slater orbitals are given. The unit of energy is 10^{-3} au. Positive values imply singlet, negative values triplet stability.

same for isosceles configurations) and $P_{13,24}$. A positive contribution implies that the singlet state is favored, a negative contribution implies triplet coupling. The unit of energy is 10^{-3} au (1 au $\approx 0.3 \times 10^6$ K). It is seen that first-order exchange favors antiparallel coupling of the cation spins, except for a region of weak ferromagnetic alignment around $\theta = 90°$. For large θ the coupling is strongly antiferromagnetic, in agreement with experiments. The angle dependence of superexchange is here due *solely* to permutation symmetry of the zeroth-order wavefunction in the non-orthogonal basis of spherically symmetric atomic functions. If non-orthogonality is neglected then the coupling appears only in higher order of the perturbation series.

Of particular interest for comparison with experiment is the narrow θ-range of weak ferromagnetic coupling. We find, in terms of a Heisenberg Hamiltonian $C - 2J_{AB}S_A \cdot S_B$, with S_A and S_B the total-spin operators for the Ni-cations, a variation in J_{AB}/k between $+3$ ($\theta = 85°$) and -7 K ($\theta = 95°$). Experimentally, it is known that the sign of J_{AB} does indeed sensitively depend on the opening θ just in this region [22].

That the non-orthogonal basis gives rise, in a Heitler–London formalism, to a θ-range of *triplet* stability can be easily explained. If we (Schmidt-) orthogonalize the cation orbitals to that of the effective electron on the anion, then at a certain θ these orbitals on the two cations become mutually orthogonal, *viz.* when for the original orbitals the (constant) cation–cation overlap is equal to the square of cation–anion overlap. We then have three orthogonal orbitals; the only first-order exchange splitting is due to direct exchange between electrons on the two cation Schmidt-orbitals, which favors parallel alignment of their spins.

3.2. 180°-Superexchange: quantitative results

On the basis of the above simple model we have evaluated [19] 180°-indirect exchange in solids of composition XMF_3 and X_2MF_4 (X = K, Rb, Tl; M = Mn, Co, Ni). The available experimental data on the exchange constants J/k and on the cation nearest-neighbor distance R enable one to make reliable estimates of the R-dependence of J/k for the same cation, as well as of the J/k-ratio for Mn^{2+}, Co^{2+} and Ni^{2+} at *fixed* R. Assuming a power-law dependence $J/k \sim R^{-n}$, the data yield $n \sim 12$ for all three cations in the experimental range and a J/k-ratio $1:3.6:7.7$ for Mn^{2+}, Co^{2+} and Ni^{2+} at $R = 4.074$ au.

To carry out the model calculations, we must determine the Slater orbital parameters for the three cations and for the anion F^-. This was accomplished as follows:

a) estimate the ratios of the cation parameters for Mn^{2+}, Co^{2+}, Ni^{2+} from SCF-calculations on the $\langle r^2 \rangle$-values of the cation electrons;
b) estimate the ratio $p_{Mn^{2+}}/p_{F^-}$ from experimental diamagnetic susceptibilities;
c) assign a value to p_{F^-} through a fit to the experimental value for J/k of Mn^{2+}. This determines all the other parameters.

Table I
Exchange constants J/k for 180°-superexchange in a number of compounds as predicted by the effective-electron model and compared with experiment. The numbers in parentheses indicate experimental accuracy; R is the cation–cation nearest-neighbor distance.

Compound	R(Å)	J/k Calculated	J/k Experimental
Cs_2MnCl_4	5.130	-5.0	$-5.0(2)$
$TlMnF_3$	4.250	-3.1	$-3.0(3)$
$KCoF_3$	4.058	-17.6	$-19.2(20)$
K_2NiF_4	3.992	-50	$-51.5(3)$
Rb_2FeF_4	4.214	-6.1	$-6.5(15)$

Proceeding in this way, the model predicts J/k-ratios of $1:3.3:7.6$ (exp. $1:3.6:7.7$). Good agreement with experiment was also found regarding the R-dependence of J/k for fixed cation. Although the model predicts an exponential R-dependence, the results for the limited R-region can indeed be closely approximated by R^{-n}, with $n \sim 12$ [19].

Additional 180°-results have in the meantime been obtained for a number of other compounds; these are given in table I above. The columns list, respectively, the compound, the cation–cation nearest-neighbor distance R(Å) and the calculated and experimental values (with accuracy) of the exchange constants J/k.

Note that in the first compound the anion is Cl^- and in the last compound the cation is Fe^{2+}.

Quantitatively, the calculated exchange constants depend, of course, strongly on the choice of the orbital parameters. Nevertheless, it is intriguing to observe that a set of parameters can be found which simultaneously yields the correct ratio of J/k-values for Mn^{2+}, Ni^{2+} and Co^{2+}, their correct R-dependence in the experimental range, as well as exchange constants themselves (including that for $Fe^{2+}-F^-$) which are in close agreement with experiment.

3.3. Superexchange in three modifications of MnS [20]

The three-center, four-electron model has also been applied to magnetic ordering in the three experimental modifications of MnS [crystal structures B1 (NaCl), B3 (sphalerite) and B4 (wurtzite)]. The difference in observed magnetic ordering (antiferromagnetic of second kind in B1, of third kind in B3 (B4)) is in the literature supposed to arise because of different covalent character of the Mn–S bond in an octahedral (B1) and a tetrahedral (B3, B4) environment. Proceeding formally in the same manner as for 180°-superexchange, we find the following results [20, 21]:

i) the stable magnetic structure in the B1-configuration is indeed antiferromagnetic of the second kind;

ii) this spin arrangement is ruled out in the B3 (B4)-modification because of the lack of (strongly antiferromagnetic) 180°-superexchange in a tetrahedral arrangement of Mn^{2+}-cations around a central S^{2-}-anion. Instead, antiferromagnetism of the third kind has the lowest magnetic energy (although the difference with antiferromagnetism of the first kind is very small on the basis of the model);

iii) covalent MnS bonding is not involved in the explanation of differences in magnetic patterns for solid MnS.

3.4. Effect of Jahn–Teller distortions on exchange coupling

An obvious extension of the analysis of magnetic interactions is to consider Cu^{2+} as the magnetic cation. Solids with very interesting magnetic behavior are e.g. those of the compounds $KCuF_3$, K_2CuF_4 and the "two-dimensional" ferromagnets $[C_nH_{2n+1}NH_3]_2 CuCl_4$ ($n = 1, 2, \ldots$) [22]. They also exhibit Jahn–Teller distortions and the question arises as to whether there is a connection between these two phenomena.

We have carried out model calculations for superexchange interactions between two Cu^{2+}-cations (called P and Q) at a distance 10 au. A Cl^--anion (A) between the cations lies at a distance x from A and y from Q (thus $x + y = 10$). Cation P is surrounded by four anions (two of these are given in the drawing) at distance y in a plane perpendicular to the line connecting P and Q, whereas Q is surrounded by four anions (including A) at the same distance y, now in the plane CPQ. This is, somewhat idealized, the arrangement in the $(CuCl_4)$-ferromagnets ($J/k \approx 20$ K).

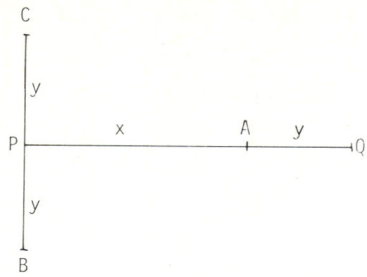

The Slater p-parameter for the cation was chosen as 1 au^{-1}, that for the anion as 0.70 au^{-1}. What interests us is, in the first place, the dependence of 180°-superexchange PAQ on x. As can be seen from the table below, this change, for x between 3 and 7 au, is only weak and no ferromagnetism can result in this way from Jahn–Teller distortions. However, if we include a *second* anion (say B), then a strong ferromagnetic shift is found when the anion A is displaced from the middle towards the cation Q. The results, for triplet-minus-singlet energies $(T - S)$ in units 10^{-6} au, are collected in the following table.

Table II
Triplet-minus-singlet $(T - S)$ energies, in units 10^{-6} au, for superexchange interactions between two Cu^{2+}-cations (P and Q) due to interaction with one anion (A), another anion (B) and both anions, as a function of the AQ- (equal to PB-) distance (see drawing).

$(T - S)$ in 10^{-6} au	PQ	PAQ	PBQ	PABQ
$x = 3$	1	795	1	792
3.5		816	1	805
4		836	1	807
4.5		851	0	786
5		857	0	729
5.5		851		622
6		836	1	466
6.5		816	0	272
7		795	3	73

In comparing the columns PAQ and PABQ, we note that a strong ferromagnetic shift results from Jahn–Teller distortions as soon as the anion A is moved towards the cation Q. Since there are several of the configurations considered for any two cations P and Q, it does not seem excluded that the total interaction may indeed become ferromagnetic. Detailed calculations are in progress.

3.4. Magnetic ordering in manganese pyrites

Neutron diffraction measurements [23] have shown that the manganese pyrites MnS_2, $MnSe_2$ and $MnTe_2$ are antiferromagnetically ordered at low temperatures. However, the usual ordering of the second kind does not occur; instead, MnS_2 is antiferromagnetic of the third kind, $MnSe_2$ shows a mixture of third and first kinds and $MnTe_2$ exhibits antiferromagnetism of the first kind. This means that from the sulfide to the telluride there is an increasing tendency for spins of *second* neighbors around a central Mn^{2+}-cation to align themselves parallel to the spin of the central cation.

In an effort to explain this behavior, we have [24] applied the effective-electron model to different Mn–X_2–Mn configurations (X = S, Se or Te), replacing the complex X_2^{2-} by two centers and two spin-paired effective electrons on a molecular orbital with bonding or anti-bonding character. This yields a four-center, four-electron model which is then analyzed on the basis of (first-order) exchange perturbation theory. The Slater parameter p for the orbital on each X^- was estimated using a value for its diamagnetic susceptibility lying between those for the X-atom and the X^{2-}-ion.

We find the following results:

1) Calculations based on a bonding type of molecular orbital for the X_2^{2-}-electrons are heavily in favor of antiferromagnetism of the second kind for all compounds and are thus rejected. Those based on an anti-bonding type of orbital show more differentiation. These results are reported in what follows;

2) Values for the exchange constants J_1/k (nearest-neighbor cations) and J_2/k (next-nearest neighbors) agree rather well with those found experimentally [25] using in part the molecular-field approximation. However, as is well known, these values do not yield the observed magnetic structures as the most stable ones;

3) There exists a narrow range of p-parameters for Mn^{2+} and S^- where antiferromagnetism of the third kind is most stable. However, the calculated Néel-temperature is too low by a factor of four;

4) In all other cases, antiferromagnetism of the second kind is the most stable magnetic ordering.

The implication of the effective-electron model as applied to this class of compounds is that the magnetic structures *cannot* be explained on the basis of localized electrons. The observed very narrow bandgaps [26] support this conclusion. Further research on the subject is in progress.

References

[1] H.A. Kramers, Physica 1 (1934) 182.
[2] J. Becquerel, W.J. de Haas and J. van den Handel, Physica 1 (1934) 383.
[3] D. de Klerk, Physica XII (1946) 513.
[4] C.G. Shull and J.S. Smart, Phys. Rev. 76 (1949) 1256.
[5] P.W. Anderson, Phys. Rev. 115 (1959) 2.
[6] N. Fuchikami, J. Phys. Soc. Japan 28 (1970) 871.
[7] F. Keffer and T. Oguchi, Phys. Rev. 115 (1959) 1428.
[8] P.W. Anderson, Sol. State Phys. 14 (1963) 99.
[9] C. Herring, Magnetism IIB (1966) 1.
[10] D.J. Newman, Proc. ICM (1976)
[11] J.B. Goodenough, Phys. Rev. 100 (1955) 564; J. Phys. Chem. Solids 6 (1958) 287.
[12] J. Kanamori, J. Phys. Chem. Solids 10 (1959) 87.
[13] See, e.g., L. Jansen and E. Lombardi, Disc. Faraday Soc. 40 (1965) 78.
L. Jansen, in Crystal Structure and Chemical Bonding in Inorganic Chemistry, C.J.M. Rooymans and A. Rabenau, eds. (North-Holland Publ. Co., Amsterdam, 1975) p. 205.
[14] E. Lombardi, R. Ritter and L. Jansen, Int. J. Quant. Chem. 7 (1973) 155.
E. Lombardi, L. Pirola, G. Tarantini, L. Jansen and R. Ritter, ibid 8 (1974) 335.
[15] E. Lombardi, G. Tarantini, L. Pirola, L. Jansen and R. Ritter, J. Chem. Phys. 61 (1974) 894.
E. Lombardi, G. Tarantini, L. Pirola and R. Ritter, J. Chem. Phys. 63 (1974) 2553.
[16] R. Block and L. Jansen, in Quantum Science (Plenum, New York, 1976) p. 123.
[17] L. Jansen, Phys. Rev. 162 (1967) 63.
W. Byers Brown, Chem. Phys. Lett. 2 (1968) 105.
D.S. Farberov, V.Ya. Mitrofanov and A.N. Men, Int. J. Quant. Chem. 6 (1972) 1057.
[18] R. Block, Physica 70 (1973) 397; 73 (1974) 312.
[19] L.J. de Jongh and R. Block, Physica 79B (1975) 568.
[20] L. Jansen, R. Ritter and E. Lombardi, Physica 71 (1974) 425.
[21] The orbitals for the effective electrons were here chosen to be of simple-Gaussian type. Later checks have shown that the results for Slater orbitals are not qualitatively different.
[22] L.J. de Jongh and A.R. Miedema, Adv. Phys. 23 (1974) 1.
[23] J.M. Hastings, N. Elliott and L.M. Corliss, Phys. Rev. 115 (1959) 13.
[24] G. van Kalkeren, R. Block and L. Jansen, submitted for publication.
[25] See, e.g., M.S. Lin and H. Hacker, Solid State Comm. 6 (1958) 687.
[26] G. Brostigen and A. Kjekshus, Acta Chem. Scand. 24 (1970) 2993.

PROSPECTS FOR *AB INITIO* CALCULATIONS AND MODELS OF SUPEREXCHANGE INTERACTIONS

D.J. NEWMAN

Department of Solid State Physics, Research School of Physical Sciences, Australian National University, Canberra A.C.T. 2600, Australia

(Invited paper)

> The distinctions between *ab initio* calculations, parametrization schemes and two types of model of superexchange interactions are made explicit. The advantage of using calculated band structures as a basis for the *ab initio* determination of superexchange parameters for transition metal ions is stressed. A detailed examination of the path model for NiO shows it to be a poor phenomenological model, and an alternative model of transition metal anisotropic superexchange is proposed. Finally, the problems in calculating lanthanide superexchange parameters are surveyed, and it is pointed out that it may be appropriate to use the superposition model in some special cases.

1. Introduction

Ab initio calculations of crystal field parameters [1, 2] for ions with both 3d and 4f open shells in insulating crystals have given insight into the mechanisms involved. In particular, they have provided a useful starting point for the development of simplified models, such as the superposition model [2]. This situation has not yet been reached in the study of superexchange, where relatively few extensive *ab initio* calculations (such as [3]) have been attempted.

A lot more work is involved in superexchange calculations than in crystal field calculations, and usually comparison with experiment is made via a single numerical result. Even when such a calculation is successful the word "fortuitous" immediately springs to mind, and it is never difficult to find processes of possible importance that have been omitted. For this reason we shall not be able to use comparisons between theoretical predictions and experimental results in the following as a test of the validity of methods of calculation. Nevertheless, we may hope that the experimental determination of complete sets of anisotropic parameters for simple systems will take place soon. These will provide an irresistible bait for theoreticians, as it is no more difficult to calculate a set of anisotropic parameters than a single isotropic parameter, and the resulting comparisons between theory and experiment should be very significant [4].

The aim of this paper is to survey the possibilities for the future development of both *ab initio* calculations and models of superexchange interactions. We shall also try to clarify the relationships between several different approaches that already exist in the literature.

2. Effective Hamiltonians

We begin by assuming that sufficient experimental evidence is available to separate superexchange effects from those due to other types of pair interaction [5]. Such effects may then be analysed at three distinct levels:

a) *Effective Hamiltonian parametrization schemes based on general considerations (such as symmetry and the number of electrons involved in the interaction) which aim to be independent of the detailed nature of the physical processes involved.* It has not been sufficiently emphasised in the literature that there is no need to base Hamiltonians of this type on detailed analyses of possible interaction processes. The most general form of anisotropic exchange interaction between an electron i on site A and an electron j on site B can be written [6, 7]:

$$\mathcal{H}(i, j) = -(\tfrac{1}{2} + 2s_i \cdot s_j) \sum_{\substack{K_1 K_2 \\ Q_1 Q_2}} \Gamma^{K_1 K_2}_{Q_1 Q_2} u^{(K_1)}_{Q_1}(A, i) u^{(K_2)}_{Q_2}(B, j), \quad (1)$$

where the unit tensor operators $u^{(K)}_Q$ are defined by their transformation properties with respect to rotations at the appropriate site. The number of non-zero parameters $\Gamma^{K_1 K_2}_{Q_1 Q_2}$ will be restricted by the symmetry of the system containing the two sites. The large number of parameters in eq. (1) make it necessary, in general, to introduce simplified models of the exchange process even before a phenomenological parametrization can be obtained. Nevertheless, some progress has been made in the determination of the general parameters [8, 9] by choosing systems in which the number of parameters in the general parametrization is severely reduced. In particular, Meltzer and Cone [8] have been able to demonstrate, at least in one case, that the two electron parameters $\Gamma_{Q_1^- Q_2^-}^{K_1 K_2}$ are the same for widely separated terms of the same configuration. That is, the two electron exchange parametrization "works", in the same sense as the one-electron crystal field parametrization.

b) *Model parametrizations, in which an attempt is made to reduce the number of phenomenological parameters in eq. (1) by making assumptions about the nature of the physical processes involved.* We shall find it convenient to distinguish two types of models according to their roles. *Heuristic* models relate exchange processes to other types of observable quantity. *Phenomenological* models give a precise reduction in the number of parameters to be fitted to experimental data. The aim of heuristic models is to provide insight into the processes involved, while phenomenological models must be "falsifiable" in the sense that it should be possible to represent a large number of observed energy levels using relatively few parameters.

c) *Full ab initio calculations, in which all parameters are calculated from the Schrödinger equation (albeit with many approximations on the way).* The most detailed of these has been carried out by Fuchikami [3].

Some background material to this paper has been provided by the recent review articles by Levy [4] and Stevens [10]. Levy is mainly concerned with the various forms of eq. (1) which are required for the analysis of experimental data. Stevens surveys possible contributions to exchange parameters using an unorthodox form of perturbation theory, hence producing a prototype *ab initio* calculation, in which numerical values have yet to be inserted.

3. Models and *ab initio* calculations

It is well known that the one-electron model of an insulating crystal containing transition metal ions (such as NiO) allows the electrons to pass freely between 3d states on different ions, and hence determines a 3d energy band. Nevertheless, the "physical" electrons are localised by the electron–electron repulsion on a given ion, which is included in its simplest form in the Hubbard Hamiltonian [11] for 3d Wannier states:

$$H = \sum_{i,j} a_i^\dagger a_j \langle i|h|j\rangle + \tfrac{1}{2} \sum_A \sum_{i,j \text{ on } A} a_i^\dagger a_j^\dagger a_j a_i U_{ij}, \qquad (2)$$

where h represents the one-electron crystal Hamiltonian and $U_{ij} = \langle ij|e^2/r_{12}|ij\rangle$ is the 3d–3d Coulomb integral between electrons in a given ion. Large values of U_{ij} ($= U$, effectively independent of i, j) ensure that all low lying states of the system have the same number of electrons in open shells. If the matrix elements of h connecting Wannier states on different ions are removed, eq. (2) provides a suitable unperturbed Hamiltonian for *ab initio* calculations of superexchange interactions as well as for the construction of models. Unfortunately, this is not always made explicit in detailed calculations [3].

In discussions of the form of effective Hamiltonians it is frequently assumed that each 3d shell contains a given number of electrons at the beginning, rather than include the Coulomb localising term explicitly, giving the unperturbed Hamiltonian as a sum of crystal field Hamiltonians as follows:

$$H_0 = \sum_A h_A = \sum_A \sum_{\text{all } ij \text{ on } A} a_i^\dagger a_j \langle i|h|j\rangle. \qquad (3)$$

Stevens ([10] page 10) objects to this form of H_0 on the grounds that, in distinguishing between the electrons on the different sites, it lowers the symmetry of the system. On the contrary, it is only necessary to assume that H_0 will always be used in relation to properly antisymmetrized many electron states in which there are the *same number* of electrons on each site. At the same time, it is true that H_0 cannot be used as the starting point of a proper perturbation formulation of superexchange, which must allow for the possibility of the virtual excitation of electrons between 3d open shells.

The most important conceptual step in superexchange theory was made by Anderson [12] who suggested that the total exchange interaction between electrons located at sites A and B

could be written as the sum of a negative (ferromagnetic) Coulomb "potential" exchange term $-\langle AB\, e^2/r_{12}|BA\rangle$ and a positive (antiferromagnetic) "kinetic" exchange term of the form $|\langle A|h|B\rangle|^2/U$ corresponding to virtual excitations of electrons between two sites. In these expressions $|A\rangle$ is used to represent an open-shell Wannier (or molecular orbital) state centred at A, h is the one-electron crystal Hamiltonian and U is the previously mentioned correlation energy localizing the electrons at the sites A and B. At the simplest level this model provides qualitative understanding of the properties of superexchange for 3d electrons, such as the Goodenough–Kanamori rules. Moreover, it can be related to *ab initio* formulations of superexchange, at least up to second-order perturbation theory using Wannier function basis states. Nevertheless, detailed *ab initio* formulations show that not all the possibly significant contributions are included in the potential and kinetic exchange terms [10, 13]. For example, important contributions may also arise due to correlated excitations from Wannier states centred on the ligands into open-shell Wannier states [14].

If atomic orbitals, rather than Wannier functions, are used as the basis set in perturbation theory, contributions to the superexchange interaction appear only in high order terms. Nevertheless, it is tempting to start with atomic orbitals as these can be determined with considerable accuracy. Most *ab initio* calculations have been formulated in this way and Fuchikami [3] makes it clear that configuration interaction calculations of this type can be related to the Anderson model. The work of Huang and Orbach [15, 16, 17] is based on the extreme simplification of considering a complex with atomic orbitals on only three ions; the two interacting paramagnetic ions with a single intervening ligand. Fuchikami [3] uses a larger complex and is able to show that the three ion result is reasonably accurate for superexchange in $KMnF_3$, but that this will not be the case in systems with stronger metal–ligand covalent bonding.

If several ligands mediate the interaction in the real system, the final result may be calculated by summing over paths of interaction (remembering that contributions to superexchange are produced by all *circular* paths including both interacting ions, not the *linear* paths between them). This suggests that it should be possible to develop quantitative *path models* in which the total interaction is expressed as a sum of products of terms corresponding to ion–ion interactions [18]. Models of this type have considerable heuristic value in that they can be related to quantities used in discussing crystal field and other metal ligand interactions [14]. We shall discuss them from another point of view in section 5.

4. Band calculations and 3d superexchange

The relationship between Anderson's kinetic exchange contribution and band structure has been discussed by Mattheiss [19] using the LCAO model. It was subsequently pointed out that the matrix elements $\langle A|h|B\rangle$ could be extracted directly from band energies by a transformation method [20], providing a viable alternative to the cluster methods for superexchange calculations. This approach is of particular interest for near-linear metal–ligand–metal arrangements, for the contribution of potential (or direct) exchange is calculated to be relatively small in this case [3].

It is instructive to compare the Fuchikami results [3] for kinetic exchange in $KMnF_3$ with those obtained for $KNiF_3$ [20] by the transformation method (see table I). Although very similar values are obtained in the two methods for t_{2g} state interactions, the band calculation gives considerably larger values for the e_g contributions. This is presumably related to the fact that the cluster calculations assume that all 3d atomic states have the same radius, while a band calculation gives more diffuse unoccupied states. In this respect the band calculation is capable of giving a more accurate representation of the electronic states and should therefore provide better values of $\langle A|h|B\rangle$. (Note, however, that this is a point of difference between Ni^{2+} and Mn^{2+}. In the Mn^{2+} ground state all orbital states have the same radial distribution, while in Ni^{2+} we expect the e_g states to be more diffuse as both the open shell holes are of this symmetry type).

These considerations lead us to suggest that much wider use should be made of band structure calculations as a means of obtaining superexchange parameters. The main problem is that the potential exchange cannot be obtained directly from a band calculation and must therefore be determined independently. Nevertheless, the transformation technique used to obtain the $\langle A|h|B \rangle$ from band energies can also be used for Coulomb integrals [21]. To obtain the various exchange integrals between Wannier functions, it is also necessary to determine a small number of Coulomb integrals between the Bloch functions which may be determined in the band calculation.

A considerable reduction in the number of anisotropic superexchange parameters can be produced simply by assuming the interaction to be factorizable [7, 22]. That is to say, we write the matrix element for an effective exchange Hamiltonian as

$$\langle AB|H_{\mathrm{eff}}|BA \rangle = \langle A|H'|B \rangle \langle B|H'|A \rangle$$

for some one electron Hamiltonian H'. This is clearly appropriate for kinetic exchange and should therefore be a good approximation in cases where kinetic exchange dominates potential exchange, as it does for the 'in line' metal–ligand–metal array. The factorizable model has one important advantage, as a phenomenological model, over path models. The matrix elements $\langle A|H'|B \rangle$ are described by a precise number of undetermined parameters. It can therefore be tested when suitably detailed experimental results on anisotropic superexchange become available, and is in this sense falsifiable.

5. Path models

Path models have a direct intuitive appeal in that they tell us which anisotropic superexchange matrix elements are likely to be large. For example, the second and third rows in table I demonstrate that Ni^{2+}–ligand–Ni^{2+} linear superexchange in both $KNiF_3$ and NiO oxide is dominated by the same processes. In fact, σ-bonding contributes to the results in the first column, while π-bonding contributes to the results in the last two columns. The second column indicates that different interaction paths are available in NiO to those in $KNiF_3$. It is of considerable interest to see whether this qualitative picture can be made quantitatively accurate.

The superposition of path contributions in superexchange is directly analogous to the superposition model [2] in crystal field theory. In the course of analysing contributions to the crystal field parameter 10Dq for Ni^{2+} in NiO, Millar [23] has broken down the transfer integrals between 3d Wannier states, centred on Ni^{2+} ions, into contributions via different paths of interaction. It was found that 13 transfer integrals required 12 path parameters to provide an accurate description. This was necessary because direct interactions between Ni^{2+} ions were found to be important even for third nearest neighbours, and Ni^{2+} ion 4s states can also mediate exchange between other Ni^{2+} ions.

This suggests that path models, especially models in which all interactions are supposed to take place via the negative ions, are bound to be crude. Tables I and II show the results of a simplified path model for (100) and $(\frac{11}{22}0)$ interactions in NiO. In this model five parameters are employed, corresponding to $2p\sigma$, $2p\pi$ and 2s ligand interactions as well as $d\sigma$ and $d\pi$ interactions between Ni^{2+} 3d states in the $(\frac{11}{22}0)$ configuration. It is seen that this model allows an accurate representation of the largest transfer integrals and a reasonable approximation to the smaller ones. Unfortunately, however, a fit of this accuracy can only be made by altering metal–ligand interaction parameters by as much as 100% from their values in the more precise fit mentioned previously. This throws doubt on the phenomenological value of such a parametrization, for the parameters can have no "intrinsic"

Table I
Values of d-state Wannier function diagonal matrix elements $\langle A|h|B \rangle$ for the linear metal–ligand–metal system derived from atomic cluster and band calculations (10^{-2} Ryd).

System	e_g states		t_{2g} states	
$KMnF_3$ cluster [3]	−1.32	−0.01	−0.04	−0.75 −0.75
$KNiF_3$ band	−2.44	−0.04	−0.04	−0.71 −0.71
NiO band [21]	−2.76	−0.14	−0.03	−0.76 −0.76
NiO path model fit	−2.76	0	0	−0.70 −0.70

Table II
Comparison between path model fit and the results of band calculation of 3d Wannier function matrix elements for the 90° metal–ligand–metal configuration in NiO (10^{-2} Ryd).

	e_g states		t_{2g} states	
Band [21]	−0.36 −0.66	0.69	0.06	−2.52
Path model fit	−0.21 −0.51	0.69	−0.01	−2.52

significance independent of the system being analyzed. This is not to suggest that path models are invalid, for we have seen that they can be directly related to configuration interaction calculations. The problem is that such models allow the introduction of too many parameters to be useful in phenomenological fits to data.

The experimental work of Johnson and Sievers [24] has recently focussed theoretical interest on the problem of distance dependence of the superexchange parameters for transition metal ions [25, 26]. However, detailed consideration of calculations such as [3] show that they are bound to produce power law exponents of the order of -10, or approximately twice the corresponding exponent for crystal field parameters. We doubt whether the existing experimental data is sufficiently accurate to determine the relative merits of different calculations or models. Nevertheless, the path model did lead Johnson and Sievers [24] to the conclusion that vacant 4s states on transition metal ions play an important role in exchange interaction, as we have already noted as a result of the theoretical analysis described above.

6. Lanthanide superexchange

Very little theoretical work has been carried out on lanthanide ion superexchange and no really convincing attempts at *ab initio* calculations of superexchange parameters for pairs of lanthanide ions in insulators have apparently been made. Most interest has centred on the exchange constants for EuO, because of its ferromagnetism. The calculations on this system carried out by Ritter, Jansen and Lombardi [27] and Falkovskaya and Sapozhnikov [28] are, to the author's knowledge, the only attempts that have been made at calculating lanthanide ion superexchange from first principles. In both cases extreme approximations are employed to evaluate isotropic exchange parameters which themselves have a considerable degree of experimental uncertainty. Ritter et al. [27] use wave functions which do not have the correct angular or radial form. Falkovskaya and Sapozhnikov [28] do not consider the possibility of the $5s^2p^6$ outer shell electrons being involved in the interaction process.

Several authors have suggested additional processes which may be important in lanthanide superexchange interactions. Smit [29] has emphasised the possible importance of the excited 5d, 6s and 6p states. This would produce an exchange process similar to the indirect exchange process known to dominate in lanthanide metals [30], but with absence of occupied 5d and 6s states. There is more convincing evidence [31, 32] that spin-polarization of the lanthanide closed $5s^2p^6$ shell plays an important role in exchange. If, following Anderson [12], we formulate these interactions in terms of Wannier functions centred on the lanthanide ions at centres A and B, the kinetic exchange term

$$\langle 4f_A | h | 4f_B \rangle \langle 4f_B | h | 4f_A \rangle / U$$

is supplemented by contributions of the form

$$\left\langle 4f_A 4f_B \left| \frac{e^2}{r_{12}} \right| 4f_B 4f_A \right\rangle,$$

which includes terms linear in the 4f/5p atomic exchange integral, and

$$\left\langle 5p_A 4f_A \left| \frac{e^2}{r_{12}} \right| 4f_A 5p_A \right\rangle \langle 5p_A | h | 5p_B \rangle \langle 5p_B | h | 5p_A \rangle$$
$$\times \left\langle 5p_B 4f_B \left| \frac{e^2}{r_{12}} \right| 4f_B 5p_B \right\rangle,$$

where energy denominators have been omitted with last expression. A configuration interaction calculation of these contributions using atomic states is feasible, although it would be necessary to allow for screening of the Coulomb integrals due to higher order effects. Band structure calculations for lanthanide insulators are not yet sufficiently reliable to be used as a basis for calculating the one electron integrals.

Cho [33] has suggested that his spin polarized band structure calculations [34] may provide a useful starting point for the determination of superexchange in EuO. Unfortunately, however, his own attempt to do this [33] is invalid, as the energy denominator U of the Anderson model is misinterpreted as being the ligand (2p)→metal (3d) excitation energy. Spin-polarized band calculations can be used to determine the Wannier function matrix elements of an effective exchange potential h_{ex}, which may be defined as follows:

$$\langle A|h_{ex}|A\rangle \equiv \sum_C \left\langle AC \left| \frac{e^2}{r_{12}} \right| CA \right\rangle,$$

$$\langle A|h_{ex}|B\rangle \equiv \sum_C \left\langle AC \left| \frac{e^2}{r_{12}} \right| CB \right\rangle.$$

Here the summations are to be taken over states as well as sites. These matrix elements represent a "molecular field", but do not determine pair interaction matrix elements of the form $\langle AB|e^2/r_{12}|BA\rangle$ and hence cannot be employed in calculations of potential exchange. It follows that spin-polarized band calculations do not provide more information than unpolarized calculations for the determination of superexchange pair interactions.

It is suggested that future effort on lanthanide superexchange calculations would be better directed at the system where relatively abundant data on the isotropic parameters exists [36], as well as the most significant work to date on anisotropic parameters [8]. This is the Gd^{3+} pair system in $LaCl_3$ and isomorphic crystals.

In the absence of reasonable ab initio calculations for the lanthanides we have no basis for discussing the relative advantages of possible models. Certainly there is no overwhelming experimental evidence in favour of a particular process. We therefore conclude by suggesting a model for systems in which exchange is manifested as a molecular field effect, rather than as a pair interaction. The outstanding example of this is interaction of lanthanide ions with the Fe^{3+} sublattice in iron garnets. The appropriate parametrization for this system has been given by Levy [36] in the form

$$\mathcal{H}_{ex} = \sum_{kq} \alpha_{kq} O_q^{(k)}(L) M_{Fe} \cdot S,$$

where S and L operate on lanthanide 4f states and M_{Fe} is the net magnetization of the iron sublattice. Several attempts have been made to determine the ten parameters α_{kq} which are nonzero in the D_2 symmetry of the lanthanide sites in the garnets [9], but without significant results. A model of $\mathcal{H}_{ex}$ would therefore help if it reduced the number of free parameters below ten.

If the ligands can be regarded as transmitting iron sublattice magnetization by a contact interaction to the 4f electrons, then it seems appropriate to use the superposition model which has been particularly successful for lanthanide crystal field parameters in the garnets [2]. In this model the observed potential is regarded as a sum of axially symmetric contributions from the ligands. In the present case this model would reduce the number of free parameters to seven, or four plus three power law exponents representing the distance dependence of the single ligand interactions. We would expect these exponents to be very similar to those already determined for the corresponding crystal field parameters. If this were so, the ratios of the α_{kq} for a given k would be similar to that of the corresponding crystal field parameters, provided that an equivalent normalization is chosen for both sets of parameters.

References

[1] T.S. Soules, J.W. Richardson and D.M. Vaught, Phys. Rev. B3 (1971) 2186.
[2] D.J. Newman, Adv. Phys. 20 (1971) 197.
[3] N. Fuchikami, J. Phys. Soc. Japan 28 (1970) 871.
[4] P.M. Levy, in: Magnetic Oxides, D.J. Craik, ed (Wiley, London, 1975) p. 181.
[5] J.M. Baker, Rep. Prog. Phys. 34 (1971) 109.
[6] R.J. Elliott and M.F. Thorpe, J Appl. Phys. 39 (1968) 802.
[7] G.M. Copland and D.J. Newman, J. Phys. C: Solid St. Phys. 5 (1972) 3253.
[8] R.S. Meltzer and R.L. Cone, Phys. Rev. B13 (1976) 2818.
[9] E. Orlich and S. Hüfner, Z. Phys. 232 (1970) 418.
[10] K.W.H. Stevens, Phys. Reps. 24 (1976) 1.
[11] J. Hubbard, Proc. Roy. Soc. A276 (1963) 238.
[12] P.W. Anderson, Phys. Rev. 115 (1959) 2.
[13] K. Gondaira and Y. Tanabe, J. Phys. Soc. Japan 21 (1966) 1527.
[14] S. Freeman, Phys. Rev. B7 (1973) 3960.
[15] N.L. Huang and R. Orbach, Phys. Rev. 154 (1967) 487.
[16] N.L. Huang, Phys. Rev. 157 (1967) 378.
[17] N.L. Huang, Phys. Rev. 164 (1967) 636.
[18] D.J. Newman, J. Phys. C: Solid St. Phys. 5 (1972) 1089.
[19] L.F. Mattheiss, Phys. Rev. B2 (1970) 3918.
[20] D.J. Newman, J. Phys. C: Solid St. Phys. 6 (1973) 2203.
[21] D.J. Newman and B.F. Lau, J. Phys. C: Solid St. Phys. 7 (1974) 2283.
[22] I. Veltruský, Czech. J. Phys. B25 (1975) 101.
[23] D.P. Millar, Research Report, Dept. of Solid State Physics, Australian National University (1976).
[24] K.C. Johnson and A.J. Sievers, Phys. Rev. B10 (1974) 1027.
[25] L.J. de Jongh and B. Block, Physica 79B (1975) 568.
[26] K.N. Shrivastava and V. Jaccarino, Phys. Rev. B13 (1976) 299.
[27] R. Ritter, L. Jansen and E. Lombardi, Phys. Rev. B8 (1973) 2139.

[28] L.D. Falkovskaya and V.A. Sapozhnikov, Phys. Stat. Sol. (b) 71 (1975) 469.
[29] J. Smit, J. Appl. Phys. 37 (1966) 1455.
[30] A.J. Freeman, in: Magnetic Properties of Rare Earth Metals, R.J. Elliott, ed (Plenum Press, London, 1972) p. 245.
[31] M.I. Bradbury and D.J. Newman, J. Phys. Chem. Solids 32 (1971) 627.
[32] G.M. Copland, Chem. Phys. Lett. 7 (1970) 175.
[33] S.J. Cho, Phys. Lett. 29A (1969) 129.
[34] S.J. Cho, Phys. Rev. B12 (1970) 4589.
[35] R.W. Cochrane and W.P. Wolf, Solid St. Comm. 9 (1971) 1997.
[36] P.M. Levy, Phys. Rev. 135 (1964) A155.

CONTRIBUTION TO THE TEMPERATURE DEPENDENCE OF THE EXCHANGE INTERACTION BY MODULATION OF THE EXCHANGE INTEGRAL IN CrBr$_3$

J.E. DRUMHELLER and C.E. ZASPEL

Department of Physics, Montana State University, Bozeman, Montana, 59715, USA

The direct modulation of the overlap integral by optical phonons through two independent paths is shown to contribute to the temperature dependence of the symmetric exchange interaction in CrBr$_3$.

The temperature dependence of the electron paramagnetic linewidth in CrBr$_3$ has been measured by Seehra and coworkers [1–3] and shows a variation of a factor of about five over a range from 50 to 500 K. These workers have attributed this variation to spin–lattice relaxation via phonon modulation of the crystalline electric field which is essentially a temperature dependence of the spin–spin correlation function. In this work we show that direct phonon modulation of the exchange integral can account for a significant portion of the temperature dependence of the linewidth.

Previously [4] we presented a model which assumed an exponential form for the exchange integral of the form $J(R) = J_0 e^{-\lambda \delta R}$ in which λ is a constant and δR is the departure from equilibrium, as well as anharmonic intermolecular potential between independent oscillators. Using bond strengths for the metal–ligand bond that were taken from known tabulated results and overlap integral that could be approximated by Slater-type orbitals we were able to get good agreement between theory and experimental results for a number of different cases. The advantage of this model is that within the assumptions made there are no adjustable parameters, although the bond strengths must be readjusted to suit the appropriate solid. No crystal symmetry appears in the model but the best agreement has occurred for layered structures where the Einstein-like optical phonon contribution is restricted to a plane. It would seem, however, that the model should work in a wide variety of situations including S-state, non-S-state systems in which an antisymmetric exchange interaction is present, and amorphous systems.

In CrBr$_3$ the superexchange has two dominant paths: intraplanar exchange where there is an intervening Br ion and interplanar where there are two Br ions to bridge the Cr [5]. The model we choose assumes that the independent oscillators contain the entire CrBr$_3$ molecule and that the exchange takes place through the anharmonic potential between CrBr$_3$ groups shown in fig. 1. Separately, one may consider the intraplanar case where the "spring" is the Cr^{3+}–Br$^-$ and for the interplanar it is a Br$^-$–Br$^-$ bond. For each case, the calculation requires an estimation of the overlap integral between the ions that form the exchange bridge, or spring, in order to find the coefficient in the exponential exchange. As before the method uses that of Slater [6] and the assumption that the Slater-type overlap has the same form as that of the exchange integral. This produces almost identical values of $3 \times 1(10)^8$ cm^{-1} for both the Br–Br and Cr–Br bridges.

To perform a thermal average, a Morse potential is assumed so that the particular states used are perturbed harmonic oscillator states and the depth of the potential is given in standard references as the metal–ligand or, for the particular case of the Br–Br bridge, the ligand–ligand bridge [7].

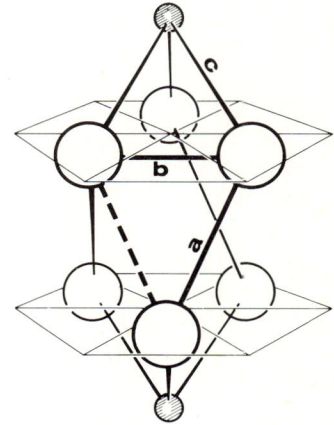

Fig. 1. The structure of CrBr$_3$ showing the two predominate exchange paths. The interplanar Br–Br' distance a is 3.56 Å, the intraplanar Br–Br distance b is 3.66 Å, and the intraplanar Cr–Br distance c is 2.59 Å. The Cr–Cr planes are separated by 6.10 Å. See ref. 5.

The question of how to readjust the bond strengths of the monomuclear and heteronuclear molecules to account for the appropriate bonds in the solid is not easily resolved. Previously [4] we used an adjustment that has its origins in Paulings method of the geometric mean [8] but that method depended more on the covalent nature of the bonds than seems to be justified by the primarily ionic character of $CrBr_3$. If one uses purely ionic radii for the Cr^--Br^- bond one finds the lengths in $CrBr_3$ to be about the same as that of the heteronuclear molecule so that the bond strength may reasonably be assumed to be of the same magnitude or about 78 kcal/mol [7]. Since superexchange is present, however, there is at least some indication that the bond has overlap of the wave functions therefore some covalent character so that this value is probably a little high. The Br^--Br^- bond is likely much weaker although the character seems to be unknown. Inasmuch as the interplanar exchange is ferromagnetic and down only a factor of $6(10)^{-2}$ of the intraplanar, it indicates at least some overlap [9]. As an upper limit the weak bond is assumed to be 50 kcal/mol for the sum of the three Br^--Br^- bonds and as a lower limit we choose 5 kcal/mol. The latter is the correct order if the bonds are primarily covalent and the method of Pauling was used.

Finally, because the exponential part of the exchange integral and the bond energy as a function of R should be the same, the Morse potential affords a method to directly relate the Slater-type overlap which in this case is λ, to the non-harmonic parameter in the Morse potential.

By inserting the reduced mass of the $CrBr^3$ complex the thermal average over the appropriate number of states can be made giving the value of the interplanar exchange J at each temperature. Finally, the exchange as a function of temperature can be compared to the linewidth through the expression $\Delta H = bM_2/2J$, where b is a constant and M_2 is the second moment of the resonance line [10], which for this calculation is considered a constant. The cross-hatch region of fig. 2 shows the calculated temperature dependence of J for the contribution of the Br^--Br^- bond over the order of magnitude of possible bond strengths discussed above. Also shown is the possible dependence

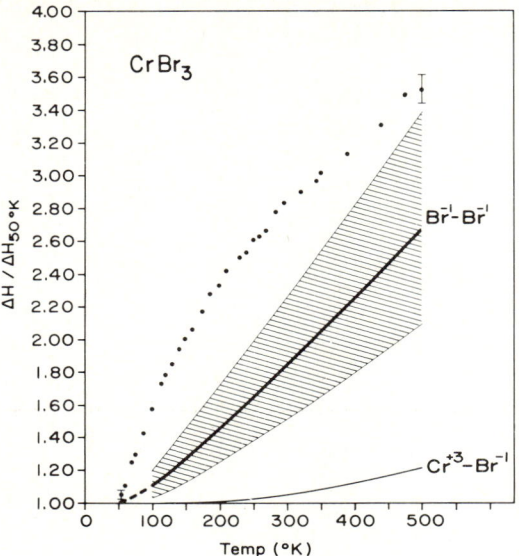

Fig. 2. Contributions to the linewidth owing to the temperature dependence of J, the exchange energy, as a function of temperature. The variation in dependence for an order of magnitude possible variation in bond strength for the interplanar Br-Br bonds is shown by the cross-hatched region. The intraplanar Cr-Br bond is shown by the smooth line. Circles are the data of Seehra and co-workers.

owing to considering only the in-plane vibrations of the $Cr^{3+}-Br^{\pm}$ bonds. It is apparent that if the above assumptions of the model are correct that direct overlap of wave functions may give a significant contribution to the temperature dependence of the exchange energy.

References

[1] M.S. Seehra and R.P. Gupta, Phys. Rev. 113 (1974) 197.
[2] M.S. Seehra and D.L. Huber, AIP Conf. Proc. 24 (1974) 261.
[3] D.L. Huber and M.S. Seehra, J. Phys. Chem. Solids 36 (1975) 723.
[4] C.E. Zaspel and J.E. Drumheller, Solid State Commun. 17 (1975) 1107. A more detailed account is given in C.E. Zaspel and J.E. Drumheller, submitted to Phys. Rev.
[5] I. Tsubokawa, J. Phys. Soc. Jap. 15 (1960) 1664.
[6] J.C. Slater, Phys. Rev. 36 (1930) 57.
[7] See, for example, Handbook of Chemistry and Physics, 51st ed. (Chemical Rubber Publishing Co., 1970–71), pp. F130–131.
[8] L. Pauling, The Nature of the Chemical Bond (Connell University Press, Ithaca, New York, 1960) p. 256, p. 93.
[9] E.J. Samuelson, R. Silberglitt, G. Shirane and J.P. Remeika, Phys. Rev. B3 (1971) 157.
[10] J.H. Van Vleck, Phys. Rev. 74 (1948) 1168.

DETERMINATION OF EXCHANGE PARAMETERS OF Cr^{3+} PAIRS IN $MgAl_2O_4$ BY OPTICAL AND ESR MEASUREMENTS

J.C.M. HENNING and H. VAN DEN BOOM
Philips Research Laboratories, Eindhoven, The Netherlands

Exchange interactions between nearest-neighbour Cr^{3+} ions ions in $MgAl_2O_4$ are studied by ESR and optical techniques. The splitting of the $(^4A_2, ^4A_2)$ ground state of the pair is described by a conventional coupling between ionic spins $S_1 = S_2 = \frac{3}{2}$. Since the first excited pair state $(^4A_2, ^2E)$ is orbitally degenerate, its exchange splitting has to be described by an orbitally dependent coupling between electronic spins $s_i = s_j = \frac{1}{2}$, summed over all occupied orbitals.

1. Introduction

Electron spin resonance (ESR), luminescence and luminescence excitation spectra of nearest-neighbour chromium pairs are studied in single crystals and powders of the spinel $MgAl_2O_4$, moderately doped with Cr (0.3–10 at.%). The objective of this investigation is to obtain information about the 90° Cr–Cr interaction. A comparison of the present results with those for the analogous system [1, 2] $ZnGa_2O_4$:Cr^{3+} with a 3% larger lattice parameter gives an insight into the dependence of exchange parameters on interionic distance.

The geometry of the nearest-neighbour Cr pair is shown in fig. 1 of ref. 3. The crystal field at the Al sites in spinel is predominantly octahedral, with a small trigonal component along a local ⟨111⟩ axis. The ground state of $Cr^{3+}(3d^3)$ in this crystal field is an orbital singlet $^4A_2(^4F)$. The lowest excited state is $^2E(^2G)$, at about 14 000 cm^{-1} above the ground state. Both the pair ground state $(^4A_2, ^4A_2)$ and first excited state $(^4A_2, ^2E)$ have been investigated.

2. ESR results

The exchange splitting of the $(^4A_2, ^4A_2)$ pair ground state can be studied in great detail by means of ESR measurements on single crystals. The pair spin hamiltonian is written

$$\mathcal{H}_p = \mathcal{H}_1 + \mathcal{H}_2 + \mathcal{H}_{ex}, \qquad (1)$$

with

$$\mathcal{H}_i = g\beta \boldsymbol{H} \cdot \boldsymbol{S}_i + D_p[S_{\zeta i}^2 - \tfrac{1}{3}S_i(S_i+1)] \\ + E_p(S_{\xi i}^2 - S_{\eta i}^2), \qquad (2)$$

and

$$\mathcal{H}_{ex} = -J\boldsymbol{S}_1 \cdot \boldsymbol{S}_2 + \boldsymbol{S}_1 \cdot \boldsymbol{J} \cdot \boldsymbol{S}_2 + j(\boldsymbol{S}_1 \cdot \boldsymbol{S}_2)^2, \qquad (3)$$

where S is the *ionic* spin $(S = \frac{3}{2})$ and $\boldsymbol{J}$ is a traceless tensor. The form eqn (3) of H_{ex} is justified since the pair ground state is orbitally nondegenerate.

Since J is larger than the microwave quantum (0.3 cm^{-1}), only transitions within the total spin multiplets $\Sigma = 3, 2, 1$ can be observed. The zero-field splittings of the multiplet $\Sigma = 2$ are solely due to $\boldsymbol{J}$ and do not depend on the single-ion anisotropy [4] (D_p, E_p) in the pair. The angular dependence of the four allowed transition within $\Sigma = 2$ can therefore be used to find the principal values and principal axes of the $\boldsymbol{J}$ tensor. The result is:

$J_{PP} = +0.1015 \pm 0.0025$ cm^{-1},

$J_{QQ} = +0.0535 \pm 0.0025$ cm^{-1},

$J_{RR} = -0.1550 \pm 0.0010$ cm^{-1},

where $P = [001]$, $Q = [1\bar{1}0]$, $R = [110]$ for a Cr–Cr pair along the R axis. Although an orthorhombic coupling is compatible with the C_{2v} pair symmetry, the physical origin of the orthorhombic term needs explanation, especially since the coupling is perfectly axial (along R) in the case of $ZnGa_2O_4$:Cr. On the assumption that the Cr–Cr separation is 1.5% larger than the Al–Al separation in $MgAl_2O_4$, the observed tensor $\boldsymbol{J}$ can be decomposed into a point dipole–dipole interaction (axial along the pair axis R) and an axial tensor along the P axis:

$J_{PP}(\text{obs.}) = +0.0695$ cm^{-1} (d–d)
$\qquad\qquad\qquad + 0.0320$ cm^{-1} (anis. exch),

$J_{QQ}(\text{obs.}) = +0.0695$ cm^{-1} (d–d)
$\qquad\qquad\qquad - 0.0160$ cm^{-1} (anis. exch.),

$J_{RR}(\text{obs.}) = -0.1390$ cm^{-1} (d–d)
$\qquad\qquad\qquad - 0.0160$ cm^{-1} (anis. exch.).

An axially symmetric anisotropic exchange along the P axis is to be expected if the two Cr^{3+} ions and the two oxygens through which the superexchange interaction takes place lie in a plane perpendicular to the P axis [5].

The values of D_p and E_p can be obtained from the angular dependence of $\Sigma = 3$ transitions. Transitions within $\Sigma = 1$ are not observed in an X-band experiment since the zero-field splitting of the $\Sigma = 1$ state exceeds the magnitude of the microwave quantum [4]. We find $D_p = +1.042 \pm 0.005$ cm^{-1}, $E_p = -0.038 \pm 0.002$ cm^{-1} which differs appreciably from the single-ion values: $D = +0.910 \pm 0.002$ cm^{-1}, $E = 0$. This points to a local lattice distortion around the Cr–Cr pair [3]. No such distortion was found in $ZnGa_2O_4:Cr^{3+}$ [1, 2].

The values of J and j are obtained from the temperature dependence of the intensities of $\Sigma = 2$ and $\Sigma = 3$ transitions. The result is $J = -28.5 \pm 2$ cm^{-1}, $j = -2.1 \pm 0.7$ cm^{-1}.

The negative sign of J accords with the negative Θ value of $MgCr_2O_4$ ($\Theta = -350$ K [6]. Using $3k\Theta = 6S(S+1)J$ it follows that 87% of the Θ value can be accounted for by nearest-neighbour interactions. The large, negative, biquadratic exchange term can be explained as being due to exchange-striction effects [7, 8].

3. Optical results

Luminescence and luminescence excitation spectra were taken from powder samples with various Cr concentrations. The $(^4A_2, ^2E) \rightarrow (^4A_2, ^4A_2)$ pair transitions appear as sharp, weak satellites of the single ion R line ($^2E \rightarrow {}^4A_2$). By studying the relative intensities as a function of Cr concentration a definite assignment to pair lines can be made. Since the excited state $(^4A_2, ^2E)$ is orbitally degenerate, the exchange interaction cannot be described by a simple coupling between ionic spins, like eq (3). Instead, a more sophisticated "orbital-dependent exchange of hamiltonian" [9–11]

$$\mathcal{H}_{ex} = -\sum_{ij} J_{ij} s_i \cdot s_j \qquad (4)$$

has to be used. Here, s_i, s_j are *electronic* spins ($s_i = s_j = \frac{1}{2}$) and the sum is over the occupied t_{2g} orbitals: $\xi = |yz\rangle$, $\eta = |zx\rangle$, $\zeta = |xy\rangle$. The local C_{2v} pair symmetry leads to four independent J_{ij} parameters. As a consequence of the interaction (4) the excited state $(^4A_2, ^2E)$ is split into eight sublevels, four with total spin $\Sigma = 1$, and four with total spin $\Sigma = 2$. The energies of these levels can be expressed in the four J_{ij} parameters and E_0, being the $^2E-{}^4A_2$ splitting for a pair (which may differ slightly from the single-ion value).

Luminescence spectra taken at $T = 77$ K show only pair lines originating from the lowest excited level. Four zero-phonon pair lines are observed, which form a Landé sequence from which the *ground state J and j* values can be deduced. The parameters obtained in this way agree with the ESR values. In addition to these zero-phonon lines some phonon sidebands of the strongest pair line (N_3) are observed.

In analogy with $ZnGa_2O_4$ much more experimental information can be expected by studying the excitation spectra with the luminescence tuned to one of the phonon sidebands of N_3. In this way pair lines from higher excited sublevels of $(^4A_2, ^2E)$ can also be observed. Since excitation measurements are essentially measurements of the absorption spectrum (detected by luminescence), the temperature dependence of the intensities gives direct information about the spins of the ground state levels. The spins of the excited state levels then follow from the selection rule $\Delta\Sigma = 0$, which has been established [2, 14] for the strongest (exchange-induced electric dipole) pair transitions.

Experimentally we have established only three of the eight excited state levels with certainty. This data apparently is insufficient to determine the exchange splitting of the excited state. The main problem with the spectrum is the presence of strong phonon sidebands which mask the zero-phonon pair lines.

References

[1] J.C.M. Henning, J.H. den Boef and G.G.P. van Gorkom, Phys. Rev. B7 (1973) 1825.
[2] G.G.P. van Gorkom, J.C.M. Henning and R.P. van Stapele, Phys. Rev. B8 (1973) 955.
[3] J.C.M. Henning and H. van den Boom, Phys. Rev. B8 (1973) 2255.
[4] J. Owen, J. Appl. Phys. 32 (1961) 213.
[5] R. W. Bené, Phys. Rev. 178 (1969) 497.
[6] G. Blasse and J.F. Fast, Philips Res. Rep. 18 (1963) 393.
[7] C. Kittel, Phys. Rev. 120 (1960) 35.
[8] J.C.M. Henning and J.P.M. Damen, Phys. Rev. B3 (1971) 3852.
[9] J. Ferguson, H. J. Guggenheim and Y. Tanabe, J. Phys. Soc. Jap. 21 (1966) 692.
[10] M.H.L. Pryce, unpublished.
[11] N.L. Huang, Phys. Rev. B1 (1970) 945.
[12] J.P. van der Ziel, Phys. Rev. B4, (1971) 2888.

TRANSFERRED HYPERFINE INTERACTIONS OF THE ^{1}H, ^{2}D AND ^{17}O NUCLEI IN A SERIES OF B.C.C. HEISENBERG FERROMAGNETS WITH THE $K_2CuCl_4 \cdot 2H_2O$ STRUCTURES

W.J. LOOYESTIJN, T.O. KLAASSEN and N.J. POULIS

Kamerlingh Onnes Laboratorium der Rijksuniversiteit, Leiden, The Netherlands

In the series of the Heisenberg ferromagnets with the $K_2CuCl_4 \cdot 2H_2O$ structure the nuclear magnetic and quadrupolar interaction parameters of the ^{1}H, ^{2}D and ^{17}O are determined. From the ^{17}O spectrum the occurence of magnetostriction effects is evident.

From many experiments it is known that the series of compounds with the $M_2CuX_4 \cdot 2H_2O$ structure (MX = KCl, NH_4Cl, RbCl, CsCl, NH_4Br and RbBr) behave like b.c.c. Heisenberg ferromagnets, with a critical temperature of about 1 K. As the coupling between the magnetic copper ions is due to superexchange interaction via two or more intermediate diamagnetic ions, it is interesting to determine the unpaired spin density on these diamagnetic ligands. Until now all hyperfine interaction parameters have been reported except those of the oxygen and hydrogen nuclei, which will be presented in this paper.

In all six compounds the ^{17}O NMR spectra have been determined in the paramagnetic and in the ferromagnetic state in zero field. Single crystals have been grown from 15 at.% ^{17}O enriched water. In the ferromagnetic state at 100 mK also in not-enriched samples these spectra could be observed as a result of the great r.f. enhancement. The crystals have tetragonal symmetry with space group $P4_2/mnm$, with two chemically equivalent formula units in the unit cell, related to each other by a $\frac{1}{2}\pi$ rotation about the c-axis. In the discussion we confine ourselves to the $(CuX_4 \cdot 2H_2O)^{2-}$ octahedron at (0, 0, 0) (fig. 1). The water molecule lies in a (110) plane, i.e. two sp^3 hybridized orbitals containing the ^{1}H ions are situated in (110), the other two (lone pair) lie in ($1\bar{1}0$). That means that the oxygen 2p orbitals point along the a-axis and the c-axis. (110) and ($1\bar{1}0$) are mirror planes, so the two 2p orbitals along the a axis are equivalent. Hence, the magnetic hyperfine interaction of the oxygen nucleus with the unpaired electron spin is expected to show axial symmetry around the c axis. From the crystal symmetry it can also be argued that the principal axes of the electric field gradient

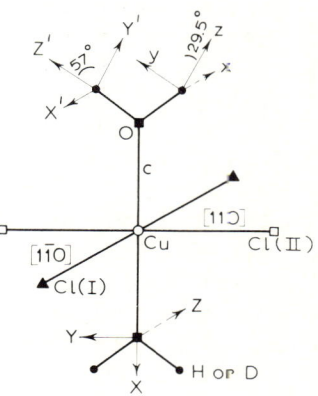

Fig. 1. Structure of the copper octahedron at (0, 0, 0), (X, Y, Z) and (X', Y', Z') are the directions of the principal axes of the EFG tensors at the oxygen and deuterium site and (x, y, z) those of the dipolar tensor at the hydrogen site.

(EFG) tensor are directed along the [110], [1$\bar{1}$0] and c-axes.

As the ^{17}O nuclear hyperfine interactions do not differ very much in these six compounds, we will restrict ourselves here, for the sake of clarity, to the presentation of the results of $Rb_2CuCl_4 \cdot 2H_2O$.

The nuclear spin hamiltonian is given by

$$\mathcal{H} = -h\mathbf{I} \cdot \left(\frac{1}{\gamma}\mathbf{H}_0 + \mathbf{A} \cdot \mathbf{g} \cdot \langle \mathbf{S} \rangle\right)$$
$$-\tfrac{1}{6}\nu_q\{3I_z^2 - I(I+1) + \tfrac{1}{2}\eta(I_+^2 + I_-^2)\},$$

where all symbols have their usual meaning.

In the paramagnetic state at $T = 1.2$ K and $H = 10$ kOe, the oxygen spectrum consists of a central line with two pairs of quadrupolar satelites, and may accurately be analysed with first and second order perturbation calculations (high-field case). The X, Y and Z axes of the EFG tensor are found to be along the c, [110] and [1$\bar{1}$0] axes, respectively. The quadrupolar splittings along the principal axis, however, are definitely not consistent with one set of quadrupole parameters ν_q and η.

In the chlorine compounds the spontaneous magnetization in the ferromagnetic state is directed along [110], so that the Y and Z spectra can be observed in the ordered state. The ferromagnetic X spectrum can be found in the bromides, as in these compounds the magnetization points along the c axis. Comparing the analogous spectra in the para- and ferromagnetic state, it can be concluded, that there are no significant differences in the X and Z spectra. The quadrupole splittings in the Y spectrum is significantly smaller in the ferromagnetic than in the paramagnetic state. As the quadrupolar interaction depends strongly on the covalent bonding, it must be concluded that the origin of the deviating value of the ^{17}O quadrupolar interaction for the magnetic field along [110] (Y spectrum) is the presence of induced and spontaneous magnetostrictive distortions of the copper octahedron. Magnetostriction effects have also been observed [1] in the Cl(II) spectra, when the magnetization is along [110].

In $Rb_2CuCl_4 \cdot 2H_2O$ we find in the paramagnetic state, after correction for the dipolar interaction, $A_{XX} = 42.16$ Mhz (c-axis), $A_{YY} = 29.34$ MHz ([110]) and $A_{ZZ} = 32.21$ Mhz ([1$\bar{1}$0]). The non-axiality of **A** is a further indication of magnetostriction. From A_{XX} and A_{ZZ}, and taking $A_{YY} = A_{ZZ}$, the unpaired spin density are calculated to be $f_s = +0.75\%$ and $f_\sigma - f_\pi = +3.3\%$. The total spin density at the oxygen ligand is approximately 4%. Again from the X and Z spectra, the values $\nu_q = 1.28$ MHz and $\eta = 0.50$ are determined. As the hydrogen (or deuterium) ion forms mainly s bonds, the magnetic interaction tensor of the hydrogen nucleus consists of an isotropic hyperfine term A_s and an anisotropic dipolar tensor D. From symmetry considerations it can easily be concluded that one principle axis of D points along [1$\bar{1}$0] and the other two lie in the (1$\bar{1}$0) plane. The same conclusions hold for the EFG tensor. In fig. 1 x, y, z are the experimentally determined directions of the principal axis of D. In $Rb_2CuCl_4 \cdot 2H_2O$ we have found $D_{xx} = -1.09$, $D_{yy} = -2.92$ and $D_{zz} = 4.11$ Mhz and $A_s = 2.82$ Mhz. The spin density at the H ion is calculated to be $f_s = -0.33\%$. In deuterated crystals the ^{2}D quadrupolar parameters have been derived from 4 to 300 K. No strong temperature dependence have been found for temperatures up to 200 K. At higher temperature the spectrum changes drastically due to tunneling of the D ions. At 20 K the experimental quadrupole parameters in $Rb_2CuCl_4 \cdot 2D_2O$ are $\nu_q = 319$ khz and $\eta = 0.11$.

An analysis of the variations in the above mentioned parameters over the series of the six compounds will give more insight in the role of the H_2O molecule in the exchange paths between the copper ions. For a detailed discussion we refer to a forthcoming paper.

The investigations are supported by Stichting 'F.O.M.'

Reference

[1] T.O. Klaassen, W.J. Looyestijn and N.J. Poulis, Physica 77 (1974) 43. and refs. therein.

HYPERFINE AND EXCHANGE INTERACTIONS IN THE EuX HEISENBERG FERROMAGNETS

Ch. SAUER and W. ZINN

Institut für Festkörperforschung der Kernforschungsanlage, Jülich, D-5170 Jülich, W. Germany

Transferred hyperfine fields measured by nuclear resonance techniques at Eu nuclei in EuX materials support theoretical models for two different transfer and exchange mechanisms: long range interactions via valence band electrons extending beyond the sixth Eu neighbours, and short range interactions due to d-orbital overlap between Eu nearest neighbours predominant in EuO.

Referring to a recent review [1] we report here on further data from ^{151}Eu and ^{153}Eu nuclear resonance experiments [2–4] in the EuX series of magnetic semiconductors and summarize the results to be related with the Heisenberg–Dirac–Van Vleck operator

$$\hat{H}_e = -2\hat{S}_i \sum_r z_r J_r \hat{S}_r \qquad (1)$$

and the attributed ordering temperature

$$\theta = [2S(S+1)/3k_B] \sum_r z_r J_r. \qquad (2)$$

Our discussion is based on the relations expected between θ and J of eq. (2) and the total transferred hyperfine (h.f.) field at an Eu nucleus at site i given by

$$B_i^{THF} = B_I - B_0 = z_1 B_1^{THF} + \sum_r^n z_r B_r^{STHF} \qquad (3)$$

as discussed in more detail elsewhere [1, 5]. B^{THF} therefore is the difference between the total measured h.f. field, B_I, and the intrinsic h.f. field, B_0, due to the Eu spin S_i at site i itself and consists of the $z_1 = 12$ transferred contributions B_1^{THF} of the Eu nearest neighbours and the sum over all supertransferred contributions B_r^{STHF} [5].

We also include in the following discussion the results on B_0 reported by K. Kojima et al. [6] from an EPR study of Eu^{2+} ion in diamagnetic SrX host crystals.

In table I all data on B_I, B_C, and B^{THF} of eq. (3) for the different ordered spin structures of the EuX members are compiled with the related contributions $z_r B_r$ to B^{THF} given in column 5.

The plots of B_I and B_0 versus the ^{151}Eu isomer shift, i.e. the s-electron density, $\rho_s(0)$, in the nuclear volume, in fig. 1(a), and of B_{NNN}^{THF} versus θ in fig. 1(b) suggest the following conclusions:

(1) A linear relation is established fairly well between EuS and EuTe for both B_I and $\rho_s(0)$ in fig. 1(a) and B^{THF} and θ in fig. 1(b). This is

Table I
Hyperfine field (B_I) with intrinsic (B_0) and totally transferred ($\Sigma z_r B_r$) hyperfine field contributions at the Eu-nuclei of the EuX-series

	Spin-structure	B_I(T)	B_0(T)	$\sum_r z_r B_r$(T)		Refs.
EuO	NNNN	-30.77 ± 0.1	-29.9	-0.9	$\left. \begin{array}{l} = 12B_1 + 6B_2 + 24B_3 + 12B_4 \\ + 24B_5 + 8B_6 + 24B_7 + 12B_8 \end{array} \right\}$	[4, 6]
EuS	NNNN	-34.44 ± 0.1	-30.0	-3.4		[4, 6]
		-28.1 ± 0.2		-5.3 ± 0.2		[4, 3]
EuSe	NNNN	-32.65 ± 0.1	-29.2	-3.4		[8, 6]
	NNS	-29.35 ± 0.1		-0.15	$= 6B_1 + 6B_3 + 6B_4 + 12B_7$	[8, 6]
	NNSS	-29.35 ± 0.1		-0.15	$= 6B_1 - 12B_5 - 12B_8$	[8, 6]
	NSNS	-26.2 ± 0.2		$+3.0$	$= -6B_2 + 12B_4 - 8B_6 + 12B_8$	[2, 6]
EuTe	NNNN	-30.6 ± 0.5	-28.3	-2.3 ± 0.5	$=$ (see above)	[2, 6]
	NSNS	-26.0 ± 0.3		$+2.3 \pm 0.3$	$=$	[2, 6]

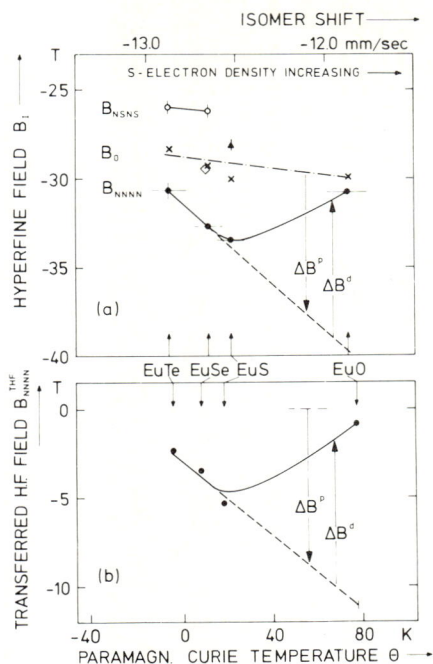

Fig. 1. (a) Total h.f. field, B_1, and intrinsic hyperfine field, B_0, as a function of the ^{151}Eu isomer shift for the ferromagnetic NNN (-●-), ferrimagnetic NNS (-◇-), and antiferromagnetic NSNS (-○-) spin structure of EuX and for Eu^{2+} ions in SrX (× from ref. 6 and ▲ from ref. 3). (b) Transferred h.f. fields, $B_{NNNN}^{THF} = B_1(NNNN) - B_0$, of the EuX-series as a function of θ. (For ΔB^p and ΔB^d see text.)

expected if exchange and transfer are due to a single mechanism only. We tend to attribute this B^{THF} contribution, $\Delta B^p = \Sigma_r z_r B_r^{STHF} = \Sigma_r z_r a^p J_r^p$, scaling with $\rho_s(0)$ and ℓ to the superexchange, J_r^p, via valence band "p"-electrons [5].

(2) For EuO a competing contribution, $\Delta B^d = z_1 B_1^{THF} = 12 a^d J_1^d$, is obvious from figs. 1(a) and 1(b). It is suggested to be related with the exchange constant, J_1^d, due to the overlap of the Eu n.n. 5d-orbitals and strongly varying with the Eu-distance [7]. This, however, is to be proved by further experiments on the Eu$_{1-x}$Sr$_x$O dilution system in preparation.

(3) Relying on the relations stated before in (1) and (2) the decrease of ΔB^p and, hence, of the superexchange $\Sigma z_r J_r^p$ with increasing covalency between EuO and EuTe is in contradiction with the Anderson molecular model for the superexchange restricted to $r = 2$ and localized p-orbitals only. The observed decrease is consistent, however, with a superexchange mechanism via delocalized valence band electrons as proposed recently by Sawatzky et al. [5]. It is expected to decrease with the density of the valence electrons and to be of long range increasing between EuO and EuTe [1].

Experimental evidence for J_r ranging at least up to the sixth Eu neighbours for EuS comes from a recent Mössbauer study of the Eu$_{1-x}$Sr$_x$S dilution series [3]. A similar long range can be concluded for EuSe from a consideration of the coefficient matrix established by the four inhomogeneous linear equations for the B_r given in table I (column 5) for the four different spin structures. For $r \leq 6$ no resolution at all is possible and with $r > 6$ the possible solutions for only four nonzero B_r contributions do not allow a meaningful conclusion. Again we conclude that all B_r up to at least $r = 6$ have to be considered.

In conclusion, the nuclear resonance studies provide experimental evidence for the existence of at least two competing exchange mechanisms. The long range mechanism dominates for EuS, EuSe, and EuTe, the competing short range mechanism for EuO. The results clearly demonstrate that the so far generally assumed restriction of the HDVV operator [eq. (1)] to the n.n. and n.n.n. exchange parameters J_1 and J_2 only, is inappropriate for the EuX family of magnetic semiconductors.

References

[1] W. Zinn, J. Magn. Magn. Mater. 3 (1976) 23.
[2] Ch. Sauer, U. Köbler, W. Zinn and G.M. Kalvius, J. Phys. Collq. 35, C6 (1974) 269.
[3] G. Crecelius, H. Maletta, H. Pink and W. Zinn, to be published in J. Magn. Magn. Mater. (1977).
[4] M. Neusser, H. Lütgemeier and W. Zinn, to be published in J. Magn. Magn. Mater. (1976).
[5] G.A. Sawatzky, W. Geertsma and C. Haas, J. Magn. Magn. Mater. 3 (1976) 37.
[6] K. Kojima, T. Komaru, T. Hihara and Y. Koi, to be published in J. Phys. Soc. Jap. (1976).
[7] T. Kasuya, IBM J. Res. Develop. 14 (1970) 214.
[8] T. Komaru, T. Hihara and Y. Koi, J. Phys. Soc. Jap. 31 (1971) 1391.

PROPERTIES OF HEISENBERG MAGNETODIELECTRICS AT HIGH PRESSURES

A.A. GALKIN, V.P. DJAKONOV, I.M. FITA and G.A. TSINTSADZE

Physico-Technical Institute, Academy of Sciences of the Ukr. SSR, 340048 Donetsk, USSR

Measurements of magnetic susceptibility of Heisenberg ferromagnets $Cu(NH_4)_2Br_4 \cdot 2H_2O$ and $CuK_2Cl_4 \cdot 2H_2O$ are carried out by the modulation method on single crystals at hydrostatic pressures up to 12 kbar in the temperature range 4.2–0.4 K. With increasing pressure the magnetic susceptibility, magnetization, Curie temperature, and the exchange constants increase for $Cu(NH_4)_2Br_4 \cdot 2H_2O$ and decrease for $CuK_2Cl_4 \cdot 2H_2O$.

Among Heisenberg ferromagnets there are hydrated crystals $Cu(NH_4)_2 \cdot Br_4 \cdot 2H_2O$ and $CuK_2Cl_4 \cdot 2H_2O$ with low Curie temperature [1–4] and rather high compressibility which allow one to observe the pressure effect on magnetic properties.

Magnetic susceptibility measurements were carried out with a low-frequency differential magnetometer. The magnetization was defined by integrating the differential susceptibility as a function of the external magnetic field. High quasi-hydrostatic pressures were produced in a cylinder-piston device and measured by the temperature values of superconducting transition of tin.

As is seen from figs. 1 and 2, isobars of magnetic susceptibilities along the easy axis below T_c are constant in these compounds but

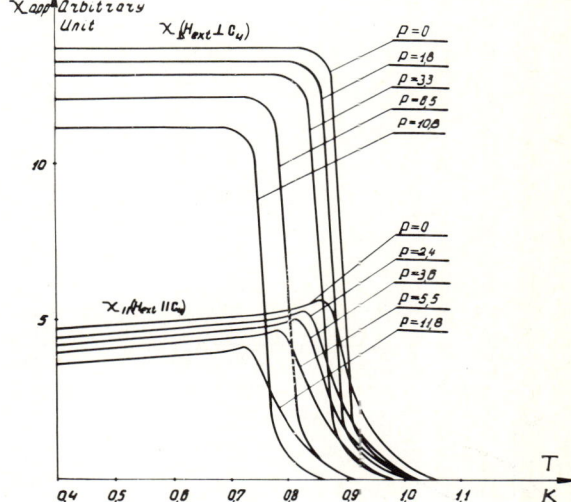

Fig. 2. Magnetic susceptibility of $CuK_2Cl_4 \cdot 2H_2O$ under the pressure. P is given in kbar.

with pressure increasing and temperature decreasing the magnitude of susceptibility anisotropy increases and that seems to be the evidence for the exchange anisotropy increasing. From experimental isobars presented the temperature dependences of phase transitions T_c on the pressure were found. As it is seen from fig. 3, T_c increases linearly for $Cu(NH_4)_2Br_4 \cdot 2H_2O$ with baric coefficient $\Delta T_c / \Delta P = +(0.033 \pm 0.003)$ deg kbar^{-1} beginning from 1.74 K and for $CuK_2Cl_4 \cdot 2H_2O$ it increases linearly with the coefficient $\Delta T_c / \Delta P = -(0.014 \pm 0.002)$ deg kbar^{-1} beginning from 0.88 K. Figs. 4 and 5 show that magnetization of $Cu(NH_4)_2Br_4 \cdot 2H_2O$ increases under the pressure, and that of $CuK_2Cl_4 \cdot 2H_2O$ decreases.

The direction of spontaneous magnetization in the Br-salt coincides with the axis C_4, and in

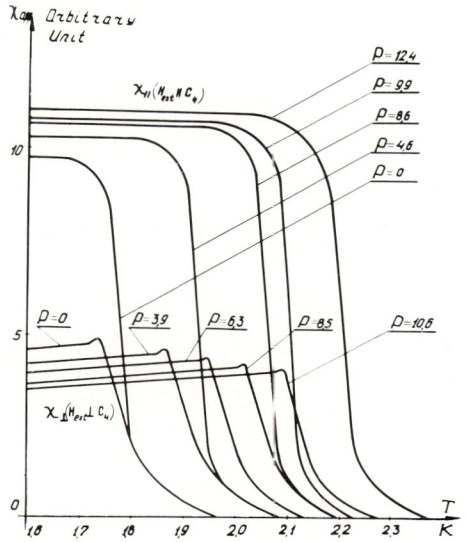

Fig. 1. Pressure effect on magnetic susceptibility of $Cu(NH_4)_2Br_4 \cdot 2H_2O$. P is given in kbar.

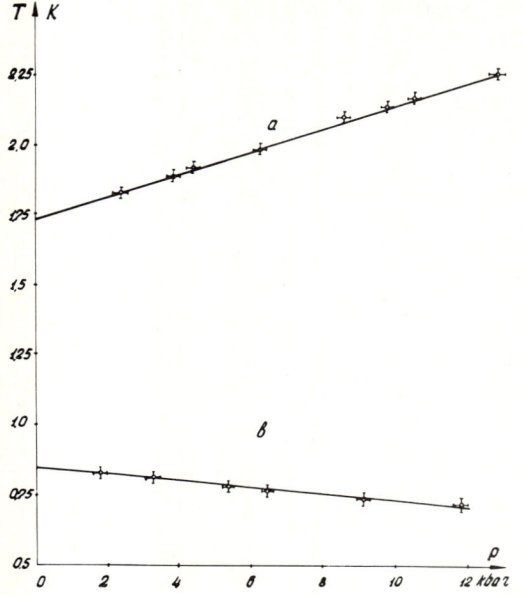

Fig. 3. Dependence of T_c on the pressure. (a) $Cu(NH_4)_2Br_4 \cdot 2H_2O$; (b) $CuK_2Cl_4 \cdot 2H_2O$.

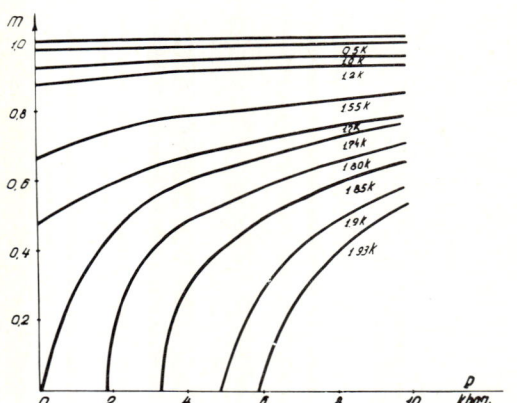

Fig. 4. Dependence of reduced magnetization of $Cu(NH_4)_2Br_4 \cdot 2H_2O$ on the pressure.

Cl-salt it is in the aa plane. In both compounds saturation is reached in higher magnetic fields. The anisotropy field of $Cu(NH_4)_2Br_4 \cdot 2H_2O$ at 0.4 K and $P = 0$ is equal to 130 Oe and at 10 kbars it is equal to 170 Oe. As for

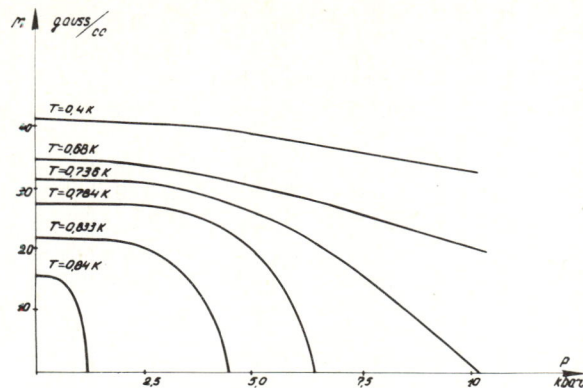

Fig. 5. Dependence of magnetization of $CuK_2Cl_4 \cdot 2H_2O$ on the pressure.

$CuK_2Cl_4 \cdot 2H_2O$ the anisotropy field at $P = 10$ kbars is equal to 60 Oe in comparison with 48 Oe at $P = 0$. The uniaxial character of the anisotropy in $Cu(NH_4)_2Br_4 \cdot 2H_2O$ is kept under pressure. The results of magnetization measurements indicate the absence of hysteresis and residual magnetization.

Data on the value and sign of $\Delta T_c/\Delta P$ allow one to determine the change in the exchange interaction between magnetic ions of Cu^{2+} under uniform compression. The nearest and next-nearest-neighbour exchange constants J_1 and J_2 calculated for $Cu(NH_4)_2Br_4 \cdot 2H_2O$ increase under uniform compression, whereas those of $CuK_2Cl_4 \cdot 2H_2O$ decrease. The behaviour of the constants J_1 and J_2 is difficult to explain but seems to be related to changes in the superexchange paths between magnetic ions of Cu^{2+}, due to changes in the tetragonal local environment of this ion.

References

[1] A.R. Miedema, R.F. Wielinga and W.J. Huiskamp, Physica 31 (1965) 1585.
[2] L. J. de Jongh, A.R. Miedema and R.F. Wielinga, Physica 46 (1970) 44.
[3] R.F. Wielinga and W.J. Huiskamp, Physica 40 (1969) 602.
[4] H. Suzuki and J. Watanabe, J. Phys. Soc. Jap. 30 (1971) 367.

A RECIPE FOR INDIRECT EXCHANGE INTERACTIONS

J.C. GILL and S.F.J. COX*

H.H. Wills Physics Laboratory, University of Bristol, Bristol BS8 1TL, England

A simple means is presented of estimating the singlet–triplet separation for pairs of magnetic ions separated by long exchange paths in insulating solids. The recipe is tested for the path Cu–O(H)–H–O(SO$_3$)–H–O(H)–Cu between Cu^{2+} ions in caesium copper sulphate.

We aim to demonstrate the usefulness of a simple recipe for calculating weak superexchange interactions between paramagnetic ions in insulating solids. The method is particularly suited to magnetically dilute materials in which the exchange paths contain more than two or three intervening ions since for such materials a proper construction of the appropriate long range orbitals, and a computation of their exchange interaction, would be extremely difficult.

For a pair of magnetic ions sufficiently far apart, the zero order singlet and triplet states will have the same energy (the direct exchange being negligible). Suppose now that the system spends fractions f_i of its time in excited states i, each with a singlet–triplet separation J_i. If the perturbation causing the admixtures is independent of, or at least symmetrical in, the two spins then the first order singlet–triplet separation which results is simply $J = \Sigma f_i J_i$. We illustrate this recipe with a calculation of J between nearest neighbour copper ions in the double sulphate Cs$_2$Cu(SO$_4$)$_2 \cdot$ 6H$_2$O along the exchange path Cu–O(H)–H–O(SO$_3$)–H–O(H)–Cu and we take the appropriate excited states to be configurations in which the two magnetic spins find themselves together on one ion of the path.

A calculation follows of the contribution to the superexchange from the configuration in which both spins find themselves on the central oxygen atom. The direct exchange in a free oxygen atom is 16 000 cm^{-1}: this is the separation of the ^{3}P and ^{1}D terms. In the crystal this energy is likely to be less because the orbitals are spread out over the surrounding atoms, particularly the sulphur atom, but the order of magnitude will be correct. Then if the two spins are together on the oxygen atom for a fraction f of the time, the contribution to J will be $\sim f \times 16\,000$ cm^{-1}. f can be estimated by considering the degree of covalency in each bond separately.

(1) The Cu–OH$_2$ bond. The fraction of time that the spin from the copper ion spends on a neighbouring oxygen atom O(1) is called the spin transfer σ_1. Owen [1] shows how it can be estimated from the magnetic resonance data for dilute copper Tutton salts. A molecular orbital is constructed from the ground state wavefunctions of the copper ion and surrounding ligands:
$\phi = \alpha \, \mathrm{d}_{x^2-y^2} - (1-\alpha^2)^{1/2}\tfrac{1}{2}(\psi_1 + \psi_4 - \psi_2 - \psi_5)$, in Owen's notation. $(\psi_1 + \psi_4 - \psi_2 - \psi_5)$ is the linear combination of the ligand orbitals allowed by symmetry, and their overlap with the copper wavefunction d$_{x^2-y^2}$, is assumed to be small. The spin spends a fraction α^2 of its time on the copper orbital, and a fraction $\tfrac{1}{4}(1-\alpha^2)$ of its time on each ligand orbital: from the experimental data on g-factors (Abragam and Pryce [2]) and hyperfine structure (Bleaney and others [3]) one obtains $\alpha^2 = 0.84$, so the spin transfer is $\sigma_1 = \tfrac{1}{4}(1 - 0.84) = 4 \times 10^{-2}$.

(2) The hydrogen bond O–H–O(SO$_3$): X-ray data at room temperature suggests that the relevant O–O bond lengths are all within a few percent of the bond length in ice, so that the spin transfer σ_2 between the two oxygens can be estimated from the theoretical predictions for ice. Pauling [4] suggests that three configurations contribute to the hydrogen bond structure; with their probabilities they are:

(a) O–HO   [covalent O(1)–H]   61%,
(b) O$^-$H$^+$O   (ionic)   34%,
(c) OH–O   [covalent H–O(2)]   5%.

The spin transfer from O(1) to H is therefore $\tfrac{1}{2} \times 0.61$, and from H to O(2) is $\tfrac{1}{2} \times 0.05$, so that $\sigma_2 = (\tfrac{1}{2})^2(0.61)(0.05) = 7.6 \times 10^{-3}$. Finally, therefore, the probability of finding both spins to-

---

* Permanent address: Rutherford Laboratory, Chilton, Didcot, Oxon, OX11 0QX, England.

gether on the central oxygen atom is $f = \sigma_1^2 \sigma_2^2 = 9.3 \times 10^{-8}$, and the contribution to the exchange from this configuration is $(-)\ 1.5 \times 10^{-3}\ cm^{-1}$ ("ferromagnetic").

Inspection of the crystal structure shows that there are two such exchange paths; the contributions from configurations in which the two spins are simultaneously on each of the constituent ions can be worked out similarly and the total exchange becomes $J = 0.011\ cm^{-1}$.

We have ignored the much smaller contributions from the additional exchange path Cu–O(H)–H–O–S(O_2)–O–H–O(H)–Cu and those from configurations in which the spins find themselves on adjacent but not orthogonal orbitals (such contributions may well be "antiferromagnetic"). As presented, the recipe is suited to exchange paths which are predominantly ionic; some predominantly covalent bonds could be accommodated as described earlier by Griffiths and others [5], taking the J_i of the recipe to be the singlet–triplet separations in the appropriate free molecules.

We have measured J in the diluted salt $Cs_2(Cu_{0.05}, Zu_{0.95})(SO_4)_2 \cdot 6H_2O$ in the manner described by Meredith and Gill [6], that is by identifying the nearly forbidden lines in the EPR spectrum which result from nearest neighbour copper pairs. A computer simulation of the pair line positions best fits the observed spectrum for a value of $J = -0.009 \pm 20\%\ cm^{-1}$, which is also consistent with the exchange specific heat, $C\alpha J^2$, measured in the undiluted salt by Benzie and others [7].

Such precise agreement between the measured and estimated values is probably fortuitous in view of the simplicity of the recipe. The order of magnitude of the result is clearly reliable, however, and we anticipate some success in explaining the variation of J throughout the copper Tutton salt series (Cs → Rb, K, NH_4) by considering the changes in the spin transfer functions with bond lengths (and also the additional hydrogen bonding in the ammonium salt). It is also possible to estimate in the same way how J is modulated by lattice strains.

References

[1] J. Owen, Proc. Roy. Soc. (London) A227 (1955) 183.
[2] A. Abragam and M.H.L. Pryce, Proc. Roy. Soc. (London) A205 (1951) 135 and A206, 164.
[3] B. Bleaney, K.D. Bowers and M.H.L. Pryce, Proc. Roy. Soc. (London) A228 (1955) 166.
[4] L. Pauling, Nature of the Chemical Bond (Oxford University Press, 1960) p. 452.
[5] J.H.E. Griffiths, J. Owen, J.G. Park and M.F. Partridge, Proc. Roy. Soc. (London) 250A (1959) 84.
[6] D.J. Meredith and J.C. Gill, Phys. Lett. 25A (1967) 429.
[7] R.J. Benzie, A.H. Cooke and S. Whitley, Proc. Roy. Soc. (London) A232 (1955) 277.

INELASTIC NEUTRON SCATTERING STUDY OF CRYSTAL FIELD LEVELS AND MAGNETIC INTERACTIONS IN HoCrO$_3$

N. SHAMIR[a], M. MELAMUD[a], H. SHAKED[a,b] and S. SHTRIKMAN[c]

[a] *Nuclear Research Center, Negev, PO Box 9001, Beer-Sheva, Israel*
[b] *Ben-Gurion University of the Negev, PO Box 2053, Beer-Sheva, Israel*
[c] *The Weizmann Institute of Science, Rehovot, Israel*

The first four excited levels of the $^5I_8$ ground multiplet of Ho^{3+} in HoCrO$_3$ are determined. The technique used is inelastic neutron scattering from powder sample. The ground and first excited level form a "doublet" whose splitting is studied as a function of temperature.

Transitions between energy levels of Ho^{3+} in HoCrO$_3$, split by electric-crystal and magnetic internal fields, were measured using inelastic neutron scattering. The measurements were performed using a triple axis neutron spectrometer and a powder sample at 4.2, 77 and 295 K. Transitions between ground (0) and first four excited (1, 2, 3, 4) levels in the ground $^5I_8$ multiplet were observed. In the 1.40 Å spectra (fig. 1), $0 \to 2$, $0 \to 3$ and $0 \to 4$ transitions were observed at 4.2 K. At 77 K, level 2 is sufficiently populated so that $2 \to 0$, 1 transitions are also observed. The $0 \to 1$ transition was observed in the 2.25 Å spectra [fig. 2(b)]. The energy values of levels 1 and 2 are in good agreement with optical data while levels 3 and 4 were not reported by other methods.

The 0, 1 levels are sufficiently separated from the higher levels and therefore can be treated as a two level system [1, 2] with energy splitting of

$$E(T) = [D^2 + P(T)^2]^{1/2}, \quad (1)$$

where D is the crystal field splitting and $P(T)$ is the magnetic splitting due to the internal magnetic field of the Cr and Ho spin systems. The Cr^{3+} ions are ordered ($T_N = 140$ K) antiferromagnetically [2, 3], while the order of the Ho^{3+} moments is induced by the field of the Cr spin system.

The energy splitting, $E(T)$, as observed in optical measurements [4], is shown in fig. 2(a) (dashed line). Molecular field calculations, using Ho^{3+} interaction parameters, which give best fit to the intensity–temperature curve (elastic neutron scattering) of a pure Ho^{3+} magnetic reflection, were made [5]. The ground doublet splitting obtained in these calculations is also shown in fig. 2(a) (solid line). Fits to spontaneous magnetization and susceptibility were also reported [2] and yield a splitting curve similar to that of the elastic neutron scattering. The positions of the $0 \to 1$, $0 \to 2$ and $1 \to 2$ lines were determined by fitting gaussians to the spectra [figs. 2(b), (c)]. The energy splittings of

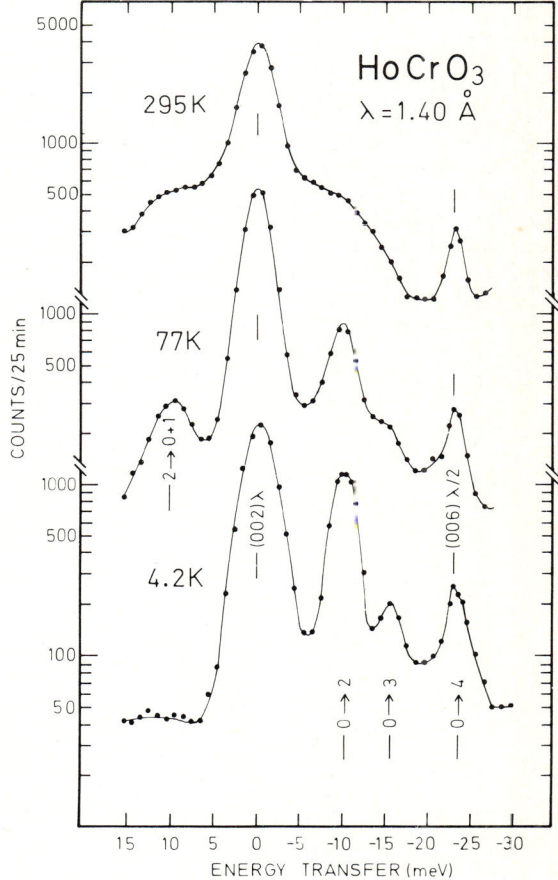

Fig. 1. Neutron (1.40 Å) energy transfer spectra.

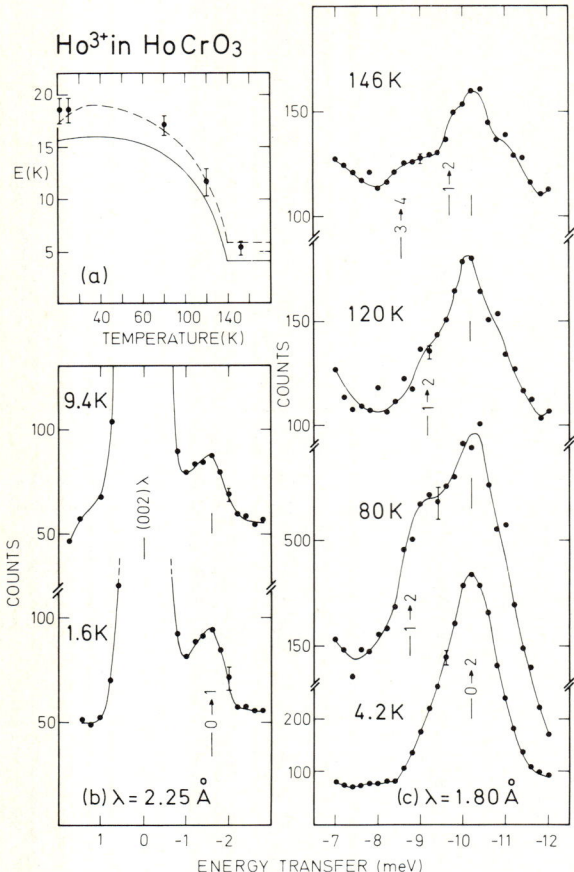

Fig. 2. (a) Ground doublet splitting as derived from: inelastic neutron scattering (experimental points, present work); optical data (dashed line [4]); and molecular field fits to neutron diffraction data (solid line, [5]). (b) Neutron (2.25 Å) energy transfer spectra. (c) Neutron (1.80 Å) energy transfer spectra. Notice that the scale of the 4.2 K spectrum is different by factor of 5 from the scales of the other spectra.

the ground doublet at 1.6, 9.4, 80, 120 and 146 K were deduced from these fits and are shown in fig. 2(a). There is a good agreement between the direct methods, i.e. inelastic neutron scattering of the present work and optics [4]. The indirect methods of molecular field fits to intensity–temperature curve [5], spontaneous magnetization and susceptibility [2] yield a consistently lower $E(T)$ curves. The small maximum observed by optics and found also by elastic neutron scattering [5], is due to the opposite but small Ho^{3+} contribution to P.

References

[1] A.P. Malozemoff and R.L. White, Solid State Commun. 8 (1970) 665.
[2] R.M. Hornreich, B.M. Wanklyn and I. Yaeger, Int. J. Magn. 2 (1972) 77.
[3] E.F. Bertaut, J. Marschal, G. de Vries, R. Aleonard, R. Pauthenet, J.P. Reboult et V. Zarubicha, IEEE Trans. Mag. 2 (1966) 453.
[4] R. Courths and S. Hüfner, Z. Phys. B24 (1976) 193.
[5] N. Shamir and H. Shaked (unpublished).

LONG-RANGE EXCHANGE INTERACTIONS

W. GEERTSMA, C. HAAS, G.A. SAWATZKY

Laboratories for Inorganic and Physical Chemistry, Materials Science Center of the University, Groningen, The Netherlands

and

G. VERTOGEN

Institute of Theoretical Physics, University of Groningen, The Netherlands

Experiment indicates that rather long-range exchange interactions and supertransferred hyperfine fields are present in some semiconductors. We discuss a model in which a d state interacts with a fully occupied band. Special band structures have been chosen to illustrate the dependence on the various parameters. The resulting exchange interaction is of a longer range than expected on grounds of perturbation theory.

There is some experimental evidence that the superexchange interactions are of fairly long range in sulfides and selenides of transition metals [1]. The high Néel temperature of $CoRh_2S_4$ has been attributed to strong superexchange via the intermediate diamagnetic Rh^{3+} ion. The spinel $Fe_{1/2}Cu_{1/2}Rh_2S_4$ has the same spin structure and approximately the same Néel temperature as α-MnS. This spin structure can only be explained by taking into account distant neighbour interactions involving exchange paths Fe–S–Rh–S–Fe and Fe–S–Rh–S–Rh–S–Fe. A phenomenon directly related to superexchange is the transferred hyperfine field at distant neighbours, which is a consequence of a long-range spin density distribution. These data are usually interpreted in terms of the superexchange theory for insulators [2]. In this theory the electronic states are taken as localized whereas solids can in most cases be better described using delocalized band states. Especially the anion valence orbitals are not localized but form broad bands. The width of the valence band for oxides is typically 4–5 eV and for sulfides and selenides 5–10 eV, as measured by photoelectron spectroscopy. The cation d states, on the other hand, often form quite narrow bands. The occupied d states of sulfides and selenides are found quite close to the top of the valence band or actually in the valence band itself. In this situation the mixing of the anion orbitals with the cation d orbitals will depend strongly on the k vector of the band state. Cluster-type calculations are no longer valid and also the strong Coulomb interaction between d electrons must be taken into account. In this paper we consider a model where localized d orbitals hybridize with band states. In an earlier paper [3] we discussed a similar model without taking into account the Coulomb interaction.

We assume that the one-electron band structure ϵ_k of the non-magnetic states is known. We now introduce the localized d state with energy E_d^0, and an on-site Coulomb interaction U. The spin-independent interaction between the d state and the band state k is V_{dk}. The hamiltonian is

$$H = \sum_{k,\sigma} \epsilon_k c_{k\sigma}^+ c_{k\sigma} + \sum_{\sigma} E_d^0 c_{d\sigma}^+ c_{d\sigma} - U n_{d\uparrow} n_{d\downarrow}$$
$$+ \sum_{k,\sigma} [V_{dk} c_{d\sigma}^+ c_{k\sigma} + V_{kd} c_{k\sigma}^+ c_{d\sigma}]. \quad (1)$$

A similar model hamiltonian has been used by Haldane and Anderson [4] to discuss the multiple charge states of transition-metal impurities in semiconductors. We adopt the Hartree–Fock approximation for the Coulomb term. The Green's function describing the d state is given by

$$G_d^\sigma(\omega) = [\omega - E_d^0 - U\langle n_{d-\sigma}\rangle - \Gamma(\omega)]^{-1}.$$

Here $\Gamma(\omega) = \sum_k |V_{dk}|^2/(\omega + is - \epsilon_k)$ and the total occupation of the d state with spin σ is given by

$$\langle n_{d\sigma}\rangle = -1/\pi \int_{\text{occupied states}} \text{Im } G_d^\sigma(\omega)\, d\omega.$$

The positions ω_σ of the localized states of spin σ, given by the poles of $G_d^\sigma(\omega)$, are

$$\omega_\sigma = E_d^0 + U\langle n_{d-\sigma}\rangle + \Gamma(\omega_\sigma). \quad (3)$$

For a more detailed discussion we take a linear array of atoms at positions $R = \nu a$ along the z-axis. We approximate the non-magnetic states by a single, fully occupied, p_z-type tight-binding band $\epsilon_k = \epsilon_0 - 2W \cos ka$. Consider the case of an extra d_{z^2} orbital at position $\nu = 0$ occupied by one spin-up electron. For the shift function we find $\Gamma(\omega) = 2(V^2/W)(\beta - \sqrt{\beta^2 - 1})$. where $\beta = (\omega - \epsilon_0)/2W$ and V is the transfer integral between the localized state and the band state Wannier function centered on a nearest neighbour of the magnetic ion. For $\langle n_{d\sigma} \rangle$ we find $\langle n_{d\uparrow} \rangle = 1$ and $\langle n_{d\downarrow} \rangle = 1 - [1 - \partial\Gamma(\omega_\downarrow)/\partial\omega_\downarrow]^{-1}$. Inserting this in eq. (3) we obtain a set of equations for $\omega_\uparrow$ and $\omega_\downarrow$. The solutions of these equations are plotted in fig. 1, with $\epsilon = (E_d^0 - \epsilon_0)/2W$. For $(V/W)^2 < 1 - \epsilon - \frac{1}{2}(U/W)$ the empty state lies in the band, for $U/W > 2(1 - \epsilon)$ the empty state is always above the band. We see that the position of the localized states outside the band is a weak function of V/W. The states which fall in the band are virtual levels for which the shift function $\Gamma(\omega)$ is complex. The level indicated in the band in fig. 1 represents the energy where the d density of states is a maximum.

The spin density for $|\nu| \geq 1$ is given by

$$S_\nu = (V/W)^2 S_d [\beta_\downarrow - \sqrt{\beta_\downarrow^2 - 1}]^{2\nu}, \quad (4)$$

where $S_d = (\langle n_{d\uparrow} \rangle - \langle n_{d\downarrow} \rangle)$ is the spin in the d state. In fig. 2 we have plotted S_1 as a function of V/W. In our model S_ν and S_d depend only on the position of the empty state relative to the band. In fig. 3 we have plotted S_d as a function of V/W. For large V/W, $S_1 \Rightarrow \frac{1}{4}$, $S_d \Rightarrow \frac{1}{2}$, and $S_{\nu>1} \Rightarrow 0$. The R dependence of S_ν is given in fig. 7 of ref. 3. If the empty state approaches the top of the band the spin density distribution becomes more long range, but the absolute magnitude of the spin density decreases.

We have carried out a calculation for two magnetic impurities separated by a distance R and interacting via the band. The resulting spin and charge redistribution, and the energy shifts of the localized states are governed by the function

$$\gamma(R, \omega) = \sum_k V_{d_1 k} V_{k d_2}/(\omega - \epsilon_k). \quad (5)$$

For large R the exchange interaction is
$$J(R) = 2[(\gamma(R, \omega_\downarrow))^2$$
$$\times [U^{-1} - \frac{1}{2}(S_d)^3(\partial^2 \gamma(R, \omega_\downarrow)/\partial\omega_\downarrow^2)]$$
$$- 2(S_d)^2 \gamma(R, \omega_\downarrow)[\partial\gamma(R, \omega_\downarrow)/\partial\omega_\downarrow], \quad (6)$$

where $\omega_\downarrow$ and S_d have values as calculated for the one-impurity case. This represents an antiferromagnetic interaction. The first term has the

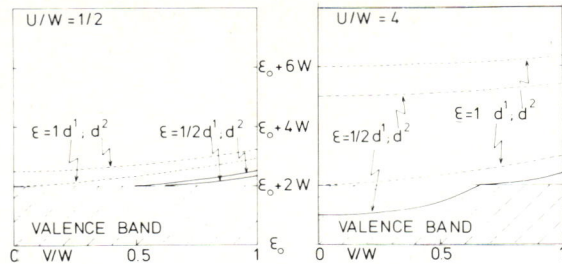

Fig. 1. Energy ω of the occupied state $\omega_\uparrow(d^1)$ and the empty localized state $\omega_\downarrow(d^2)$ as a function of V/W.

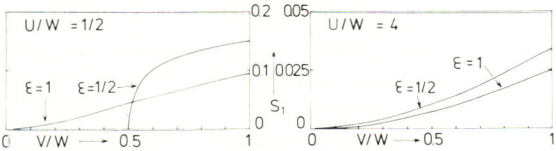

Fig. 2. Spin density S_1 on the nearest neighbour ν of the magnetic ion as a function of V/W.

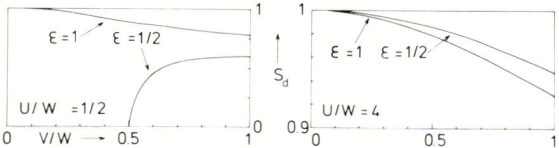

Fig. 3. Spin S_d in the d state as a function of V/W.

same R dependence as the spin density. For a free-electron-like band $\epsilon_k = \epsilon_0 - \hbar^2 k^2/2m^*$ the first term is $J(R) \propto R^{-2} \exp(-2k_0 R)$ with $k_0^2 = (\omega_\downarrow - \epsilon_0)(2m^*/\hbar^2)$. This is comparable with the perturbation result $J(R) \propto (k_0 R)^{-2} \exp(-2k_0 R)$ [5]. The last term of (6) gives $J(R) \propto R^{-1} \exp(-2k_0 R)$ which gives a larger range of the exchange interaction than the first term.

We are of the opinion that many of the transition-metal chalcogenides should be described in the above way rather than using a localized-cluster approximation. More detailed calculations using the actual crystal structures and the one-electron band structures will have to be done to check the validity of this model.

References
[1] R. Plumier and F.K. Lotgering, Solid State Commun. 8 (1970) 477.
[2] P.W. Anderson, Solid State Phys. 14 (1963) 99.
[3] G.A. Sawatzky, W. Geertsma and C. Haas, J. Magn. Magn. Mater. 3 (1976) 37.
[4] F.D.M. Haldane and P.W. Anderson, Phys. Rev. B6 (1976) 2553.
[5] C.E.T. Gonçalves da Silva and L.M. Falicov, J. Phys. C: Solid State Phys. 5 (1972) 63.

ANISOTROPIC EXCHANGE IN ANTIFERROMAGNETIC RbCoF$_3$

T.M. HOLDEN, W.J.L. BUYERS and E.C. SVENSSON
Atomic Energy of Canada Limited, Chalk River, Ontario, Canada K0J 1J0

The dispersion relations for the three lowest branches of magnetic excitations in antiferromagnetic RbCoF$_3$ have been determined by neutron inelastic scattering. A Hamiltonian that includes anisotropic exchange gives a better description of the results than one with only isotropic exchange but is still inadequate.

Because the ground state of the Co^{2+} ion in an octahedral crystal field is an orbital triplet, cobalt salts exhibit more complex behaviour than materials containing ions with no orbital angular momentum. In the presence of spin–orbit coupling and an exchange field the orbital triplet in RbCoF$_3$ splits into a manifold of twelve levels spanning approximately 35 THz. We have studied the transitions between the ground state and the first three excited states (S_-, S_+ and the Davydov split S_z modes, respectively). The orbital angular momentum of the Co^{2+} ion is associated with an anisotropic spatial distribution of electrons which leads, in principle, to anisotropic exchange between Co^{2+} ions. The results are analysed with the pseudoboson model [1] with crystal-field, spin–orbit and isotropic-exchange parameters and also with anisotropic-exchange interactions [2]. RbCoF$_3$ has the cubic perovskite structure in the paramagnetic phase and becomes a type G antiferromagnet [3] below 98 ± 5 K with a small tetragonal distortion.

Measurements of the dispersion relations of RbCoF$_3$ were made at the NRU reactor, Chalk River with a triple axis crystal spectrometer operated in the constant-momentum-transfer mode. Two large crystals (total volume 9 cm^3) were aligned so as to have a common orientation with the [1$\bar{1}$0] axis vertical and were maintained at 4.2 K throughout the experiment. A preliminary account of some of the measurements has appeared elsewhere [4]. In Fig. 1 are shown the results for the first three branches of magnetic excitations and for several phonon branches. Only the modes which were predominantly magnetic were analysed with the models described later. Excitations observed at approximately 1.0, 10.1 and 12.6 THz in light scattering experiments [5] agree with our zone-centre frequencies, but, except for the lowest

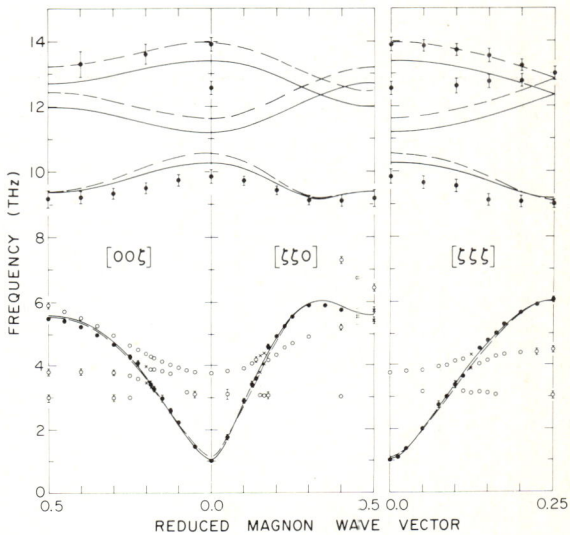

Fig. 1. Dispersion relations for magnons and phonons in RbCoF$_3$ at 4.2 K. Modes which are predominantly magnetic are shown as solid circles, and phonons are shown by open circles. Modes of mixed character are designated by crosses. The symmetry directions [00ζ] and [$\zeta\zeta$0] apply only for the magnons. The curves are the predictions of the spin wave model for isotropic (solid curves) and anisotropic exchange (dashed curves).

mode, the assignments are not consistent with our observations.

The hamiltonian which is conventionally used [1] to describe the properties of the lowest multiplet is

$$H = \sum_{j,k} [-\alpha\lambda \mathbf{l}(jk) \cdot \mathbf{S}(jk) + c l_z^2(jk)]$$
$$+ \sum_{jk}\sum_{j'k'} J\binom{jj'}{kk'} \mathbf{S}(jk) \cdot \mathbf{S}(j'k'),$$

where c represents a tetragonal distortion and $J\binom{jj'}{kk'}$ represents the isotropic exchange between spins j and j' on sublattices k and

k', respectively. The true angular momentum L has been replaced by an effective operator l acting within the orbital triplet and the effective spin–orbit parameter is $\alpha\lambda$. α includes a small correction for mixing to states of the next multiplet and has the value 1.421 [6] in the similar compound $KCoF_3$. The best least-squares fit to the results of the many-level spin-wave model [1] is shown by the solid curves in fig. 1. The model describes the lowest branch accurately, but the fitted second branch is too high, the fitted third branch is too low, and the splitting of the third branch is overestimated. The model parameters obtained were $\lambda = -4.48 \pm 0.05$ THz, $c = -0.30 \pm 0.03$ THz and $J = 0.271 \pm 0.002$ THz for the nearest-neighbour isotropic exchange. Note that λ is 30% lower than the free-ion value, -5.35 THz, and only about a third of this difference can be attributed to covalency [6].

Copland and Levy [2] have suggested that the anisotropic exchange between near-neighbour Co^{2+} ions has the form

$$\sum_{jk}\sum_{j'k'} \{J + K_1 l(jk) \cdot l(j'k') + K_2 C_0^{[2]}$$
$$\times [l(jk) \cdot l(j'k') - 3l_z(jk)l_z(j'k')] + \ldots\}$$
$$\times \{\tfrac{1}{4} + S(jk) \cdot S(j'k')\}.$$

The terms in K_1 and K_2 depend on the orbital state of the ions and K_2 gives rise to anisotropy with respect to bond axes. Because of the spherical harmonic $C_0^{[2]}$ the K_2 term is absent for $[\zeta\zeta\zeta]$ wavevectors where the most complete results were available, and it was not included in the analysis. Including the anisotropic exchange K_1 leads to a better overall description of the results and, as predicted by Copland and Levy [2], the spin–orbit parameter is no longer very different from the free-ion value. A fit where λ was constrained equal to the free-ion value (dashed curve in fig. 1) gave $J = 0.284 \pm 0.002$ THz and a relatively large anisotropic coupling $K_1 = 0.12 \pm 0.02$ THz. Although the zone-boundary frequencies are well described, the Davydov-splitting and the bandwidth of the S_+ branch remain seriously overestimated.

We conclude that the anisotropic exchange interactions considered here are not by themselves sufficient to give a completely satisfactory description of the magnons. However, we have only considered direct exchange as proposed by Copland and Levy and only their leading order term; more general forms of exchange are certainly possible. It is interesting to note that isotropic exchange gave a good description of the spin-wave spectrum of $KCoF_3$ [1] where only the lowest two branches were observed. Inclusion of the third branch as in the present experiment provides a much more stringent test of the theory and demonstrates the need for more complex interactions.

References

[1] W.J.L. Buyers, T.M. Holden, E.C. Svensson, R.A. Cowley and M.T. Hutchings, J. Phys. C4 (1971) 2139.
[2] G.M. Copland and P.M. Levy, Phys. Rev. B1 (1970) 3043.
[3] Y. Allain, J. Denis, A. Herpin, J. Lecomte, P. Meriel, F. Plique and A. Zarembovitch, J. Phys. (Paris) 32C 1 (1970) 611.
[4] E.C. Svensson, T. M. Holden and W.J.L. Buyers, Proc. 14th Int. Conf. on Low Temperature Physics, Vol. 3, M. Krusius and M. Vuorio, eds. (North-Holland, Amsterdam, 1975) p. 184.
[5] J. Nouet, D.J. Toms and J.F. Scott, Phys. Rev. B7 (1973) 4874.
G.H. Johnson, Ph.D. Thesis, Cornell University (1975).
[6] J.H.M. Thornley, C.G. Windsor and J. Owen, Proc. Roy. Soc. A284 (1965) 252.

BIQUADRATIC EXCHANGE INTERACTIONS IN DySb*

J.S. KOUVEL,† T.O. BRUN and F.W. KORTY†

Argonne National Laboratory, Argonne, Illinois 60439, USA and University of Illinois, Chicago, Illinois 60680, USA

From a comparison of magnetization data on a DySb crystal above its Néel point with magnetization calculations based on its crystal-field states, it is shown that the exchange interactions have a substantial biquadratic component, as suggested by the field-induced state with orthogonal sublattice moments.

At the Néel temperature ($T_N \approx 9.5$ K) in zero magnetic field, the cubic NaCl-structured compound DySb transforms abruptly into a type-II antiferromagnet in which the Dy moments in adjacent ferromagnetic (111) planes are antiparallel [1, 2]. The first-order transition is accompanied by a nearly tetragonal lattice distortion and the ordered moments ($\sim 10\ \mu_B$/Dy-atom) lie along the tetragonal [001] axis. Moreover, magnetic measurements [3] on a DySb crystal at 1.5 K have shown that the compound undergoes a rapid field-induced transition from the antiferromagnetic state to a state whose magnetization corresponds to an alignment of Dy moments in equal numbers along each of two orthogonal ⟨100⟩ directions. It was suggested [3] that the magnetic structure of this quadrature-spin (Q) state of DySb was that of HoP, in which the moments in adjacent ferromagnetic (111) planes are aligned orthogonally, and this has been borne out by neutron diffraction measurements on a DySb crystal in high fields [4]. The latter work [4] also included a magnetization study which gave a determination of the magnetic phase diagrams for the applied field (H) along ⟨100⟩ and ⟨110⟩. These are shown in fig. 1, together with the phase diagram for H along ⟨111⟩, which we recently determined from a similar study of the same DySb crystal. The nature and positions of the phase boundaries, plus the occurrence of tricritical points [5], will be discussed in a separate report.

For our present purpose, the magnetic phase diagrams for DySb serve to emphasize the high-field existence of the Q state, which cannot be stabilized by bilinear exchange interactions alone but requires higher-order interactions in

*Work performed under the auspices of the US Energy Research and Development Administration.
†Supported in part by the National Science Foundation and the Office of Naval Research.

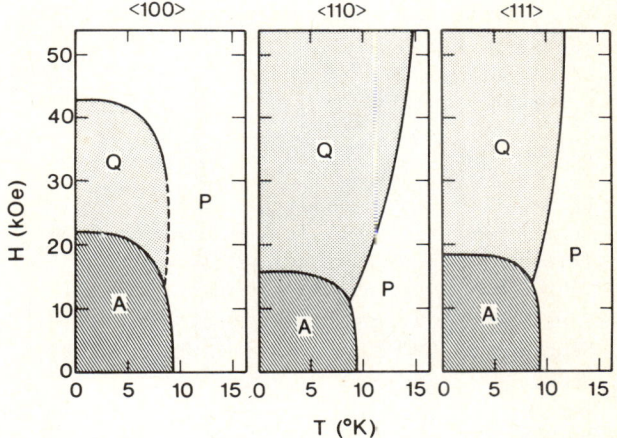

Fig. 1. Magnetic phase diagrams of field vs. temperature for DySb, indicating the antiferromagnetic (A), quadrature-spin (Q), and paramagnetic (P) regimes.

addition, as was pointed out earlier [4]. Specifically, if the higher-order interactions have a simple biquadratic form, the exchange hamiltonian (of the ith spin) can be written as

$$\mathcal{H}_i = -\sum_j [J_{ij} \mathbf{S}_i \cdot \mathbf{S}_j + J'_{ij}(\mathbf{S}_i \cdot \mathbf{S}_j)^2]. \quad (1)$$

The existence of the Q state would then suggest that the inter-sublattice component of the biquadratic exchange coefficient J'_{ij} is negative, thus favoring quadrature (orthogonal) over collinear alignment of the sublattice moments. The importance of biquadratic interactions in DySb has also been deduced previously from a mean-field analysis of elastic and magnetothermal data, and it was concluded that J'_{ij} is effectively positive [6, 7]. Although a positive J'_{ij} provides a plausible rationale for the first-order transition of DySb at T_N in zero field [7], it was shown more recently that the abruptness of this transition can be justified solely on the basis of magnetic symmetry [8].

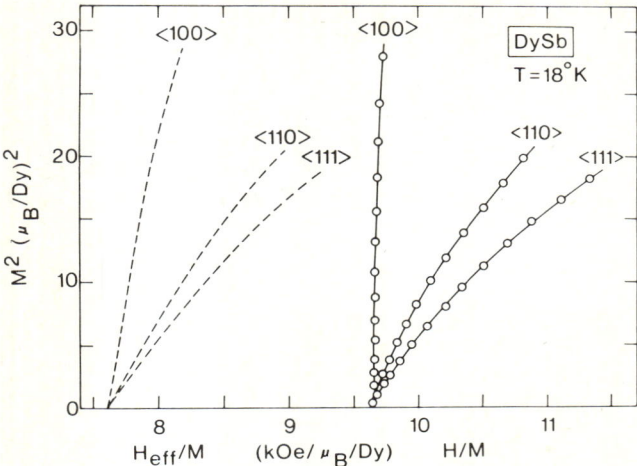

Fig. 2. Calculated (dashed) curves of M^2 vs. H_{eff}/M and experimental data of M^2 vs. H/M for DySb at 18 K with the effective field (H_{eff}), applied field (H), and magnetization (M) along various crystallographic directions.

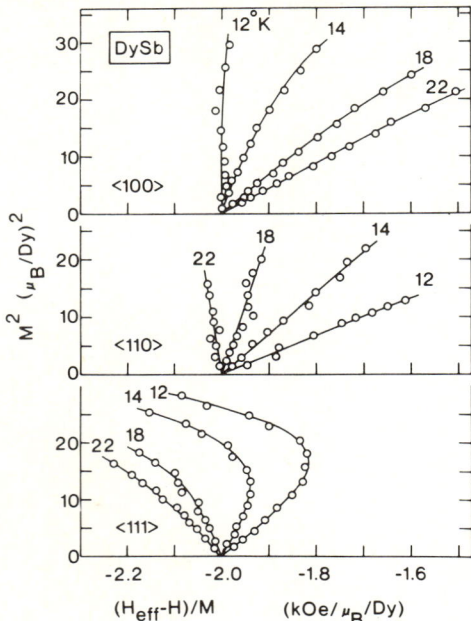

Fig. 3. M^2 vs. $(H_{eff} - H)/M$ for DySb at various temperatures (in K) for different directions of the effective field (H_{eff}), applied field (H), and magnetization (M).

In order to reveal explicitly any biquadratic coupling in DySb, we have followed a procedure recently used for the compound PrAg [9]. First, we had to determine the magnetization (M) vs. effective field (H_{eff}) behavior of DySb from its crystal-field states. The latter are presently available from interpolations between the neutron scattering results on various other isomorphic rare-earth antimonides [10]. We could thus establish quite reliably that the crystal-field coefficients for DySb are $A_4^0 \langle r^4 \rangle = 60$ K and $A_6^0 \langle r^6 \rangle = 2$ K, from which we deduced that relative to the lowest lying Γ_6 doublet the three Γ_8 quartets are at 8.8, 96.0 and 120.2 K, and the Γ_7 doublet is at 60.4 K [11]. From this crystal-field scheme, we calculated M vs. H_{eff} at various temperatures for H_{eff} along each of the principal crystallographic directions; the results for 18 K are presented as M^2 vs. H_{eff}/M in fig. 2. Also shown in this figure are our experimental magnetization results for DySb at 18 K in fields up to 56 kOe applied along the same three directions, which are plotted analogously as M^2 vs. H/M. These and similar experimental curves obtained at other temperatures within the paramagnetic regime were each compared against the corresponding calculated M^2 vs. H_{eff}/M isotherm; $(H_{eff} - H)/M$ was then evaluated at each experimental value of M^2.

Our results are plotted as M^2 vs. $(H_{eff} - H)/M$ in fig. 3. They can be compared meaningfully with an expression for the total exchange field,

$$H_{exch} = H_{eff} - H = \lambda M + \lambda' M^3, \qquad (2)$$

which has been derived [9] from spin hamiltonian (1) within a mean-field framework, where λ and λ' are proportional to the j-summations of J_{ij} and J'_{ij}, respectively, for the paramagnetic state. Thus, if λ and λ' were constant, M^2 vs. $(H_{eff} - H)/M$ would describe a straight line and, if λ' were zero, the line would be vertical. The pronounced deviations of the curves in fig. 3 from verticality can therefore be taken as evidence for substantial biquadratic interactions in DySb. Furthermore, these interactions vary significantly with temperature and with the direction (and, in some cases, the magnitude) of the magnetization. A detailed analysis of these results will be presented in a future report.

References

[1] E. Bucher, R.J. Birgeneau, J.P. Maita, G.P. Felcher and T.O. Brun, Phys. Rev. Lett. 28 (1972) 746.
[2] G.P. Felcher, T.O. Brun, R.J. Gambino and M. Kuznietz, Phys. Rev. B8 (1973) 260.
[3] G. Busch and O. Vogt, J. Appl. Phys. 39 (1968) 1334.

[4] T.O. Brun, G.H. Lander, F.W. Korty and J.S. Kouvel, AIP Conf. Proc. 24 (1974) 244.
[5] P. Streit, G.E. Everett and A.W. Lawson, Phys. Lett. 50A (1974) 199.
[6] T.J. Moran, R.L. Thomas, P.M. Levy and H.H. Chen, Phys. Rev. B7 (1973) 3238.
[7] L.F. Uffer, P.M. Levy and H.H. Chen, AIP Conf. Proc. 10 (1973) 553.
[8] P. Bak, S. Krinsky and D. Mukamel, Phys. Rev. Lett. 36 (1976) 52.
[9] T.O. Brun, J.S. Kouvel and G.H. Lander, Phys. Rev. B13 (1976) 5007.
[10] R.J. Birgeneau, E. Bucher, J.P. Maita, L. Passell and K.C. Turberfield, Phys. Rev. B8 (1973) 5345.
[11] These crystal-field coefficients and energy levels for DySb differ somewhat from those deduced in ref. 1 from more limited spectroscopic information.